Symmetry, Broken Symmetry, and Topology in Modern Physics

A First Course

Written for use in teaching and for self-study, this book provides a comprehensive and pedagogical introduction to groups, algebras, geometry, and topology. It assimilates modern applications of these concepts, assuming only an advanced undergraduate preparation in physics. It provides a balanced view of group theory, Lie algebras, and topological concepts, while emphasizing a broad range of modern applications such as Lorentz and Poincaré invariance, coherent states, quantum phase transitions, the quantum Hall effect, topological matter, and Chern numbers, among many others. An example-based approach is adopted from the outset, and the book includes worked examples and informational boxes to illustrate and expand on key concepts. 344 homework problems are included, with full solutions available to instructors, and a subset of 172 of these problems have full solutions available to students.

Mike Guidry is Professor in Physics and Astronomy at the University of Tennessee. He is the author of more than 125 journal articles and six published textbooks. He has been the Lead Educational Technology Developer for several major college textbooks in introductory physics, astronomy, biology, genetics, and microbiology. During his career, he has won multiple teaching awards and has taken the lead in a variety of science outreach initiatives.

Yang Sun gained his Ph.D. at the Technical University of Munich and has many years of experience teaching undergraduate courses, ranging from introductory physics to quantum mechanics. He is the author of more than 250 journal articles, mainly in the field of nuclear many-body theory, but also in other correlated fermionic systems. He was awarded the Wu Youxun Prize by the Chinese Physical Society for his research achievements.

Symmetry, Broken Symmetry, and Topology in Modern Physics

A First Course

MIKE GUIDRY
University of Tennessee, Knoxville, TN

YANG SUN
Shanghai Jiao Tong University, Shanghai

CAMBRIDGE
UNIVERSITY PRESS

CAMBRIDGE
UNIVERSITY PRESS

University Printing House, Cambridge CB2 8BS, United Kingdom

One Liberty Plaza, 20th Floor, New York, NY 10006, USA

477 Williamstown Road, Port Melbourne, VIC 3207, Australia

314–321, 3rd Floor, Plot 3, Splendor Forum, Jasola District Centre, New Delhi – 110025, India

103 Penang Road, #05–06/07, Visioncrest Commercial, Singapore 238467

Cambridge University Press is part of the University of Cambridge.

It furthers the University's mission by disseminating knowledge in the pursuit of education, learning, and research at the highest international levels of excellence.

www.cambridge.org
Information on this title: www.cambridge.org/highereducation/isbn/9781316518618
DOI: 10.1017/9781009000949

First published 2022

Printed in the United Kingdom by TJ Books Limited, Padstow Cornwall 2022

A catalogue record for this publication is available from the British Library.

ISBN 978-1-316-51861-8 Hardback

Additional resources for this publication at www.cambridge.org/guidry-sun

Yang Sun dedicates this book to his wife Ping Zheng and his son Yan Sun.
Mike Guidry dedicates this book to his late father, Clifford Guidry, for instilling the virtues of critical thinking and innate curiosity.

Brief Contents

Part II Broken Symmetry

Part III Topology and Geometry

Part IV A Variety of Physical Applications

Contents

Part III Topology and Geometry

Preface

Undergraduate physics majors today take classes and participate in research experiences that expose them increasingly to more advanced mathematical topics that they will encounter in many research fields if they continue on to graduate school. For example, at the end of the twentieth century a significant introduction to general relativity was uncommon at the undergraduate level. Now the rudimentary mathematics underlying general relativity (differential geometry and associated tensor calculus) and the basics of physical applications in gravitational physics are taught routinely at the advanced undergraduate level in many universities.

However, there is one set of topics and associated mathematics of increasing importance for various research disciplines that undergraduates (and even many graduate students, truth be told) are seldom exposed to in any systematic fashion. We will organize these topics loosely under the rubric of applying group theory, Lie algebras, geometry, and topological concepts to physical systems. In fact, we would argue that in the landscape of undergraduate mathematical preparation for research at the next level these topics are the most important omission in the mathematical and conceptual preparation for many of our students.

Why is this so? Is it an essential problem generated by the topics being inherently too difficult to teach effectively before graduate school, or is it more a problem associated with the lack of pedagogical materials to enable them to be taught easily to advanced undergraduate physics majors? Although some mathematical concepts in these areas are not trivial, we would argue that – assuming students to have had a solid one-year course in quantum mechanics (and ideally a similar course in electromagnetism), as routinely taught at the advanced undergraduate level – they are no more difficult than the mathematics of general relativity, which also involves concepts and mathematical techniques to which many undergraduates have had little prior exposure. Therefore, we would argue that the major impediment to being able to teach these concepts systematically to advanced undergraduates is not that it is too difficult for the students, but rather it is the scarcity of pedagogical material for instructors that would make teaching such a course feasible within the usual undergraduate curriculum.

To be sure, there are many good book and journal article resources available on the general topics of groups, algebras, topology, and geometry as applied in modern physics (many in the bibliography of this book). However, few if any address these topics in a coherent and cross-disciplinary fashion that is deliberately pitched at an audience having advanced undergraduate preparation and broad scientific interests. This book is a step toward filling that gap. It is adapted from lectures taught initially over a number of years to graduate students, and from many of our own research articles on applying Lie algebras and Lie groups to a variety of many-body systems. However, the material has been systematically reorganized, rewritten, and supplemented by additional pedagogical material to make it intelligible to advanced undergraduate students.

It may be noted that there is historical precedent for this idea, already alluded to above. In our opinion, the reason that we now teach general relativity systematically to undergraduates was primarily the advent in the early 2000s of textbook and resource material that demonstrated explicitly a blueprint for teaching undergraduate general relativity (aided greatly by inherent student interest generated by remarkable discoveries in modern gravitational physics). Thus, one may view the present book as an attempt to do something similar to enable systematic teaching of groups, algebras, topology, geometry, and their modern physics applications to new generations of undergraduate students.

Student Preparation: The reader of this book is expected to have physics experience commensurate with that of a third or fourth year undergraduate physics major in a U.S. university, and to be familiar with the material typically covered in an advanced undergraduate quantum mechanics course, including quantum mechanics expressed in second-quantized Dirac notation (bras, kets, and creation and annihilation operators), and to be aware of basic concepts from special relativity as typically taught in introductory modern physics. While some familiarity with the Dirac equation and with relativistic quantum field theory is useful, we do not assume such preparation, and introduce these concepts as part of the presentation where needed. It is desirable that the student have an advanced undergraduate understanding of electromagnetism (in particular Maxwell's equations and gauge transformations), but this is not essential for the diligent student as we introduce the required concepts as an integral part of the presentation. We assume the reader to be familiar with basic algebra, geometry, calculus, differential equations, and linear algebra, but to have minimal prior knowledge of group theory, Lie algebras, differential geometry, and topology.

Examples and Boxes: To aid in comprehension, many worked examples and supplementary information boxes are scattered throughout each chapter. These serve two general functions: to illustrate how to do some essential tasks, like showing how to write a cyclic group as a direct product of smaller cyclic groups, or to set in context and provide broader perspective, as in a discussion of the general concept of equivalence classes as background for understanding the partition of group elements into conjugacy classes.

Problems and Solutions: A total of 344 problems of varying complexity and difficulty may be found at the ends of the chapters, each chosen to familiarize the reader with basic concepts, illustrate important points, fill in details, or prove assertions made in the text. The solutions for all 344 problems are available from the publisher as a PDF file in typeset book format for instructors, and a subset of 172 problem solutions is available to students in the same format. Those problems with solutions available to students are marked by the symbol *** at the end of the problem.

Conventions: A broad variety of conventions may be found in books and journal literature related to the subject matter of this book. In choosing our conventions we have been guided by two main principles: (1) this is not a mathematics textbook but rather one about physics applications, and (2) a central goal is to equip advanced undergraduate students to use the available physics literature in fields covered by this book.

While parts of the book employ standard MKS or CGS units, professionals in many fields touched on by these topics routinely use natural units that are defined such that fundamental constants like the speed of light or Planck's constant take unit value. One of the purposes of the present material is to address the significance of these topics for cutting-edge research in a variety of fields, and to encourage students to use and explore the corresponding literature. Thus we have not shied away from using and explaining natural units where appropriate.

Mathematically one should distinguish groups from associated Lie algebras symbolically. But in the kinds of practical physics applications emphasized here it is usually clear from the context whether one means a group or an algebra, and it can become rather pedantic to distinguish formally. Therefore, we have generally used the same symbols for a Lie algebra and associated Lie groups, stating in words whether the symbol stands for a group or algebra if there is any chance of confusion.

Likewise, it is common in introductory quantum mechanics to distinguish operators from non-operator quantities by special notation, such as a special font or placing a caret over operators. Since the reader is assumed to understand enough quantum mechanics to know the difference between operator and non-operator quantities, in the interest of clean and compact notation we will usually not use special fonts or symbols to denote operators, letting the context dictate which quantities are operators. We deviate from this practice only if confusion might otherwise result.

Because it is common in many fields, we often use the Einstein convention for implied summation on repeated indices. Mathematically, one should distinguish upper and lower indices and employ them consistently, requiring summation on precisely one upper and one lower repeated index. However, the practical distinction between upper and lower tensor indices is important primarily for non-euclidean metrics and, since many physics problems are formulated explicitly or implicitly in euclidean manifolds, one finds a broad variety of adherence or non-adherence to this rule in the physics literature. Therefore we have adopted a loose overall summation convention that any repeated index implies a summation on that index, irrespective of vertical position unless stated otherwise, but have adhered to more rigorous mathematical conventions for non-euclidean metrics like for Minkowski space, where we distinguish explicitly vectors from dual vectors, and accordingly require clear distinction between upper and lower tensor indices in notation and summation convention. While less than desirable formally, we believe that this hybrid approach is of practical utility in preparing students to engage with real-world physics literature.

Resources for Teaching: For those wishing to teach from this book, two free resources are available from Cambridge University Press.

1. *Instructor Solutions Manual,* which is a PDF file typeset in the format of the book that presents the full solutions for all 344 problems at the ends of chapters. This manual is available only to instructors.
2. *Student Solutions Manual,* which is a PDF file typeset in the format of the book that contains the full solutions for a subset of 172 of the 344 problems at the ends of chapters.

This manual is available to students, instructors, and general readers. The problems contained in this solutions manual for students are marked by *** at the end of the problem in the text.

These resources may be found at the Cambridge University Press website for this book.

Sample Courses: For those teaching from this book, unless one has the luxury of a two-semester timeframe the material contained here is too much for a single course. This book should be viewed as an integrated resource from which one can tailor courses of varying length and emphasis. We give some examples to suggest possible directions to go, assuming a single-semester course.

1. *Overview Course:* A broad introduction to groups, algebras, topology, and geometry can be constructed from Chs. 2–6, 8–10, 12–18, 24, and 27–29 (add Ch. 7 if you wish to cover formal classification of Lie groups). Worked problems in the *Instructor Solutions Manual* 244. Worked problems in the *Student Solutions Manual* 121. Supplemental special topics: Chs. 30–34.
2. *Symmetry and Broken Symmetry Course:* Chs. 2–23. Worked problems in the *Instructor Solutions Manual* 273. Worked problems in the *Student Solutions Manual* 135. Supplemental special topics: Chs. 30–34.
3. *Topology in Physics Course* (including basic group theory, Bloch theorem and Brillouin zone, Dirac, Weyl, and Majorana equations, and Lorentz/Poincaré and gauge symmetry as background): Chs. 2–5, 13–16, 24–29. Worked problems in the *Instructor Solutions Manual* 187. Worked problems in the *Student Solutions Manual* 95.

Note that the 344 problems with complete solutions in the *Instructor Solutions Manual* are a resource that can be used to broaden and deepen the coverage of particular topics in these chapters.

Summary: This book provides a unified and pedagogical discussion of groups, algebras, geometry, and topology in modern physics that is tailored specifically for advanced undergraduate students, with integrated and comprehensive support material enabling the aspiring instructor to teach these topics at that level. It is our hope that this material will facilitate a systematic introduction to these concepts at the advanced undergraduate level, enabling our students to enter graduate school armed with the tools and understanding to begin participating immediately in many of the most interesting and challenging research topics in modern physics.

Acknowledgments: We would like to extend our thanks to the many students and colleagues whose questions and comments sharpened this presentation, to our long-time research collaborators Cheng-Li Wu, Da Hsuan Feng, Lian-Ao Wu, and Weimin Zhang in the development of dynamical symmetries for many-body systems that have been incorporated in this book, and to Nicolas Gibbons, Sarah Armstrong, Nicola Chapman, and Christian Green at Cambridge University Press, and Franklin Mathews Jabaraj at Straive for all their help in bringing this book to publication. This book was completed during the Covid-19 pandemic of 2020–2021. We are particularly grateful to our families, professional colleagues, support staff, and myriad food, medical, and service workers whose sacrifices and willingness to brave the virus kept us well, fed, as sane as could be expected, and reasonably functional during trying times.

PART I

SYMMETRY GROUPS AND ALGEBRAS

1 Introduction

Symmetry principles and geometrical/topological concepts are central to many of the most interesting developments in modern physics. Sophisticated mathematical advances in applications of groups, algebras, geometry, and topology to physical systems are now in routine use by theoretical physicists, and symmetry principles and concepts pervade the language that we use in our physical descriptions; but this was not always so.

Groups were invented by mathematicians who rejoiced that they had finally discovered something that was of no practical use to the natural scientists [72].[1] Lie algebras were invented largely by the Norwegian mathematician Sophus Lie and the German mathematician Wilhelm Killing in the period ~1873–1890. Just as differential geometry was developed by mathematicians more than a half century before Einstein (with the help of his mathematician friend Marcel Grossmann) adapted it to the description of gravity in his theory of general relativity, Lie algebras and Lie groups were developed mathematically decades before they began to find serious applications in physics. Likewise, just as the development of general relativity was the impetus for the introduction of differential geometry into physics applications in the period 1912–1915, the realization that Lie algebras and Lie groups might be important for physics was largely an outgrowth of the invention of quantum mechanics in 1925–1926.

In general relativity it was Einstein's realization that gravity must be associated with curved spacetime that necessitated a mathematical description of curvature that was intrinsic to the four-dimensional spacetime manifold. That description (differential geometry) had been invented by Riemann (building on the work of Gauss) more than half a century before. In quantum theory, sets of operators that close under commutation belong to a finite-dimensional Lie algebra and transformations are described by a finite number of continuous parameters belonging to a Lie group. Thus, Lie algebras and Lie groups provided a natural language to describe physical problems in the new quantum mechanics. However, many were slow to grasp this insight. Eugene Wigner, who would win a Nobel Prize for his applications of symmetry principles to nuclear and elementary particle physics, was advised more than once by giants of the field that the use of group

[1] Not the only time that mathematicians have erred in this expectation. Physicists have shown themselves to be quite clever in co-opting pure mathematics for their own purposes!

theory in quantum mechanics was a passing fad, best put aside in favor of less abstract approaches based on more traditional mathematics.[2]

Today symmetry concepts are integral to much of our thinking and discussion in many areas of physics, either explicitly or implicitly; so much so that it is now possible to teach angular momentum theory or the significance of conservation laws in quantum mechanics without much thought for the underlying Lie groups and Lie algebras that are ultimately responsible for these concepts. As a consequence, modern physics undergraduate students often gain a vague appreciation for, but generally do not get a systematic and mathematical grounding in, the role of symmetry in modern physics research.

Furthermore, in recent decades there has been an explosion of ideas built on geometrical and topological concepts in various fields of physics. Initially these concepts were pervasive in disciplines like elementary particle physics because they most often appeared as solutions of relativistic quantum field theories. But now geometrical and topological themes are proliferating in such unexpected research domains as materials science and condensed matter physics, which are built on non-relativistic quantum fields. These developments have significant implications for fundamental physics, but also may be of large importance for practical applications in quantum information, quantum computing, and related disciplines. Today, even experimentalists in elementary particle physics, broad ranges of condensed matter physics and material science, and quantum information and computing must at least know the language and concepts of topological quantum theories. Needless to say, most undergraduate physics students are even less prepared to deal with issues in topology and advanced geometry than they are to deal with those from sophisticated applications of groups and algebras in modern physics.

This book addresses the issues described above by providing a unified and pedagogical discussion of groups, algebras, geometry, and topology in modern physics. It is tailored specifically for physicists and students with at least an advanced undergraduate preparation, but without a specific background in these areas.

[2] Though of course the matrices employed in some formulations of quantum mechanics were themselves mathematics that had not been much used in physics prior to the advent of quantum theory (and that were, in fact, closely related to Lie groups). Schrödinger reputedly advised a young Wigner that the idea that Lie groups were relevant to quantum physics would be largely forgotten within five years. Not the first time that a scientist of considerable renown was in serious error about the future of a field!

2 Some Properties of Groups

Our goal in this book is to examine basic principles of symmetry, topology, and geometry in the context of modern research in physics. We begin with symmetry and the mathematical concept of a *group*. In this chapter some fundamental definitions and terminology will be introduced, using as illustration a few simple groups that often have transparent geometrical or combinatorial interpretations.

2.1 Invariance and Conservation Laws

The essence of a symmetry principle is that some quantity is fundamentally unobservable and that this implies an invariance under a related mathematical transformation. For a geometrical object, this may be illustrated in terms of its (geometrical) *covering operations*: the set of (1) rotations, (2) inversions, and (3) reflections that leave it indistinguishable from the object before the transformation. For example, a sphere has a high degree of symmetry and it looks the same after any rotation, reflection, or inversion. A square has lower symmetry, but it is unchanged after rotation by $\frac{\pi}{2}$ or after various reflections, and so on.

This fundamental symmetry principle also may be implemented in more abstract terms. In particular, we may consider mathematical operations that are not essentially geometrical in nature such as the various transformations important in quantum mechanics. Suppose that a symmetry operation is implemented by the quantum-mechanical operator U. If the Hamiltonian operator H is invariant under this symmetry transformation,[1]

$$H = UHU^{-1}, \tag{2.1}$$

where the inverse U^{-1} is defined through $UU^{-1} = 1$. Operating from the right with U,

$$HU = UHU^{-1}U = UH \quad \rightarrow \quad [U, H] = 0, \tag{2.2}$$

[1] In quantum theory operators are often distinguished from non-operator quantities by the use of a different font or by a special notation such as placing a hat over operators: $\hat{H}$. In the interest of clean and concise notation, we will let the context dictate which quantities are operators and not use a separate notation for them except in special circumstances.

Table 2.1. Some symmetries and associated conservation laws

Non-observable	Transformation	Conservation law
Absolute position	Space translation: $x \to x + \delta x$	Momentum
Absolute time	Time translation: $t \to t + \delta t$	Energy
Absolute direction	Rotation: $\theta \to \theta + \delta\theta$	Angular momentum

where the *commutator* of two operators A and B is defined by

$$[A, B] \equiv AB - BA. \tag{2.3}$$

But in quantum mechanics the Heisenberg equation of motion for an operator U is

$$\frac{dU}{dt} = \frac{\partial U}{\partial t} + \frac{i}{\hbar}[U, H], \tag{2.4}$$

and if U has no explicit time dependence it is a *constant of motion* if it commutes with H. For most (not all) cases the operators U may be chosen to be unitary and written

$$U = e^{i\alpha A}, \tag{2.5}$$

where α is a real number and the unitarity of U, defined by $U^\dagger = U^{-1}$, implies that A is hermitian ($A = A^\dagger$, where $A^\dagger$ indicates hermitian conjugation of A: in a matrix representation, complex conjugate each element and interchange rows and columns). But in quantum mechanics hermitian operators imply observables (their real eigenvalues) and we conclude that invariance of a Hamiltonian operator with respect to some unitary transformation leads to a conservation law since, if U is a constant of motion, A will be also. Generally, the eigenvalues of A will be conserved quantum numbers and the operator conjugate to A in the sense of the uncertainty principle will be associated with an inherently unobservable quantity. Table 2.1 displays the relations among some well-known symmetries, conservation laws, and corresponding non-observables, as illustrated in Examples 2.1 and 2.2.

Example 2.1 The impossibility of determining absolute spatial position (homogeneity of space) implies an invariance under spatial translations. Conservation of linear momentum follows directly from this invariance.

Example 2.2 Inability to distinguish absolute orientation (isotropy of space) implies an invariance under rotations, which leads directly to conservation of angular momentum.

The natural framework for analysis of such symmetries is that of group theory and the algebras of its associated operators. Let us now turn to a consideration of those mathematical techniques, with groups to be introduced in this chapter and algebras in Ch. 3.

2.2 Definition of a Group

A *set* is a collection of distinct objects having the properties reviewed in Box 2.1. A *group* is a set $G = \{x, y, \ldots\}$ for which a binary operation $a \cdot b$ called *group multiplication* (or, more mathematically, *composition*) is defined, which has the following properties.

1. *Associativity:* Multiplication is associative, $(x \cdot y) \cdot z = x \cdot (y \cdot z)$ for each x, y, and z in G.
2. *Closure:* If x and y are elements of G, then $x \cdot y$ is an element of G also.
3. *Unique identity:* An identity element e exists such that $e \cdot x = x \cdot e = x$ for any x in G.
4. *Unique inverse:* Each group element x has an inverse x^{-1} such that $xx^{-1} = x^{-1}x = e$.

Box 2.1 **Properties of Sets**

A *set* is a collection of distinct objects considered as an entity. The members of a set are called *elements* and they can be almost anything (including other sets). The number of members is called the *cardinality* of the set.

Defining elements of a set: A set may be populated with elements in two ways.

1. List the elements explicitly: $A = \{\text{red, yellow, green}\}$.
2. Specify a rule giving all elements: "the negative integers greater than -4." A common rule is to specify a property $P(x)$ that all members x of a set have using *set-builder notation*: $\{x|P(x)\}$ means "the set of all x such that the property $P(x)$ is true." *Example*: If $P(x)$ is the property "x is a prime number," then $\{x|P(x)\} = \{x|\ x \text{ is a prime number}\}$ is the set of all prime numbers.

Set membership: Writing $x \in A$ indicates that x is a member of set A and writing $x \notin A$ means that it is not. These may be read "x is in A" and "x is not in A," respectively.

Equivalence: Sets are *equivalent* iff (which means "if and only if") they have precisely the same elements. We assume each element to be listed only once, so $\{a, a, b, c\}$ and $\{a, b, c\}$ are equivalent sets. Likewise, the order of elements does not matter, so $\{a, b, c\}$ and $\{b, a, c\}$ are equivalent sets.

Null set: The *null set* is the set having no elements, commonly denoted by $\emptyset$ or $\{\}$.

Subsets: Set B is a *subset* of set A if each element of B is in A. The notation $B \subset A$ indicates that B is a subset of A. Equivalently $A \supset B$ indicates that A is a superset of B (contains B as a subset). Every set has itself and the null set as subsets. If $B \subset A$ but A and B are not equivalent and $B \neq \{\emptyset\}$, B is a *proper subset* of A.

Union and intersection: The *union* $A \cup B$ of two sets A and B is the set containing all elements that are in A or in B, or in both A and B. The *intersection* $A \cap B$ of sets A and B is the set of only those elements contained in both A and B. The union and intersection of two sets are conveniently represented in terms of *Venn diagrams*.

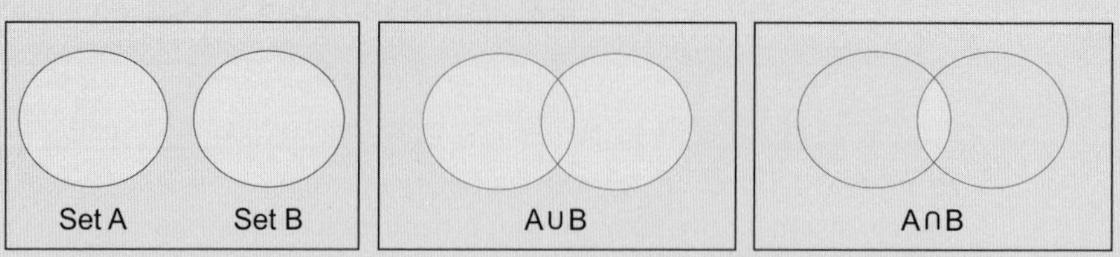

Cartesian product: A cartesian product $X \times Y$ of sets X and Y is the set of ordered pairs (x, y) with $x \in X$ and $y \in Y$. The prototype is $\mathbb{R}^2 = \mathbb{R}^1 \times \mathbb{R}^1$, where $\mathbb{R}^1$ is the real number line and $\mathbb{R}^2$ is the 2D cartesian plane. If X is of dimension m and Y is of dimension n, the cartesian product $X \times Y$ is a set of dimension $m \times n$.

More advanced properties such as open and closed sets, or equivalence classes for sets, will be addressed as they are needed (for example, in Box 24.1).

Group "multiplication" is a law of binary combination that can be much more abstract than ordinary arithmetic multiplication. The group definition requires associativity but not commutivity. If a group has commutative elements so that $ab = ba$ for all a and b, the group is *abelian*; if it has non-commuting elements ($ab \neq ba$) the group is *non-abelian*.

2.3 Examples of Groups

The preceding requirements are relatively easy to satisfy and many sets can form groups under some law of binary combination. Let us give a few simple examples.

2.3.1 Additive Group of Integers

The set of all integers $G = \{\ldots, -3, -2, -1, 0, 1, 2, 3, \ldots\}$ forms a group under the binary operation of ordinary arithmetic addition.

- The sum of any two integers is an integer.
- A unique identity element $e = 0$ exists.
- A unique inverse exists for each element since $-n + n = e = 0$.
- Arithmetic addition is associative.

This is called the *additive group of integers* Z; it is abelian since addition commutes. Whether a set forms a group under some binary operation depends both on the set members and the binary operation.

Example 2.3 The integers are a group under addition but the positive integers are not (elements have no inverse). The real numbers are a group under addition but not under multiplication (zero has no inverse).

Box 2.2

The Two-Element Group

The group with two elements may be defined by listing the elements $\{e, a\}$, where e is the identity, and specifying the results of the multiplication operation. One way to do this is to give the *multiplication table* for elements of the group.

The Multiplication Table

From the properties of the identity we must have $ee = e$ and $ea = ae = a$ (for compactness the multiplication symbol often will be omitted: $ae \equiv a \cdot e$). Thus, only the product aa is required to complete the multiplication table. Either $aa = a$, or $aa = e$, if the closure property is to be satisfied. The first is impossible because multiplying the equation from both sides by a^{-1} gives $a = e$, which must be false if this is a two-element group.[a] So a is its own inverse, $aa = e$, and the *multiplication table* is

C_2	e	a
e	e	a
a	a	e

As indicated, this group is commonly denoted C_2, the *cyclic group* of order 2. Equivalently, it is the group Z_2 of integers 0 and 1 under addition modulo 2 (Problem 2.31).

Important Lessons

Two important properties of groups have been illustrated by this simple example.

1. Only *one* abstract group with two elements exists, since any two-element group must have the multiplication table just constructed.
2. Finite groups may be specified by their multiplication table, since this tells us both the elements of the set and the nature of the multiplication operation.

As you are asked to show in Problem 2.4, the group with three elements is also unique because the multiplication table is determined by general group properties independent of the specific group elements or nature of the binary combination law. The two-element and three-element groups illustrate the abstract nature of groups, since their properties are fixed by the group postulates, independent of specifics.

[a] This proof can be generalized to the useful rule that in a group multiplication table each group element must appear exactly once in each row and exactly once in each column of multiplication results. This may be viewed as a special case of the *rearrangement lemma* discussed in Section 6.2.5.

The additive group of integers has an infinity of discrete elements. There also are groups where the elements form an uncountable continuum (*continuous groups*), and groups with a finite number of elements (*finite groups*). An example of a finite group is given in Box 2.2.

2.3.2 Rotation and Translation Groups

As another example of a group, consider rotations. A natural multiplication may be defined by taking the product of two operations to be the resultant of applying first one operation and then the other to an arbitrary physical system. The product of two transformations A

and B may be denoted by $AB \equiv A \cdot B$, with the convention that the operator B is to be applied first and then the operator A. For rotations the following is clear physically.

1. Two successive rotations on a physical system are equivalent to some other rotation.
2. Rotation of a system through the null angle corresponds to an identity operation.
3. A system may be rotated through an angle and then rotated back to the original orientation. The second operation defines the unique inverse of the first.
4. If three successive rotations are performed, the same result is obtained if we perform the first and then the resultant of the other two, or if we perform the resultant of the first two and then the third rotation. Thus, rotation is associative.
5. For rotations in three dimensions, the result of two successive rotations about different axes is not necessarily equivalent to the result obtained by performing these rotations in reverse order. Thus, 3D rotations do not generally commute.

Therefore, 3D rotations form a *rotation group* that is non-abelian. This is a *continuous group* since the possible angles for a rotational transformation are continuous. Such groups of continuous transformations will be among the most important that we will examine. As another example of a continuous group, it is easily verified that the set of translations form a group under the natural definition that multiplication of two translations corresponds to applying first one and then the other translation operation to a physical system. The group of rotations and the group of translations are members of a special class of continuous groups called *Lie groups* that will occupy a prominent place in our discussion.

Sets of rotations and sets of translations each define continuous groups sharing many features, but they differ in two important aspects. The first is that translation groups are abelian but the 3D rotation group is non-abelian. The second concerns the group parameter space. Rotational parameters are real angles and the space is *bounded* (rotation by 2π about an axis returns us to the starting point). However, the parameter space for translations is *unbounded.* A continuous group for which the parameter space is closed and bounded is called *compact*; otherwise it is termed *non-compact.* As will be seen, the mathematical structure and the associated physical implications of compact and non-compact groups differ considerably; accordingly they have different roles to play in physical applications.

2.3.3 Parameterization of Continuous Groups

The number of elements is the *order* of a group, which can be finite or infinite. For finite groups of low order the group elements may simply be listed. This is inappropriate for continuous groups such as the rotation group, for which it is customary to specify a set of continuously varying parameters that describe the transformation: $\boldsymbol{\alpha} = (\alpha_1, \alpha_2, \ldots, \alpha_n)$. The continuous group associated with this parameter set is called an *n-parameter group.* The 3D rotations are a three-parameter group since there are three axes of rotation, each described by an independent angle. *Do not confuse the order of a group and the number of continuous parameters that define it.* For continuous groups the order is infinite but the number of parameters is usually finite. Our standard notation for continuous groups will be $G(\alpha_1, \alpha_2, \ldots, \alpha_n)$, where the n parameters α_n all vary continuously over some range.

2.3.4 Permutation Groups

Another example of a relatively simple group is the *permutation group* S_n. This group consists of the set of all permutations on n objects, with a multiplication law defined by the rule that $A \cdot B$ means the transformation obtained by first making the permutation B and then applying the permutation A to the result. The order matters because S_n is not abelian if $n \geq 3$. This group is particularly significant for our discussion because symmetry under particle exchange is of fundamental importance in quantum mechanics. *Be certain that you understand what constitutes the group here!*

> The elements of a permutation group are not the objects being permuted but rather the permutation operations on those objects. Also, do not confuse a permutation with a group multiplication. Permutation interchanges objects; group multiplication is the application of two successive such permutations.

An arbitrary permutation of n objects may be specified by[2]

$$p = \begin{pmatrix} 1 & 2 & 3 & \cdots & n \\ p_1 & p_2 & p_3 & \cdots & p_n \end{pmatrix}, \tag{2.6}$$

where the notation means that each entry in the first row is to be replaced by the corresponding entry in the second row. A more compact notation may be introduced by classifying permutations according to their *cycle structure.* To illustrate, consider the permutation

$$\begin{pmatrix} 1 & 2 & 3 & 4 & 5 & 6 \\ 3 & 5 & 4 & 1 & 2 & 6 \end{pmatrix} = \begin{pmatrix} 1 & 3 & 4 & 2 & 5 & 6 \\ 3 & 4 & 1 & 5 & 2 & 6 \end{pmatrix} \tag{2.7}$$

on six objects, where in the second expression invariance under column interchange has been exploited. In this permutation 1 is replaced by 3, which is replaced by 4, which is then replaced by 1, which was the starting point. This is called a *three-cycle*, which can be represented by (134). In a similar fashion we see that interchanges on objects 2 and 5 form a *two-cycle* (25), and object 6 participates in a *one-cycle* (6). In *cycle notation* the permutation (2.7) is specified as (134)(25)(6), which may be abbreviated to (134)(25) since one-cycles are commonly omitted in the notation. The cycles contain no elements in common so they commute and the order in the list does not matter: (134)(25) = (25)(134). Giving the cycle structure of any permutation completely defines that permutation.

The permutations on n objects form a group of order $n!$ because (1) the product of two permutations is also a permutation, (2) products of permutations are associative, (3) the identity is the symbol representing no permutations [$n = p_n$ for all n in Eq. (2.6)], and (4) interchanging the two rows in Eq. (2.6) defines an inverse. Example 2.4 illustrates.

[2] The order of swaps is irrelevant so the permutation symbol (2.6) is invariant under interchange of columns.

Table 2.2. Multiplication table $A \cdot B$ for S_3[†]

$A\backslash B$	e	(12)	(23)	(13)	(123)	(321)
e	e	(12)	(23)	(13)	(123)	(321)
(12)	(12)	e	(321)	(123)	(13)	(23)
(23)	(23)	(123)	e	(321)	(12)	(13)
(13)	(13)	(321)	(123)	e	(23)	(12)
(123)	(123)	(23)	(13)	(12)	(321)	e
(321)	(321)	(13)	(12)	(23)	e	(123)

[†]Entries $A \cdot B$, where the left column gives A and the top row gives B.

Example 2.4 Consider the group S_3 of permutations on three objects. There are $3! = 6$ independent ways that a set of objects $\{a, b, c\}$ may be rearranged (see Section 4.1 for a more extensive discussion of the notation).

(e) Do nothing: $(abc) \to (abc)$.
(12) Swap the objects in positions 1 and 2: $(abc) \to (bac)$.
(23) Swap the objects in positions 2 and 3: $(abc) \to (acb)$.
(13) Swap the objects in positions 1 and 3: $(abc) \to (cba)$.
(321) Perform a cyclic permutation: $(abc) \to (cab)$.
(123) Perform an anticyclic permutation: $(abc) \to (bca)$.

Applying the multiplication rule for these transformations, we find as an example that $(12) \cdot (13) = (123)$, because (13) generates $(abc) \to (cba)$ and subsequent application of (12) produces $(cba) \to (bca)$, which is equivalent to direct application of (123) to (abc). The full group multiplication table is given in Table 2.2 (see Problem 2.2).[3]

We see that Table 2.2 is closed under multiplication, defining a finite group that is non-abelian because generally $A \cdot B \neq B \cdot A$.

2.4 Subgroups

One or more subsets H of elements within a group G may satisfy the group requirements among themselves, subject to the same law of multiplication as for the whole group. Such a subset H is termed a *subgroup* of the larger group G; symbolically $G \supset H$. A subgroup is not just a subset of group elements; it must embody the *same multiplication rule* as for the full group. The following example illustrates.

[3] Our convention will be that for $A \cdot B$ the operation A is listed in the left column and the operation B is listed in the top row of a multiplication table. For example, from Table 2.2 we obtain $(12) \cdot (13) = (123)$ but $(13) \cdot (12) = (321)$. The ordering convention matters since S_3 is non-abelian. We will consider only a few simple finite groups and it is often convenient to specify them by multiplication tables. For larger groups or many cases this can be tedious and there are more efficient ways to represent the content of a multiplication table (see Problem 2.28). We will not need them but Jones [125] may be consulted for examples.

Example 2.5 Consider a group G consisting of the rational numbers under addition, and a second group H consisting of the positive rational numbers under multiplication. The positive rational numbers are a *subset* of the rational numbers, but H is *not a subgroup* of G because the law of binary combination is different for G and for H.

Note that any subgroup must contain the identity element of the full group. Otherwise either the subgroup would not satisfy the group conditions, or the full group would not have a unique identity. Subgroups may be classified as (1) *proper* or (2) *improper.* A group always contains as improper subgroups the group itself and the subgroup consisting only of the identity element; the remaining subgroups constitute the proper subgroups.

Example 2.6 The 2D spatial rotations around a single axis constitute a proper subgroup of the full 3D rotation group. Furthermore, although the full rotation group is non-abelian the subgroup of rotations about a single axis is abelian, since the order of two successive rotations about the same axis is immaterial.

Example 2.7 As may be verified from Table 2.2, the permutation group S_3 has proper subgroups S_2 corresponding to sets of permutations on only two objects, and a proper subgroup A_3 consisting of only the even permutations $\{1, (123), (321)\}$ on three objects.

Unless otherwise noted, subgroups will be assumed to be proper subgroups in our discussion. Examples 2.6 and 2.7 illustrate two common ways to obtain transformation subgroups.

1. Restrict the objects on which the transformations can act (for example, the subgroup S_2 of permutations on only two objects).
2. Restrict the action of the transformation operator on all objects (for example, the subgroup A_3 of only even permutations on three objects).

As illustrated in Examples 2.6 and 2.7, for actual problems there will often be a simple physical interpretation of the mathematical relationship between a group and a subgroup.

2.5 Homomorphism and Isomorphism

Frequently we will invoke a *map,* which generalizes a function and associates an object in one set with an object in the same or another set, as reviewed in Box 2.3.[4] Of crucial importance for physics applications, often a map $f : A \to B$ can be defined between the elements of a group A and the elements of a group B that preserves the group multiplication

[4] The usage of "map" and "function" varies among fields and authors. Some use "map" and "function" to mean essentially the same thing; others reserve "function" for a map where the sets being mapped are real or complex numbers, and use "map" to include more general possibilities.

Box 2.3

Maps

A map assigns to each object in a set one or more objects in another (or the same) set. *Example:* If A is the set of all integers, "multiply objects in set A by two" defines a map from the set of integers to the set of even integers. The *domain* is the set of all objects the map can operate on (integers in this example); the *codomain* is the set into which all output of the map falls; the *image* is the set of all outputs created by the map (even integers in this example), which is generally a subset of the codomain. The *inverse image* of a subset S of the codomain is the set of all elements of the domain that map to members of S. A map is single valued, so it can connect more than one object in the domain to a single object in the image, but it cannot connect an object in the domain to more than one object in the image.

Mapping Notation

If $f(x)$ depends on a single variable, $f : \mathbb{R} \longrightarrow \mathbb{R}$ indicates that $f(x)$ provides a map from real numbers $\mathbb{R}$ to real numbers $\mathbb{R}$. The notation can also indicate the map more explicitly. For a function of two variables, $f(x, y) = x + y$, we can write that the function maps the xy plane $\mathbb{R}^2$ to the real number line $\mathbb{R}$, $f : \mathbb{R}^2 \longrightarrow \mathbb{R}$, or indicate the map more explicitly as $f : (x, y) \mapsto x + y$. The inverse image $f^{-1}(S)$ of a subset S of the codomain is indicated by $f^{-1}(S) = \{x \in X : f(x) = S\}$, for a domain X.

Homomorphism and Isomorphism

The maps illustrated schematically in Fig. 2.1 are between group (set) elements. In the case of isomorphism exactly one object in one set is mapped into exactly one object in another set. In the case of homomorphism one or more objects in the first set are mapped into single objects in the second set.

General Mapping Relationships

The general nature of the mapping between two sets is often described using a terminology that is illustrated in the following figure.

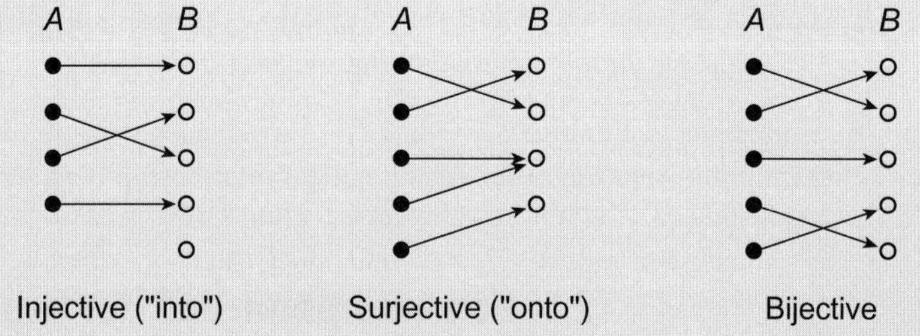

(1) A map from A to B is *injective* ("into") if an element in B is related to no more than one element in A, but B can have unmatched elements. (2) A map is *surjective* ("onto") if at least one element of A is associated with each element of B. (3) A map is *bijective* if each element of A has a matching element in B and each element in B has a matching element in A; a bijective map is *invertible*. Isomorphism is bijective.

structure. Such a mapping is called a *homomorphism* and the two groups are said to be *homomorphic*. For the special case where the groups are of the same order and the mapping is one to one, the homomorphism is termed an *isomorphism* (also called a *bijective map*; see Box 2.3) and the two groups are said to be *isomorphic*. Homomorphism and isomorphism

Fig. 2.1 Schematic examples of a 2:1 homomorphism and an isomorphism between two groups.

are illustrated in Fig. 2.1. If a group D is homomorphic to a group G, the group D may be said to form a *representation* of G. A representation requires that every $g \in G$ has a corresponding operator $D(g)$, with a mapping that preserves the group multiplication law:

$$D(x) \cdot D(y) = D(x \cdot y), \tag{2.8}$$

where x and y are arbitrary elements of the group G. Example 2.8 illustrates a representation giving the effect of a transformation of coordinates on a function of those coordinates.

Example 2.8 Suppose a space S of functions $\psi(\boldsymbol{r})$ that is invariant under a group of coordinate transformations G_i in the sense that if $\psi(\boldsymbol{r}) \in S$, then $\psi(G_i^{-1}\boldsymbol{r}) \in S$. Then a representation T may be defined in the function space by the requirement that

$$T(G_i)\psi(\boldsymbol{r}) = \psi(G_i^{-1}\boldsymbol{r}). \tag{2.9}$$

As shown in Problem 2.35, the operators $T(G_i)$ of Eq. (2.9) satisfy Eq. (2.8); hence they constitute a valid representation of G.[5] Notice that we are dealing with group transformations on two different spaces here: (1) transformations by G in the *space of coordinates* $\boldsymbol{r}$, and (2) transformations by the representation $T(G)$ of G in the *space of functions* $\psi(\boldsymbol{r})$.

If a homomorphic mapping is of the group to itself, it is termed an *endomorphism*. If the mapping of the group to itself is also isomorphic, the endomorphism is termed an *automorphism*. Two examples of automorphisms that we shall encounter are

1. the mapping of a group to itself by conjugation of its elements (Section 2.11), and
2. the mapping of Lie group generators to their complex conjugates (Section 8.4).

Endomorphisms and automorphisms are important in the fundamental analysis of groups, but for physics applications we shall most often be interested in homomorphisms and isomorphisms that are between different groups, as in Fig. 2.1.

2.6 Matrix Representations

A set of matrices that forms a group under matrix multiplication is termed a *matrix group*. If a matrix group is homomorphic to some abstract group, it forms a representation of that group. Let us illustrate for the permutation group S_3.

[5] Transformation of a function caused by transformation of a variable on which the function depends, as in Eq. (2.9), is termed an *induced transformation*.

2.6.1 A Matrix Representation of S_3

It is readily verified that the set of six matrices D specified by

$$D(e) = \begin{pmatrix} 1 & 0 & 0 \\ 0 & 1 & 0 \\ 0 & 0 & 1 \end{pmatrix} \quad D(12) = \begin{pmatrix} 0 & 1 & 0 \\ 1 & 0 & 0 \\ 0 & 0 & 1 \end{pmatrix} \quad D(13) = \begin{pmatrix} 0 & 0 & 1 \\ 0 & 1 & 0 \\ 1 & 0 & 0 \end{pmatrix},$$
$$D(23) = \begin{pmatrix} 1 & 0 & 0 \\ 0 & 0 & 1 \\ 0 & 1 & 0 \end{pmatrix} \quad D(123) = \begin{pmatrix} 0 & 1 & 0 \\ 0 & 0 & 1 \\ 1 & 0 & 0 \end{pmatrix} \quad D(321) = \begin{pmatrix} 0 & 0 & 1 \\ 1 & 0 & 0 \\ 0 & 1 & 0 \end{pmatrix}, \tag{2.10}$$

forms a group under the binary operation of ordinary matrix multiplication, and that the multiplication table is in one to one correspondence with that of the permutation group on three objects, $S_3 = \{e, (12), (23), (13), (123), (321)\}$, given in Table 2.2.

Example 2.9 If the arrangement of objects $\{a, b, c\}$ is specified by a column vector, then

$$D(12)\begin{pmatrix} a \\ b \\ c \end{pmatrix} = \begin{pmatrix} 0 & 1 & 0 \\ 1 & 0 & 0 \\ 0 & 0 & 1 \end{pmatrix}\begin{pmatrix} a \\ b \\ c \end{pmatrix} = \begin{pmatrix} b \\ a \\ c \end{pmatrix}$$

and, comparing with the group operations for S_3 given in Section 2.3.4, the matrix $D(12)$ is mapped isomorphically to the permutation (12). Symbolically, $D(12) \leftrightarrow (12)$, where (12) is an element of the group S_3 and $D(12)$ is an element of the matrix group (2.10). By explicit matrix multiplication we find that the matrices D have a multiplication table in one to one correspondence with that in Table 2.2 for the group S_3. For example, the S_3 multiplication $(13)(12) = (321)$ is mapped to the matrix multiplication

$$D(13)D(12) = \begin{pmatrix} 0 & 0 & 1 \\ 0 & 1 & 0 \\ 1 & 0 & 0 \end{pmatrix}\begin{pmatrix} 0 & 1 & 0 \\ 1 & 0 & 0 \\ 0 & 0 & 1 \end{pmatrix} = \begin{pmatrix} 0 & 0 & 1 \\ 1 & 0 & 0 \\ 0 & 1 & 0 \end{pmatrix} = D((13)(12)) = D(321),$$

as may be verified by comparison of Table 2.2 and Eq. (2.10). This is a specific illustration of the condition (2.8) that a representation must satisfy.

Therefore, the group of permutations S_3 on three objects and the group of matrices D given in Eq. (2.10) are isomorphic and D constitutes a *matrix representation* of the permutation group S_3. When a matrix group is isomorphic to another group G, the matrices are said to be a *faithful representation* of G. Thus the matrices (2.10) are a faithful representation of S_3. The physical importance of matrix representations is discussed in Box 2.4.

2.6.2 Dimensionality of Matrix Representations

The *dimension* (or *degree*) of a matrix representation is the dimension of each matrix in the representation. A representation in terms of $N \times N$ matrices is termed an *N-dimensional*

Box 2.4 Abstract Groups and Concrete Physics

Isomorphisms with matrix groups provide much of the practical linkage between abstract groups and concrete physics. We may consider a group to be a multiplication table[a] satisfying the postulates of group theory and a representation to be a tangible realization of that multiplication. For many cases of interest in quantum physics the representations will be in terms of unitary matrices. The power of group theory is that it is possible to determine many properties of any representation from the abstract properties of the group. On the other hand, in physical problems it is often easier to think in terms of some concrete representation rather than in terms of the abstract group itself. We will often be working with representations of groups in terms of matrices. The foregoing tells us that this is adequate, since the matrix representation has the same multiplication table as the abstract group.

[a] Or the set of structure constants for a Lie group, which will act as multiplication table entries for those continuous groups (see Section 3.2).

representation of the group. The matrices D employed in a representation must be non-singular ($\det D \neq 0$) because the requirement that each group element have an inverse demands that the matrices be invertible.[6] The representation (2.10) is a three-dimensional matrix representation of the group S_3. The *fundamental matrix representation* of a group is the lowest-dimensional faithful matrix representation. Some groups have more than one fundamental representation. For example, SU(2) has one fundamental representation but SU(3) has two. There may be many representations of a group having dimensionality different from that of the fundamental representation(s).

2.6.3 Linear Operators and Matrix Representations

Quantum mechanics is formulated in terms of linear vector spaces (*Hilbert spaces*). Therefore, it is often convenient to view representations of quantum-mechanical groups in terms of abstract linear operators that have concrete manifestation in a given basis as a matrix. If $g \in G$ for a group G and $|i\rangle$ is a complete orthonormal basis for the space in which $D(g)$ acts as a linear operator, then by inserting a complete set of states using $1 = \sum_j |j\rangle\langle j|$ from Eq. (A.6),

$$D(g)\,|i\rangle = \sum_j |j\rangle\langle j|\,D(g)\,|i\rangle \equiv \sum_j |j\rangle\,[D(g)]_{ji}, \tag{2.11}$$

which provides the connection between a linear operator $D(g)$ and the elements of its matrix realization $[D(g)]_{ji} = \langle j|\,D(g)\,|i\rangle$.

[6] Not all $N \times N$ matrices are invertible, but those that are form a group under matrix multiplication. Some examples are given in Section 2.9.

2.7 Reducible and Irreducible Representations

A matrix representation of a group is not unique because we are free to make linear transformations to express the representation in a new basis. Representations D and D' of a group are deemed *equivalent* if they are related by a *similarity transformation,*

$$D'(x) = SD(x)S^{-1}, \tag{2.12}$$

with the same operator S for each matrix $D(x)$ in the representation. Thus D and D' are sets of matrices that represent the *same linear operator* expressed in two different bases. A representation is *reducible* if it has an invariant subspace [the action of any $D(g)$ on a vector in the subspace yields a vector in that subspace]. A representation that is not reducible is *irreducible*. A representation is *completely reducible* if its matrices can be similarity transformed to *block-diagonal form*:

$$D'(x) = SD(x)S^{-1} = \begin{pmatrix} D'_1(x) & 0 & 0 & 0 \\ 0 & D'_2(x) & 0 & 0 \\ 0 & 0 & \ddots & \vdots \\ 0 & 0 & \cdots & D'_k(x) \end{pmatrix}, \tag{2.13}$$

where matrices on the diagonal of Eq. (2.13) are *irreducible representations (irreps)*. For finite or compact continuous groups, it may be shown that if a representation is reducible it is completely reducible and we shall use reducible to mean completely reducible. The representation $D'(x)$ is termed the *direct sum* of the irreps $D'_1, D'_2, \ldots, D'_k$, which may be symbolized by

$$D' = D'_1 \oplus D'_2 \oplus \cdots \oplus D'_k, \tag{2.14}$$

where the meaning of the direct sum is precisely that the matrix D' is related to the matrices D'_i though the block-diagonal structure (2.13).

2.8 Degenerate Multiplet Structure

We now show that the irreducible representations of Eq. (2.13) imply degenerate multiplets for the eigenfunctions when the Hamiltonian commutes with symmetry operators [see Eq. (2.2)]. The full set of operators that commute with the Hamiltonian form a symmetry group that we may call the *Schrödinger group,* and eigenvalues of these operators may be used to label states of conserved energy. Let us apply a symmetry operator U to the Schrödinger equation, $UH\psi_n = UE_n\psi_n$, and since H and U commute, $H(U\psi_n) = E_n(U\psi_n)$. Thus $U\psi_n$ is an eigenfunction with the same energy E_n as the original state ψ_n. Other eigenfunctions with the same energy may be obtained by applying all symmetry operators that commute with H to these eigenfunctions. The full set of eigenfunctions created in this way corresponds to a multiplet of states degenerate in energy as a direct consequence of symmetry. Conversely, any state that happens to have energy E_n that cannot be obtained by actions of the symmetry generators is said to be *accidentally degenerate.*

Now, how is this degenerate multiplet structure related to the representations of the symmetry? Let us exclude accidental degeneracy and assume that the eigenstates are ℓ-fold degenerate because of the symmetry, implying a basis of ℓ eigenfunctions, each having energy E_n. This represents a subspace of the full Hilbert space that is invariant under all operations of the Schrödinger group. From the preceding considerations, application of those symmetry operators $U_\alpha = \{U_A, U_B, \ldots\}$ that commute with H to any wavefunction $\psi_i^{(n)}$ in the subspace must produce a linear combination of the basis vectors,

$$U_\alpha \psi_i^{(n)} = \sum_{i=1}^{\ell} \Gamma^{(n)}(\alpha)_{ij} \psi_j^{(n)}. \tag{2.15}$$

The matrices $\Gamma^{(n)}(\alpha)$ form a representation of the Schrödinger symmetry group since

$$\begin{aligned} U_{\alpha\beta}\psi_i \equiv U_\alpha U_\beta \psi_i &= U_\alpha \sum_j \psi_j \Gamma(\beta)_{ji} = \sum_j (U_\alpha \psi_j)\Gamma(\beta)_{ji} \\ &= \sum_{jk} \psi_k \Gamma(\alpha)_{kj} \Gamma(\beta)_{ji} = \sum_k \psi_k (\Gamma(\alpha)\Gamma(\beta))_{ki}, \end{aligned}$$

and this along with $U_{\alpha\beta}\psi_i = \sum_k \psi_k \Gamma(\alpha\beta)_{ki}$ implies the matrix equation

$$\Gamma(\alpha\beta) = \Gamma(\alpha)\Gamma(\beta), \tag{2.16}$$

which is the condition (2.8) for the matrices Γ to form a representation. Furthermore, the representation is irreducible because it was generated by symmetry operations within a single subspace.

> The ℓ eigenfunctions of energy E_n are basis functions for an ℓ-dimensional irrep $\Gamma^{(n)}$ of the group of symmetry operations that commute with the Hamiltonian.

Thus, the space acted on by the transformed matrix $D'(x)$ in Eq. (2.13) breaks up into k orthogonal subspaces, each mapped to itself by all operators $D'(x)$, as illustrated in Fig. 2.2, with the resulting energy spectrum exhibiting degenerate multiplet structure.

Example 2.10 In Ch. 3 the quantum theory of angular momentum will be formulated in terms of the special unitary group SU(2). The matrices defining the angular momentum may be brought to the form (2.13) by a unitary transformation, with each subspace D_J corresponding to a definite total angular momentum J [irrep of SU(2)]. The matrices within a subspace are of dimension $(2J + 1) \times (2J + 1)$ and they operate on state vectors with $2J + 1$ components, which is the magnetic-substate degeneracy for angular momentum J.

In Example 2.10, the block-diagonal form of Eq. (2.13) will ensure conservation of angular momentum because multiplets of different J correspond to different irreps and cannot mix.

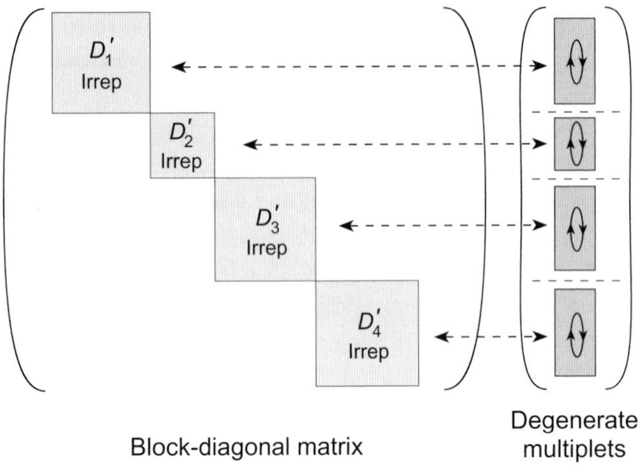

Fig. 2.2 Block-diagonal matrix and the corresponding degenerate multiplet structure. The matrices along the diagonal are irreducible representations (irreps).

2.9 Some Examples of Matrix Groups

Since matrix groups are pivotal in many applications of group theory to physics, it is useful to make a general classification of some important matrix groups according to the nature of their constituent matrices. In this section we list some of the more consequential ones.

2.9.1 General Linear Groups

The most general matrix group is the *complex general linear group*, $\mathrm{GL}(n, \mathrm{C})$, consisting of the regular, invertible, complex (the designation C), $n \times n$ matrices. Each such matrix has n^2 complex entries, so there are $2n^2$ real parameters and $\mathrm{GL}(n, \mathrm{C})$ is $2n^2$-dimensional. Restriction to real matrices (denoted by R) yields a subgroup $\mathrm{GL}(n, \mathrm{R})$,

$$\mathrm{GL}(n, \mathrm{C}) \supset \mathrm{GL}(n, \mathrm{R}), \tag{2.17}$$

with n^2 real parameters. *Special linear groups* are subgroups of $\mathrm{GL}(n, \mathrm{C})$ that restrict to matrices with unit determinant,

$$\mathrm{GL}(n, \mathrm{C}) \supset \mathrm{SL}(n, \mathrm{C}) \supset \mathrm{SL}(n, \mathrm{R}) \qquad \mathrm{GL}(n, \mathrm{R}) \supset \mathrm{SL}(n, \mathrm{R}), \tag{2.18}$$

where $\mathrm{SL}(n, \mathrm{C})$ denotes the *complex* special linear group with $2(n^2 - 1)$ parameters and $\mathrm{SL}(n, \mathrm{R})$ denotes the *real* special linear group with $n^2 - 1$ parameters.

2.9.2 Unitary Groups

The group $\mathrm{GL}(n, \mathrm{C})$ has many other subgroups, which can be classified by specifying the algebraic form that they leave invariant. For example, the $n \times n$ *unitary matrices* form elements of an n^2-parameter *unitary group* $\mathrm{U}(n)$ that leaves invariant the form $\sum_i z_i z_i^*$, where z_i is a complex number. If consideration is restricted further to unitary matrices with unit determinant, we obtain the *special unitary group* SU(n), which has $n^2 - 1$ parameters.

2.9.3 Orthogonal Groups

The group $\mathrm{O}(n, \mathrm{C})$ consists of the *complex orthogonal matrices of degree n*. It is an $n(n-1)$-parameter group and, since the orthogonality requirement $A^{\mathrm{T}}A = 1$ implies $\det A = \pm 1$, the group $\mathrm{O}(n, \mathrm{C})$ decomposes into two disconnected pieces. The subgroup $\mathrm{SO}(n, \mathrm{C})$ results from restricting the orthogonal matrices to those with unit determinant. This *special complex orthogonal group* $\mathrm{SO}(n, \mathrm{C})$ leaves invariant the quadratic form $\sum_i z_i^2$, and has $n(n-1)$ parameters. Obviously, it is also possible to form the real orthogonal and special real orthogonal groups $\mathrm{O}(n, \mathrm{R})$ and $\mathrm{SO}(n, \mathrm{R})$ respectively. [Note that we shall often drop the field labels R or C; for example, using abbreviations like SO(n) for $\mathrm{SO}(n, \mathrm{R})$.]

2.9.4 Symplectic Groups

Orthogonal matrices R satisfy $R^{\mathrm{T}}gR = g$ for a matrix g with elements $g_{ij} = \delta_{ij}$ called the *metric tensor*, implying that orthogonal transformations preserve symmetric bilinear forms. *Symplectic matrices* of dimension $2n \times 2n$ satisfy instead $R^{\mathrm{T}}gR = g$ for a metric tensor

$$g = \begin{pmatrix} 0 & I_n \\ -I_n & 0 \end{pmatrix}, \tag{2.19}$$

where I_n is the $n \times n$ unit matrix and 0 is an $n \times n$ matrix of zeros, which implies that symplectic transformations preserve antisymmetric bilinear forms. These matrices form the $2n(2n+1)$-parameter complex *symplectic group* $\mathrm{Sp}(2n, \mathrm{C})$, which is a subgroup of the general linear group,

$$\mathrm{GL}(n, \mathrm{C}) \supset \mathrm{Sp}(2n, \mathrm{C}). \tag{2.20}$$

We may also define a real symplectic group $\mathrm{Sp}(2n, \mathrm{R})$ by restricting to real matrices, and a unitary symplectic group $\mathrm{Sp}(2n) = \mathrm{USp}(2n) \equiv \mathrm{U}(2n) \cap \mathrm{Sp}(2n, \mathrm{C})$.

2.10 Group Generators

The minimal number of distinct group operations that in various combinations and powers gives all of the group operations is called the set of *generators* for the group. Conventionally, the generators are defined somewhat differently for discrete groups and for continuous groups. The following examples illustrate identifying generators for permutation groups and translation groups.

Example 2.11 For the discrete permutation groups S_n it is easy to show that all permutations on n objects can be written as products of operations that interchange two objects (the transpositions) and their inverses. In general, the permutation groups S_n are generated by the $n-1$ adjacent interchanges $P_{i,i+1}$, because it can be shown that any cycle (and hence any permutation) can be written as a set of transpositions.

Example 2.12 Consider the operator $U(a)$ that translates a 1D quantum system with wavefunction $\psi(x)$ continuously by a distance a. Expanding in a Taylor series about x,

$$\begin{aligned}\psi(x+a) &= U(a)\psi(x) \\ &= \psi(x) + a\frac{d\psi(x)}{dx} + \frac{a^2}{2!}\frac{d^2\psi(x)}{dx^2} + \cdots \\ &= \sum_{n=0}^{\infty}\frac{a^n}{n!}\frac{d^n}{dx^n}\psi(x) \\ &= \exp\left[ia\left(-i\frac{d}{dx}\right)\right]\psi(x).\end{aligned}$$

Thus, the operator for 1D translations in x is

$$U(a) = \exp\left(\frac{i}{\hbar}aP_x\right) \qquad P_x \equiv \frac{\hbar}{i}\frac{d}{dx},$$

where P_x is the momentum operator. Since all translations can be built from a sequence of infinitesimal ones, the operator $U(a)$ generates one-dimensional translations. However, for continuous groups such as this where operators can be parameterized in the form $U = e^{iaA}$, with U and A matrices, it is common to refer to A as the generator rather than U.

The advantage of viewing P_x rather than $U(a) = \exp[(i/\hbar)aP_x]$ as the generator of translations is that translational groups are Lie groups, with local properties determined by behavior of the group elements near the origin of parameter space (see Ch. 3). So for small a

$$\psi(x+a) \sim \psi(x) + \frac{i}{\hbar}aP_x\psi(x) \qquad U(a) \sim 1 + \frac{i}{\hbar}aP_x.$$

Then attention can be focused on the single operator P_x, rather than on the infinity of operators $U(a)$ corresponding to different values of the continuous parameter a.

2.11 Conjugate Classes

Elements of a group may be partitioned into two useful categories: *conjugate classes* and *cosets*. We examine conjugate classes here and cosets in Section 2.14. If $a, b, g \in G$ and

$$b = gag^{-1} \tag{2.21}$$

then element b is conjugate to element a.[7] Denoting conjugation by $\sim$, it is easy to show that (Problem 2.8)

[7] Conjugation is a special example of an *automorphism* (a mapping of a group to itself; see Section 2.5).

Box 2.5 **Equivalence Classes**

An abstract but very useful concept for a set is that of an *equivalence class*, which is predicated on the idea of an *equivalence relation* defined between set members.

For example, if for the set of all balls of a solid color we define an equivalence relation "having the same color," then the subset of all yellow balls would be one equivalence class of the set of all balls of a solid color, and the subset of all black balls would be another.

When a set S has an equivalence relation defined on it the set may be partitioned naturally into disjoint (non-overlapping) equivalence classes, with elements a and b belonging to the same equivalence class if and only if they are equivalent. Formally, equivalence is a binary relation $\sim$ between members of a set S exhibiting the following properties.

1. *Reflexivity*: For each $a \in \mathrm{S}$, we have $a \sim a$.
2. *Symmetry*: For each $a, b \in \mathrm{S}$, if $a \sim b$, then $b \sim a$.
3. *Transitivity*: For each $a, b, c \in \mathrm{S}$, if $a \sim b$ and $b \sim c$, then $a \sim c$.

Since groups are sets with added structure, equivalence classes are defined in a similar way for groups. An important equivalence relation for groups is the conjugation operation (2.21), which partitions groups uniquely into conjugacy classes.

- $a \sim a$;
- if $a \sim b$, then $b \sim a$;
- if $a \sim b$ and $b \sim c$, then $a \sim c$.

Group elements that are conjugate to each other form a *conjugacy class* (often just termed a *class*). A conjugacy class is a particular example of an *equivalence class,* with the equivalence operation being conjugation as in Eq. (2.21); see Box 2.5. For matrix groups this means that all elements in the same class are related by similarity transformations (2.12).

Example 2.13 From Table 2.2 for the group S_3, utilizing that the transpositions are their own inverses,

$$(23)(12)(23)^{-1} = (13) \qquad (12)(23)(12)^{-1} = (13) \qquad (12)(123)(12)^{-1} = (321).$$

Furthermore $aea^{-1} = aa^{-1} = e$ for any group element a. Therefore, the transpositions (12), (13), and (23) are in one class, the cyclic and anticyclic permutations (123) and (321) are in another class, and the identity e is in its own class.

Generally each group element belongs to one and only one class, and the identity is necessarily in a class by itself. A class is specified uniquely by giving a single element, since all others may be obtained from it by similarity transforms. As we found in

Example 2.13, in physical applications the elements in a class often share some obvious characteristic. For example, rotations about different axes but of the same magnitude are in the same class of the rotation group (Ch. 6). For a matrix representation the *character* χ of an element is the trace (sum of diagonal elements) of the corresponding matrix

$$\chi(D) = \operatorname{Tr} D = \sum_i D_{ii}. \tag{2.22}$$

An essential property of matrices is that the trace is invariant under a similarity transformation.[8] Therefore, *elements in the same class have the same characters* $\chi(D)$.

Example 2.14 From the matrix representation (2.10) for S_3 we deduce the following.

1. The identity (e) forms a class by itself, with character $\chi(e) = 3$.
2. The three transpositions form a second class, with $\chi(12) = \chi(13) = \chi(23) = 1$.
3. The cyclic and anticyclic permutations form a third class, with $\chi(321) = \chi(123) = 0$.

This is consistent with the result found in Example 2.13.

The preceding example generalizes to permutation groups of arbitrary order, where it is found that permutations exhibiting equivalent cycle structure are in the same class.

2.12 Invariant Subgroups

Suppose that $g \in G$, and that H is a subgroup of G with elements h. Then the group H' with the elements

$$H' = \{ghg^{-1}\,;\; h \in H,\; g \in G\} \tag{2.23}$$

is also a subgroup of G that is termed the *conjugate subgroup* to H. The subgroups H and H' have the same number of elements and either coincide exactly, or have only the identity in common. If H and H' have the same elements,

$$gHg^{-1} = H \qquad g \in G, \tag{2.24}$$

then H is an *invariant subgroup* of G (also called a *normal subgroup*). It follows that a subgroup composed of whole classes is an invariant subgroup. Every group has two trivial invariant subgroups, the identity e and the full group G, but our primary interest here will lie in non-trivial invariant subgroups.

Before proceeding, be certain that you understand the meaning of the invariant subgroup condition (2.24). Multiplying it from the right by g gives $gH = Hg$. This is *not* a statement that each element of H commutes with each element of G, which would mean that $gh_1 =$

[8] The only invariants of a matrix are its eigenvalues, but it is useful to define combinations of the eigenvalues that also are invariant. Two important ones are the *character* (sum of the eigenvalues) and the *determinant* (product of the eigenvalues).

$h_1 g$, $gh_2 = h_2 g$, and so on. Rather, it is a statement that for $h_1 \in H$ and $g \in G$, there is always an element $h_2 \in H$ such that $gh_1g^{-1} = h_2$, with h_1 and h_2 not necessarily the same elements of H. That is, if H is an invariant subgroup of G then, *taken as an entity,* H commutes with G. If in fact each element of H commutes with each element of G, then for that special case we term H an *abelian invariant subgroup.*

Example 2.15 Consider the cyclic group of order four, $C_4 = \{e, a, a^2, a^3\}$, where $e = a^4$ defines the cyclic condition,[9] with a multiplication table

C_4	e	a	a^2	a^3
e	e	a	a^2	a^3
a	a	a^2	a^3	e
a^2	a^2	a^3	e	a
a^3	a^3	e	a	a^2

As shown in Problem 2.5, the subgroup $H = \{e, a^2\}$ is identical to all of its conjugate subgroups $H' = \{ghg^{-1};\ g \in G,\ h \in H\}$, and is an (abelian) invariant subgroup of C_4.

The result of Example 2.15 could have been obtained immediately by noting that each element of an abelian group is in a class by itself since $gHg^{-1} = gg^{-1}H = H$.

Every subgroup is invariant for an abelian group because each subgroup of an abelian group is composed of whole classes.

More generally, all cyclic groups C_n are abelian so all subgroups of C_n are (abelian) invariant subgroups.

2.13 Simple and Semisimple Groups

Invariant subgroups and abelian invariant subgroups provide the basis for an important classification of groups. As shown below in Section 2.14.2, if a group contains a proper invariant subgroup it can be factored, suggesting that products of groups without invariant subgroups can be used to build larger groups. This leads to two useful definitions.[10]

1. A *simple group* contains no invariant subgroup other than the identity.
2. A *semisimple group* contains no *abelian* invariant subgroup other than the identity.

[9] We encountered the cyclic group C_2 in Box 2.2. Generally, the cyclic group of order n has the elements $C_n = \{e, a, a^2, a^3, \dots, a^{n-1}\}$, with n any positive integer and a closure condition $e = a^n$. Geometrically, C_n is isomorphic to the group of rotations preserving a regular polygon with n sides. All cyclic groups are abelian.

[10] For Lie groups we will not use these definitions directly but will instead introduce simple and semisimple Lie groups in terms of properties for the Lie algebra of their generators; see Section 3.2.

Thus simple groups are a subset of semisimple groups and a simple group is also semisimple, but a semisimple group need not be simple. *Example:* S_3 is neither simple nor semisimple because it contains the abelian invariant subgroup $\{e, (123), (321)\}$ (Problem 2.6).

2.14 Cosets and Factor Groups

If $H = \{h_1, h_2, \dots\}$ is a subgroup of G and g is an element of G, a *left coset* gH and a *right coset* Hg are defined by[11]

$$gH \equiv \{gh_1, gh_2, \dots\} \qquad Hg \equiv \{h_1 g, h_2 g, \dots\}. \tag{2.25}$$

Cosets contain the same number of elements as H. They are *not subgroups* of G because they do not contain the identity e (except when $g = e$); *cosets are sets,* not groups.

2.14.1 Left and Right Coset Decompositions

For a finite group the algorithm for constructing a left-coset decomposition of $G \supset H$ is as follows.

1. Choose an element $g_1 \in G$ and construct the coset $g_1 H$.
2. If the coset $g_1 H$ does not exhaust the elements of G, choose another element $g_2 \in G$ and form the coset $g_2 H$.
3. Continue in this manner until the sum of the cosets exhausts the content of G,[12]

$$g_1 H + g_2 H + \cdots + g_\ell H = G. \tag{2.26}$$

This defines a left coset decomposition of G with respect to the subgroup H.

A right coset decomposition is formed in similar manner. In the decomposition (2.26), we can see the following.

1. Different coset terms have no elements in common, so cosets provide a *unique decomposition of G.*
2. The elements g_i are called the *coset representatives.*
3. The number ℓ of distinct cosets is called the *index of H in G.*

Left and right coset decompositions may have different elements but they have the same index. If n is the order of a finite group G and m is the order of a subgroup H, then $n = m\ell$

[11] Some reverse this definition and call gH a right coset and Hg a left coset. Nothing of consequence depends on this choice, if used consistently. If H is an *invariant subgroup,* left and right cosets are equivalent.
[12] The sum in Eq. (2.26) is over sets; it is equal to the set of objects contained in at least one of the summed sets. Equality in Eq. (2.26) is equality of sets: equal sets have the same members, up to permutation of list order.

for the coset decomposition of G with respect to H, since each coset has the same number of elements as H (because of the *rearrangement lemma* described in Section 6.2.5).[13]

Example 2.16 Let us use the group multiplication Table 2.2 to construct left and right coset decompositions of $G = S_3 = \{e, (12), (23), (13), (123), (321)\}$, with respect to its subgroup $H = \{e, (12)\}$. For the left cosets, choosing $g_1 = e$,

$$g_1 H = eH = \{e, (12)\},$$

choosing $g_2 = (23)$ gives

$$g_2 H = (23)H = \{(23)e, (23)(12)\} = \{(23), (123)\},$$

and choosing $g_3 = (13)$ gives

$$g_3 H = (13)H = \{(13)e, (13)(12)\} = \{(13), (321)\}.$$

Summing these sets gives the left coset decomposition of G,

$$eH + g_2 H + g_3 H = \{e, (12), (23), (123), (13), (321)\} = G.$$

Likewise, for the right cosets,

$$\begin{aligned} Hg_1 &= He = \{e, (12)\}, \\ Hg_2 &= H(23) = \{e(23), (12)(23)\} = \{(23), (321)\}, \\ Hg_3 &= H(13) = \{e(13), (12)(13)\} = \{(13), (123)\}, \end{aligned}$$

and summing these sets gives the right coset decomposition of G,

$$He + Hg_2 + Hg_3 = \{e, (12), (23), (321), (13), (123)\} = G.$$

The left or right cosets give a unique decomposition of G and the index of H in G for both cases is $\ell = n/m = 6/2 = 3$, where $n = 6$ is the order of G and $m = 2$ is the order of H.

Example 2.16 exhibits the general properties that are expected for coset decompositions.

1. The number of elements in a coset equals the number of elements in the subgroup H.
2. The cosets are sets, not groups, because they do not contain the group identity (except for eH and He).
3. The cosets in either the left or right coset decomposition have no elements in common.

If H is an invariant subgroup the left and right cosets are equivalent, since $gHg^{-1} = H$ implies that $gH = Hg$. Whence, for invariant subgroups one may speak simply of cosets. Problems 2.17 and 2.18 illustrate the use of coset decompositions to prove some important general properties of finite groups.

[13] Thus the order of any subgroup H of a finite group G is a divisor of the order of G. This is the content of a famous result known as *Lagrange's theorem*.

2.14.2 Factor Groups

The partitioning of a group G into cosets is unique, so it is natural to consider a factorization of G based on this partitioning. When H is an invariant subgroup of G the cosets form a group called the *quotient group* or *factor group*, denoted formally by G/H. The elements of G/H are the cosets of H and composition (group multiplication) within the quotient group is defined in terms of products of cosets. That is, the *law of coset multiplication* is

$$pHqH = (pq)H, \tag{2.27}$$

for the product of cosets pH and qH. *Factor groups are not subgroups*: elements of a subgroup are elements of the group whereas elements of a factor group are *cosets* of the group. If a group G has an invariant subgroup H there is a natural homomorphism from G to the factor group G/H that preserves group multiplication: $g \in G \rightarrow gH \in G/H$. This rather abstract discussion is made more tangible by applying it to our old standby, the group S_3.

Example 2.17 In Problem 2.6 you are asked to show that $H = \{e, (123), (321)\}$ is an invariant subgroup of S_3. Using the multiplication Table 2.2 to form the left cosets,

$$e\{e, (123), (321)\} = \{e, (123), (321)\} \qquad (23)\{e, (123), (321)\} = \{(23), (12), (13)\},$$

which accounts for all elements of S_3. Therefore, there are two independent cosets and $G/H = S_3/H = H + M$, where

$$H = \{e, (123), (321)\} = eH \qquad M = \{(23), (12), (13)\} = (23)H.$$

The coset multiplication law $pHqH = (pq)H$ of Eq. (2.27) gives the products

$$H^2 = eHeH = (ee)H = H \qquad MH = (23)HH = (23)H = M,$$
$$HM = eH(23)H = (23)H = M \qquad M^2 = (23)H(23)H = (23)(23)H = H.$$

This corresponds to the multiplication table in Fig. 2.3(a), which is isomorphic to that for C_2 or Z_2 (see Box 2.2). The mapping from S_3 to the factor group is shown in Fig. 2.3(b).

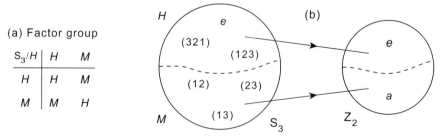

(a) Factor group

S_3/H	H	M
H	H	M
M	M	H

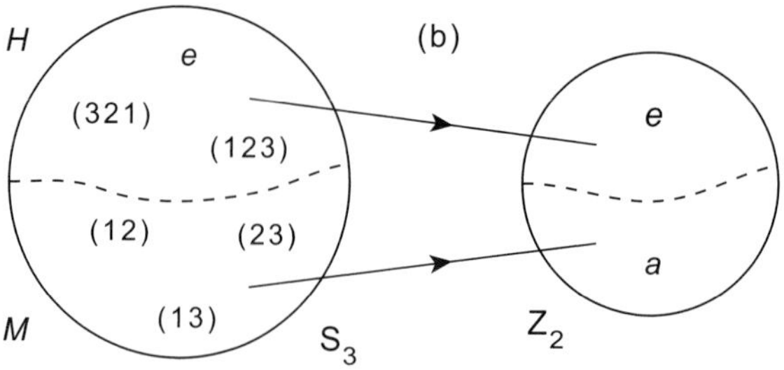

Fig. 2.3 (a) Multiplication table for the factor group $S_3/H = C_2$ constructed in Example 2.17. (b) Homomorphism of S_3 to the factor group $Z_2 \sim C_2$ [199].

Table 2.3. Multiplication table for C_6

C_6	e	a	a^2	a^3	a^4	a^5
e	e	a	a^2	a^3	a^4	a^5
a	a	a^2	a^3	a^4	a^5	e
a^2	a^2	a^3	a^4	a^5	e	a
a^3	a^3	a^4	a^5	e	a	a^2
a^4	a^4	a^5	e	a	a^2	a^3
a^5	a^5	e	a	a^2	a^3	a^4

For a homomorphism between groups A and B defined by $f : A \to B$, the *kernel of the homomorphism* [denoted by $\ker f$ or $f^{-1}(e_B)$, where e_B is the identity of B] is the subset of elements in A mapped into the identity element of B. Generally the kernel is an invariant subgroup of A and the homomorphism maps the kernel of A into the identity of B, and entire cosets of A into single elements of B. This is seen to be the case in Fig. 2.3(b). The kernel $\{e, (123), (321)\}$ constitutes an invariant subgroup of S_3 (see Problem 2.6) and is mapped to the identity e of Z_2, while the entire coset $\{(23), (12), (13)\}$ of S_3 found in Example 2.17 is mapped into the single element a of Z_2.

2.15 Direct Product Groups

Suppose a group G to have subgroups H_1 and H_2 that have the following properties.

1. Each element of H_1 commutes with each element of H_2.
2. Each element g of G can be written uniquely as $g = h_1h_2$, where $h_1 \in H_1$ and $h_2 \in H_2$.

Then G is the *direct product* of H_1 and H_2, written symbolically as $G = H_1 \times H_2$. An illustration of such a direct product group is given in Example 2.18.

Example 2.18 Let us show that the cyclic group C_6 can be written as the direct product of two subgroups. The multiplication table for C_6 is given in Table 2.3,[14] with the proviso that closure requires the cyclic condition $a^6 = e$. Clearly $H_1 \equiv \{e, a^3\}$ and $H_2 \equiv \{e, a^2, a^4\}$ are subgroups, each element $h_1 \in H_1$ commutes with each element $h_2 \in H_2$ (cyclic groups are abelian), and each element of C_6 can be written uniquely as a product h_1h_2:

$$e = ee \qquad a = a^3a^4 \qquad a^2 = ea^2 \qquad a^3 = a^3e \qquad a^4 = ea^4 \qquad a^5 = a^3a^2.$$

From the multiplication table, $H_1 = C_2$ and $H_2 = C_3$, so we have shown that C_6 can be written as a direct product $C_6 = C_2 \times C_3$ of two smaller cyclic groups.

[14] Notice that for cyclic groups each row or column of the multiplication table is a cyclic permutation of the preceding one, and that for abelian groups like C_6 the table is reflection-symmetric about the main diagonal.

Box 2.6 Physical Meaning of a Direct Product Group

A physical meaning can be attached to the direct product structure of the C_6 group worked out in Example 2.18 if the cyclic groups are interpreted as groups of rotations by angles that are integer divisors of the circle.

Realization of C_6 by Finite Rotations

The group C_6 may be realized physically in terms of a set of rotations $R(\theta)$ by integer multiples of 60°:

$$C_6 : \{R(0^\circ), R(60^\circ), R(120^\circ), R(180^\circ), R(240^\circ), R(300^\circ)\},$$

where $e = R(0^\circ), a = R(60^\circ), a^2 = R(120^\circ)$, and so on. These clearly have the same multiplication table as C_6 because $a^6 = R(360^\circ) = e$ imposes the cyclic condition. Likewise, C_3 and C_2 may be expressed in terms of finite rotations by multiples of 120° and 180°, respectively:

$$C_3 : \{R(0^\circ), R(120^\circ), R(240^\circ)\} \qquad C_2 : \{R(0^\circ), R(180^\circ)\}.$$

Now each element of C_3 commutes with each element of C_2, and each rotation in C_6 may be expressed uniquely as a product of a C_2 rotation and a C_3 rotation. For example, the product of a C_2 rotation by 180° and a C_3 rotation by 240° is

$$R(240^\circ)R(180^\circ) = R(420^\circ) = R(60^\circ)$$

which is the second element of C_6. The following table summarizes combinations of C_2 and C_3 rotations (in either order) that reproduce uniquely all C_6 rotations.

Element of C_2	Element of C_3	Element of C_6
$R(0^\circ)$	$R(0^\circ)$	$e = R(0^\circ)$
$R(180^\circ)$	$R(240^\circ)$	$a = R(60^\circ)$
$R(0^\circ)$	$R(120^\circ)$	$a^2 = R(120^\circ)$
$R(180^\circ)$	$R(0^\circ)$	$a^3 = R(180^\circ)$
$R(0^\circ)$	$R(240^\circ)$	$a^4 = R(240^\circ)$
$R(180^\circ)$	$R(120^\circ)$	$a^5 = R(300^\circ)$

Hence the physical meaning of $C_2 \times C_3 = C_3 \times C_2 = C_6$ is that all C_6 rotations factor into *two independent rotations,* one an element of the subgroup C_2 and one an element of the subgroup C_3.

Direct Products and Independent Subgroups

A similar interpretation as for the C_6 example given above applies generally to direct product groups. In physical terms a direct product structure implies that the full group separates into independent pieces associated with symmetries of subgroups, and that these subgroups can be analyzed separately.

Often factorization of a group into direct products of smaller groups has a clear physical interpretation, as examined further in Box 2.6. Finally we note that in a direct product $G = A \times B$ the groups A and B are invariant subgroups of G, as proved in Problem 2.32.

2.16 Direct Product of Representations

We can also define the direct product of two representations for the same group. For a matrix group G, if A is an $m \times m$ matrix representation with elements A_{ij} and B is an $n \times n$ matrix representation with elements B_{kl}, their direct product (or *Kronecker product*) $A \otimes B$ is an $mn \times mn$ matrix representation C of G,

$$A \otimes B = C \qquad C_{ik,jl} \equiv A_{ij} B_{kl}. \tag{2.28}$$

Example 2.19 illustrates for 2×2 matrices.

Example 2.19 From Eq. (2.28) we have for the direct product of 2×2 matrix representations A and B,

$$A \otimes B = \begin{pmatrix} A_{11}B & A_{12}B \\ A_{21}B & A_{22}B \end{pmatrix} = \begin{pmatrix} A_{11}B_{11} & A_{11}B_{12} & A_{12}B_{11} & A_{12}B_{12} \\ A_{11}B_{21} & A_{11}B_{22} & A_{12}B_{21} & A_{12}B_{22} \\ A_{21}B_{11} & A_{21}B_{12} & A_{22}B_{11} & A_{22}B_{12} \\ A_{21}B_{21} & A_{21}B_{22} & A_{22}B_{21} & A_{22}B_{22} \end{pmatrix}.$$

Thus, each element of A is multiplied by each element of B, giving a 4×4 matrix.

The space on which $A \otimes B$ operates is called the *direct product space.* Such spaces occur in quantum mechanics when a system has two independent degrees of freedom that are described by the same symmetry group. An example is afforded by the coupling of two angular momenta to a good total angular momentum in a *Clebsch–Gordan series*, which will be introduced in Section 3.3.4. Representations of direct product groups exhibit an important property: *the irreducible representations of a direct product group are equivalent to the direct product of the irreducible representations of the two groups in the product.* This means that a direct product group can be analyzed in terms of the independent group factors and their irreps, and the results combined at the end of the analysis.

2.17 Characters of Representations

For finite groups, various orthogonality and completeness conditions are obeyed by the irreps that are of central importance in their analysis. For example [199],

$$\frac{n_\mu}{n_G} \sum_g D^\dagger_{(\mu)}(g)^k_i D^{(\nu)}(g)^j_l = \delta^\nu_\mu \delta^j_i \delta^k_l, \tag{2.29a}$$

$$\sum_\mu n_\mu^2 = n_G, \tag{2.29b}$$

$$\sum_{\mu,l,k} \frac{n_\mu}{n_G} D^{(\mu)}(g)^l_k D^\dagger_\mu(g')^k_l = \delta_{gg'}, \tag{2.29c}$$

where $D^{(\mu)}$ is a representation matrix for the irreducible representation μ of dimension n_μ, $n_{\rm G}$ is the order of the group, g stands for the group elements, and the hermitian conjugate $D^\dagger$ of a matrix D is obtained by transposing the rows and columns and replacing each element of the resulting matrix by its complex conjugate. However, the representation matrices are basis dependent and it is useful to formulate these important results in terms of the group characters (traces of the matrices), which are invariant under similarity transformation between different bases.

2.17.1 Character Theorems

Equations (2.29) imply that traces χ of the representation matrices satisfy the orthonormality and completeness relations [199]

$$\sum_i \frac{n_i}{n_{\rm G}} \chi^*_\mu(i)\chi^\nu(i) = \delta^\nu_\mu \qquad \frac{n_i}{n_{\rm G}} \sum_\mu \chi^\mu(i)\chi^{*j}_\mu(i) = \delta^j_i, \tag{2.30}$$

where the index i ranges over classes of the representation, n_i is the number of class members, and the greek indices label irreducible representations. This has a corollary for finite groups that *the number of irreps is equal to the number of distinct classes.* But not all representations are irreducible. Two trace theorems allow us to decide whether a representation is reducible and to find the number of times each irrep occurs if it is reducible. First, a representation of a finite group is irreducible only if

$$\sum_i n_i \, |\chi_i|^2 = n_{\rm G}, \tag{2.31}$$

where the sum is over classes, n_i is the number of members in class i, and $n_{\rm G}$ is the order of the group. Conversely, if the representation is reducible the sum in Eq. (2.31) will be greater than $n_{\rm G}$. Second, if the representation is reducible, it will be a direct sum of irreducible representations for the group. Since the number of irreps equals the number of classes, it is finite for a finite group and a reducible representation $U(g)$ can be written as a finite direct sum of these irreps. The number of times a_ν that the irrep ν occurs in this reduction is

$$a_\nu = \sum_i \frac{n_i}{n_{\rm G}} \chi^*_\nu(i)\chi(i), \tag{2.32}$$

where i labels classes, and $\chi_\nu(i)$ and $\chi(i)$ are characters for class i in the irrep ν and the reducible representation $U(g)$, respectively. The following examples illustrate [180].

Example 2.20 Consider the six representations $\Gamma^{(i)}$ of S_3 displayed in Fig. 2.4. Which of these correspond to irreps? The order of S_3 is $3! = 6$, so from Eq. (2.31) the representations that are irreps must satisfy $\sum_j \left|{\rm Tr}\, A_j\right|^2 = 6$ for the matrices A_j in the representation. By inspection $\Gamma^{(1)}$, $\Gamma^{(2)}$, and $\Gamma^{(3)}$ fulfill this condition, so they are S_3 irreducible representations. We could examine the remaining representations in the same way but there is no need! From Eq. (2.29b) the sum of the squares of the irrep dimensions for a finite group must be equal to the order of the group, so the 1D irreps $\Gamma^{(1)}$ and $\Gamma^{(2)}$, and the 2D irrep

	$\Gamma^{(1)}$	$\Gamma^{(2)}$	$\Gamma^{(3)}$	$\Gamma^{(4)}$	$\Gamma^{(5)}$	$\Gamma^{(6)}$
(1)	1	1	$\begin{pmatrix} 1 & 0 \\ 0 & 1 \end{pmatrix}$	$\begin{pmatrix} 1 & 0 & 0 \\ 0 & 1 & 0 \\ 0 & 0 & 1 \end{pmatrix}$	$\begin{pmatrix} 1 & 0 & 0 \\ 0 & 1 & 0 \\ 0 & 0 & 1 \end{pmatrix}$	$\begin{pmatrix} 1 & 0 & 0 & 0 & 0 & 0 \\ 0 & 1 & 0 & 0 & 0 & 0 \\ 0 & 0 & 1 & 0 & 0 & 0 \\ 0 & 0 & 0 & 1 & 0 & 0 \\ 0 & 0 & 0 & 0 & 1 & 0 \\ 0 & 0 & 0 & 0 & 0 & 1 \end{pmatrix}$
(123)	1	1	$\begin{pmatrix} -1/2 & \sqrt{3}/2 \\ -\sqrt{3}/2 & -1/2 \end{pmatrix}$	$\begin{pmatrix} 1 & 0 & 0 \\ 0 & -1/2 & \sqrt{3}/2 \\ 0 & -\sqrt{3}/2 & -1/2 \end{pmatrix}$	$\begin{pmatrix} 0 & 0 & 1 \\ 1 & 0 & 0 \\ 0 & 1 & 0 \end{pmatrix}$	$\begin{pmatrix} 0 & 0 & 1 & 0 & 0 & 0 \\ 1 & 0 & 0 & 0 & 0 & 0 \\ 0 & 1 & 0 & 0 & 0 & 0 \\ 0 & 0 & 0 & 0 & 1 & 0 \\ 0 & 0 & 0 & 0 & 0 & 1 \\ 0 & 0 & 0 & 1 & 0 & 0 \end{pmatrix}$
(321)	1	1	$\begin{pmatrix} -1/2 & -\sqrt{3}/2 \\ \sqrt{3}/2 & -1/2 \end{pmatrix}$	$\begin{pmatrix} 1 & 0 & 0 \\ 0 & -1/2 & -\sqrt{3}/2 \\ 0 & \sqrt{3}/2 & -1/2 \end{pmatrix}$	$\begin{pmatrix} 0 & 1 & 0 \\ 0 & 0 & 1 \\ 1 & 0 & 0 \end{pmatrix}$	$\begin{pmatrix} 0 & 1 & 0 & 0 & 0 & 0 \\ 0 & 0 & 1 & 0 & 0 & 0 \\ 1 & 0 & 0 & 0 & 0 & 0 \\ 0 & 0 & 0 & 0 & 0 & 1 \\ 0 & 0 & 0 & 1 & 0 & 0 \\ 0 & 0 & 0 & 0 & 1 & 0 \end{pmatrix}$
(23)	1	−1	$\begin{pmatrix} -1 & 0 \\ 0 & 1 \end{pmatrix}$	$\begin{pmatrix} 1 & 0 & 0 \\ 0 & -1 & 0 \\ 0 & 0 & 1 \end{pmatrix}$	$\begin{pmatrix} 1 & 0 & 0 \\ 0 & 0 & 1 \\ 0 & 1 & 0 \end{pmatrix}$	$\begin{pmatrix} 0 & 0 & 0 & 1 & 0 & 0 \\ 0 & 0 & 0 & 0 & 1 & 0 \\ 0 & 0 & 0 & 0 & 0 & 1 \\ 1 & 0 & 0 & 0 & 0 & 0 \\ 0 & 1 & 0 & 0 & 0 & 0 \\ 0 & 0 & 1 & 0 & 0 & 0 \end{pmatrix}$
(13)	1	−1	$\begin{pmatrix} 1/2 & -\sqrt{3}/2 \\ -\sqrt{3}/2 & -1/2 \end{pmatrix}$	$\begin{pmatrix} 1 & 0 & 0 \\ 0 & 1/2 & -\sqrt{3}/2 \\ 0 & -\sqrt{3}/2 & -1/2 \end{pmatrix}$	$\begin{pmatrix} 0 & 0 & 1 \\ 0 & 1 & 0 \\ 1 & 0 & 0 \end{pmatrix}$	$\begin{pmatrix} 0 & 0 & 0 & 0 & 1 & 0 \\ 0 & 0 & 0 & 0 & 0 & 1 \\ 0 & 0 & 0 & 1 & 0 & 0 \\ 0 & 0 & 1 & 0 & 0 & 0 \\ 1 & 0 & 0 & 0 & 0 & 0 \\ 0 & 1 & 0 & 0 & 0 & 0 \end{pmatrix}$
(12)	1	−1	$\begin{pmatrix} 1/2 & \sqrt{3}/2 \\ \sqrt{3}/2 & -1/2 \end{pmatrix}$	$\begin{pmatrix} 1 & 0 & 0 \\ 0 & 1/2 & \sqrt{3}/2 \\ 0 & \sqrt{3}/2 & -1/2 \end{pmatrix}$	$\begin{pmatrix} 0 & 1 & 0 \\ 1 & 0 & 0 \\ 0 & 0 & 1 \end{pmatrix}$	$\begin{pmatrix} 0 & 0 & 0 & 0 & 0 & 1 \\ 0 & 0 & 0 & 1 & 0 & 0 \\ 0 & 0 & 0 & 0 & 1 & 0 \\ 0 & 1 & 0 & 0 & 0 & 0 \\ 0 & 0 & 1 & 0 & 0 & 0 \\ 1 & 0 & 0 & 0 & 0 & 0 \end{pmatrix}$

Fig. 2.4 Some matrix representations of the group S_3 [180]. The left column labels group elements and other columns give matrices for six representations $\Gamma^{(i)}$. The identity is labeled (1).

$\Gamma^{(3)}$, exhaust the independent possibilities for the irreps of S_3.[15] Therefore, the remaining representations in Fig. 2.4 must be direct sums of $\Gamma^{(1)}$, $\Gamma^{(2)}$, and $\Gamma^{(3)}$.

Example 2.21 Let us use the representation $\Gamma^{(6)}$ in Fig. 2.4 to illustrate finding the irrep content of one of the reducible representations. This example is simple because the only

[15] This is consistent with the previous observations that S_3 has three classes and that the number of classes for a finite group is equal to the number of irreducible representations.

Table 2.4. Characters for the group S_3

Irrep\Class	$e(1)$	$P_{ijk}(2)$	$P_{ij}(3)$
$\Gamma^{(1)}$	1	1	1
$\Gamma^{(2)}$	1	1	−1
$\Gamma^{(3)}$	2	−1	0

matrix in the representation $\Gamma^{(6)}$ with a finite trace is the identity $\Gamma^{(6)}((1))$ in the first row of Fig. 2.4. From Eq. (2.32), the number of times a_1 that the irrep $\Gamma^{(1)}$ appears is given by

$$
\begin{aligned}
a_1 &= \frac{1}{6}\Big[\underbrace{\mathrm{Tr}\,1\cdot\mathrm{Tr}\,\Gamma^{(6)}((1))}_{\text{class }\{(1)\}} + 2\,\underbrace{\mathrm{Tr}\,1\cdot\mathrm{Tr}\,\Gamma^{(6)}((123))}_{\text{class }\{(123),(321)\}} + 3\,\underbrace{\mathrm{Tr}\,1\cdot\mathrm{Tr}\,\Gamma^{(6)}((23))}_{\text{class }\{(23),(13),(12)\}}\Big]\\
&= \frac{1}{6}[(1)(6) + 2(1)(0) + 3(1)(0)]\\
&= \frac{1}{6}(1)(6) = 1,
\end{aligned}
$$

while for the irrep $\Gamma^{(2)}$ we have $a_2 = \frac{1}{6}(1)(6) = 1$, and finally for $\Gamma^{(3)}$

$$
a_3 = \frac{1}{6}\,\mathrm{Tr}\begin{pmatrix}1 & 0\\ 0 & 1\end{pmatrix}\,\mathrm{Tr}\begin{pmatrix}1&0&0&0&0&0\\0&1&0&0&0&0\\0&0&1&0&0&0\\0&0&0&1&0&0\\0&0&0&0&1&0\\0&0&0&0&0&1\end{pmatrix} = \frac{1}{6}(2)(6) = 2.
$$

Thus the representation $\Gamma^{(6)}$ reduces to the direct sum $\Gamma^{(6)} = \Gamma^{(1)} \oplus \Gamma^{(2)} \oplus 2\,\Gamma^{(3)}$. The meaning of this reduction is that the space spanned by any set of six functions that transform according to $\Gamma^{(6)}$ under any group isomorphic to S_3 may be reduced to four invariant and irreducible subspaces: two of dimension one and two of dimension two (see Fig. 2.2).

Example 2.22 The sums of the squares for the characters of both $\Gamma^{(4)}$ and $\Gamma^{(5)}$ in Fig. 2.4 are equal to 12, so they are reducible, but further inspection shows that their characters are equal so $\Gamma^{(4)}$ and $\Gamma^{(5)}$ are also *equivalent* (related by a similarity transform). The irrep content of these equivalent reducible representations can be determined by the method of Example 2.21, which gives $\Gamma^{(4)} = \Gamma^{(5)} = \Gamma^{(1)} \oplus \Gamma^{(3)}$ (see Problem 2.15).

2.17.2 Character Tables

Finite groups have a finite number of independent irreducible representations, characters may be used to determine most properties of interest, and characters are invariant under similarity transformations (change of basis). Thus it is convenient to tabulate once and for all the characters for each class and irrep in a *character table* for the group. In such tables the irreps typically are arrayed vertically and the classes horizontally, with the number of members for each class given in parentheses. Table 2.4 illustrates for the permutation group S_3. Note that a character table displays the dimensionality of the irreps because

$$\text{Character of identity } e = \text{Dimensionality of irrep } \Gamma, \tag{2.33}$$

since the identity is the unit matrix and the trace of an $n \times n$ unit matrix is n.

Example 2.23 As Table 2.4 indicates, the group S_3 has three irreducible representations $\Gamma^{(1)}$, $\Gamma^{(2)}$, and $\Gamma^{(3)}$, and three classes: the identity e with one member, the cyclic permutations P_{ijk} with two members, and the transpositions P_{ij} with three members. The table entries give the characters for each class in each irrep. For example, the character of the class of transpositions P_{ij} in the irrep $\Gamma^{(2)}$ is -1. From Eq. (2.33), the irreps $\Gamma^{(1)}$ and $\Gamma^{(2)}$ are one-dimensional and the irrep $\Gamma^{(3)}$ is two-dimensional.

Character tables for many finite groups may be found by searching online, and are included in many books; for example, Refs. [108, 196, 208].

Background and Further Reading

Introductions to the basic concepts of group theory for physics with varying tradeoffs between sophistication and clarity may be found in Elliott and Dawber [56], Georgi [68], Hamermesh [104], Heine [108], Ludwig and Falter [143], Ma [145], Ramond [169], Schensted [180], Stephenson [184], Tinkham [196], Tung [199] (with solutions for all problems given in Aivazis [8]), Wigner [211], Wybourne [224], and Zee [228].

Problems

2.1 Prove that for a group the inverse of each group element and the identity are unique. Prove that the inverse of the product of two group elements is given by $(a \cdot b)^{-1} = b^{-1}a^{-1}$. Check this against the product $a \cdot b = (12) \cdot (23)$ from Table 2.2. ***

2.2 Write out the multiplication table for all possible products of elements in the group S_3 (permutations on three objects). Use this to demonstrate explicitly that S_3 is a group, that it is non-abelian, and that it has two proper subgroups: the group S_2 of permutations on two objects, and a group (called the alternating group) $A_3 \equiv \{e, (123), (321)\}$. Show that for an operator c_3 that rotates a system by $\frac{2\pi}{3}$ about a given axis, the set $C_3 \equiv \{1, c_3, c_3^2\}$ also constitutes an abelian group and it is isomorphic to A_3. ***

2.3 If the operator c_4 rotates a system by $\frac{\pi}{2}$ about a specified axis, demonstrate that the operator set $C_4 = \{e, c_4, c_4^2, c_4^3\}$ constitutes an abelian group with the group multiplication defined by application of two successive rotations.

2.4 Show that the group $G = \{e, a, b\}$ with e the identity has one possible multiplication table; thus there is only one finite group of order three. *Hint*: See Box 2.2. ***

2.5 Demonstrate that for the cyclic group C_4 with multiplication table given in Example 2.15, the subgroup $H = \{e, a^2\}$ is an abelian invariant subgroup. ***

2.6 Prove that $\{e, (123), (321)\}$ is an invariant subgroup of S_3 but $\{e, (12)\}$ is not. ***

2.7 Show that the cyclic group C_4 is neither simple nor semisimple.

2.8 Prove that if a, b, and c are elements of a group and class conjugation is indicated by $\sim$, then (1) $a \sim a$, (2) if $a \sim b$, then $b \sim a$, and (3) if $a \sim b$ and $b \sim c$, then $a \sim c$. Thus conjugacy is an equivalence relation (see Box 2.5). ***

2.9 Show that the group $\{e, a, b, c\}$ with multiplication table (b) below, is in one to one correspondence with the geometrical symmetry operations on figure (a) below

(a)

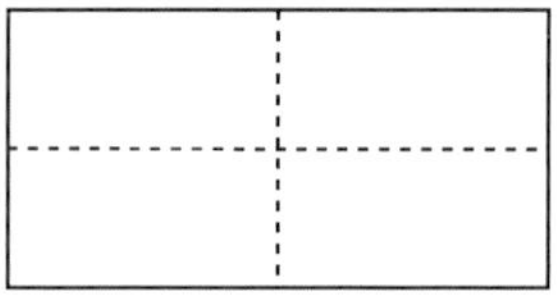

(b)

D_2	e	a	b	c
e	e	a	b	c
a	a	e	c	b
b	b	c	e	a
c	c	b	a	e

This is called the *4-group* or *dihedral group* D_2. Show that D_2 has three subgroups, $\{e, a\}$, $\{e, b\}$, and $\{e, c\}$, each isomorphic to the cyclic group C_2.

2.10 Demonstrate that the identity, reflections about the three symmetry axes (dashed lines), and rotations by $\frac{2\pi}{3}$ and $\frac{4\pi}{3}$ about the center of the equilateral triangle

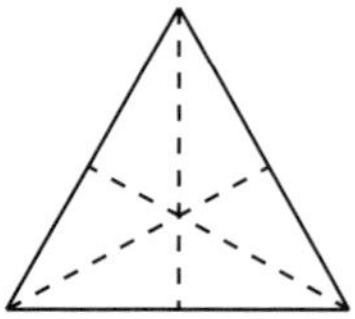

form a group of order six (the dihedral group, D_3) that is isomorphic to the permutation group S_3. Show that D_3 has four distinct subgroups. (This result is a special case of *Cayley's theorem*: every group of order n is isomorphic to a subgroup of S_n.)

2.11 Show that the quotient group of the *4-group* D_2 defined in Problem 2.9 is C_2. ***

2.12 Prove that the angular momentum operator L_z generates rotations around the z-axis.

2.13 Show that for real numbers α, β, and δ the matrices

$$G = \begin{pmatrix} 1 & \alpha & \delta \\ 0 & 1 & \beta \\ 0 & 0 & 1 \end{pmatrix}$$

form a group under matrix multiplication. Show that the matrices G with $\alpha = \beta = 0$ form an invariant subgroup of G.

2.14 Prove that the direct product of two representations is a representation, and that the character of the direct product is the product of characters for the representations.

2.15 Determine the irrep content for the equivalent reducible S_3 representations $\Gamma^{(4)}$ and $\Gamma^{(5)}$ of Fig. 2.4. ***

2.16 Demonstrate explicitly that for invariant subgroups the cosets form a group under the coset multiplication law (2.27). ***

2.17 Prove that for a finite group G with an invariant subgroup $G \supset H$ and $g_i \in G$, two cosets $g_i H$ and $g_j H$ have no elements in common if $i \neq j$. ***

2.18 Use cosets to show that a finite group with an order that is a prime number can have no proper subgroups. Show that a finite group with order equal to a prime number is isomorphic to a cyclic group. *Hint*: Show that the group has a cyclic subgroup. ***

2.19 Show that the set of matrices $\{a, b, c, d\}$ given by

$$a = \begin{pmatrix} 1 & 0 \\ 0 & 1 \end{pmatrix} \qquad b = \begin{pmatrix} 1 & 0 \\ 0 & -1 \end{pmatrix} \qquad c = \begin{pmatrix} -1 & 0 \\ 0 & 1 \end{pmatrix} \qquad d = \begin{pmatrix} -1 & 0 \\ 0 & -1 \end{pmatrix}$$

closes under multiplication and is a representation of the group D_2 in Problem 2.9.

2.20 Show that the set of functions $\{f_1(x) = x,\ f_2(x) = -x,\ f_3(x) = x^{-1},\ f_4(x) = -x^{-1}\}$ forms a group under the binary operation of substitution of one function into another. Show that the group is isomorphic to the matrix group of Problem 2.19. ***

2.21 The real numbers form a group under the binary operation of arithmetic addition. Show that for real numbers v the matrices

$$D(v) = \begin{pmatrix} 1 & 0 \\ v & 1 \end{pmatrix}$$

form a 2D representation of this additive group of real numbers. Show that transformation by this matrix corresponds to the Galilean transformations of classical physics, $x' = x + vt$ and $t' = t$, relating time and coordinate (t, x) for one observer to time and coordinate (t', x') for an observer with relative velocity v along the x-axis.

2.22 Show that for a matrix representation $D(x)$ satisfying Eq. (2.8), a new set of matrices formed by performing the same similarity transform $S^{-1}D(x)S$ for a fixed matrix S on all matrices $D(x)$ is also a representation.

2.23 Consider a cartesian (xyz) coordinate system and define the following operations: R = rotation by π in the x–y plane, E = do nothing, I = inversion of all three axes, and σ = reflection in the x–y plane. Show that these operations form a group under the product of transformations. Show that this group contains subgroups $S_2 = \{E, I\}$ (*inversion subgroup*) and $C_2 = \{E, R\}$ (*cyclic subgroup*), and that the full group can be written as a direct product of these subgroups.

2.24 Verify that the mapping $e \to 1$ and $a \to -1$ gives a representation of the cyclic group C_2 described in Box 2.2 that preserves the group multiplication, as does the trivial mapping $e \to 1$ and $a \to 1$. Show that these two representations are irreducible and that they are in fact the only irreps for C_2, up to possible isomorphisms.

2.25 Show that the matrices

$$t_1 = \begin{pmatrix} 1 & 0 \\ 0 & 1 \end{pmatrix} \qquad t_2 = \begin{pmatrix} 0 & 1 \\ 1 & 0 \end{pmatrix}$$

constitute a representation of the group C_2 described in Box 2.2. Diagonalize this set of matrices and show that the 2D representation (t_1, t_2) is reducible to a direct sum of C_2 irreps. *Hint*: The irreps for C_2 are given in Problem 2.24.

2.26 Prove the trigonometric identities

$$\cos(\phi + \theta) = \cos\phi\cos\theta - \sin\phi\sin\theta$$
$$\sin(\phi + \theta) = \cos\phi\sin\theta + \sin\phi\cos\theta$$

by the following group-theoretical means.

1. Show that the complex numbers of unit modulus $c = x + iy$ with $|x|^2 + |y|^2 = 1$ form a group under multiplication, and that $e^{i\phi}$ with $-\pi \le \phi \le \pi$ is a faithful representation: $\cos\phi + i\sin\phi = e^{i\phi}$.
2. Use the representation $e^{i\phi}$ and group multiplication to prove the identities.

This is a simple example of a more general idea: the special functions of mathematical physics often are representations of some group, and standard identities can be obtained by appropriate operations on the representations of that group. ***

2.27 (a) Divide the integers up into four equivalence classes (see Box 2.5),

$$e \equiv \{0, 4, -4, 8, -8, \ldots\} \qquad a \equiv \{1, 5, -3, 9, -7, \ldots\}$$
$$b \equiv \{2, 6, -2, 10, -6, \ldots\} \qquad c \equiv \{3, 7, -1, 11, -5, \ldots\}.$$

Show that $\{e, a, b, c\}$ form a group (Z_4), under addition modulo 4 by constructing the multiplication table. *Hint*: Recall that in addition modulo N two integers are added normally and then an integer multiple of N is added or subtracted to bring the result into the range $-N$ to $+N$. For example, $3 + 2 \bmod 4 = 1$.

(b) Show that the matrices

$$e = \begin{pmatrix} 1 & 0 \\ 0 & 1 \end{pmatrix} \qquad a = \begin{pmatrix} 0 & -1 \\ 1 & 0 \end{pmatrix} \qquad b = \begin{pmatrix} -1 & 0 \\ 0 & -1 \end{pmatrix} \qquad c = \begin{pmatrix} 0 & 1 \\ -1 & 0 \end{pmatrix}$$

are a representation of Z_4 under matrix multiplication.

(c) Define a homomorphism between Z_4 and the cyclic group C_4 of rotation operations in a plane $\{e, a, b, c\}$, with $e \equiv$ rotate by an integer multiple of 2π, $a \equiv$ rotate counterclockwise by $\frac{\pi}{2}$, $b \equiv$ rotate counterclockwise by π, and $c \equiv$ rotate counterclockwise by $\frac{3\pi}{2}$, by showing that Z_4 and C_4 have equivalent multiplication tables.

(d) Elements of Z_N may be represented by $z_i = \exp(2\pi i n/N)$ with $n = 0, 1, \ldots, N-1$. Construct this representation for Z_4 and show that it has the same multiplication table as that obtained above for Z_4 and C_4.***

2.28 Show that the multiplication table for the 4-group $\{I, a, b, c\}$ given in Problem 2.9 follows from the algebraic requirements $a^2 = b^2 = I$ and $ab = ba = c$.

2.29 For finite groups each group element a must give the identity e when raised to some finite power: $a^p = e$. The integer p is called the *order of the element* a. Show that two elements in the same conjugacy class have the same order p.

2.30 Prove that the group identity e is always in a conjugacy class of its own, and that no group element can be in two different conjugacy classes. ***

2.31 Show that the group Z_2 of integers under addition modulo 2 is isomorphic to the cyclic group C_2 described in Box 2.2. Show that the set $\{1, -1\}$ is isomorphic to C_2 under arithmetic multiplication.

2.32 Show that for a direct product group $G = A \times B$, the groups A and B are invariant subgroups of G. ***

2.33 From the solution of Problem 2.5, $H = \{e, a^2\}$ is an abelian invariant subgroup of the group C_4 described in Example 2.15. Perform a left-coset decomposition of C_4 with respect to the abelian invariant subgroup H and show that the factor group C_4/H is isomorphic to the group C_2. *Hint*: See Examples 2.16 and 2.17.

2.34 Show that the special linear group SL(2, C) of 2×2 matrices with complex entries and unit determinant [see Eq. (2.18)] satisfies the group postulates of Section 2.2.

2.35 Verify that the representation T given by Eq. (2.9) obeys $T(G_i)T(G_j) = T(G_iG_j)$, so it preserves the group multiplication law for G and is a valid representation. ***

3 Introduction to Lie Groups

Chapter 2 introduced some basic concepts relevant to an understanding of group theory and its application in physics. Often these concepts have been illustrated with finite groups, although most apply with suitable modification both to finite and to continuous groups. This chapter continues our introductory survey but now the emphasis will be on continuous groups, in particular on a certain kind of continuous group called a *Lie group*. We have already met some examples of these groups in the preceding chapter, but now their properties will be considered more systematically.

3.1 Lie Groups

Lie groups are continuous groups with elements labeled by a finite number of parameters and a multiplication law that depends smoothly on those parameters. As discussed in Ch. 2, a continuous one-parameter group $G(\alpha)$ satisfies [see Eq. (2.8)]

$$G(a) \cdot G(b) = G(c), \tag{3.1}$$

where a, b, and c are particular values of the continuous parameter α. This continuous group is also a (one-parameter) Lie group if it obeys the additional restriction that c is an analytical function of a and b. Thus, Lie groups are sometimes termed *continuous analytical groups*. Likewise, N-parameter Lie groups may be defined by generalizing α to a vector, $\boldsymbol{\alpha} = (\alpha_1, \alpha_2, \ldots, \alpha_N)$. The conventional notation for Lie group elements $U(\boldsymbol{\alpha})$ is

$$U(\alpha_1, \alpha_2, \ldots, \alpha_N) = e^{i\alpha_1 X_1 + i\alpha_2 X_2 + \cdots + i\alpha_N X_N} = e^{i\sum_a \alpha_a X_a} \equiv e^{i\alpha_a X_a}, \tag{3.2}$$

where an *Einstein summation convention* has been introduced.[1]

> There is an implied summation over any index (either upper or lower, unless stated otherwise) that appears twice on the same side of an equation.

The operators X_a in Eq. (3.2) were encountered previously in Section 2.10 and are called the *generators* of the Lie group. The generators X_a have several important properties.

1. They are hermitian ($X_a = X_a^\dagger$) if U is unitary.
2. They are linearly independent.

[1] In some cases, like in Ch. 13, we will require also that one index must be upper and one must be lower.

Box 3.1

Linear Vector Spaces

Linear vector spaces will be of paramount importance in the discussion of groups in general, and Lie groups in particular. Let us review their basic properties.

Definition of a Vector Space

A *vector space* V over a field K (typically the real or complex numbers) is a set with two operations defined, (1) addition of vectors (a map $V \times V \to V$ of the vector space to itself) and (2) multiplication by scalars (a map $C \times V \to V$, where C is the set of complex numbers), which satisfies the following axioms [152]:

1. $\mathrm{u} + \mathrm{v} = \mathrm{v} + \mathrm{u}$;
2. $(\mathrm{u} + \mathrm{v}) + \mathrm{w} = \mathrm{u} + (\mathrm{v} + \mathrm{w})$;
3. a zero vector 0 exists, such that $\mathrm{u} + 0 = \mathrm{u}$;
4. for each u there is a corresponding $-\mathrm{u}$ such that $\mathrm{u} + (-\mathrm{u}) = 0$;
5. $\alpha(\mathrm{u} + \mathrm{v}) = \alpha\mathrm{u} + \alpha\mathrm{v}$;
6. $(\alpha + \beta)\mathrm{u} = \alpha\mathrm{u} + \beta\mathrm{u}$;
7. $(\alpha\beta)\mathrm{u} = \alpha(\beta\mathrm{u})$;
8. $1\mathrm{u} = \mathrm{u}$.

The vectors $\mathrm{u}, \mathrm{v}, \mathrm{w} \in V$, the scalars $\alpha, \beta \in K$, and 1 is the unit element of K. From a more intuitive perspective, we may think of a linear vector space as a set of objects (the vectors) that can be multiplied by numbers and added together in a linear way, while exhibiting *closure*: any such operations on elements of the set give back a linear combination of elements. For arbitrary vectors A and B, and arbitrary scalars a and b, one expects then that expressions like

$$(a + b)(A + B) = aA + aB + bA + bB$$

should be satisfied. These rules are not very restrictive and many sets can be turned easily into vector spaces. Vector spaces employed in physics often have additional structure defined beyond the minimal requirements listed above, like an inner product and a norm.

Basis Vectors

A *basis* for a vector space is a set of vectors that *span the space* (any vector is a linear combination of basis vectors) and that are *linearly independent* (no basis vector is a linear combination of other basis vectors). The number of basis vectors is equal to the *dimensionality* of the vector space.

3. They form a basis in a linear vector space (see Box 3.1).
4. They form an algebra under commutation called a *Lie algebra* (Section 3.2 below).

The N-dimensional vector space spanned by the group generators should not be confused with the Hilbert space in which each generator acts as a quantum operator.

Example 3.1 The rotation group in 3D has three continuous parameters $(\theta_1, \theta_2, \theta_3)$, and three generators of angular momentum (L_1, L_2, L_3). The generators form a basis for a 3D linear vector space associated with the group structure, but each generator also acts as a quantum operator in a Hilbert space parameterized by the continuous angles $(\theta_1, \theta_2, \theta_3)$.

Equation (3.2) makes clear that the number of generators equals the number of parameters for a Lie group. For many Lie groups – specifically for the compact ones, see Section 6.2.4 – any representation is equivalent to one in terms of unitary matrices.

3.2 Lie Algebras

The generators X_a of a Lie group form a closed commutator algebra called a *Lie algebra*:

$$[X_a, X_b] = i f_{abc} X_c \qquad [A, B] \equiv AB - BA, \tag{3.3}$$

where the quantities $f_{abc} = -f_{bac}$ are *structure constants* of the Lie algebra.[2] The structure constants are not unique because they depend on the basis chosen for the generators.

Because the generators of a Lie algebra form a linear vector space, independent linear combinations of these generators also are generators of the Lie algebra.

Each new set of generators formed by taking linear combinations implies a different set of structure constants, which defines the *same Lie algebra* but in a *different basis.* Lie groups result from exponentiating the generators of Eq. (3.3), according to Eq. (3.2). A subset of generators satisfying the commutator (3.3) forms a Lie subalgebra, and exponentiating them as in Eq. (3.2) generates a subgroup of the original Lie group.

Box 3.2 Two Conventions for Lie Algebras

We will most often use Eq. (3.3) to specify Lie algebras but at times use another convention that eliminates the explicit i displayed in Eq. (3.3) by defining the generators to have an additional factor of i. For example, consider the angular momentum algebra defined in Example 3.2, $[L_i, L_j] = i\epsilon_{ijk} L_k$. If the L_i are mapped to a new set of generators by $L_i \to iX_i$, then $[L_i, L_j] = i\epsilon_{ijk} L_k$ becomes

$$[X_i, X_j] = c_{ij}^k X_k$$

(implied sum on k), where the $c_{ij}^k = -c_{ji}^k$ are also called structure constants. Then

$$[X_1, X_2] = X_3 \qquad [X_2, X_3] = X_1 \qquad [X_3, X_1] = X_2$$

is the explicit form of the angular momentum algebra with these new generators.

[2] An alternative convention for Eq. (3.3) is given in Box 3.2. Rigorously, a Lie algebra is a vector space equipped with a binary law of combination $[A, B]$ called a *Lie bracket* that maps vectors A and B to a third vector, while satisfying $[A, B] = -[B, A]$ and the Jacobi identity (3.6). For applications in this book, A and B will be linear operators or matrices and the Lie bracket may be taken to be the commutator $[A, B] \equiv AB - BA$.

Example 3.2 The Lie algebra of the angular momentum operators $L_i\ (i = 1, 2, 3)$ is $[L_i, L_j] = i\epsilon_{ijk}L_k$, where ϵ_{ijk} is the completely antisymmetric rank-3 tensor, with

1. $\epsilon_{ijk} = +1$ if the indices i, j, k vary cyclically (for example, $\epsilon_{123} = 1$),
2. $\epsilon_{ijk} = -1$ if the indices are anticyclic (for example, $\epsilon_{321} = -1$),
3. $\epsilon_{ijk} = 0$ if any two indices are the same.

This Lie algebra is called SU(2); later it will be found to be associated with *two Lie groups*, SU(2) and SO(3). We can form a new set of operators from linear combinations of the old ones, $L_\pm = L_1 \pm iL_2$ and L_3, and from the original commutators these have the Lie algebra

$$[L_3, L_\pm] = \pm L_\pm \qquad [L_+, L_-] = 2L_3.$$

The structure constants differ for the generator sets (L_+, L_-, L_3) and (L_1, L_2, L_3) but this is still the SU(2) Lie algebra; the generators have just been expressed in a different basis.

We will show later that the local properties of a Lie group may be defined in terms of its associated Lie algebra. In fact, we will find (as in Example 3.2) that a particular Lie algebra usually determines the local structure of more than one Lie group.

The structure constants f_{abc} in a Lie group are analogous to a multiplication table for a finite group, since in a certain sense commutation is the quantum-mechanical equivalent of classical multiplication.

It will often be convenient to define Lie groups in terms of Lie algebras: *a Lie group is any continuous group that has generators X_a satisfying Eq. (3.3)*. Therefore we will often go back and forth between Lie algebras and Lie groups, with the context indicating whether a group or its associated algebra are under consideration.[3] Rigorously, different symbols should be used for algebras and groups. For example, lower case can be used for an algebra, su(2), and upper case for a group, SU(2). However, in typical physics applications one learns quickly to tell from context whether SU(2) means the algebra or the group, and it can become rather pedantic to distinguish formally. Therefore, we shall often use the same symbol for an algebra and a corresponding group, depending upon context or explicit statements to tell whether an algebra or a group is meant.

3.2.1 Invariant Subalgebras

A subset F of group generators G might be closed when commuted with the entire set. That is, for the members of the subset F, commutation with any member of the whole set G gives a linear combination of generators lying only in the subset F. In symbols,

[3] However, as already noted Lie groups and Lie algebras are not in one to one correspondence because a Lie algebra is typically associated with more than one Lie group. We shall elaborate on this later.

$$[X_i, X_j] = i f_{ijk} X_k \qquad (i, k \in F; j \in G). \tag{3.4}$$

Such an algebra is called an *invariant subalgebra* or an *ideal*. When exponentiated as in Eq. (3.2), an invariant subalgebra generates an invariant subgroup. Every algebra has two trivial improper invariant subalgebras: (1) the set containing only the identity element and (2) the whole algebra. If the invariant subalgebra F commutes with all members of G,

$$[X_i, X_j] = 0 \qquad (i \in F; j \in G), \tag{3.5}$$

F is termed an *abelian invariant subalgebra*, a *maximal ideal*, or a *center of the algebra*.

A non-abelian algebra having no proper invariant subalgebras is a *simple algebra,* and the corresponding group is a *simple group*. A *semisimple Lie algebra* contains no proper abelian invariant subalgebras. Semisimple Lie algebras give rise to *semisimple Lie groups,* which may be constructed from direct products of simple groups.[4] Large portions of our discussion will involve semisimple algebras and groups. The structure constants carry little information for abelian invariant subalgebras, since they all vanish. In contrast, the f_{ijk} are rich in content for semisimple algebras because many are non-zero.

3.2.2 Adjoint Representation of the Algebra

Commutators satisfy two conditions that place restrictions on the structure constants of Lie algebras. The first is obvious from the definition of a commutator, $[A, B] = -[B, A]$. The second is called the *Jacobi identity:* $[[A, B], C]$ + cyclic permutations $= 0$, or explicitly

$$[[A, B], C] + [[B, C], A] + [[C, A], B] = 0, \tag{3.6}$$

which may be verified directly by expanding the nested commutators. Combining Eq. (3.6) and $[A, B] = -[B, A]$ with Eq. (3.3) indicates that the structure constants must satisfy

$$f_{abc} = -f_{bac}, \tag{3.7a}$$

$$f_{abd} f_{cde} + f_{bcd} f_{ade} + f_{cad} f_{bde} = 0, \tag{3.7b}$$

where now both (3.6) and (3.7b) will be termed the Jacobi identity. These properties of the structure constants are cornerstones of the theory of Lie algebras.

A significant consequence of Eq. (3.7b) is that *the structure constants themselves generate a representation of the algebra.* That is, if we define a matrix T_a having matrix elements

$$(T_a)_{bc} = \langle X_b | \, T_a \, | X_c \rangle \equiv -i f_{abc}, \tag{3.8}$$

[4] Note from these definitions that a simple algebra is necessarily semisimple, but a semisimple algebra need not be simple. These definitions of simple and semisimple in terms of Lie algebras differ somewhat from those given in Section 2.13 in terms of group properties for non-Lie groups. A formal procedure that uses the structure constants to determine whether a Lie algebra is semisimple will be given in Section 7.2.2.

then it follows from the Jacobi identity that

$$[T_a, T_b] = if_{abc}T_c. \tag{3.9}$$

The representation (3.8) generated by the structure constants of a Lie algebra is unique and termed the *adjoint representation* or the *regular representation*. It is pivotal in defining the structure of a Lie group and in the uses of Lie groups in physics. Since the dimension of a matrix representation is the dimension of the vector space on which it acts, *the adjoint representation has a dimensionality equal to the number of group generators.* This is, in turn, equal to the number of real, continuous variables that parameterize the group.

Example 3.3 Consider the rotation group in three dimensions, which has three parameters, three generators, and thus a unique 3D adjoint representation. We will see shortly that it is associated with multiplets of angular momentum $J = 1$. As another example, SU(3) has eight generators, implying a unique eight-dimensional adjoint representation.

Let us now elaborate on the preceding ideas by considering a specific Lie algebra and associated Lie groups – the algebra SU(2) and the associated groups SU(2) and SO(3) – first in their role as the symmetry of angular momentum in quantum mechanics, and then as symmetries associated with more abstract physical concepts such as isospin.

3.3 Angular Momentum and the Group SU(2)

The SU(2) group is especially well suited to the pedagogical intentions of this chapter for several reasons. The algebra is semisimple and non-abelian. Thus, it is not too difficult to deal with but still exhibits many of the features found in more complicated groups. In addition, you likely already have a basic knowledge of the Lie algebra and group theory of SU(2) in the guise of angular momentum theory in quantum mechanics. Therefore, our task is simplified because various aspects of the discussion will consist of applying new terminology and new ways of thinking to some familiar concepts.

3.3.1 Fundamental Representation of SU(2)

The fundamental representation of a group was defined in Section 2.6.2 as the lowest-dimensional faithful matrix representation. For SU(2) this is in terms of 2×2 matrices U that operate on a two-component column vector χ (*fundamental doublet* or *Pauli spinor*). General transformations of these doublets will take the form $\chi' = U\chi$, where

$$\chi = \begin{pmatrix} \chi_1 \\ \chi_2 \end{pmatrix} \qquad U = \begin{pmatrix} u_{11} & u_{12} \\ u_{21} & u_{22} \end{pmatrix},$$

Table 3.1. Properties of the Pauli matrices σ_i

$$\sigma_1 = \begin{pmatrix} 0 & 1 \\ 1 & 0 \end{pmatrix} \qquad \sigma_2 = \begin{pmatrix} 0 & -i \\ i & 0 \end{pmatrix} \qquad \sigma_3 = \begin{pmatrix} 1 & 0 \\ 0 & -1 \end{pmatrix}$$

$$[\tfrac{\sigma_i}{2}, \tfrac{\sigma_j}{2}] = i\epsilon_{ijk}\left(\tfrac{\sigma_k}{2}\right) \qquad \{\sigma_i, \sigma_j\} = 2\delta_{ij} \qquad \mathrm{Tr}\,\sigma_i = 0$$

$$\mathrm{Tr}\,(\sigma_i\sigma_j) = 2\delta_{ij} \qquad \sigma_i\sigma_j = \delta_{ij} + i\epsilon_{ijk}\sigma_k \qquad \sigma_i^2 = 1$$

$$\text{Completeness: } \sum_i (\sigma_i)_{ab}(\sigma_i)_{cd} = 2(\delta_{bc}\delta_{ad} - \tfrac{1}{2}\delta_{ab}\delta_{cd})$$

Table 3.2. Completely antisymmetric rank-3 tensor ϵ_{ijk}

$\epsilon_{123} = \epsilon_{231} = \epsilon_{312} = 1$	$\epsilon_{132} = \epsilon_{213} = \epsilon_{321} = -1$	All other $\epsilon_{ijk} = 0$
$\epsilon_{ipq}\epsilon_{jpq} = 2\delta_{ij}$	$\epsilon_{ijk}\epsilon_{ijk} = 6$	$\epsilon_{ijk}\epsilon_{pqk} = \delta_{ip}\delta_{jq} - \delta_{iq}\delta_{jp}$

with the conditions $U^\dagger U = UU^\dagger = 1$ and $\det U = +1$. For a matrix A we have the useful relation proved in Problem 3.11(d) that

$$\det e^A = e^{\mathrm{Tr}\,A}. \tag{3.10}$$

The SU(2) unitary matrices are parameterized conventionally as $U = e^{\frac{i}{2}\theta_j\sigma_j}$, so that the generators are $\frac{1}{2}\sigma_j$. The unit determinant condition $\det U = \exp[\mathrm{Tr}\,(\frac{i}{2}\theta_j\sigma_j)] = 1$ requires that $\mathrm{Tr}\,(\sigma_j) = 0$. The inverse and hermitian conjugate of U are

$$U^{-1} = e^{-\frac{i}{2}\theta_j\sigma_j} \qquad U^\dagger = e^{-\frac{i}{2}\theta_j\sigma_j^\dagger},$$

respectively, but unitarity implies $U^\dagger = U^{-1}$ and the matrices σ are necessarily hermitian: $\sigma_j = \sigma_j^\dagger$. These results generalize to all special unitary groups SU(N) and we find that the special unitary generators are $N \times N$ matrices that are *traceless* and *hermitian.* The three generators for SU(2) are usually chosen to be $\frac{1}{2}\sigma_i$, where σ_i denotes the *Pauli matrices,*

$$\sigma_1 = \begin{pmatrix} 0 & 1 \\ 1 & 0 \end{pmatrix} \qquad \sigma_2 = \begin{pmatrix} 0 & -i \\ i & 0 \end{pmatrix} \qquad \sigma_3 = \begin{pmatrix} 1 & 0 \\ 0 & -1 \end{pmatrix}, \tag{3.11}$$

which obey the SU(2) Lie algebra

$$\left[\frac{\sigma_i}{2}, \frac{\sigma_j}{2}\right] = i\epsilon_{ijk}\left(\frac{\sigma_k}{2}\right). \tag{3.12}$$

The basic properties of the Pauli matrices are summarized in Table 3.1 and the properties of the completely antisymmetric rank-3 tensor ϵ_{ijk} (Levi-Civita symbol) are summarized in Table 3.2. More generally, N-dimensional representations of SU(2) may be constructed from $N \times N$ matrices J that satisfy the algebra

$$[J_i, J_j] = i\epsilon_{ijk}J_k, \tag{3.13}$$

and operate on wavefunctions that are N-component multiplets.

3.3.2 The Cartan–Dynkin Method

Let us now apply the SU(2) algebra to analysis of angular momentum in quantum mechanics. The approach to be used is termed the *Cartan–Dynkin method.* Many concepts will be familiar from the elementary quantum mechanics of angular momentum but the same technique forms the basis of a powerful and systematic method to analyze the structure of more complicated Lie groups that will be developed in Ch. 7. The essence of the method is to divide the generators of the group into two sets by taking expeditious linear combinations.

1. The first set consists of hermitian operators that can be diagonalized to give quantum numbers suitable for labeling members of irreducible multiplets.
2. The second set consists of *stepping operators* that permit moving through an irreducible multiplet in such a manner that from any member of the multiplet it is possible to reach any other member by successive stepping operations.

A systematic procedure to accomplish this may be outlined as follows.

Casimir Operators: For non-abelian groups the generators do not generally commute with each other but it is possible to find operators that are non-linear functions of generators that do commute with all the generators; these are called *Casimir operators.* A semisimple Lie algebra has l independent Casimir operators, where l is termed the *rank of the algebra.* The rank can usually be expressed in terms of simple formulas. For example, the SU(N) groups are of rank $l = N - 1$. For semisimple algebras the lowest-order Casimir is proportional to the sum of the squares of the group generators, with higher-order Casimirs involving higher powers of the generators. In physical applications, the lowest-order Casimir is often dominant. The significance of the Casimir operators is expressed by *Schur's lemma.*[5]

Schur's Lemma: A matrix that commutes with every matrix of an irreducible representation is a multiple of the unit matrix.

Thus Casimir operators give *eigenvalue equations* and *quantum numbers* that label the irreps of a group. Whether quantum numbers associated with the Casimir operators are sufficient to completely specify an irrep depends on factors that will be taken up later.

Cartan Subalgebra: The generators X_a are basis vectors of a linear vector space and we construct linear combinations in such a way that as many operators as possible are (1) hermitian, (2) mutually commuting, and (3) diagonal. *The maximum number of such operators is equal to l,* the rank of the algebra, and this maximal set of commuting operators H_i ($i = 1, 2, \ldots, l$) is termed the *Cartan subalgebra.* The l diagonal generators of the Cartan subalgebra provide l eigenvalue equations

$$H_i \,|jm_i\rangle = m_i \,|jm_i\rangle \qquad (i = 1, 2, \ldots, l), \tag{3.14}$$

[5] We shall refer simply to this statement as *Schur's lemma.* However, it is also called variously Schur's first lemma, Schur's second lemma, or a corollary of Schur's lemma in the physics literature.

where j denotes all labels necessary to specify the representation and m_i is termed the *weight* of $|jm_i\rangle$. These eigenvalue equations give additive quantum numbers, corresponding to components of vectors, while the Casimir operators give quantum numbers that are not simply additive. For example, we will see that the Casimir eigenvalue for the SU(2) algebra in a state of angular momentum J is proportional to $J(J+1)$, while the single quantum number coming from the Cartan subalgebra is the magnetic projection J_3.

Weight Vectors and Stepping Operators: The weights may be assembled into a *weight vector,* $m = (m_1, m_2, \ldots, m_l)$. The components of m span the *weight space* and the l-dimensional plot of all the weights for a representation is called the *weight diagram.* The weights for the adjoint representation are of unique importance in the analysis of Lie groups (see Ch. 7), so weights in the adjoint representation get a special name: *roots.* The *multiplicity* or *degeneracy* of a weight is the number of eigenvectors in a representation having that weight. The remaining generators not in the Cartan subalgebra may be combined to give the stepping operators that allow moving from state to state in the representation.

3.3.3 Cartan–Dynkin Analysis of SU(2)

Let us now apply the Cartan–Dynkin analysis to the specific case of SU(2). In general, for the special unitary groups SU(N),

1. SU(N) has $N^2 - 1$ generators, and of these
2. $N-1$ may be diagonalized simultaneously (Cartan subalgebra), so the rank of SU(N) is $l = N-1$ and it has $N-1$ Casimir operators,
3. which leaves $N^2 - 1 - (N-1) = N^2 - N$ stepping operators.

Thus SU(2) has three generators, one weight, one Casimir, and two stepping operators.

Casimir Operator and the Cartan Subalgebra: The angular momentum algebra is $[J_i, J_j] = i\epsilon_{ijk} J_k$ and the Cartan subalgebra of SU(2) may be chosen to be $H_1 = J_3$, corresponding to an eigenvalue equation (in $\hbar = c = 1$ units)[6]

$$\hat{J}_3 \, |J, J_3\rangle = J_3 \, |J, J_3\rangle \,, \tag{3.15}$$

where the single weight is $m = J_3$. The Casimir operator C and its eigenvalue equation are

$$C = J^2 = J_1^2 + J_2^2 + J_3^2 \qquad C\psi^{(J)} = J(J+1)\psi^{(J)}. \tag{3.16}$$

SU(2) is rank-1 so this is the only Casimir operator and J, which we know labels the total angular momentum, is seen also to be the quantum number labeling different SU(2)

[6] We will work often in units where $\hbar$ (Planck's constant divided by 2π) and the speed of light c are set to unity: $\hbar = c = 1$. Then neither $\hbar$ nor c need be displayed in equations and proper "engineering units" can be restored by dimensional analysis, if needed. Such *natural units* are common for many fields of modern physics. They are discussed further in Appendix B, and Problems 3.19 through 3.22 give some practice using them.

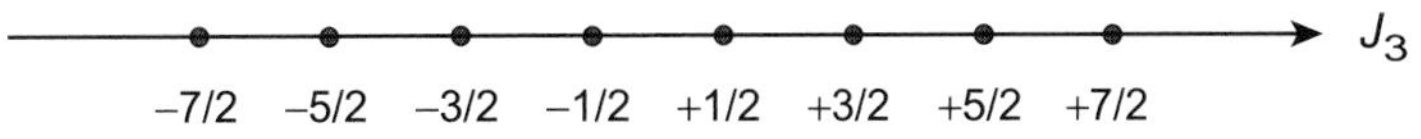

Fig. 3.1 Weight space for a $J = \frac{7}{2}$ irreducible representation of SU(2).

irreducible representations. There are $(2J+1)$ weights $m = J_3$ corresponding to an angular momentum J, implying that the irreps of SU(2) are $(2J + 1)$-dimensional.

Stepping Operators: The two stepping operators that may be constructed from generators not in the Cartan subalgebra are familiar from elementary quantum physics:

$$J_+ \equiv J_1 + iJ_2 \qquad J_- \equiv J_1 - iJ_2, \tag{3.17}$$

as are the algebraic properties

$$\begin{aligned} [\, J_3, J_\pm \,] &= \pm J_\pm \qquad J^2 = \frac{1}{2}\,(J_+J_- + J_-J_+) + J_3^2 \\ [\, J_+, J_- \,] &= 2J_3 \qquad [\, J^2, J_i \,] = 0. \end{aligned} \tag{3.18}$$

As will be elaborated in Chs. 6 and 7, these algebraic properties imply that

$$J_\pm \left| J, J_3 \right\rangle \propto \left| J, J_3 \pm 1 \right\rangle, \tag{3.19}$$

except that

$$J_+ \left| J_3 = J \right\rangle = 0 \qquad J_- \left| J_3 = -J \right\rangle = 0. \tag{3.20}$$

Therefore, we find the following.

1. Successive application of stepping operators takes us between all members of a $(2J+1)$-dimensional multiplet corresponding to an irrep of SU(2), but not out of the irreducible representation.
2. All states accessible to the stepping operators are labeled by the Casimir eigenvalue of the state from which we start.

> Thus the SU(2) stepping operators $J_\pm$ may be used to reach all members of an irreducible representation starting from any member, without ever reaching a state that is not in the irrep.

Weight Space: The multiplicity of each weight in an SU(2) irrep is one, since each eigenvector has a unique value of J_3 within an angular momentum J multiplet. Thus the weight diagram for an SU(2) irrep may be represented by a line with sites that are singly occupied at weight values J_3 from $-J$ to $+J$, and otherwise unoccupied. Figure 3.1 illustrates for the $J = \frac{7}{2}$ irrep of SU(2). Lest we conclude that weight diagrams are always this simple, let us take a quick look at the weights for an irrep of the somewhat more complicated group SU(3) that is summarized in Box 3.3, and will be discussed more extensively in Ch. 8.

Box 3.3 **Weight Space for SU(3)**

From the general formula for SU(N) the group SU(3) has $N^2 - 1 = 8$ generators, with an eight-dimensional adjoint representation and two distinct three-dimensional fundamental representations.

Cartan Subalgebra

It follows that there are $N - 1 = 2$ operators to diagonalize in the Cartan subalgebra and SU(3) is a rank-2 group, with

- two independent Casimir operators,
- six stepping operators, and
- a two-dimensional weight space.

That is, SU(3) has two weight quantum numbers in place of the single magnetic quantum number in SU(2), and six stepping operators instead of the two for SU(2).

Weight Space

The following figure shows the weight diagram for a 27-dimensional SU(3) irrep, with dots indicating single degeneracy and circles denoting additional degeneracy.

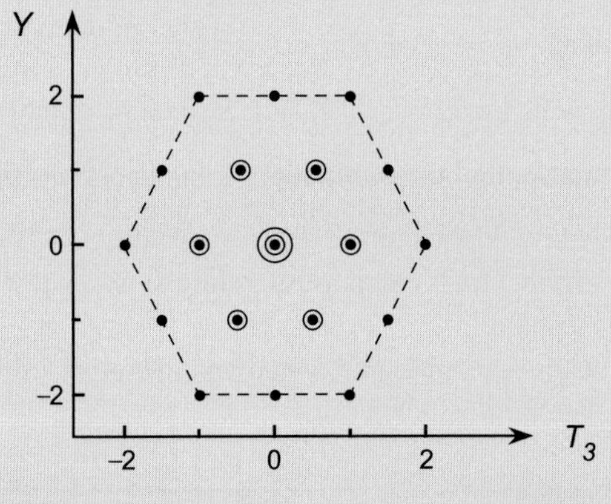

We see that the sites of the weight diagram for the SU(3) irrep may be multiply occupied, and that the stepping operators must be more numerous than for SU(2) because it is necessary to range through the 2D (T_3, Y) weight space to reach all multiplet members. The SU(3) weight space will be examined further in Ch. 8. For rank-3 and higher the weights are harder to visualize and we will turn to more abstract ways to characterize the weight space in Ch. 7.

3.3.4 The Clebsch–Gordan Series for SU(2)

Let us now consider the weight space for the direct product of two SU(2) representations. It is well known from angular momentum theory that the product of two spherical harmonics can be written as a sum over spherical harmonics. This *spherical harmonic addition theorem* is just a special case of a more general result from group theory.

The direct product of representations D_1 and D_2 for a group can be written as a direct sum over irreducible representations $D^{(i)}$ of that group

$$D_1 \otimes D_2 = \sum_i \Gamma_i D^{(i)}. \tag{3.21}$$

This extremely important result is called the *Clebsch–Gordan series.*

Equation (3.21) leads to the spherical harmonic addition theorem because we shall show in Ch. 6 that spherical harmonics are irreducible representations of SU(2).

Multiplicity of Irreps in the Direct Sum: The coefficient Γ_i in Eq. (3.21) is called the *multiplicity of the irrep* in the direct sum. Groups for which the multiplicities are all either zero or unity are called *simply reducible*, since each irrep occurs no more than once in the Clebsch–Gordan series. The group SU(2) is simply reducible but some other groups that we will consider are not. This will require additional labels to distinguish the irreps occurring more than once in the Clebsch–Gordan series.

Highest-Weight Algorithm: The direct product of two matrix representations will generally lead to a matrix that must be similarity transformed to exhibit explicitly the direct sum of irreps implied by Eq. (3.21) and Fig. 2.2. Let us now demonstrate how to decompose an SU(2) direct product into a sum of irreps with weight sites no more than singly occupied. This is most easily implemented in terms of the *highest-weight algorithm*, which rests on the following assertions, valid for any compact, semisimple group.

1. For each irrep there is a unique weight corresponding to a singly occupied site that can by some consistent prescription be designated the "highest weight."
2. Two equivalent irreps have the same highest weight; two irreps with the same highest weight are equivalent.

Example 3.4 illustrates the assignment of weights to a (generally reducible) representation resulting from a direct product of SU(2) irreps.

Example 3.4 Consider the direct product of SU(2) representations labeled by the Casimir eigenvalues $J(J+1)$ with $J = 1$ and $J = \frac{3}{2}$: $\left(J = \frac{3}{2}\right) \otimes (J = 1) = \mathbf{4} \otimes \mathbf{3}$. Here a common notation for the special unitary groups is introduced where an irrep is specified by giving the dimension [$2J + 1$ for SU(2)] of the representation as a boldface number. For SU(2) this gives a unique labeling of irreps since J^2 is the only Casimir operator. For SU(N) with $N > 2$ this labeling is not unique, but it is still commonly used because the ambiguity is not large, at least in low-dimensional representations. The direct product $\left(J = \frac{3}{2}\right) \otimes (J = 1)$ gives the weight diagram shown in Fig. 3.2. Site occupations can be deduced by counting the ways that the allowed weights (magnetic quantum numbers) for the two representations can combine additively to give a particular weight in the direct product space. For example, consider the possible ways to make $J_3 = \frac{1}{2}$ in Fig. 3.2: $J_3 = \frac{1}{2} = \left\{\frac{3}{2} - 1,\ -\frac{1}{2} + 1,\ \frac{1}{2} + 0\right\}$, so $J_3 = \frac{1}{2}$ has a multiplicity of three. But, there is only one way to add the $J = \frac{3}{2}$ and $J = 1$ weights to get $J_3 = \frac{5}{2}$, so that site is singly occupied.

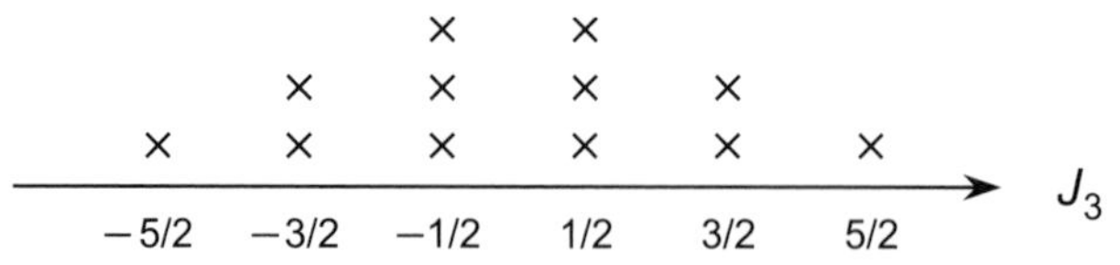

Fig. 3.2 Weight diagram for the SU(2) direct product $(J = \frac{3}{2}) \otimes (J = 1)$ in Example 3.4, with each cross denoting occupation of the corresponding site. Some sites are multiply occupied because this is a reducible representation of SU(2).

Decomposition into a Direct Sum of Irreps: The reducible weight diagram of Fig. 3.2 can now be decomposed into a direct sum of irreps using the highest-weight algorithm.

1. The weight $\frac{5}{2}$ is singly occupied and corresponds to the unique highest weight of an irrep with $J = \frac{5}{2}$. Using the stepping operator J_- with properties (3.19)–(3.20), we may start from this highest-weight state and step through the multiplet, removing states with
$$J_3 = \left\{\frac{5}{2}, \frac{3}{2}, \frac{1}{2}, -\frac{1}{2}, -\frac{3}{2}, -\frac{5}{2}\right\},$$
since the sequence terminates after $J_3 = -\frac{5}{2}$ because of the stepping operator algebra. This corresponds to the $2J + 1$ weights for a $J = \frac{5}{2}$ state. We have now removed the bottom row of crosses in Fig. 3.2, leaving a weight diagram with $J_3 = \frac{3}{2}$ singly occupied.
2. Now $J_3 = \frac{3}{2}$ is the highest weight state in a multiplet for which $J = \frac{3}{2}$. Applying the J_- operator sequentially as before, we may remove from the diagram
$$J_3 = \left\{\frac{3}{2}, \frac{1}{2}, -\frac{1}{2}, -\frac{3}{2}\right\},$$
which corresponds to weights for a $J = \frac{3}{2}$ state.
3. This removes the second row of crosses from Fig. 3.2, leaving a diagram with only the $J_3 = \pm\frac{1}{2}$ sites occupied; obviously this corresponds to a $J = \frac{1}{2}$ irrep of SU(2).

Thus the direct product of a $J = 1$ and $J = \frac{3}{2}$ representation for SU(2) has been decomposed into a $J = \frac{5}{2}$, a $J = \frac{3}{2}$, and a $J = \frac{1}{2}$ irrep. Labeling the representations by their dimensions, this can be written symbolically as $\mathbf{4} \otimes \mathbf{3} = \mathbf{6} \oplus \mathbf{4} \oplus \mathbf{2}$. In accordance with previous assertions, the direct product of these SU(2) representations is seen to be simply reducible into a direct sum in which no irrep occurs more than once, and the weight space sites are no more than singly occupied in each irrep.

3.3.5 SU(2) Adjoint Representation

The adjoint representation was introduced previously as the unique representation equal in dimension to the number of group generators. Therefore, the SU(2) adjoint representation is three-dimensional. It is easily shown by matrix multiplication that the set

$$J_1 = \frac{1}{\sqrt{2}}\begin{pmatrix} 0 & -1 & 0 \\ -1 & 0 & 1 \\ 0 & 1 & 0 \end{pmatrix} \quad J_2 = \frac{i}{\sqrt{2}}\begin{pmatrix} 0 & 1 & 0 \\ -1 & 0 & -1 \\ 0 & 1 & 0 \end{pmatrix} \quad J_3 = \begin{pmatrix} 1 & 0 & 0 \\ 0 & 0 & 0 \\ 0 & 0 & -1 \end{pmatrix}, \tag{3.22}$$

obeys the SU(2) Lie algebra and is an adjoint matrix representation.[7] From the previous discussion it is clear that the adjoint representation for SU(2) must correspond to angular momentum $J = 1$. For this reason, the SU(2) adjoint representation is sometimes called the *vector representation*. The adjoint representation may be constructed in two ways.

1. Use Eq. (3.8) to build it using the structure constants of the group.
2. Utilize a property of semisimple groups that all representations can be built from direct products of the fundamental representations with themselves.

Construction of the adjoint representation using the first method is illustrated in Problem 3.6, while Example 3.5 illustrates use of the second method.

Example 3.5 The product of the SU(2) fundamental representation **2** with itself gives the direct sum $\mathbf{2} \otimes \mathbf{2} = \mathbf{1} \oplus \mathbf{3}$, where the **3** is the adjoint representation. Properties of the **3** may then be inferred from the properties of the **2**.

Example 3.5 corresponds to the well-known result that two spin-$\frac{1}{2}$ electrons [fundamental representation of SU(2)] can couple to a resultant spin of $J = 0$ or 1 [singlet and triplet states, with the triplet state corresponding to the adjoint representation of SU(2)].

3.4 Isospin

If symmetries like SU(2) were applicable only to angular momentum, Lie algebras and Lie groups would be only a sidelight in physics (though a technically quite important one). The proliferation in application of Lie groups to modern physical problems stems from the realization that there are many objects of fundamental significance in quantum mechanics other than angular momentum operators that may close under commutation to form Lie algebras of varying complexities. In particular, the constraints imposed by commutation or anticommutation relations that creation and annihilation operators are required to obey often result in closed Lie algebras for physical systems. The history of this fruitful idea dates from the realization that neutrons and protons can be described approximately as two different "spin states" in an abstract space, which implies a group structure isomorphic to SU(2). "Rotations" in this abstract space interconvert neutrons and protons and the resulting formalism is called *isotopic spin* or *isobaric spin*, which we shorten to *isospin*.

3.4.1 The Neutron–Proton System

Introduce a set of creation and annihilation operators for neutrons $N_i^\dagger$ and N_i, respectively, and a corresponding set for protons $P_i^\dagger$ and P_i, with i denoting the quantum numbers specifying the state. These operators must obey the fermion anticommutation relations

[7] As for any representation, the SU(2) adjoint representation can be expressed in terms of many other equivalent sets of matrices by making the same similarity transform on all matrices in Eq. (3.22).

$$\{P_i, P_j^\dagger\} = \{N_i, N_j^\dagger\} = \delta_{ij} \qquad \{N_i, N_j\} = \{P_i, P_j\} = 0,$$
$$\{N_i^\dagger, N_j^\dagger\} = \{P_i^\dagger, P_j^\dagger\} = 0, \tag{3.23}$$

where the *anticommutator* is defined by $\{A, B\} \equiv AB + BA$. The bilinear combinations of these operators that conserve the total number of nucleons (neutrons plus protons) are

$$P_i^\dagger N_i \qquad N_i^\dagger P_i \qquad P_i^\dagger P_i \qquad N_i^\dagger N_i$$

(sum on repeated indices), from which it is convenient to form the linear combinations

$$B = P_i^\dagger P_i + N_i^\dagger N_i \qquad T_+ = P_i^\dagger N_i \qquad T_- = N_i^\dagger P_i,$$
$$T_3 = \frac{1}{2}\left(P_i^\dagger P_i - N_i^\dagger N_i\right) = Q - \frac{1}{2}B, \tag{3.24}$$

where the total charge operator $Q \equiv P_i^\dagger P_i$ counts the number of protons.

3.4.2 Algebraic Structure for Isospin

It is straightforward to calculate the commutators among the operators (3.24) using Eq. (3.23), and that neutron and proton operators anticommute. For example,

$$\begin{aligned}
[T_+, T_-] &= P_i^\dagger N_i N_j^\dagger P_j - N_j^\dagger P_j P_i^\dagger N_i \\
&= P_i^\dagger P_j N_i N_j^\dagger - P_j P_i^\dagger N_j^\dagger N_i \\
&= P_i^\dagger P_j \left(\delta_{ij} - N_j^\dagger N_i\right) - \left(\delta_{ij} - P_i^\dagger P_j\right) N_j^\dagger N_i \\
&= P_i^\dagger P_i - N_i^\dagger N_i \\
&= 2T_3.
\end{aligned}$$

Calculating all the commutators by a similar procedure gives the algebra

$$[T_+, T_-] = 2T_3 \qquad [T_3, T_\pm] = \pm T_\pm \qquad [B, T_\pm] = [B, T_3] = 0. \tag{3.25}$$

Therefore, the operator sets B and $(T_3, T_\pm)$ are separately closed under commutation and the group structure is isomorphic to a direct product of two Lie groups. Comparing with Eqs. (3.17)–(3.18), we conclude that the operators T_3 and $T_\pm$ constitute a complete set of generators for the group SU(2), while the operator B generates a one-parameter continuous abelian group isomorphic to the unitary group U(1). It is readily verified that the matrices

$$\tau_0 = \begin{pmatrix} 1 & 0 \\ 0 & 1 \end{pmatrix} \qquad \tau_1 = \begin{pmatrix} 0 & 1 \\ 1 & 0 \end{pmatrix} \qquad \tau_2 = \begin{pmatrix} 0 & -i \\ i & 0 \end{pmatrix} \qquad \tau_3 = \begin{pmatrix} 1 & 0 \\ 0 & -1 \end{pmatrix} \tag{3.26}$$

are a representation of the algebra (3.25), because they obey the commutator algebra

$$\left[\frac{\tau_i}{2}, \frac{\tau_j}{2}\right] = i\epsilon_{ijk}\left(\frac{\tau_k}{2}\right) \qquad [\tau_0, \tau_i] = 0. \tag{3.27}$$

The matrices τ_i $(i = 1, 2, 3)$ are just the Pauli matrices (3.11), but it is conventional to use the symbol τ for them when they represent generators of isospin rotations. Alternatively, we may take linear combinations to define the equivalent set of generators

$$\tau_+ = \frac{1}{\sqrt{2}}(\tau_1 + i\tau_2) = \frac{2}{\sqrt{2}}\begin{pmatrix}0 & 1\\ 0 & 0\end{pmatrix} \qquad \tau_- = \frac{1}{\sqrt{2}}(\tau_1 - i\tau_2) = \frac{2}{\sqrt{2}}\begin{pmatrix}0 & 0\\ 1 & 0\end{pmatrix},$$
$$\tau_3 = \begin{pmatrix}1 & 0\\ 0 & -1\end{pmatrix} \qquad \tau_0 = \begin{pmatrix}1 & 0\\ 0 & 1\end{pmatrix}, \tag{3.28}$$

which exhibit the commutator algebra

$$[\,\tau_0, \tau_\pm\,] = [\,\tau_0, \tau_3\,] = 0 \qquad [\,\tau_+, \tau_-\,] = 2\tau_3 \qquad [\,\tau_3, \tau_\pm\,] = \pm\tau_\pm. \tag{3.29}$$

The group elements $U = e^{\frac{i}{2}\tau_i\theta_i}$ are unitary because all the matrices (3.26) are hermitian. However, the unitary matrix associated with the generator τ_0 is not *special* unitary because

$$\det U_0 = \det e^{\frac{i}{2}\theta_0\tau_0} = e^{\frac{i}{2}\theta_0 \mathrm{Tr}\,\tau_0} \neq 1,$$

where Eq. (3.10) was used. We conclude that the overall group structure is U(2), the set of unitary, invertible, 2×2 matrices under matrix multiplication.[8]

3.4.3 The U(1) and SU(2) Subgroups of U(2)

The group associated with the abelian algebra for B in (3.25) is U(1), the group of unitary 1×1 matrices $e^{i\alpha N}$, with the group multiplication law

$$e^{i\alpha N}e^{i\alpha' N} = e^{i(\alpha+\alpha')N}. \tag{3.30}$$

This U(1) group is generated by τ_0 and we may write[9]

$$U(\alpha) = e^{i\alpha\tau_0} = \exp\left[i\alpha\begin{pmatrix}1 & 0\\ 0 & 1\end{pmatrix}\right] \simeq 1 + \begin{pmatrix}i\alpha & 0\\ 0 & i\alpha\end{pmatrix} = \begin{pmatrix}1+i\alpha & 0\\ 0 & 1+i\alpha\end{pmatrix}. \tag{3.31}$$

All abelian groups have 1D irreps, so this representation of U(1) is seen to be a direct sum of two 1D representations, as will be elaborated in Section 6.2.2. The isospin group structure is then defined by the homomorphism $\mathrm{U}(2) \to \mathrm{U}(1) \times \mathrm{SU}(2)$. In general there is a homomorphism $\mathrm{U}(N) \to \mathrm{U}(1) \times \mathrm{SU}(N)$ relating unitary and special unitary groups. Because of the direct product structure the U(1) and SU(N) symmetries are independent and U(N) may be examined using SU(N) if we keep track of the U(1) factor separately.

The U(1) factors in U(N) constitute an abelian invariant subalgebra. In most applications of U(N) to internal symmetries such as isospin the physical meaning of these U(1) factors is that some quantity related to a particle number is conserved. For example, the operator B is the baryon number operator for a system consisting only of neutrons and protons, and the U(1) symmetry in this case is a statement that the total number of baryons is conserved.

8 The group U(2) has two diagonal generators [τ_0 and τ_3 in the representation (3.26)], so it is a rank-2 group. In general the group U(N) has N^2 generators and the restriction to unit determinant reduces this to the $N^2 - 1$ generators of SU(N), with $N - 1$ of them diagonal.

9 Do not confuse the *generator of the Lie algebra* τ_0 with the *Lie group element* $U(\alpha) = \exp(i\alpha\tau_0)$. Both may be written as 2×2 matrices for this example, but in different spaces.

3.4.4 Analogy between Angular Momentum and Isospin

It is now clear that isospin may be analyzed in terms of the group SU(2), provided that we work in a system of definite baryon number. But this means that a mathematical formalism for isospin can be taken over intact from the application of SU(2) to angular momentum. The Lie algebra and Lie group theory are exactly the same, the only difference being the physical interpretation of the SU(2) generators. The normal convention, although some nuclear physics authors reverse this, is to associate isospin up with the proton and isospin down with the neutron, so that the fundamental nucleon doublet is

$$|p\rangle = \begin{pmatrix} 1 \\ 0 \end{pmatrix} \qquad |n\rangle = \begin{pmatrix} 0 \\ 1 \end{pmatrix}. \tag{3.32}$$

The fundamental doublet consists of "spin-up" and "spin-down" nucleons in the isospace. For example, from Eq. (3.32) and the matrix representation (3.28),

$$\frac{1}{2}\tau_3 \, |p\rangle = \frac{1}{2}\begin{pmatrix} 1 & 0 \\ 0 & -1 \end{pmatrix}\begin{pmatrix} 1 \\ 0 \end{pmatrix} = \frac{1}{2}\begin{pmatrix} 1 \\ 0 \end{pmatrix} = \frac{1}{2}\, |p\rangle,$$

$$\frac{1}{2}\tau_3 \, |n\rangle = \frac{1}{2}\begin{pmatrix} 1 & 0 \\ 0 & -1 \end{pmatrix}\begin{pmatrix} 0 \\ 1 \end{pmatrix} = -\frac{1}{2}\begin{pmatrix} 0 \\ 1 \end{pmatrix} = -\frac{1}{2}\, |n\rangle.$$

[Recall that the SU(2) generators are $\frac{1}{2}\tau_i$; see Eq. (3.27).] The actions of the raising and lowering operators defined in Eq. (3.28) on the states (3.32) are

$$\tau_- \, |p\rangle = \frac{2}{\sqrt{2}}\, |n\rangle \qquad \tau_+ \, |p\rangle = 0 \qquad \tau_- \, |n\rangle = 0 \qquad \tau_+ \, |n\rangle = \frac{2}{\sqrt{2}}\, |p\rangle.$$

From Eq. (3.26), the single Casimir operator is

$$T^2 = \left(\frac{\tau_1}{2}\right)^2 + \left(\frac{\tau_2}{2}\right)^2 + \left(\frac{\tau_3}{2}\right)^2 = \frac{3}{4}\begin{pmatrix} 1 & 0 \\ 0 & 1 \end{pmatrix}, \tag{3.33}$$

which we recognize as an example of Schur's lemma since T^2 commutes with all generators (see Section 3.3.2). Applying the Casimir operator to proton and neutron states gives

$$T^2 \, |p\rangle = T(T+1)\, |p\rangle = \frac{3}{4}\, |p\rangle \qquad T^2 \, |n\rangle = T(T+1)\, |n\rangle = \frac{3}{4}\, |n\rangle.$$

These considerations indicate that the neutron and proton behave in all respects under isospin SU(2) as if they were members of a $T = \frac{1}{2}$ angular momentum state with two magnetic substates, one corresponding to the proton and one to the neutron. In that sense, neutrons and protons are two different orientations of the same vector in an abstract isospace.

By analogy with angular momentum, isospin symmetry implies rotational invariance in isospace. This is a fancy way of saying that for an isospin invariant nucleus the energy is unchanged if we substitute neutrons for protons or protons for neutrons, subject to constraints of the Pauli principle. In nuclear physics this observation is called the *charge independence hypothesis* – the strong nuclear interaction between two nucleons is independent of their electrical charge. This symmetry is not respected by the electromagnetic interaction between nucleons, which certainly does care about the charges. However, in nuclear and elementary particle physics the strong interactions often dominate and isospin symmetry is good enough to be useful, with typical Hamiltonians for the strong interactions having isospin-violating terms contributing only at the few percent level. Isospin is thus an *approximate phenomenological symmetry* of the strong interactions, as will be discussed further in Box 19.2. As for angular momentum, isospin symmetry implies degenerate multiplet structure, with irreps (isospin multiplets) characterized by a total isospin T with

$$T^2 = T_1^2 + T_2^2 + T_3^2 = \frac{1}{2}(T_+T_- + T_-T_+) + T_3^2,$$
$$T^2 \left|\psi\right\rangle = T(T+1)\left|\psi\right\rangle,$$

and having multiplicity $2T+1$. Multiplet components can be labeled by the third component of isospin T_3 that is specified by the SU(2) Cartan subalgebra, which is related to the total charge Q and baryon number B by $T_3 = Q - \frac{1}{2}B$.

Example 3.6 For SU(2) the direct product of two spin-$\frac{1}{2}$ irreps gives the Clebsch–Gordan series $\mathbf{2}\otimes\mathbf{2} = \mathbf{1}\oplus\mathbf{3}$ (compare Examples 3.4 and 3.5), implying that the two-nucleon system has the isospin singlet and triplet states illustrated in Fig. 3.3. Identifying spin-up and spin-down isospin projections with the proton and the neutron, respectively, the $|T, T_3\rangle = |1, -1\rangle$ state may be associated with nn (dineutron) and the $|1, +1\rangle$ state with pp (diproton), but the np state (deuteron or ^{2}H) can exist in either a $T = 0$ isospin singlet or a $T = 1$ isospin triplet state:

$$\mathbf{1}:\ |00\rangle = \frac{1}{\sqrt{2}}\left(\,|p\rangle\,|n\rangle - |n\rangle\,|p\rangle\,\right) \qquad \mathbf{3}:\ \begin{cases} |11\rangle = |p\rangle\,|p\rangle \\ |10\rangle = \dfrac{1}{\sqrt{2}}\left(\,|p\rangle\,|n\rangle + |n\rangle\,|p\rangle\,\right) \\ |1-1\rangle = |n\rangle\,|n\rangle\,. \end{cases}$$

Near degeneracy of isospin multiplets implies approximate isospin invariance, just as exact degeneracy of an angular momentum multiplet implies exact rotational invariance.

Experimentally the two-nucleon $T = 0$ isosinglet has lower mass than the $T = 1$ isotriplet, as indicated in Fig. 3.3. This is not explained by isospin symmetry. To understand it we must either go beyond symmetry considerations, or expand the group structure to include

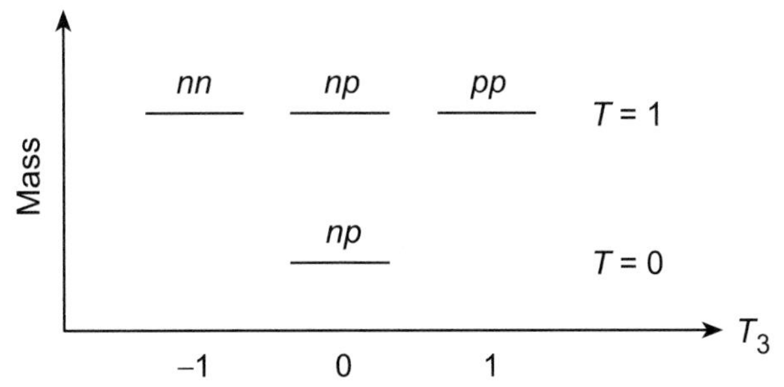

Fig. 3.3 Two-nucleon isospin states $|T, T_3\rangle$, where p labels protons and n labels neutrons.

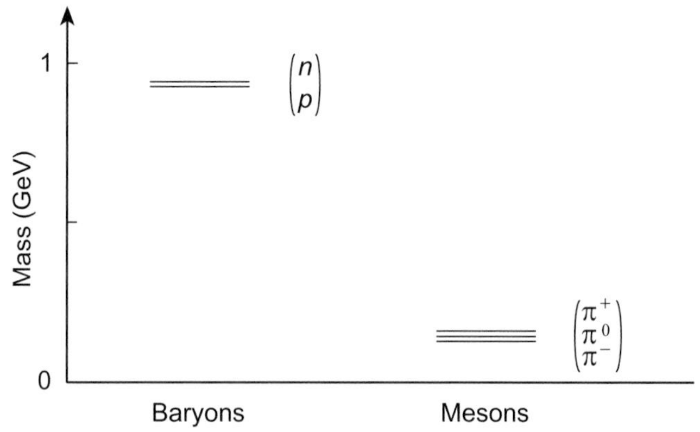

Fig. 3.4 The nucleon $T = \frac{1}{2}$ doublet and the pion $T = 1$ triplet in the hadronic mass spectrum. The near degeneracy of the members in each multiplet reflects approximate isospin symmetry.

more physics than the present simple isospin picture. Models for nuclear structure built on larger groups with more generators that incorporate additional physics will be considered in later chapters.

3.4.5 The Adjoint Representation of Isospin

For angular momentum SU(2), a $J = 1$ adjoint representation arises from the direct product of fundamental representations for a two-electron system (Example 3.5). However, a single particle might also transform as the $J = 1$ representation; for example, the intermediate vector bosons ($W^{\pm}, Z^0$) described in Section 19.1.5 are fundamental spin-1 particles. In the two-nucleon system an isospin adjoint representation can be obtained from the product of fundamental representations, but could a single strongly interacting particle correspond to an adjoint isospin representation? The candidate $T = 1$ particle must exhibit three mass-degenerate states that differ only in charge. The nearly degenerate pion triplet having $m_\pi c^2 \sim 140$ MeV displayed in Fig. 3.4 corresponds to such a $T = 1$ isospin multiplet.[10]

3.5 The Importance of Lie Groups in Physics

Lie groups are of fundamental importance in applications of symmetry to physics. This status derives from several theorems and observations that we summarize without proof [147]. For a set of operators satisfying a Lie algebra $[L_i, L_j] = i f_{ijk} L_k$, one may show the following.

[10] This discussion has treated pions, neutrons, and protons as "fundamental" particles. They are not, since they are tightly bound composites of quarks and antiquarks. However, at low energy this internal structure is invisible and π, n, and p behave effectively as fundamental particles with approximate isospin symmetry.

1. There exist one or more Lie groups with these operators as generators.
2. Group multiplets are *uniquely determined* by the structure constants f_{ijk}.
3. Matrix elements between members of a multiplet are *uniquely determined* by the f_{ijk}.
4. Matrix elements of operators that are constructed from sums of Casimir operators (i) are *uniquely determined* by the f_{ijk} and (ii) vanish for transitions out of the multiplet.
5. In many groups of interest the generators may have direct physical significance.

The importance of simple and semisimple Lie groups rests on the following assertions.

1. A finite number of compact semisimple Lie groups of a given rank exist. For example, there are only three of rank two.
2. Each finite-dimensional, invariant subspace of a semisimple Lie group decomposes into non-overlapping irreducible representations.
3. Simple Lie groups of rank ℓ have ℓ *fundamental irreducible representations* (faithful lowest-dimensional irreps), and every representation can be obtained from products of fundamental irreps.
4. Every representation of a compact, semisimple Lie group is equivalent to a representation by unitary matrices.

These properties will have important implications for many aspects of our discussion.

3.6 Symmetry and Dynamics

Symmetries in physics often are associated with conservation laws such as those for angular momentum or global charge. However, perhaps the most important development in applications of symmetry to physical problems in recent decades has been the use of Lie algebras to determine the *dynamics* of physical systems. Such applications are much more powerful than those that merely impose global conservation laws.

> Conservation laws deriving from global symmetries tell us what is permitted; symmetries associated with dynamics can tell us what actually happens.

Two broad categories of symmetry applications have dynamical implications: (1) imposition of a *local gauge symmetry* through Lie algebras with generators depending on local spacetime coordinates, and (2) the use of Lie algebras to construct *dynamical symmetries* that describe emergent states in quantum many-body systems. We give a brief introduction to these in the next two subsections, and will discuss them extensively in later chapters.

3.6.1 Local Gauge Theories

Local gauge invariance requires that charges of some kind (generally defined through integrals of particle currents) be conserved *locally,* not just globally. For abelian gauge symmetries this implies *local conservation of electric charge* and leads to quantum electrodynamics. For non-abelian gauge symmetries this leads to Yang–Mills field theories and the Standard Model of particle physics, which are built on local conservation of more abstract charges.

> Processes at different spacetime points that separately violate charge conservation but that offset each other so that charge is conserved globally are allowed by global gauge symmetry, but forbidden by local gauge symmetry.

Local gauge invariance has *dynamical implications* that are especially stringent when the gauge group is non-abelian because the commutator algebra imposes non-trivial conditions. We shall have much more to say about local gauge invariance and dynamics in Chs. 16–19.

3.6.2 Dynamical Symmetries

The second category of symmetry considerations having dynamical implications is termed *dynamical symmetry,* which results when the Hamiltonian for a quantum system can be expressed as a polynomial in the invariants of a chain of subgroups deriving from some highest symmetry. We provide an introduction to this concept in Ch. 20 and discuss ambitious applications of dynamical symmetries in Section 20.2 and Chs. 31 and 32. As will be seen, dynamical symmetry is really a statement about how some highest symmetry is *broken* in a physical system. Therefore, it is closely related to many concepts to be discussed in Part III, which deals in some depth with the concept of broken symmetry.

Background and Further Reading

Introductions to Lie groups in physical applications may be found in Close [42], Gasiorowicz [66], Georgi [68], Gilmore [72, 73], Hamermesh [104], Iachello [115], Jeevanjee [124], Lichtenberg [141], Lipkin [142], McVoy [147], O'Raifeartaigh [158], Ramond [169], Schensted [180], Wybourne [224], and Zee [228]. Comprehensive treatments of angular momentum are contained in Brink and Satchler [29], Edmonds [51], and Rose [173]. For isospin see Elliott and Dawber [56], Georgi [68], Lichtenberg [141], and Lipkin [142]. The physical importance of Lie groups is discussed concisely by McVoy [147]. For an introduction to local gauge theories that does not assume a background in particle physics, see Guidry [85]. Comprehensive reviews of many-body dynamical symmetries are given in Guidry, Sun, Wu, and Wu [98], Iachello and Arima [116], and Wu, Feng, and Guidry [218].

Problems

3.1 Use the highest-weight algorithm to show that $\mathbf{2} \otimes \mathbf{2} \otimes \mathbf{2} = \mathbf{4} \oplus \mathbf{2} \oplus \mathbf{2}$, for the product of three fundamental SU(2) representations. ***

3.2 Suppose that $D^{(\alpha)}$ and $D^{(\beta)}$ are irreps of a compact simple group with dimensions n_α and n_β, respectively, and basis vectors x_i and y_i, respectively. If the basis vectors

$\boldsymbol{x} = (x_1, x_2, \ldots)$ and $\boldsymbol{y} = (y_1, y_2, \ldots)$ describe independent degrees of freedom for the system, investigate the transformation properties of the two-index quantities $x_n y_n$ and show that they transform as the matrix $C_{ik,jl}$ in Eq. (2.28).

3.3 By expanding the group elements written in the canonical exponential form (3.2) about the origin, derive the Lie algebra $[\, X_a, X_b \,] = i f_{abc} X_c$. *Hint*: A commutator can be defined by taking the difference between the product of two group elements and the product in reverse order. ***

3.4 Show that the most general form for 2×2 unitary matrices of unit determinant is

$$U = \begin{pmatrix} a & b \\ -b^* & a^* \end{pmatrix} \qquad |a|^2 + |b|^2 = 1,$$

where a and b are complex numbers.

3.5 Prove that if matrices T_a with matrix elements $(T_a)_{bc}$ proportional to the structure constants f_{abc} are defined as in Eq. (3.8), these matrices satisfy the Lie algebra (3.9). Thus, show that the structure constants generate a representation of the algebra with dimension equal to the number of generators. *Hint*: Use Eqs. (3.3) and (3.7). ***

3.6 Use that the adjoint representation for SU(2) is three-dimensional and that generators of the adjoint representation are the structure constants of the group to construct a 3D matrix representation of SU(2). Verify explicitly that the resulting matrices satisfy the SU(2) algebra. Using the standard methods of matrix algebra, transform this set of matrices to a new set T_1, T_2, T_3, where T_3 is diagonal. ***

3.7 Prove *Schur's lemma*: a matrix that commutes with all generators of an irrep is a multiple of the unit matrix. *Hint*: Assume that $[\, T_a, M \,] = 0$ for all a, where T_a is a group generator and M is some matrix. Show that there is a contradiction unless every member of the irrep has the same eigenvalue with respect to M. ***

3.8 Show that for a four-dimensional cartesian space (x, y, z, t) the operators

$$M_1 = z\frac{\partial}{\partial y} - y\frac{\partial}{\partial z} \qquad M_2 = x\frac{\partial}{\partial z} - z\frac{\partial}{\partial x} \qquad M_3 = y\frac{\partial}{\partial x} - x\frac{\partial}{\partial y},$$

$$N_1 = x\frac{\partial}{\partial t} - t\frac{\partial}{\partial x} \qquad N_2 = y\frac{\partial}{\partial t} - t\frac{\partial}{\partial y} \qquad N_3 = z\frac{\partial}{\partial t} - t\frac{\partial}{\partial z},$$

obey the Lie algebra

$$[\, M_i, M_j \,] = \epsilon_{ijk} M_k \qquad [\, M_i, N_j \,] = \epsilon_{ijk} N_k \qquad [\, N_i, N_j \,] = \epsilon_{ijk} M_k,$$

which is the algebra associated with the group SO(4). Show that this SO(4) is locally isomorphic to SU(2) $\times$ SU(2) by showing that the new operator set,

$$J_i \equiv \frac{1}{2}(M_i + N_i) \qquad K_i \equiv \frac{1}{2}(M_i - N_i) \qquad (i = 1, 2, 3),$$

satisfies the Lie algebra

$$[J_i, J_j] = \epsilon_{ijk} J_k \qquad [K_i, K_j] = \epsilon_{ijk} K_k \qquad [K_i, J_j] = 0,$$

which corresponds to the product of two independent SU(2) algebras. ***

3.9 Use the cyclic property of the trace [see Problem 3.11(f)], and that the generators of compact, semisimple algebras may be normalized $\mathrm{Tr}\,(T_a T_b) = \lambda \delta_{ab}$ with λ a constant, to show that

$$f_{abc} = f_{bca} = f_{cab} = -f_{bac} = -f_{cba} = -f_{acb}$$

for the structure constants in Eq. (3.3). *Hint*: Show that $\mathrm{Tr}\,([T_a, T_b] T_c) = i\lambda f_{abc}$.

3.10 Show that the structure constants f_{abc} appearing in Eq. (3.3) are real if the generators of the Lie algebra are hermitian.

3.11 Matrices are central to representation theory. If A, B, C, and D are square, invertible, $n \times n$ matrices with complex entries, prove the following useful properties.

a. The trace of a matrix is invariant under similarity transforms, $\mathrm{Tr}\,(BAB^{-1}) = \mathrm{Tr}\,A$. *Hint*: Written out with indices, $(BAB^{-1})_{ij} = B_{i\ell} A_{\ell k} B^{-1}_{kj}$.
b. Products of exponentiated matrices behave in the expected way, provided that the matrices commute with each other: $e^A e^B = e^{A+B}$ if $[A, B] = 0$. (See the BCH formula quoted in Problem 3.15 for the case $[A, B] \neq 0$.) *Hint*: We may expand

$$e^A = \sum_{m=0}^{\infty} \frac{1}{m!} A^m = 1 + A + \frac{1}{2!} A^2 + \frac{1}{3!} A^3 + \cdots$$

for a matrix A.
c. The exponential e^A has eigenvalues $e^{\lambda_1}, e^{\lambda_2}, \ldots, e^{\lambda_n}$ if the matrix A has eigenvalues $\lambda_1, \lambda_2, \ldots, \lambda_n$.
d. The determinant of an exponentiated matrix is the exponentiated trace of the matrix, $\det e^A = e^{\mathrm{Tr}\,A}$. *Hint*: The determinant of a matrix is the product of its eigenvalues and the trace of a matrix is the sum of its eigenvalues.
e. The trace of a product of two square matrices is independent of the order of multiplication: $\mathrm{Tr}\,(AB) = \mathrm{Tr}\,(BA)$.
f. *Cyclic property of the trace:* The trace of a matrix product is invariant under cyclic permutation of the factors in the product. *Hint*: Use the results of Problem 3.11(e) and that matrix multiplication is associative, $ABC = (AB)C$.
g. The hermitian conjugate of a matrix product is the product of the hermitian conjugates for each matrix, in reverse order. For example, $(ABCD)^\dagger = D^\dagger C^\dagger B^\dagger A^\dagger$. *Hint*: The hermitian conjugate of a matrix is the transpose of its complex conjugate, so the matrix elements for $M^\dagger$ are given by $(M^\dagger)_{ij} = (M_{ji})^*$.
h. The inverse of a matrix product is the product of the inverses for each matrix, in reverse order. For example, $(ABCD)^{-1} = D^{-1}C^{-1}B^{-1}A^{-1}$.

i. Quantum-mechanical matrix elements $\langle a|\, A\, |b\rangle$ obey

$$\sum_{ab} |\, \langle a|\, A\, |b\rangle\, |^2 \equiv \sum_{ab} \langle a|\, A\, |b\rangle^* \, \langle a|\, A\, |b\rangle = \mathrm{Tr}\,(AA^\dagger) = \mathrm{Tr}\,(A^\dagger A),$$

for a matrix A. *Hint*: $\langle a|\, A\, |b\rangle^* = \langle b|\, A^\dagger\, |a\rangle$ and $\sum_a |a\rangle \langle a| = 1$.***

3.12 Show that for a continuous parameter θ the set of matrices

$$G = \begin{pmatrix} \cos\theta & \sin\theta \\ -\sin\theta & \cos\theta \end{pmatrix}$$

forms a one-parameter abelian Lie group under matrix multiplication, and that if the matrices G operate on a 2D cartesian space (x_1, x_2) they leave $x_1^2 + x_2^2$ invariant. ***

3.13 Show that the Pauli matrices (3.11) generate a matrix representation of Lie group elements having the form

$$U = e^{\frac{i}{2}(\sigma_1\theta_1 + \sigma_2\theta_2 + \sigma_3\theta_3)} \simeq \begin{pmatrix} a & b \\ -b^* & a* \end{pmatrix},$$

as required by Problem 3.4.

3.14 Prove that if the Lie group G_1, with generators $J_i(1)$, is isomorphic to a group G_2, with generators $J_i(2)$, and $J_i(1)$ and $J_i(2)$ operate on independent degrees of freedom, then the sum operators $J_i \equiv J_i(1) + J_i(2)$ obey the same Lie algebra as the generators of G_1 and G_2. Thus the J_i generate a group G that is isomorphic to G_1 and G_2.

3.15 Show that the group-element commutator $R_y(\delta\theta)R_x(\delta\theta)R_y^{-1}(\delta\theta)R_x^{-1}(\delta\theta)$, is related to the generator commutator $[\, J_x, J_y\,]$ by

$$[\, J_x, J_y\,] = \frac{i}{\delta\theta}\left[1 - R_y(\delta\theta)R_x(\delta\theta)R_y^{-1}(\delta\theta)R_x^{-1}(\delta\theta)\right],$$

where the group elements and generators are related by $R_x(\theta) = \exp(iJ_x\theta)$ and $R_y(\theta) = \exp(iJ_y\theta)$ for a rotation angle θ. *Hint*: Keep the first three terms of

$$e^A e^B = \exp\left(A + B + \frac{1}{2}[\, A, B\,] + \frac{1}{12}[\, A, [\, A, B\,]] + \frac{1}{12}[\, B, [\, B, A\,]] + \cdots\right),$$

which is called the *Baker–Campbell–Hausdorff (BCH) formula.*

3.16 (a) Confirm the validity of the Jacobi identity given in Eq. (3.6). (b) Use Eq. (3.6) to confirm the validity of Eq. (3.7b).

3.17 If a local density operator is expressed by $\rho(r) = \sum_i \delta(r - r_i)$, where the sum is over particles and $\delta(r - r_i)$ is the Dirac delta function, what is its second-quantized form? *Hint*: See Appendix A and note that the sum is an integral in the continuum limit.

3.18 Given a single-particle operator $F = \sum_i f(r_i, p_i)$, the second-quantized form is

$$F = \sum_{ij} \langle i|\, f\, |j\rangle\, a_i^\dagger a_j \qquad \langle i|\, f\, |j\rangle \equiv f_{ij} \equiv \int_r \psi_i^* f(r,p)\psi_j(r)$$

(see Appendix A). Consider the spin operators $S^a = \frac{1}{2}\sigma^a$, with the representation of the $\sigma^a (a = 1, 2, 3)$ given in Eq. (3.11). Evaluate the second-quantized form of the three spin operators (S^1, S^2, S^3) assuming a basis

$$|\uparrow\rangle = \begin{pmatrix}1\\0\end{pmatrix} \quad \text{(spin up)} \qquad |\downarrow\rangle = \begin{pmatrix}0\\1\end{pmatrix} \quad \text{(spin down)}$$

to show that

$$S^1 = \frac{1}{2}(a_\uparrow^\dagger a_\downarrow + a_\downarrow^\dagger a_\uparrow) \qquad S^2 = \frac{1}{2i}(a_\uparrow^\dagger a_\downarrow - a_\downarrow^\dagger a_\uparrow) \qquad S^3 = \frac{1}{2}(a_\uparrow^\dagger a_\uparrow - a_\downarrow^\dagger a_\downarrow)$$

for spin-$\frac{1}{2}$ electrons.

3.19 In $\hbar = c = 1$ natural units a particular hadronic cross section σ is estimated to be

$$\sigma \sim \frac{1}{(M_\pi)^2} \sim \frac{1}{(140)^2}\,\text{MeV}^{-2},$$

where the mass of the pion is $M_\pi \sim 140$ MeV. What is this cross section in more standard units of barns (b), where 1 b $\equiv 10^{-24}$ cm^2? *Hint*: See Appendix B and use dimensional analysis. ***

3.20 In special relativity it is common to use units where the speed of light c is set to one. The world record in the 100 meter dash is about 9.6 seconds. What is this time expressed in $c = 1$ units? What is the physical meaning of your result? ***

3.21 In natural ($\hbar = c = 1$) units the mean life for the decay $\Sigma^0 \to \Lambda + \gamma$, where Σ^0 and Λ are elementary particles and γ is a photon, is

$$\tau \simeq \frac{\pi(M_\Lambda + M_\Sigma)^2}{e^2 E_\gamma^3},$$

where M_Σ and M_Λ are the masses of the elementary particles, E_γ is the energy of the photon, and $\pi/e^2 \sim 137/4$. What is the lifetime in seconds for this decay if $(M_\Lambda + M_\Sigma) = 2307$ MeV and $E_\gamma = 74.5$ MeV? ***

3.22 What is one Joule in $c = 1$ units? What is one atmosphere (10^5 N m^{-2}) of pressure expressed in $c = 1$ units? ***

3.23 Using the commutators for J_i and K_i given in Problem 3.8, argue that the SO(4) Lie algebra is semisimple, but not simple. Argue that SO(4) can be written as a direct product of two simple groups, which can be analyzed independently.

4 Permutation Groups

In Ch. 2 the group S_3 of permutations on three objects was used to illustrate some important group-theoretical concepts. More generally, the symmetric or permutation groups S_n of permutations on n objects are important in group theory for several reasons.

1. A finite group of order n is isomorphic to a subgroup of S_n (*Cayley's theorem*).
2. The permutation groups are of direct physical relevance in the description of symmetry under particle exchange for systems of identical particles.
3. Through the methods of tensor analysis, the irreps of S_n provide a powerful tool to analyze the irreps for groups of continuous transformations such as U(N) and SU(N).

In this chapter we will introduce analysis of the symmetric group through the method of *Young diagrams*. In subsequent chapters we will see how Young diagrams of the permutation groups may be used in the analysis of other groups.

4.1 Young Diagrams

As an example of notation, let us label three distinguishable single-particle states by a, b, and c, and three identical particles by 1, 2, and 3.[1] Consider the three-particle wavefunction $\psi(123) \equiv a(1)b(2)c(3)$, where the notation means that particle 1 is in state a, particle 2 is in state b, and particle 3 is in state c. This may be specified more compactly by writing in order the single-particle state for particles 1, 2, and 3: $\psi(123) \equiv abc$. We wish to consider the effect of a permutation (2.6) on this state, and more generally to investigate the permutation symmetry for n objects distributed over m available quantum states. To that end we introduce a diagrammatic formalism in which *a box stands for a particle* and the *Young diagram* for n particles corresponds to a pattern of n boxes arranged according to rules that will be explained shortly. In these diagrams a row implies *symmetrization* of the particles (boxes) in that row under exchange of the coordinates, while a column implies *antisymmetrization* of the particles (boxes) in that column under coordinate exchange.

[1] It is *essential* to keep track of what is labeling particles (numbers $1, 2, 3, \ldots$ in this example) and what is labeling states available to those particles (letters $a, b, c, \ldots$ for this example).

4.1.1 Two-Particle Young Diagrams

The boxes for a two-particle system may be arranged in two ways,

$\psi_S =$ 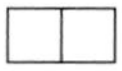 $\psi_A =$  . (4.1)

A Young diagram having only a row of boxes denotes a completely symmetric wavefunction (ψ_S) and one having only a column of boxes denotes a completely antisymmetric wavefunction (ψ_A). For more than two particles the situation will get more complicated.

4.1.2 Many-Particle Young Diagrams

Consider n particles. The rule for allowed Young diagrams that prevents double counting is that the n boxes may be put together in any fashion for which *no row is longer than the row above it.* For three particles then, there are three allowed diagrams,

$\psi_S =$ 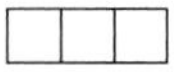 $\psi_M =$ 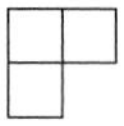 $\psi_A =$ 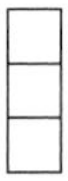 . (4.2)

The first and last diagrams correspond to completely symmetric and completely antisymmetric combinations, respectively, that are analogous to those in Eq. (4.1), but the middle diagram is new. It is called a diagram of *mixed symmetry*, which indicates a wavefunction symmetric with respect to exchange of two particles, but antisymmetric with respect to subsequent exchange of one of those particles with the third particle.

What is the meaning of mixed-symmetry diagrams? All known elementary particles are either fermions or bosons. Collections of indistinguishable fermions must be completely antisymmetric with respect to exchange of the particles, while systems of indistinguishable bosons must be completely symmetric with respect to exchange of the particles. However, these statements apply to simultaneous exchange of *all* coordinates for the particles; in some instances it is useful to consider the symmetry of a many-body system under exchange of only some of the coordinates of the particles. In that case we may encounter states of mixed symmetry under that exchange, even though they are either totally symmetric or totally antisymmetric under the exchange of all particle coordinates.[2]

4.1.3 A Compact Notation

For systems of many particles it is convenient to introduce a compact notation for designating Young diagrams. Several possibilities are in common use. We may illustrate two of them by considering the following 15-box diagram for a 15-particle system,

[2] By all coordinates for a particle we mean variables describing any internal degrees of freedom such as spin or isospin, in addition to the spacetime coordinates.

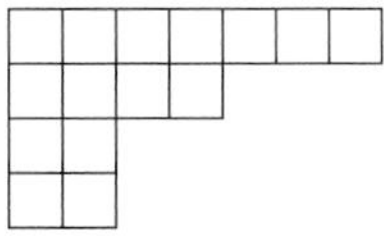

A *partition* of an integer n is a set of positive integers that sum to n. A Young diagram can be labeled by a partition specifying the number of boxes in each row of the diagram. For example, the diagram shown above is specified by the partition $\lambda = (\lambda_1\lambda_2\lambda_3\lambda_4) = (7422)$, which can be abbreviated as (742^2).

A valid arrangement of a Young diagram for an n-particle system corresponds to all possible partitions of n, subject to the rule that in the resulting diagram *no row is longer than the row above it* ($\lambda_i \geq \lambda_{i+1}$, where i labels the row).

Another useful way of denoting a Young diagram is by employing an ordered set of integers that gives the *difference* in the number of boxes in successive rows. If we define $p_i = \lambda_i - \lambda_{i+1}$, the diagram shown above can be specified by $p = (p_1\, p_2\, p_3\, p_4) = (3202)$.

4.2 Standard Arrangement of Young Tableaux

The $n!$ possible permutations on $\psi(123\ldots n)$ generally form *reducible representations* of the permutation group S_n. The irreducible representations of S_n are the linear combinations of these permutations having definite permutational symmetry properties (that is, they correspond to a specific Young diagram). Generally, one finds a single totally symmetric and a single totally antisymmetric linear combination, each corresponding to a one-dimensional irrep of S_n. All other irreducible representations of S_n are of mixed symmetry and are multidimensional. Young diagrams can be used to find the dimensionality of the irreps of S_n if we define a *Young tableau* and a *standard arrangement* of tableaux. A Young diagram with numbers in the boxes as specified below is, by strict usage, termed a *Young tableau*. A standard arrangement of a Young tableau is defined to be a Young diagram in which no row is longer than the row above it, with a set of positive integers placed in the boxes subject to the following restrictions.

1. The numbers do not decrease in reading from left to right in any row.
2. The numbers increase from top to bottom in any column.
3. If m states are available to a particle, the numbers j appearing in boxes satisfy $1 \leq j \leq m$.

In applying these rules, do not confuse the *number of particles* (the number of boxes in the diagram) with the *number of states* available to those particles (the maximum value of an integer that can appear in a box).

The following example illustrates application of these rules for the case of three available states that can be populated by two identical particles.

Example 4.1 The standard arrangements of Young tableaux for two identical particles distributed over three available states are

$$\text{Symmetric:}\quad \begin{array}{|c|c|}\hline 1&1\\\hline\end{array}\quad \begin{array}{|c|c|}\hline 1&2\\\hline\end{array}\quad \begin{array}{|c|c|}\hline 1&3\\\hline\end{array}\quad \begin{array}{|c|c|}\hline 2&2\\\hline\end{array}\quad \begin{array}{|c|c|}\hline 2&3\\\hline\end{array}\quad \begin{array}{|c|c|}\hline 3&3\\\hline\end{array}$$

$$\text{Antisymmetric:}\quad \begin{array}{|c|}\hline 1\\\hline 2\\\hline\end{array}\quad \begin{array}{|c|}\hline 1\\\hline 3\\\hline\end{array}\quad \begin{array}{|c|}\hline 2\\\hline 3\\\hline\end{array}$$

where each diagram has two boxes since there are two particles, and each box contains an integer from 1 to 3 (subject to the rules given above), since three states are available.

In standard arrangements such as in Example 4.1 the same number can occur twice in the same row but *never twice in the same column,* because a column is an injunction to antisymmetrize. A number appearing twice in the same column means that two particles are in the same state, which implies a wavefunction impossible to antisymmetrize.

4.3 Irreducible Representations

In the discussion of the permutation groups in Ch. 2, it was noted that all permutations in S_n with the same cycle structure belong to the same class. This result can be used to establish a one to one correspondence between partitions of n and the irreps of S_n, because each tableau corresponds to a conjugacy class and thus to an irreducible representation.

4.3.1 Counting Standard Arrangements

To determine the dimensionality of an S_n irrep we must count the number of *standard arrangements of Young tableaux that have each particle in a different state.* This requires that we further restrict the standard arrangement of tableaux so that the same number does not appear twice. The following example illustrates.

Example 4.2 Let us determine the dimensionality of S_3 irreps. This reduces to counting standard arrangements with three identical particles distributed on three states, but with no two particles in the same state. The standard arrangements satisfying these conditions are

$$\begin{array}{|c|c|c|}\hline 1&2&3\\\hline\end{array}\qquad \begin{array}{|c|}\hline 1\\\hline 2\\\hline 3\\\hline\end{array}\qquad \begin{array}{|c|c}\hline 1&\multicolumn{1}{c|}{2}\\\hline 3\\\cline{1-1}\end{array}\qquad \begin{array}{|c|c}\hline 1&\multicolumn{1}{c|}{3}\\\hline 2\\\cline{1-1}\end{array}$$

The first two diagrams correspond to 1D irreps of S_3. The next two have equivalent Young patterns and constitute a 2D irreducible representation. We conclude that S_3 has a completely symmetric 1D irrep, a completely antisymmetric 1D irrep, and a 2D irrep of mixed symmetry. From Eq. (2.29b) the sum of the squares of the dimensions of the irreps must equal the dimension of the group, so this exhausts the possible irreps for S_3.

A similar technique can be used to find the irrep dimensionality of any symmetric group S_n. Problem 4.4 illustrates for S_4. However, for larger n this method becomes tedious. We now describe a dimensionality recipe that is much more efficient for many-particle states.

4.3.2 The Hook Rule

First we define a *hook number* for a box that is the number of boxes that a line passes through if it enters from the far right of the row of the box, goes to the center of the box, makes a 90° downward turn, and then exits from the bottom of the diagram. For example,

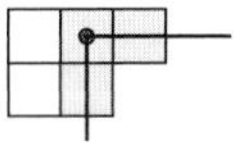

h_i = hook length = 3.

Then, for the group S_n the dimension Dim (S_n) of an irrep specified by a Young diagram is given by the formula

$$\text{Dim}\,(S_n) = \frac{n!}{\prod_i h_i}, \tag{4.3}$$

where the denominator product is over the hook numbers for all boxes in the diagram. To illustrate, let us consider the irrep corresponding to the partition (421) for S_7. Drawing the diagram and placing the hook length in each box (do not confuse these hook numbers with the ones we have been placing in boxes to label particle states),

$$\begin{array}{|c|c|c|c|}\hline 6 & 4 & 2 & 1 \\ \hline 3 & 1 \\ \cline{1-2} 1 \\ \cline{1-1}\end{array} \longrightarrow \quad \text{Dim} = \frac{7!}{6\cdot 4\cdot 3\cdot 2\cdot 1\cdot 1\cdot 1} = 35.$$

This is a *much faster way* to calculate irrep dimensionality than enumerating the standard tableaux for cases like this one where there are more than a few boxes (particles).

4.4 Basis Vectors

Let us now consider the construction of basis vectors for wavefunctions corresponding to the irreducible representations of S_n. To facilitate this we first define some useful symmetrizing and antisymmetrizing operators [180]. Suppose a given tableau is designated by τ, and that we introduce a *row permutation* p_τ that interchanges numbers in a row of the diagram. Then the *row symmetrizer* $\mathbb{P}_\tau$ is defined by the sum over all row permutations

$$\mathbb{P}_\tau = \sum p_\tau. \tag{4.4}$$

Likewise, we may introduce a *column permutation* q_τ and define a *column antisymmetrizer*

$$\mathbb{Q}_\tau = \sum \eta_{q_\tau} q_\tau \qquad \eta_{q_\tau} = \begin{cases} +1 & \text{if } q_\tau \text{ is even} \\ -1 & \text{if } q_\tau \text{ is odd,} \end{cases} \tag{4.5}$$

where the sum is over all column permutations. The *Young operator* $\mathbb{Y}_\tau$ is then defined by

$$\mathbb{Y}_\tau = \mathbb{Q}_\tau \mathbb{P}_\tau = \sum_{q_\tau} \sum_{p_\tau} \eta_{q_\tau} q_\tau p_\tau . \tag{4.6}$$

Suitably normalized, $\mathbb{Y}$ is a *primitive idempotent* of the group algebra. This means that it is equal to its square, and this means that it is a *projection operator* that extracts states of definite permutation symmetry (irreps of S_n) when applied to an arbitrary wavefunction.

Example 4.3 Consider the S_3 Young diagram $\boxed{1}\boxed{2}\boxed{3}$, for which

$$\mathbb{P}_\tau = 1 + (12) + (23) + (13) + (123) + (321)$$

and $\mathbb{Q}_\tau = 1$. Therefore,

$$\mathbb{Y}\left(\boxed{1}\boxed{2}\boxed{3}\right) = \mathbb{Q}_\tau \mathbb{P}_\tau = 1 + (12) + (23) + (13) + (123) + (321).$$

Applying this operator to the three-particle wavefunction $\psi(123) \equiv abc$ gives

$$\psi_S = \mathbb{Y}\left(\boxed{1}\boxed{2}\boxed{3}\right) abc = abc + bac + acb + cba + bca + cab,$$

for the S_3 irrep basis vector. Similar examples for S_4 are addressed in Problem 4.6.

Thus the Young operator (4.6) serves as a projector that may be used to construct basis vectors for irreducible representations of the permutation group S_n.

4.5 Products of Representations

Two important products may be defined for representations of the permutation groups.

The Direct Product: The direct product is a product of different representations for the *same number of particles* (the same group S_n). It is useful when the functions being multiplied refer to *different coordinates for the same set of particles*. An important application occurs for the multiplication of a spin function by a spatial function for some set of particles in quantum mechanics. The direct product is described in Section 4.5.1.

The Outer Product: The outer product occurs when the symmetry states for two *different* sets of particles are combined. That is, it is a product of representations for two *different groups,* S_n and $S_{n'}$. This product is important when considering the permutation symmetry of $n + n'$ particles, given the symmetry of the n particles and n' particles separately. The outer product is described in Section 4.5.2.

4.5.1 Direct Products

The direct product of representations for finite or compact, semisimple groups generally leads to a direct sum of irreps (Section 2.7). The invariance of the characters determines the irrep content of the direct product in the following way. The Clebsch–Gordan series,

$$D^{(i)} \otimes D^{(j)} = \sum_{k\oplus} c_k D^{(k)},$$

where $\otimes$ indicates the direct product and $k\oplus$ indicates a direct sum over irreps as in Eq. (3.21), implies a corresponding relation for the characters of the representations

$$\chi_\alpha^{(i)} \chi_\alpha^{(j)} = \sum_k c_k \chi_\alpha^{(k)}, \tag{4.7}$$

where α is a class index. This relation and the group character table may be used with Eq. (2.32) to deduce the irrep content for direct product of S_n representations.

Example 4.4 Consider the direct product of the S_3 irrep $\Gamma^{(3)}$ given in Fig. 2.4 with itself. From Table 2.4 the character of the direct product is $\chi^{(3\otimes 3)} = (4, 1, 0)$, where the three numbers refer to the product of characters within the e, P_{ij}, and P_{ijk} classes of Table 2.4, respectively. Applying Eq. (2.32) for instances a_ν of the irreps $\Gamma^{(\nu)}$ in the direct product,

$$a_1 = \frac{1}{6}(1)(4) + \frac{2}{6}(1)(1) + \frac{3}{6}(1)(0) = 1 \qquad a_2 = \frac{1}{6}(1)(4) + \frac{2}{6}(1)(1) + 0 = 1$$

$$a_3 = \frac{1}{6}(2)(4) + \frac{2}{6}(-1)(1) + 0 = 1,$$

so that the direct product of the S_3 irrep $\Gamma^{(3)}$ with itself is $\Gamma^{(3)} \otimes \Gamma^{(3)} = \Gamma^{(1)} \oplus \Gamma^{(2)} \oplus \Gamma^{(3)}$.

Since only completely symmetric (boson) or completely antisymmetric (fermion) irreps of S_n occur in nature, it is important to know when the direct product of two irreps of S_n contains a completely symmetric or completely antisymmetric representation.

1. The completely symmetric irrep occurs only in the direct product of a representation with itself.
2. The completely antisymmetric irrep occurs only in the direct product of a representation with its *associate representation*.[3]

If a representation of S_n is described by a Young diagram, the associate representation is described by the corresponding Young diagram in which rows are interchanged for columns (reflection about the main diagonal). For example, consider the S_4 diagrams

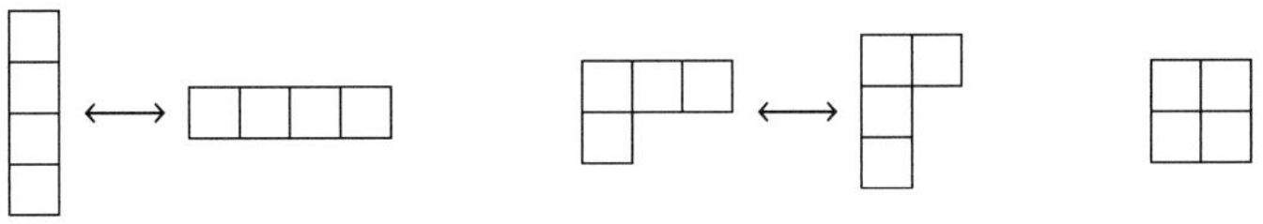

The first two pairs are associate, while the last diagram is its own associate (*self-associate*). Note a simple rule to obtain an associate diagram: rotate clockwise by 90 degrees and reflect left–right.

[3] Associate representations are called *conjugate representations* or *adjoint representations* by some authors.

4.5.2 Outer Products

The outer product (denoted by ×) generates a representation of $S_{n+n'}$ from a representation $D^{(\alpha)}$ of S_n and a representation $D^{(\alpha')}$ of $S_{n'}$. It can be used to build wavefunctions of definite permutation symmetry for $n + n'$ particles, if the permutation symmetries of the systems of n and n' particles are known. Its most important application for us will be in constructing the direct product of unitary symmetries, which will be elaborated in Section 8.9. There is a simple diagrammatic procedure for finding the outer product of two S_n representations.

1. Draw the diagrams for the two representations and label the second diagram with numbers in the boxes denoting the *row in which the box appears.* For example,

   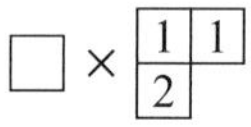

   Break the second diagram up into individual boxes and attach the boxes (with their numbers) to the first diagram in all possible ways subject to the following restrictions.

   a. All diagrams are proper diagrams (rows do not increase in length down the diagram).
   b. On a path passing through each row from right to left and moving from top to bottom, at all points on the path the number of boxes encountered containing the number i is less than or equal to the number of boxes encountered containing the number $i - 1$.
   c. The numbers in the boxes do not decrease from left to right in a row.
   d. The numbers in each column increase from top to bottom.

2. Each diagram constructed by this procedure corresponds to an irrep of $S_{n+n'}$.

Let us illustrate for the outer product of an S_4 irrep and an S_2 irrep.

Example 4.5 Consider the outer product of the [31] irrep of S_4 and the [2] irrep of S_2:

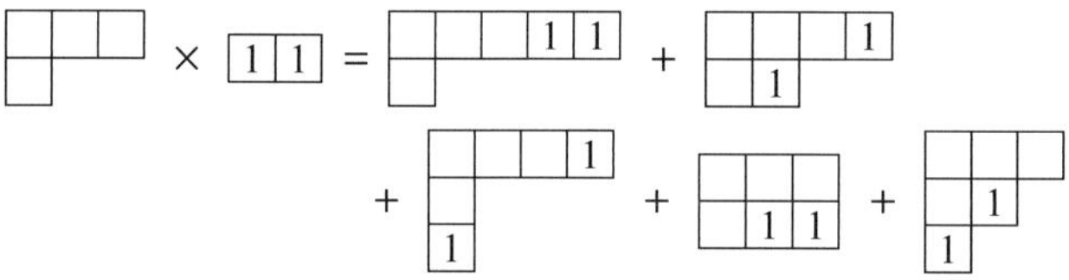

(4.8)

or equivalently, [31] × [2] = [51] + [42] + [411] + [33] + [321], in terms of partitions.

An outer product construction may be checked by counting dimensionalities. In the outer product of a representation $[f]$ of dimension d for S_k and a representation $[f']$ of dimension d' for $S_{k'}$, the number of states in the outer product representation of $S_{k+k'}$ is [106]

$$\text{Dim}\ ([f] \times [f']) = \frac{(k + k')!}{k!\, k'!}\, dd'. \tag{4.9}$$

In Example 4.5, $k = 4$ and $k' = 2$, and from Eq. (4.3) $d = 3$ and $d' = 1$, which gives Dim ([31] × [2]) = 45 from Eq. (4.9). Counting dimensionalities on the right side of Eq. (4.8) using Eq. (4.3) gives the same result: 5 + 9 + 10 + 5 + 16 = 45.

Background and Further Reading

Permutation groups are discussed clearly in Elliott and Dawber [56], Georgi [68], Hamermesh [104], and Schensted [180].

Problems

4.1 Show that the set of permutations on n objects forms a group with a multiplication operation defined as the application of two successive permutation operations.

4.2 Verify that the mapping $\{e, (123), (321)\} \rightarrow 1$ and $\{(12), (23), (31)\} \rightarrow -1$ preserves the group multiplication operation for the permutation group S_3. ***

4.3 Sketch the Young tableaux for the permutation group S_5 and find their dimensionalities using the hook rule.

4.4 Use counting of Young tableaux to find the dimensionalities of the irreps for the group S_4. ***

4.5 For the irreps $\Gamma^{(1)}$, $\Gamma^{(2)}$, and $\Gamma^{(3)}$ of the group S_3 (Fig. 2.4 and Table 2.4), determine the irrep content of the nine direct products $\Gamma^{(n)} \otimes \Gamma^{(m)}$.

4.6 Write the Young operators for the S_4 tableaux

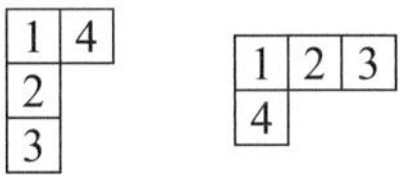

and construct explicit wavefunctions having these permutational symmetries. ***

4.7 Construct the outer product [21] × [21] for the partition [21] of S_3. Check the dimensionalities using Eq. (4.9). *Hint*: See Example 4.5. ***

4.8 Construct the outer product [4] × [3] for the partitions [4] of S_4 and [3] of S_3. Check the dimensionalities using Eq. (4.9). *Hint*: See Example 4.5.

4.9 Construct the wavefunction for the S_3 irrep  using the method of Example 4.3.

4.10 Determine the irrep content of

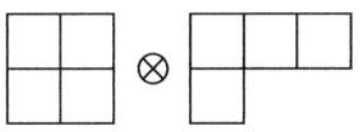

for S_4 representations. *Hint*: The character table for S_4 is given by [180]

Irrep\ Class	$e(1)$	$A(8)$	$B(3)$	$C(6)$	$D(6)$
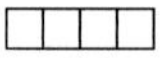	1	1	1	1	1
	1	1	1	–1	–1
	2	–1	2	0	0
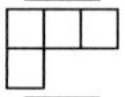	3	0	–1	1	–1
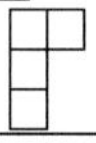	3	0	–1	–1	1

where e is the identity, A, B, C, and D are convenient class labels, and the number of elements for each class is shown in parentheses. ***

5 Electrons on Periodic Lattices

Many systems form crystalline phases over significant ranges of temperature and density. A *perfect crystal* may be thought of as a lattice with a periodically repeated basic structural unit – which may itself have a non-trivial structure – filling all of space. The electronic properties of the solid state are strongly influenced by such periodic crystalline lattices. Periodicity implies symmetry, with significant implications for the physics of condensed matter. Unlike for many other applications tending to emphasize symmetries that are continuous, symmetries important in condensed matter often involve combinations of discrete and continuous symmetries because of the propagation of waves through discrete lattice structures. For every crystal structure there are two lattices of physical significance.

1. The spatial lattice, often called the *real-space lattice* or *direct lattice.*
2. A momentum-space lattice obtained from the direct lattice by Fourier transform that is termed the *reciprocal lattice.*

Sometimes the direct lattice is termed *r-space* and the reciprocal lattice *k-space.* Vectors defined on the direct lattice and on the reciprocal lattice have a relationship similar to the duality between vectors and dual vectors described in Section 13.1.1. As is generally the case in quantum mechanics, such real-space and momentum-space representations are equivalent descriptions if no approximations are made, but one may be more useful than the other in a particular context and both can be important for a full physical picture of a lattice problem.

5.1 The Direct Lattice

Let us begin with the direct lattice. The repeating structural unit of a perfect crystal is called a *unit cell* and a unit cell of minimum possible volume is termed a *primitive unit cell.*

5.1.1 Brevais Lattices

A *Brevais lattice* is a mathematical collection of points forming a periodic array in which any lattice point corresponds to a position vector $\boldsymbol{R}$ having the form (in three dimensions)

$$\boldsymbol{R} = n_1\boldsymbol{a}_1 + n_2\boldsymbol{a}_2 + n_3\boldsymbol{a}_3, \tag{5.1}$$

where the coefficients n_i can take all integer values and where the vectors $\boldsymbol{a}_i$ are called *primitive vectors.* Thus, any point in the lattice may be specified by an integral combination

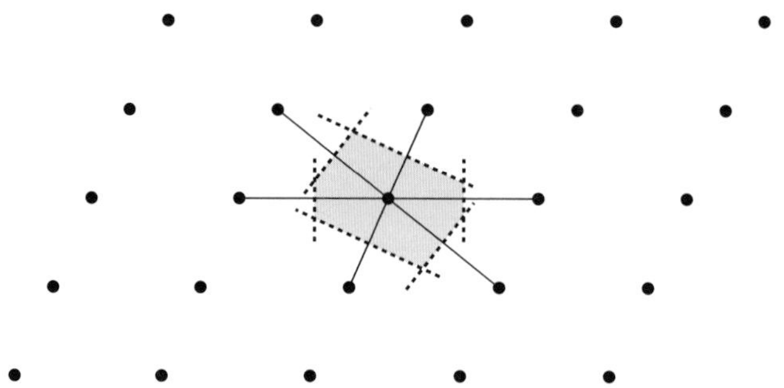

Fig. 5.1 Construction of a Wigner–Seitz unit cell around a site for a two-dimensional lattice. Each dotted line is a perpendicular bisector of a line joining the site to a nearest-neighbor site.

of primitive lattice vectors $\boldsymbol{a}_i$. For an actual physical lattice the vectors $\boldsymbol{a}_i$ have the dimension of length and the corresponding real-space lattice is termed the *direct lattice.*[1] In three dimensions, 14 distinct Brevais lattices are possible. A *translation vector* or *lattice vector* $\boldsymbol{T}$ connects two points on the lattice, $\boldsymbol{T} = \boldsymbol{R}_i - \boldsymbol{R}_j$. If $\boldsymbol{T}_1$ and $\boldsymbol{T}_2$ are lattice vectors, then their negatives, their sums, and their differences are lattice vectors too.

5.1.2 Wigner–Seitz Cells

There is no unique way to select a primitive unit cell for a Brevais lattice but a common choice is a *Wigner–Seitz cell,* which is that region of space bounded by the perpendicular bisectors for all lines connecting a site to nearest-neighbor sites, as illustrated in Fig. 5.1.Since a Wigner–Seitz cell is a primitive unit cell, the entire lattice can be constructed by translating copies of it through all possible lattice vectors. The complete description of a perfect crystal requires specifying both a Brevais lattice and the distribution of mass (atoms) within each unit cell. If the unit cell contains more than one atom, the positions of these atoms with respect to the center of the cell is called the *basis*. Thus a crystal structure is characterized by its Brevais lattice and its basis.

5.2 The Reciprocal Lattice

Consider a real-space Brevais lattice specified by a set of points $\boldsymbol{R}$. A plane wave $e^{i\boldsymbol{k}\cdot\boldsymbol{r}}$ will have the periodicity of this lattice only for particular choices of the wavevector $\boldsymbol{k}$. The set of all wavevectors $\boldsymbol{K}$ that give plane waves having the periodicity of a real-space (direct)

[1] Equivalently, a Brevais lattice is an array of points that appears exactly the same when viewed from any position in the lattice, implying that the lattice must be of infinite extent. Real crystals are finite but they contain so many atoms that the fiction of infinite extent is useful. Formally, applying infinite-lattice assumptions to a real crystal means either that surface effects are neglected, or that periodic boundary conditions are imposed such that a pattern of finite extent is repeated to fill all of space.

lattice is called the *reciprocal lattice* associated with the direct lattice. A wavevector $\boldsymbol{K}$ satisfies the periodicity condition if

$$e^{i\boldsymbol{K}\cdot(\boldsymbol{r}+\boldsymbol{R})} = e^{i\boldsymbol{K}\cdot\boldsymbol{r}} e^{i\boldsymbol{K}\cdot\boldsymbol{R}} = e^{i\boldsymbol{K}\cdot\boldsymbol{r}},$$

for all $\boldsymbol{r}$ and $\boldsymbol{R}$ on the real-space Brevais lattice. Thus, factoring out the common exponential, the reciprocal lattice corresponds to the wavevectors $\boldsymbol{K}$ satisfying

$$e^{i\boldsymbol{K}\cdot\boldsymbol{R}} = 1, \tag{5.2}$$

for all $\boldsymbol{R}$ on the real-space Brevais lattice. A reciprocal lattice is defined with respect to a *particular* direct Brevais lattice. It may be shown that

1. the reciprocal lattice of a direct Brevais lattice is itself a Brevais lattice in the k-space,
2. the reciprocal of the reciprocal lattice is the original direct Brevais lattice, and
3. the point group symmetry (see Section 5.6) of the reciprocal lattice is the same as the point group symmetry of the direct lattice.

Primitive vectors $\boldsymbol{b}_i$ for the reciprocal lattice have a dimension of inverse length and are given in terms of the primitive vectors $\boldsymbol{a}_i$ for the corresponding direct lattice by

$$\boldsymbol{b}_1 = 2\pi\frac{\boldsymbol{a}_2 \times \boldsymbol{a}_3}{\boldsymbol{a}_1 \cdot (\boldsymbol{a}_2 \times \boldsymbol{a}_3)} \qquad \boldsymbol{b}_2 = 2\pi\frac{\boldsymbol{a}_3 \times \boldsymbol{a}_1}{\boldsymbol{a}_1 \cdot (\boldsymbol{a}_2 \times \boldsymbol{a}_3)} \qquad \boldsymbol{b}_3 = 2\pi\frac{\boldsymbol{a}_1 \times \boldsymbol{a}_2}{\boldsymbol{a}_1 \cdot (\boldsymbol{a}_2 \times \boldsymbol{a}_3)}, \tag{5.3}$$

and the reciprocal lattice is a Brevais lattice corresponding to those points $\boldsymbol{K}$ satisfying[2]

$$\boldsymbol{K} = n_1\boldsymbol{b}_1 + n_2\boldsymbol{b}_2 + n_3\boldsymbol{b}_3, \tag{5.4}$$

for arbitrary integers n_i. From the definitions (5.3),

1. the vectors $\boldsymbol{a}_i$ and $\boldsymbol{b}_j$ obey $\boldsymbol{a}_i \cdot \boldsymbol{b}_j = 2\pi\delta_{ij}$,
2. the volume of a unit cell for the reciprocal lattice is inversely proportional to the volume of a unit cell for the direct lattice, and
3. the vector $\boldsymbol{K}$ in Eq. (5.4) satisfies the periodicity condition (5.2),

as may be proved using basic vector algebra.

5.3 Brillouin Zones

Bragg planes of the reciprocal lattice correspond to perpendicular bisectors of lines joining a point to nearest-neighbor points, next-nearest-neighbor points, and so on. Some examples are shown in Fig. 5.2. From the definitions of Bragg planes and Wigner–Seitz primitive unit cells, a Wigner–Seitz unit cell for the reciprocal lattice is the region that can be reached from the origin without crossing a Bragg plane. This Wigner–Seitz primitive unit cell for the reciprocal lattice is called the *first Brillouin zone.* Wigner–Seitz cells can be defined in any space but Brillouin zones are always defined in k-space: although it is common to

[2] Some omit the factors of 2π in the definitions (5.3), instead multiplying the right side of Eq. (5.4) by 2π.

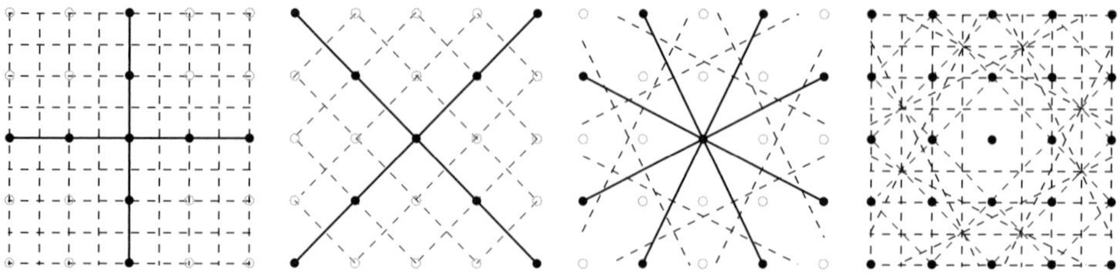

Fig. 5.2 Some Bragg planes for a square 2D reciprocal lattice. The left three diagrams show separate sets of Bragg planes relative to the central point as dashed lines and the rightmost diagram shows them all superposed. Solid black lines indicate the perpendiculars to each set of Bragg planes.

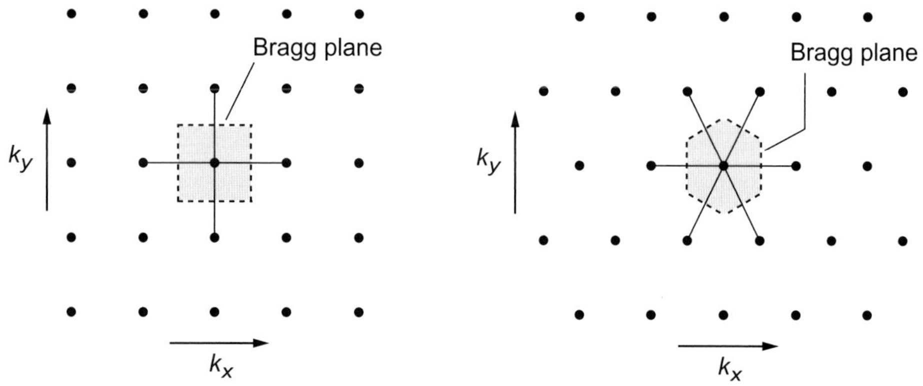

Fig. 5.3 First Brillouin zones for square and hexagonal 2D lattices. Central points are assumed to lie at the origin of k-space. Dashed lines indicate the bounding Bragg planes.

refer loosely to Brillouin zones of a particular real-space lattice, this is really shorthand for Wigner–Seitz cells of the corresponding reciprocal lattice.

> The first Brillouin zone defines a *primitive cell in reciprocal space.* Just as a real-space lattice is divided up into Wigner–Seitz unit cells, we may view the reciprocal lattice as being divided up into Brillouin zones.

The first Brillouin zones for square and hexagonal two-dimensional lattices are illustrated in Fig. 5.3. We may define the *second Brillouin zone* to be that region of the reciprocal lattice that can be reached from the first Brillouin zone by crossing only one Bragg plane, and generally the nth Brillouin zone is the region that can be reached from the origin of k-space by crossing exactly $n - 1$ Bragg planes. The first several Brillouin zones for a square reciprocal lattice are illustrated in Fig. 5.4.

As a consequence of Bloch's theorem (described in Section 5.4), *all* lattice wavefunctions may be classified by a wavevector $\boldsymbol{k}$ lying in a single Brillouin zone, which we choose conventionally to be the first Brillouin zone and henceforth term simply "the Brillouin zone." The Brillouin zone is central to understanding electronic structure on a lattice because it corresponds to the set of all possible momenta $\boldsymbol{k}$ for electrons in a crystal.

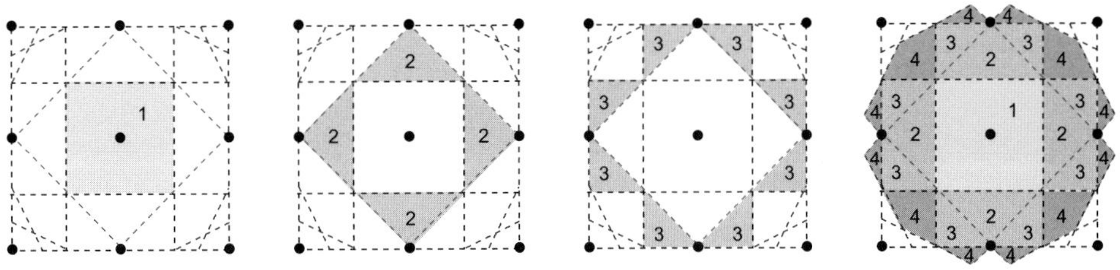

Fig. 5.4 The first several Brillouin zones for a square reciprocal lattice. Unshaded outer portions of the rightmost figure are segments of zones 5 and 6.

5.4 Bloch's Theorem

A translationally invariant lattice can have excitations that break the invariance. For example, the lattice can undergo quantized vibrations (phonon states), electrons can move on the lattice (electronic states), spins can be excited on localized atoms (spin waves), and so on. The key to understanding this more complex behavior is that each of these possibilities is described dynamically by an equation that is *invariant under lattice translations.* For example, electronic states may be described by a Schrödinger equation of the form

$$\left(-\frac{\hbar^2}{2m}\nabla^2 + V(\boldsymbol{r}) - E\right)\psi = 0, \tag{5.5}$$

for which $V(\boldsymbol{r}+\boldsymbol{R}) = V(\boldsymbol{r})$ for all $\boldsymbol{R}$, since the periodicity of the lattice implies a periodicity in the potential seen by an electron. Thus Eq. (5.5) is invariant under $\boldsymbol{r} \to \boldsymbol{r} + \boldsymbol{R}$.

Suppose translations by lattice vectors to be the only symmetries of a lattice. The most general translation by a lattice vector $\boldsymbol{R}$ of Eq. (5.1) is a product of three translations: $T(n_1\boldsymbol{a}_1)T(n_2\boldsymbol{a}_2)T(n_3\boldsymbol{a}_3)$, where the action of a translation operator on a wavefunction is

$$T(\boldsymbol{s})\psi(\boldsymbol{r}) = \psi(\boldsymbol{r} - \boldsymbol{s}). \tag{5.6}$$

For infinite crystals the product of two translations is another translation and translations commute with each other, so the translations form a group T that is a direct product of abelian groups corresponding to translation in the three orthogonal directions,

$$T = T_1 \times T_2 \times T_3. \tag{5.7}$$

To deal mathematically with the finite nature of real crystals, it is useful to introduce a periodic boundary condition in each direction such that after some number of lattice translations N the next translation returns to the origin. For simplicity, we will assume the same large N for each translation direction. Because Eq. (5.7) is a direct product, translations in each orthogonal direction may be considered independently.

In each direction the periodic boundary condition implies that the translation group T becomes a cyclic group of order N. As shown in Problems 5.14 and 5.15, this implies that for wavefunctions defined on a periodic lattice translational invariance requires that

$$\psi_k(\boldsymbol{r}+\boldsymbol{T}) = e^{i\boldsymbol{k}\cdot\boldsymbol{T}}\psi_k(\boldsymbol{r}), \tag{5.8}$$

where $\boldsymbol{T}$ is a lattice translation vector, or equivalently,

$$\psi_{\boldsymbol{k}}(\boldsymbol{r}) = u_{\boldsymbol{k}}(\boldsymbol{r})e^{i\boldsymbol{k}\cdot\boldsymbol{r}} \qquad u_{\boldsymbol{k}}(\boldsymbol{r}) = u_{\boldsymbol{k}}(\boldsymbol{r}+\boldsymbol{T}). \tag{5.9}$$

Equations (5.8) or (5.9) define *Bloch's theorem.*

Bloch's Theorem: Because of translational invariance, wavefunctions on a periodic lattice take the form (5.9) of a plane wave modulated by a function having the periodicity of the lattice.

Wavefunctions $\psi_k(\boldsymbol{r})$ satisfying (5.9) or (5.8) are called *Bloch functions.* The vector $\boldsymbol{k}$ is the wavevector for Bloch waves propagating through the lattice, which is called the *crystal momentum*. The Bloch theorem is a powerful and quite general result, since it was derived using only the translational symmetry and thus follows solely from group theory and the assumed periodicity of the lattice.

5.5 Electronic Band Structure

For atoms arranged in a crystal lattice, electrons from neighboring atoms hybridize and the orbitals of individual atoms form bands of states, each with a finite range of energies depending on the momentum of the electrons. The states of such bands correspond to electrons shared among many atoms of the crystal, but are still subject to Bloch's theorem because of the lattice periodicity and translational invariance. This band structure is indicated schematically in Fig. 5.5, and Box 5.1 and Fig. 5.6 outline how these bands provide an elementary understanding of insulating and conducting materials. As will be discussed in Box 28.5, the periodicity deriving from Bloch's theorem and boundary

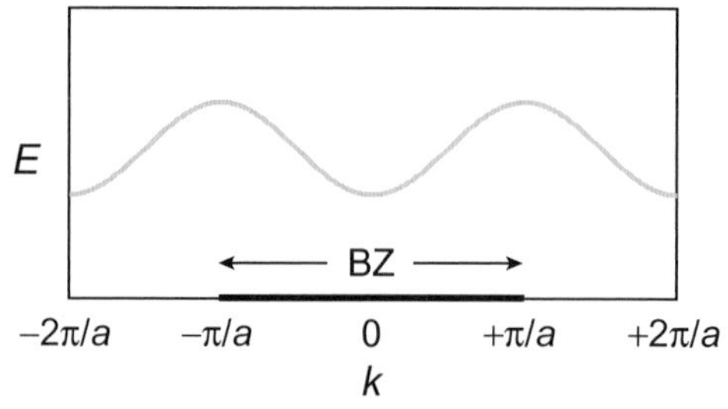

Fig. 5.5 Schematic energy E versus momentum k for an electronic band in a uniform crystal. The lattice spacing is a and the Brillouin zone (BZ) is indicated.

Box 5.1

Band Theory of Electronic Conduction

Bands of electronic states form in crystals because of orbital hybridization between adjacent atoms, as illustrated in Fig. 5.6. *Band theory* gives a simple explanation for why some materials are electrical insulators and some are conductors (metals).

Band Insulators and Conductors

In band theory the electrons occupy energy bands as a consequence of the Bloch theorem and the ground state is formed by filling states from the bottom up, subject to the Pauli principle. Two qualitatively different pictures may be identified,

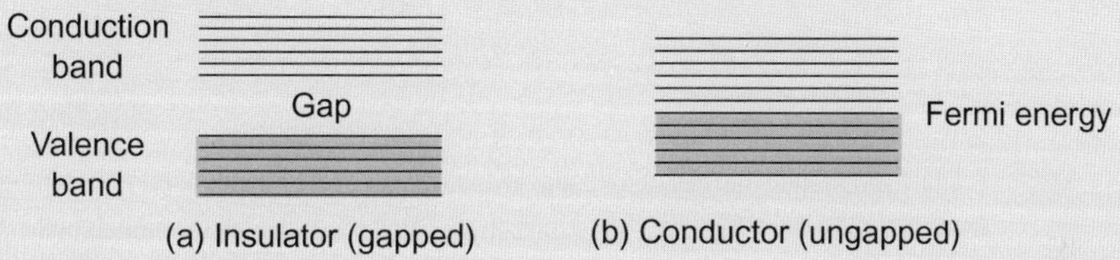

where filled levels are shaded. *Band insulators* have *no partially filled bands:* lower-energy, completely filled bands are separated from higher-energy, completely empty bands by a *bandgap,* and charge transport is inhibited because it requires excitation across the bandgap. In contrast, a metal has a *Fermi surface* (momentum-space energy surface separating occupied and unoccupied levels) that lies in a partially filled band, as indicated by the Fermi energy in the diagram above. This permits charge transport at a cost of arbitrarily small energy. Band theory explains a lot, but it is incomplete. For example, it does not explain Mott insulators (Box 32.3) and band theory can be modified in surprising ways by topological considerations (Ch. 29). Nevertheless, it is foundational for understanding the solid state.

Gapped and Ungapped Phases

Band theory leads to some standard terminology in which phases are classified as *gapped,* meaning that they are insulators because of a bandgap above the highest occupied state, or *ungapped,* meaning that they are metals because there is no energy gap above the highest occupied state. Gapped phases are sometimes termed *incompressible,* for reasons explained in Box 28.2.

A More Sophisticated Classification

More precisely it is useful to divide materials into (1) *metals,* (2) *semimetals,* (3) *semiconductors,* and (4) *insulators.* Insulators and metals are as described above. Semiconductors have a valence–conduction bandgap, but it is small enough that thermal promotion of electrons can convert insulators to conductors. Semimetals have a small overlap between the valence and conduction bands. Thus, in metals and semimetals the Fermi energy lies within one or more bands, but in insulators and semiconductors it lies in a bandgap. The density of states at the Fermi energy is high in metals but low in semimetals. Hence a semimetal behaves as a metal, but a poor one. Undoped graphene is an extreme example (see Section 20.2.1).

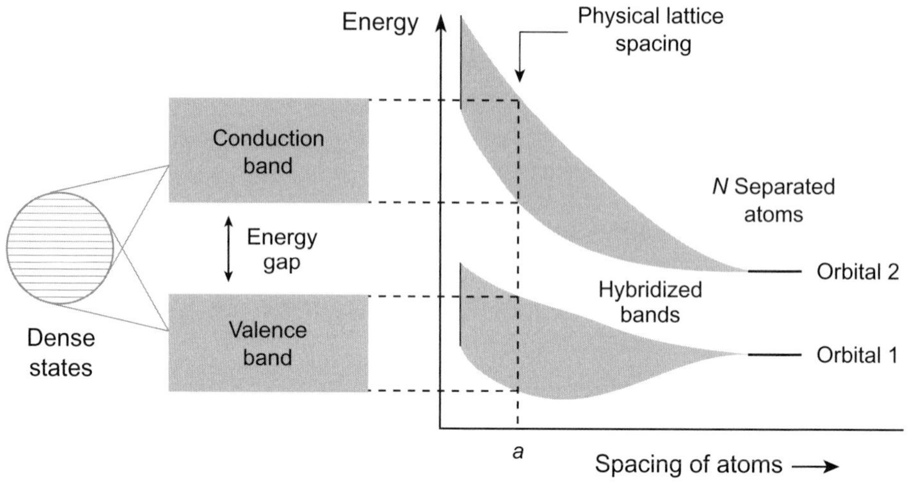

Fig. 5.6 Orbital hybridization into electronic bands in a solid for N identical atoms. Far right corresponds to two energy levels in well-separated atoms. As spacing between atoms is decreased the orbitals in the N atoms hybridize, forming dense bands of states. At the physical lattice spacing a of the corresponding crystal we can project a valence band and a conduction band, with an energy gap between them. Such bands form the basis of the band theory of conduction described in Box 5.1. The bands are actually dense sets of energy levels, but if the number of atoms in the crystal N is of the order of Avogadro's number, we may estimate that the average spacing between states in each band is $\Delta E \sim 10^{-22}$ eV and the bands essentially define a continuum.

conditions imply that states in the Brillouin zone can exhibit non-trivial topology. This requires a more nuanced description than the simple band-theory distinction between insulators and metals described in Box 5.1. This *topological view of matter* will be developed in Chs. 28 and 29.

5.6 Point Groups

Lattice translations are not the only symmetry operations for crystals. Various rotations and reflections about lattice points may also leave the crystal indistinguishable from its state before the operation. Recall from Section 2.1 that the set of all operations that leave a geometrical object unchanged is called the set of *covering operations* for the object. The complete set of such covering operations for a crystal is called the *space group* of the crystal. If we delete from the set of covering operations all translations, the remaining set of rotation, reflection, and inversion operations is called the *point group* of the crystal.[3]

[3] The set of distance-preserving transformations on an object can be built up from three basic types: (1) rotations, (2) reflections, and (3) translations, with translation a possible symmetry only if the body is of infinite extent. If translations are omitted, it can be shown that the set of geometrical symmetry operations on an object must leave at least one point invariant. This is the origin of the terminology *point group*.

For crystals the requirement of translational invariance restricts the number of allowed point groups to 32. These point groups that leave a spatial lattice invariant are called the *crystallographic point groups,* and will be the ones of primary concern here. The corresponding maximum number of space groups for crystal structure is 230. In this section we develop the theory of point groups for crystals. In section 5.9 we shall describe briefly the extension to the full set of space groups.

5.6.1 Point Group Operations

Point groups are groups of rotations and/or reflections about a point: any group of such transformations that take a finite body into itself while preserving the center of mass defines a point group. Point group covering operations fall into three categories.

1. *Rotations about axes through the origin.* These are through discrete angles and thus are subgroups of the full rotation group. The rotations by all multiples of $360°/n$, where n is an integer, define an *n-fold axis of (rotational) symmetry.* We shall see below that in an infinite crystal only $n = 1, 2, 3, 4$, and 6 are possible for rotational symmetry axes.
2. *Reflections through planes that contain the origin.* Rotation combined with reflection is an *improper rotation*; pure rotations are *proper rotations.* Improper rotations involve an *n-fold roto-reflection axis* and correspond to rotation by multiples of $360°/n$, followed by reflection through a plane containing the origin that is orthogonal to the rotation axis.
3. *Inversions* $\boldsymbol{r} \to -\boldsymbol{r}$ *through the origin.* These are equivalent to rotation by π, followed by reflection in the plane perpendicular to the rotation axis, so they are not independent of the first two categories. But it is useful conceptually to distinguish them.

Two common methods of labeling crystallographic point group operations are *Schoenflies notation* and the *international system,* which are described in Box 5.2.

5.6.2 The Crystallographic Point Groups

The 32 point groups that exhaust the possibilities for crystals occur in two general categories: (1) the *simple rotations,* for which there is one rotation axis that is of higher symmetry than any others, and (2) the *groups of higher symmetry,* which have no unique axis of highest symmetry but instead have more than one three-fold or greater axis. A detailed discussion of the classification may be found in Ref. [196]. These 32 crystallographic groups can be tabulated with respect to the properties of the translational unit cell that is compatible with the given point group symmetry. Table 5.1 gives this classification in terms of the unit-cell lengths and angles displayed in Fig. 5.7. The names of the groups listed in Table 5.1 are given in both the *Schoenflies system* and the *international system*, and the number of elements in the group is given in the last column.

Box 5.2

Schoenflies Notation

One standard method of labeling crystallographic point group operations is the *Schoenflies notation* [196].

1. E is the identity.
2. c_n is a rotation by $2\pi/n$, where in crystals n is restricted to the values 1, 2, 3, 4, and 6.
3. σ_h is a reflection through a plane perpendicular to the rotation axis of highest symmetry and passing through the origin (termed a *horizontal plane*).
4. σ_v is a reflection through a plane containing the rotation axis of highest symmetry (termed a *vertical plane*).
5. σ_d (a special case of σ_v) is a reflection through a plane that contains the symmetry axis and bisects the angle between the twofold axes perpendicular to the symmetry axis (termed a *diagonal plane*).
6. $\mathbb{S}_n$ is an improper rotation through $2\pi/n$.
7. i is inversion through the origin of the coordinate system. This is equivalent to rotation by π and then a σ_h reflection, so it is the same as $\mathbb{S}_2$.

The 32 allowed crystallographic point groups then correspond to the independent ways in which such operations can be combined into sets that satisfy the group criteria, and that are consistent with translational invariance for crystals. Another standard labeling method for point groups is the *Hermann–Mauguin (H–M) system,* which is often termed the *international system.* The particulars of this system are not important for our purposes but we will use it for labels in some tables.

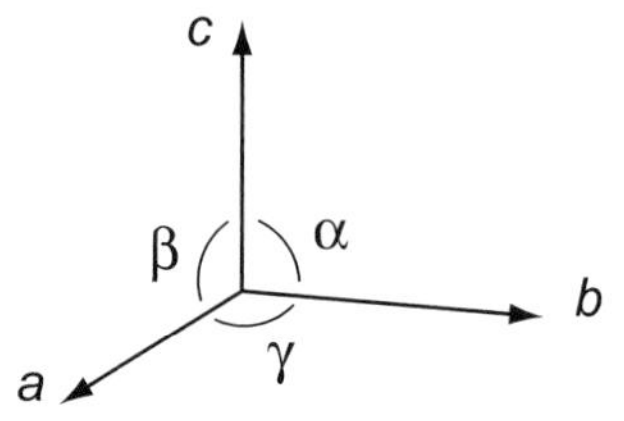

Fig. 5.7 Unit-cell parameters for the crystal systems listed in Table 5.1

5.7 Example: The Ammonia Molecule

To illustrate the properties of point groups, let us analyze the symmetry operations that can be performed on the ammonia molecule, which has the geometry illustrated in Fig. 5.8.

5.7.1 Symmetry Operations

For symmetry purposes, we may view ammonia as an equilateral triangle of hydrogen atoms in a plane with the nitrogen atom on an axis through the center of the triangle and normal to the plane. The symmetry operations that leave Fig. 5.8 invariant are the following.

Table 5.1. Crystal systems and their associated point groups [104, 196]

System	Unit cell†	Schoenflies	International	Elements
Triclinic	$a \neq b \neq c$	C_1	1	1
	$\alpha \neq \beta \neq \gamma$	$\mathbb{S}_2$	$\bar{1}$	2
Monoclinic	$a \neq b \neq c$	C_{1h}	m	2
	$\alpha = \gamma = \pi/2 \neq \beta$	C_2	2	2
		C_{2h}	$2/m$	4
Orthorhombic	$a \neq b \neq c$	C_{2v}	$2mm$	4
	$\alpha = \beta = \gamma = \pi/2$	D_2	222	4
		D_{2h}	mmm	8
Tetragonal	$a = b \neq c$	C_4	4	4
	$\alpha = \beta = \gamma = \pi/2$	$\mathbb{S}_4$	$\bar{4}$	4
		C_{4h}	$4/m$	8
		D_{2d}	$\bar{4}2m$	8
		C_{4v}	$4mm$	8
		D_4	422	8
		D_{4h}	$4/mmm$	16
Rhombohedral	$a = b = c$	C_3	3	3
	$\alpha = \beta = \gamma < 2\pi/3 \neq \pi/2$	$\mathbb{S}_6$	$\bar{3}$	6
		C_{3v}	$3m$	6
		D_3	32	6
		D_{3d}	$\bar{3}m$	12
Hexagonal	$a = b \neq c$	C_{3h}	$\bar{6}$	6
	$\alpha = \beta = \pi/2, \gamma = 2\pi/3$	C_6	6	6
		C_{6h}	$6/m$	12
		D_{3h}	$\bar{6}m2$	12
		C_{6v}	$6mm$	12
		D_6	622	12
		D_{6h}	$6/mmm$	24
Cubic	$a = b = c$	$\mathbb{T}$	23	12
	$\alpha = \beta = \gamma = \pi/2$	$\mathbb{T}_h$	$m3$	24
		$\mathbb{T}_d$	$\bar{4}3m$	24
		$\mathbb{O}$	432	24
		$\mathbb{O}_h$	$m3m$	48

†See Fig. 5.7 for the coordinate system.

1. The identity e, which does not change the figure.
2. The operation c_3, which is a rotation in the plane of the hydrogen atoms counterclockwise by an angle $2\pi/3$.
3. Rotation in the plane of the hydrogen atoms counterclockwise by an angle $4\pi/3$, which is denoted by $c_3^2 = c_3 c_3$. Note that c_3^3 is equivalent to the identity.
4. Reflection through the plane oad, which is denoted σ_a.
5. Reflection through the plane obd, which is denoted σ_b.
6. Reflection through the plane ocd, which is denoted σ_c.

Table 5.2. Multiplication table $A \cdot B$ for C_{3v}

$A \backslash B$	e	c_3	c_3^2	σ_a	σ_b	σ_c
e	e	c_3	c_3^2	σ_a	σ_b	σ_c
c_3	c_3	c_3^2	e	σ_c	σ_a	σ_b
c_3^2	c_3^2	e	c_3	σ_b	σ_c	σ_a
σ_a	σ_a	σ_b	σ_c	e	c_3	c_3^2
σ_b	σ_b	σ_c	σ_a	c_3^2	e	c_3
σ_c	σ_c	σ_a	σ_b	c_3	c_3^2	e

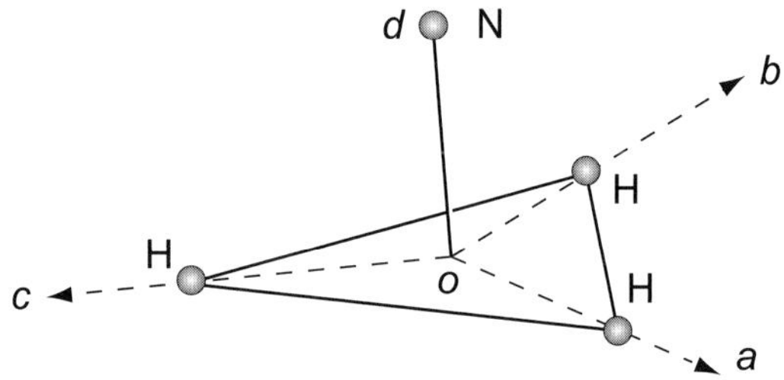

Fig. 5.8 Geometry of the ammonia molecule NH_3. The three hydrogen atoms (H) lie in a plane and the nitrogen atom (N) is on an axis perpendicular to this plane.

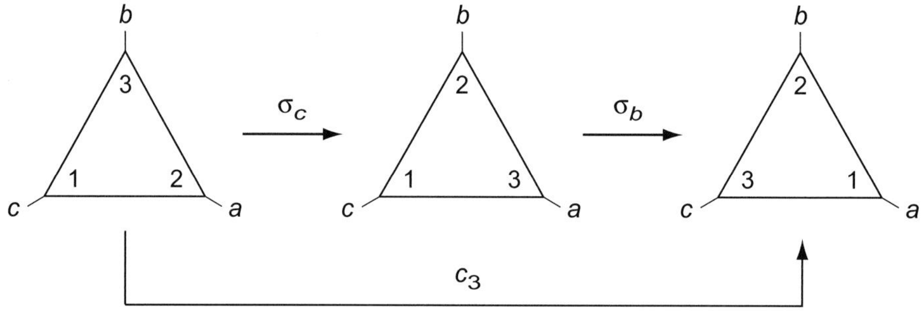

Fig. 5.9 Product of symmetry operations on the ammonia molecule of Fig. 5.8 with axes labeled by letters and vertices labeled by numbers, demonstrating geometrically that $\sigma_b \sigma_c = c_3$.

A multiplication table may be formed from products of operations AB, where B and then A are performed on Fig. 5.8. The results are visualized easily if the locations of the hydrogen atoms are labeled with numbers, as in Fig. 5.9. In this example we see that the application of σ_c, followed by the application of σ_b, is indistinguishable from the application of c_3 to the original figure. Thus, $\sigma_b \sigma_c = c_3$. By similar considerations all 36 possible products of the six symmetry operations may be evaluated, giving Table 5.2.This multiplication table defines the group C_{3v} in Schoenflies notation ($3m$ in international notation); see Table 5.1. The group is non-abelian since from Table 5.2 we see that AB is not always equal to BA.

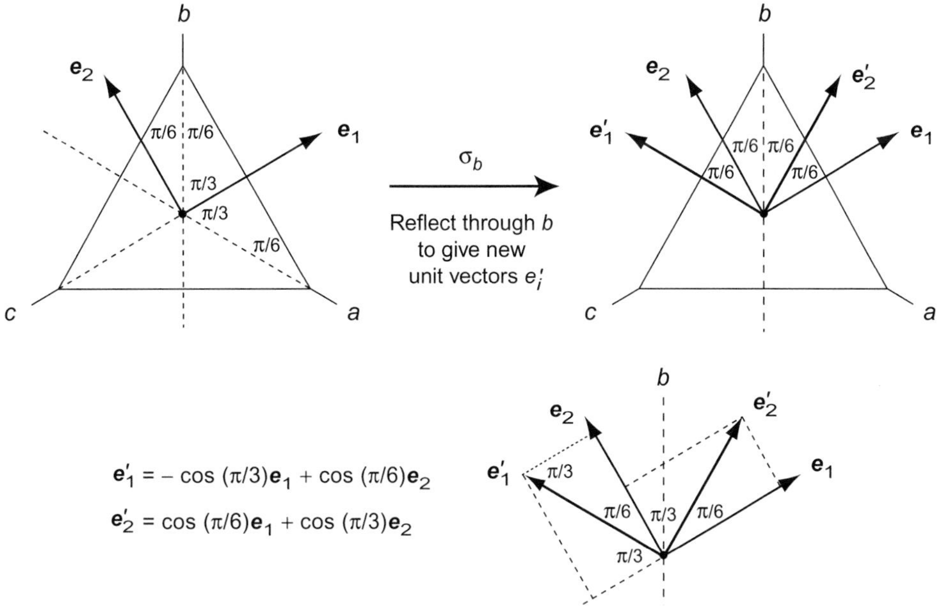

Fig. 5.10 The reflection σ_b acting on orthogonal unit vectors $\boldsymbol{e}_1$ and $\boldsymbol{e}_2$ (with $\boldsymbol{e}_1$ chosen coincident with the c-axis) gives new unit vectors $\boldsymbol{e}_1'$ and $\boldsymbol{e}_2'$. The lower diagram illustrates the geometrical relationship between the original unit vectors and the reflected ones. The two equations for $\boldsymbol{e}_1'$ and $\boldsymbol{e}_2'$ are equivalent to the matrix relation (5.10).

5.7.2 A Matrix Representation

Suppose that we attach orthogonal unit vectors $\boldsymbol{e}_1$ and $\boldsymbol{e}_2$ to the center of the triangles in Fig. 5.9. The C_{3v} symmetry operations will then transform this set of orthogonal unit vectors into a new set $(\boldsymbol{e}_1', \boldsymbol{e}_2')$. For a particular choice of initial $(\boldsymbol{e}_1, \boldsymbol{e}_2)$, Fig. 5.10 illustrates the action of the reflection σ_b. By taking the projections of the new axes on the old axes as illustrated in the lower portion of Fig. 5.10, one sees that under the symmetry operation σ_b a new set of unit vectors is obtained that is related to the old set by the matrix equation

$$\begin{pmatrix} \boldsymbol{e}_1' \\ \boldsymbol{e}_2' \end{pmatrix}_{\sigma_b} = \begin{pmatrix} -\cos(\frac{\pi}{3}) & \cos(\frac{\pi}{6}) \\ \cos(\frac{\pi}{6}) & \cos(\frac{\pi}{3}) \end{pmatrix} \begin{pmatrix} \boldsymbol{e}_1 \\ \boldsymbol{e}_2 \end{pmatrix} = \begin{pmatrix} -\frac{1}{2} & \frac{\sqrt{3}}{2} \\ \frac{\sqrt{3}}{2} & \frac{1}{2} \end{pmatrix} \begin{pmatrix} \boldsymbol{e}_1 \\ \boldsymbol{e}_2 \end{pmatrix}, \tag{5.10}$$

where the subscript σ_b indicates explicitly that the transformation of the unit vectors is by the symmetry operation corresponding to σ_b. By similar considerations, the action of σ_a on the same unit vectors is given by (Problem 5.12)

$$\begin{pmatrix} \boldsymbol{e}_1' \\ \boldsymbol{e}_2' \end{pmatrix}_{\sigma_a} = \begin{pmatrix} -\cos(\frac{\pi}{3}) & -\cos(\frac{\pi}{6}) \\ -\cos(\frac{\pi}{6}) & \cos(\frac{\pi}{3}) \end{pmatrix} \begin{pmatrix} \boldsymbol{e}_1 \\ \boldsymbol{e}_2 \end{pmatrix} = \begin{pmatrix} -\frac{1}{2} & -\frac{\sqrt{3}}{2} \\ -\frac{\sqrt{3}}{2} & \frac{1}{2} \end{pmatrix} \begin{pmatrix} \boldsymbol{e}_1 \\ \boldsymbol{e}_2 \end{pmatrix}, \tag{5.11}$$

and the action of σ_c is simply

$$\begin{pmatrix} \boldsymbol{e}_1' \\ \boldsymbol{e}_2' \end{pmatrix}_{\sigma_c} = \begin{pmatrix} 1 & 0 \\ 0 & -1 \end{pmatrix} \begin{pmatrix} \boldsymbol{e}_1 \\ \boldsymbol{e}_2 \end{pmatrix}. \tag{5.12}$$

Transformations of unit vectors may be constructed by inspection for simple cases like Fig. 5.10. However, we may systematize the procedure as follows. Assume an orthonormal basis $(\boldsymbol{e}_1, \boldsymbol{e}_2, \ldots, \boldsymbol{e}_n)$, and that the transformation of this basis to a new basis $(\boldsymbol{e}_1', \boldsymbol{e}_2', \ldots, \boldsymbol{e}_n')$ is implemented by a matrix T with $\boldsymbol{e}_i' = \sum_k T_{ki}\boldsymbol{e}_k$. Now consider the scalar product taken between a basis vector in the original basis and a basis vector in the transformed basis,

$$\boldsymbol{e}_j \cdot \boldsymbol{e}_i' = \boldsymbol{e}_j \cdot (T\boldsymbol{e}_i) = \sum_k T_{ki}\boldsymbol{e}_j \cdot \boldsymbol{e}_k = \sum_k T_{ki}\delta_{jk} = T_{ji}.$$

Therefore, for orthonormal basis vectors the elements of the transformation matrix between the bases are given by the scalar product of old and new basis vectors,

$$T_{ji} = \boldsymbol{e}_j \cdot \boldsymbol{e}_i'. \tag{5.13}$$

Example 5.1 illustrates the use of this method.

Example 5.1 Consider the reflection transformation σ_b in Fig. 5.10. Denoting the matrix implementing this reflection by $T(\sigma_b)$, from Eq. (5.13) the elements of this matrix are given by $T(\sigma_b)_{ji} = \boldsymbol{e}_j \cdot \boldsymbol{e}_i'$ and from the right triangles that are displayed in Fig. 5.10,

$$T(\sigma_b)_{11} = \boldsymbol{e}_1 \cdot \boldsymbol{e}_1' = -\cos\left(\frac{\pi}{3}\right) \qquad T(\sigma_b)_{12} = \boldsymbol{e}_1 \cdot \boldsymbol{e}_2' = \cos\left(\frac{\pi}{6}\right),$$

$$T(\sigma_b)_{21} = \boldsymbol{e}_2 \cdot \boldsymbol{e}_1' = \cos\left(\frac{\pi}{6}\right) \qquad T(\sigma_b)_{22} = \boldsymbol{e}_2 \cdot \boldsymbol{e}_2' = \cos\left(\frac{\pi}{3}\right),$$

which implies that

$$T(\sigma_b) = \begin{pmatrix} -\cos(\frac{\pi}{3}) & \cos(\frac{\pi}{6}) \\ \cos(\frac{\pi}{6}) & \cos(\frac{\pi}{3}) \end{pmatrix} = \begin{pmatrix} -\frac{1}{2} & \frac{\sqrt{3}}{2} \\ \frac{\sqrt{3}}{2} & \frac{1}{2} \end{pmatrix},$$

is the transformation matrix corresponding to σ_b, in agreement with Eq. (5.10).

Applying the same approach to the other transformations utilizing Fig. 5.11 for the rotations gives the matrices (Problem 5.2)

$$\begin{gathered} T(\sigma_a) = \begin{pmatrix} -\frac{1}{2} & -\frac{\sqrt{3}}{2} \\ -\frac{\sqrt{3}}{2} & \frac{1}{2} \end{pmatrix} \quad T(\sigma_b) = \begin{pmatrix} -\frac{1}{2} & \frac{\sqrt{3}}{2} \\ \frac{\sqrt{3}}{2} & \frac{1}{2} \end{pmatrix} \quad T(\sigma_c) = \begin{pmatrix} 1 & 0 \\ 0 & -1 \end{pmatrix}, \\ T(c_3) = \begin{pmatrix} -\frac{1}{2} & -\frac{\sqrt{3}}{2} \\ \frac{\sqrt{3}}{2} & -\frac{1}{2} \end{pmatrix} \quad T\left(c_3^2\right) = \begin{pmatrix} -\frac{1}{2} & \frac{\sqrt{3}}{2} \\ -\frac{\sqrt{3}}{2} & -\frac{1}{2} \end{pmatrix} \quad T(e) = \begin{pmatrix} 1 & 0 \\ 0 & 1 \end{pmatrix}. \end{gathered} \tag{5.14}$$

You should verify that the set of six matrices given in Eq. (5.14) is closed under ordinary matrix multiplication (Problem 5.3). The complete multiplication table for the matrices of Eq. (5.14) is given in Table 5.3. Comparison of Table 5.3 with Table 5.2 shows that the matrices (5.14) have the same multiplication table as the group C_{3v} with the correspondence $T(A) \leftrightarrow A$ for a C_{3v} symmetry operation A and the corresponding matrix $T(A)$. Thus Eq. (5.14) defines a matrix representation of the group C_{3v}. The representation

Table 5.3. Multiplication table for the matrices T in Eq. (5.14)

C_{3v}	$T(e)$	$T(c_3)$	$T(c_3^2)$	$T(\sigma_a)$	$T(\sigma_b)$	$T(\sigma_c)$
$T(e)$	$T(e)$	$T(c_3)$	$T(c_3^2)$	$T(\sigma_a)$	$T(\sigma_b)$	$T(\sigma_c)$
$T(c_3)$	$T(c_3)$	$T(c_3^2)$	$T(e)$	$T(\sigma_c)$	$T(\sigma_a)$	$T(\sigma_b)$
$T(c_3^2)$	$T(c_3^2)$	$T(e)$	$T(c_3)$	$T(\sigma_b)$	$T(\sigma_c)$	$T(\sigma_a)$
$T(\sigma_a)$	$T(\sigma_a)$	$T(\sigma_b)$	$T(\sigma_c)$	$T(e)$	$T(c_3)$	$T(c_3^2)$
$T(\sigma_b)$	$T(\sigma_b)$	$T(\sigma_c)$	$T(\sigma_a)$	$T(c_3^2)$	$T(e)$	$T(c_3)$
$T(\sigma_c)$	$T(\sigma_c)$	$T(\sigma_a)$	$T(\sigma_b)$	$T(c_3)$	$T(c_3^2)$	$T(e)$

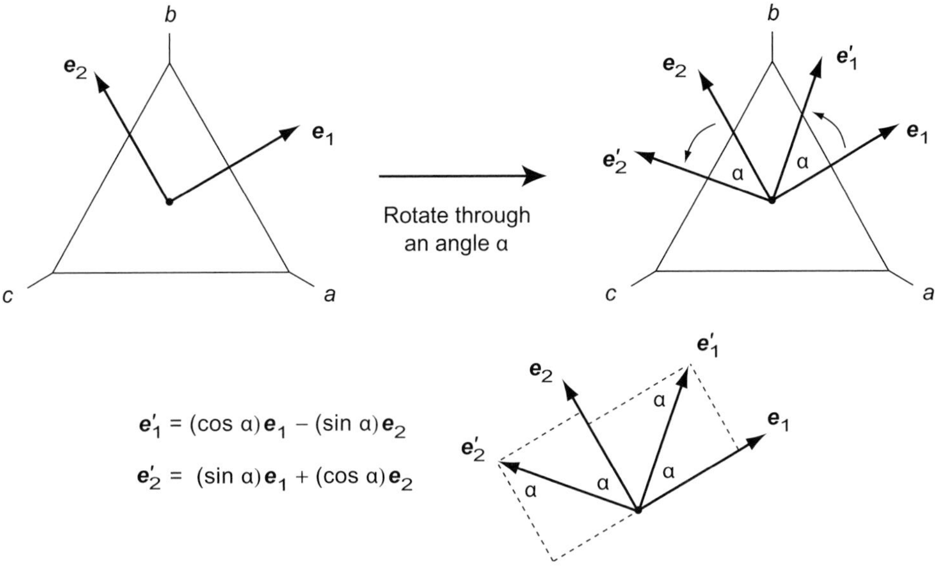

Fig. 5.11 Rotation counterclockwise by an angle α on orthogonal unit vectors $\boldsymbol{e}_1$ and $\boldsymbol{e}_2$ gives new orthogonal unit vectors $\boldsymbol{e}'_1$ and $\boldsymbol{e}'_2$. The lower diagram illustrates the geometrical relationship between the original unit vectors and the rotated ones.

(5.14) is not unique since a similarity transformation on this set of matrices yields a new set that is also a representation of C_{3v}. For example, in Problem 5.10 you are asked to show that

$$
\begin{array}{lll}
T'(\sigma_a) = \begin{pmatrix} \frac{1}{2} & -\frac{\sqrt{3}}{2} \\ -\frac{\sqrt{3}}{2} & -\frac{1}{2} \end{pmatrix} & T'(\sigma_b) = \begin{pmatrix} -1 & 0 \\ 0 & 1 \end{pmatrix} & T'(\sigma_c) = \begin{pmatrix} \frac{1}{2} & \frac{\sqrt{3}}{2} \\ \frac{\sqrt{3}}{2} & -\frac{1}{2} \end{pmatrix} \\
T'(c_3) = \begin{pmatrix} -\frac{1}{2} & -\frac{\sqrt{3}}{2} \\ \frac{\sqrt{3}}{2} & -\frac{1}{2} \end{pmatrix} & T'\left(c_3^2\right) = \begin{pmatrix} -\frac{1}{2} & \frac{\sqrt{3}}{2} \\ -\frac{\sqrt{3}}{2} & -\frac{1}{2} \end{pmatrix} & T'(e) = \begin{pmatrix} 1 & 0 \\ 0 & 1 \end{pmatrix}
\end{array}
\qquad (5.15)
$$

is also a matrix representation of C_{3v}. It is easy to verify that these matrices have a multiplication table in one to one correspondence with those in Tables 5.2 and 5.3.

Example 5.2 By direct matrix multiplication using the matrices in Eq. (5.15),

$$T'(c_3)T'(\sigma_b) = \begin{pmatrix} -\frac{1}{2} & -\frac{\sqrt{3}}{2} \\ \frac{\sqrt{3}}{2} & -\frac{1}{2} \end{pmatrix}\begin{pmatrix} -1 & 0 \\ 0 & 1 \end{pmatrix} = \begin{pmatrix} \frac{1}{2} & -\frac{\sqrt{3}}{2} \\ -\frac{\sqrt{3}}{2} & -\frac{1}{2} \end{pmatrix} = T'(\sigma_a),$$

which maps to $c_3\sigma_b = \sigma_a$ from Table 5.2 and to $T(c_3)T(\sigma_b) = T(\sigma_a)$ from Table 5.3.

Generally the correspondence is $A \leftrightarrow T(A) \leftrightarrow T'(A)$ for the group operations A in C_{3v} and the matrix representations (5.14) and (5.15).

5.7.3 Class Structure

The matrices $T'(A)$ in Eq. (5.15) and $T(A)$ in Eq. (5.14) are related by a similarity transformation, so they are matrix representations of the same group. We see by explicit construction a physical statement of their similarity: each represents a set of 2×2 matrix transformations implementing C_{3v} symmetry, but they are expressed in coordinate systems that are rotated by $\frac{\pi}{6}$ with respect to each other (see Problem 5.10). Notice that the traces of the matrices (characters) are preserved by the similarity transformation between Eqs. (5.14) and (5.15), as they must be since the trace is a matrix invariant. By inspection,

1. $T(\sigma_a)$, $T(\sigma_b)$, and $T(\sigma_c)$, as well as $T'(\sigma_a)$, $T'(\sigma_b)$, and $T'(\sigma_c)$, all have a trace of 0,
2. $T(c_3)$ and $T(c_3^2)$, as well as $T'(c_3)$ and $T'(c_3^2)$, all have a trace of -1, and
3. $T(e)$ and $T'(e)$ both have a trace of 2.

Thus, from the discussion in Section 2.11 there are three classes: (1) the three reflections, (2) the two rotations, and (3) the identity.

Example 5.3 The class structure just found for C_{3v} could be verified in two other ways.

1. *Physically:* Group operations in the same class typically share similar characteristics. Thus the three reflections are essentially the same operation since the choice of axes is arbitrary and they constitute a class, the two rotations can be viewed as being by the same angle but in the opposite sense[4] (because c_3^2 is equivalent to c_3^{-1}), so they are in the same class, which leaves the identity in a class by itself.
2. *Formally:* The members of a class may be inferred as in Section 2.11 by forming for each group element q the conjugate elements $g_i^{-1}qg_i$. You are asked to verify the C_{3v} classes by this more formal approach in Problem 5.4.

The group C_{3v} has irreducible representations in addition to the ones considered above; in the next section we shall set about finding them.

[4] Unless there is an obvious choice for sense of rotation within a group, rotations about the same axis by the same angle but in opposite directions will belong to the same class.

Box 5.3

Representation Labels for Point Groups

There is a rather opaque but often used notation for labeling irreducible representations of point groups.

1. A one-dimensional representation having character $+1$ under the principal rotation (c_3 for C_{3v}, for example) is labeled A.
2. A one-dimensional representation with character -1 under the principal rotation is labeled B.
3. A two-dimensional representation is labeled E.
4. A three-dimensional representation is labeled T.

If there is more than one representation of a given type, they are distinguished by subscripts, for example, A_1 and A_2. If inversion symmetry is present, it is also standard to label a representation that is

1. even with respect to inversion by a subscript g (German *gerade*), and
2. odd with respect to inversion by a subscript u (German *ungerade*).

For example, A_{1g} labels a one-dimensional representation with character $+1$ under the principal rotation that is even under inversion.

5.7.4 Other Irreducible Representations

Let us introduce a common labeling convention for representations of point groups that is summarized in Box 5.3. By that scheme the 2D irrep (5.14) or (5.15) that we have found is labeled E. A group always has a one-dimensional representation corresponding to mapping all group elements to the one-dimensional unit matrix:

$$T(e) = T(c_3) = T(c_3^2) = T(\sigma_a) = T(\sigma_b) = T(\sigma_c) = +1, \tag{5.16}$$

since this satisfies Table 5.3 trivially. Using the naming scheme in Box 5.3, this irrep may be labeled A_1. Physically, we may think of the irrep A_1 in the following way. The irrep E is seen from the preceding discussion to be the set of two-dimensional matrices that transform the unit vectors $\boldsymbol{e}_1$ and $\boldsymbol{e}_2$ into linear combinations of each other under the six symmetry operations of the group. The irrep A_1 can be thought of as the corresponding set of matrices that transform the unit vector $\boldsymbol{e}_3$ perpendicular to the plane of the triangle in Fig. 5.8 (pointed toward the nitrogen atom in the physical example of an ammonia molecule). Since none of the six symmetry operations of C_{3v} alters this vector, these matrices are all one-dimensional, with value unity corresponding to the identity operation.

From Eq. (2.29b), for a finite group the sum of the squares of the dimensions of the irreps must sum to the order of the group, so there must be an additional one-dimensional representation (giving $1^2 + 1^2 + 2^2 = 6$ for the case we are examining). It can be specified by the mapping

$$T(e) = T(c_3) = T(c_3^2) = +1 \qquad T(\sigma_a) = T(\sigma_b) = T(\sigma_c) = -1, \tag{5.17}$$

Table 5.4. Characters for irreps of the group C_{3v}

Irrep\Class	$e(1)$	$c_3(2)$	$\sigma_v(3)$	Basis
A_1	1	1	1	$\boldsymbol{e}_3$
A_2	1	1	-1	$\boldsymbol{e}_1 \times \boldsymbol{e}_2$
E	2	-1	0	$(\boldsymbol{e}_1, \boldsymbol{e}_2)$

which satisfies the C_{3v} multiplication table of Table 5.3.[5] Using the terminology of Box 5.3, this irrep may be labeled A_2. It may be interpreted physically as follows. We have seen that E is associated with transformations that mix $\boldsymbol{e}_1$ and $\boldsymbol{e}_2$ in the plane of the triangle in Fig. 5.8, and that A_1 is associated with group operations on the unit vector $\boldsymbol{e}_3$ perpendicular to the plane of the triangle for C_{3v}. Consider now the *pseudovector* defined by $\boldsymbol{e}_1 \times \boldsymbol{e}_2$. Under the identity or the two rotation operations of C_{3v} the pseudovector is unchanged, so those operations are mapped to 1 in Eq. (5.17). However, under the three reflection operations in the plane the pseudovector $\boldsymbol{e}_1 \times \boldsymbol{e}_2$ changes sign. Thus the reflection operations are mapped to -1 in Eq. (5.17). In summary then, C_{3v} has

1. one 2D irrep E that characterizes the transformations among the orthogonal unit vectors $\boldsymbol{e}_1$ and $\boldsymbol{e}_2$ in the plane of the triangle,
2. one 1D irrep A_1 that characterizes the effect of transformations on a unit vector $\boldsymbol{e}_3$ perpendicular to the plane of the triangle, and
3. one 1D irrep A_2 that characterizes actions of the group on the pseudovector $\boldsymbol{e}_1 \times \boldsymbol{e}_2$ (which is perpendicular to the plane of the triangle but without a definite sense).

This information about the class structure and irreps of C_{3v} is summarized concisely in the character table displayed in Table 5.4.

5.8 General Lattice Symmetry Classifications

If all symmetries of the lattice were ignored except for those associated with translations, the full symmetry classification of electronic states would be given by Bloch's theorem (5.9). Conversely, if the translational symmetry were ignored, cataloging electronic states would reduce to point group classifications about individual atomic sites. Such approximations may be justified in specific cases such as when there is weak overlap of the atomic wavefunctions on different sites, but generally we require classifications that reflect both the translational symmetries of the lattice and the rotational and reflection symmetries

[5] The representation (5.14) is faithful (see Section 2.6.1), since there is a one to one correspondence between group elements and matrices, but the representations (5.16) and (5.17) are not faithful because they have a many-to-one relationship between group elements and matrices. Note also that the representation (5.17) corresponds to the set of determinants of the matrices in the representation (5.14). This is no accident, as you will find if you work Problem 5.9.

about individual lattice sites.[6] This is a daunting problem but significant understanding results if we make some simplifying assumptions that still retain many realistic features.

5.9 Space Groups

The full set of geometrical symmetry operations on a crystal consists of rotations and translations (and inversions and reflections in the case of improper rotations). This full set of transformations forms a group called the *space group* of the crystal. A complete discussion is rather involved. We will be content with giving a basic introduction and some references for the interested reader to pursue on their own.

5.9.1 Elements of the Space Group

The most general space group element may be expressed as $\{\boldsymbol{R}|\boldsymbol{t}\}$, where $\boldsymbol{R}$ denotes a proper or improper rotation and $\boldsymbol{t}$ denotes a translation. The corresponding coordinate transformation is $\boldsymbol{x}' = \boldsymbol{R}\boldsymbol{x} + \boldsymbol{t}$ (see Section 12.4.2), and the group multiplication law is

$$\{\boldsymbol{R}|\boldsymbol{t}\}\{\boldsymbol{R}'|\boldsymbol{t}'\} = \{\boldsymbol{R}\boldsymbol{R}'|\boldsymbol{R}\boldsymbol{t}' + \boldsymbol{t}\}. \tag{5.18}$$

The rotational parts of these operators are obtained by setting $\boldsymbol{t} = 0$ and form a subgroup corresponding to the point group symmetry. The translation operators alone (the elements $\{\boldsymbol{E}|\boldsymbol{t}\}$, where $\boldsymbol{E}$ is the identity), form an abelian invariant translation subgroup, provided that the crystal is of infinite extent or that we impose periodic boundary conditions so that the translations close on themselves.

5.9.2 Symmorphic Space Groups

The full range of possibilities for space group symmetries is large and intricate. For simplicity, many discussions restrict to space groups for which all operators obtained by taking the direct product of the translational subgroup with the point group lie in the space group. These are termed *symmorphic* space groups, and there are 73 of them. For the special case of symmorphic space groups, the entire point group is a subgroup of the space group. The 157 of 230 total possible space groups excluded by this simplifying choice contain (1) *glide planes* and (2) *screw axes*, which combine reflection or rotation with translation [196]. We shall omit further discussion, referring the reader to the more specialized literature (for example, Lax [138]).

[6] In Sections 20.2 and 32.3 we shall become even more ambitious and ask whether there are additional symmetries of the lattice wavefunctions that embody *internal properties of the system Hamiltonian.* Such symmetries will have dynamical consequences, not just the geometrical ones emphasized in this chapter.

Background and Further Reading

Ashcroft and Mermin [15], Kittel [131], and Ziman [232] provide good general introductions to condensed matter physics. More advanced treatments of various topics in this field are given by Chaikin and Lubensky [36], El-Batanouny [54], and Girvin and Yang [75]. Introductions to point groups, space groups, and general symmetry issues in condensed matter may be found in Butler [30], Elliott and Dawber [56], Hamermesh [104], Heine [108], Inui, Tanabe, and Onodera [119], Lax [138], Tinkham [196], and Wherrett [208].

Problems

5.1 (a) Verify the entries given in Table 5.2 for the multiplication table of C_{3v}. (b) Show that the multiplication table for C_{3v} can be put into one to one correspondence with that in Table 2.2 for S_3.

5.2 Apply Fig. 5.11 and the scalar product method of Eq. (5.13) to construction of the matrix representation in Eq. (5.14). ***

5.3 Verify that the set of matrices (5.14) is closed under ordinary matrix multiplication.

5.4 Find the classes and their members for C_{3v} as in Section 2.11 by forming for each group element q the conjugate elements $g_i^{-1} q g_i$ for all elements g_i of the group.

5.5 The group D_3 in Schoenflies notation (32 in international notation, which is read "three-two"; see Table 5.1) consists of the proper (those not reflections or inversions) covering operations on an equilateral triangle. Label the vertices of a triangle as

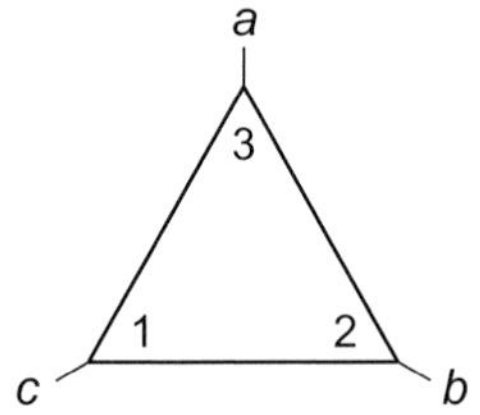

and show that there are six proper symmetry operations and find the corresponding multiplication table. Show from this that the groups D_3 and C_{3v} are isomorphic. ***

5.6 Derive the two-dimensional matrix representation

$$T(c_{2a}) = \begin{pmatrix} 1 & 0 \\ 0 & -1 \end{pmatrix} \qquad T(c_{2b}) = \begin{pmatrix} -\frac{1}{2} & \frac{\sqrt{3}}{2} \\ \frac{\sqrt{3}}{2} & \frac{1}{2} \end{pmatrix} \qquad T(c_{2c}) = \begin{pmatrix} -\frac{1}{2} & -\frac{\sqrt{3}}{2} \\ -\frac{\sqrt{3}}{2} & \frac{1}{2} \end{pmatrix}$$

$$T(c_3) = \begin{pmatrix} -\frac{1}{2} & -\frac{\sqrt{3}}{2} \\ \frac{\sqrt{3}}{2} & -\frac{1}{2} \end{pmatrix} \qquad T\left(c_3^2\right) = \begin{pmatrix} -\frac{1}{2} & \frac{\sqrt{3}}{2} \\ -\frac{\sqrt{3}}{2} & -\frac{1}{2} \end{pmatrix} \qquad T(e) = \begin{pmatrix} 1 & 0 \\ 0 & 1 \end{pmatrix}$$

for the group D_3, using the basis $(\boldsymbol{e}_1, \boldsymbol{e}_2)$ defined in the following figure.

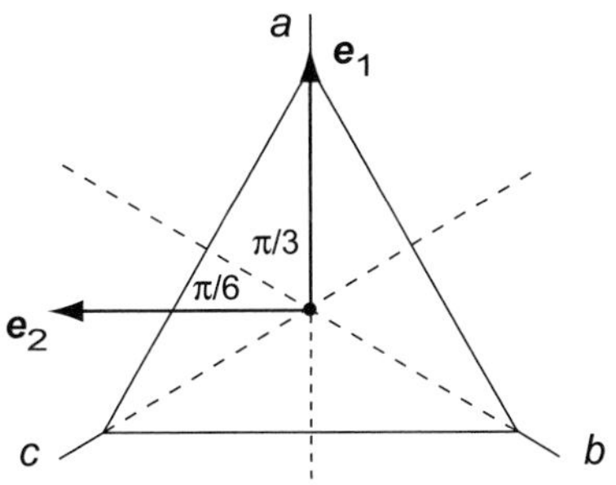

Hint: See construction of matrix representations for C_{3v} in Section 5.7.2. ***

5.7 Prove that the matrix representation of D_3 worked out in Problem 5.6 is irreducible.

5.8 Show that the group D_3 has two 1D irreps in addition to the 2D irrep found in Problem 5.6, and construct the character table. ***

5.9 Demonstrate that if x labels elements of a group and the set $D(x)$ is a matrix representation of this group such that Eq. (2.8) is obeyed, $D(a) \cdot D(b) = D(a \cdot b)$, then the set of determinants $\det(D(x))$ also forms a representation of the group. *Hint*: Recall that $\det A \cdot \det B = \det(A \cdot B)$ for matrices A and B.

5.10 The matrix representation of C_{3v} given in Eq. (5.14) was constructed with respect to the particular coordinate system defined by the unit vectors $\boldsymbol{e}_1$ and $\boldsymbol{e}_2$ in Fig. 5.10. Show that a corresponding matrix representation for basis vectors rotated clockwise by $\frac{\pi}{6}$ relative to those in Fig. 5.10 is given by Eq. (5.15). *Hint*: We could proceed geometrically as in Fig. 5.10, but a more elegant approach is to exploit rotational symmetry and use the rotation operator of Eq. (6.3) to perform a similarity transformation (2.12) on the original matrices. ***

5.11 Show that the groups C_{3v} and D_3 have equivalent characters, but the basis functions corresponding to their irreps are different. ***

5.12 Prove that the action of the symmetry operations σ_b and σ_c on the basis vectors $\boldsymbol{e}_1$ and $\boldsymbol{e}_2$ in Fig. 5.10 are given by the matrix equations (5.11) and (5.12). ***

5.13 Consider the following figure

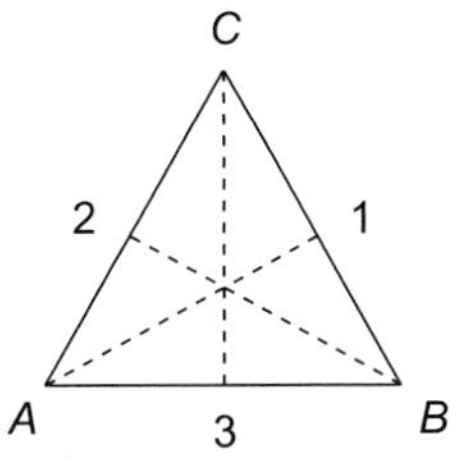

Define an operator c that rotates in the plane of the triangle by $-2\pi/3$ such that $c^2 = -4\pi/3$, operators r_1, r_2, and r_3 that rotate by π around the axes labeled 1, 2, and 3, respectively, and an identity e.

(a) Show that the set $\{e, c, c^2, r_1, r_2, r_3\}$ forms a group (the group D_3 listed in Table 5.1) under a multiplication operation defined as two successive transformations by constructing the multiplication table using geometrical arguments.

(b) Show that D_3 is isomorphic to the permutation group S_3, and has an abelian invariant subgroup $H = \{e, c, c^2\}$.

(c) Partition the group D_3 into conjugacy classes.

(d) Construct the cosets with respect to the abelian invariant subgroup $H = \{e, c, c^2\}$ and show that the factor group D_3/H is isomorphic to the cyclic group C_2.

5.14 If cyclic boundary conditions are imposed on a periodic 1D lattice by identifying the two ends with N cycles between the boundaries, the translation group becomes a cyclic group of order N. Show that the irreps of this group are of the form

$$\Gamma^{(p)}(C) = e^{2\pi i p/N} \quad (p = 1, 2, 3, \ldots, N),$$

where C denotes group elements. *Hint*: Elements of the cyclic group of order N are $C_1 = c$, $C_2 = c^2$, $C_3 = c^3, \ldots, C_N = c^N = 1$, because of the closure condition $C_N = 1$, where 1 is the group identity. Thus the group is abelian and irreps are 1D and labeled by complex numbers. ***

5.15 As shown in Problem 5.14, the symmetry group for a finite 1D periodic lattice having cyclic boundary conditions with N periods between boundaries is the cyclic group of order N, with irreps of the form

$$\Gamma^{(p)} = e^{2\pi i p/N} \quad (p = 1, 2, 3, \ldots, N).$$

Show that if the total length of the finite 1D lattice is $L = aN$, wavefunctions defined on the lattice obey $\psi_k(x + a) = e^{ika}\psi_k(x)$, or equivalently $\psi_k(x) = u_k(x)$, where $u_k(x)$ is periodic, $u_k(x) = u_k(x + a)$, and $k \equiv 2\pi p/L$ is called the *crystal momentum.* This result is a 1D version of Bloch's theorem, described in Section 5.4. ***

6 The Rotation Group

In this chapter we consider the group SO(3) of continuous rotations in 3D space and the closely related special unitary group SU(2). These groups are of practical significance because of the importance of angular momentum in quantum mechanics, and they serve as examples of techniques that may be adapted to the analysis of more complicated groups. As part of this discussion we will investigate the relationship between the groups SO(3) and SU(2). They will be found to obey the same Lie algebra, so they are locally identical but differ in the global structure of the group manifold. Hence, we will also introduce in this chapter a distinction between the local and global properties of Lie groups.

6.1 Three-Dimensional Rotations

Rotations of vectors in 3D euclidean space may be implemented by $\boldsymbol{x}' = R\boldsymbol{x}$, where $\boldsymbol{x}$ is the initial vector, $\boldsymbol{x}'$ is the rotated vector, and R is a 3×3 matrix. Let $\boldsymbol{x}$ be an arbitrary vector expanded in a basis, $\boldsymbol{x} = \hat{e}_i x^i$, where $\hat{e}_i$ is a unit cartesian vector (implied summation on repeated indices). Then, under a rotation R the components transform as $x_i' = R_i^j x_j$. Physically, a rotational transformation on a vector leaves its *squared length unchanged*: $\boldsymbol{x}^2 = \boldsymbol{x}'^2$. The scalar product $\boldsymbol{x}^2 \equiv \boldsymbol{x} \cdot \boldsymbol{x}$ written in matrix notation is $\boldsymbol{x}^2 = x^{\mathrm{T}} x$, where x^{T} denotes the *transpose* of x (interchange rows and columns). Therefore, $\boldsymbol{x}^2 = \boldsymbol{x}'^2$ requires that $(x^{\mathrm{T}} x)^2 = (x^{\mathrm{T}} R^{\mathrm{T}} R x)^2$, which is true only if $R^{\mathrm{T}} R = 1$, or equivalently $R^{\mathrm{T}} = R^{-1}$. But this is the condition that *R is orthogonal,* so rotational transformations are implemented by orthogonal matrices.

Furthermore, the determinant of a matrix product is the product of determinants for each matrix: $\det(A \cdot B) = \det A \cdot \det B$. Thus $R^{\mathrm{T}} R = 1$ implies that $\det R^{\mathrm{T}} \det R = 1$. But the determinant of a matrix is equal to the determinant of its transpose, $\det R^{\mathrm{T}} = \det R$, so

$$\det R = \pm 1, \tag{6.1}$$

for orthogonal matrices. Rotations are continuous and all physical rotations may be reached from the identity, which corresponds to no rotation, by a series of infinitesimal rotations. But the identity is a unit matrix and therefore has determinant +1. Since the determinant of a matrix cannot change from +1 to −1 under continuous infinitesimal rotations, we conclude that the orthogonal matrices R that describe physical rotations must be further limited to those having a determinant of +1. The adjective “special” and the prefix “S” are commonly used to indicate matrices restricted to those with determinant +1. Thus the

Box 6.1 Relationship of the Groups O(3) and SO(3)

The group O(3) corresponds to the normal rotations plus a discrete space reflection in all axes (parity transformation) that is implemented by the matrix

$$\begin{pmatrix} -1 & 0 & 0 \\ 0 & -1 & 0 \\ 0 & 0 & -1 \end{pmatrix}.$$

It has two disjoint pieces corresponding to the opposite signs for det R in Eq. (6.1). The subgroup O(3) $\supset$ SO(3) may be identified with the det $R = +1$ piece of O(3), which contains the identity. It follows that the det $R = -1$ portion of O(3) is not a group because it does not contain the identity.

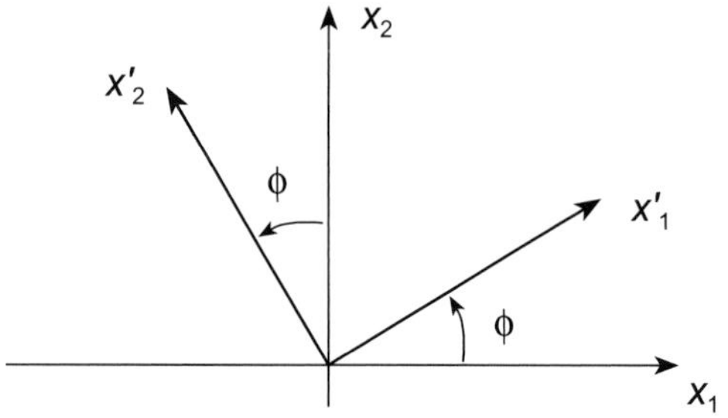

Fig. 6.1 Rotation of a two-dimensional cartesian coordinate system (x_1, x_2) counterclockwise by an angle ϕ into a new coordinate system (x_1', x_2').

rotation matrices are *special orthogonal*, and the three-dimensional rotation group is the SO(3) subgroup of the full orthogonal group O(3), as discussed further in Box 6.1.

6.2 The SO(2) Group

It is obvious that rotations around a single axis commute and satisfy the group postulates, so they form an abelian SO(2) subgroup of the full rotation group in three dimensions. We will analyze this subgroup first, before tackling the full SO(3) group.

6.2.1 Generators of SO(2) Rotations

Consider the rotation of an orthogonal two-dimensional coordinate system (x_1, x_2) by an angle ϕ, as depicted in Fig. 6.1. The equation defining this rotation is

$$\begin{pmatrix} x_1' \\ x_2' \end{pmatrix} = R \begin{pmatrix} x_1 \\ x_2 \end{pmatrix}, \tag{6.2}$$

where the rotation matrix R can be parameterized (see Problem 6.1)

$$R(\phi) = \begin{pmatrix} \cos\phi & -\sin\phi \\ \sin\phi & \cos\phi \end{pmatrix}. \tag{6.3}$$

These matrices form a rotational subgroup SO(2) of the full rotation group, O(3) $\supset$ SO(3) $\supset$ SO(2), with an abelian law of composition (multiplication law)

$$R(\phi_1)R(\phi_2) = R(\phi_1 + \phi_2), \tag{6.4}$$

but with the proviso that

$$R(\phi) = R(\phi \pm 2\pi), \tag{6.5}$$

if $\phi_1 + \phi_2$ is not in the range $(0, 2\pi)$.

Because SO(2) is a Lie group, we may determine all of its local properties from the behavior of the group near the origin (the identity, corresponding to rotation by $\phi = 0$). An infinitesimal rotation through an angle $d\phi$ corresponds to

$$R(d\phi) = 1 - iJd\phi, \tag{6.6}$$

where the factor i is conventional and J is independent of $d\phi$. We may write

$$R(\phi + d\phi) = R(\phi) + \frac{dR(\phi)}{d\phi}\, d\phi, \tag{6.7}$$

and from (6.4) and (6.6)

$$R(\phi + d\phi) = R(\phi)R(d\phi) = R(\phi) - iJR(\phi)d\phi. \tag{6.8}$$

Comparing Eqs. (6.7) and (6.8) implies that $dR/d\phi = -iJR(\phi)$, which has the solution

$$R(\phi) = e^{-iJ\phi}, \tag{6.9}$$

if $R(0) = 1$ is used as the boundary condition. The quantity J is a group generator and we note that the group multiplication rule (6.4) is satisfied automatically by Eq. (6.9):

$$R(\phi_1)R(\phi_2) = e^{-iJ\phi_1}e^{-iJ\phi_2} = e^{-iJ(\phi_1+\phi_2)} = R(\phi_1 + \phi_2).$$

A matrix form of the infinitesimal generator J can be determined by examining the matrix R in Eq. (6.3) for infinitesimal rotations $d\phi$. This gives (Problem 6.2),

$$J = \begin{pmatrix} 0 & -i \\ i & 0 \end{pmatrix}, \tag{6.10}$$

for the single SO(2) generator J in the basis implied by Eq. (6.3).

6.2.2 SO(2) Irreducible Representations

The rotation matrix (6.3) is a 2D representation of SO(2) but it is *not an irreducible representation* because SO(2) is abelian so its irreps must all be 1D. As shown in Problem 6.21, the rotation matrix (6.3) may be converted by similarity transformation to

$$R(\phi) = \begin{pmatrix} e^{i\phi} & 0 \\ 0 & e^{-i\phi} \end{pmatrix}, \tag{6.11}$$

which operates on a transformed basis

$$\hat{e}_+ = \frac{1}{\sqrt{2}}(\hat{e}_1 + i\hat{e}_2) \qquad \hat{e}_- = \frac{1}{\sqrt{2}}(\hat{e}_1 - i\hat{e}_2), \tag{6.12}$$

and J is brought by the same transformation to

$$J = \begin{pmatrix} 1 & 0 \\ 0 & -1 \end{pmatrix}.$$

Therefore, $e^{i\phi}$ and $e^{-i\phi}$ are irreducible representations of SO(2) and the matrix (6.3) is a reducible representation that may be transformed to a direct sum of these two irreps.

We have looked at the symmetry under transformations of the SO(2) rotor in the parameter space ϕ. The quantum problem for the SO(2) rotor is formulated in a linear vector space (Hilbert space) in terms of state vectors that depend on the parameter ϕ. Therefore, we wish to understand how SO(2) transformations in the parameter space induce corresponding SO(2) transformations in the Hilbert space. The basic procedure is given in Example 2.8 and Section 2.8, and was worked out for the 1D translation group in Problems 5.14 and 5.15. The 1D translation group and SO(2) are both abelian so the steps here are similar, except that we must account for the constraint (6.5) on SO(2) transformations.

Let the Hilbert space operator $U(\phi)$ correspond to the coordinate space operator $R(\phi)$. By the discussion in Example 2.8, $U(\phi)$ will generate an SO(2) representation in the vector space if it preserves the SO(2) multiplication law from the coordinate space,

$$U(\phi_1)U(\phi_2) = U(\phi_1 + \phi_2), \tag{6.13}$$

along with the continuity condition $U(\phi) = U(\phi \pm 2\pi)$. Note that we are now dealing with two spaces: the *group space* where the transformations $R(\phi)$ live, and the *representation space* where the transformations $U(\theta)$ live. Repeating the previous arguments in Eqs. (6.6)–(6.9), we obtain $U(d\phi) = 1 - iJd\phi$ and [199]

$$U(\phi) = e^{-i\phi J}, \tag{6.14}$$

which is similar formally to Eq. (6.9) but now $U(\phi)$ and J are *operators* acting on state vectors $|\alpha\rangle$ in the Hilbert space. By the same arguments as for the translation group in Section 5.4 (see also the solutions of Problems 5.14 and 5.15), for a vector $|\alpha\rangle$ in a minimal subspace invariant under SO(2),

$$J\,|\alpha\rangle = \alpha\,|\alpha\rangle \qquad U(\phi)\,|\alpha\rangle = e^{-i\alpha\phi} \tag{6.15}$$

[see Eq. (2.15)], where we have used the results of Problem 3.11(c) to evaluate the eigenvalue of the exponentiated operator J in (6.14). This satisfies the group multiplication rule (6.13) for any α, but it must also satisfy the global continuity requirement $U(\phi) = U(\phi \pm 2\pi)$. Inserting this condition in Eq. (6.15) gives $e^{\pm 2\pi i\alpha} = 1$, which restricts α to integer values. Therefore, the irreducible representations of SO(2) are of the general form

$$U_m(\phi) = e^{-im\phi}, \tag{6.16}$$

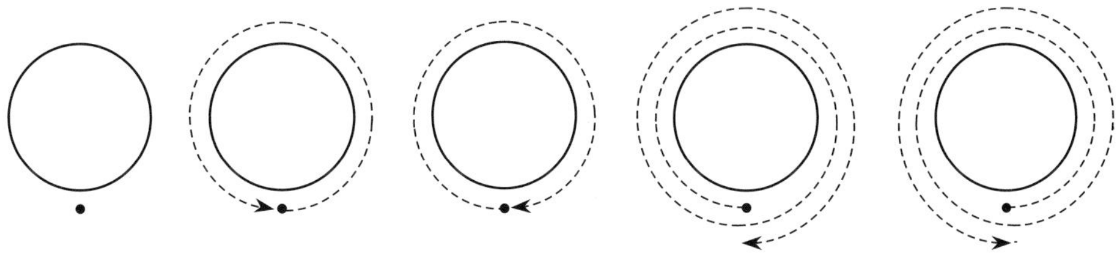

Fig. 6.2 Some winding numbers characterizing the topology of the SO(2) manifold S^1. Intuitively the winding number is the number of times the dashed curve is wrapped around the solid circle, clockwise (+) or counterclockwise (−). A more rigorous discussion of winding numbers as topological invariants will be given in Section 24.4.

where m is a positive or negative integer. The defining equation (6.3) for 2D rotations $R(\phi)$ is then, by virtue of Eq. (6.11), a 2D reducible SO(2) representation corresponding to a direct sum of the $m = \pm 1$ irreps.

6.2.3 Connectedness of the Manifold

From Eq. (6.16) the topology of the SO(2) manifold is that of the unit circle S^1, which is illustrated in Fig. 6.2.[1]

1. If $m = 0$, we obtain the identity representation, $R(\phi) \to U_0(\phi) = 1$.
2. If $m = -1$, the isomorphic mapping is $R(\phi) \to U_{-1}(\phi) = e^{i\phi}$ and as R ranges over group space, in representation space U_{-1} covers the unit circle once counterclockwise.
3. If $m = +1$, the isomorphic mapping is $R(\phi) \to U_1(\phi) = e^{-i\phi}$ and as R ranges over the group space, in representation space $U_1(\phi)$ covers the unit circle once clockwise.
4. If $m = \pm 2$, the mapping $R(\phi) \to U_{\mp 2}(\phi) = e^{\pm 2i\phi}$ is no longer isomorphic because the unit circle is covered *twice* (clockwise or counterclockwise).

Likewise, higher $|m|$ defines mappings that wind more times around the unit circle. Thus, only the $m = \pm 1$ irreps are faithful representations of SO(2).

Homotopic Paths: The *connectedness* of a group manifold is specified by the nature of the closed paths in the manifold. If two paths with the same endpoints can be deformed continuously one to the other, the paths are said to be *homotopic*. A manifold in which all paths can be deformed continuously to a single path is said to be *simply connected*, as illustrated in Fig. 6.3(a). A manifold in which there are paths between the same endpoints that cannot be deformed continuously into each other is said to be *multiply connected*. Figure 6.3(b) illustrates a multiply connected space because the paths 1 and 2 are not homotopic. From these considerations we conclude that the SO(2) manifold is *infinitely connected*, because an infinity of paths exists corresponding to differing numbers of times wrapped around the unit circle for different values of m, and these cannot be deformed continuously one into the other.

[1] In this chapter we will deal qualitatively with two central concepts from topology: *connectedness* and *compactness*. These issues will be taken up more mathematically beginning in Ch. 24.

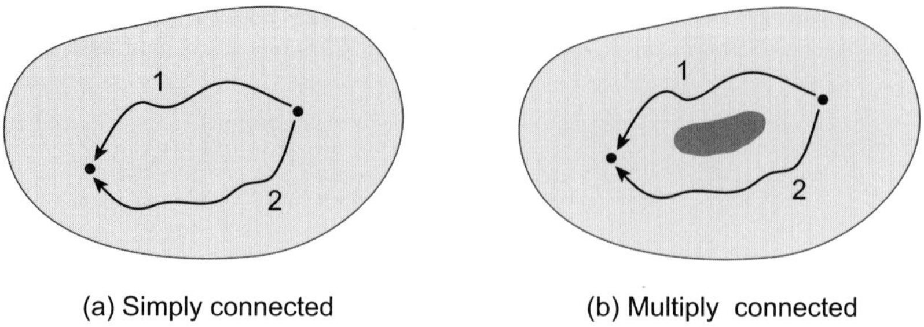

Fig. 6.3 Simply connected and multiply connected manifolds. (a) Paths 1 and 2 can be continuously deformed into each other, so this manifold is simply connected. (b) Paths 1 and 2 cannot be deformed continuously into each other because the dark shaded region is not part of the manifold, so this manifold is multiply connected.

Universal Covering Group: Generally more than one Lie group shares a given Lie algebra but *only one of them will be simply connected.* This unique group is called the *(universal) covering group* of the algebra. The other groups having the same Lie algebra are homomorphic to the covering group. For example, we will see that the groups SO(3) and SU(2) have the same Lie algebra but they differ topologically. The group SU(2) is simply connected and is the covering group for the algebra, while SO(3) is doubly connected and related to the covering group SU(2) by a homomorphism. For SO(2) the universal covering group is the additive group of real numbers, which is simply connected.

6.2.4 Compactness of the Manifold

A compact Lie group is characterized by a parameter space that is both *closed* and *bounded*. (This definition is adequate for most physics applications but a mathematically more rigorous one will be given in Section 24.2.3.) A set is bounded if no member of the set exceeds a certain positive number in absolute value. A set is closed if the limit of every convergent sequence of points in the set is also in the set. These definitions are made more tangible by considering a few examples.

1. The real number line between 0 and 1 is bounded but not closed, because it does not contain the endpoints of the interval.
2. The real number line from 0 to 1 including the endpoints is both closed and bounded.
3. Rotations in any number of dimensions are bounded and closed because around any axis the rotations lie in a closed interval $0 - 2\pi$.
4. One-dimensional translations are not compact because they are not bounded (if they are bounded, they do not form a group).
5. The group of Lorentz transformations discussed in Ch. 13 is not compact because closure is not satisfied: the velocity v is bounded by the speed of light c, but the limit point $v = c$ is not in the set because the Lorentz boost transformation (13.11) is infinite at $v = c$.

Since the manifold of SO(2) illustrated in Fig. 6.2 is the unit circle S^1, it is *compact* and *multiply connected.*

6.2.5 Invariant Group Integration

In formal proofs of the relations for representations and characters in finite groups such as Eqs. (2.29)–(2.32), the following kinds of equalities are used repeatedly:

$$\sum_j f(g_j) = \sum_j f(hg_j) = \sum_j f(g_j h), \tag{6.17}$$

where f is an arbitrary function of the group elements g_j and h is any element of the group. This relation is called the *rearrangement lemma*,[2] because the product of two group elements is another group element and the multiplication of g_j and h just rearranges terms in the sum, without changing the value of the sum itself (see Problem 6.3).

Integration Measure: Whether theorems involving summations that we have used for finite groups can be extended to continuous groups depends on whether a convergent integral with a consistent integration measure can be defined that generalizes summations on representations and characters. Such a measure exists if the parameter space is compact, so that the group volume is finite and the integrals are well behaved. Schematically, Eq. (6.17) can be written $\sum_g = \sum_{g'}$, where g is a group element and $g' = hg$, where h is some other group element. Thus, a reasonable requirement for an invariant group integration measure dg is that for some function $f(g)$ of the group elements

$$\int dg f(g) = \int dg f(g'g), \tag{6.18}$$

where g' is an arbitrary group element. That is, the integral must be invariant under the replacement $g \to g'g$ in the function, because the product $g'g$ is also an element of the group and this replacement just corresponds to a reshuffling of the elements in the group space and does not change the number of points being integrated (summed) over. Let $\xi = (\xi_1, \xi_2, \ldots, \xi_n)$ label the group elements and define a *weight function* $J(\xi)$ such that

$$\int dg f(g) = \int d\xi_1 d\xi_2 \cdots d\xi_n J(\xi) f[g(\xi)]. \tag{6.19}$$

Then we may hope to satisfy (6.18) by a suitable choice of $J(\xi)$. An invariant integration measure satisfying Eq. (6.18) exists if the group manifold is compact (see Section 24.2.3), but it also exists if the manifold satisfies a less stringent condition called *local compactness.*We skip the details of defining local compactness but remark that *not all Lie groups are compact, but all are locally compact.* Thus for Lie group manifolds an integration measure satisfying Eq. (6.18) may be constructed. In mathematics the corresponding integration measure is called the *Haar measure.*

Let us illustrate these ideas with a one-parameter group; from Eq. (2.8) the Lie group multiplication must satisfy $g[\alpha(\beta,\gamma)] = g(\beta)g(\gamma)$, for specific values α, β, and γ of the parameters. Combining the preceding equations gives for a one-parameter group

$$\int d\beta J(\beta) f[g(\beta)] = \int d\beta J(\beta) f[g(\alpha(\beta,\gamma))], \tag{6.20}$$

[2] As noted in Box 2.2, the rearrangement lemma is the reason that in the multiplication table for a finite group each group element appears exactly once in each row and exactly once in each column.

which defines an invariant integration measure. The weight function $J(\beta)$ can be determined by examining behavior near the origin. Problem 6.4 illustrates, where we find that

$$dg_{\mathrm{SO(2)}} = \frac{1}{2\pi} d\phi \tag{6.21}$$

for the invariant SO(2) integration measure. Similar methods may be used to find the invariant integration measure for other compact Lie groups [199].

Orthogonality and Completeness: Using Eq. (6.21), the SO(2) orthogonality and completeness relations may be written as

$$\frac{1}{2\pi}\int_0^{2\pi} U_n^\dagger(\phi)U_m(\phi)d\phi = \delta_{mn} \qquad \sum_n U_n(\phi)U_n^\dagger(\phi') = \delta(\phi - \phi'), \tag{6.22}$$

where $U_m = e^{-im\phi}$ and $\delta(\phi - \phi')$ is the Dirac δ-function. These represent extensions of the orthogonality and completeness conditions (2.29) to the case of a continuous group.

Characters: Irreducible representations of the abelian group SO(2) are one-dimensional and characters of the irreps are just the representations $\chi_m(\phi) = e^{-im\phi}$, which are continuous functions of the parameters. Invariant integration over characters for SO(2) takes the form

$$\int_0^{2\pi} \chi_m(\phi)\chi_{m'}^*(\phi)d\phi = \int_0^{2\pi} e^{i\phi(m-m')}d\phi = 2\pi\delta_{mm'}, \tag{6.23}$$

which is analogous to the second of Eqs. (2.30) for finite groups.

Expansion of Arbitrary Functions: Arbitrary functions may be expanded as a sum over a finite number of irreps for finite groups. For continuous groups the dimension of each irrep is finite but the characters become continuous functions, implying an infinite number of classes and thus irreps. Hence the expansion of an arbitrary function on the irreps for a continuous group usually will involve an infinite number of terms.

Example 6.1 For SO(2) this expansion in terms of irreps is just the Fourier series

$$f(\phi) = \sum_{m=-\infty}^{+\infty} c_m e^{im\phi}, \tag{6.24}$$

for a function of a single angular variable ϕ.

Example 6.2 Shortly we will encounter the expansion of functions defined on a 2D surface in terms of spherical harmonics $Y_{lm}(\theta, \phi)$,

$$f(\theta, \phi) = \sum_{lm} a_{lm} Y_{lm}(\theta, \phi), \tag{6.25}$$

where $Y_{lm}(\theta, \phi)$ will be found to be a $(2l + 1)$-dimensional representation of SO(3).

6.3 The SO(3) Group

We turn now to the full 3D rotation group SO(3). Our experience with SO(2) will be put to good use because all SO(3) rotations may be built from a succession of infinitesimal rotations about each of three independent axes.

6.3.1 Generators of SO(3)

The first task is to determine the Lie algebra obeyed by the 3D rotation group. Consider an infinitesimal rotation about an axis; the effect of this rotation on a function $f(x,y,z)$ may be found either by rotating the function in one direction (an *active rotation*), or by rotating the coordinate system in the opposite direction (a *passive rotation*). From Example 2.8, the general result is

$$O_R f(p) = f\left(R^{-1}p\right), \tag{6.26}$$

where R is a mapping of 3D space onto itself, $p = (x,y,z)$ is a point in that space, and O_R is the operator implementing the mapping R in the function space. Thus, after being acted upon by O_R the function is equal to the same function but evaluated at the point $R^{-1}p$. For rotations $R(\phi)$ about a single axis the inverse is $R^{-1}(\phi) = R(-\phi)$, and from Eq. (6.3) for infinitesimal rotations by an angle $-d\phi$ about the z-axis in an (r,θ,ϕ) coordinate system,

$$\begin{pmatrix} x' \\ y' \end{pmatrix} = R^{-1}(d\phi)\begin{pmatrix} x \\ y \end{pmatrix} = R(-d\phi)\begin{pmatrix} x \\ y \end{pmatrix} = \begin{pmatrix} 1 & d\phi \\ -d\phi & 1 \end{pmatrix}\begin{pmatrix} x \\ y \end{pmatrix},$$

and we may write for a small rotation of an arbitrary function $f(x,y,z)$ about the z-axis,

$$\begin{aligned} O_R f(x,y,z) &= f\left(R^{-1}(x,y,z)\right) \\ &= f(x+dx,\ y+dy, z) \\ &= f(x + y\,d\phi, y - x\,d\phi, z). \end{aligned} \tag{6.27}$$

As shown in Problem 6.26, expanding in a Taylor series and repeating for rotations around the other axes then leads to

$$J_x = -i\left(y\frac{\partial}{\partial z} - z\frac{\partial}{\partial y}\right) \quad J_y = -i\left(z\frac{\partial}{\partial x} - x\frac{\partial}{\partial z}\right) \quad J_z = i\left(y\frac{\partial}{\partial x} - x\frac{\partial}{\partial y}\right), \tag{6.28}$$

for the generators of 3D rotations. From this we find

$$[\,J_i, J_j\,] = i\epsilon_{ijk}J_k, \tag{6.29}$$

where $x = 1$, $y = 2$, and $z = 3$. Comparing with Eq. (3.13), the angular momentum operators J_i are the generators of an SU(2) Lie algebra.

6.3.2 Matrix Elements of the Rotation Operator

We wish to evaluate the quantum matrix elements of the rotation operator in the SO(3) basis.[3] To do so it is useful to introduce the *Euler angle parameterization* for a general

[3] Be warned! Different conventions are followed by various authors for angular momentum matrix elements. Here we follow the presentation of Brink and Satchler [29].

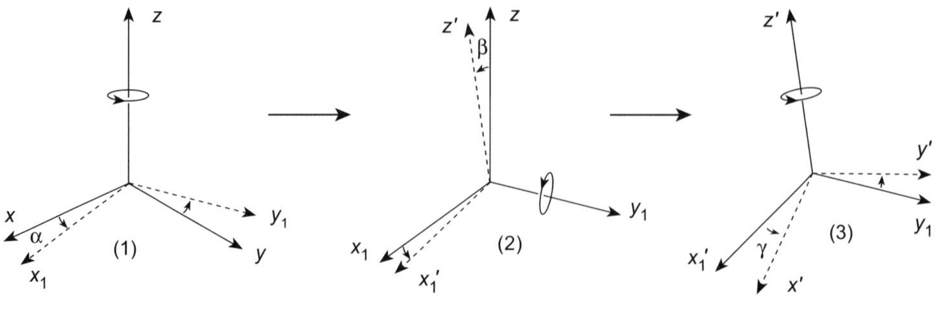

Fig. 6.4 Euler angles for 3D rotations. Positive rotations are defined by the right-hand screw rule.

3D rotation. An arbitrary rotation may be decomposed into three successive Euler angle rotations, as depicted in Fig. 6.4: (1) rotate by an angle α about the z-axis, then (2) rotate by an angle β around the new y_1-axis, and finally (3) rotate by an angle γ around the new z'-axis. But the same transformation can be accomplished by another sequence of rotations: (1) rotate by γ about the original z-axis, then (2) rotate by β about the original y-axis, and finally (3) rotate by α about the original z-axis, which is implemented by the operator

$$D(\alpha, \beta, \gamma) = e^{-i\alpha J_z} e^{-i\beta J_y} e^{-i\gamma J_z}. \tag{6.30}$$

Let us evaluate matrix elements of (6.30) in a basis corresponding to an irreducible representation of SO(3). Since the irrep is associated with an SU(2) algebra, it is labeled by a total angular momentum J, it is $(2J + 1)$-dimensional, and it has a Cartan subalgebra for which $J_z \, |JM\rangle = M \, |JM\rangle$. The matrix elements of the operator (6.30) in this basis are (Problem 6.22)[4]

$$D^J_{M'M}(\alpha, \beta, \gamma) \equiv \langle JM'| \, D(\alpha, \beta, \gamma) \, |JM\rangle = e^{-i(\alpha M' + \gamma M)} d^J_{M'M}(\beta), \tag{6.31a}$$

$$d^J_{M'M}(\beta) \equiv \langle JM'| \, e^{-i\beta J_y} \, |JM\rangle. \tag{6.31b}$$

The matrix element (6.31b) is not diagonal in J_y because only one generator can be chosen diagonal in the SU(2) algebra and we have already selected J_z for that honor.

Phases for the D-matrices depend on the Euler angle convention and the phase choice for the matrix elements of $\boldsymbol{J}$. As is common, we follow the *Condon–Shortley phase convention* alluded to in Section 6.3.7. Then $d^j_{mn}(\beta)$ is a *real orthogonal matrix* given by

$$d^j_{mn}(\beta) = \sum_t (-1)^t \left(\frac{[(j+m)! \, (j-m)! \, (j+n)! \, (j-n)! \,]^{1/2}}{(j+m-t)! \, (j-n-t)! \, t! \, (t+n-m)!} \right) \times \left(\cos \frac{\beta}{2} \right)^{2j+m-n-2t} \left(\sin \frac{\beta}{2} \right)^{2t+n-m}, \tag{6.32}$$

where the sum is over all values of t that give non-negative factorials. (Other phase conventions typically give differing phase factors in off-diagonal elements of the representation

[4] Among practitioners it is common to refer to the matrix element defined in Eq. (6.31a) as the "(Wigner) D-matrix" and to the matrix element defined in Eq. (6.31b) as the "(Wigner) little d-matrix."

matrices.) For low values of angular momenta it is not hard to construct the D-matrices, as illustrated in the following example.

Example 6.3 Let us work out the D-matrix for $J = \frac{1}{2}$. Since SO(3) and SU(2) share the same Lie algebra the $J = \frac{1}{2}$ representation is in terms of the Pauli matrices (3.11) with

$$J_1 = \frac{1}{2}\begin{pmatrix} 0 & 1 \\ 1 & 0 \end{pmatrix} \qquad J_2 = \frac{1}{2}\begin{pmatrix} 0 & -i \\ i & 0 \end{pmatrix} \qquad J_3 = \frac{1}{2}\begin{pmatrix} 1 & 0 \\ 0 & -1 \end{pmatrix},$$

from which we find that

$$d^{1/2} = e^{-\frac{i}{2}\beta\sigma_2} = 1 - i\left(\frac{\beta}{2}\right)\sigma_2 - \frac{1}{2}\left(\frac{\beta}{2}\right)^2 \sigma_2^2 + \frac{i}{3!}\left(\frac{\beta}{2}\right)^3 \sigma_2^3 - \cdots$$

But for the Pauli matrices, $\sigma_2^2 = 1$, $\sigma_2^3 = \sigma_2$, $\sigma_2^4 = 1, \ldots$, and

$$\begin{aligned} d^{1/2}(\beta) &= \left[1 - \frac{1}{2}\left(\frac{\beta}{2}\right)^2 + \frac{1}{4!}\left(\frac{\beta}{2}\right)^4 - \cdots\right] - i\sigma_2\left[\frac{\beta}{2} - \frac{1}{3!}\left(\frac{\beta}{2}\right)^3 + \cdots\right] \\ &= \cos\frac{\beta}{2} - i\sigma_2 \sin\frac{\beta}{2} = \begin{pmatrix} \cos\frac{\beta}{2} & -\sin\frac{\beta}{2} \\ \sin\frac{\beta}{2} & \cos\frac{\beta}{2} \end{pmatrix}, \end{aligned} \tag{6.33}$$

where we have omitted explicit display of 2×2 unit-matrix factors. Therefore,

$$\begin{aligned} D^{1/2}_{MM'}(\alpha, \beta, \gamma) &= e^{-i(\alpha M + \gamma M')} d^{1/2}_{MM'} \\ &= \begin{pmatrix} e^{-i\alpha/2}\cos\left(\frac{\beta}{2}\right)e^{-i\gamma/2} & -e^{-i\alpha/2}\sin\left(\frac{\beta}{2}\right)e^{i\gamma/2} \\ e^{i\alpha/2}\sin\left(\frac{\beta}{2}\right)e^{-i\gamma/2} & e^{i\alpha/2}\cos\left(\frac{\beta}{2}\right)e^{i\gamma/2} \end{pmatrix}, \end{aligned} \tag{6.34}$$

for the D-matrix corresponding to $J = \frac{1}{2}$.

Elements of d-matrices for low angular momenta are given in Table C.1 of Appendix C.

6.3.3 Properties of *D*-Matrices

The d-matrices and D-matrices have a number of important properties.

1. The $D^J_{M'M}$ are special unitary, by construction

$$D^\dagger(\alpha, \beta, \gamma) = D^{-1}(\alpha, \beta, \gamma) = D(-\gamma, -\beta, -\alpha), \tag{6.35a}$$

$$D^J_{MN}(\alpha, \beta\gamma)^* = (-1)^{M-N} D^J_{-M-N}(\alpha, \beta, \gamma), \tag{6.35b}$$

$$D^J_{MN}(\alpha, \beta, \gamma)^* = D^J_{NM}(-\gamma, -\beta, -\alpha) \qquad \det D(\alpha, \beta, \gamma) = 1. \tag{6.35c}$$

$$\sum_{M'} D^J_{M'N}(\Omega)^* D^J_{M'M}(\Omega) = \delta_{MN} \qquad \sum_{M'} D^J_{MM'}(\Omega) D^J_{NM'}(\Omega)^* = \delta_{MN}, \tag{6.35d}$$

   where $\Omega \equiv (\alpha, \beta, \gamma)$.
2. The functions $d^J_{MM'}$ are real orthogonal in the Condon–Shortley phase convention.

3. The functions $d^I_{MN}(\beta)$ have the properties

$$d^I_{MN}(\pi) = (-1)^{I+M}\delta_{M,-N} \qquad d^I_{MN}(2\pi) = (-1)^{2I} d^I_{MN}(0)$$
$$d^I_{MN}(2\pi) = (-1)^{2I}\delta_{MN}. \tag{6.36}$$

4. The invariant integration measures for SO(3) and SU(2) are

$$d\tau_{\mathrm{so}(3)} = \frac{\sin\beta}{8\pi^2}\, d\alpha\, d\beta\, d\gamma \qquad d\tau_{\mathrm{su}(2)} = \frac{1}{2} d\tau_{\mathrm{so}(3)}. \tag{6.37}$$

5. Defining $d\Omega \equiv \sin\beta d\beta d\alpha d\gamma$, the D-functions obey the orthogonality relation

$$\int_0^{2\pi}\int_0^{2\pi}\int_0^{\pi} \sin\beta\, d\beta\, d\alpha\, d\gamma\, D^I_{MM'}(\Omega)^* D^J_{NN'}(\Omega) = \frac{8\pi^2}{2I+1}\delta_{MN}\delta_{M'N'}\delta_{IJ}. \tag{6.38}$$

6. The D-functions are complete

$$f(\alpha,\beta,\gamma) = \sum_{JMM'} f^J_{MM'} D^J_{MM'}(\alpha,\beta,\gamma), \tag{6.39}$$

where $f(\alpha,\beta,\gamma)$ is an arbitrary function of the Euler angles and where,

$$f^J_{MM'} = \frac{2J+1}{8\pi^2}\int_0^{2\pi}\int_0^{2\pi}\int_0^{\pi} \sin\beta\, d\beta\, d\alpha\, d\gamma\, D^J_{MM'}(\alpha,\beta,\gamma)^* f(\alpha,\beta,\gamma),$$

by orthogonality.

7. The spherical harmonics $Y_{LM}(\theta,\phi)$ are related to the D-functions by

$$Y^*_{LM}(\theta,\phi) = \sqrt{\frac{2L+1}{4\pi}}\, D^L_{M0}(\phi,\theta,0). \tag{6.40}$$

Some low-L spherical harmonics are tabulated in Table C.2 of Appendix C.

Proofs of these relations may be found in Refs. [29, 51, 173, 201].

6.3.4 Characters for SO(3)

Since all rotations by the same angle around any axis are in the same class, the character is a continuous function of the rotation angle α about any axis and we may choose any axis to evaluate it. We already know that around the z-axis

$$\chi^{(j)}_\alpha = \sum_{m=-j}^{+j} e^{-im\alpha}. \tag{6.41}$$

Using basic trigonometric identities we find (Problem 6.8)

$$\chi^{(j)}_\alpha = \frac{\sin\left(j+\frac{1}{2}\right)\alpha}{\sin\left(\frac{1}{2}\alpha\right)}, \tag{6.42}$$

from which it may be concluded that for the identity element

$$\lim_{\alpha\to 0}\chi^{(j)}_\alpha = 2j+1, \tag{6.43}$$

and the character of the identity element is the irrep dimensionality, just as for finite groups.

6.3.5 Direct Products of SO(3) Representations

The direct product of SO(3) representations $D^{(j)}$ leads to the Clebsch–Gordan series

$$D^{(j_1)} \otimes D^{(j_2)} = \sum_J c_J D^{(J)} \tag{6.44}$$

(see Section 3.3.4). One way to evaluate the coefficients c_J in Eq. (6.44) is to use the characters of the SO(3) representations. As illustrated in Problem 6.27, all the c_J are equal to zero or one so the group SO(3) is *simply reducible* and we obtain

$$D^{(j_1)} \otimes D^{(j_2)} = D^{(J=j_1+j_2)} \oplus D^{(J=j_1+j_2-1)} \oplus \cdots \oplus D^{(J=|j_1-j_2|)} \tag{6.45}$$

for the SO(3) Clebsch–Gordan series.

6.3.6 SO(3) Vector-Coupling Coefficients

The most common physical application of the Clebsch–Gordan series for SO(3) occurs for a system with two independent angular momenta. The direct product of representations corresponding to the two angular momenta will not generally yield matrices in the direct-sum, block-diagonal form of Eq. (2.13), but they can be put in that form by a unitary transformation that corresponds to a coupling of the independent angular momenta of the system to a resultant angular momentum that is a conserved quantum number. The elements of this transformation are called the *vector-coupling* or *Clebsch–Gordan coefficients*, which we shall now construct for the group SO(3).

> The essential property that will be employed in constructing Clebsch–Gordan coefficients is that the generators of the direct product representation are the sums of the corresponding generators for the product representations.

Suppose that $D^{(j_1)}$ and $D^{(j_2)}$ are irreps of SO(3) with eigenvectors $|j_1m_1\rangle$ and $|j_2m_2\rangle$, respectively, and assume that the total angular momentum is the sum of the two angular momenta, $\boldsymbol{J} = \boldsymbol{J}_1 + \boldsymbol{J}_2$. If the interaction between the two components leaves the individual angular momenta and their z-components constants of motion, a complete set of commuting operators consists of $\{H, \boldsymbol{J}_1^2, J_{1z}, \boldsymbol{J}_2^2, J_{2z}\}$, where H is the Hamiltonian. The eigenfunctions can be written as a product, $|j_1m_1j_2m_2\rangle \equiv |j_1m_1\rangle\,|j_2m_2\rangle$, and the eigenvalue equations are

$$J_1^2\,|j_1m_1j_2m_2\rangle = j_1(j_1+1)\,|j_1m_1j_2m_2\rangle \qquad J_{1z}\,|j_1m_1j_2m_2\rangle = m_1\,|j_1m_1j_2m_2\rangle,$$
$$J_2^2\,|j_1m_1j_2m_2\rangle = j_2(j_2+1)\,|j_1m_1j_2m_2\rangle \qquad J_{2z}\,|j_1m_1j_2m_2\rangle = m_2\,|j_1m_1j_2m_2\rangle.$$

But alternatively we could choose $\{H, \boldsymbol{J}_1^2, \boldsymbol{J}_2^2, \boldsymbol{J}^2 = (\boldsymbol{J}_1+\boldsymbol{J}_2)^2, J_z = J_{1z}+J_{2z}\}$ as a mutually commuting set, with eigenfunctions $|j_1j_2JM\rangle$ and eigenvalue equations

$$J^2\,|j_1j_2JM\rangle = J(J+1)\,|j_1j_2JM\rangle \qquad J_z\,|j_1j_2JM\rangle = M\,|j_1j_2JM\rangle,$$
$$J_1^2\,|j_1j_2JM\rangle = j_1(j_1+1)\,|j_1j_2JM\rangle \qquad J_2^2\,|j_1j_2JM\rangle = j_2(j_2+1)\,|j_1j_2JM\rangle.$$

These labeling schemes describe the same state so they must be related by a unitary transformation:

$$|j_1 j_2 JM\rangle = \sum_{m_1 m_2} |j_1 m_1 j_2 m_2\rangle \langle j_1 m_1 j_2 m_2 |\, j_1 j_2 JM\rangle , \tag{6.46a}$$

$$|j_1 j_2 m_1 m_2\rangle = \sum_{JM} |j_1 j_2 JM\rangle \langle j_1 j_2 JM |\, j_1 m_1 j_2 m_2\rangle , \tag{6.46b}$$

where the elements $\langle j_1 j_2 JM |\, j_1 m_1 j_2 m_2\rangle$ of the unitary transformation matrix are called *Clebsch–Gordan coefficients, vector-coupling coefficients,* or *Wigner coefficients*. The above labeling is somewhat redundant and we will often suppress explicit display of the labels j_1 and j_2 in $\langle j_1 j_2 JM|$, writing $\langle JM|\, j_1 m_1 j_2 m_2\rangle$ to mean $\langle j_1 j_2 JM|\, j_1 m_1 j_2 m_2\rangle$, for example.

The Clebsch–Gordan coefficients may be regarded as forming a unitary matrix of dimensions $(2j_1+1)(2j_2+1)$, with the rows labeled by the indices JM and the columns by the indices $m_1 m_2$. By transformation with the matrix of Clebsch–Gordan coefficients, the generally reducible representation $D^{(j_1)} \otimes D^{(j_2)}$ is brought to reduced form with the irreps $D^{(J)}$ along the diagonal. Schematically (see Fig. 2.2),

$$\mathbb{S}^{-1} D^{(j_1)} \otimes D^{(j_2)} \mathbb{S} = \begin{pmatrix} D^{(j_1+j_2)} & & & \\ & D^{(j_1+j_2-1)} & & \\ & & \ddots & \\ & & & D^{(|j_1-j_2|)} \end{pmatrix} , \tag{6.47}$$

where blank matrix entries are zeros and $\mathbb{S}$ is the matrix of Clebsch–Gordan coefficients.

6.3.7 Properties of SO(3) Clebsch–Gordan Coefficients

The SO(3) Clebsch–Gordan coefficients may be chosen real and satisfy the relations

$$\langle JM|\, j_1 m_1 j_2 m_2\rangle = \langle j_1 m_1 j_2 m_2|\, JM\rangle^* , \tag{6.48a}$$

$$\sum_{m_1 m_2} \langle JM|\, j_1 m_1 j_2 m_2\rangle \langle j_1\, m_1\, j_2\, m_2|\, J'\, M'\rangle = \delta_{JJ'} \delta_{MM'}, \tag{6.48b}$$

$$\sum_{JM} \langle j_1 m_1 j_2 m_2|\, JM\rangle \langle JM|\, j_1 m_1' j_2 m_2'\rangle = \delta_{m_1 m_1'} \delta_{m_2 m_2'}, \tag{6.48c}$$

by virtue of the unitarity of the transformations (6.46). They also obey the restrictions that $\langle j_1\, m_1\, j_2\, m_2|\, J\, M\rangle = 0$ unless $m_1 + m_2 = M$,[5] and $|j_1 - j_2| \leq J \leq j_1 + j_2$ (*triangle inequality*), which follow from additivity of the generators. There is considerable latitude in phase choice but the standard one is called the *Condon–Shortley convention*. Clebsch–Gordan coefficients are unsymmetric under exchange of angular momentum labels:

[5] Thus the sum over M in Eq. (6.48) is formal since the coefficients vanish unless $M = m_1 + m_2$.

$$\langle j_1\, m_1\, j_2\, m_2 |\, J\, M \rangle = (-1)^{j_1+j_2-J} \langle j_2 m_2 j_1 m_1 |\, J M \rangle, \tag{6.49a}$$

$$\langle j_1\, m_1\, j_2\, m_2 |\, J\, M \rangle = (-1)^{j_1-m_1} \sqrt{\frac{2J+1}{2j_2+1}} \langle j_1 m_1 J - M |\, j_2 - m_2 \rangle, \tag{6.49b}$$

$$\langle j_1\, m_1\, j_2\, m_2 |\, J\, M \rangle = (-1)^{j_2+m_2} \sqrt{\frac{2J+1}{2j_1+1}} \langle J - M j_2 m_2 |\, j_1 - m_1 \rangle. \tag{6.49c}$$

Because of this asymmetry an alternative vector-coupling coefficient called the *3J symbol* is often employed.

6.3.8 3*J* Symbols

A more symmetric vector-coupling coefficient may be obtained if the angular momentum coupling is viewed as coupling the three angular momenta j_1, j_2, and J to a resultant of zero. This leads to the 3*J symbol*, which is related to the Clebsch–Gordan coefficient by

$$\begin{pmatrix} j_1 & j_2 & J \\ m_1 & m_2 & -M \end{pmatrix} = \frac{(-1)^{j_1-j_2+M}}{\sqrt{2J+1}} \langle j_1\, m_1\, j_2\, m_2 |\, J\, M \rangle. \tag{6.50}$$

The $3J$ symbol is

1. invariant under an even permutation of columns,
2. multiplied by a phase $(-1)^{j_1+j_2+J}$ under an odd permutation of columns, and
3. multiplied by a phase $(-1)^{j_1+j_2+J}$ if all signs in the bottom row are changed.

These symmetries of the $3J$ symbols correspond to the relations (6.49) for the Clebsch–Gordan coefficients.

6.3.9 Construction of SO(3) Irreducible Multiplets

We may construct SO(3) multiplets and corresponding Clebsch–Gordan coefficients using the highest-weight algorithm discussed in Section 3.3.4. For the coupling of two angular momenta $\boldsymbol{J} = \boldsymbol{j} + \boldsymbol{j}'$, when new linear combinations are taken to get a good $|JM\rangle$ basis there is only one combination of the old basis vectors $|jm\rangle\, |j'm'\rangle$ that can contribute to the unique "maximally stretched" vector $|JM\rangle = |J = j + j', M = j + j'\rangle$,

$$|J = j + j', M = j + j'\rangle = |j, m = j\rangle\, |j', m' = j'\rangle \equiv |j, m = j, j', m' = j'\rangle.$$

Multiplying from the left with $\langle j, m = j, j', m' = j'|$ gives the Clebsch–Gordan coefficient

$$\langle j, m = j, j', m' = j' |\, J = j + j', M = j + j' \rangle = 1,$$

if we assume that $|jm\rangle$ is normalized to unity. We now have constructed the highest-weight member of this multiplet and associated Clebsch–Gordan coefficient . The remaining members may be generated by successive applications of the lowering operator J_-, where

$$J_-\, |JM\rangle = \sqrt{J(J+1) - M(M-1)}\ \ |J, M-1\rangle \qquad J_- = J_1 - iJ_2 = j_- + j'_-,$$

with the last equation following because the generators of the direct product are additive. Operating on the coupled basis with J_- gives

$$J_-|J = j + j', M = j + j'\rangle = \sqrt{2(j + j')}\ |J = j + j', M = j + j' - 1\rangle,$$

where $M = J = j + j'$ has been used. Operating on the uncoupled basis with J_- gives

$$\begin{aligned} J_-|m = j, m' = j'\rangle &= (j_- + j'_-)\,|m = j, m' = j'\rangle \\ &= \sqrt{2j}\,|m = j - 1, m' = j'\rangle + \sqrt{2j'}\,|m = j, m' = j' - 1\rangle, \end{aligned}$$

where we have suppressed j and j' labels. Equating the preceding two expressions leads to

$$\begin{aligned} |J = j + j', M = j + j' - 1\rangle &= \left(\frac{j}{j + j'}\right)^{1/2} |m = j - 1, m' = j'\rangle \\ &\quad + \left(\frac{j'}{j + j'}\right)^{1/2} |m = j, m' = j' - 1\rangle. \end{aligned}$$

The Clebsch–Gordan coefficients are obtained by multiplying this expression from the left by $\langle j, m = j - 1, j', m' = j'|$ and $\langle j, m = j, j', m' = j' - 1|$, respectively, to give

$$\begin{aligned} \langle j, m = j - 1, j', m' = j'|\, J = j + j', M = j + j' - 1\rangle &= \sqrt{j/(j + j')}, \\ \langle J = j + j', M = j + j' - 1|\, j, m = j, j', m' = j' - 1\rangle &= \sqrt{j'/(j + j')}. \end{aligned}$$

Continuing in this fashion, we can construct all the $2J + 1$ members of this irreducible representation and the corresponding Clebsch–Gordan coefficients. This defines an invariant subspace labeled by $J = j + j'$. The highest remaining weight is $M = j + j' - 1$. There are two such states, and one is already included in the $J = j + j'$ representation just constructed. Therefore the remaining one (which must be orthogonal to the first) is the highest weight of an irrep with $J = j + j' - 1$, and another invariant subspace can be generated by repeated application of J_- to this state. This whole procedure can be repeated until completing the subspace with $J = |j - j'|$, which exhausts all the states.

By the algorithm just outlined, all of the states and the corresponding Clebsch–Gordan coefficients may be constructed for the direct product of SO(3) representations. However, real-world calculations often require the use of many Clebsch–Gordan coefficients and it is not very efficient to work them out by hand each time they are needed. Many tables of SO(3) Clebsch–Gordan coefficients or $3J$ symbols have been published and there are computer subroutines to calculate them. Tables C.3 and C.4 in Appendix C give explicit expressions for some frequently encountered SO(3) vector coupling coefficients.

6.4 Tensor Operators under Group Transformations

Previous discussion has alluded to the utility of tensor methods for analyzing group representations. We now begin a systematic introduction to tensors for representation theory. Angular momentum groups will be used to illustrate but similar methods will prove useful for more complicated groups to be discussed in later chapters.

Let $|\Lambda\lambda\rangle$ define basis vectors for the representation of some group G. The representation is labeled by Λ, which may stand for more than one quantum number, and the individual components of the representation are labeled by λ, which also may represent more than one quantum number. Generally, the quantities Λ may be related to eigenvalues of Casimir operators and the quantum numbers λ are the weights of the representation.[6] The matrix elements of a linear operator R in this space will be denoted by $\langle\Lambda\lambda'|\,R\,|\Lambda\lambda\rangle$. The effect of R on a basis vector is to produce a linear combination of basis vectors since, by completeness (that is, $\sum_i |i\rangle\langle i| = 1$),

$$R\,|\Lambda\lambda\rangle = \sum_{\lambda'} |\Lambda\lambda'\rangle\langle\Lambda\lambda'|\,R\,|\Lambda\lambda\rangle. \tag{6.51}$$

A set $\boldsymbol{T}(\Lambda)$ of linearly independent operators $T(\Lambda\lambda)$ forms a *tensor operator* belonging to the representation Λ of the group G if it transforms as [224]

$$RT(\Lambda\lambda)R^{-1} = \sum_{\lambda'} \langle\Lambda\lambda'|\,R\,|\Lambda\lambda\rangle\,T(\Lambda\lambda'). \tag{6.52}$$

A tensor operator $\boldsymbol{T}(\Lambda)$ shares the adjectives *reducible*, *irreducible*, or *equivalent* with the representation Λ.

Example 6.4 A spherical harmonic $Y_{LM}(\theta,\phi)$ is a tensor by the preceding definition, for it is well known that they transform under rotations as

$$C_{LM}(\theta',\phi') = RC_{LM}(\theta,\phi)R^{-1} = \sum_N D^L_{NM}(\Omega)C_{LN}(\theta,\phi), \tag{6.53}$$

where $C_{LM}(\theta,\phi)$ has been defined through

$$Y_{LM}(\theta,\phi) = \sqrt{\frac{2L+1}{4\pi}}\,C_{LM}(\theta,\phi), \tag{6.54}$$

(θ,ϕ) and (θ',ϕ') define angles before and after rotation by Euler angles Ω, respectively, and the $D^L_{NM}(\Omega)$ are the matrix elements of the rotation operator (represented by $\langle\Lambda\lambda'|\,R\,|\Lambda\lambda\rangle$ in the preceding notation).

By considering infinitesimal transformations, the definition (6.52) for a tensor operator may also be expressed in terms of a commutator with the infinitesimal generators X_μ,

$$[\,X_\mu, T(\Lambda\lambda)\,] = \sum_{\lambda'} \langle\Lambda\lambda'|\,X_\mu\,|\Lambda\lambda\rangle\,T(\Lambda\lambda'). \tag{6.55}$$

Either Eq. (6.52) or Eq. (6.55) may be taken as definition of a tensor operator $T(\Lambda\lambda)$.

[6] If there are degeneracies in the weight space additional quantum numbers α may also be required to distinguish states. We will assume that such quantum numbers are supplied if needed, and will often suppress the corresponding quantities in the notation unless they are required for the discussion.

6.5 Tensors for the Rotation Group

The preceding discussion has been a general one. Let us now specialize to SO(3) [or SU(2)]. If $\boldsymbol{T}(k)$ is an irreducible SO(3) tensor operator transforming according to the irrep $D^{(k)}$, then the $2k+1$ components T_{kq} $(q = -k, -k+1, \ldots, +k)$ must satisfy (Problem 6.10)

$$[\, J_i, T_{kq} \,] = \sum_{q'} \langle kq' |\, J_i \,| kq \rangle\, T_{kq'}, \tag{6.56}$$

or equivalently the relations

$$[\, J_3, T_{kq} \,] = qT_{kq} \qquad [\, J_\pm, T_{kq} \,] = \sqrt{k(k+1) - q(q \pm 1)}\; T_{k,q\pm1}, \tag{6.57}$$

where

$$J_\pm \,| JM \rangle = \sqrt{J(J+1) - M(M \pm 1)}\; | J, M \pm 1 \rangle \qquad J_3 \,| JM \rangle = M \;| JM \rangle. \tag{6.58}$$

These SO(3) operators are said to be of *rank k* and are often called *spherical tensors.*

Example 6.5 A rank-zero (scalar) operator commutes with the angular momentum generators. The Casimir operator J^2 is an SO(3) scalar, since $[\, J_\pm, J^2 \,] = [\, J_3, J^2 \,] = 0$.

Example 6.6 The angular momentum operators are SO(3) tensors, for

$$\langle b | J_a^{(1)} | c \rangle = -i\epsilon_{abc},$$

because the structure constants generate the adjoint $(J = 1)$ representation [see Eq. (3.8), where this matrix element is denoted $(T_a)_{bc}$], and

$$[\, J_j, J_i \,] = \langle k | J_j^{(1)} | i \rangle J_k,$$

which should be compared with Eq. (6.56). Therefore the angular momentum operators transform as the *vector (or adjoint) representation* $(J = 1)$ of SO(3).

Example 6.7 The position coordinate r is an SO(3) tensor. As shown in Problem 6.11(a), the angular momentum operator L_a satisfies

$$[\, L_a, r_b \,] = \langle b | L_a^{(1)} | c \rangle r_c$$

and position r_b is a vector under SO(3) (as is the momentum; see Problem 6.11).

More generally, any three-component quantity A_l $(l = 1, 2, 3)$ that satisfies

$$[\, J_k, A_l \,] = i\epsilon_{klm} A_m, \tag{6.59}$$

when commuted with an angular momentum operator J_k, is a vector under SO(3).

6.6 SO(3) Tensor Products

If T_{kq} and $T_{k'q'}$ are irreducible tensors of rank k and k' respectively, the $(2k+1)(2k'+1)$ products $T_{kq}T_{k'q'}$ form a tensor transforming under $D^{(k)} \otimes D^{(k')}$ of the rotation group. This is reducible using Clebsch–Gordan coefficients and the techniques outlined in Section 6.3. The reduction gives the irreducible tensors

$$T_{KQ}(kk') = \sum_{qq'} \langle k\, q\, k'\, q' |\, K\, Q \rangle\, T_{kq} T_{k'q'}, \tag{6.60}$$

where $K = k+k', k+k'-1, \ldots, |k-k'|$, and $Q = q+q'$.

Example 6.8 Consider the most general rank-2 cartesian tensor constructed from the nine products of the components of two vectors, $T_{ik} = A_i B_k$ $(i, k = 1, 2, 3)$. This product is irreducible under the general linear group of matrices (Section 2.9.1). However, it must be reducible under the SO(3) subgroup of linear transformations because the components A_i and B_k transform as vector irreps of SO(3) and, from the SO(3) Clebsch–Gordan series, $D^{(1)} \otimes D^{(1)} = D^{(0)} \oplus D^{(1)} \oplus D^{(2)}$. Therefore, the rank-2 tensor T_{ik} may be decomposed into pieces transforming as angular momentum 0, 1, and 2 irreps of the SO(3) subgroup.

Let us use the SO(3) Clebsch–Gordan coefficients to form the irreducible tensors $D^{(0)}$, $D^{(1)}$, and $D^{(2)}$ deduced in Example 6.8. The scalar component $D^{(0)}$ is

$$T(AB)_{00} = \sum_{MN} \langle 1\, M\, 1\, N |\, 0\, 0 \rangle\, A_M B_N.$$

From Problem 6.5 the Clebsch–Gordan coefficient evaluates to

$$\langle j_1\, m_1\, j_2\, m_2 |\, 0\, 0 \rangle = \frac{(-1)^{j_1 - m_1}}{\sqrt{2j_1+1}}\, \delta_{j_1 j_2} \delta_{m_1 - m_2},$$

and we obtain

$$T(AB)_{00} = \frac{1}{\sqrt{3}} \left(A_{-1}B_1 + A_1 B_{-1} - A_0 B_0 \right), \tag{6.61}$$

where A_N and B_N are spherical components of the vectors $\boldsymbol{A}$ and $\boldsymbol{B}$, respectively. This takes a form perhaps more familiar under a transformation to cartesian components using

$$V_{-1} = \frac{1}{\sqrt{2}} \left(V_x - iV_y \right) \qquad V_1 = \frac{1}{\sqrt{2}} \left(V_x + iV_y \right) \qquad V_0 = V_z, \tag{6.62}$$

which is valid for any vector operator $\boldsymbol{V}$. From this we find that

$$T(AB)_{00} = -\frac{1}{\sqrt{3}} \left(A_x B_x + A_y B_y + A_z B_z \right) = -\frac{1}{\sqrt{3}}\, \boldsymbol{A} \cdot \boldsymbol{B}, \tag{6.63}$$

which expresses the rotational invariance of the scalar product of vectors. The components of the vector representation $D^{(1)}$ are obtained using Table C.3 as

$$T(AB)_{11} = \langle 1\,0\,1\,1 |\, 1\,1\rangle\, A_0B_1 + \langle 1\,1\,1\,0 |\, 1\,1\rangle\, A_1B_0$$
$$= \frac{1}{\sqrt{2}}\,(A_1B_0 - A_0B_1)\,, \tag{6.64}$$

$$T(AB)_{1-1} = \langle 1\,0\,1\,-1 |\, 1\,-1\rangle\, A_0B_{-1} + \langle 1\,-1\,1\,0 |\, 1\,-1\rangle\, A_{-1}B_0$$
$$= \frac{1}{\sqrt{2}}\,(A_0B_{-1} - A_{-1}B_0)\,, \tag{6.65}$$

$$T(AB)_{10} = \langle 1\,1\,1\,-1 |\, 1\,0\rangle\, A_1B_{-1} + \langle 1\,-1\,1\,1 |\, 1\,0\rangle\, A_{-1}B_1$$
$$= \frac{1}{\sqrt{2}}\,(A_1B_{-1} - A_{-1}B_1)\,. \tag{6.66}$$

The five components of the rank-2 tensor that can be constructed from the two vectors are

$$T(AB)_{2Q} = \sum_{MN} \langle 1\,M\,1\,N |\, 2\,Q\rangle\, A_M B_N. \tag{6.67}$$

It is left as an exercise (see Problem 6.12) to construct the five components $T(AB)_{2Q}$ of $D^{(2)}$. For example, you should find that

$$T(AB)_{22} = A_1B_1 \qquad T(AB)_{21} = \frac{1}{\sqrt{2}}\,(A_0B_1 + A_1B_0)\,,$$
$$T(AB)_{20} = \frac{1}{\sqrt{6}}\,(A_{-1}B_1 + A_1B_{-1} + 2A_0B_0)\,. \tag{6.68}$$

Thus we have shown explicitly how to construct the irreducible SO(3) representations corresponding to a direct product of vectors such as $D^{(1)} \otimes D^{(1)}$ in Example 6.8.

6.7 The Wigner–Eckart Theorem

If operators are expressed as spherical tensors, the calculation of matrix elements reduces to evaluating quantities of the general form $\langle \Lambda_1\lambda_1 |\, T(\Lambda\lambda)\, |\Lambda_2\lambda_2\rangle$.

Wigner–Eckart Theorem: Matrix elements $\langle \Lambda_1\lambda_1 |\, T(\Lambda\lambda)\, |\Lambda_2\lambda_2\rangle$ can be factored in the form [224],

$$\langle \Lambda_1\lambda_1 |\, T(\Lambda\lambda)\, |\Lambda_2\lambda_2\rangle = \sum_{\alpha} \langle \alpha\Lambda_1\lambda_1 |\, \lambda\lambda_2\rangle^* \langle \alpha\Lambda_1 \,\|\, T(\Lambda) \,\|\, \alpha\Lambda_2\rangle, \tag{6.69}$$

where $\langle \alpha\Lambda_1 \,\|\, T(\Lambda) \,\|\, \alpha\Lambda_2\rangle$ is termed the *reduced matrix element.*

The Wigner–Eckart factorization (6.69) has the following properties.

1. The coupling coefficient $\langle \Lambda_2\lambda_2\Lambda\lambda |\, \Lambda_1\lambda_1\alpha\rangle$ contains all the dependence of the matrix elements on the weights λ, but no details of the internal structure.

2. The label α represents any quantum numbers needed to distinguish representations. For example, a sum over α is required if the group is not simply reducible.
3. The reduced matrix element is independent of the weights λ_i.

The coupling coefficients are elements of a unitary transformation and the expression for the reduced matrix element may be inverted to give an equation defining the reduced matrix element $\langle \alpha\Lambda_1 \,\|\, T(\Lambda) \,\|\, \alpha\Lambda_2 \rangle$ in terms of the full matrix element $\langle \Lambda_1\lambda_1 | \, T(\Lambda\lambda) \, | \Lambda_2\lambda_2 \rangle$,

$$\langle \alpha\Lambda_1 \,\|\, T(\Lambda) \,\|\, \alpha\Lambda_2 \rangle = \sum_{\lambda_1\lambda_2} \langle \lambda\lambda_2 | \, \alpha\Lambda_1\lambda_1 \rangle \langle \Lambda_1\lambda_1 | \, T(\Lambda\lambda) \, | \Lambda_2\lambda_2 \rangle . \tag{6.70}$$

Evaluation of the reduced matrix element typically proceeds by using Eq. (6.70) for a simple case, with the full matrix element on the right side determined by theory or by relating it to a measurable quantity.

6.8 The Wigner–Eckart Theorem for SO(3)

The Wigner–Eckart theorem is quite general, but let us illustrate it for the group SO(3) [or SU(2), which has the same Lie algebra]. The SO(3) Wigner–Eckart theorem takes a simple form because the group is simply reducible so there is no summation over α, and the coupling coefficient is just a Clebsch–Gordan coefficient. Then Eq. (6.69) becomes

$$\langle \alpha JM | \, T_{kq} \, | \alpha' J' M' \rangle = (-1)^{2k} \langle J' \, M' \, k \, q | \, J \, M \rangle \langle \alpha J \,\|\, \boldsymbol{T}_k \,\|\, \alpha' J' \rangle, \tag{6.71}$$

where the phase $(-1)^{2k}$ was introduced for convenience, Eq. (6.70) becomes

$$\langle \alpha J \,\|\, \boldsymbol{T}_k \,\|\, \alpha' J' \rangle = (-1)^{2k} \sum_{M'q} \langle J' \, M' \, k \, q | \, J \, M \rangle \langle \alpha JM | \, T_{kq} \, | \alpha' J' M' \rangle , \tag{6.72}$$

and we continue to follow the conventions of Brink and Satchler [29].

6.8.1 Reduced Matrix Elements

The calculation of reduced matrix elements will now be illustrated with two examples from angular momentum theory.

Example 6.9 We first evaluate the reduced matrix elements of the angular momentum operator $\boldsymbol{J}$, which is a spherical tensor of rank one. Choosing the generator J_z to be diagonal and also invoking Eq. (6.71) gives two relations,

$$\langle JM | \, J_z \, | J' M' \rangle = M \, \delta_{JJ'} \delta_{MM'} \qquad \langle JM | \, J_z \, | J' M' \rangle = \langle J' \, M' \, 1 \, 0 | \, J \, M \rangle \langle J \,\|\, \boldsymbol{J} \,\|\, J' \rangle,$$

and setting these equations equal gives

$$\langle J \,\|\, \boldsymbol{J} \,\|\, J' \rangle = \langle J' \, M' \, 1 \, 0 | \, J \, M \rangle^{-1} \, M \, \delta_{JJ'} \delta_{MM'}.$$

Utilizing Eq. (6.50), symmetries of the $3J$ coefficient given in Section 6.3.8, and Table C.4, the Clebsch–Gordan coefficient is $M(J(J+1))^{-1/2}$ and

$$\langle J' \| \boldsymbol{J} \| J \rangle = \sqrt{J(J+1)}\, \delta_{JJ'}. \tag{6.73}$$

As expected, the reduced matrix element has no M dependence.

Example 6.10 Consider the reduced matrix element of a spherical harmonic evaluated for states of definite angular momentum. From the Wigner–Eckart theorem (6.71),

$$\langle lm | Y_{LM} | l'm' \rangle = \langle l'\, m'\, L\, M | \, l\, m \rangle \langle l \| \boldsymbol{Y}_L \| l' \rangle.$$

As you are asked to demonstrate in Problem 6.14, this equation and the tensor and orthonormality properties of the Y_{LM} may be employed to show that

$$\langle l \| \boldsymbol{Y}_L \| l' \rangle = \sqrt{\frac{(2L+1)(2l'+1)}{4\pi(2l+1)}} \, \langle L\, 0\, l'\, 0 | \, l\, 0 \rangle, \tag{6.74}$$

for the reduced matrix element.

Often we can construct basis states described by a chain of subgroups $A \supset B \supset C \supset \cdots \supset Z$, and the Wigner–Eckart theorem can be applied systematically through the chain. In this way matrix elements may be factored into products of coupling coefficients for the groups in the chain. Examples of this approach will be discussed in Section 11.4.

6.8.2 Selection Rules

The structure of the factored matrix element in the SO(3) Wigner–Eckart theorem implies several properties of matrix elements that are independent of details contained in the reduced matrix element.

1. The Clebsch–Gordan coefficients contain the "directional information" (magnetic quantum numbers) for SO(3), so any process not involving a specific direction depends only on the reduced matrix element.
2. The Clebsch–Gordan coefficient vanishes unless $M' = q + M$ in Eq. (6.71).
3. The triangle inequality $|J - J'| \leq k \leq J + J'$ must be satisfied if the Clebsch–Gordan coefficient is not to vanish.

These last two conditions imply general selection rules in the matrix elements (6.71) that depend only on the SO(3) group structure of the problem; there may be additional selection rules associated with the reduced matrix element.

Example 6.11 In the expression (6.74) for the reduced matrix element of $Y_{LM}(\theta, \phi)$ there is a Clebsch–Gordan coefficient $\langle L\, 0\, l'\, 0 | \, l\, 0 \rangle$ that vanishes unless $l + l' + L$ is even. Since the parity of a spherical harmonic Y_{LM} is $(-1)^L$, this implies a parity-conserving matrix element.

6.9 Relationship of SO(3) and SU(2)

We have at various times noted that SU(2) and SO(3) share the same Lie algebra, which means that they behave in the same way near the origin of group space. We conclude this chapter by elaborating on the exact relationship between these groups. We will see that they have the same infinitesimal generators, but their global structures differ because the topology of the SO(3) manifold is distinct from that of the SU(2) manifold.

6.9.1 SO(3) and SU(2) Group Manifolds

Let us examine the connectedness properties of the SO(3) and SU(2) manifolds.

Topology of the SO(3) Manifold: The SO(3) manifold may be visualized using the *axis–angle parameterization* of Fig. 6.5(a), where the direction of a vector represents the direction of the rotation axis and the length of the vector denotes the angle of rotation.[7] From this parameterization the manifold is the *volume* of a sphere of radius π, which is compact, but *antipodal points are the same points* and must be identified because rotations by π and by $-\pi$ around the same axis are equivalent. Thus the manifold of SO(3) is compact and *doubly-connected,* because the identification of antipodal points implies that there are *two kinds of closed paths* in the parameter space, as illustrated in Fig. 6.5(b).

Topology of the SU(2) Manifold: A two-dimensional U(2) matrix may be parameterized

$$U = e^{i\lambda}\begin{pmatrix} a & b \\ -b^* & a^* \end{pmatrix} = e^{i\lambda}\begin{pmatrix} \cos\theta e^{i\xi} & -\sin\theta e^{i\eta} \\ \sin\theta e^{-i\eta} & \cos\theta e^{-i\xi} \end{pmatrix}, \tag{6.75}$$

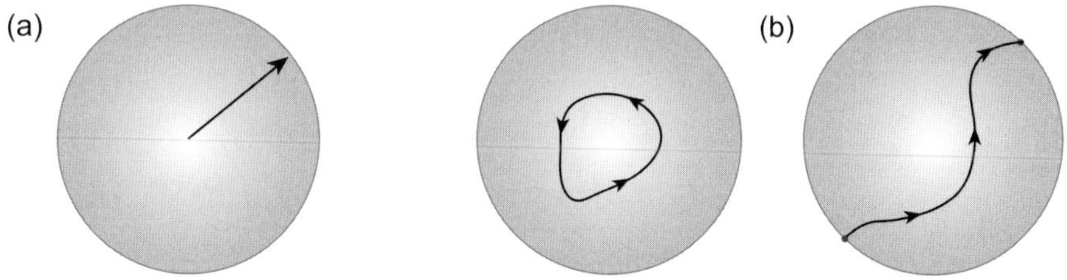

Fig. 6.5 (a) Axis–angle parameterization of the SO(3) manifold as the *volume* of a sphere of radius π, with *antipodal points identified.* The direction of the arrow is the direction of the rotation axis and its length indicates the angle of rotation. (b) Two kinds of closed paths; these cannot be deformed continuously into each other so the SO(3) manifold is doubly connected.

[7] All rotations by the same angle lie in the same class, irrespective of the direction of the rotation axis. Thus different classes correspond to concentric spheres with different radii in the axis–angle parameterization.

where $|a|^2 + |b|^2 = 1$. The special unitary subgroup $U(2) \supset SU(2)$ having unit determinant results if we restrict to $\lambda = 0$ (see Problem 3.4). The SU(2) matrix (6.75) with $\lambda = 0$ is equivalent to the SO(3) matrix $D^{1/2}(\alpha, \beta, \gamma)$ of Eq. (6.34) if the identifications

$$\theta = \frac{1}{2}\beta \qquad \xi = \frac{1}{2}(-\alpha - \gamma) \qquad \eta = \frac{1}{2}(-\alpha + \gamma)$$

are made. However, there are *two SU(2) matrices for each SO(3) matrix*: rotation by 2π returns to the original SO(3) matrix but rotation by 2π gives the *negative* of the original SU(2) matrix; only upon rotation by 4π does the SU(2) matrix return to the original one. Thus, SU(2) matrices form a *double-valued representation* of SO(3) because two SU(2) matrices correspond to the same SO(3) rotation (2:1 homomorphism). The topology of the SU(2) manifold may be displayed by parameterizing an SU(2) group element as

$$U = \begin{pmatrix} a & b \\ -b^* & a^* \end{pmatrix} = \begin{pmatrix} r_0 - ir_3 & -r_2 - ir_1 \\ r_2 - ir_1 & r_0 + ir_3 \end{pmatrix}, \tag{6.76}$$

where a and b are complex numbers and the r_i are real. Then the SU(2) requirement $\det U = 1$ implies the condition

$$r_0^2 + r_1^2 + r_2^2 + r_3^2 = 1, \tag{6.77}$$

which is the equation for a unit 3-sphere, S^3. Spheres S^n with $n \geq 2$ are known to be simply-connected, closed, and bounded, so SU(2) is a compact and simply-connected group.[8]

Homomorphism of SU(2) and SO(3): We conclude that SU(2) and SO(3) share the same Lie algebra and are *locally isomorphic,* but are 2:1 homomorphic in the large. The structure constants describe the group near the identity but do not specify the global properties of the manifold. One consequence is that SO(3) has only integer single-valued representations but SU(2) has single-valued integer and half-integer representations (the half-integer representations are termed *Pauli spinor representations*). Hence, SU(2) contains all possible angular momentum states in single-valued representations, and all properties of SO(3) representations can be deduced from single-valued representations of SU(2). The converse is not true; SO(3) spinor representations are double valued.

6.9.2 Universal Covering Group of the SU(2) Algebra

As we saw in Section 6.2.3, when several Lie groups share the same Lie algebra only one of them, the *universal covering group,* will be compact and simply-connected. The *Lie group* SU(2) is the universal covering group of the *Lie algebra* SU(2). Other groups sharing the same algebra are related to the covering group by a homomorphism. The group SU(2) has an abelian invariant subgroup Z_2 with elements

$$e = \begin{pmatrix} 1 & 0 \\ 0 & 1 \end{pmatrix} \qquad a = \begin{pmatrix} -1 & 0 \\ 0 & -1 \end{pmatrix}$$

[8] That S^n with $n = 2$ is simply connected can be stated in the picturesque phrase "you can't lasso a basketball" [43]: any closed path on a 2D sphere can be deformed continuously to a point, so the space is simply-connected. The same is true for $n > 2$. However, we have seen that S^1 (the unit circle) is infinitely connected.

(see Box 2.2), and SO(3) is then isomorphic to the factor group $SU(2)/Z_2$ (see Section 2.14.2). The SU(2) $\to$ SO(3) homomorphism is not one to one because the kernel of the mapping $K = \{e, a\}$ has two SU(2) elements that map to the identity of SO(3) (see the discussion in Section 2.14.2). Further insight into the relationship between SU(2) and SO(3) may be found in the solution of Problem 6.17. More generally, Z_n is the group of the nth roots of unity (n of them) multiplied by the $n \times n$ unit matrix, and Z_n is an invariant subgroup of SU(n). The factor group SU(n)/Z_n then contains only some of the representations of SU(n) as single-valued representations, but all representations of SU(n) are included if multiple-valued representations of SU(n)/Z_n are admitted [141].

Background and Further Reading

Comprehensive discussions of angular momentum theory are contained in Brink and Satchler [29], Edmonds [51], Rose [173], and Varshalovich, Moskalev, and Khersonskii [201]. Our treatment of SO(2) has been guided by the presentation in Tung [199]. Invariant group integration is explained in Creutz [45], Hamermesh [104], and Tung [199]. Tensor operators for SO(3) and the associated angular momentum algebra are discussed extensively in Brink and Satchler [29], Edmonds [51], Rose [173], Wybourne [224], and Varshalovich, Moskalev, and Khersonskii [201]. Authors in this list may use different phase and notational conventions, so care must be taken in comparing formulas between different sources.

Problems

6.1 Show that an arbitrary 2×2 matrix with real entries that is orthogonal and has unit determinant can always be parameterized as in Eq. (6.3). Thus any SO(2) matrix can be interpreted as a rotation in some plane. ***

6.2 (a) Demonstrate that Eq. (6.10) defines a generator of SO(2) by examining the 2D rotation matrix (6.3) for an infinitesimal rotation $d\phi$. (b) Show that Eqs. (6.3) and (6.9) are equivalent by expanding the exponential in Eq. (6.9) to all orders.

6.3 Verify explicitly the validity of the rearrangement lemma (6.17) for the three-element group with multiplication table given in the solution of Problem 2.4, and the dihedral group D_2 with multiplication table given in Problem 2.9.

6.4 Use Eq. (6.20) to show that Eq. (6.21) defines the SO(2) integration measure. ***

6.5 Show that the SO(3) Clebsch–Gordan coefficients $\langle j_1\, m_1\, j_2\, m_2 |\, J\, M\rangle$ evaluate to

$$\langle jmj'm'|\, 00\rangle = \frac{(-1)^{j-m}}{\sqrt{2j+1}}\, \delta_{jj'}\delta_{m,-m'},$$

for the special case $J = M = 0$.

6.6 Consider a two-electron state in $L--S$ coupling,

$$\boldsymbol{L} = \boldsymbol{l}_1 + \boldsymbol{l}_2 \qquad \boldsymbol{S} = \boldsymbol{s}_1 + \boldsymbol{s}_2 \qquad \boldsymbol{J} = \boldsymbol{L} + \boldsymbol{S},$$

where $\boldsymbol{L}$ is the orbital angular momentum and $\boldsymbol{S}$ the spin angular momentum operator.

a. Use Clebsch–Gordan coefficients to show that the singlet and triplet spin wavefunctions ${}^{2S+1}\sigma_{m_S}$ are given by

$$ {}^1\sigma_0 = \frac{1}{\sqrt{2}}\left[\alpha(1)\beta(2) - \beta(1)\alpha(2)\right] \qquad {}^3\sigma_1 = \alpha(1)\alpha(2), $$
$$ {}^3\sigma_0 = \frac{1}{\sqrt{2}}\left[\alpha(1)\beta(2) + \alpha(2)\beta(1)\right] \qquad {}^3\sigma_{-1} = \beta(1)\beta(2), $$

where $\alpha(n)$ denotes spin up and $\beta(n)$ spin down for electron n.

b. Show that the total angular momentum wavefunction ψ^M_{JSL} for $J = S = 1$, $L = 2$, and $M = 1$ is given by

$$ \psi^1_{112} = \frac{1}{\sqrt{10}}\,{}^3\sigma_1 Y_{20}(\theta,\phi) - \sqrt{\frac{3}{10}}\,{}^3\sigma_0 Y_{21}(\theta,\phi) + \sqrt{\frac{3}{5}}\,{}^3\sigma_{-1} Y_{22}(\theta,\phi), $$

where the $Y_{\ell m}(\theta,\phi)$ are spherical harmonics. *Hint*: You will need some Clebsch–Gordan coefficients for larger angular momenta than those found in Table C.3. One way to obtain them is to use Table C.4, as illustrated in Problem 6.9.

6.7 Use the method of Section 6.3.9 with stepping operators to find the Clebsch–Gordan coefficients for coupling $j = 1$ and $j' = \frac{1}{2}$ to good total angular momentum J.

6.8 Prove Eqs. (6.42) and (6.43), thus showing that the dimension of the SO(3) irrep labeled by j is the character $2j + 1$ of the identity. *Hint*: Rearrange the sum and use

$$ \sum_{n=0}^{2j} e^{in\alpha} = \frac{\sin\left(j + \frac{1}{2}\right)\alpha}{\sin\left(\frac{1}{2}\alpha\right)}\, e^{ij\alpha}. $$

Then take the limit $\alpha \to 0$ to define the identity element. ***

6.9 Use the relation (6.50) between Clebsch–Gordan coefficients and $3J$ symbols, symmetries under column permutation of the $3J$ symbols summarized in Section 6.3.8, and Table C.4 to find values of the SO(3) Clebsch–Gordan coefficients $\langle 2011|\,11\rangle$, $\langle 2110|\,11\rangle$, and $\langle 221-1|\,11\rangle$.

6.10 Derive Eqs. (6.55) and (6.57) for the commutators of tensor operators, beginning from the tensor transformation law given in Eq. (6.52). ***

6.11 (a) Prove that the position coordinate r transforms as a vector under 3D rotations; that is, show that it is an SO(3) tensor of rank one. *Hint*: Begin by noting that the orbital angular momentum may be written in the form $L_a = \epsilon_{abc} r_b p_c$, where ϵ_{abc} is the completely antisymmetric rank-3 tensor, r_b is the position component, and p_c

is the momentum component. (b) Show that the linear momentum transforms as an SO(3) vector. ***

6.12 Derive the forms for the spherical tensor components given in Eq. (6.68). ***

6.13 Prove the spherical harmonic addition theorem

$$\sum_m (-1)^m Y_{lm}(\theta_1, \phi_1) Y_{l-m}(\theta_2, \phi_2) = \frac{2l+1}{4\pi} P_l(\cos\theta_{12})$$

(where $\theta_{12} \equiv \theta_1 - \theta_2$), by coupling two spherical harmonics to an SO(3) scalar and invoking the invariance of that scalar under rotations.

6.14 Use tensor methods to evaluate the reduced matrix element of the spherical harmonic $Y_{LM}(\theta, \phi)$ between states of good angular momentum $|JM\rangle$. ***

6.15 Define a quadrupole operator $Q_{20} = r^2 Y_{20}(\theta, \phi)$. The quadrupole moment Q for a state of good angular momentum $|jm\rangle$ is conventionally defined as the expectation value of this operator in the substate $|j, m = j\rangle$, multiplied by a factor $\sqrt{16\pi/5}$,

$$Q \equiv \sqrt{\frac{16\pi}{5}} \langle j, m = j | \, Q_{20} \, | j, m = j \rangle .$$

Use the Wigner–Eckart theorem to show that

$$\sqrt{\frac{16\pi}{5}} \langle jm | \, Q_{20} \, | jm \rangle = \frac{3m^2 - j(j+1))}{j(2j-1)} Q.$$

Hint: First express the reduced matrix element (6.72) in terms of Q. ***

6.16 Prove that the operator $a^\dagger_{jm}$ that acts on the vacuum as $a^\dagger_{jm} |0\rangle = |jm\rangle$ to create a fermion with angular momentum j and magnetic quantum number m transforms as a spherical tensor of rank j. *Hint*: The second-quantized form of the angular momentum operator is

$$J_a = \sum_{\alpha\beta} \langle \alpha | \, J_a \left| \beta \right\rangle a^\dagger_\alpha a_\beta ,$$

where α and β label single-particle states, and fermion operators obey $\{ a_i, a^\dagger_j \} = \delta_{ij}$ and $\{ a^\dagger_i, a^\dagger_j \} = \{ a_i, a_j \} = 0$.

6.17 Demonstrate the relationship between the groups SO(3) and SU(2) as follows.

(a) Associate each 3D euclidean coordinate $\boldsymbol{x} = (x_1, x_2, x_3)$ with a 2×2, traceless, hermitian matrix X through the map $X = \sigma_i x_i$, where the σ_i are the Pauli matrices of Eq. (3.11). Evaluate the matrix X explicitly and show that $\det X = -|\boldsymbol{x}|^2$.

(b) Let U be an arbitrary SU(2) matrix that induces a linear transformation $X \to X' = UXU^{-1}$ on X. Prove that if (as assumed) X is hermitian and traceless, so is X'.

(c) Argue that because of the properties derived above, X' is defined by the expansion $X' = \sigma_i x'_i$, and that $\det X' = \det X$.

(d) Use results (a)–(c) to show that the SU(2) transformation $X \to X' = UXU^{-1}$ induces an SO(3) transformation in 3D euclidean space. *Hint*: Preservation of squared euclidean length $|\boldsymbol{x}^2|$ is a defining characteristic of an SO(3) transformation.

(e) Argue that the mapping of $U \in$ SU(2) to an element of SO(3) established above is 2:1 because two different SU(2) transformations U can give the same X'. *Hint*: Is the sign of U determined by the equations that you have derived? ***

6.18 (a) Show that the most general 2×2 unitary matrix with unit determinant can be parameterized as in Eqs. (6.76) and (6.77). *Hint*: See Problem 3.4. (b) Take the group identity element $U(1,0,0,0)$ to correspond to $r_1 = r_2 = r_3 = 0$ and expand around the identity to show that $U \simeq 1 - idr_i\sigma_i$, where σ_i is a Pauli matrix.

6.19 Consider collisions of pions with nucleons. View the pions as a $T=1$ isospin triplet $\pi = (\pi^+, \pi^0, \pi^-)$, and the nucleon as a $T = \frac{1}{2}$ isospin doublet, $N = (p, n)$. The combined system may be coupled to a total isospin $T = \frac{3}{2}$ or $\frac{1}{2}$. Write the wavefunctions for all possible states $|TM_T\rangle$ of the coupled $\pi - N$ system in terms of the uncoupled products $|\pi N\rangle \equiv |\pi\rangle\,|N\rangle$. Use unitarity of the Clebsch–Gordan coefficients to invert these expressions and show that

$$|\pi^- p\rangle = \frac{1}{\sqrt{3}}\left|T = \tfrac{3}{2}, M_T = -\tfrac{1}{2}\right\rangle - \sqrt{\frac{2}{3}}\left|T = \tfrac{1}{2}, M_T = -\tfrac{1}{2}\right\rangle$$

for the uncoupled wavefunction $|\pi^-\rangle\,|p\rangle \equiv |\pi^- p\rangle$.

6.20 Derive the Clebsch–Gordan coefficients for the SU(2) direct product $\mathbf{2} \otimes \mathbf{2}$.

6.21 The 2D rotation matrix $R(\phi)$ defined in Eq. (6.3) is a reducible representation of SO(2). Diagonalize $R(\phi)$ to give the eigenvalues $\lambda_\pm = e^{\pm i\phi}$ and Eq. (6.11). Show that the basis vectors in the new basis after diagonalization are given by Eq. (6.12). Find the form of the generator J given by Eq. (6.10) in the new basis. Find the operators C and C^{-1} that perform the similarity transformation between the original basis and the diagonalized basis, $CR(\phi)C^{-1} = R'$. *Hint*: The solution of Problem 14.14 gives much of the required math. ***

6.22 Prove the result of Eq. (6.31a) for the matrix element of the $D(\alpha, \beta, \gamma)$ operator. *Hint*: Use the form (6.30) and insert complete sets of states using $1 = \sum_n |n\rangle\,\langle n|$.

6.23 Construct the rotation matrix $d^l_{mm'}$ for $l = 1$. Check your results against the entries in Table C.1 of Appendix C. *Hint*: You can save time by considering the product $d^{1/2} \otimes d^{1/2}$ and using the expression for $d^{1/2}$ already constructed in Eq. (6.33). ***

6.24 Show that if x and z are positive real numbers and y is an arbitrary real number, the matrices

$$\begin{pmatrix} x & y \\ 0 & z \end{pmatrix}$$

form a group under matrix multiplication but the naive group integration measure $dg = dx\,dy\,dz$ is not invariant under left multiplication of the group elements; that is, if $f(g)$ is a function of the group elements g,

$$\int f(g_0 g)d(g_0 g) \neq \int f(g_0 g)dg,$$

where g_0 is some group element. However, show that a new integration measure,

$$dg_{\rm L} = \frac{dx\,dy\,dz}{x^2 z},$$

is left invariant: $d(g_0 g) = dg$. Likewise, show that $dg = dx\,dy\,dz$ is not invariant under multiplication from the right by a group element g_0, but a new measure,

$$dg_{\rm R} = \frac{dx\,dy\,dz}{xz^2},$$

is right invariant. ***

6.25 Clebsch–Gordan coefficients not listed in Table C.3 often can be computed from Eq. (6.50) and the $3J$ symbols given in Table C.4. Show that

$$\left\langle \tfrac{3}{2}\, -\tfrac{3}{2}\, \tfrac{3}{2}\, \tfrac{3}{2} \right| 2\,0\rangle = \left\langle \tfrac{3}{2}\, -\tfrac{1}{2}\, \tfrac{3}{2}\, \tfrac{1}{2} \right| 2\,0\rangle = -\tfrac{1}{2},$$

$$\left\langle \tfrac{3}{2}\, \tfrac{1}{2}\, \tfrac{3}{2}\, -\tfrac{1}{2} \right| 2\,0\rangle = \left\langle \tfrac{3}{2}\, \tfrac{3}{2}\, \tfrac{3}{2}\, -\tfrac{3}{2} \right| 2\,0\rangle = \tfrac{1}{2},$$

starting from the formulas for $3J$ symbols given in Table C.4.

6.26 Starting from Eq. (6.27) and similar expressions for rotations around the x and y axes, show that the generators of 3D rotations are given by Eq. (6.28). *Hint*: Assume group elements to be parameterized as in Eq. (3.2) with generators $J_a = X_a$, expand expressions like Eq. (6.27) in a Taylor series, and compare with Eq. (6.6). ***

6.27 Use SO(3) group characters to show that in Eq. (6.44) the coefficients c_J are all zero or one [so SO(3) is simply reducible], which leads to the SO(3) Clebsch–Gordan series (6.45). *Hint*: Use the results of Section 6.3.4, and that the SO(3) characters $\chi_\alpha^{(j)}$ obey a relation $\chi_\alpha^{(j_1)}\chi_\alpha^{(j_2)} = \sum_J c_J \chi_\alpha^{(J)}$ that is analogous to Eq. (6.44). ***

7 Classification of Lie Algebras

We would like to generalize methods developed in preceding chapters for angular momentum to larger algebras and their associated Lie groups, with an eye toward more ambitious physics applications. As a first step, we consider methods that permit us to classify the possible Lie algebras. The key point is that the generators of a Lie algebra form a basis for a linear vector space, so any linearly independent combination of generators is itself a set of generators. This freedom of linear transformation among sets of generators may be used to simplify the analysis of an algebra by reducing the number of non-zero structure constants (recall that the values of the structure constants depend on the representation). This chapter describes a generalization of the standard treatment of angular momentum discussed in Ch. 3 called the *Cartan–Dynkin analysis* that leads to such a simplification, and provides a framework for systematic classification of Lie algebras.

7.1 Adjoint Representations

The adjoint representation that was introduced in Section 3.2.2 is singularly important because (1) states in the adjoint representation are in one to one correspondence with the generators themselves, and (2) the adjoint representation distinguishes between different Lie algebras, so it may be used to implement a unique classification of all semisimple Lie algebras.

7.1.1 The Cartan Subalgebra

As discussed in Section 3.3.2, the set of mutually commuting generators for a rank-ℓ Lie algebra is called the *Cartan subalgebra.* The generators H_i $(i = 1, \ldots, \ell)$ of the Cartan subalgebra are hermitian ($H_i = H_i^\dagger$), mutually commuting ($[H_i, H_j] = 0$), and form a basis for a linear vector space. It is convenient to choose a normalization

$$\mathrm{Tr}\,(H_i H_j) = \lambda\delta_{ij} \qquad (i, j = 1, 2, \ldots, \ell), \tag{7.1}$$

where λ is a positive constant and i and j range from 1 to ℓ. Upon diagonalizing the Cartan generators, $H_i\,|m, \kappa, D\rangle = m_i\,|m, \kappa, D\rangle$, where $m = (m_1, m_2, \ldots, m_\ell)$ is the *weight vector,* κ denotes any additional quantum numbers required to specify a state, and D labels the representation. *Roots* were defined in Section 3.3.2 as the weights for states in the adjoint representation. The dimensionality of the adjoint representation is equal to the number of group generators, and from the matrix (3.8) defining the adjoint representation the states of the adjoint representation are in direct correspondence with the generators. We shall denote a state corresponding to a generator X_a by $|X_a\rangle$.[1] Linear combinations of these states then correspond to linear combinations of generators, which may be denoted by

$$\alpha_a |X_a\rangle \equiv \sum_a \alpha_a |X_a\rangle = |\alpha_a X_a\rangle = |\alpha_1 X_1 + \alpha_2 X_2, + \cdots\rangle, \tag{7.2}$$

with implied summation on the repeated index a. As shown in Problem 7.7, the definition (3.8) of the adjoint representation matrices and Eq. (7.2) may be used to compute the action of a generator X_a on a state $|X_b\rangle$ [68],

$$X_a |X_b\rangle = if_{abc} |X_c\rangle = |if_{abc} X_c\rangle = |[X_a, X_b]\rangle. \tag{7.3}$$

Because the roots are weights in the adjoint representation and the Cartan subalgebra is commutative, Eq. (7.3) implies that

$$H_i|H_j\rangle = |\left[H_i, H_j\right]\rangle = 0, \tag{7.4}$$

and the states in the adjoint representation corresponding to generators of the Cartan subalgebra have zero weight vectors. Let us define a scalar product $\langle X_a| X_b\rangle$ through

$$\langle X_a| X_b\rangle \equiv \frac{1}{\lambda}\mathrm{Tr}\,(X_a^\dagger X_b). \tag{7.5}$$

Then from Eqs. (7.5) and (7.1), and that $H_i = H_i^\dagger$,

$$\langle H_i|H_j\rangle = \frac{1}{\lambda}\mathrm{Tr}\,\langle H_i H_j\rangle = \delta_{ij}, \tag{7.6}$$

so the states of the Cartan subalgebra are orthonormal.

7.1.2 Raising and Lowering Operators

The adjoint representation states $|E_\alpha\rangle$ associated with generators E_α that are not in the Cartan subalgebra satisfy [68]

$$H_i |E_\alpha\rangle = \alpha_i |E_\alpha\rangle, \tag{7.7}$$

where the weights α_i are generally non-zero. But from Eq. (7.3)

$$H_i |E_\alpha\rangle = |[H_i, E_\alpha]\rangle, \tag{7.8}$$

and from Eqs. (7.7) and (7.8) the generators satisfy

$$[H_i, E_\alpha] = \alpha_i E_\alpha, \tag{7.9}$$

[1] The $|X_a\rangle$ are vectors in a linear vector space and the vector space can be associated with an algebraic structure for corresponding linear operators. For example, the generators X_a of the algebra are associated with the vectors $|X_a\rangle$ through a map $X_a \to |X_a\rangle$. This implies that a combinatorial operation in the algebraic structure such as commutation is associated with a linear transformation in the associated vector space. For a Lie algebra, we will see in Eq. (7.3) that a mapping $A|B\rangle = |[B, A]\rangle$ associates vectors in the vector space with other vectors defined naturally through the combinatorial operation. See Ch. 7 of Gilmore [72] for further discussion.

which is an analog of the usual quantum eigenvalue equation $A|f\rangle = a|f\rangle$, since a commutator plays the role of multiplication in the Lie algebra.[2] The E_α associated with the roots not in the Cartan subalgebra satisfy

$$[A, E_\alpha] = \alpha E_\alpha, \tag{7.10}$$

for a general linear combination of generators $A = a_\mu X_\mu$.

The operators E_α are not hermitian. Taking the hermitian adjoint of Eq. (7.9),

$$[H_i, E_\alpha]^\dagger = E_\alpha^\dagger H_i^\dagger - H_i^\dagger E_\alpha^\dagger = \alpha_i^\dagger E_\alpha^\dagger.$$

But H_i and α_i are hermitian, implying that

$$[H_i, E_\alpha^\dagger] = -\alpha_i E_\alpha^\dagger, \tag{7.11}$$

which is satisfied by the choice $E_\alpha^\dagger = E_{-\alpha}$. States with different weights are orthogonal and we may choose a normalization for adjoint states

$$\langle E_\alpha | E_\beta \rangle = \frac{1}{\lambda}\mathrm{Tr}\,(E_\alpha^\dagger E_\beta) = \frac{1}{\lambda}\mathrm{Tr}\,(E_{-\alpha}E_\beta) = \delta_{\alpha\beta}, \tag{7.12}$$

where $\delta_{\alpha\beta} \equiv \Pi_i \delta_{\alpha_i\beta_i}$. As shown in Problem 7.10,

$$H_i\,(E_{\pm\alpha}|mD\rangle) = (m \pm \alpha)_i\, E_{\pm\alpha}|mD\rangle, \tag{7.13}$$

so application of $E_{\pm\alpha}$ to a state changes the weight quantum number of the state by $\pm\alpha$ and the $E_{\pm\alpha}$ act as raising and lowering operators in the algebra, analogous to the role of $J_\pm$ for SU(2) in Section 3.3.3.

Consider $E_\alpha|E_{-\alpha}\rangle$ for the particular case of the adjoint representation. Since by Eq. (7.13) this has weight $\alpha - \alpha = 0$, in the adjoint representation it must be a linear combination of Cartan subalgebra generators (which are the only generators with zero weight):

$$E_\alpha|E_{-\alpha}\rangle = \beta_i|H_i\rangle = \left|\beta_i H_i\right\rangle \equiv \left|\beta \cdot H\right\rangle = |[E_\alpha, E_{-\alpha}]\rangle, \tag{7.14}$$

where (7.2) and (7.3) have been used. Equations (7.14) and (7.3) may then be used to prove that $\beta_i = \alpha_i$ (Problem 7.8). Therefore, from Eq. (7.14),

$$\left|\beta_i H_i\right\rangle = |\alpha_i H_i\rangle \equiv |\alpha \cdot H\rangle = |[E_\alpha, E_{-\alpha}]\rangle,$$

and we conclude that

$$[E_\alpha, E_{-\alpha}] = \alpha_i H_i \equiv \alpha \cdot H. \tag{7.15}$$

From Eq. (7.13), $E_\alpha\left|E_\beta\right\rangle \propto \left|E_{\alpha+\beta}\right\rangle$, and from Eq. (7.3) we obtain

$$[E_\alpha, E_\beta] = N_{\alpha\beta}E_{\alpha+\beta}, \tag{7.16}$$

if $\alpha + \beta$ is a non-zero root; otherwise $[E_\alpha, E_\beta] = 0$, where $N_{\alpha\beta}$ depends only on α and β, and can be zero. Later we will show that $N_{\alpha\beta}$ can be evaluated from the roots.

[2] An operator A obeying an equation of the form $[H, A] = \lambda A$ is said to be an *eigenoperator* of the matrix H, with eigenvalue λ.

7.2 The Cartan–Weyl Basis

Summarizing results to this point, from Eqs. (7.4), (7.9), (7.15), and (7.16),

$$[H_i, H_j] = 0 \qquad (i, j = 1, 2, \ldots, \ell), \tag{7.17a}$$

$$[H_i, E_\alpha] = \alpha_i E_\alpha, \tag{7.17b}$$

$$[E_\alpha, E_{-\alpha}] = \alpha_i H_i, \tag{7.17c}$$

$$[E_\alpha, E_\beta] = N_{\alpha\beta} E_{\alpha+\beta} \quad (\alpha + \beta \neq 0). \tag{7.17d}$$

These equations define the *Cartan–Weyl basis*. Other bases are sometimes convenient but we shall restrict the present discussion to the basis (7.17).

Example 7.1 The Cartan–Weyl generators for SU(2) and their commutators are

$$H_1 = J_3 \qquad \sqrt{2}\,E_1 = J_+ \equiv J_1 + iJ_2 \qquad \sqrt{2}\,E_{-1} = J_- \equiv J_1 - iJ_2,$$

$$[H_1, E_1] = E_1 \qquad [H_1, E_{-1}] = -E_{-1} \qquad [E_1, E_{-1}] = H_1.$$

No values of $N_{\alpha\beta}$ are required since there is only one positive root.

Example 7.1 is (up to normalization) just a reformulation of the SU(2) angular momentum algebra of Eqs. (3.17)–(3.18). This suggests that the Cartan–Weyl basis can be used to generalize the Cartan–Dynkin analysis of SU(2) in Section 3.3 to other algebras.

7.2.1 Semisimple Algebras

Much of our discussion will focus on semisimple algebras and the corresponding groups (Section 3.2.1). The Cartan–Weyl basis can be used to classify systematically all semisimple algebras because it separates the generators of such algebras into two sets.

1. The ℓ generators H_i of the maximally commuting algebra (Cartan subalgebra).
2. The remaining generators E_α, which obey Eq. (7.17b) with Cartan generators $H = (H_1, H_2, \ldots, H_\ell)$ and eigenvalue $\alpha = (\alpha_1, \alpha_2, \ldots, \alpha_\ell)$.

Each non-zero root α is in one to one correspondence with an eigenvector E_α, and there is a one to one correspondence between the semisimple algebras of rank ℓ and the sets of roots in the ℓ-dimensional root space. The next section describes a systematic way to ascertain whether an algebra is semisimple.

7.2.2 Metric Tensor, Semisimplicity, and Compactness

Whether an algebra is semisimple may be determined directly from its structure constants. For a set of generators X_i, assume a Lie algebra (sum on repeated indices)

$$[X_i, X_j] = c_{ij}^k X_k, \tag{7.18}$$

where $c_{ij}^k = -c_{ji}^k$, and where Box 3.2 and Problem 7.4 illustrate the relationship of this form of the Lie algebra to our standard form (3.3). A *metric tensor* or *Cartan–Killing form* g_{ij} may be defined for any Lie algebra or Lie group by

$$g_{ij} = g_{ji} \equiv c_{ik}^{\ell} c_{j\ell}^{k}. \tag{7.19}$$

We then make the following assertions [224].

1. A Lie algebra is semisimple if and only if the *Cartan condition* $\det g \neq 0$ is satisfied.
2. A Lie algebra defined on the field of real numbers is *compact* if g_{ij} is negative definite.
3. This implies, by virtue of $\det g \neq 0$, that *compact algebras are necessarily semisimple.*
4. The quadratic Casimir operators for semisimple algebras are given by $C \equiv g_{ij} X^i X^j$.

Thus, whether an algebra is compact and semisimple can be investigated by constructing its metric tensor from the structure constants, as illustrated in Problems 7.6, 12.6, and 12.7.

7.3 Structure of the Root Space

Let us now investigate the structure of the root space in more detail, using as outline the presentation in Refs. [68, 72]. We begin by introducing the definitions

$$E_{\pm} \equiv \frac{E_{\pm\alpha}}{|\alpha|} \qquad E_3 \equiv \frac{\alpha \cdot H}{\alpha^2}. \tag{7.20}$$

As shown in Problem 7.9, these operators obey the commutators

$$[\, E_+, E_- \,] = E_3 \qquad [\, E_3, E_{\pm} \,] = \pm E_{\pm}, \tag{7.21}$$

and comparison with Example 3.2 indicates that this is the SU(2) Lie algebra, up to normalization.

7.3.1 Root Space Restrictions

For a state $|m, x, D\rangle$ of representation D,

$$E_3 \,|m, x, D\rangle = \frac{\alpha \cdot H}{\alpha^2} \,|m, x, D\rangle = \frac{\alpha \cdot m}{\alpha^2} \,|m, x, D\rangle . \tag{7.22}$$

But from Eq. (7.21), E_3 is a generator of an SU(2) Cartan subalgebra and its eigenvalues $\alpha \cdot m/\alpha^2$ must be integer or half-integer, which implies that $2\alpha \cdot m/\alpha^2$ is an integer. The state $|m, x, D\rangle$ can be written as a linear combination of states transforming according to definite irreducible representations of the SU(2) algebra (7.21). Furthermore, the operators $E_{\pm}$ behave as SU(2) raising and lowering operators, stepping between states in a representation. Suppose an integer $p \geq 0$ such that

$$(E_+)^p \,|m, x, D\rangle \neq 0 \qquad (E_+)^{(p+1)} \,|m, x, D\rangle = 0. \tag{7.23}$$

Then $(E_+)^p |m, x, D\rangle$ is the *highest-E_3 state* of the corresponding SU(2) representation and it has SU(2) weight $m + p\alpha$ since, by virtue of (7.20) and (7.13), each application of E_+ changes m by α. The E_3 value of the state generated by $(E_+)^p |m, x, D\rangle$ is [see Eq. (7.22)]

$$E_3 |m + p\alpha, x, D\rangle = \frac{\alpha \cdot (m + p\alpha)}{\alpha^2} = \frac{\alpha \cdot m}{\alpha^2} + p = j, \tag{7.24}$$

where j is the quantum number labeling this SU(2) irrep. By similar reasoning, suppose an integer $q \geq 0$ such that

$$(E_-)^q |m, x, D\rangle \neq 0 \qquad (E_-)^{(q+1)} |m, x, D\rangle = 0. \tag{7.25}$$

Then $(E_-)^q |m, x, D\rangle$ is the *lowest-E_3 state* of the SU(2) representation, with weight $m - q\alpha$. The E_3 value for the state generated by $(E_-)^q |m, x, D\rangle$ in Eq. (7.25) is then

$$E_3 |m - q\alpha, x, D\rangle = \frac{\alpha \cdot (m - q\alpha)}{\alpha^2} = \frac{\alpha \cdot m}{\alpha^2} - q = -j. \tag{7.26}$$

Adding (7.24) and (7.26) gives

$$n \equiv 2\frac{\alpha \cdot m}{\alpha^2} = q - p. \tag{7.27}$$

As we shall now see, because q, p, and n must be integers, these results lead to a systematic classification of Lie algebras.

7.3.2 Lengths and Angles for Root Vectors

Consider a pair of distinct roots α and β, and assume the SU(2) algebra (7.21) to be defined in terms of $E_{\pm\alpha}$, as in Eq. (7.20). Then Eq. (7.27) implies that

$$n = 2\frac{\alpha \cdot \beta}{\alpha^2} = q - p. \tag{7.28}$$

But we could equally well define the SU(2) algebra (7.21) using $E_{\pm\beta}$, in which case the equivalent of Eq. (7.27) implies that

$$n' \equiv 2\frac{\beta \cdot \alpha}{\beta^2} = q' - p'. \tag{7.29}$$

Multiplying (7.28) and (7.29) gives

$$\left(\frac{\alpha \cdot \beta}{\alpha\beta}\right)^2 = \cos^2\theta_{\alpha\beta} = \frac{(p - q)(p' - q')}{4} = \frac{nn'}{4}, \tag{7.30}$$

where the euclidean nature of the root-space metric has been used to express the scalar product in terms of the angle $\theta_{\alpha\beta}$ between the roots α and β. Since from the preceding derivation $(p - q)(p' - q')$ must be a non-negative integer, only a limited number of possibilities for the angle $\theta_{\alpha\beta}$ permit a solution of Eq. (7.30), as summarized in Table 7.1. Furthermore, the relative lengths of the root vectors α and β are constrained by this construction because dividing (7.29) by (7.28) gives

$$\frac{n'}{n} = \frac{2\beta \cdot \alpha/\beta^2}{2\alpha \cdot \beta/\alpha^2} = \frac{\alpha^2}{\beta^2} = \frac{q' - p'}{q - p}, \tag{7.31}$$

Table 7.1. Possible relative root vector angles and lengths [72]

$\cos^2(\alpha,\beta)$	$\theta(\alpha,\beta)^\dagger$	$n = 2\dfrac{\alpha\cdot\beta}{\alpha\cdot\alpha}$	$n' = 2\dfrac{\alpha\cdot\beta}{\beta\cdot\beta}$	$\dfrac{n'}{n} = \dfrac{\alpha\cdot\alpha}{\beta\cdot\beta}$
1	0°, 180°	±2	±2	1
$\frac{3}{4}$	30°, 150°	±3	±1	$\frac{1}{3}$
		±1	±3	3
$\frac{1}{2}$	45°, 135°	±2	±1	$\frac{1}{2}$
		±1	±2	2
$\frac{1}{4}$	60°, 120°	±1	±1	1
0	90°	0	0	indeterminate

†180° is redundant by Theorem 7.1 and 0° is excluded by uniqueness of roots.

Table 7.2. Allowed combinations for $nn' = 2$

n	n'	nn'	$\cos^2\theta_{\alpha\beta}$	n'/n
1	2	2	$\frac{1}{2}$	2
−1	−2	2	$\frac{1}{2}$	2
2	1	2	$\frac{1}{2}$	$\frac{1}{2}$
−2	−1	2	$\frac{1}{2}$	$\frac{1}{2}$

which fixes the relationship between lengths of root vectors. The integers n and n', and the length ratios n'/n, are also tabulated in Table 7.1 for possible solutions of Eq. (7.30).

Example 7.2 Consider the possibility $nn' = 2$. The allowed combinations are given in Table 7.2, which accounts for the entries $\theta_{\alpha\beta} = 45°, 135°$ in Table 7.1.

Other entries, and the failure of possibilities not listed in Table 7.1 to satisfy Eq. (7.30), may be verified in similar fashion.

7.4 Construction of Root Diagrams

The information in Table 7.1 may be used to construct systematically the root diagrams for semisimple Lie algebras, when supplemented by the following theorems [141, 224].

Theorem 7.1 *If α is a non-vanishing root, then $-\alpha$ is a root also.*

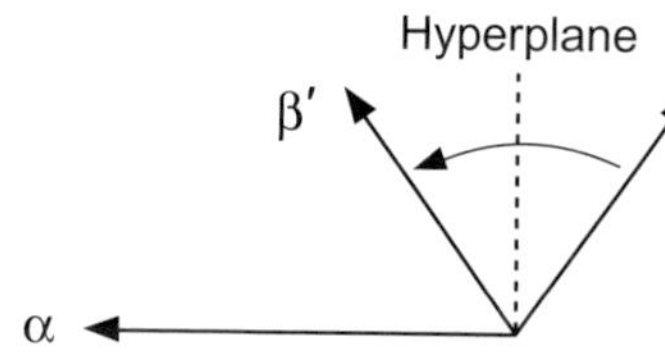

Fig. 7.1 Geometrical interpretation of Theorem 7.2. For two-dimensional root spaces the dashed line represents a line perpendicular to the root vector α. In higher-dimensional root spaces the dashed line represents a hyperplane orthogonal to the root vector α.

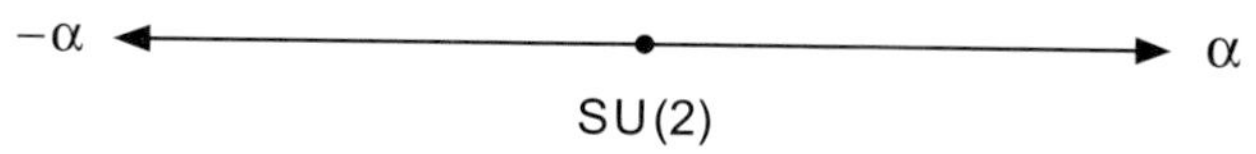

Fig. 7.2 Root diagram for the rank-1 algebra SU(2) in physics notation, or A_1 in Cartan notation.

Theorem 7.2 *If α and β are roots, then $2\alpha \cdot \beta/\alpha \cdot \alpha$ and $2\alpha \cdot \beta/\beta \cdot \beta$ are integers and $\beta' = \beta - 2\alpha(\alpha \cdot \beta)/\alpha \cdot \alpha$ is a root also.*[3]

Theorem 7.3 *The only integer multiples of a root α are 0 and $\pm\alpha$.*

Theorem 7.4 *There is only a finite number of compact Lie algebras of a given rank ℓ.*

Let us illustrate by constructing root diagrams for rank-1 and rank-2 compact Lie algebras.

7.4.1 Rank-1 and Rank-2 Compact Lie Algebras

There is only one independent rank-1 algebra, isomorphic to SU(2). The root diagram is illustrated in Fig. 7.2. If α is a root then so is $-\alpha$, by Theorem 7.1. This algebra corresponds to $\theta_{\alpha\beta} = 180°$ in Table 7.1. Notice that $0°$ implies $\alpha = \beta$, which is excluded by uniqueness of the roots, and that $180°$ is redundant because of Theorem 7.1. There are three independent rank-2 compact algebras, corresponding to $\cos^2\theta_{\alpha\beta} = \frac{1}{4}, \frac{1}{2}$, and $\frac{3}{4}$, respectively. The following example illustrates construction of the root diagram for the $\cos^2\theta_{\alpha\beta} = \frac{3}{4}$ case.

Example 7.3 From Table 7.1, we take two roots separated by $30°$ with lengths in the ratio $1/\sqrt{3}$. Choose the shorter root vector to be $\alpha = (1, 0)$ and the longer root β to be rotated by $30°$ counterclockwise relative to α. Then $\beta = (\frac{3}{2}, \frac{\sqrt{3}}{2})$, as illustrated in Fig. 7.3(a). Next, invoke Theorem 7.2 to construct the reflected root vectors α' and β', also illustrated in

[3] A geometrical interpretation of Theorem 7.2 is given in Fig. 7.1. The root β' is obtained by reflecting the root β through the hyperplane perpendicular to the root α. These reflection planes are called *Weyl hyperplanes* and the reflections are called *Weyl reflections*. The reflections and their products form a group termed the *Weyl group* that explains the high degree of symmetry found in the weight diagrams for semisimple algebras.

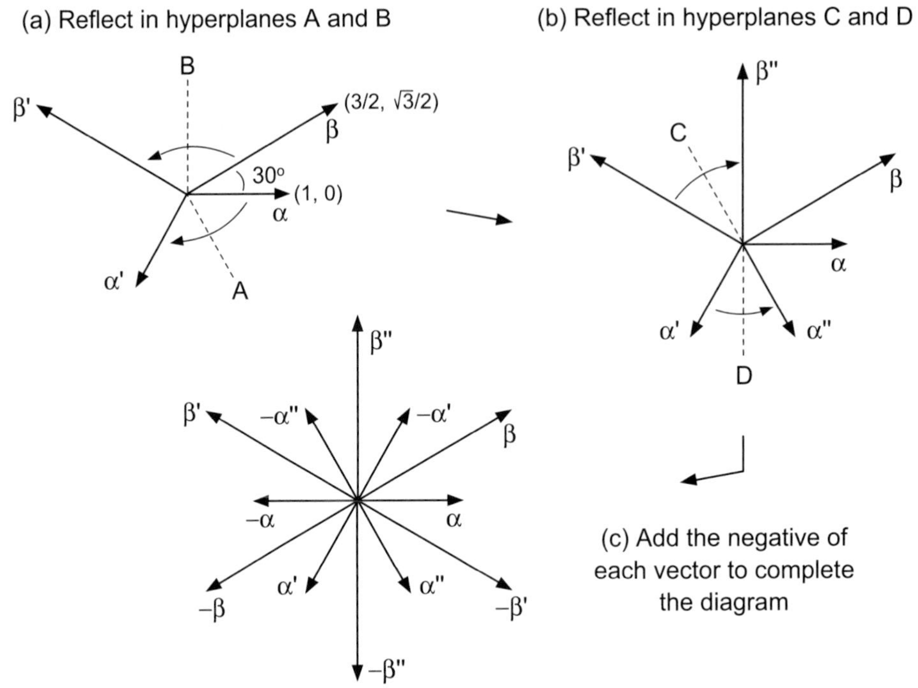

Fig. 7.3 Construction of the root diagram for G_2 from root vectors α and β in Example 7.3.

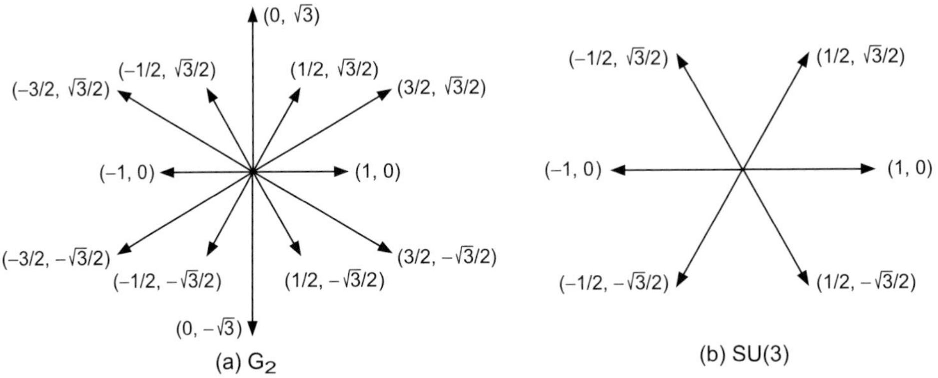

Fig. 7.4 (a) G_2 root diagram with weights labeled. (b) SU(3) root diagram with weights labeled. Signs for weights are chosen according to the convention discussed in Section 7.4.2.

Fig. 7.3(a). Invoke Theorem 7.2 again to add the root vectors α'' and β'', as in Fig. 7.3(b), and invoke Theorem 7.1 to add the negative of all root vectors displayed in Fig. 7.3(b), giving Fig. 7.3(c). Figure 7.4(a) displays the final root diagram, with each root labeled with its components. The corresponding algebra is called G_2 in the Cartan classification.

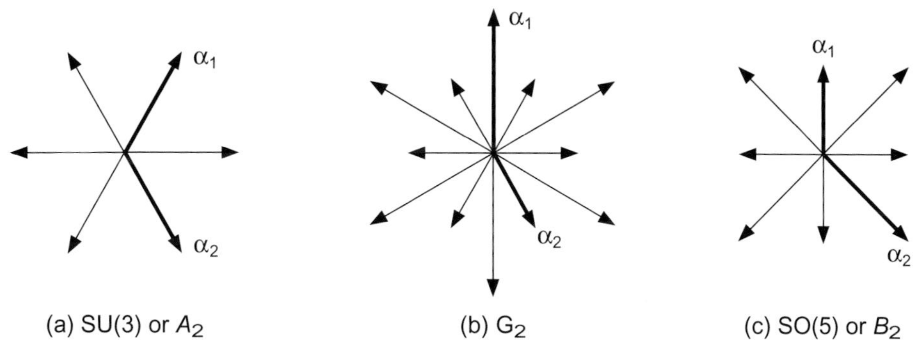

Fig. 7.5 Root diagrams for the compact rank-2 Lie algebras. Simple roots are indicated by heavier lines labeled α_1 and α_2. All roots may be constructed from vector sums of the simple roots, as explained in Section 7.5.

The other two rank-2 algebras correspond to the choices $\theta_{\alpha\beta} = 60°, 120°$ and $\theta_{\alpha\beta} = 45°, 135°$ in Table 7.1. Their root diagrams may be constructed in a similar way as for the G_2 example just given. The corresponding rank-2 compact algebras are called SU(3) (A_2 in the Cartan classification) and SO(5) (B_2 in the Cartan classification). A root diagram for SU(3) labeled explicitly with root components is shown in Fig. 7.4(b) and the root diagrams of all three rank-2 compact algebras are summarized in Fig. 7.5.

7.4.2 An Ordering Prescription for Weights

To extend the Cartan–Dynkin method for SU(2) to algebras with higher-dimensional weight spaces requires a definition of positivity for weights. This will allow specifying (1) a direction for stepping operators (raising or lowering), and (2) the concept of a highest weight in a representation. Any consistent prescription will do; the definition itself is not very significant but the root classification that it enables is. We shall adopt the following convention: *a weight is positive if its first non-zero component is positive; otherwise, it is negative.*

Example 7.4 In Fig. 7.4(a) all roots in the right half of the figure are positive, as is $(0, \sqrt{3})$, while all roots in the left half of the diagram are negative, as is $(0, -\sqrt{3})$ (because the first non-zero component is the second one, which is negative).

This definition allows the roots to be divided into three classes: (1) positive roots, (2) negative roots, and (3) elements of the Cartan subalgebra (with roots of zero length), which enables an ordering. *A weight m is greater than a weight n if $m - n > 0$.* Within the adjoint representation positive roots act like raising operators and negative roots act like lowering operators, in analogy with the $J_\pm$ raising and lowering operators of the SU(2) algebra.

7.5 Simple Roots

Some roots in diagrams may be recognized as vector sums of other roots. For example, from Fig. 7.4(a) for G_2, the root $\beta = 2\beta'' + 3\alpha''$, where we have used the labeling in Fig. 7.3(c). This suggests that root diagrams may be specified in terms of a small number of basic roots from which all others may be built. Because of Theorem 7.1, we need only consider positive roots. A *simple positive root* (which we abbreviate to *simple root*) cannot be expressed as a sum of two other positive roots. Several theorems permit the full root space and the corresponding Lie algebra to be constructed from the simple roots [224].

Theorem 7.5 *The number of simple roots is equal to the rank of the algebra.*

Theorem 7.6 *If α and β are simple roots, then $\alpha - \beta$ is not a simple root.*

Theorem 7.7 *The simple roots are linearly independent.*

Theorem 7.8 *For simple roots α and β, there may be a set of roots called an α-string,*

$$\beta + \alpha, \beta + 2\alpha, \ldots, \beta + n\alpha \qquad \text{(alpha-string)}, \tag{7.32}$$

with the value of n that terminates this sequence given by

$$n = -2\,\frac{\alpha \cdot \beta}{\alpha \cdot \alpha}. \tag{7.33}$$

Likewise, there may be a set of roots called a β-string,

$$\alpha + \beta, \alpha + 2\beta, \ldots, \alpha + n'\beta \qquad \text{(beta-string)}, \tag{7.34}$$

with the value of n' given by

$$n' = -2\,\frac{\alpha \cdot \beta}{\beta \cdot \beta} \tag{7.35}$$

terminating this sequence.

These theorems allow the full set of roots to be constructed from the simple roots by virtue of Theorems 7.9 and 7.10.

Theorem 7.9 *Any positive root ϕ can be written as a linear combination of simple roots α with non-negative integer coefficients*

$$\phi = \sum_\alpha k_\alpha \alpha. \tag{7.36}$$

The positive roots and their coefficients may be determined by forming all allowed strings (7.32) and (7.34) from the simple roots. The level *of a root is defined by*

$$k = \text{level} \equiv \sum_\alpha k_\alpha, \tag{7.37}$$

where the k_α are the coefficients appearing in Eq. (7.36). The roots corresponding to a given level are formed by action of the simple roots on roots at the previous level.

Theorem 7.10 *All roots, and the entire Lie algebra up to a phase convention, may be determined from the simple roots of the algebra.*

Let us illustrate the use of these theorems by finding the simple roots for the SU(3) algebra.

Example 7.5 Since we know SU(3) to be a rank-2 algebra, by Theorem 7.5 there are two simple roots. From Fig. 7.4(b) the positive roots are

$$\alpha_1 = \left(\frac{1}{2}, \frac{\sqrt{3}}{2}\right) \qquad \alpha_2 = \left(\frac{1}{2}, -\frac{\sqrt{3}}{2}\right) \qquad \alpha_3 = (1, 0).$$

Obviously, $\alpha_3 = \alpha_1 + \alpha_2$, so α_1 and α_2 are the simple roots for SU(3), as we have already indicated graphically in Fig. 7.5(a).

By Theorem 7.6, if α and β are simple roots then $\alpha - \beta$ is not a root, implying that

$$E_{-\alpha} \left|E_\beta\right\rangle = E_{-\beta} \left|E_\alpha\right\rangle = 0. \tag{7.38}$$

Therefore, if α and β are simple roots, $q = 0$ in Eq. (7.28),

$$\frac{\alpha \cdot \beta}{\alpha^2} = \frac{1}{2}(q - p) = -\frac{p}{2}, \tag{7.39}$$

and $q' = 0$ in Eq. (7.29),

$$\frac{\beta \cdot \alpha}{\beta^2} = \frac{1}{2}(q' - p') = -\frac{p'}{2}. \tag{7.40}$$

Taking the product of (7.39) and (7.40), and using Eq. (7.30) gives

$$\cos \theta_{\alpha\beta} = -\frac{1}{2}\sqrt{pp'}, \tag{7.41}$$

for the angle between simple roots. Taking their ratio yields

$$\frac{\beta^2}{\alpha^2} = \frac{p}{p'}, \tag{7.42}$$

which defines the relative length of the two simple roots. From Eq. (7.39), $2\alpha \cdot \beta/\alpha^2 = -p \leq 0$, which implies that for the angle between simple roots $\theta_{\alpha\beta} \geq 90°$, and because we are considering only positive roots we also require that $\theta_{\alpha\beta} \leq 180°$. Since the angles that solve Eq. (7.41) are restricted by the requirement that p and p' be integers and that $90° \leq \theta_{\alpha\beta} \leq 180°$, comparison with Table 7.1 indicates that the allowed angles for simple roots are $\{150°, 135°, 120°, 90°\}$. Because of these restrictions on simple roots, they may be represented in a simple diagrammatic notation that we discuss next.

7.6 Dynkin Diagrams

In a *Dynkin diagram,* each simple root is indicated by an open circle and the circles are connected by a set of lines, with the number of lines indicating the angle between the two

Table 7.3. Dynkin diagrams

Angle between roots	Dynkin diagram
150°	
135°	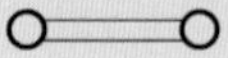
120°	
90°	

circles (roots). The standard notation is indicated in Table 7.3. For the angle between two simple roots: (1) no lines indicates 90°, (2) one line indicates 120°, (3) two lines indicates 135°, and (4) three lines indicates 150°. Dynkin diagrams may be labeled with the root for each circle, and a standard convention uses filled circles for shorter simple roots and open circles for longer simple roots.[4] The weight space for a rank-ℓ algebra is difficult to visualize for $\ell > 2$. Conversely, Dynkin diagrams exhibit all the information required to obtain the complete set of root vectors, including all root length and angle information, in a concise 2D graph for any ℓ.

Example 7.6 Let us construct the Dynkin diagram for an algebra having the simple roots $\alpha_1 = (0, \sqrt{3})$ and $\alpha_2 = \left(\frac{1}{2}, -\frac{\sqrt{3}}{2}\right)$. From these roots

$$\begin{gathered}\alpha_1 \cdot \alpha_1 = 0 + 3 = 3 \qquad \alpha_2 \cdot \alpha_2 = \frac{1}{4} + \frac{3}{4} = 1 \qquad \alpha_1 \cdot \alpha_2 = 0 - \frac{3}{2} = -\frac{3}{2} \\ p = -2\,\frac{\alpha_1 \cdot \alpha_2}{\alpha_1^2} = 1 \qquad p' = -2\,\frac{\alpha_1 \cdot \alpha_2}{\alpha_2^2} = 3.\end{gathered} \tag{7.43}$$

Then from Eq. (7.41), the angle between the two simple roots is

$$\theta_{\alpha\beta} = \cos^{-1}\left(-\frac{1}{2}\sqrt{pp'}\right) = \cos^{-1}\left(-\frac{\sqrt{3}}{2}\right) = 150°,$$

while from Eq. (7.42), $\alpha_2/\alpha_1 = 1/\sqrt{3}$. From Table 7.3 the Dynkin diagram is , with the filled circle indicating the shorter root α_2; this is the algebra G_2 in Fig. 7.4(a).

7.6.1 The Cartan Matrix

The only invariants for the simple roots are their lengths and scalar products. This may be expressed concisely through the *Cartan matrix* A, which has elements

[4] This convention is unambiguous because the set of simple roots never has more than two distinct lengths. Some authors use an arrow on the lines pointing from shorter to longer rather than filled and open circles to distinguish root lengths in Dynkin diagrams.

$$A_{ij} = \frac{2\alpha_i \cdot \alpha_j}{\alpha_i \cdot \alpha_i}, \tag{7.44}$$

for two simple roots α_i and α_j.

Example 7.7 The SU(3) Dynkin diagram is , implying from Table 7.3 and Table 7.1 that there are two simple roots of equal length, with an angle of 120° between them. Therefore, $\alpha_1 \cdot \alpha_1 = \alpha_2 \cdot \alpha_2 = 1$ and $\alpha_1 \cdot \alpha_2 = \cos 120° = -\frac{1}{2}$, and from Eq. (7.44),

$$A_{11} = \frac{2\alpha_1 \cdot \alpha_1}{\alpha_1 \cdot \alpha_1} = 2 \qquad A_{12} = \frac{2\alpha_1 \cdot \alpha_2}{\alpha_1 \cdot \alpha_1} = -1,$$
$$A_{21} = \frac{2\alpha_2 \cdot \alpha_1}{\alpha_2 \cdot \alpha_2} = -1 \qquad A_{22} = \frac{2\alpha_2 \cdot \alpha_2}{\alpha_2 \cdot \alpha_2} = 2,$$

so that the Cartan matrix is

$$A_{\mathrm{SU(3)}} = \begin{pmatrix} A_{11} & A_{12} \\ A_{21} & A_{22} \end{pmatrix} = \begin{pmatrix} 2 & -1 \\ -1 & 2 \end{pmatrix}$$

for the SU(3) Lie algebra.

A related quantity called the *Coxeter matrix* is sometimes used, which may be obtained by replacing the denominator of Eq. (7.44) with $|\alpha_i|\,|\alpha_j|$.

7.6.2 Constructing All Roots from Dynkin Diagrams

Example 7.6 illustrated how to construct a Dynkin diagram from simple roots. Let us now consider the inverse process: using a Dynkin diagram to deduce the simple roots and then using the simple roots to construct the full set of roots. Since many of the details have already been worked out, the algebra G_2 will be used again as an example. The Dynkin diagram is , which indicates that there are two simple roots with an angle of 150° between them. As illustrated in the solution of Problem 7.11, this information may be used to recover the full root diagram of Fig. 7.4(a) for the algebra G_2.

7.6.3 Constructing the Algebra from the Roots

We have shown how the full set of roots may be constructed from the simple roots, and that these are in turn specified by the Dynkin diagram. Let us now illustrate a procedure to construct the algebra from the roots, thereby demonstrating that the set of simple roots (and thus the Dynkin diagram) determines both the root structure and the Lie algebra [224]. Consider the Lie algebra A_2, or in physicist-speak, SU(3), and choose the normalization of roots implied in Fig. 7.6, where the simple roots are

$$\alpha = \left(\frac{1}{2}, \frac{\sqrt{3}}{2}\right) \qquad \beta = \left(\frac{1}{2}, -\frac{\sqrt{3}}{2}\right)$$

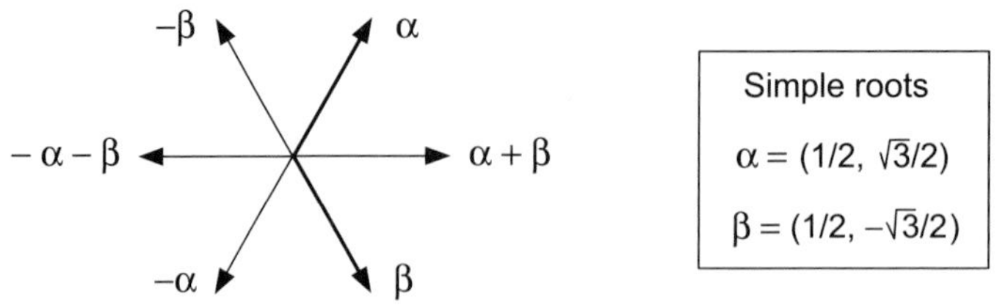

Fig. 7.6 Root basis for constructing the SU(3) Lie algebra. Heavy lines indicate the simple roots.

and the full set of roots is given by

$$\pm\alpha = \pm\left(\frac{1}{2}, \frac{\sqrt{3}}{2}\right) \qquad \pm\beta = \pm\left(\frac{1}{2}, -\frac{\sqrt{3}}{2}\right) \qquad \pm(\alpha+\beta) = \pm(1,0).$$

Thus, we must find the commutation algebra associated with the operators $E_{\pm\alpha}$, $E_{\pm\beta}$, and $E_{\pm(\alpha+\beta)}$, and the two generators of the Cartan subalgebra, designated H_1 and H_2.

To construct all possible commutators among these operators, assume the Cartan–Weyl basis (7.17) and the root space of Fig. 7.6. First, the Cartan subalgebra is commutative, by definition: $[H_i, H_j] = 0$. Then from Eq. (7.17b)

$$[H_1, E_{\pm\alpha}] = \pm\frac{1}{2}E_{\pm\alpha} \qquad [H_1, E_{\pm\beta}] = \pm\frac{1}{2}E_{\pm\beta},$$
$$[H_2, E_{\pm\alpha}] = \pm\frac{\sqrt{3}}{2}E_{\pm\alpha} \qquad [H_2, E_{\pm\beta}] = \mp\frac{\sqrt{3}}{2}E_{\pm\beta},$$

while from Eq. (7.17c),

$$[E_\alpha, E_{-\alpha}] = \frac{1}{2}H_1 + \frac{\sqrt{3}}{2}H_2 \qquad [E_\beta, E_{-\beta}] = \frac{1}{2}H_1 - \frac{\sqrt{3}}{2}H_2.$$

Since the weights are additive, from Eq. (7.17b)

$$[H_i, E_{\alpha+\beta}] = (\alpha+\beta)_i E_{\alpha+\beta},$$

which gives

$$[H_1, E_{\pm(\alpha+\beta)}] = \pm\left(\frac{1}{2}+\frac{1}{2}\right)E_{\pm(\alpha+\beta)} = \pm E_{\pm(\alpha+\beta)},$$
$$[H_2, E_{\pm(\alpha+\beta)}] = \pm\left(\frac{\sqrt{3}}{2}-\frac{\sqrt{3}}{2}\right)E_{\pm(\alpha+\beta)} = 0.$$

From Eq. (7.17c), we have

$$[E_{\alpha+\beta}, E_{-\alpha-\beta}] = (\alpha+\beta)_i H_i = \left(\frac{1}{2}+\frac{1}{2}\right)H_1 + \left(\frac{\sqrt{3}}{2}-\frac{\sqrt{3}}{2}\right)H_2 = H_1.$$

In the Cartan–Weyl basis the commutators $[E_\alpha, E_\beta]$ are given by Eq. (7.17d), which requires the constants $N_{\alpha\beta}$. These can be evaluated directly from the properties of the root space as follows. For root strings of the form

$$\beta + j\,\alpha, \beta + (j-1)\,\alpha, \ldots, \beta, \ldots, \beta - k\,\alpha,$$

where $\beta - (k+1)\alpha$ and $\beta + (j+1)\alpha$ are not roots, the $N_{\alpha\beta}$ exhibit the symmetries [72]

$$N_{\alpha\beta} = -N_{\beta\alpha} = -N_{-\alpha,-\beta} = N_{\beta,-\alpha-\beta} = N_{-\alpha-\beta,\alpha}, \tag{7.45}$$

and are given explicitly by

$$N_{\alpha\beta}^2 = \frac{1}{2} j(k+1)\, \alpha \cdot \alpha. \tag{7.46}$$

For SU(3) the positive roots are α, β, and $\alpha + \beta$. Then

$$[\, E_\alpha, E_\beta \,] = N_{\alpha\beta} E_{\alpha+\beta} = \frac{1}{\sqrt{2}} E_{\alpha+\beta},$$

since $\alpha + \beta$ is a root and in Eqs. (7.45)–(7.46) we have $j = 1$, $k = 0$, and $\alpha \cdot \alpha = 1$, which imply that $N_{\alpha\beta} = \frac{1}{\sqrt{2}}$. Likewise, Eq. (7.17d) implies that

$$[\, E_\alpha, E_{\alpha+\beta} \,] = 0 \qquad [\, E_\beta, E_{\alpha+\beta} \,] = 0$$

where the first commutator vanishes because $\alpha + \alpha + \beta$ is not a root and the second vanishes because $\beta + \alpha + \beta$ is not a root. Finally,

$$[\, E_\alpha, E_{-(\alpha+\beta)} \,] = N_{\alpha,-\alpha-\beta} E_{\alpha-\alpha-\beta} = N_{\alpha,-\alpha-\beta} E_{-\beta} = -\frac{1}{\sqrt{2}} E_{-\beta},$$

$$[\, E_\beta, E_{-(\alpha+\beta)} \,] = N_{\beta,-\alpha-\beta} E_{\beta-\alpha-\beta} = N_{\beta,-\alpha-\beta} E_{-\alpha} = \frac{1}{\sqrt{2}} E_{-\alpha},$$

where Eqs. (7.45) and (7.46) have been used.

7.7 Dynkin Diagrams and the Simple Algebras

The Dynkin diagrams introduced in Table 7.3 may be extended to higher-rank groups by combining circles and lines subject to a restrictive set of rules for allowed diagrams (see Gilmore [72], Ch. 8). These rules permit all simple Lie algebras to be classified because there is a one to one correspondence between an allowed diagram and a simple Lie algebra. The full Dynkin classification is shown in Appendix D, and gives rise to four series of *classical algebras,* A_ℓ, B_ℓ, C_ℓ, and D_ℓ, and to five *exceptional algebras*, G_2, F_4, E_6, E_7, and E_8. It is remarkable that these are the *only* possibilities for allowed Dynkin diagrams and therefore the only possibilities for simple Lie algebras. This suggests that for physical systems having a few collective degrees of freedom [for example, a superconductor (Ch. 32), a deformed atomic nucleus (Ch. 31), or a sheet of graphene in a strong magnetic field (Ch. 20)], only a limited number of Lie algebras are candidates for the physical description of their collective (emergent) modes. We shall exploit this "quantization" of allowed low-dimensional algebras for compact groups at various places in this book.

Background and Further Reading

This chapter draws from Georgi [68], Wybourne [224], Lichtenberg [141], and Gilmore [72]. See also Elliott and Dawber [56], Greiner and Müller [82], and O'Raifeartaigh [158].

Problems

7.1 Use Eqs. (7.27)–(7.31) to verify the entries in Table 7.1.

7.2 Use Table 7.1 and Theorems 7.1–7.2 to construct root diagrams for the rank-2 compact algebras SU(3) and SO(5).

7.3 Find the simple roots for the rank-2 Lie algebras G_2 and SO(5). ***

7.4 Show that the angular momentum Lie algebra $[J_i, J_j] = i\epsilon_{ijk} J_k$ can be put in the form

$$[X_1, X_2] = X_3 \qquad [X_2, X_3] = X_1 \qquad [X_3, X_1] = X_2,$$

by substituting $J_i \to iX_i$, which is the form $[X_i, X_j] = c_{ij}^k X_k$ described in Box 3.2 and assumed for Eq. (7.18).

7.5 Beginning from the Dynkin diagram  for the SU(3) algebra, construct the complete root diagram. ***

7.6 For the group SO(3), find the metric tensor (7.19) and show that SO(3) is compact and semisimple. Use the metric tensor to construct the Casimir operator. *Hint*: The SO(3) algebra has been put in the form (7.18) in Problem 7.4. ***

7.7 Use the definition of the adjoint representation matrices (3.8), to compute the action of a generator X_a on a state $|X_b\rangle$ given in Eq. (7.3).

7.8 Use Eqs. (7.14) and (7.5) to prove that $\beta_i = \alpha_i$. *Hint*: You will need the cyclic property of the trace; see Problem 3.11(f). ***

7.9 Show that the operators given in Eq. (7.20) have the SU(2) commutators (7.21).

7.10 Prove the result of Eq. (7.13) that the $E_{\pm\alpha}$ act as raising and lowering operators within the weight space. ***

7.11 Use the Dynkin diagram  to construct the simple roots, and from those all roots, for the algebra G_2. ***

7.12 From Problem 7.11, suitably normalized simple roots for the algebra G_2 are $\alpha_1 = \left(0, \sqrt{3}\right)$ and $\alpha_2 = \left(\frac{1}{2}, -\frac{\sqrt{3}}{2}\right)$. What is the corresponding Cartan matrix?

7.13 For an operator $A = a_\mu X_\mu$ corresponding to a linear combination of generators X_μ for a Lie algebra, use Eq. (7.10) and the Jacobi identity (3.6) to prove that

$$[A, [E_\alpha, E_\beta]] = (\alpha + \beta)[E_\alpha, E_\beta],$$

where E_α and E_β are generators that are not in the Cartan subalgebra.

8 Unitary and Special Unitary Groups

The continuous symmetries discussed so far have emphasized spatial rotations under the groups SO(2), SO(3), and SU(2). However, as shown in Ch. 7 there are more complicated Lie groups. As an example of more sophisticated symmetries, we now consider the *unitary groups* U(N) with $N > 2$ and their *special unitary subgroups,* SU(N), which are crucial in the formulation of quantum mechanics because unitary operators conserve probabilities. We will often illustrate the properties of unitary groups using the group SU(3) that we found in Ch. 7 to be associated with the algebra labeled A_2 by Cartan.

8.1 Generators and Commutators for SU(3)

SU(N) is of rank $l = N - 1$, with $N^2 - 1$ generators. Thus SU(3) is rank-2, with eight generators that may be represented by 3×3, traceless, hermitian matrices, two of which may be diagonalized simultaneously. The group elements may be parameterized

$$U = e^{\frac{1}{2}i\alpha_k\lambda_k} \qquad (k = 1, 2, \ldots, 8), \tag{8.1}$$

where the generators are $\frac{1}{2}\lambda_k$, with the λ_k being the eight *Gell-Mann matrices,*

$$\begin{aligned} \lambda_1 &= \begin{pmatrix} 0 & 1 & 0 \\ 1 & 0 & 0 \\ 0 & 0 & 0 \end{pmatrix} \quad \lambda_2 = \begin{pmatrix} 0 & -i & 0 \\ i & 0 & 0 \\ 0 & 0 & 0 \end{pmatrix} \quad \lambda_3 = \begin{pmatrix} 1 & 0 & 0 \\ 0 & -1 & 0 \\ 0 & 0 & 0 \end{pmatrix}, \\ \lambda_4 &= \begin{pmatrix} 0 & 0 & 1 \\ 0 & 0 & 0 \\ 1 & 0 & 0 \end{pmatrix} \quad \lambda_5 = \begin{pmatrix} 0 & 0 & -i \\ 0 & 0 & 0 \\ i & 0 & 0 \end{pmatrix} \quad \lambda_6 = \begin{pmatrix} 0 & 0 & 0 \\ 0 & 0 & 1 \\ 0 & 1 & 0 \end{pmatrix}, \\ \lambda_7 &= \begin{pmatrix} 0 & 0 & 0 \\ 0 & 0 & -i \\ 0 & i & 0 \end{pmatrix} \quad \lambda_8 = \frac{1}{\sqrt{3}}\begin{pmatrix} 1 & 0 & 0 \\ 0 & 1 & 0 \\ 0 & 0 & -2 \end{pmatrix}, \end{aligned} \tag{8.2}$$

with λ_3 and λ_8 chosen as the two diagonal generators. The Lie algebra is

$$\left[\frac{\lambda_i}{2}, \frac{\lambda_j}{2}\right] = i f_{ijk}\left(\frac{\lambda_k}{2}\right), \tag{8.3}$$

where f_{ijk} is antisymmetric under exchange of any two indices, with non-zero values

$$\begin{aligned} f_{123} &= 1 \qquad f_{147} = f_{246} = f_{257} = f_{345} = \frac{1}{2}, \\ f_{156} &= f_{367} = -\frac{1}{2} \qquad f_{458} = f_{678} = \frac{\sqrt{3}}{2}, \end{aligned} \tag{8.4}$$

and permutations of these indices. A compact group with generators τ_i may be normalized

$$\mathrm{Tr}\,(\tau_i \tau_j) = \mathrm{constant} \times \delta_{ij}. \tag{8.5}$$

The traceless SU(3) generators $\frac{1}{2}\lambda_k$ have been normalized such that

$$\mathrm{Tr}\left(\frac{\lambda_i}{2} \cdot \frac{\lambda_j}{2}\right) = \frac{1}{2}\delta_{ij}. \tag{8.6}$$

Just as for SU(2), it is convenient to form new operators from the generators with the maximal number diagonal (two in this case), and the remaining six linear combinations giving stepping operators in the two-dimensional SU(3) weight space. To facilitate compact notation let us define $F_k = \frac{1}{2}\lambda_k$ and use these to form two diagonal operators

$$T_3 = F_3 \qquad Y = \frac{2}{\sqrt{3}} F_8, \tag{8.7}$$

and six additional operators,

$$T_\pm = F_1 \pm iF_2 \qquad U_\pm = F_6 \pm iF_7 \qquad V_\pm = F_4 \pm iF_5. \tag{8.8}$$

The commutation relations obeyed by these operators are summarized in Table 8.1.

Table 8.1. Some SU(3) commutators [66]

$$[T_3, T_\pm] = \pm T_\pm \qquad [Y, T_\pm] = 0$$
$$[T_3, U_\pm] = \mp\tfrac{1}{2} U_\pm \qquad [Y, U_\pm] = \pm U_\pm$$
$$[T_3, V_\pm] = \pm\tfrac{1}{2} V_\pm \qquad [Y, V_\pm] = \pm V_\pm$$

$$[T_+, T_-] = 2T_3$$
$$[U_+, U_-] = \tfrac{3}{2}Y - T_3 \equiv 2U_3$$
$$[V_+, V_-] = \tfrac{3}{2}Y + T_3 \equiv 2V_3$$
$$[T_+, V_+] = [T_+, U_-] = [U_+, V_+] = 0$$

$$[T_+, V_-] = -U_- \qquad [T_+, U_+] = V_+$$
$$[U_+, V_-] = T_- \qquad [T_3, Y] = 0$$

Unlisted commutation relations may be obtained from

$$T_+ = (T_-)^\dagger \qquad U_+ = (U_-)^\dagger \qquad V_+ = (V_-)^\dagger$$

Table 8.2. Non-zero values of d_{ijk} for the SU(3) Lie algebra

$d_{118} = d_{228} = d_{338} = -d_{888} = \frac{1}{\sqrt{3}}$	$d_{146} = d_{157} = d_{256} = d_{344} = d_{355} = \frac{1}{2}$
$d_{247} = d_{366} = d_{377} = -\frac{1}{2}$	$d_{448} = d_{558} = d_{668} = d_{778} = \frac{-1}{2\sqrt{3}}$

Other values follow from noting that d_{ijk} is completely symmetric in its indices.

8.2 SU(3) Casimir Operators

The quadratic SU(3) Casimir operator that is the analog of J^2 for SU(2) is

$$C_2 = F^2 \equiv \sum_{i=1}^{8} F_i F_i, \tag{8.9}$$

and since SU(3) is rank-2 there is a second Casimir operator, cubic in the generators,

$$C_3 = 8 \sum_{ijk} d_{ijk} F_i F_j F_k, \tag{8.10}$$

with the d_{ijk} values defined in Eq. (11.17) and listed in Table 8.2. The Casimir operators C_2 and C_3 commute with all SU(3) generators.

8.3 SU(3) Weight Space

Since SU(3) is rank-2 the weight space is two-dimensional. It is conventional to use the additive quantum numbers T_3 (x-axis) and Y (y-axis) to label points in the weight space. The notation derives from applications of SU(3) in elementary particle physics where the quantum number Y is the hypercharge and T_3 is the third component of isospin. We will employ this notation for convenience but the discussion in this chapter will depend only on group properties and not on any particular physical interpretation of the generators.

8.3.1 SU(3) Raising and Lowering Operators

By examining the commutation relations of the operators $U_\pm$, $V_\pm$, and $T_\pm$ in Table 8.1, it may be concluded that these are the raising and lowering generators of three separate SU(2) subgroups (sometimes termed U-spin, V-spin, and T-spin, respectively), and that they step in the (Y, T_3) plane in the manner depicted in Fig. 8.1. For example, the operator U_+ increases Y by one and decreases T_3 by $\frac{1}{2}$, when applied to a member of an SU(3) multiplet, unless this action would lead out of the irrep, in which case it gives zero. For SU(2) we analyzed the weight space by using the properties of the stepping operators and the concept of a state of maximal weight within each irreducible multiplet. This procedure may be generalized to SU(3). Since the operators T_+, U_-, and V_+ all increase the value of

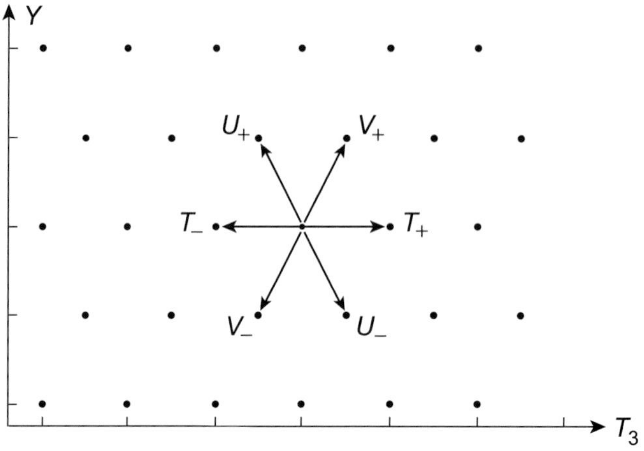

Fig. 8.1 The weight space for SU(3) and action of the six stepping operators $U_\pm$, $V_\pm$, and $T_\pm$. Allowed sites and the action of the operators follow from the relations in Table 8.1. The horizontal spacing between points is one unit in T_3 and the vertical spacing is one unit in Y.

T_3, for SU(3) representations there must exist a *maximally stretched state* ϕ_{max} satisfying $T_+\phi_{\text{max}} = U_-\phi_{\text{max}} = V_+\phi_{\text{max}} = 0$. As for SU(2) representations, other members of an SU(3) irrep may be generated by sequential application of the stepping operators to ϕ_{max}.

8.3.2 SU(3) Irreducible Representations

The weight diagram for SU(3) consists of points in a two-dimensional plane, and by connecting these points with lines we obtain geometrical figures that uniquely specify the irreducible representations. For SU(3) irreps the outer boundary of these figures is always convex, which suggests a systematic way to construct and label such diagrams [66]. Repeated application of V_- to ϕ_{max} eventually will reach a state where the next application of V_- will give zero, because it would lead to a state not in the representation. This defines a corner of the diagram and an integer p that is the number of V_- applications required to reach this corner, $(V_-)^{p+1}\phi_{\text{max}} = 0$. Then we may operate repeatedly with T_- (q times), until another corner is reached and $(T_-)^{q+1}(V_-)^p\phi_{\text{max}} = 0$. The SU(3) representation and corresponding weight diagram are uniquely specified by the positive integers (p, q) because the algebra in Table 8.1 implies that the diagrams must be symmetric under reflection about $T_3 = 0$, and symmetric about axes perpendicular to the $U_\pm$ and $V_\pm$ axes.

8.3.3 Dimensionality of SU(3) Irreps

For example, there exists a 27-dimensional irreducible representation of SU(3) that has the weight diagram shown in Fig. 8.2. From the shape of the outer boundary, the ordered numbers $(p, q) = (2, 2)$ may be used to characterize this irrep. It is also common to specify the irreps by giving the dimensionality in bold numbers (**27**), even though the dimensionality may not uniquely specify the SU(3) irrep as it does for the SU(2) irreps.

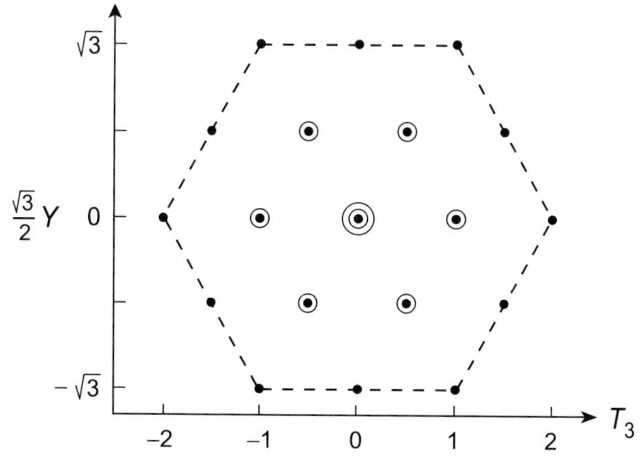

Fig. 8.2 Weight diagram for the SU(3) irrep (2, 2), or **27**. Circles around a point denote additional degeneracy. Thus the central point $(T_3, Y) = (0, 0)$ is triply degenerate.

For SU(3) weight diagrams as in Fig. 8.2, we adopt the standard convention that (1) a point stands for a singly occupied site and (2) each circle about a point implies an additional degeneracy of one. Unlike for SU(2), sites in the SU(3) weight space may be multiply occupied, even in irreducible representations. The general rules for the multiplicity at each site are as follows.

1. Sites on the outer boundary of an SU(3) irrep diagram are singly occupied.
2. Site degeneracy increases by one for each layer inward in the diagram, except that inside a triangular layer the degeneracy remains constant at the value for the triangular layer.

For the $(2, 2)$ irrep in Fig. 8.2 this gives single occupation for points in the outer layer, double occupation for points in the next layer inward, a triply occupied central site, and a representation dimensionality of **27**, by summing site occupations. Table 8.3 lists some irreps of SU(3), labeled both by (p, q) and by dimensionality of the representation D, which is related to (p, q) by

$$D = \frac{1}{2}(p + 1)(p + q + 2)(q + 1), \tag{8.11}$$

and the expectation value of the quadratic Casimir (8.9), which is in terms of the (p, q),

$$C_2 = F^2 = \frac{1}{3}(p^2 + pq + q^2) + p + q. \tag{8.12}$$

The weights of the state ϕ_{max} of maximal T_3 can also be expressed as

$$T_3 = \frac{1}{2}(p + q) \qquad Y = \frac{1}{3}(p - q), \tag{8.13}$$

in terms of the irrep labels p and q.

Table 8.3. Some SU(3) representations [85]

Diagram†	Dim D	(p, q)	$\langle C_2 \rangle$
	1	(0, 0)	0
	3	(1, 0)	$\frac{4}{3}$
	$\overline{\mathbf{3}}$	(0, 1)	$\frac{4}{3}$
	8	(1, 1)	3
	6	(2, 0)	$\frac{10}{3}$
	$\overline{\mathbf{6}}$	(0, 2)	$\frac{10}{3}$
	10	(3, 0)	6
	$\overline{\mathbf{10}}$	(0, 3)	6
	27	(2, 2)	8

†SU(3) Young diagrams are described in Section 8.7.

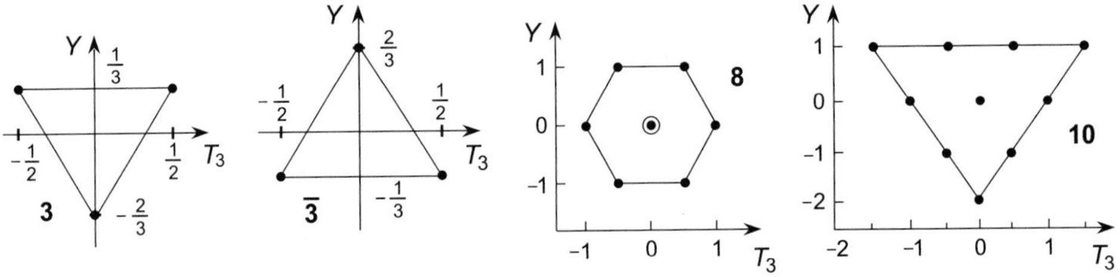

Fig. 8.3 Weight diagrams (Y, T_3) for some SU(3) irreducible representations in Table 8.3, labeled by their dimensionality in bold numbers.

8.3.4 Construction of SU(3) Weight Diagrams

Weight diagrams for the representations listed in Table 8.3 may be constructed starting from the state of maximal T_3. A few cases are shown in Fig. 8.3. The following example illustrates construction of the weight diagram for the **10** of Fig. 8.3.

Example 8.1 For the 10-dimensional SU(3) irreducible representation $(3, 0)$ in Fig. 8.3, from Eq. (8.13) the state of maximal T_3 has $T_3 = \frac{3}{2}$ and $Y = 1$. Starting from this state, $p = 3$ applications of V_- take us to a boundary, then $q = 0$ applications of T_- leave us at the same point, which is $T_3 = 0$ and $Y = -2$. The SU(3) diagrams are symmetric about the vertical axis, so the outer boundary of the $(3, 0)$ must be a triangle. The horizontal sites must be separated by one unit in T_3 because of the isospin SU(2) subgroups in horizontal rows, so the top row has four occupied sites and $T_3 = Y = 0$ is the only occupied interior site.

The outer boundary is singly occupied and triangular, so the multiplicity of the single interior site is one and the irrep dimensionality is 10, by counting occupied sites.

Weight diagrams for other irreps in Table 8.3 may be constructed in a similar fashion.

8.4 Complex Conjugate Representations

Suppose a Lie algebra $[T_i, T_j] = if_{ijk}T_k$. Complex conjugating both sides of this equation,

$$[T_i^*, T_j^*] = [-T_i^*, -T_j^*] = if_{ijk}(-T_k^*), \tag{8.14}$$

since the f_{ijk} are real. Therefore, if the matrices T_i satisfy a Lie algebra the matrices $-T_i^*$ satisfy the same Lie algebra. The representation generated by the matrices $-T_i^*$ is said to be the (*complex*) *conjugate representation* of the one generated by the matrices T_i.

SU(2) Conjugate Representation: For SU(2) the effect of $T_i \to -T_i^*$ is $\sigma_3 \to -\sigma_3$ and $\sigma_\pm \to -(\sigma_1 \mp i\sigma_2)$, which changes the sign of the magnetic quantum number m and switches the actions of σ_+ and σ_-. But SU(2) weight diagrams are symmetric in $\pm m$, so the conjugate representation, symbolized by $\bar{\mathbf{2}}$ or $\mathbf{2}^*$, is equivalent to the fundamental representation **2**.

SU(3) Conjugate Representation: For SU(3) the situation is different. The replacement $T_i \to -T_i^*$ has three effects on the weight space: (1) $T_3 \to -T_3$, (2) $Y \to -Y$, and (3) raising and lowering roles are interchanged for the operator pairs $V_\pm$, $U_\pm$, and $T_\pm$. These changes correspond to a reflection of the weight diagram in both the T_3 and Y axes, which interchanges the representation labels p and q. If we denote the SU(3) representations by $D^{(pq)}$, then the conjugate representation is $\bar{D}^{(pq)} = D^{(qp)}$. Example 8.2 illustrates for the fundamental representation of SU(3).

Example 8.2 The conjugate of the SU(3) fundamental representation **3**, corresponding to $(p, q) = (1, 0)$, is $\bar{\mathbf{3}}$, which corresponds to $(p, q) = (0, 1)$.

From the SU(3) weight diagrams for the **3** and $\bar{\mathbf{3}}$ of SU(3) in Fig. 8.3, it is evident geometrically that they are not equivalent representations.

8.5 Real and Complex Representations

A representation is *complex* if it is not equivalent to its complex conjugate representation; if it is equivalent to its complex conjugate representation it is *real* or *self-conjugate*.[1] If H_i denotes the Cartan diagonal generators of a representation, $-H_i^*$ are the Cartan diagonal

[1] Mapping of generators to their complex conjugates is an example of *automorphism*; see Section 2.5.

generators of the conjugate representation. But the eigenvalues of H_i^* and H_i are equivalent (H_i is hermitian), so if m is a weight of the representation, $-m$ is a weight of the conjugate representation. Since a representation is uniquely specified by its highest or lowest weight, this gives a prescription for determining whether a representation is complex.

A representation is real if its lowest weight is the negative of the highest weight; otherwise, it is complex.

For SU(2) this condition is always fulfilled and all SU(2) representations are real. For SU(3), the **3** does not satisfy this condition (Fig. 8.3), so $\overline{\mathbf{3}} \neq \mathbf{3}$. However, irreps of SU(3) with $p = q$ like $\mathbf{8} = (1, 1)$ are self-conjugate (real). They correspond to weight diagrams that are symmetric under reflection in both Y and T_3 (see the **8** in Fig. 8.3).

8.6 Unitary Symmetry and Young Diagrams

Following Elliott and Dawber [56], let $\phi_1, \phi_2, \ldots, \phi_N$ denote basis vectors in an N-dimensional space. A unitary group U(N) is defined by the set of unitary transformations on this space

$$\phi_j' = U\phi_j = \sum_{i=1}^{N} U_{ij}\phi_i, \tag{8.15}$$

where U is an $N \times N$ unitary matrix. We may define a product space

$$\Phi = \phi_i(1)\phi_j(2)\phi_k(3)\cdots\phi_p(n), \tag{8.16}$$

where the numbers $(1, 2, \ldots, n)$ in parentheses label *particles* and subscripts $(i, j, k, \ldots, N)$ denote *states*. Thus, Φ corresponds to an n-particle state, with each of the n particles in one of the N single-particle states of the fundamental representation of U(N). In general, there are N^n possible independent products Φ. Let us denote by $T(U)$ the operator that performs the unitary operation (8.15) on all particles in the product state given by Eq. (8.16),

$$\begin{aligned} T(U)\Phi &= \phi_i'(1)\phi_j'(2)\phi_k'(3)\cdots\phi_p'(n) \\ &= \sum_{i'j'k'\cdots p'} U_{i'i}U_{j'j}U_{k'k}\cdots U_{p'p}\phi_{i'}(1)\phi_{j'}(2)\phi_{k'}(3)\cdots\phi_{p'}(n). \end{aligned} \tag{8.17}$$

The N^n-dimensional space $\mathscr{L}$ of the products Φ is invariant under $T(U)$ and corresponds to a U(N) representation T of U(N) given by the direct product $T = U \otimes U \otimes U \otimes \cdots \otimes U$ of n matrices U, each of dimension $N \times N$. However, T is *not generally an irreducible representation of U(N).*

Our task is to rearrange the space $\mathscr{L}$ into invariant subspaces and reduce the representation T to a direct sum of U(N) irreps. This may be accomplished rather elegantly by analyzing *the permutation group* S_n. This remarkable property follows because if projection operators (Section 4.4) are used to construct subspaces $\mathscr{L}^{(\alpha)}$ that are invariant with respect to S_n, the subspace $\mathscr{L}^{(\alpha)}$ is also invariant under the product group $S_n \times U(N)$ because *permutations commute with unitary transformations.*

> If an n-particle state is an S_n irrep and it is constructed from single-particle states that are basis vectors of an N-dimensional irrep of U(N), then the state is an irreducible U(N) tensor. Therefore, the dimensionality of a U(N) irrep is the number of standard Young tableaux, and the tableaux define basis vectors of the irreps. This is also true for the general linear group of which U(N) is a subgroup, and for the subgroup SU(N).

It follows that Young diagrams may be used to deduce the irreps of U(N). However, before considering examples we note the connection between U(N) and its subgroup SU(N) described in Box 8.1. From that discussion we see that for most considerations it will be adequate to specialize from U(N) to the subgroup SU(N), which has simpler diagrams.

8.7 Young Diagrams for SU(N)

For SU(N) Young diagrams the restrictions given in Section 4.2 that numbers in the boxes (state labels) must increase down a column, and that they must be less than or equal to the number of states N, means that irreps of SU(N) correspond to diagrams with *no more than N rows.* A column with more than N boxes denotes a null SU(N) tensor because a column implies antisymmetrization and N boxes (particles) cannot be antisymmetrized with fewer than N states [the N states available in the fundamental representation of SU(N)]. But for SU(N) any columns with N boxes are irrelevant for dimensionalities because an N-box column has only one standard arrangement. In practical terms, this means that for SU(N) we require only Young diagrams with up to $N-1$ rows. If the multiplication of SU(N) representations yields Young diagrams with N or more rows, we may simply remove all columns from a diagram containing N boxes, as far as determining the dimensionality of representations is concerned. For example, in SU(3)

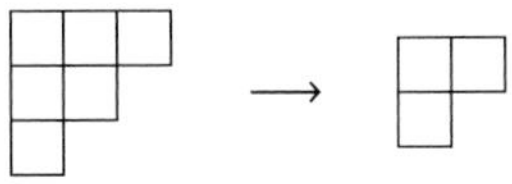

The first diagram corresponds to a six-particle SU(3) state and the second diagram to a three-particle state, but they denote representations of equivalent dimensionality.

In Ch. 4, we labeled N-row Young diagrams by the N numbers $(p_1 p_2 p_3 \cdots p_N)$, where the number of boxes in row i is l_i and $p_i \equiv l_i - l_{i-1}$. Therefore, for SU(N) the Young diagrams corresponding to irreducible representations may be specified by the set of numbers $(p_1 p_2 \cdots p_{N-1})$. Irreps of SU(2) correspond to one-row diagrams, labeled by a number p that is the number of boxes in the diagram. Irreps of SU(3) correspond to two-row diagrams, labeled by the numbers $(p_1, p_2) \equiv (\lambda, \mu)$, the irreducible representations of

Box 8.1 **Relationship between U(N) and SU(N)**

As noted in Section 3.4.3, U(N) may be written as the direct product U(N) = U(1) × SU(N), where U(1) is the group of $N \times N$ matrices of the form $\mathrm{I}e^{-i\phi}$, with I the $N \times N$ unit matrix (usually we will suppress the explicit unit matrix in the notation). From the matrix property det (AB) = det A det B, any unitary matrix U can be written

$$U = e^{-i\psi}\tilde{U},$$

where $\tilde{U}$ is a matrix with det $\tilde{U} = 1$. If U is an arbitrary element of the unitary group U(N), the matrices $\tilde{U}$ form the group SU(N) under matrix multiplication.

Irreps of the Unitary Group

The direct product of two groups has the general property that the irreducible representations of the direct product are direct products of the irreducible representations of the two groups. Therefore, the irreps of U(N) take the form

$$U^\alpha = e^{-in\phi}\tilde{U}^\alpha,$$

where n is an integer that measures particle number,[a] and $\tilde{U}^\alpha$ denotes the irreps of the subgroup SU(N). The irreps $e^{-in\phi}$ of U(1) are one-dimensional so the unitary group U(N) and the special unitary group SU(N) have *representations of the same dimensionality*. We say that U(N) does not reduce (in dimensionality) upon restriction to its subgroup SU(N). Thus, if a multiplet constitutes an invariant subspace under U(N), the full multiplet is still an invariant subspace under SU(N).

Role of the Abelian Invariant Subalgebra

This behavior is not generally the case upon restriction to a subgroup. For example, under SO(3) ⊃ SO(2) a $(2J+1)$-dimensional invariant subspace of SO(3) breaks into $2J+1$ one-dimensional invariant subspaces under SO(2). The failure of U(N) to reduce upon restriction to SU(N) is associated with the presence of a U(1) abelian invariant subalgebra in U(N).

Identification of U(N) and SU(N) Young Diagrams

It follows that the *same Young diagrams* may be used for U(N) and SU(N) if we keep track separately of the phase $e^{-in\phi}$, which is typically associated with conservation of particle number or a quantity related to it like electrical charge. For example, in Section 3.4.3 isospin was discussed as a U(2) symmetry but the only information lost upon restriction to SU(2) was the baryon number associated with the U(1) factor. Since the baryon number is usually assumed to be conserved, we typically discuss isospin as an SU(2) symmetry rather than as a full U(2) symmetry.

[a] Hence n is the number of boxes in the U(N) Young diagram. This follows from the expression for the n-particle wavefunction Φ, which has n factors U.

SU(4) are three-row diagrams specified by three integers, and so on. Now in our discussion of the weight spaces for SU(2) and SU(3) we saw that the irreps could be denoted uniquely by the quantum number J in the former case and the quantum numbers (p, q) in the second case. Therefore, these quantum numbers arising in the weight space analysis must be related to the quantum numbers p and (λ, μ) arising in the Young diagram analysis of these groups. In fact, the relationships are quite direct: $J = \frac{1}{2}p$ for SU(2) and $(p, q) = (\lambda, \mu)$ for SU(3), as may be verified by comparing equivalent expressions in the two labeling schemes. Similar identifications between quantum numbers appearing in the weight space and in the Young diagrams may be made for the higher unitary symmetries. Let us now consider some examples that will make these ideas a little more concrete [56].

8.7.1 Two Particles in Two States

Suppose that two particles are placed in a two-dimensional space (n = particles = 2; N = states = 2). This corresponds to two-particle states with U(2) symmetry, but in light of preceding considerations we may specialize to SU(2). There are $N^n = 4$ product functions, $|11\rangle$, $|12\rangle$, $|21\rangle$, and $|22\rangle$, where we adopt the compact notation

$$|ijk \cdots p\rangle \equiv \phi_i(1)\phi_j(2)\phi_k(3) \cdots \phi_p(n).$$

The irreps of the permutation group S_2 may be used to construct the irreducible representations of SU(2) states in this space. The allowed Young diagrams are

$$[2] : \begin{array}{|c|c|}\hline 1 & 1 \\ \hline\end{array} \quad \begin{array}{|c|c|}\hline 1 & 2 \\ \hline\end{array} \quad \begin{array}{|c|c|}\hline 2 & 2 \\ \hline\end{array} \qquad [11] : \begin{array}{|c|}\hline 1 \\ \hline 2 \\ \hline\end{array}\,.$$

Applying the row symmetrizer for the symmetric diagrams, the column antisymmetrizer for the antisymmetric diagrams, and normalizing (see Section 4.4), we may construct the basis vectors for the two subspaces $L^{[\alpha]}$, where α denotes the partition,

$$L^{[2]} : \left(|11\rangle, \frac{1}{\sqrt{2}}(\,|12\rangle + |21\rangle\,), |22\rangle \right) \qquad L^{[11]} : \frac{1}{\sqrt{2}}(\,|12\rangle - |21\rangle\,).$$

Thus the partition [2] is a **3** under SU(2) but it is a 1D representation of S_2, because the dimensionality of S_n is determined by counting only standard tableaux with each particle in a different state (Section 4.3.1). The partition [11] is a 1D irrep for both SU(2) and S_2.

Example 8.3 The physical content of this result may be appreciated by considering SU(2) explicitly as an angular momentum symmetry. For two spin-$\frac{1}{2}$ electrons [fundamental doublet of SU(2)], there are two states available, corresponding to spin projections $m_s = \pm\frac{1}{2}$. With respect to permutation symmetry, two electrons having only spin internal degrees of freedom can form either a symmetric or an antisymmetric state, one way each. Hence there are two S_2 irreps, each of dimension one. With respect to the SU(2) symmetry, the two electrons can couple to angular momentum $S = 1$ or $S = 0$. The first case corresponds to a $2S + 1 = 3$-dimensional representation and the second to a $2S + 1 = 1$-dimensional representation of SU(2). Under permutation each of the three wavefunctions in the $S = 1$ representation $L^{[2]}$ is symmetric, while the wavefunction of the $S = 0$ representation $L^{[11]}$ is antisymmetric.

8.7.2 Two Particles in Three States

As a second example, consider two particles, but with three states available ($n = 2$; $N = 3$). The symmetry is U(3) $\supset$ SU(3) and we restrict to the SU(3) subgroup by the previous arguments. There are now $N^n = 3^2 = 9$ two-particle states in the product basis,

$$|11\rangle \quad |12\rangle \quad |13\rangle \quad |21\rangle \quad |22\rangle \quad |23\rangle \quad |31\rangle \quad |32\rangle \quad |33\rangle .$$

The allowed Young diagrams correspond to a six-dimensional symmetric representation and a three-dimensional antisymmetric representation of SU(3),

$$\text{Symmetric:} \quad \begin{array}{|c|c|}\hline 1&1\\\hline\end{array} \quad \begin{array}{|c|c|}\hline 1&2\\\hline\end{array} \quad \begin{array}{|c|c|}\hline 1&3\\\hline\end{array} \quad \begin{array}{|c|c|}\hline 2&2\\\hline\end{array} \quad \begin{array}{|c|c|}\hline 2&3\\\hline\end{array} \quad \begin{array}{|c|c|}\hline 3&3\\\hline\end{array},$$

$$\text{Antisymmetric:} \quad \begin{array}{|c|}\hline 1\\\hline 2\\\hline\end{array} \quad \begin{array}{|c|}\hline 1\\\hline 3\\\hline\end{array} \quad \begin{array}{|c|}\hline 2\\\hline 3\\\hline\end{array}.$$

The corresponding basis vectors are

$$L^{[2]}: \quad \left\{ \begin{array}{l} |11\rangle, \ \dfrac{1}{\sqrt{2}}(\,|12\rangle + |21\rangle\,), \ \dfrac{1}{\sqrt{2}}(\,|13\rangle + |31\rangle\,), \\ |22\rangle, \ \dfrac{1}{\sqrt{2}}(\,|23\rangle + |32\rangle\,), \ |33\rangle \end{array} \right.$$

$$L^{[11]}: \quad \left(\frac{1}{\sqrt{2}}(\,|12\rangle - |21\rangle\,), \ \frac{1}{\sqrt{2}}(\,|13\rangle - |31\rangle\,), \ \frac{1}{\sqrt{2}}(\,|23\rangle - |32\rangle\,) \right).$$

Therefore, the symmetric **3** of SU(2) becomes a symmetric **6** in SU(3), and the antisymmetric SU(2) singlet becomes a **3** in SU(3).

8.7.3 Fundamental and Conjugate Representations

Fundamental and conjugate SU(N) representations are denoted by a single box and a column of boxes with $N - 1$ rows, respectively,

$$\text{SU}(N)\ \text{fundamental:} \ \begin{array}{|c|}\hline \\\hline\end{array} \qquad \text{SU}(N)\ \text{conjugate:} \ \begin{array}{|c|}\hline 1\\\hline 2\\\hline \vdots\\\hline N-1\\\hline\end{array}$$

For example, the fundamental and conjugate representations are equivalent, $\begin{array}{|c|}\hline \\\hline\end{array} = \mathbf{2} = \bar{\mathbf{2}}$, for SU(2), but the fundamental and conjugate representations,

$$\begin{array}{|c|}\hline \\\hline\end{array} = \mathbf{3} \qquad \begin{array}{|c|}\hline \\\hline \\\hline\end{array} = \bar{\mathbf{3}}$$

are distinguishable for SU(3).

8.8 Dimensionality of SU(N) Representations

The dimensionality of an SU(N) representation is determined by counting the number of distinct standard Young diagrams that can be drawn subject to the limitation that

- no numbers larger than N appear in the boxes, and
- there are no columns with more than $N - 1$ boxes.

In the simple examples considered above the dimensionality followed easily from counting. For more complicated diagrams counting can become tedious. We now give without proof a faster dimensionality algorithm. The dimensionality of an SU(N) representation corresponding to a particular Young diagram may be determined by the following procedure [42].

1. For an SU(N) diagram, insert the number N in each box along the main diagonal. Then insert the numbers $N + 1$, $N + 2$, ... along each successive diagonal to the right of the main diagonal, and the numbers $N - 1$, $N - 2$, ... along each successive diagonal to the left of the main diagonal. This is continued until each box has a number. For example,

N	N+1	N+2
$N-1$	N	N+1
$N-2$	$N-1$	N
$N-3$		

Now define an integer $\mathscr{N}$ that is the product of all numbers in the boxes of the diagram,

$$\mathscr{N} = \prod_{\text{boxes}} (\text{numbers in the boxes}). \tag{8.18}$$

2. For each box in the diagram, define a hook number h_i, as described in Section 4.3.2, and define an integer $\mathscr{D}$ that is the product of all the hook numbers in the boxes,

$$\mathscr{D} = \prod_{\text{boxes}} (\text{hook numbers}). \tag{8.19}$$

3. The dimensionality of the SU(N) representation is then given by the ratio

$$\text{Dim (rep)} = \frac{\mathscr{N}}{\mathscr{D}}. \tag{8.20}$$

Example 8.4 illustrates application of this procedure to determine the dimensionality of irreps for special unitary groups.

Example 8.4 In a notation inspired by the preceding discussion,

$$\text{Dim}\left(\square\square\right) = \frac{\mathscr{N}}{\mathscr{D}} = \frac{\boxed{N}\boxed{N{+}1}}{\boxed{2}\boxed{1}} = \frac{N \times (N+1)}{2 \times 1} = \frac{N(N+1)}{2},$$

so that $\square\square$ = **3** for SU(2) but $\square\square$ = **6** for SU(3), in agreement with our previous results.

8.9 Direct Products of SU(N) Representations

In physical applications of SU(N) symmetries we must be able to construct the Clebsch–Gordan series (3.21), which expresses the direct product of representations as a direct sum over irreducible representations. There is an easy way to do this based upon the close connection between permutation groups and unitary groups.

> Finding the Clebsch–Gordan series corresponding to direct products of U(N) representations is equivalent to determining the irrep content of the *outer product* for the permutation group, which we demonstrated in Section 4.5.2.

In this section we give a prescription for decomposing the direct product of SU(N) representations that exploits this relationship with the permutation groups. The Clebsch–Gordan series may be constructed by multiplying Young diagrams, subject to the following rules [85].

1. Draw the diagrams for the two representations to be multiplied and label the second diagram with numbers in the boxes giving the row in which the box occurs. For example,

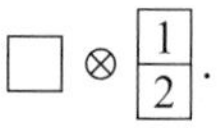

   Break the second diagram up into individual boxes and attach each box of the second diagram to the first diagram in all possible ways such that the following are satisfied.

   a. All diagrams are proper (rows do not increase in length from top to bottom).
   b. No column has more than N boxes for SU(N).
   c. On a path through each row from right to left and from top to bottom through the diagram, at each point the number of boxes encountered containing the number i is less than or equal to the number of boxes encountered containing the number $i - 1$.
   d. The numbers in the boxes do not decrease from left to right in a row.
   e. The numbers in a column increase from top to bottom.

2. Each diagram constructed in this manner corresponds to an irrep of SU(N) occurring in the direct product.

The following example illustrates this procedure.

Example 8.5 Let us construct the Clebsch–Gordan series for the product $\mathbf{8} \otimes \mathbf{8}$ in SU(3). The $\mathbf{8}$ corresponds to $(\lambda, \mu) = (1, 1)$, so by the prescription given above,

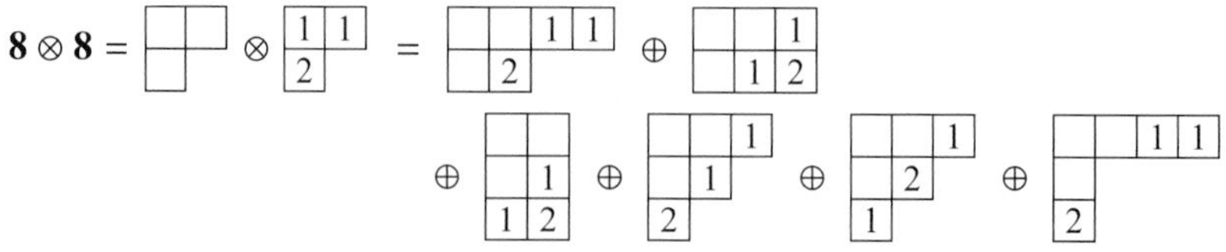

For representation dimensionality in SU(3) the 3-box columns may be omitted from each diagram and the Clebsch–Gordan series in terms of Young diagrams is

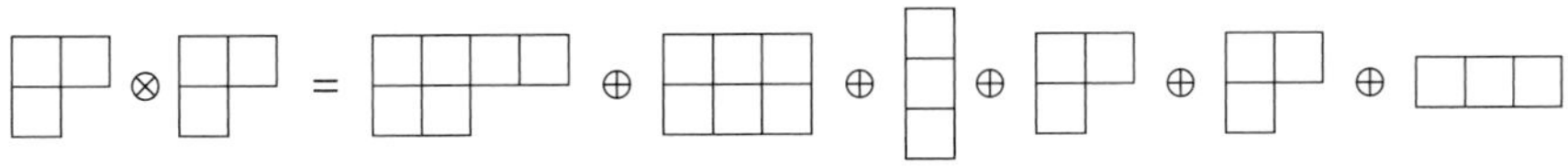

This may also be written in terms of dimensionalities,

$$\mathbf{8} \otimes \mathbf{8} = \mathbf{27} \oplus \overline{\mathbf{10}} \oplus \mathbf{1} \oplus \mathbf{8} \oplus \mathbf{8} \oplus \mathbf{10},$$

or as the series

$$(1,1) \otimes (1,1) = (2,2) \oplus (0,3) \oplus (0,0) \oplus (1,1) \oplus (1,1) \oplus (3,0),$$

in terms of the ordered pair of numbers (λ, μ) labeling irreps.

Direct products of other unitary representations may be constructed by a procedure similar to that of Example 8.5.

8.10 Weights from Young Diagrams

Because weights are additive quantum numbers, Young diagrams allow us to deduce weight diagrams for multiparticle states. For example, consider the two-particle SU(3) states resulting from $\mathbf{3} \otimes \mathbf{3} = \mathbf{6} \oplus \bar{\mathbf{3}}$. The weights $(T_3, \sqrt{3}Y/2)$ of the fundamental representation are

$$m\left(\boxed{1}\right) = \left(\frac{1}{2}, \frac{1}{2\sqrt{3}}\right) = \left(T_3, \frac{\sqrt{3}}{2}Y\right),$$

$$m\left(\boxed{2}\right) = \left(-\frac{1}{2}, \frac{1}{2\sqrt{3}}\right) \qquad m\left(\boxed{3}\right) = \left(0, -\frac{1}{\sqrt{3}}\right).$$

Then the weights of the symmetric $\mathbf{6}$ are sums of weights for individual particles

$$m\left(\boxed{1}\boxed{1}\right) = \left(1, \frac{1}{\sqrt{3}}\right) \quad m\left(\boxed{1}\boxed{2}\right) = \left(0, \frac{1}{\sqrt{3}}\right) \quad m\left(\boxed{2}\boxed{2}\right) = \left(-1, \frac{1}{\sqrt{3}}\right),$$

$$m\left(\boxed{1}\boxed{3}\right) = \left(\frac{1}{2}, -\frac{1}{2\sqrt{3}}\right) \quad m\left(\boxed{2}\boxed{3}\right) = \left(-\frac{1}{2}, -\frac{1}{2\sqrt{3}}\right) \quad m\left(\boxed{3}\boxed{3}\right) = \left(0, -\frac{2}{\sqrt{3}}\right),$$

and for the antisymmetric $\bar{\mathbf{3}}$,

$$m\left(\begin{array}{|c|}\hline 1 \\ \hline 2 \\ \hline\end{array}\right) = \left(0, \frac{1}{\sqrt{3}}\right) \quad m\left(\begin{array}{|c|}\hline 1 \\ \hline 3 \\ \hline\end{array}\right) = \left(\frac{1}{2}, -\frac{1}{2\sqrt{3}}\right) \quad m\left(\begin{array}{|c|}\hline 2 \\ \hline 3 \\ \hline\end{array}\right) = \left(-\frac{1}{2}, -\frac{1}{2\sqrt{3}}\right).$$

These results are displayed in the SU(3) weight plane in Fig. 8.4.

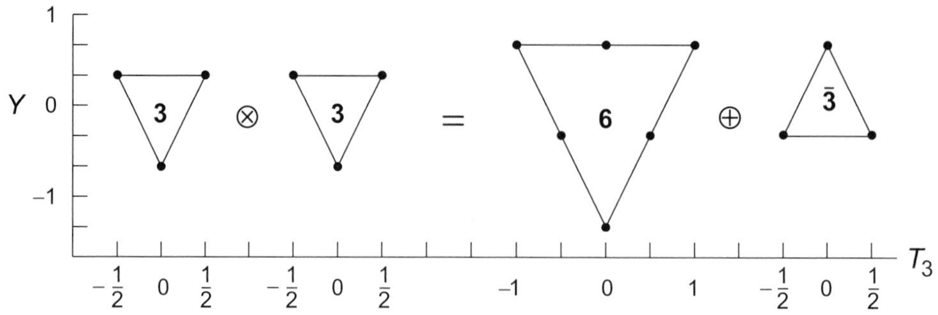

Fig. 8.4 Weight space diagrams for the direct product of fundamental SU(3) representations constructed using Young diagrams.

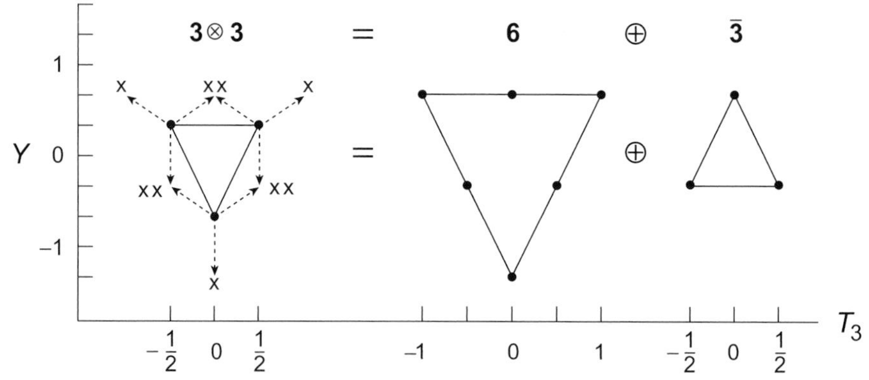

Fig. 8.5 Graphical construction of the SU(3) direct product $\mathbf{3} \otimes \mathbf{3} = \mathbf{6} \oplus \bar{\mathbf{3}}$. Each occupation of a weight site in the product is indicated by an x.

8.11 Graphical Construction of Direct Products

Additivity of the weights also suggests a graphical means for the construction of direct products that is illustrated in Fig. 8.5. The occupied sites in the weight diagram for the direct product may be obtained by placing at each occupied site of the first representation a set of vectors defining the shape of the second representation. Then each vector tip implies an occupation of the corresponding site in the direct product. For simple cases it is then easy to see how to decompose these occupations uniquely into a sum of irreducible representations, as illustrated in Fig. 8.5.

Background and Further Reading

Elliott and Dawber [56] present a general discussion of unitary symmetries. The SU(3) group is analyzed with an emphasis on physical applications in Close [42], Gasiorowicz [66], and Lichtenberg [141].

Problems

8.1 Use the SU(3) algebra to prove that $T_\pm$, $V_\pm$, and $U_\pm$ have the raising and lowering properties in the (T_3, Y) plane that we have ascribed to them. Prove that the allowed values of Y and T_3 are indicated by the dots shown in the following diagram

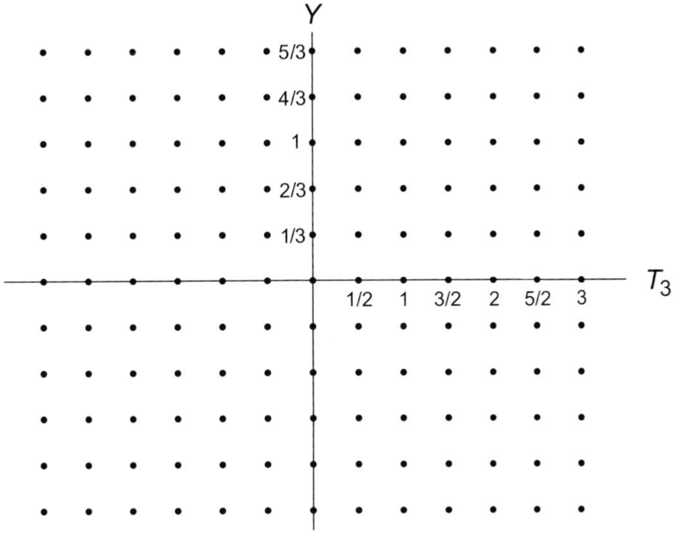

but that the basic principles of representation theory require that the possible occupied sites in the SU(3) irreps be further restricted to those marked by a heavy dot in the following diagram.

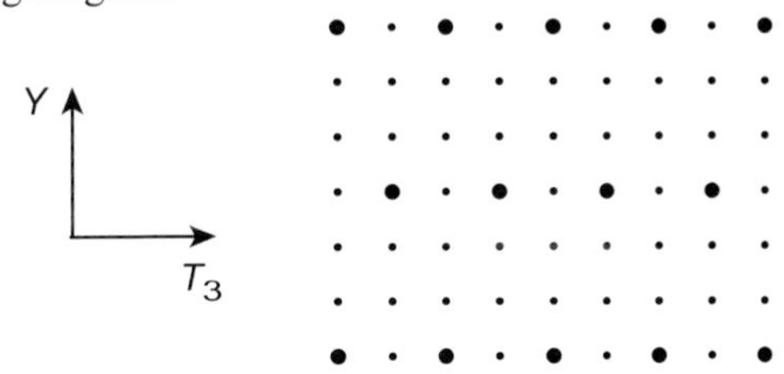

Hint: One way to proceed is to generalize the proof of the analogous properties of the SU(2) generators $J_\pm$. ***

8.2 Prove that for SU(2) symmetry $\mathbf{2} \otimes \mathbf{2} \otimes \mathbf{2} = \mathbf{4} \oplus \mathbf{2} \oplus \mathbf{2}$, while for SU(3) symmetry

$$\mathbf{3} \otimes \mathbf{3} \otimes \mathbf{3} = \mathbf{10} \oplus \mathbf{8} \oplus \mathbf{8} \oplus \mathbf{1} \qquad \bar{\mathbf{3}} \otimes \mathbf{3} = \mathbf{8} \oplus \mathbf{1} \qquad \mathbf{3} \otimes \mathbf{3} = \mathbf{6} \oplus \mathbf{3}.$$

What is the irrep content of $\mathbf{8} \otimes \mathbf{8} \otimes \mathbf{8}$ in SU(3)?

8.3 Determine the SU(3) irreps appearing in the direct product $(2,1) \otimes (3,0)$. Label them by their (λ, μ) quantum numbers and dimensionality. ***

8.4 Use the method of Young diagrams to find the irrep content of $\mathbf{6} \otimes \mathbf{10}$ for SU(3).

8.5 Show that the SU(3) quadratic Casimir operator (8.9) can be written as

$$\langle F^2 \rangle = \frac{1}{3}(p^2 + pq + q^2) + p + q,$$

where the integers p and q labeling representations are defined in Section 8.3.2.

8.6 A representation is said to be complex if it is equivalent to its complex conjugate representation. Show that the **2** and $\bar{\mathbf{2}}$ representations are equivalent for SU(2) but $\mathbf{3} \neq \bar{\mathbf{3}}$ for SU(3). *Hint*: Look at the weight space.

8.7 Use the graphical method described in Section 8.11 to find the direct product $\mathbf{3} \otimes \bar{\mathbf{3}}$ for SU(3).

9 SU(3) Flavor Symmetry

The rank-2 Lie algebra and associated group SU(3) were introduced in Ch. 8. Although the algebra and group structure of SU(3) can be developed without reference to physics, it is easiest to illustrate its features through concrete applications to important physical problems. In this chapter we examine the pivotal contribution of SU(3) symmetry in imposing order on the phenomenology of strongly interacting elementary particles.

9.1 Symmetry in Particle Physics

The historical motivation for application of SU(3) in high-energy physics was the proliferation of known elementary particles in the 1950s and 1960s, and the associated difficulty of constructing a rational scheme to account for why there should be so many "elementary" particles. One approach was to view the expanding zoo of new particles that were being discovered as a manifestation of some underlying symmetry. In particular, the success of isospin in nuclear and particle physics motivated an attempt to extend isospin symmetry, with its single additive quantum number T_3, to a larger symmetry with two additive quantum numbers (that is, a two-element Cartan subalgebra), with isospin as a subgroup.

9.1.1 SU(3) Phenomenology and Quarks

The rationale for a group with two additive quantum numbers was that both T_3 and the hypercharge Y (alternatively the strangeness S) were observed to be rather good quantum numbers for the hadrons that were known in the 1960s. Two additive quantum numbers require a rank-2 algebra and in the simplest case the algebra can be expected to be compact and semisimple. Three Lie algebras satisfy these requirements (see the classification of the rank-2 algebras in Ch. 7), but of these only SU(3) was consistent with the detailed properties of known particles. This symmetry was called *flavor SU(3).*

Historically flavor SU(3) symmetry was proposed as a phenomenology that allowed progress until a more detailed microscopic theory was available. The foundations of that more microscopic theory were laid with the proposal that the SU(3) hadronic symmetry could be built from fundamental triplets of unobserved elementary particles called *quarks* and *antiquarks,* with the quarks transforming as the **3** and the antiquarks as the $\bar{\mathbf{3}}$ representation of flavor SU(3). Although free quarks have never been seen, that proposal gained increasing credibility until the discovery of the J/ψ particle in the mid-1970s, which implied the existence of a fourth flavor quantum number called *charm* for quarks, converted almost all of particle physics to a belief in the essential correctness of the quark hypothesis.

9.1.2 Non-Abelian Gauge Symmetries

The development of non-abelian gauge theories provided a dynamical scheme in which the original SU(3) flavor symmetries may be understood. From *quantum chromodynamics* (QCD), the gauge theory of the strong interactions discussed in Section 19.2, the SU(3) flavor symmetries are now understood to be a kind of "accident" resulting from the similar effective masses of the lightest quarks. Nevertheless, the SU(3) flavor symmetries that we discuss in this chapter retain phenomenological importance in the study of the strong interactions, and are of considerable utility in understanding the weak interactions.

It is important to appreciate that the flavor model discussed here, and the QCD theory to be presented in Section 19.2, both exhibit SU(3) symmetry but *in different spaces.* Quantum chromodynamics operates on a *color* degree of freedom for the quarks, while the theory introduced in this chapter involves a separate *flavor* degree of freedom. There is historical precedent for such dual roles for a symmetry group. The proton and neutron have an SU(2) symmetry associated with their spin and a *different* SU(2) symmetry associated with their isospin. These symmetries correspond mathematically to the same Lie group but physically they have nothing to do with each other: generators of spin SU(2) and generators of isospin SU(2) involve degrees of freedom that are physically distinct. Likewise flavor SU(3) and color SU(3) are isomorphic mathematically but are unconnected physically.

9.2 Fundamental SU(3) Quark Representations

We assume that the lighter hadrons may be constructed from a fundamental triplet of quarks q and a fundamental triplet of antiquarks $\bar{q}$, denoted by

$$q = \begin{pmatrix} u \\ d \\ s \end{pmatrix} = \begin{pmatrix} \text{up} \\ \text{down} \\ \text{strange} \end{pmatrix} \qquad \bar{q} = \begin{pmatrix} \bar{u} \\ \bar{d} \\ \bar{s} \end{pmatrix} = \begin{pmatrix} \text{anti-up} \\ \text{anti-down} \\ \text{anti-strange} \end{pmatrix}. \tag{9.1}$$

Table 9.1 lists the quantum numbers that must be assigned to these quarks and antiquarks if this hypothesis is to be consistent with known properties of the hadrons (we also list the attributes of the charmed quark c, the bottom quark b, and the top quark t, which enter into subsequent discussions but will be ignored in this chapter).[1] The labels (u, d, s) will be termed *flavor indices.* The empirical relationships among charge Q, baryon number B, hypercharge Y, strangeness S, and the third component of isospin T_3 are

[1] Specifically, the present chapter will restrict attention to hadrons with masses below the mass of the J/ψ particle at about 3 GeV, so that charm and heavier quark flavors are not important in the valence structure.

Table 9.1. Quantum number assignments for quarks [85]

	Up	Down	Strange	Charm	Bottom	Top
Symbol	u	d	s	c	b	t
Baryon number (B)	$\frac{1}{3}$	$\frac{1}{3}$	$\frac{1}{3}$	$\frac{1}{3}$	$\frac{1}{3}$	$\frac{1}{3}$
Spin	$\frac{1}{2}$	$\frac{1}{2}$	$\frac{1}{2}$	$\frac{1}{2}$	$\frac{1}{2}$	$\frac{1}{2}$
Charge (Q)	$\frac{2}{3}$	$-\frac{1}{3}$	$-\frac{1}{3}$	$\frac{2}{3}$	$-\frac{1}{3}$	$\frac{2}{3}$
Isospin (T)	$\frac{1}{2}$	$\frac{1}{2}$	0	0	0	0
T_3	$\frac{1}{2}$	$-\frac{1}{2}$	0	0	0	0
Strangeness number (S)	0	0	−1	0	0	0
Charm number (c)	0	0	0	1	0	0
Bottom number (b)	0	0	0	0	−1	0
Top number (t)	0	0	0	0	0	1

The additive quantum numbers Q, T_3, S, c, B, b, and t of the corresponding antiquarks are the negative of those for the quarks. The charge is given by $Q = T_3 + \frac{1}{2}(B + S + c + b + t)$.

$$Q = T_3 + \frac{B+S}{2} = T_3 + \frac{Y}{2} \qquad Y \equiv B + S. \tag{9.2}$$

We will associate the quantum numbers T_3 and Y with the eigenvalues of the two diagonal generators for the SU(3) matrices (8.7)–(8.8). From Table 9.1, the operator $\hat{T}_3$ applied to the fundamental triplet (9.1) gives a quantum number T_3, while the operator $\hat{Y}$ has the hypercharge Y as an eigenvalue for the fundamental triplet of quarks. Therefore,

$$Q = \frac{\lambda_3}{2} + \frac{\lambda_8}{2\sqrt{3}}, \tag{9.3}$$

is the SU(3) charge operator.

9.3 SU(3) Flavor Multiplets

Suppose that the known low-mass hadrons are plotted as points in the Y–T_3 plane, grouped according to baryon number and spin, as in Fig. 9.1. We have seen such diagrams before: they have the familiar shapes of some of the SU(3) representations discussed in Ch. 8. This suggests that SU(3) might be a useful symmetry for classifying hadrons, provided that the mass splittings within these multiplets is not too large, since in the symmetry limit we would expect all members of an SU(3) flavor multiplet to have the same mass (energy).

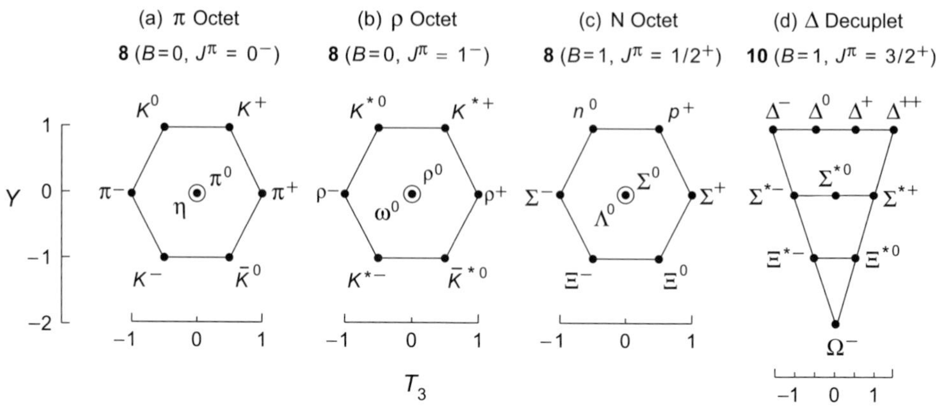

Fig. 9.1 Some SU(3) flavor multiplets labeled by baryon number B, spin J, and parity π [85]. The Σ^* is now called $\Sigma(1385)$ and Ξ^* is now $\Xi(1530)$. Reproduced with permission from Wiley Interscience: *Gauge Field Theories – An Introduction with Applications,* M. Guidry (1991).

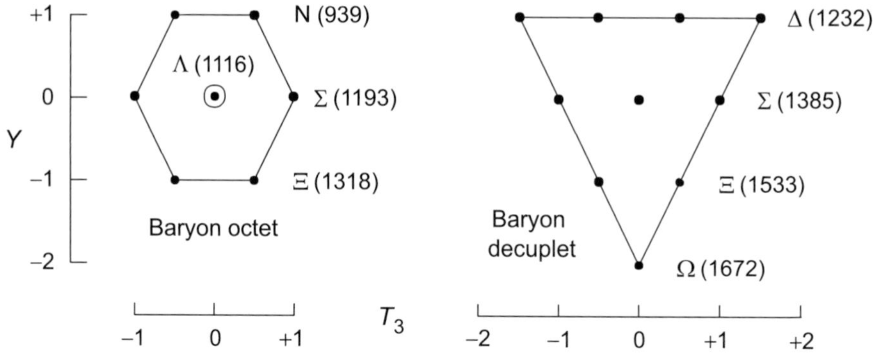

Fig. 9.2 Average mass of isospin multiplets in the $J^\pi = \frac{1}{2}^+$ baryon **8** and the $J^\pi = \frac{3}{2}^+$ baryon **10**.

9.3.1 Mass Splittings in SU(3) Multiplets

In fact, masses within such multiplets are often similar. Figure 9.2 displays the $J = \frac{1}{2}$ baryon octet (**8**) and the $J = \frac{3}{2}$ baryon decuplet (**10**). Horizontal rows are associated with different values of T_3 connected by SU(2) raising and lowering operators $T_\pm$; thus, they correspond to isospin multiplets. For example, the top row of the baryon octet is the nucleon isospin doublet and the middle row corresponds to two isospin multiplets: the Σ triplet and the Λ singlet. The average mass of the isospin multiplet is given for each horizontal row. The mass deviation between members of each isospin multiplet is usually less than a few percent. The average mass of each SU(2) isospin multiplet typically differs from the average mass of the entire SU(3) multiplet by 10–20%. Although this is considerably larger than the mass deviation within isospin multiplets, it is a sufficiently small difference to suggest that SU(3) flavor is a useful approximate symmetry of the strong interactions, representing a generalization of the isospin symmetry to encompass the hypercharge degree of freedom.

9.3.2 Quark Structure for Mesons and Baryons

The hadrons may be grouped into *mesons,* with baryon number zero (Fig. 9.1(a,b)), and *baryons,* with non-zero baryon number (Fig. 9.1(c,d)). The systematics of the SU(3) classification require **8**s and **10**s for baryon representations and **8**s and **1**s for the meson representations. In Ch. 8 we found that $\mathbf{3}\otimes\mathbf{3}\otimes\mathbf{3}=\mathbf{10}\oplus\mathbf{8}\oplus\mathbf{8}$ and $\bar{\mathbf{3}}\otimes\mathbf{3}=\mathbf{8}\oplus\mathbf{1}$ for SU(3). Therefore, identification of the basic triplet of quarks with the fundamental representation and the triplet of antiquarks with the conjugate representation suggests a valence quark structure qqq for baryons and $q\bar{q}$ for mesons, where q and $\bar{q}$ stand for the three flavors of quark or antiquark, respectively. The quark and antiquark triplets have the weight diagrams displayed in Fig. 9.3 (see also Table 9.1), and the resulting quark content of some hadrons is displayed in Fig. 9.4. Let us illustrate construction of the middle diagram in Fig. 9.4 for the nucleon **8**. The allowed Young diagrams are

$$\begin{array}{|c|c|}\hline 1&1\\ \hline 2\\ \cline{1-1}\end{array}\quad \begin{array}{|c|c|}\hline 1&2\\ \hline 2\\ \cline{1-1}\end{array}\quad \begin{array}{|c|c|}\hline 1&3\\ \hline 2\\ \cline{1-1}\end{array}\quad \begin{array}{|c|c|}\hline 1&1\\ \hline 3\\ \cline{1-1}\end{array}\quad \begin{array}{|c|c|}\hline 1&2\\ \hline 3\\ \cline{1-1}\end{array}\quad \begin{array}{|c|c|}\hline 1&3\\ \hline 3\\ \cline{1-1}\end{array}\quad \begin{array}{|c|c|}\hline 2&2\\ \hline 3\\ \cline{1-1}\end{array}\quad \begin{array}{|c|c|}\hline 2&3\\ \hline 3\\ \cline{1-1}\end{array}$$

and the weights $m=(T_3,Y)$ are additive with

$$m(\boxed{1})=\left(\frac{1}{2},\frac{1}{3}\right)\qquad m(\boxed{2})=\left(-\frac{1}{2},\frac{1}{3}\right)\qquad m(\boxed{3})=\left(0,-\frac{2}{3}\right).$$

Hence the weights for the baryon octet are those given in Table 9.2, where the quark assignments come from the identifications

$$m(\boxed{1})\longleftrightarrow u\qquad m(\boxed{2})\longleftrightarrow d\qquad m(\boxed{3})\longleftrightarrow s,$$

between the weights m and the quarks u, d, and s.

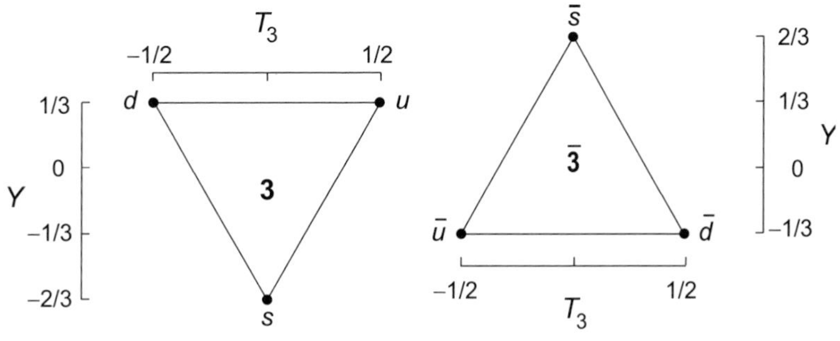

Fig. 9.3 Fundamental and conjugate quark representations for SU(3) flavor.

Table 9.2. Weights for the baryon octet

Quarks	Diagram	Weights (T_3, Y)
uud	$\begin{smallmatrix}1&1\\2&\end{smallmatrix}$	$2m(\boxed{1}) + m(\boxed{2}) = \left(\frac{1}{2}, 1\right)$
udd	$\begin{smallmatrix}1&2\\2&\end{smallmatrix}$	$m(\boxed{1}) + 2m(\boxed{2}) = \left(-\frac{1}{2}, 1\right)$
uds	$\begin{smallmatrix}1&3\\2&\end{smallmatrix}$	$m(\boxed{1}) + m(\boxed{2}) + m(\boxed{3}) = (0, 0)$
uus	$\begin{smallmatrix}1&1\\3&\end{smallmatrix}$	$2m(\boxed{1}) + m(\boxed{3}) = (1, 0)$
uds	$\begin{smallmatrix}1&2\\3&\end{smallmatrix}$	$m(\boxed{1}) + m(\boxed{2}) + m(\boxed{3}) = (0, 0)$
uss	$\begin{smallmatrix}1&3\\3&\end{smallmatrix}$	$m(\boxed{1}) + 2m(\boxed{3}) = \left(\frac{1}{2}, -1\right)$
dds	$\begin{smallmatrix}2&2\\3&\end{smallmatrix}$	$2m(\boxed{2}) + m(\boxed{3}) = (-1, 0)$
dss	$\begin{smallmatrix}2&3\\3&\end{smallmatrix}$	$m(\boxed{2}) + 2m(\boxed{3}) = \left(-\frac{1}{2}, -1\right)$

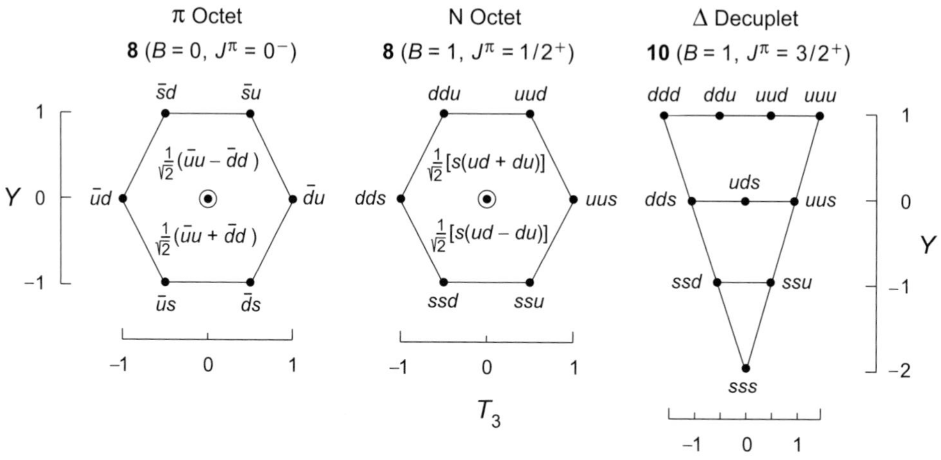

Fig. 9.4 Quark content of the SU(3) **8** and **10** [85]. The notation is shorthand for properly symmetrized expressions (see Ch. 4 and Problem 9.4). For example, *udd* stands for the symmetric flavor wavefunction $(udd + ddu + dud)/\sqrt{3}$. Reproduced with permission from Wiley Interscience: *Gauge Field Theories – An Introduction with Applications,* M. Guidry (1991).

9.4 Isospin Subgroups of SU(3)

The analysis of subgroups is important for several reasons.

1. Mathematically, subgroup chains can be useful in the elucidation of the group structure. For example, a chain of subgroups provides quantum numbers for labeling states.

2. If subgroups are chosen judiciously (except for the simplest groups, there is usually more than one subgroup chain), they may attain a direct physical significance.

One method of obtaining a subgroup is to restrict the set of generators. For example, the subgroup SO(2) is generated by any one of the three SO(3) generators J_1, J_2, or J_3. If we wish to determine an intermediate subgroup G_x in a chain $G_a \supset \cdots \supset G_x \supset \cdots \supset G_b$, our task is to find a subset of generators for G_a that generates a group G_x and, at the same time, contains as a subset the full complement of generators for the group G_b. A specific example occurs for systems in which angular momentum is conserved. Then the physical group chains must contain an SO(3) group generated by operators that can be identified with the physical angular momentum, and the parent group and subsequent subgroups must number these angular momentum generators among their generators if angular momentum is conserved by all states.

Example 9.1 Consider the angular momentum subgroup chain

$$\underset{J,\,M}{\mathrm{SO}(3)} \supset \underset{M}{\mathrm{SO}(2)},$$

where the quantum numbers that label the irreducible representations of the group are listed below the groups. The chain provides quantum numbers (J, M) that both label the states and provide a mathematical description of the group properties. In addition, the subgroup quantum number M attains a palpable meaning if a preferred direction in space is established, such as by a magnetic field. The physical significance is that at the SO(3) level there are no preferred directions (isotropy of 3D space), but at the SO(2) level only symmetry under continuous rotations about a single axis remains (isotropy only in a plane).[2]

Example 9.2 Consider the isospin subgroup chain (see Section 3.4),

$$\underset{B,T,T_3}{\mathrm{U}(2)} \supset \underset{B,T,T_3}{\mathrm{U}(1)_B \times \mathrm{SU}(2)_T} \supset \underset{B,T_3}{\mathrm{U}(1)_\mathrm{B} \times \mathrm{U}(1)_\mathrm{Q}}$$

where B, T, and T_3 denote the baryon number, isospin, and third component of isospin, respectively. At the U(2) level and at the intermediate $\mathrm{U}(1)_B \times \mathrm{SU}(2)_T$ level, B, T, and T_3 are conserved, but at the final $\mathrm{U}(1)_B \times \mathrm{U}(1)_{T_3}$ level only the baryon number B and third component of isospin (or equivalently charge Q, since $T_3 = Q - \frac{1}{2}B$) remain conserved.

9.4.1 Subgroup Analysis Using Weight Diagrams

Inspection of the SU(3) commutation relations in Table 8.1 reveals several SU(2) subgroups that are associated with the generator sets $(U_\pm, U_3)$, $(V_\pm, V_3)$, and $(T_\pm, T_3)$, respectively. We have seen evidence of the mathematical significance associated with these

[2] The subgroup decomposition might be carried further by restricting to a finite subgroup of SO(2) consisting of rotation through discrete angles (for example, the cyclic groups C_2, C_3, and C_6 studied in Example 2.18). This is of physical relevance for an atom embedded in a crystal of finite point-group symmetry, where even the symmetry for continuous rotations about a single axis might be lost because of crystal-field anisotropy and only symmetry under rotation by certain discrete angles retained.

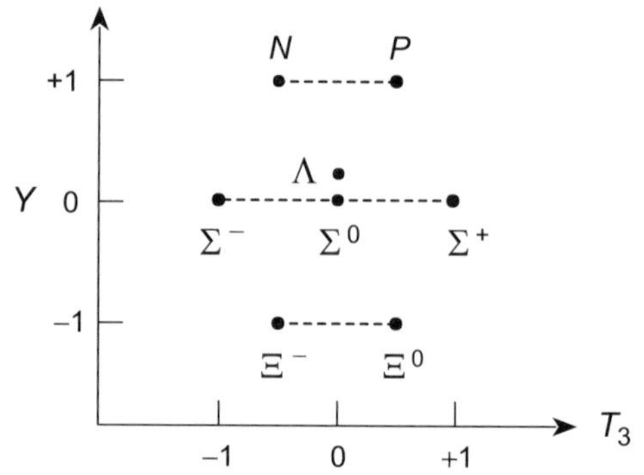

Fig. 9.5 Isospin subgroups $\{N, P\}$, $\{\Lambda\}$, $\{\Sigma^-, \Sigma^0, \Sigma^+\}$, and $\{\Xi^-, \Xi^0\}$ of the SU(3) flavor octet.

subgroups. For example, in Ch. 8 these subgroups were used to determine the symmetry of SU(3) representations under reflection about various axes in the weight space. This is of great utility in constructing weight diagrams for SU(3) representations. In applications of SU(3) to flavor symmetries, these SU(2) subgroups attain physical significance as well, because of their connection to the physical quantum numbers for isospin and hypercharge. Let us consider the isospin subgroups of SU(3) that are generated by $(T_\pm, T_3)$. We may find these easily by consulting the weight diagrams. Since restricting the full complement of SU(3) generators to the set $(T_\pm, T_3)$ leads to SU(2) weight spaces that are lines parallel to the T_3-axis, SU(2) isospin multiplets may be read directly from the sites in horizontal rows of flavor SU(3) diagrams.

Example 9.3 The octet representation of flavor SU(3) is displayed in Fig. 9.5. From the algebraic properties of the operators $T_\pm$ (recall the discussion in Section 3.3), we find immediately that the top and bottom rows correspond to isospin doublets ($T = \frac{1}{2}$ and $T_3 = \pm\frac{1}{2}$). Since the middle row contains a multiply occupied site, it cannot represent a single irrep of SU(2). It clearly corresponds to the direct sum of an SU(2) scalar ($T = 0$) and an SU(2) triplet ($T = 1$), as we could show formally by applying the highest-weight algorithm for SU(2) to the horizontal rows. Hence, the SU(3) irrep **8** contains the SU(2) irreps **1**, **2** (twice), and **3** associated with the isospin subgroup.

9.4.2 Subgroup Analysis Using Young Diagrams

Subgroup analysis of unitary groups may also be carried out in terms of Young diagrams. The use of Young diagrams has the advantage that it generalizes readily to all the unitary groups, whereas the method just discussed in Section 9.4.1 requires detailed knowledge of the weight space for each particular group. Determination of the SU($N-1$) content of an SU(N) diagram hinges on the observation that the Young diagrams for the subgroup SU($N-1$) cannot contain a box with the number N. The rule is to place Ns in the boxes consistent with the standard arrangements from no Ns up to the maximum allowed number. Removing boxes containing Ns from the resulting diagrams leaves the diagrams corresponding to the SU($N-1$) subgroup of SU(N). For example,

$$
\begin{array}{lcccc}
\mathrm{SU}(N): & \begin{array}{|c|c|}\hline \; & \; \\ \hline \; & \multicolumn{1}{c}{} \\ \cline{1-1}\end{array} & \begin{array}{|c|c|}\hline \; & N \\ \hline \; & \multicolumn{1}{c}{} \\ \cline{1-1}\end{array} & \begin{array}{|c|c|}\hline \; & \; \\ \hline N & \multicolumn{1}{c}{} \\ \cline{1-1}\end{array} & \begin{array}{|c|c|}\hline \; & N \\ \hline N & \multicolumn{1}{c}{} \\ \cline{1-1}\end{array} \\
 & \Downarrow & \Downarrow & \Downarrow & \Downarrow \\
\mathrm{SU}(N-1): & \begin{array}{|c|c|}\hline \; & \; \\ \hline \; & \multicolumn{1}{c}{} \\ \cline{1-1}\end{array} & \begin{array}{|c|}\hline \; \\ \hline \; \\ \hline\end{array} & \begin{array}{|c|c|}\hline \; & \; \\ \hline\end{array} & \begin{array}{|c|}\hline \; \\ \hline\end{array}
\end{array}
\tag{9.4}
$$

upon implementing this prescription.

Example 9.4 Specializing to SU(3), Eq. (9.4) is just the octet example analyzed previously using the weight space. After removing all columns containing two boxes from the bottom row of Eq. (9.4) because the SU(2) subgroup has Young diagrams with only one row of boxes,

$$\square + (\mathbf{1}) + \square\square + \square = \mathbf{2} + \mathbf{1} + \mathbf{3} + \mathbf{2}.$$

Thus, we have shown that the same dimensionalities are obtained for the SU(3) octet upon reduction to SU(2) by weight space analysis and by using Young diagrams.

We still must show that the isospin multiplets obtained in Example 9.4 are those of the horizontal rows of the SU(3) weight diagram in Fig. 9.5. This is done easily using additivity of the weights. For example, in the reduction

$$\begin{array}{|c|c|}\hline \; & \; \\ \hline \; & \multicolumn{1}{c}{} \\ \cline{1-1}\end{array} \longrightarrow \square = \mathbf{2},$$

the allowed SU(3) diagrams are those in which no Ns ($N = 3$ in this case) appear. From Table 9.2, the octet diagrams satisfying this condition and their weights (T_3, Y) are

$$\begin{array}{|c|c|}\hline 1 & 1 \\ \hline 2 & \multicolumn{1}{c}{} \\ \cline{1-1}\end{array} \qquad m = 2m(\boxed{1}) + m(\boxed{2}) = \left(\frac{1}{2}, 1\right),$$

$$\begin{array}{|c|c|}\hline 1 & 2 \\ \hline 2 & \multicolumn{1}{c}{} \\ \cline{1-1}\end{array} \qquad m = m(\boxed{1}) + 2m(\boxed{2}) = \left(-\frac{1}{2}, 1\right).$$

Therefore, this is a $Y = 1, T = \frac{1}{2}$ isospin doublet, corresponding to the top row of the SU(3) octet diagram in Fig. 9.5. In the reduction

$$\begin{array}{|c|c|}\hline \; & \; \\ \hline 3 & \multicolumn{1}{c}{} \\ \cline{1-1}\end{array} \longrightarrow \square\square$$

only the bottom box can contain a 3 and from Table 9.2 the allowed weights are

$$\begin{array}{|c|c|}\hline 1 & 2 \\ \hline 3 & \multicolumn{1}{c}{} \\ \cline{1-1}\end{array} \longrightarrow \quad m = m(\boxed{1}) + m(\boxed{2}) + m(\boxed{3}) = (0, 0),$$

$$\begin{array}{|c|c|}\hline 1 & 1 \\ \hline 3 & \multicolumn{1}{c}{} \\ \cline{1-1}\end{array} \longrightarrow \quad m = 2m(\boxed{1}) + m(\boxed{3}) = (1, 0),$$

$$\begin{array}{|c|c|}\hline 2 & 2 \\ \hline 3 & \multicolumn{1}{c}{} \\ \cline{1-1}\end{array} \longrightarrow \quad m = 2m(\boxed{2}) + m(\boxed{3}) = (-1, 0).$$

Therefore, this is a $Y = 0$ isovector ($T = 1$), corresponding to the middle row of the SU(3) octet in Fig. 9.5. Proceeding in this manner, upon restriction to the isospin subgroup the diagrams

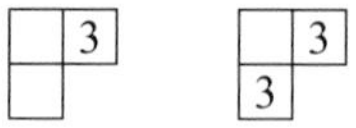

yield, respectively, the isoscalar with weight $(0, 0)$ and the $Y = -1$ isodoublet in the bottom row of the octet diagram. Hence, we have used Young diagrams to obtain the same SU(2) content of the SU(3) octet as that found previously by inspection of the weight space.

9.5 Extensions of Flavor SU(3) Symmetry

Isospin is an excellent phenomenological symmetry of the strong interactions, good to a few percent typically in particle masses for the lighter hadrons (those involving only the up and down quarks and antiquarks). Likewise, for hadrons involving only up, down, and strange quarks, flavor SU(3) symmetry is useful, with mass symmetry breaking typically 20% or less. It is natural to ask whether this approach can be extended to describe a broader range of hadrons and their properties. Two general directions suggest themselves: the first is to ask whether flavor SU(3) can be extended to a flavor symmetry of rank higher than two, thereby incorporating more additive quantum numbers in the description of heavier hadronic states; the second is to extend the flavor model by attempting to incorporate both flavor and spin within the same group structure.

9.5.1 Higher-Rank Flavor Symmetries

From Table 9.1 there are three quarks (and corresponding antiquarks) known in addition to the up, down, and strange quarks discussed to this point: charm (c), bottom (b), and top (t). These quarks are associated with additional additive quantum numbers, suggesting an extension of flavor SU(3) to higher-rank algebras. The simplest possibility is to expand flavor SU(3) to the rank-3 algebra SU(4), thereby incorporating the charm quantum number in addition to isospin and strangeness in the description of hadrons.

However, this is not a very fruitful approach because of the relative effective mass scales for the known quarks. Isospin is a good symmetry because of the near equivalence of the effective u and d quark masses. Flavor SU(3) is still a reasonably good symmetry because the effective mass of the strange quark is larger than that of the up and down quarks, but not so much larger. However, the effective mass of the charm quark is considerably greater than that of the strange quark. Therefore, a flavor SU(4) symmetry turns out not to be very useful because the corresponding symmetry breaking is large. Extensions to higher-rank algebras to incorporate additive quantum numbers associated with bottom and top quark degrees of freedom would be even worse because their mass scales are larger still.

9.5.2 SU(6) Flavor–Spin Symmetry

A (somewhat) more promising extension of flavor SU(3) is motivated by noting that the spin-$\frac{1}{2}$ baryon octet and the spin-$\frac{3}{2}$ baryon decuplet displayed in Fig. 9.1 differ in average mass by an amount that is comparable to the splitting between isospin multiplets within each of these SU(3) multiplets. This suggests that perhaps these two SU(3) multiplets – which transform as different irreducible representations of SU(3) – might be part of the *same irreducible representation* of a larger group. The baryon **8** and **10** differ in one fundamental characteristic: the former consists of spin-$\frac{1}{2}$ fermions and the latter of spin-$\frac{3}{2}$ fermions. Spin (real spin, not isospin!) is not a generator of SU(3) but is a conserved quantum number in low-energy interactions. This invites us to consider a larger group structure containing both flavor SU(3) and spin SU(2) as subgroups. The simplest possibility is $\mathrm{SU}(6) \supset \mathrm{SU}(3)_{\mathrm{flavor}} \times \mathrm{SU}(2)_{\mathrm{spin}}$, with the SU(6) symmetry having a six-dimensional fundamental representation of u, d, and s quarks, each with spin-up and spin-down components (recall from Table 9.1 that the quarks are all spin-$\frac{1}{2}$ fermions).

Thus, in flavor SU(3) the fundamental representation is a **3**, representing three quark flavors with no spin degrees of freedom [the spin degree of freedom for the quarks in the flavor SU(3) model is not a part of the flavor symmetry and the full flavor–spin symmetry is a direct product, $\mathrm{SU}(3)_{\mathrm{flavor}} \times \mathrm{SU}(2)_{\mathrm{spin}}$]. But in the flavor–spin SU(6) symmetry the fundamental representation is a **6**, representing three quarks, each with two spin states. The direct product of flavor and spin is a *subgroup* of SU(6) for which spin and flavor generators commute, but the full SU(6) group incorporates flavor and spin degrees of freedom on the same footing (a non-abelian flavor–spin algebra). This extension of SU(3) to include the spin degree of freedom was proposed as a generalization of the Wigner supermultiplet theory discussed in Section 10.3.2 to include the strangeness degree of freedom, thereby enlarging the algebraic structure from SU(4) to SU(6) [101] .

9.5.3 Baryons and Mesons under SU(6) Symmetry

The quark model requires that baryons be composed of three valence quarks, suggesting that a baryonic state under SU(6) is a direct product of three SU(6) fundamental representations. Using the Young diagram methods introduced in Ch. 8 for the Clebsch–Gordan series associated with direct products of unitary representations, we find that for SU(6)

$$\mathbf{6} \otimes \mathbf{6} \otimes \mathbf{6} = \mathbf{56} \oplus \mathbf{70} \oplus \mathbf{70} \oplus \mathbf{20}. \tag{9.5}$$

The fully symmetric **56** is exactly the right size to accommodate the baryon octet and decuplet representations alluded to above: each octet member has $2s + 1 = 2\left(\frac{1}{2}\right) + 1 = 2$ possible spin states and each decuplet member $2\left(\frac{3}{2}\right) + 1 = 4$ possible spin states, giving a total degeneracy of $2 \times 8 + 4 \times 10 = 56$ for these two SU(3) flavor multiplets. This decomposition of the **56** of SU(6) with respect to its SU(3) flavor and SU(2) spin subgroups may be represented compactly as

$$\mathbf{56} \to (\mathbf{8}, \mathbf{2}) \oplus (\mathbf{10}, \mathbf{4}), \tag{9.6}$$

where the dimensionality on the left side is for the SU(6) representation and within each set of parentheses on the right side the first number is the SU(3) flavor dimensionality and the second number is the SU(2) spin dimensionality.

Example 9.5 The $(\mathbf{10}, \mathbf{4})$ representation on the right side of Eq. (9.6) means the 10-dimensional SU(3) flavor representation, for which each state is of spin $s = \frac{3}{2}$ and therefore corresponds to an SU(2) spin representation of dimension $2s + 1 = 2(\frac{3}{2}) + 1 = 4$.

Example 9.6 Meson states in the quark model have a quark–antiquark $(q\bar{q})$ structure, which suggests that a basic meson state in the SU(6) model corresponds to

$$\mathbf{6} \otimes \overline{\mathbf{6}} = \mathbf{35} \oplus \mathbf{1}. \tag{9.7}$$

The low-lying meson states in the SU(3) flavor model fit nicely into the SU(6) representation **35** of this decomposition.

The SU(6) model does a reasonably good job of describing lighter hadrons, with perhaps its most notable success being a description of magnetic moments for baryons. This success may be fortuitous because in a fully relativistic theory it is not possible to separate spin and orbital degrees of freedom and only the total angular momentum of a particle is conserved. Hence, flavor–spin SU(6) symmetry is inconsistent with Lorentz invariance (Chs. 13–14), and should be viewed as only a low-energy approximation to a more correct theory.

Background and Further Reading

Flavor SU(3) symmetry is discussed in Close [42], Gasiorowicz [66], and Lichtenberg [141]. Treatments of SU(6) flavor–spin symmetry may be found in Close [42].

Problems

9.1 Use the fundamental quark triplet and Young diagrams to construct the quark content of the Δ decuplet illustrated in Fig. 9.4. ***

9.2 Use the Young diagram method to deduce the SU(2) isospin content of the SU(3) flavor representations **6** and **27**.

9.3 Use Young diagrams to deduce the isospin subgroup irrep content of the Δ decuplet illustrated in Fig. 9.4.

9.4 Use Young diagrams and the methods of Ch. 4 to construct SU(3) flavor wavefunctions for the Δ decuplet (baryon **10**) of Fig. 9.4 having the appropriate permutation symmetry. ***

9.5 Use Young diagrams to deduce the SU(2) isospin irrep content of the flavor SU(3) irrep $(\lambda, \mu) = (2, 1)$. ***

9.6 Use Young diagrams to find the SU(2) isospin content of the flavor SU(3) irreducible representation $(1, 2)$.

9.7 Use the methods of Ch. 4 to construct a proton flavor–spin wavefunction that is symmetric with respect to flavor–spin exchange. *Hint*: See Problem 9.11. ***

9.8 If we assume a non-relativistic model, the magnetic moment of a point quark is given by $\mu_i = q_i/2m_i$, where q_i is the charge and m_i the effective mass of quark i. Assume the magnetic moment of the proton to be given by the sum over valence quark contributions

$$\mu_p = \sum_{i=1}^{3} \langle p_{1/2}|\, \mu_i \sigma_3^i \,|p_{1/2}\rangle ,$$

where $|p_{1/2}\rangle$ denotes a proton in the $M_J = J = \frac{1}{2}$ angular momentum state, and σ_3^i is a Pauli matrix operating on the spin wavefunction of the ith quark. Use the proton wavefunction constructed in Problem 9.7 to show that the proton magnetic moment is given by $\mu_p = \frac{4}{3}\mu_u - \frac{1}{3}u_d$, where μ_u and μ_d are magnetic moments of the up and down quarks, respectively, and the u and d quark masses are assumed equal.

9.9 Construct a quark wavefunction for the neutron and proton, and use the results of Problem 9.8 to show that for the ratio of magnetic moments for protons and neutrons, $\mu_p/\mu_n = -\frac{3}{2}$, if we approximate $m_u = m_d$.

9.10 Use Young diagrams to obtain the SU(6) Clebsch–Gordan series $\mathbf{6} \otimes \mathbf{6} \otimes \mathbf{6} = \mathbf{56} \oplus \mathbf{70} \oplus \mathbf{70} \oplus \mathbf{20}$ of Eq. (9.5) that is relevant for baryons, and the Clebsch–Gordan series $\mathbf{6} \otimes \mathbf{6} \otimes \bar{\mathbf{6}} = \mathbf{35} \oplus \mathbf{1}$ of Eq. (9.7) that is relevant for mesons.

9.11 Show that the if space, spin, and flavor are assumed to be the operative degrees of freedom for particles in the baryonic **10** of Fig. 9.1(d), the ground states for the spin-$\frac{3}{2}$ baryons are in conflict with the Pauli principle (which requires a fermionic wavefunction to be totally antisymmetric). Assume an additional degree of freedom (color) associated with a new SU(3) symmetry of the quarks that is independent of flavor SU(3), with the quarks transforming as fundamental representations under the color SU(3) symmetry. Show that the Pauli principle can now be satisfied if the particles of the baryon decuplet transform as a **10** with respect to flavor SU(3) but as a **1** with respect to color SU(3). ***

10 Harmonic Oscillators and SU(3)

The three-dimensional harmonic oscillator is important in many areas of physics and it is well known that the 3D quantum oscillator exhibits an unusually high level of degeneracy. We have learned that non-accidental degeneracy is typical of a Hamiltonian invariant with respect to some symmetry. The degeneracy of the oscillator exceeds that deriving from rotational invariance, so it may be expected that the group associated with this symmetry is larger than the SO(3) of angular momentum. In this chapter we demonstrate that the 3D isotropic harmonic oscillator has an SU(3) symmetry. As one example of exploiting this insight, this symmetry will be used to simplify the discussion of collective motion in light atomic nuclei.

10.1 The 3D Quantum Oscillator

The Hamiltonian of the three-dimensional isotropic harmonic oscillator may be expressed as

$$H = \frac{1}{2}\left(\boldsymbol{p}^2 + \boldsymbol{r}^2\right), \tag{10.1}$$

where the coordinate is $\boldsymbol{r}$, the momentum is $\boldsymbol{p}$, and natural units $M = \hbar = \omega = 1$, with M the effective mass and ω the oscillator frequency, have been used.

10.1.1 Eigenvalues

It is convenient to introduce the operator $\boldsymbol{a}$ and its hermitian adjoint $\boldsymbol{a}^\dagger$ through the relations

$$\boldsymbol{a}^\dagger \equiv \frac{1}{\sqrt{2}}(\boldsymbol{r} - i\boldsymbol{p}) \qquad \boldsymbol{a} \equiv \frac{1}{\sqrt{2}}(\boldsymbol{r} + i\boldsymbol{p}). \tag{10.2}$$

From the usual commutation rules for the coordinates and momenta, it is easy to show that the operators $\boldsymbol{a}^\dagger$ and $\boldsymbol{a}$ satisfy the bosonic commutation relations

$$[a_i, a_j^\dagger] = \delta_{ij} \qquad [a_i, a_j] = [a_i^\dagger, a_j^\dagger] = 0, \tag{10.3}$$

where indices refer to the cartesian axes for the oscillator. The Hamiltonian is then

$$H = \boldsymbol{a}^\dagger \cdot \boldsymbol{a} + \frac{3}{2}, \tag{10.4}$$

which has energy eigenvalues $E = n + \frac{3}{2}$, where $n = 0, 1, 2, \ldots, \infty$.

10.1.2 Wavefunctions

Commutators for components of $\boldsymbol{a}^\dagger$ and $\boldsymbol{a}$ with the Hamiltonian are given by

$$[\,a_q, H\,] = a_q \qquad [\,a_q^\dagger, H\,] = -a_q^\dagger. \tag{10.5}$$

These may be proved easily using Eq. (10.3), as illustrated in Example 10.1.

Example 10.1 Utilizing Eqs. (10.3) and (10.4)

$$\begin{aligned}
[\,a_q^\dagger, H\,] &= a_q^\dagger \sum_p a_p^\dagger a_p - \sum_p a_p^\dagger a_p a_q^\dagger \\
&= \sum_p a_p^\dagger a_q^\dagger a_p - \sum_p a_p^\dagger(\delta_{pq} + a_q^\dagger a_p) \\
&= \sum_p a_p^\dagger a_q^\dagger a_p - a_q^\dagger - \sum_p a_p^\dagger a_q^\dagger a_p \\
&= -a_q^\dagger.
\end{aligned}$$

By an analogous proof, $[\,a_q, H\,] = a_q$.

From the commutators (10.5), we conclude in Problem 10.1 that, if $H\psi = E\psi$, then

$$H(a_q\psi) = (E-1)(a_q\psi) \qquad H(a_q^\dagger\psi) = (E+1)(a_q^\dagger\psi).$$

Therefore, $a_q^\dagger$ raises the oscillator energy by one unit, through creation of an oscillator quantum along the q-axis, and a_q lowers the energy by one oscillator unit, through annihilation of an oscillator quantum along the q-axis. These operators are termed *creation and annihilation operators* because of this property. The normalized wavefunctions $|n_1 n_2 n_3\rangle$ may be expressed using the creation operators as [224]

$$|n_1 n_2 n_3\rangle = \prod_{i=1}^{3} \frac{(a_i^\dagger)^{n_i}}{\sqrt{n_i!}} \, |000\rangle\,, \tag{10.6}$$

where the vacuum state $|000\rangle$ satisfies $a_i\,|000\rangle = 0$, and where the total number of oscillator quanta is $N = n_1 + n_2 + n_3$. This implies a high degeneracy for the 3D oscillator, since various combinations of n_1, n_2, and n_3 can give the same total N defining an energy state.

10.1.3 Unitary Symmetry

To find the symmetry associated with the high level of oscillator degeneracy it is useful to define nine *shift operators* $\tilde{A}_{ij}$ by the anticommutation relations

$$\tilde{A}_{ij} \equiv \frac{1}{2}\{a_i^\dagger, a_j\} = \frac{1}{2}(a_i^\dagger a_j + a_j a_i^\dagger). \tag{10.7}$$

The $\tilde{A}_{ij}$ are called *shift operators* because they act to annihilate an oscillator quantum along one axis and to create one along another axis. They satisfy

$$[\,\tilde{A}_{ij}, \tilde{A}_{kl}\,] = \delta_{jk}\tilde{A}_{il} - \delta_{il}\tilde{A}_{kj}, \tag{10.8}$$

so they are closed under commutation and form a Lie algebra. The algebra is found to be U(3), which is not semisimple because the diagonal operators $\tilde{A}_{ii}$ form a self-commuting set and $H \equiv \tilde{A}_{11} + \tilde{A}_{22} + \tilde{A}_{33}$ commutes with all other operators, $[H, \tilde{A}_{kl}] = 0$. If H is omitted from the set of nine operators the algebra is restricted to the special unitary subgroup, U(3) ⊃ SU(3). The omitted generator implements unitary transformations of the form $U_0 = e^{i\lambda H}$, which modify the basis functions by a phase and are not of physical significance if the total energy of the system (number of oscillator quanta) is fixed. Therefore, SU(3) will be taken as the 3D oscillator symmetry.

That we are dealing with a U(3) or SU(3) symmetry also could be inferred from the general arguments about many-particle states and single-particle unitary multiplets used in Ch. 8. For the 3D isotropic oscillator there are three degenerate fundamental states, each corresponding to a single oscillator quantum along one of the three axes, from which more complicated states can be built. If n quanta are distributed on these $N = 3$ states the situation is as discussed in Section 8.6 and Box 8.1, and we expect a U(3) ⊃ SU(3) symmetry for the corresponding many-body problem. Collecting the preceding observations, the generators of SU(3) may be taken to be

$$A_{ij} \equiv \tilde{A}_{ij} - \frac{H}{3}\delta_{ij} \qquad (i, j = 1, 2, 3), \tag{10.9}$$

with the understanding that only eight are linearly independent. We may then write

$$C = \sum_{ij} A_{ij} A_{ji} \tag{10.10}$$

for the quadratic SU(3) Casimir operator C.

10.1.4 Angular Momentum Subgroup

The orbital angular momentum operator is $\boldsymbol{L} = \boldsymbol{r} \times \boldsymbol{p} = i\boldsymbol{a} \times \boldsymbol{a}^\dagger$, with components

$$L_k = i\epsilon_{ijk} a_i a_j^\dagger, \tag{10.11}$$

where the second-quantized form follows from the definitions (10.2) and the completely antisymmetric rank-3 tensor ϵ_{ijk} has the properties summarized in Table 3.2. Therefore, the operators for cartesian components of the angular momentum are

$$\begin{aligned} L_x &= i(A_{32} - A_{23}) = i\left(a_y a_z^\dagger - a_z a_y^\dagger\right), \\ L_y &= i(A_{13} - A_{31}) = i\left(a_z a_x^\dagger - a_x a_z^\dagger\right), \\ L_z &= i(A_{21} - A_{12}) = i\left(a_x a_y^\dagger - a_y a_x^\dagger\right). \end{aligned} \tag{10.12}$$

These angular momentum components form an SO(3) algebra under commutation and they commute with the oscillator Hamiltonian (see below). Therefore the isotropic harmonic oscillator in three dimensions is rotationally invariant and its states may be classified in terms of an SO(3) subgroup of SU(3). This means that we can form linear combinations of the SU(3) generators A_{ij} that simultaneously have definite transformation properties under both SU(3) and the angular momentum SO(3) subgroup (they will be spherical tensors).

10.1.5 SO(3) Transformation Properties

Using the commutation relations (10.3), it is easy to show that

$$[L_i, a_j^\dagger] = i\epsilon_{ijk} a_k^\dagger, \tag{10.13}$$

with a similar relation for a_i. Comparing with Eq. (6.59), we see that the oscillator creation and annihilation operators transform as rank-1 tensors (vector operators) under SO(3), and the nine bilinear products $a_i^\dagger a_j$ transform under the orbital angular momentum subgroup as the SO(3) direct product

$$D^{(1)} \otimes D^{(1)} = D^{(0)} \oplus D^{(1)} \oplus D^{(2)}. \tag{10.14}$$

Thus, linear combinations of SU(3) generators may be constructed that transform as SO(3) spherical tensors of rank zero, one, and two. The SO(3) tensor of rank zero is essentially the Hamiltonian $H = \sum_q a_q^\dagger a_q + \frac{3}{2}$. The rank-1 tensor is just the angular momentum, which has spherical components

$$\begin{aligned} L_{+1} &= \frac{1}{\sqrt{2}}(A_{13} - A_{31}) + \frac{i}{\sqrt{2}}(A_{23} - A_{32}), \\ L_{-1} &= \frac{1}{\sqrt{2}}(A_{13} - A_{31}) - \frac{i}{\sqrt{2}}(A_{23} - A_{32}) \qquad L_0 = i(A_{21} - A_{12}), \end{aligned} \tag{10.15}$$

the rank-2 or quadrupole tensor has the spherical components

$$\begin{aligned} Q_1 &= -\frac{\sqrt{6}}{2}(A_{13} + A_{31}) - i\frac{\sqrt{6}}{2}(A_{23} + A_{32}), \\ Q_{-1} &= \frac{\sqrt{6}}{2}(A_{13} + A_{31}) - i\frac{\sqrt{6}}{2}(A_{23} + A_{32}), \\ Q_2 &= \frac{\sqrt{6}}{2}(A_{11} - A_{22}) + i\frac{\sqrt{6}}{2}(A_{21} + A_{12}), \\ Q_{-2} &= \frac{\sqrt{6}}{2}(A_{11} - A_{22}) - i\frac{\sqrt{6}}{2}(A_{21} + A_{12}), \\ Q_0 &= -A_{11} - A_{22} + 2A_{33}, \end{aligned} \tag{10.16}$$

and the additional generator for U(3) is the SO(3) scalar

$$H_0 = A_{11} + A_{22} + A_{33}. \tag{10.17}$$

Using Eq. (10.8), these spherical tensors are found to satisfy the commutation relations

$$\begin{aligned} &[L_k, L_q] = -\sqrt{2}\langle 1k1q|\,1(k+q)\rangle L_{k+q} \qquad [Q_k, L_q] = -\sqrt{6}\langle 2k1q|\,2(k+q)\rangle Q_{k+q}, \\ &[Q_k, Q_q] = 3\sqrt{10}\langle 2k2q|\,1(k+q)\rangle L_{k+q} \qquad [H, L_k] = [H, Q_k] = 0. \end{aligned} \tag{10.18}$$

In terms of the spherical tensors $\boldsymbol{L}$ and $\boldsymbol{Q}$, Eq. (10.10) may be used to write

$$C_{\mathrm{SU(3)}} = \frac{1}{6}\boldsymbol{Q}\cdot\boldsymbol{Q}, + \frac{1}{2}\boldsymbol{L}\cdot\boldsymbol{L}, \tag{10.19}$$

for the quadratic SU(3) Casimir operator.

10.1.6 Group Structure

The operators $(L_0, L_\pm)$ form an SO(3) subgroup and L_0 may be chosen as the generator of an SO(2) subgroup of SO(3), so the group structure for the 3D isotropic oscillator is

$$\mathrm{U}(3) \supset \mathrm{SU}(3) \supset \mathrm{SO}(3) \supset \mathrm{SO}(2), \tag{10.20}$$

where the SO(2) subgroup is relevant only if spatial isotropy is broken by a field. By virtue of this subchain decomposition, harmonic oscillator states may be labeled by a total oscillator quantum number N associated with U(3), the quantum numbers (λ, μ) associated with the SU(3) subgroup, and the angular momentum quantum numbers (L, M) associated with the SO(3) subgroup. Additional quantum numbers may come from symmetries that are direct products with the U(3) symmetry. For example, parity will be a good quantum number for oscillator states because space inversion commutes with the oscillator generators (see Box 22.6). These additional quantum numbers will not be displayed in the present discussion unless specifically required.

10.1.7 Many-Body Operators

An important application of the preceding discussion is to the motion of a particle moving in a 3D potential with a linear restoring force in the three cartesian directions. If the "spring constant" is assumed to be the same in all three directions, we refer to such a potential as an *isotropic oscillator potential.* Since such linear restoring terms appear in the low-order expansion of arbitrary well-behaved potentials about a minimum, the oscillator potential is a useful first approximation in many quantum-mechanical problems involving the motion of particles in potential wells. The simplest example involves the motion of a single particle in such a potential, with the particle carrying one oscillator quantum along one of the three independent axes. In more complicated situations a single particle may carry more than one oscillator quantum and there may be more than one particle in the system. The results just discussed generalize in a straightforward fashion to many-body operators that are sums of the corresponding single-particle operators. For an A-particle system,

$$H = \sum_{t=1}^{A} H(t) \qquad A_{ij} = \sum_{t=1}^{A} A_{ij}(t) \qquad L_q = \sum_{t=1}^{A} L_q(t) \qquad Q_q = \sum_{t=1}^{A} Q_q(t). \tag{10.21}$$

The single-particle operators on the right sides and the many-body operators on the left sides of these equations both obey the previous equations for the oscillator group structure.

10.2 SU(3) and the Nuclear Shell Model

The nuclear many-body problem is greatly simplified by the shell model (mean field) approximation in which the nucleons are described by a Hamiltonian

$$H = T_0 + V_0 + U_R \qquad U_R = \sum_i V_i + \sum_{i<j} V_{ij}, \tag{10.22}$$

where T_0 is the kinetic energy, V_0 is the average single-particle potential, and U_R is the *residual interaction*, simplified here to contain only one-body terms V_i and two-body terms V_{ij}.[1] For light nuclei (say atomic mass number $A < 30$), the potential V_0 is well approximated by a 3D isotropic harmonic oscillator potential $V_0 \simeq \hbar\omega_0(x^2 + y^2 + z^2) = \hbar\omega_0 r^2$. This approximation fails for heavier nuclei because the radial shape of the potential at the nuclear surface deviates substantially from the harmonic-oscillator form, and (most importantly) because a spin–orbit contribution to the potential can no longer be neglected relative to the spatial part. Figure 10.1(a) compares the radial shape of a realistic single-particle nuclear potential with the harmonic approximation; Fig. 10.1(b) compares the single-particle energy levels in the harmonic approximation with more realistic ones that include a more correct shape for the radial potential and a spin–orbit interaction.

The residual interaction $U_R \simeq V_P + V_Q$ is dominated by short-range *pairing correlations* V_P and longer-range *quadrupole correlations* V_Q. Pairing correlations are of *particle–particle* or *hole–hole* type (created by $a^\dagger a^\dagger$ or aa).[2] They favor spherical symmetry, dominate U_R near closed shells, and are responsible for effects such as the appearance of a pairing gap in the single-particle spectrum and large modifications of moments of inertia from that expected for independent particles. Quadrupole correlations are of particle–hole type (created by $a^\dagger a$) and tend to dominate U_R further from the closed shells. They abet the formation of a deformed intrinsic mean field and are responsible for the low-lying collective vibrations and rotations that are seen in many nuclei.

10.3 SU(3) Classification of SD Shell States

If the nuclear single-particle potential is approximated by an isotropic oscillator, we may expect that the states can be classified according to SU(3) oscillator symmetry. Figure 10.1(b) suggests that the most fruitful area for applying this idea is in the shell containing the 2s and 1d harmonic oscillator levels. In nuclear physics this is called the *SD shell*; it is the valence shell being filled for both neutrons and protons between ^{16}O and ^{40}Ca in the chart of isotopes. For the lower 1p shell the application of SU(3) symmetry yields little new insight, while for heavier shells spin–orbit interaction alters the level scheme from that of an oscillator potential, invalidating the conditions for application of the symmetry. Accordingly, we consider the nuclear shell model in harmonic oscillator approximation for the SD shell in the examples to be addressed in this chapter. As discussed in Box 10.1,

[1] The restriction of U_R to no more than two-body forces is motivated by the average separation of nucleons implied by the density of nuclear matter being sufficiently large compared with the range of the residual interaction to make three-body or higher interactions less likely.

[2] The operator $a^\dagger$ creates a particle and the operator a destroys a particle, or equivalently creates a "hole" (absence of a particle). Thus the "particle–particle operator" $a^\dagger a^\dagger$ creates two particles, the "hole–hole operator" aa creates two holes, and the "particle–hole operator" $a^\dagger a$ creates a particle and a hole.

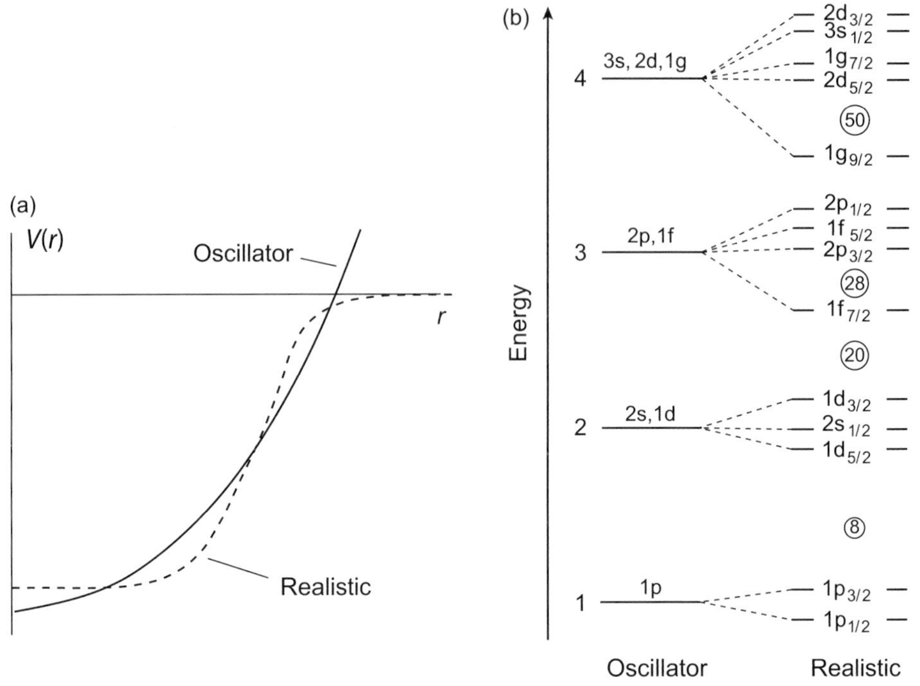

Fig. 10.1 (a) Comparison of a spherical oscillator potential with a realistic shell model potential. (b) Energy levels in light nuclei for an oscillator potential (left) and for a realistic potential including spin–orbit interaction (right). Oscillator levels are labeled by the oscillator shell number and letters s, p, d, f, g denoting orbital angular momenta 0, 1, 2, 3, 4, respectively. Levels in the realistic potential are labeled by letters for orbital angular momentum and a subscript giving total angular momentum. Shell gaps are indicated by circled nucleon numbers. A more realistic shell model extending to heavier nuclei will be given in Fig. 31.1.

we shall consider symmetries built on more general nuclear degrees of freedom that are applicable to a broader range of nuclear structure problems in later chapters.

10.3.1 Classification Strategy

Accounting for the $(2L + 1)$ orbital angular momentum degeneracy of the orbitals, the SD shell in harmonic approximation consists of six degenerate orbitals: five associated with the 1d level and one with the 2s level in Fig. 10.1(b). By the considerations of Ch. 8, with six orbitals available the orbital symmetry of the shell is expected to be $U(6) \supset SU(6)$, where restriction to the special unitary group follows from the now familiar procedure of neglecting an overall phase in the transformations. Since we assume the particles to move in an oscillator potential, it must also be possible to classify states in the SD shell according to the SU(3) oscillator chain (10.20). Therefore, the oscillator symmetry SU(3) must be a subgroup of the orbital SU(6) symmetry (see Harvey [106], Appendix E), and we will classify states according to

$$SU(6) \supset SU(3) \supset SO(3) \supset SO(2). \tag{10.23}$$

Box 10.1 Groups, Algebras, and Nuclear Structure

In this book we will discuss several ways in which symmetries have been used to simplify the nuclear many-body problem. The *Elliott Model* discussed in this chapter assumes that the quadrupole portion of the residual interaction is dominant. The *Quasispin Model* introduced in the problems of Ch. 31 assumes that the nucleon pairing dominates the residual interaction. The *Fermion Dynamical Symmetry Model* described in Ch. 31 assumes that both the pairing and quadrupole interactions are important in the residual interaction. All of these are attempts to solve the nuclear shell model problem by algebraic means (use of Lie algebras and Lie groups to compute matrix elements of observables) assuming the nucleons to be fermions. A fourth algebraic approach that we will mention for nuclear structure but not discuss in depth is the *Interacting Boson Model,* which replaces the fermion problem by one that is less realistic but easier to solve: interacting pairs of nucleons that are assumed to obey boson statistics rather than the fermion statistics of the individual nucleons. In the remainder of this chapter we describe the Elliott model, with these other methods discussed in the chapters indicated above.

Our procedure will be to classify according to the permutation group S_n for n particles in the SD shell, since this is equivalent to a classification with respect to the SU(6) orbital symmetry. Then we will classify with respect to the permutation group S_N, where N is the number of oscillator quanta carried by the particles, since this is equivalent to a classification with respect to the SU(3) oscillator symmetry. Finally, we will classify these states according to their orbital angular momentum content.

10.3.2 Orbital and Spin–Isospin Symmetry

The orbital symmetry expected in light nuclei may be deduced from the net attractive nature of the nucleon–nucleon interaction at the relevant energies and the generalized Pauli exclusion principle, which states in this context that if the total wavefunction is written as a product of a space (orbital) part and a spin–isospin part, $\psi = \psi_{\text{orbital}} \cdot \psi_{\text{spin–isospin}}$, then the product of the permutation symmetries for the orbital and spin–isospin symmetries must be antisymmetric. Considerable insight may be obtained by assuming the nucleon–nucleon interaction to be (1) charge independent (see Section 3.4.4) and (2) spin independent. The first assumption is rather good; the second is not as good but not unreasonably so for the light nuclei. A single nucleon with ($T = \frac{1}{2}, S = \frac{1}{2}$) can be in any of four states,

$$\left|\tfrac{1}{2}, \tfrac{1}{2}\right\rangle \qquad \left|\tfrac{1}{2}, -\tfrac{1}{2}\right\rangle \qquad \left|-\tfrac{1}{2}, \tfrac{1}{2}\right\rangle \qquad \left|-\tfrac{1}{2}, -\tfrac{1}{2}\right\rangle,$$

conveniently labeled by the third components of spin and isospin $|T_3, S_3\rangle$. If the nucleon–nucleon interaction is independent of both spin and isospin, these will be degenerate in energy and form the basis of a four-dimensional fundamental representation of SU(4) [by analogy with the SU(6) flavor–spin symmetry discussed in Section 9.5]. This spin–isospin unitary symmetry is termed the *Wigner supermultiplet theory* [209]. Thus, at this level of approximation the spin–isospin symmetry is SU(4) and the orbital symmetry is SU(6).

10.3.3 Permutation Symmetry

Orbital and spin–isospin variables represent independent degrees of freedom for the *same particle,* so the total wavefunction will transform as the direct product of S_n representations associated with SU(4) and SU(6), as discussed in Section 4.5.1. But the antisymmetry requirement for the total wavefunction and the rules for forming the direct product of permutation group representations permit this only if the two S_n representations are *associate* (related by interchanging rows and columns in their Young diagrams). Since SU(4) diagrams can have no more than four boxes in each column, the associate SU(6) orbital symmetry diagrams are restricted to having no more than four boxes in any row, if the total wavefunction is to be antisymmetric. The general result for wavefunctions where the effective interaction is attractive between particles is that the lowest-energy orbital states are the ones that are maximally symmetric, subject to required overall antisymmetry for the total wavefunction. This is because symmetric many-body orbital states tend to maximize the time that particles spend near each other, which is optimal if the net interaction is attractive.[3] Therefore, for spin–isospin independent forces the lowest-energy states will have orbital symmetry corresponding to the most symmetric Young diagrams, but antisymmetry restricts these to having no more than four boxes per row. Example 10.2 illustrates.

Example 10.2 For six particles in the SD shell

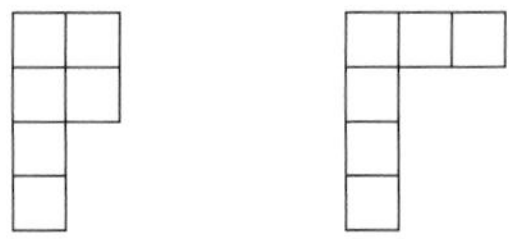

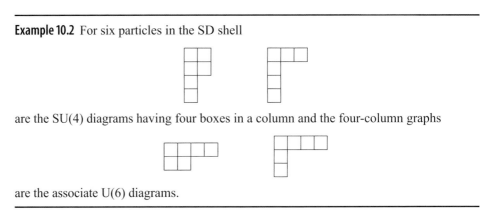

are the SU(4) diagrams having four boxes in a column and the four-column graphs

are the associate U(6) diagrams.

Since only the product of associate diagrams gives completely antisymmetric wavefunctions and the attractive nucleon–nucleon interaction favors symmetric orbital configurations, SU(6) irreps corresponding to partitions [42] and [411] are expected to lie lowest in energy for six particles in the SD shell. We shall see that this is the observed situation.

10.3.4 Example: Two Particles in the SD Shell

As an example of the general classification scheme, let us consider the states resulting from placing two particles in the SD shell.

[3] In atomic physics similar considerations lead to *Hund's rules,* but in that case the electrons interact by a net *repulsive* effective interaction and prefer antisymmetric orbital states that *minimize* the time particles spend near each other.

SU(6) Classification: The classification with respect to the SU(6) symmetry is equivalent to the classification for irreps of the permutation group S_2 for the two particles distributed on the six available orbitals (Section 8.6). Since this is a two-particle state, the SU(6) dimensionality of the allowed Young diagrams is (Section 8.8)

Dim ([2]) = Dim (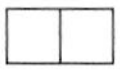) = 21 Dim ([11]) = Dim () = 15.

The total dimensionality is 21 + 15 = 36, corresponding to the $6^2 = 36$ independent states that may be formed by placing two particles in six orbitals. From the preceding considerations, we expect the symmetric partition [2] to lie lower in energy than the [11] partition.

SU(3) Classification: Now let us classify according to the SU(3) oscillator symmetry for these two-particle states. For a single-particle state with $N = 0, 1, 2, \ldots$ oscillator quanta, the corresponding SU(3) representations are $(\lambda, \mu) = (0, 0), (1, 0), (2, 0), \ldots$, as may be seen in the following way. The SU(3) symmetry refers to the *oscillator quanta,* so the boxes in the Young diagrams for SU(3) denote the *number of oscillator quanta, not the number of particles* as was the case for the SU(6) orbital symmetry. For the SD shell $N = 2$, so *each particle* in the 2s or 1d orbital carries two oscillator quanta and we expect that single-particle states correspond to two-box SU(3) diagrams,

However, the second diagram is not permitted because *a one-particle state cannot be antisymmetrized,* as illustrated explicitly in the following example.

Example 10.3 Suppose a state with one oscillator quantum in the x direction and one in the y direction. The symmetric one-particle state is $\psi_S \simeq a_x^\dagger a_y^\dagger$ but the antisymmetric state is $\psi_A \simeq a_x^\dagger a_y^\dagger - a_x^\dagger a_y^\dagger = 0$. Only the symmetric combination is non-zero.

Hence, for the $N = 2$ oscillator shell the one-particle SU(3) state is the symmetric two-oscillator quantum state  $= (2, 0)$. Now n-particle SU(3) states can be constructed from the direct product of single-particle states, and they are generally reducible to SU(3) irreps by exploiting the outer product of the permutation groups, as described in Section 8.9. Since each particle in an $N\hbar\omega$ shell carries N oscillator quanta, an n-particle state corresponds to $(n \times N)$-box Young diagrams. Specifically, for two particles ($n = 2$) in the SD shell ($N = 2$) the SU(3) Clebsch–Gordan series is, in several alternative notations,

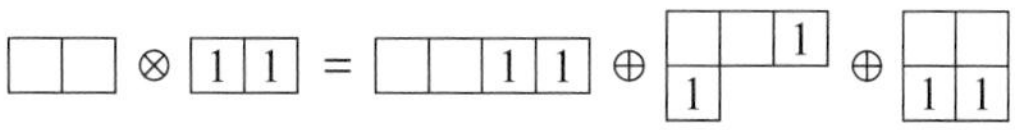

$$(2,0) \otimes (2,0) = (4,0) \oplus (2,1) \oplus (0,2), \tag{10.24}$$

$$\mathbf{6} \otimes \mathbf{6} = \mathbf{15} \oplus \mathbf{15} \oplus \mathbf{6}.$$

Because from Eq. (10.23) the oscillator symmetry SU(3) is a subgroup of the orbital symmetry SU(6), these SU(3) representations must fit into the SU(6) representations **15** and **21** in such a way that each SU(3) representation is contained entirely in a single SU(6) representation. For simple cases these assignments can often be done by inspection.

(a)

[f]	(4, 0)	(2, 1)	(0, 2)
		1	
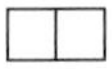	1		1

(b)

N	(λ, μ)	L
1	(1, 0)	1
2	(2, 0)	0, 2
2	(0, 1)	1

Fig. 10.2 Classification for two particles in the SD shell. (a) Classification with respect to SU(6) ⊃ SU(3). (b) Classification with respect to the SO(3) orbital angular momentum subgroup of SU(3).

Example 10.4 In Eq. (10.24) the $(4, 0)$ representation consists of a horizontal row of boxes, so it must be symmetric and must correspond to the symmetric SU(6) partition [2]. This accounts for 15 of the 21 states transforming as the [2]. The remaining six states of the [2] must correspond to the SU(3) representation $(0, 2)$, leaving the 15-dimensional $(2, 1)$ representation of SU(3) to be assigned to the 15-dimensional SU(6) partition [11].

In summary, for two particles in the SD shell the complete classification with respect to the group chain SU(6) ⊃ SU(3) is given in Fig. 10.2(a). By a similar procedure we may classify all SD shell n-particle states. Such classifications may be found in Refs. [53, 55].

SO(3) Classification: The n-particle states of the SD shell have been classified according to the symmetry SU(6) ⊃ SU(3). The classification with respect to the SO(3) orbital angular momentum subgroup of SU(3) is summarized in Fig. 10.2(b), where N represents the number of oscillator quanta, (λ, μ) specifies the SU(3) representation, and L labels the SO(3) representation. Thus, a one-oscillator quantum state transforms as the $L = 1$ representation of SO(3), while a two-oscillator quantum state can correspond to either the $(2, 0)$ or the $(0, 1)$ representation of SU(3), with $L = 0, 2$ in the former case and $L = 1$ in the latter case (which is possible only for multiparticle states since the diagram is antisymmetric).

These results may be obtained in several ways. One notes that the dimensionalities of the SU(3) representations $(1, 0)$, $(2, 0)$, and $(0, 1)$ are 3, 6, and 3, respectively, and that the dimensionalities of the SO(3) representations are $2L + 1$, which requires the assignments in Fig. 10.2(b) if the representations of the SO(3) subgroup are to fit completely within the SU(3) representations. Also, one notes that the operators $a_q^\dagger$ and a_q that create and annihilate the oscillator quanta transform as vector operators under SO(3), so single-quantum states must correspond to $L = 1$ and states with two oscillator quanta correspond to $L = 0, 1, 2$, with $L = 0, 2$ belonging to the symmetric representation and $L = 1$ to the antisymmetric representation, by dimensionality arguments. Proceeding in this way, a general rule may be obtained for the orbital angular momentum quantum numbers that occur for an SU(3) representation (λ, μ) under the reduction SU(3) ⊃ SO(3),

$$
\begin{aligned}
L &= K, K+1, \ldots, K + \max(\lambda, \mu) \qquad (K \neq 0), \\
L &= \max(\lambda, \mu), \max(\lambda, \mu) - 2, \ldots, 1 \text{ or } 0 \qquad (K = 0), \\
K &\equiv \min(\lambda, \mu), \min(\lambda, \mu) - 2, \ldots, 1 \text{ or } 0,
\end{aligned}
\tag{10.25}
$$

with $\min(\lambda, \mu)$ the minimum and $\max(\lambda, \mu)$ the maximum value of the pair λ and μ.

Example 10.5 If there are four particles in the SD shell an SU(3) representation $(4, 2)$ occurs in the reduction of the orbital symmetry. Applying Eq. (10.25), we find that $K = 0, 2$, with $L = 0, 2, 4$ for $K = 0$ and $L = 2, 3, 4, 5, 6$ for $K = 2$.

The reduction under $\mathrm{SO}(3) \supset \mathrm{SO}(2)$ is obvious, since each value of L that labels a $(2L+1)$-dimensional SO(3) representation implies $2L + 1$ one-dimensional irreps of the subgroup SO(2), each labeled by the magnetic quantum number M_L. Therefore, we have outlined a method of labeling n-particle states for the SD shell under the subgroup chain

$$\mathrm{SU}(6)_{\text{orbital}} \supset \mathrm{SU}(3)_{\text{oscillator}} \supset \mathrm{SO}(3)_L \supset \mathrm{SO}(2)_{M_L}. \tag{10.26}$$

As will be seen, this symmetry and the corresponding labeling of states permits a spectrum to be obtained analytically.

10.4 SU(2) Subgroups and Intrinsic States

In our discussion of SU(3) flavor symmetries in Ch. 9, several SU(2) subgroups that are of mathematical and sometimes physical importance were considered. The SU(3) flavor symmetries do not contain orbital angular momentum as a subgroup, since the flavor symmetries are assumed to act on internal (not spacetime) degrees of freedom of the strongly interacting particles.[4] If a group chain is constructed in the oscillator model that is the analog of the isospin decomposition of flavor SU(3), the chain does not contain the angular momentum SO(3) subgroup and the corresponding states do not have a definite value of the physical angular momentum. In many-body physics, a state that does not have a definite value of the total angular momentum (or some other normally conserved quantum number) is called an *intrinsic state.* Such states are useful because states with good quantum numbers can be recovered from intrinsic states using projection integrals (see Ch. 22), and because intrinsic states acquire a physical meaning in the classical limit. Such states are examples of *spontaneous symmetry breaking,* which will be discussed extensively later.

10.4.1 Weight Space Operators and Diagrams

Let us establish a connection between oscillator SU(3) operators and the flavor SU(3) operators of Ch. 8. As shown in Problem 10.7, from Eqs. (8.2) and (8.7)–(8.8), we find relations like

$$\begin{aligned} T_3 &\equiv \frac{1}{2}(N_x - N_y) \qquad & T_+ &= a_x^\dagger a_y, \\ T_- &= a_y^\dagger a_x \qquad & Y &\equiv \frac{1}{3}(N_x + N_y - 2N_z), \end{aligned} \tag{10.27}$$

[4] The Lorentz spacetime symmetries, including rotational invariance, would be treated as a direct product with the SU(3) flavor symmetry. Flavor SU(3) has subgroups with the angular momentum SU(2) algebra but the interpretation of the generators for these subgroups is not that of physical angular momentum [see the discussion of SU(6) flavor–spin symmetry in Ch. 9].

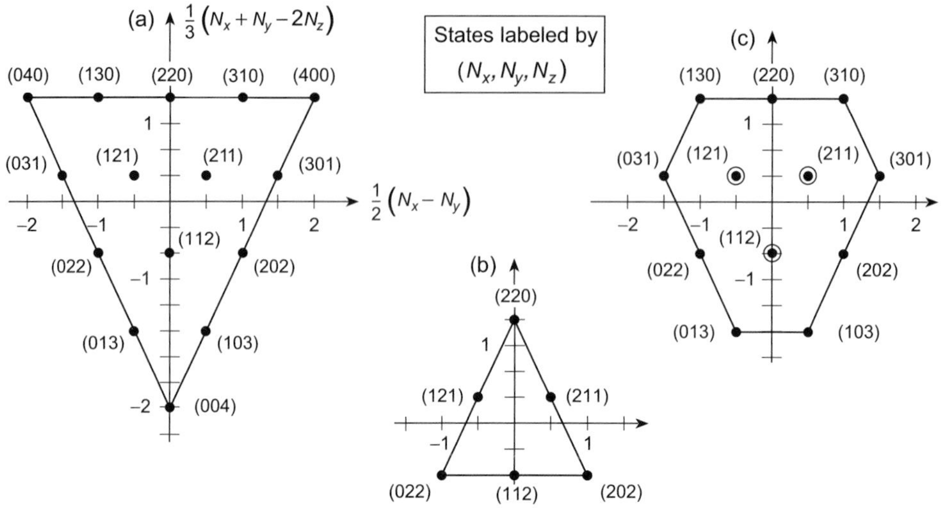

Fig. 10.3 Weight diagrams for two particles in the nuclear SD shell. These correspond to (a) $(\lambda, \mu) = (4, 0)$, (b) $(\lambda, \mu) = (0, 2)$, and (c) $(\lambda, \mu) = (2, 1)$ representations of SU(3).

where T_3 and Y label the states and the $T_\pm$ are stepping operators moving horizontally in the SU(3) oscillator representations by one unit of T_3, analogous to the isospin raising and lowering operators in flavor SU(3). As an example of constructing weight diagrams, consider two particles in the SD shell, which leads to the SU(3) representations $(4, 0)$, $(0, 2)$, and $(2, 1)$. The corresponding weight diagrams are shown in Fig. 10.3, with each site in the weight diagram labeled by the integers (N_x, N_y, N_z). These diagrams have been constructed using the methods in Ch. 8, and the identifications in Eq. (10.27) and Problem 10.7 [56, 193].

Example 10.6 In the $(4, 0)$ representation of Fig. 10.3(a) the maximal weight state is given by Eq. (8.13): $T_3 = 2$ and $Y = \frac{4}{3}$, and since the number of oscillator quanta is $N = 4$ for two particles in the SD shell and $N_x + N_y + N_z = N$, the upper right occupied site corresponds to $(N_x, N_y, N_z) = (400)$. Using the properties of the stepping operators worked out in Problem 10.7 (for example, application of T_- decreases N_x by one and increases N_y by one), Fig. 10.3(a) is easily constructed.

The weight diagram of Fig. 10.3(b) for the $(0, 2)$ representation and of Fig. 10.3(c) for the $(2, 1)$ representations can be constructed in a similar way as for Example 10.6.

10.4.2 Angular Momentum Content of Multiplets

The angular momentum content for states in each of these multiplets may be determined as follows [56]. The angular momentum operator $L_z = L_3$ is

$$
\begin{aligned}
L_z &= -i\left(a_x^\dagger a_y - a_y^\dagger a_x\right) \\
&= -i\,(T_+ - T_-) \\
&= -i(T_x + iT_y - T_x + iT_y) \\
&= 2T_y.
\end{aligned}
\tag{10.28}
$$

Hence, within a representation (λ, μ) eigenvalues of L_z equal those of $2T_y$, and this set of eigenvalues is the same as the set for $2T_3$. Thus eigenvalues for L_z are the same as those for $2T_3 = N_x - N_y$, which may be read from the weight diagrams. Once L_z values are enumerated, the SO(3) highest-weight algorithm gives the values of L in the SU(3) irrep.

Example 10.7 Occupied sites for the $(4, 0)$ irrep in Fig. 10.3(a) correspond to $L_z = \{4, 3, 2, 2, 1, 1, 0, 0, 0, -1, -1, -2, -2, -3, -4\}$, so by the highest-weight algorithm of Section 3.3.4 we obtain $L = 0, 2$, and 4. Likewise, $L = 0, 2$ for the $(0, 2)$ representation of Fig. 10.3(b) and $L = 1, 2, 3$ for the $(2, 1)$ representation of Fig. 10.3(c). These agree with Eq. (10.25).

We emphasize that the angular momentum decomposition for the oscillator wavefunctions described here is *not* the same process as the restriction from flavor SU(3) to isospin SU(2) described in Section 9.4. The generators of angular momentum SO(3) and those of isospin SU(2) have the same Lie algebra, but the microscopic structure and corresponding physical interpretation of their generators differ fundamentally.

10.5 Collective Motion in the Nuclear SD Shell

Certain atomic nuclei having valence particles in the SD shell exhibit a rotational spectrum $E \simeq J(J+1)$ for states of angular momentum J, with electromagnetic transitions between these states strongly enhanced relative to that for uncorrelated single-particle motion. A major triumph of the Elliott model was a microscopic explanation of these collective modes using the SU(3) oscillator symmetry. In this section we consider an approximate solution of the shell model Hamiltonian for the SD shell that exploits the symmetries we have discussed to provide a description of the low-lying states in the SD shell.

10.5.1 Hamiltonian

The residual interaction U_R may be approximated in the SD shell by

$$
U_\mathrm{R} = aV_\mathrm{M} + \frac{1}{4}b\,\boldsymbol{Q}\cdot\boldsymbol{Q}, \tag{10.29}
$$

where a and b are parameters and V_M is a *Majorana potential* of the form

$$
V_\mathrm{M} = \frac{1}{4}V_0(1 + \boldsymbol{\sigma}_1\cdot\boldsymbol{\sigma}_2)(1 + \boldsymbol{\tau}_1\cdot\boldsymbol{\tau}_2). \tag{10.30}
$$

In this expression, σ denotes Pauli spin operators and τ denotes the corresponding Pauli isospin operator. Thus V_{M} is both a spin and isospin scalar. The second term in Eq. (10.29) is the quadrupole–quadrupole interaction, with

$$Q \cdot Q = \sum_{i=1}^{A} Q(i) \cdot Q(i) + \sum_{i \neq j} Q(i) \cdot Q(j). \tag{10.31}$$

The first term contributes to the effective single-particle energy and the second term is the two-body quadrupole–quadrupole interaction, which is the most important part of the long-range residual nucleon–nucleon interaction. Absorbing the first term into the single-particle energy H_0 and utilizing Eq. (10.19) for the SU(3) quadratic Casimir operator $C_{\mathrm{SU(3)}}$

$$Q \cdot Q = \sum_{i \neq j} Q(i) \cdot Q(j) = 6C_{\mathrm{SU(3)}} - 3L \cdot L.$$

Therefore, an approximate shell model Hamiltonian for the SD shell is

$$H = H_0 + U_{\mathrm{R}} = H_0 + aV_{\mathrm{M}} + \frac{3}{2}b\, C_{\mathrm{SU(3)}} - \frac{3}{4}\, b\, L \cdot L, \tag{10.32}$$

where H_0 is a single-particle energy contribution.

10.5.2 Group-Theoretical Solution

The Hamiltonian (10.32) is composed of scalars with respect to the group chain (10.23). This means that it is diagonal in the basis labeled by the quantum numbers of the chain and eigenvalues can be found analytically if H_0 and the coefficients a and b are specified. First, consider the Majorana operator (10.30). For Young diagrams with no more than four columns (the maximum allowed for total antisymmetry if the forces are spin and isospin independent) a general formula can be derived for its expectation value in a state of definite permutation symmetry [49]

$$\langle V_{\mathrm{M}} \rangle = n_2 + 2n_3 + 3n_4 - \frac{1}{2}\sum_{i=1}^{4} n_i(n_i - 1), \tag{10.33}$$

where n_i is the number of boxes in the column i.

Example 10.8 For two particles in the SD shell the expectation values of the Majorana operator in the irreps labeled by the partitions [2] and [11] are $\langle V_{\mathrm{M}} \rangle_{[2]} = 1$ and $\langle V_M \rangle_{[11]} = -1$. Here, as for the general case, the Majorana term favors the more symmetric state if the coefficient a in (10.32) is negative, as the data require.

The expectation value of the SU(3) quadratic Casimir operator in the state (λ, μ) is

$$\langle C \rangle_{\lambda\mu} = 4\left(\lambda^2 + \mu^2 + \lambda\mu + 3\lambda + 3\mu\right). \tag{10.34}$$

Because the constant b is required by data to be negative, SU(3) configurations that maximize $\langle C \rangle$ are expected to lie lowest in energy. Finally, for states of good orbital angular momentum $\langle \boldsymbol{L} \cdot \boldsymbol{L} \rangle = L(L+1)$ and the energy of an SD shell state may be written in this approximation as

$$E = E_0 + a\langle V_{\mathrm{M}} \rangle + \frac{3}{2} b \langle C \rangle_{\lambda\mu} - \frac{3}{4} bL(L+1), \tag{10.35}$$

where $\langle V_{\mathrm{M}} \rangle$ is given by Eq. (10.33) and $\langle C \rangle_{\lambda\mu}$ is given by Eq. (10.34).

10.5.3 The Theoretical Spectrum

Figure 10.4 illustrates a typical calculation retaining only the lowest representations from the spectrum (10.35). In the left column the Majorana term splits the orbital SU(6) degeneracy, favoring the more symmetric orbital representations. In the second column the addition of the SU(3) Casimir term splits the SU(6) multiplets, with the representations of SU(3) within a given SU(6) multiplet that maximize the expectation value of the Casimir operator lying lowest in energy. Finally, the third column adds the splitting of the degeneracy within an SU(3) multiplet by the $\langle L^2 \rangle \sim L(L+1)$ term. The right column shows the experimental low-energy spectrum observed in ^{20}Ne. The calculation is seen to be in qualitative agreement with observations. In particular, a rotational spectrum is produced and further calculations within the model demonstrate that these are collective states, with enhanced electromagnetic transition probabilities similar to those observed experimentally (see Fig. 11.6). Calculations for higher-lying states that include additional representations exhibit qualitative agreement with data but reproduce details less well.

The procedure outlined in this chapter for an approximate solution of the nuclear shell model is powerful, and is a prototype for even more ambitious applications of groups and algebras to the many-body problem that will be discussed in subsequent chapters. As for the relatively simple example discussed here, if we can classify states with respect to some

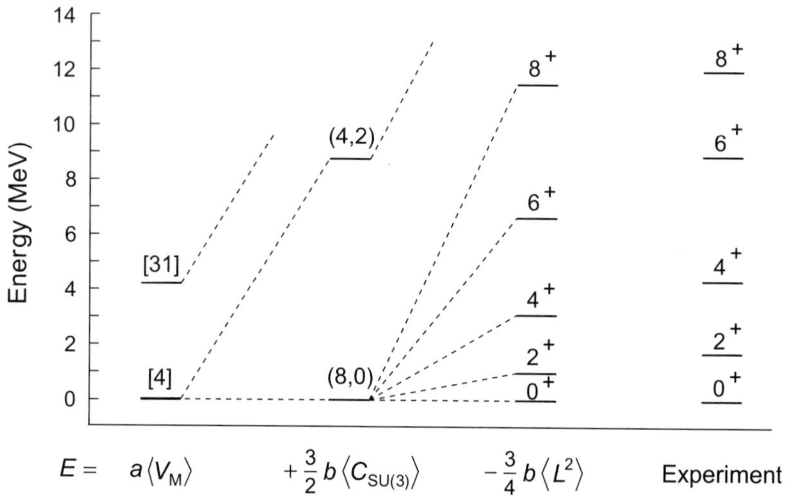

Fig. 10.4 Low-energy states for the nucleus ^{20}Ne, calculated using Eq. (10.35). The parameter b is negative. States in the last two columns are labeled by angular momentum and parity ($\pm$).

group chain and can construct an effective Hamiltonian entirely from invariant operators of that group chain, the resulting eigenvalue problem can be solved analytically. We shall see that by expanding to larger groups that encompass more physics, a highly satisfactory agreement with data is often possible for many-body problems in a variety of fields.

Background and Further Reading

Much of this chapter derives from the seminal work of Elliott [55], which is discussed well in the review article by Harvey [106], and in the books by Wybourne [224], and by Eisenberg and Greiner [53]. The papers by Wigner [209] on symmetry properties of the nuclear Hamiltonian, and by Elliott [55] on SU(3) symmetries in light nuclei, broke fertile new ground in the application of algebraic and group-theoretical techniques to physical problems. The influence of that work extends far beyond its origins in specific nuclear physics applications.

Problems

10.1 Prove that because of the commutation relations (10.5), $a_q^\dagger$ may be interpreted as a creation operator and a_q as an annihilation operator for oscillator quanta. ***

10.2 Demonstrate that the shift operators $\tilde{A}_{ij}$ obey the commutation relations (10.8). ***

10.3 Prove that the 3D harmonic oscillator orbital angular momentum operators are given by $\boldsymbol{L} = i\boldsymbol{a} \times \boldsymbol{a}^\dagger$. Show that the components L_k obey the commutator (10.13).

10.4 For three particles in the nuclear SD shell, classify the lowest states according to the irreps of SU(6) ⊃ SU(3) ⊃ SO(3) ⊃ SO(2). For this problem, assume that the forces are spin and isospin independent, and consider only the lowest two orbital states.

10.5 Find the relationship of the eight SU(3) operators $T_\pm$, $V_\pm$, $U_\pm$, T_3, and Y defined in Eqs. (8.2) and (8.7)–(8.8), and the nine oscillator operators $(A_i^j)_{kl} \equiv \delta_{ik}\delta_{jl} - \frac{1}{3}\delta_{kl}$, with a constraint $A_1^1 + A_2^2 + A_3^3 = 0$. ***

10.6 For four particles in the nuclear SD shell the lowest-energy orbital symmetry partition and its SU(3) (λ, μ) content are [4] : $(8,0), (4,2), (0,4), (2,0)$. Deduce the angular momentum content of these representations. Assuming an attractive $\boldsymbol{Q} \cdot \boldsymbol{Q}$ quadrupole interaction, order these SU(3) irreps with respect to energy.

10.7 Establish the connections of Eq. (10.27) by comparing the oscillator operators of Ch. 10 with the flavor matrix operators in Eqs. (8.2) and (8.7)–(8.8). ***

10.8 For four particles in the SD shell the SU(3) representation $(\lambda, \mu) = (8, 0)$ occurs in the Elliott model. What is the angular momentum content of this representation?

10.9 In Fig. 10.1(b) the $N = 3$ oscillator shell consists of the 2p and 1f orbitals (this is called the fp shell in nuclear physics). Use Young diagrams to classify the Elliott model orbital and SU(3) states for two particles in this oscillator shell. ***

11 SU(3) Matrix Elements

Chapter 6 described how to use the technology of Clebsch–Gordan coefficients and the Wigner–Eckart theorem to calculate matrix elements for the groups SU(2) and SO(3). In this chapter we wish to extend those methods to a more complicated group and illustrate some general means for calculating matrix elements when a symmetry and associated group structure of physical interest can be attached to a problem. We choose the group SU(3) for this purpose because it is complicated enough to illustrate general principles, but simple enough so as not to obscure those principles by overly complicated mathematics, and because there are varied and important physical applications of SU(3) symmetries in many fields of physics.

11.1 Clebsch–Gordan Coefficients for SU(3)

The Clebsch–Gordan series for the group SU(3) is of the form

$$(\lambda_1\mu_1) \otimes (\lambda_2\mu_2) = \sum_{(\lambda\mu)\oplus} M_{\lambda\mu} \cdot (\lambda\mu), \tag{11.1}$$

where $M_{\lambda\mu}$ is the multiplicity associated with a representation $(\lambda\mu)$ in the series and $\oplus$ reminds us that the sum on the right side is a direct sum of irreducible representations.

> The SU(3) Clebsch–Gordan coefficients relate basis vectors in the "uncoupled representations" on the left side to those in the "coupled representations" on the right side of Eq. (11.1). This generalizes the SU(2) Clebsch–Gordan series (3.21), which relates a product of two angular momenta defined in an uncoupled basis to a sum of states in a coupled basis of good total angular momentum.

An SU(3) state requires more labels to specify it than are required for SU(2). Let us illustrate using members of the SU(3) octet representation displayed in Fig. 11.1(a).

1. First, two SU(3) quantum numbers are needed to specify the irrep. A convenient choice is (λ, μ), but we will often use a single Greek letter for compact notation.
2. Next, individual members of the irrep (λ, μ) must be labeled. From Fig. 11.1(a), the weights Y and T_3 are not sufficient because the center site is degenerate and an additional quantum number is required. One convenient choice for flavor SU(3) is to give the SU(2) isospin T as well as Y and T_3. This distinguishes the two states at the central site of the **8** because one belongs to a $T = 0$ and one to a $T = 1$ isospin multiplet.

Using a more complete set of quantum numbers, the SU(3) Clebsch–Gordan series (11.1) becomes

$$\left|\lambda_1\mu_1T_1(T_3)_1y_1\right\rangle \otimes \left|\lambda_2\mu_2T_2(T_3)_2y_2\right\rangle = \sum_{\oplus} M_{\lambda\mu}\left|\lambda\mu TT_3y\right\rangle, \tag{11.2}$$

which may be inverted using unitarity to express the coupled representation as

$$\begin{aligned}\left|\lambda\mu TT_3y\right\rangle = \sum_{q_1}\sum_{q_2}\sum_{\rho}\left\langle\lambda_1\mu_1T_1(T_3)_1y_1;\lambda_2\mu_2T_2(T_3)_2y_2\right|\left.\lambda\mu TT_3y\rho\right\rangle \\ \times\left|\lambda_1\mu_1T_1(T_3)_1y_1\right\rangle\left|\lambda_2\mu_2T_2(T_3)_2y_2\right\rangle,\end{aligned} \tag{11.3}$$

where the symbols $q_n \equiv \{\lambda_n, \mu_n, T_n, (T_3)_n, y_n\}$ denote composite summation labels, ρ distinguishes representations with multiplicity greater than one, and the first factor is the SU(3) Clebsch–Gordan coefficient. Equations like (11.3) contain a lot of indices, so a more concise notation is sometimes useful. For example, consider the product $\mathbf{3}\otimes\bar{\mathbf{3}} = \mathbf{8}\oplus\mathbf{1}$. For the $\mathbf{8}$ we may write in highly schematic notation $|\mathbf{8}\rangle = \left\langle\mathbf{3},\bar{\mathbf{3}}\right|\left.\mathbf{8}\right\rangle|\mathbf{3}\rangle\left|\bar{\mathbf{3}}\right\rangle$, where $\left\langle\mathbf{3},\bar{\mathbf{3}}\right|\left.\mathbf{8}\right\rangle$ is the SU(3) Clebsch–Gordan coefficient in a rather terse form.

11.2 Constructing SU(3) Clebsch–Gordan Coefficients

The wavefunctions and Clebsch–Gordan coefficients for SU(3) representations may be determined by a process analogous to that considered previously for SO(3) [141].

Example 11.1 Consider the product of SU(3) fundamental representations,

$$\mathbf{3}\otimes\mathbf{3} = \mathbf{6}\oplus\bar{\mathbf{3}} =$$ 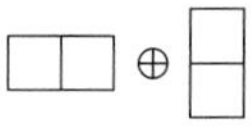

The basis states corresponding to the $\mathbf{3}$ and $\bar{\mathbf{3}}$ representations of SU(3) are illustrated in Fig. 11.1(b) and the corresponding quantum numbers are given in Table 11.1. For the

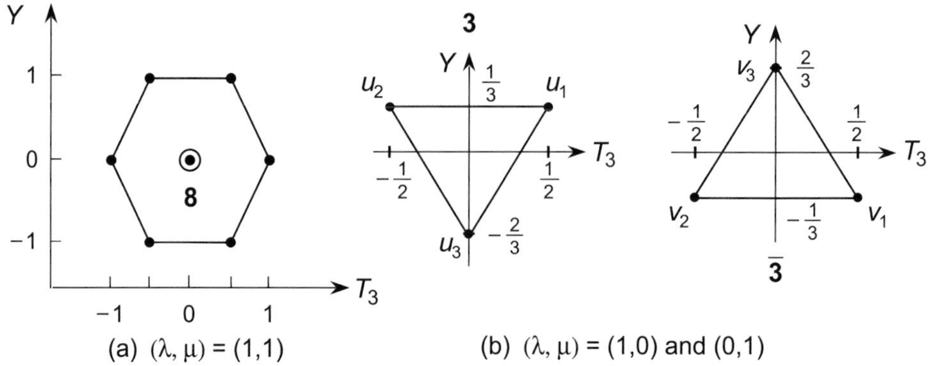

Fig. 11.1 (a) The SU(3) representation (1, 1) or $\mathbf{8}$. (b) The $\mathbf{3}$ and $\bar{\mathbf{3}}$ representations for SU(3).

Table 11.1. The **3** and $\mathbf{\bar{3}}$ basis states for SU(3)

u_i	t	t_3	y	v_i	t	t_3	y
u_1	$\frac{1}{2}$	$\frac{1}{2}$	$\frac{1}{3}$	v_1	$\frac{1}{2}$	$\frac{1}{2}$	$-\frac{1}{3}$
u_2	$\frac{1}{2}$	$-\frac{1}{2}$	$\frac{1}{3}$	v_2	$\frac{1}{2}$	$-\frac{1}{2}$	$-\frac{1}{3}$
u_3	0	0	$-\frac{2}{3}$	v_3	0	0	$\frac{2}{3}$

product $\mathbf{3} \otimes \mathbf{3}$ there are nine two-particle states $u_i u_j (i, j = 1, 2, 3)$. For the symmetric **6** occurring in the product, use of the Young projectors (Section 4.4) and normalizing gives

$$\begin{array}{|c|c|}\hline 1&1\\\hline\end{array} = u_1 u_1 \equiv \psi_1^{(6)} \qquad \begin{array}{|c|c|}\hline 1&2\\\hline\end{array} = \frac{1}{\sqrt{2}}(u_1 u_2 + u_2 u_1) \equiv \psi_2^{(6)},$$

$$\begin{array}{|c|c|}\hline 2&2\\\hline\end{array} = u_2 u_2 \equiv \psi_3^{(6)} \qquad \begin{array}{|c|c|}\hline 1&3\\\hline\end{array} = \frac{1}{\sqrt{2}}(u_1 u_3 + u_3 u_1) \equiv \psi_4^{(6)},$$

$$\begin{array}{|c|c|}\hline 2&3\\\hline\end{array} = \frac{1}{\sqrt{2}}(u_2 u_3 + u_3 u_2) \equiv \psi_5^{(6)} \qquad \begin{array}{|c|c|}\hline 3&3\\\hline\end{array} = u_3 u_3 \equiv \psi_6^{(6)},$$

and for the antisymmetric $\mathbf{\bar{3}}$,

$$\begin{array}{|c|}\hline 1\\\hline 2\\\hline\end{array} = \frac{1}{\sqrt{2}}(u_1 u_2 - u_2 u_1) \equiv \psi_1^{(\bar{3})} \qquad \begin{array}{|c|}\hline 1\\\hline 3\\\hline\end{array} = \frac{1}{\sqrt{2}}(u_1 u_3 - u_3 u_1) \equiv \psi_2^{(3)},$$

$$\begin{array}{|c|}\hline 2\\\hline 3\\\hline\end{array} = \frac{1}{\sqrt{2}}(u_2 u_3 - u_3 u_2) \equiv \psi_3^{(3)}.$$

The corresponding SU(3) Clebsch–Gordan coefficients are just the coefficients of terms in the preceding nine equations, which may be extracted by taking overlaps with wavefunctions. For example, we obtain the SU(3) Clebsch–Gordan coefficient

$$\left\langle \lambda_1 \mu_1 T_1 (T_3)_1 y_1; \lambda_2 \mu_2 T_2 (T_3)_2 y_2 \middle| \lambda \mu T T_3 y \right\rangle = \left\langle 10\tfrac{1}{2}\tfrac{1}{2}\tfrac{1}{3}; 10\tfrac{1}{2} - \tfrac{1}{2}\tfrac{1}{3} \middle| 2010\tfrac{2}{3} \right\rangle = \frac{1}{\sqrt{2}},$$

from the expression given above for $\psi_2^{(6)}$.

Example 11.2 Consider the SU(3) wavefunctions for $\mathbf{3} \otimes \mathbf{\bar{3}} = \mathbf{8} \oplus \mathbf{1}$. Using the graphical method of Section 8.10 (see Problem 8.7) and the basis vectors of Fig. 11.1(b) gives the results of Fig. 11.2. The states of the octet and singlet wavefunctions resulting from $\mathbf{3} \otimes \mathbf{\bar{3}} = \mathbf{8} \oplus \mathbf{1}$ are displayed in Fig. 11.3 (see Problem 11.3),and Clebsch–Gordan coefficients again can be obtained from overlaps with these wavefunctions. For the center states in Fig. 11.3,

1. $\psi_5^{(8)}$ is an isosinglet ($T = 0$),
2. $\psi_4^{(8)}$ is part of a $T = 1$ isotriplet that also contains $\psi_3^{(8)}$ and $\psi_6^{(8)}$, and
3. $\psi^{(1)}$ is the single state for the **1**,

and these states are all mutually orthogonal.

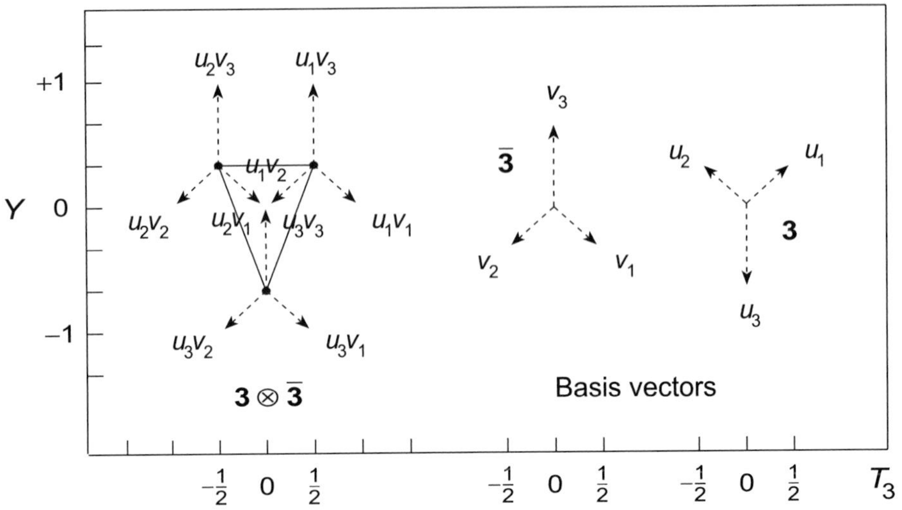

Fig. 11.2 Graphical construction of $\mathbf{3}\otimes\overline{\mathbf{3}}$ for SU(3). Tips of arrows in the left diagram correspond to weights for the nine states of the product $\mathbf{3}\otimes\overline{\mathbf{3}}=\mathbf{8}\oplus\mathbf{1}$. Center states in the product $[(T_3, Y)=(0,0)]$ are triply degenerate, with two states belonging to the $\mathbf{8}$ and one to the $\mathbf{1}$.

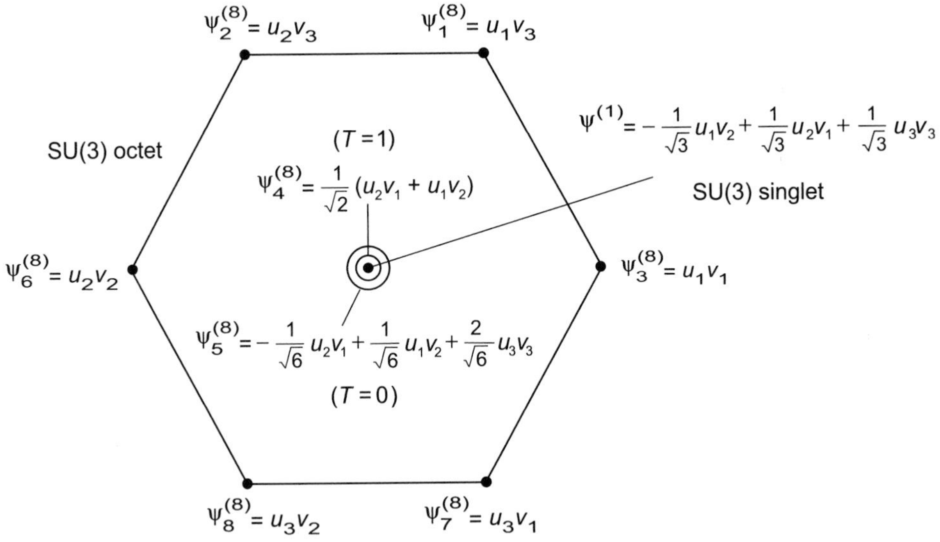

Fig. 11.3 SU(3) wavefunctions for $\mathbf{3}\otimes\overline{\mathbf{3}}=\mathbf{8}\oplus\mathbf{1}$. Center states $\psi_4^{(8)}$, $\psi_5^{(8)}$, and $\psi^{(1)}$ are mutually orthogonal. Coefficients in the equations are the Clebsch–Gordan coefficients.

Wavefunctions and Clebsch–Gordan coefficients for other representations may be obtained in a similar way, as illustrated in Problems 11.1–11.3 and 11.9. For example, Fig. 11.4 displays the wavefunctions for the **10** in the product $\mathbf{6}\otimes\mathbf{3}=\mathbf{10}\oplus\mathbf{8}$. Lichtenberg [141] may be consulted for further discussion.

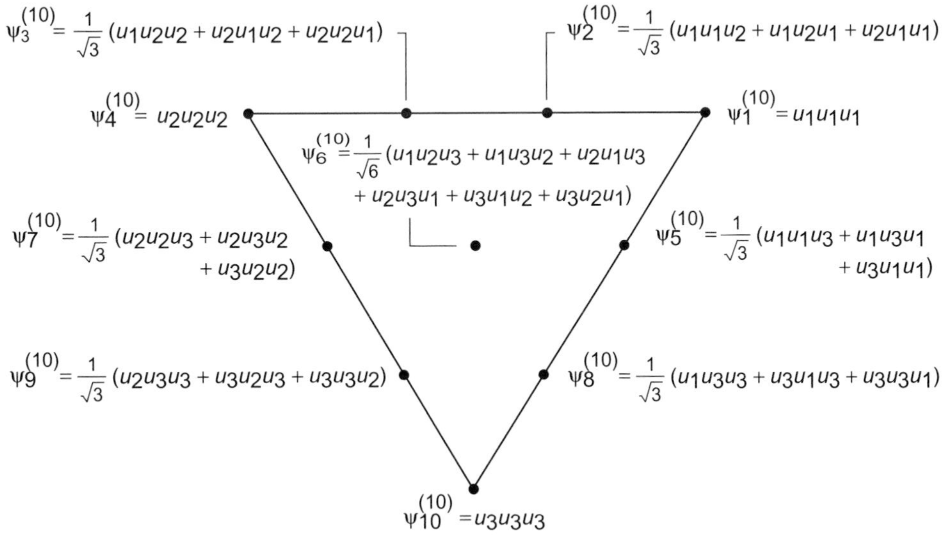

Fig. 11.4 Display in the weight space of the SU(3) wavefunctions for the **10** in the product $\mathbf{6} \otimes \mathbf{3} = \mathbf{10} \oplus \mathbf{8}$. The coefficients in the equations are the Clebsch–Gordan coefficients.

11.3 Matrix Elements of Generators

The results obtained in preceding sections may be used to calculate the matrix elements of generators within a representation. For example, using the state labeling in Fig. 11.3,

$$\left\langle \psi_5^{(8)} \middle| V_- \middle| \psi_1^{(8)} \right\rangle = \left\langle \psi_5^{(8)} \middle| V_- \middle| u_1 v_3 \right\rangle = \left\langle \psi_5^{(8)} \middle| u_3 v_3 \right\rangle + \left\langle \psi_5^{(8)} \middle| u_1 v_2 \right\rangle$$
$$= \frac{2}{\sqrt{6}} \langle u_3 v_3 | \, u_3 v_3 \rangle + \frac{1}{\sqrt{6}} \langle u_1 v_2 | \, u_1 v_2 \rangle = \sqrt{\frac{3}{2}}, \tag{11.4}$$

where the action of V_- from Fig. 8.1 was used on Fig. 11.1(b). Matrix elements for the **8** and **10** are summarized in Fig. 11.5 (Problems 11.4 and 11.5). Matrix elements of generators are important because transition operators must be proportional to generators if they are not to break the symmetry dynamically by causing transitions out of a representation.

11.4 Isoscalar Factors

Let us label the members of an SU(3) multiplet by $\psi_{y\,t\,t_3}^{(\alpha)}$, where α denotes the representation and y, t, and t_3 denote the hypercharge, isospin, and third component of isospin, respectively, obtained from the group chain $\mathrm{SU(3)}_{\mathrm{flavor}} \supset \mathrm{SU(2)}_{\mathrm{isospin}}$. Now consider direct products of SU(3) irreps. We can form states of good isospin from the product tensors,

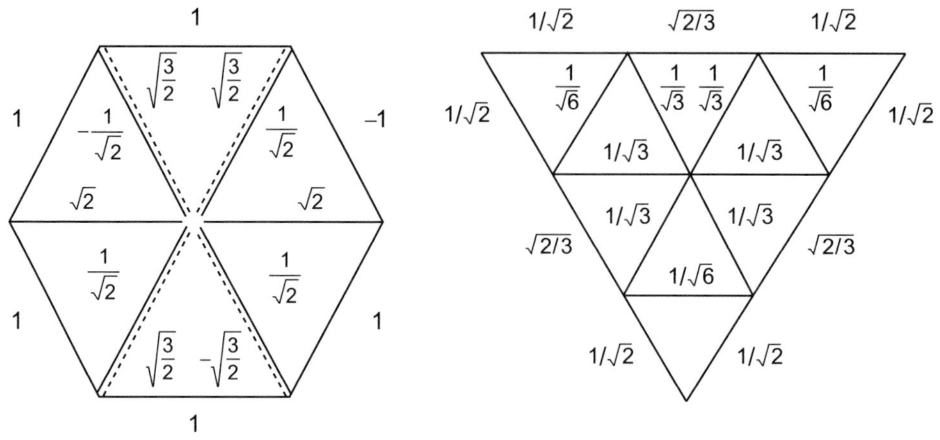

Fig. 11.5 Matrix elements of the SU(3) shift operators $T_\pm$, $V_\pm$, and $U_\pm$ in the **8** (left) and **10** (right) representations. Dashed lines in the left figure are for transitions to the isosinglet ($T = 0$ state) of the SU(3) octet representation.

$$\phi^{(T)}_{yty't'} = \sum_{t_3 t'_3} \left\langle tt_3 t' t'_3 \middle| TT_3 \right\rangle \psi^{(\alpha)}_{ytt_3} \psi^{(\beta)}_{y't't'_3}, \tag{11.5}$$

where $\left\langle tt_3 t' t'_3 \middle| TT_3 \right\rangle$ is an SU(2) Clebsch–Gordan coefficient. This is an eigenstate of SU(2) but not of SU(3). Eigenstates of SU(3) may be formed by taking linear combinations of these isospin functions:

$$\psi^{(\gamma,\rho)}_{YTT_3} = \sum_{yy'tt'} \langle \alpha yt; \beta y't' \| \gamma \rho YT \rangle \phi^{(T)}_{yty't'}, \tag{11.6}$$

where α, β, and γ are irrep labels and ρ is an additional label required to distinguish representations when two equivalent ones occur in the SU(3) direct product. The coefficient $\langle \alpha yt; \beta y't' \| \gamma \rho YT \rangle$ is termed the *isoscalar factor.* But we may also write

$$\psi^{(\gamma,\rho)}_{YTT_3} = \sum_{ytt_3} \sum_{y't't'_3} \left\langle \alpha ytt_3; \beta y't't'_3 \middle| \gamma \rho YTT_3 \right\rangle \psi^{(\alpha)}_{ytt_3} \psi^{(\beta)}_{y't't'_3}, \tag{11.7}$$

where the expansion coefficient is an SU(3) Clebsch–Gordan coefficient. Comparing with the preceding expressions, we see that the SU(3) Clebsch–Gordan coefficient may be written as a product of an SU(2) Clebsch–Gordan coefficient and an isoscalar factor:

$$\left\langle \alpha ytt_3; \beta y't't'_3 \middle| \gamma \rho YTT_3 \right\rangle = \langle \alpha yt; \beta y't' \| \gamma \rho YT \rangle \left\langle tt_3 t' t'_3 \middle| TT_3 \right\rangle. \tag{11.8}$$

The isoscalar factor does not depend on the weights for the SU(2) isospin subgroup of SU(3); thus, the isoscalar factors for all states in an isospin multiplet [horizontal row in the SU(3) weight space] are equivalent. This is the origin of the adjective "isoscalar." Isoscalar factors are elements of a unitary transformation and obey the usual orthogonality and symmetry conditions associated with such transformations; see Wybourne [224] for particulars.

11.4.1 Racah Factorization Lemma

The preceding factorization of the SU(3) Clebsch–Gordan coefficient into a Clebsch–Gordan coefficient for its SU(2) subgroup and an isoscalar factor independent of SU(2) weights is a specific example of the *Racah factorization lemma.*

> ***Racah Factorization Lemma:*** If $G_1 \supset G_2$, then $C_1 = fC_2$, where f is the $G_1 \supset G_2$ isoscalar factor, C_1 is the Clebsch–Gordan coefficient for G_1, and C_2 is the Clebsch–Gordan coefficient for G_2.

Generally, for a nested chain of subgroups, $G_1 \supset G_2 \supset G_3 \supset \cdots \supset G_n$, the Clebsch–Gordan coefficient of G_1 may be factored as $C_1 = f_2 C_2 = f_2 f_3 C_3 = f_2 f_3 \cdots f_n C_n$. Hence, evaluation of a matrix element in a subgroup chain reduces to determination of three distinct quantities: (1) the reduced matrix element, (2) the Clebsch–Gordan coefficients for the last group, and (3) the isoscalar factors for each step in the chain.

11.4.2 Evaluating and Using Isoscalar Factors

The relationship among the SU(3) Clebsch–Gordan coefficients C_1, the SU(2) Clebsch–Gordan coefficients C_2, and the corresponding isoscalar factor f is of the schematic form $C_1 = fC_2$. It follows that we may construct the isoscalar factor from this definition, if we know the SU(2) and SU(3) Clebsch–Gordan coefficients. The solution of Problem 11.8 illustrates for the states of the **8** in the SU(3) product $\mathbf{3} \otimes \bar{\mathbf{3}} = \mathbf{8} \oplus \mathbf{1}$. Conversely, given the isoscalar factors, SU(3) Clebsch–Gordan coefficients may be constructed from SU(2) Clebsch–Gordan coefficients. Various tabulations and computer codes give the isoscalar factors for $G_1 \supset G_2$, where the groups G_1 and G_2 are of physical interest. For SU(3) ⊃ SU(2) or SU(3) ⊃ SO(3), Refs. [192, 202] give useful compilations.

11.5 SU(3) ⊃ SO(3) Tensor Operators

In an obvious generalization of SO(3) spherical tensors, irreducible tensors $T(\alpha, \nu)$ under SU(3) ⊃ SO(3) may be defined through the commutation requirement [cf. Eq. (6.55)]

$$[F_i, T(\alpha,\nu)] = \sum_\sigma T(\alpha,\sigma)\langle \alpha\sigma | F_i | \alpha\nu\rangle, \tag{11.9}$$

where F_i is an SU(3) generator, the index α labels irreps, and ν labels states of an irrep. In terms of the SU(3) operators L_ν and Q_ρ introduced in Section 10.1.5, this takes the form [202]

$$[L_\nu, T^{\lambda\mu}_{\kappa LM}] = \sum_{\kappa' L'} \Big\langle (\lambda\mu)\kappa' L'(M+\nu) \Big| L_\nu \Big| (\lambda\mu)\kappa LM \Big\rangle T^{\lambda\mu}_{\kappa' L' M+\nu}, \tag{11.10}$$

$$[Q_\rho, T^{\lambda\mu}_{\kappa LM}] = \sum_{\kappa' L'} \Big\langle (\lambda\mu)\kappa' L'(M+\rho) \Big| Q_\rho \Big| (\lambda\mu)\kappa LM \Big\rangle T^{\lambda\mu}_{\kappa' L' M+\rho}, \tag{11.11}$$

where κ is an additional quantum number introduced to distinguish SU(3) states uniquely. From these definitions the generators themselves are tensors transforming according to the adjoint or $(1,1)$ representation. By analogy with the corresponding situation in SO(3) which is discussed in Section 6.5, SU(3) operators transforming according to the adjoint irrep are termed *vector operators*. The explicit $(1,1)$ tensors $T^{\lambda\mu}_{LM}$ may be chosen as

$$T^{11}_{11} = -\frac{1}{\sqrt{2}}L_+ = L_1 \qquad T^{11}_{10} = L_0,$$
$$T^{11}_{1-1} = \frac{1}{\sqrt{2}}L_- = L_{-1} \qquad T^{11}_{2\rho} = -\frac{1}{\sqrt{3}}Q_\rho, \tag{11.12}$$

where we have defined

$$L_\pm = L_x \pm iL_y \qquad L_1 = -\frac{1}{\sqrt{2}}L_+ \qquad L_{-1} = \frac{1}{\sqrt{2}}L_-, \tag{11.13}$$

and where phases are chosen so that the $T^{\lambda\mu}_{1\nu}$ coincide with the normal definitions of angular momentum tensors given in Section 6.5.

Example 11.3 Consider the commutation of L_0 with T^{11}_{11}. For $\nu = 0$,

$$\left\langle (\lambda\mu)\kappa' L'(M+\nu)\right| L_\nu \left|(\lambda\mu)\kappa LM\right\rangle = M\delta_{\kappa\kappa'}\delta_{LL'}.$$

For $L = M = 1$ and $\nu = 0$, Eq. (11.10) gives $[\,L_0, T^{11}_{LM}\,] = MT^{11}_{LM} = T^{11}_{11}$, but from Eq. (10.18),

$$[\,L_0, T^{11}_{11}\,] = [\,L_0, L_1\,] = -\sqrt{2}\,\langle 1011|\,11\rangle\, L_1 = L_1 = T^{11}_{11}.$$

Therefore, T^{11}_{11} satisfies Eq. (11.10) for $\nu = 0$.

The commutation with other components of L may be checked in a similar way and all satisfy Eq. (11.10). Therefore, T^{11}_{11} is a vector operator (Problem 11.6).

11.6 The SU(3) Wigner–Eckart Theorem

As for SU(2), it is often useful to factor SU(3) matrix elements into a part containing the dynamics and a factor that is essentially "geometrical" in nature. Thus, we wish to write a Wigner–Eckart theorem for SU(3) and, by analogy, for other more complicated groups. This is a relatively straightforward formal generalization. For example, consider a set of states labeled by the chain $SU(3)_{\text{flavor}} \supset SU(2)_{\text{isospin}}$. For an SU(3) tensor operator $T^\gamma_{YT_3}$ the corresponding SU(3) Wigner–Eckart theorem takes the form [compare Eq. (6.69)]

$$\langle \alpha y t t_3 |\, T^\gamma_{YT_3} \left|\beta y' t' t'_3\right\rangle = \sum_\rho \left\langle \alpha y t t_3; \beta y' t' t'_3 \right|\, \gamma\rho Y T T_3 \right\rangle \langle \alpha t \parallel T^\gamma \parallel \beta t'\rangle_\rho, \tag{11.14}$$

where ρ is an index labeling different occurrences of the same irrep if the product is not simply reducible. Just as SU(2) reduced matrix elements are independent of SU(2) weights, the SU(3) reduced matrix element $\langle \alpha t \,\|\, T^{\gamma} \,\|\, \beta t' \rangle_{\rho}$ is independent of the weight quantum numbers y and t_3. Equation (11.14) also may be expressed in terms of isoscalar factors $\langle \alpha yt; \beta y't' \| \gamma \rho YT \rangle$ and SU(2) Clebsch–Gordan coefficients $\langle tt_3 t't'_3 | \, TT_3 \rangle$,

$$\langle \alpha y t t_3 | \, T^{\gamma}_{YT_3} \, | \beta y' t' t'_3 \rangle = \sum_{\rho} \langle \alpha yt; \beta y't' \| \gamma \rho YT \rangle \langle tt_3 t't'_3 | \, TT_3 \rangle \langle \alpha t \,\|\, T^{\gamma} \,\|\, \beta t' \rangle_{\rho}. \quad (11.15)$$

As was the case for SU(2), if the sum over ρ in (11.14) or (11.15) involves a single term the reduced matrix element may be inferred by (1) determining the matrix element appearing on the left side of these expressions, either for the simplest theoretical situation or from a measurement, and then (2) inverting the equation using the unitarity of the coefficients to solve for the reduced matrix element. If the direct product is not simply reducible so that the sum over ρ involves more than one term, this procedure leads instead to a set of equations that must be solved simultaneously for the reduced matrix elements.

11.7 Structure of SU(3) Matrix Elements

The preceding considerations suggest that the evaluation of a matrix element for a system described by some subgroup chain reduces to the determination of three quantities:

1. the reduced matrix element,
2. the Clebsch–Gordan coefficients for the last group in the group chain, and
3. the isoscalar factors for each step in the group chain.

We know how to determine each of these, but before proceeding let us give some consideration to the general structure of the matrix elements to be calculated. This will often simplify the calculations, on the one hand, and provide useful physical insight on the other.

Suppose that the state $|A\rangle$ transforms as the irrep D_A, the state $|B\rangle$ transforms as the irrep D_B, and the state $|C\rangle$ transforms as the irrep D_C, all of some group G. The matrix element $\langle A | \, B \, | C \rangle$ is a number, so it transforms as a group scalar. This implies that the number of independent reduced matrix elements contributing to $\langle A | \, B \, | C \rangle$ is the number of times that the scalar irrep **1** appears in the direct product $\bar{D}_A \otimes D_B \otimes D_C$, where the bar denotes the conjugate representation (because $\langle A|$ transforms as the conjugate of $|A\rangle$). Alternatively, it is the number of times that D_A is contained in the direct product $D_B \otimes D_C$, since $\bar{D}_A \otimes D_A$ contains a single **1**. For SU(2) this product is simply reducible and D_A can appear no more than once in the product; thus, the summation over α in (6.69) has only a single term, leading to Eq. (6.71). For a group that is not simply reducible things are more complicated because an irrep can appear more than once in the direct product. Examples 11.4 and 11.5 illustrate.

Example 11.4 Consider an SU(3) matrix element of the form

$$\langle (\lambda_1 \mu_1) | \, T^{\lambda\mu} \, | (\lambda_2 \mu_2) \rangle = \langle (1,1) | \, T^{(11)} \, | (1,1) \rangle .$$

That is, the two states and the operator in the matrix element each transform as the 8-dimensional $(\lambda, \mu) = (1, 1)$ adjoint representation of SU(3). Since from Example 8.5,

$$(1,1) \otimes (1,1) = (2,2) \oplus (0,3) \oplus (3,0) \oplus (1,1) \oplus (1,1) \oplus (0,0),$$

there will be two distinct reduced matrix elements, corresponding to the two octets $(1, 1)$ appearing in the direct product.

Example 11.5 Consider matrix elements $\langle (3,0)| T^{(11)} |(3,0)\rangle$ of $T^{(11)}$ between states of the 10-dimensional SU(3) irrep $(3, 0)$. Since

$$(1,1) \otimes (3,0) = (4,1) \oplus (2,2) \oplus (3,0) \oplus (1,1),$$

$\langle (3,0)| T^{(11)} |(3,0)\rangle$ receives contributions from only a single reduced matrix element.

The possibility of more than one reduced matrix element for particular representations arises because there are *two* independent sets of SU(3) tensor operators that satisfy Eq. (11.9) for vector operators. One set consists of the generators F_i [just as the generator J_i is a vector operator for SU(2)]. The additional set is composed of the operators D_i,

$$D_i \equiv \frac{2}{3} \sum_{jk} d_{ijk} F_j F_k, \tag{11.16}$$

with the coefficients d_{ijk} defined through the anticommutator

$$\{ \lambda_i, \lambda_j \} = \frac{4}{3} \delta_{ij} \mathrm{I} + 2 d_{ijk} \lambda_k, \tag{11.17}$$

where $F_i = \frac{1}{2}\lambda_i$, the Gell-Mann matrices λ_i are given in Eq. (8.2), and I is the unit matrix. The d_{ijk} are completely symmetric in their indices; non-zero values were listed in Table 8.2.

The SU(2) Wigner–Eckart theorem implies that the matrix elements of a vector operator between states of good angular momentum are proportional to a matrix element of a generator. For example, $\langle JM'| \mathbf{r} |JM\rangle = \alpha(J) \langle JM'| \mathbf{J} |JM\rangle$, where $\alpha(J)$ is a constant depending on the representation but not on the SU(2) weight quantum number M. The generalization of this result to SU(3) is that the matrix elements of a vector operator (one transforming like the adjoint representation) taken between good SU(3) states has the form

$$\langle \alpha' \nu| T_i^{11} |\alpha\nu\rangle = C_1(\alpha) \langle \alpha' \nu| F_i |\alpha\nu\rangle + C_2(\alpha) \langle \alpha' \nu| D_i |\alpha\nu\rangle \tag{11.18}$$

where α is the irrep index, ν labels members of an irrep, and the constants C_1 and C_2 depend on the representation.

11.8 The Gell-Mann, Okubo Mass Formula

The masses within hadronic SU(3) flavor multiplets exhibit deviations from the average mass of the multiplet as large as 20%, indicating considerable symmetry breaking (see Fig. 9.2). The Gell-Mann, Okubo hypothesis is that the mass operator in an SU(3) flavor representation is of the form $M = M_0 + M'$, where M_0 transforms as an SU(3) scalar and M' is a perturbation breaking the full SU(3) symmetry that transforms as the eighth component of an SU(3) octet; that is, M' transforms like the hypercharge operator $F_8 = \frac{\sqrt{3}}{2}Y$ of Eq. (8.7). This assumption is motivated by noting that mass splittings in hadronic multiplets are small within isospin multiplets, but increase approximately linearly with the hypercharge quantum number.

Let us illustrate application of the Gell-Mann, Okubo hypothesis by considering the masses in the SU(3) irrep **8**, which will be given by

$$M_\nu = \left\langle \psi_\nu^{11} \middle| M \middle| \psi_\nu^{11} \right\rangle = M_\nu^{(0)} + \left\langle \psi_\nu^{11} \middle| M' \middle| \psi_\nu^{11} \right\rangle. \tag{11.19}$$

Now M' transforms as a component of an SU(3) tensor $T^{(11)}$ (that is, it transforms as a generator). We have seen in Section 11.7 that *two* reduced matrix elements will contribute when the Wigner–Eckart theorem is applied to the second term for the **8**, and the octet mass formula takes the form

$$M_\nu = M_\nu^{(0)} + C_1 \left\langle \psi_\nu^{11} \middle| F_8 \middle| \psi_\nu^{11} \right\rangle + C_2 \left\langle \psi_\nu^{11} \middle| D_8 \middle| \psi_\nu^{11} \right\rangle. \tag{11.20}$$

Explicitly, we have from Eqs. (8.7) and (11.16),

$$F_8 = \frac{\sqrt{3}}{2}Y \qquad D_8 = \frac{1}{\sqrt{3}}\left(T^2 - \frac{1}{3}F^2 - \frac{1}{4}Y^2\right),$$

$$T^2 = \sum_{i=1}^{3} F_i F_i \qquad F^2 = \sum_{i=1}^{8} F_i F_i = C_{\mathrm{SU}(3)},$$

as shown in Problem 11.7. Thus, the matrix elements are diagonal in the SU(3) ⊃ SU(2) basis and the mass formula for the octet takes the general form

$$M_\nu = a + bY + c\left(T(T+1) - \frac{1}{4}Y^2\right), \tag{11.21}$$

where a, b, and c are empirical constants for a given representation. Equation (11.21) implies testable relations among masses within an octet, as illustrated in the Example 11.6.

Example 11.6 Application of Eq. (11.21) to the hadrons in the nucleon octet of Figs. 9.2 and 9.5 leads to predictions like $\frac{1}{2}(M_N + M_\Xi) = \frac{3}{4}M_\Lambda + \frac{1}{4}M_\Sigma$, where the masses are the averages for the isospin multiplets. Experimentally it is found that [141]

$$\frac{1}{2}M_N + \frac{1}{2}M_{\Xi^0} = 1127.2 \pm 0.7\,\mathrm{MeV} \qquad \frac{3}{4}M_\Lambda + \frac{1}{4}M_{\Sigma^0} = 1134.8 \pm 0.2\,\mathrm{MeV},$$

which is in rather good agreement with this prediction of Eq. (11.21).

As another illustration, you are asked to show in Problem 11.13 that the mass formula for the Δ decuplet of Fig. 9.1 is

$$M_\nu = M_\nu^{(0)} + C_1 \left\langle \psi_\nu^{30} \middle| F_8 \middle| \psi_\nu^{30} \right\rangle = a + bY, \tag{11.22}$$

Box 11.1 **Prediction of the Ω^- Particle**

When the SU(3) flavor model was proposed the Ω^- at the bottom of the **10** in Fig. 9.1(d) had not yet been discovered. The mass formula of Eq. (11.22) indicates that there should be constant spacing between isospin multiplets of the **10**, and the known masses of the other members suggest that this spacing between isospin multiplets is about 150 MeV. The mass of the Ξ^* is about 1530 MeV, so Gell-Mann

1. predicted a new particle at $1530 + 150 = 1680$ MeV, and
2. predicted that its dominant decay mode would be by weak interactions because of its mass and quantum numbers.

The subsequent discovery of the Ω^- at 1672 MeV and its decay by the predicted modes was spectacular vindication of the SU(3) phenomenology. This led to broad acceptance of the SU(3) flavor model, which in turn motivated the quark hypothesis and eventually the gauge theory of strong interactions, quantum chromodynamics.

where a and b are parameters to be determined. The implications of this result are of some historical importance, as described in Box 11.1.

11.9 SU(3) Oscillator Reduced Matrix Elements

As another illustration of tensor methods for SU(3) matrix elements, let us consider the construction of reduced matrix elements for operators in the SU(3) $\supset$ SO(3) oscillator symmetry discussed in Ch. 10. We begin with the reduced matrix element for the creation operator $a^\dagger$ and its adjoint destruction operator a. Since other operators of interest can be formed from combinations of $a^\dagger$ and a, this will provide a means to construct reduced matrix elements for a variety of cases.

11.9.1 Spherical Operators

Let us switch from the cartesian representation employed in much of Ch. 10 to spherical operators defined by

$$a_{\pm 1} = \mp \frac{1}{\sqrt{2}}(a_x \pm i a_y) \qquad a_0 = a_z \qquad a^\dagger_{\pm 1} = \mp \frac{1}{\sqrt{2}}(a^\dagger_x \pm i a^\dagger_y) \qquad a^\dagger_0 = a^\dagger_z \tag{11.23}$$

[compare Eqs. (11.12) and (11.13)], which obey the commutation relations (Problem 11.10)

$$[\, a_p, a^\dagger_q \,] = (-1)^p \delta_{p,-q} \qquad [\, a_p, a_q \,] = [\, a^\dagger_p, a^\dagger_q \,] = 0, \tag{11.24}$$

for p and q equal to $(0, \pm 1)$. As shown in Problem 11.11, these new operators transform as vectors under SU(2). The SU(3) operators that were introduced in Ch. 10 may be rewritten in terms of these components, as illustrated in Table 11.2.

Table 11.2. SU(3) ⊃ SO(3) tensors

Operator	Tensor notation	Cartesian components	Spherical components
$a_x^\dagger$	$-\frac{1}{\sqrt{2}}(T_{11}^{10} - T_{1-1}^{10})$	$a_x^\dagger$	$-\frac{1}{\sqrt{2}}(a_1^\dagger - a_{-1}^\dagger)$
a_x	$-\frac{1}{\sqrt{2}}(T_{11}^{01} - T_{1-1}^{01})$	a_x	$-\frac{1}{\sqrt{2}}(a_1 - a_{-1})$
$a_y^\dagger$	$\frac{i}{\sqrt{2}}(T_{11}^{10} + T_{1-1}^{10})$	$a_y^\dagger$	$\frac{i}{\sqrt{2}}(a_1^\dagger + a_{-1}^\dagger)$
a_y	$\frac{i}{\sqrt{2}}(T_{11}^{01} + T_{1-1}^{01})$	a_y	$\frac{i}{\sqrt{2}}(a_1 + a_{-1})$
$a_z^\dagger$	T_{10}^{10}	$a_z^\dagger$	$a_0^\dagger$
a_z	T_{10}^{01}	a_z	a_0
$a_{\pm 1}^\dagger$	$T_{1\pm 1}^{10}$	$\mp\frac{1}{\sqrt{2}}(a_x^\dagger \pm i a_y^\dagger)$	$a_{\pm 1}^\dagger$
$a_0^\dagger$	T_{10}^{10}	$a_z^\dagger$	$a_0^\dagger$
$a_{\pm 1}$	$T_{1\pm 1}^{01}$	$\mp\frac{1}{\sqrt{2}}(a_x \pm i a_y)$	$a_{\pm 1}$
a_0	T_{10}^{01}	a_z	a_0
$\frac{-1}{\sqrt{3}}\left(H - \frac{3}{2}\right)$	T_{00}^{00}	$\frac{-1}{\sqrt{3}}\left(a_x^\dagger a_x + a_y^\dagger a_y + a_z^\dagger a_z\right)$	$\frac{1}{\sqrt{3}}\left(a_{-1}^\dagger a_1 + a_1^\dagger a_{-1} - a_0^\dagger a_0\right)$
L_0	T_{10}^{11}	$i(a_x a_y^\dagger - a_y a_x^\dagger)$	$a_1 a_{-1}^\dagger - a_{-1} a_1^\dagger$
$L_{\pm 1}$	$T_{1\pm 1}^{11}$	$\frac{1}{\sqrt{2}}(a_x^\dagger a_z - a_z^\dagger a_x)$ $\pm\frac{i}{\sqrt{2}}(a_y^\dagger a_z - a_z^\dagger a_y$	$\mp(a_{\pm 1}^\dagger a_0 - a_0^\dagger a_{\pm 1})$
Q_0	T_{20}^{11}	$-a_x^\dagger a_x - a_y^\dagger a_y + 2a_z^\dagger a_z$	$2a_0^\dagger a_0 + a_1^\dagger a_{-1} + a_{-1}^\dagger a_1$
$Q_{\pm 1}$	$T_{2\pm 1}^{11}$	$\pm\frac{\sqrt{6}}{2}(a_x^\dagger a_z + a_z^\dagger a_x)$ $-\frac{i\sqrt{6}}{2}(a_y^\dagger a_z + a_z^\dagger a_y)$	$-\sqrt{3}(a_{\pm 1}^\dagger a_0 + a_0^\dagger a_{\pm 1})$
$Q_{\pm 2}$	$T_{2\pm 2}^{11}$	$\frac{\sqrt{6}}{2}(a_x^\dagger a_x - a_y^\dagger a_y)$ $\pm\frac{i\sqrt{6}}{2}(a_y^\dagger a_x - a_x^\dagger a_y)$	$\sqrt{6}\,(a_{\pm 1}^\dagger a_{\pm 1})$

11.9.2 Matrix Elements for Creation and Annihilation Operators

The general SO(3) scalar that can be constructed from the product of two vectors is given by Eq. (6.61). If this is to be an SU(3) scalar too, it must be of the form $a^\dagger a$ or $aa^\dagger$, since

$$a^\dagger \simeq \mathbf{3} \qquad a \simeq \overline{\mathbf{3}} \qquad \mathbf{3} \otimes \overline{\mathbf{3}} = \mathbf{8} \oplus \mathbf{1},$$

but neither $\mathbf{3} \otimes \mathbf{3}$ nor $\overline{\mathbf{3}} \otimes \overline{\mathbf{3}}$ yields a scalar (the irrep $\mathbf{1}$). Thus,

$$T_{00}^{00} = \frac{1}{\sqrt{3}}\left(a_{-1}^\dagger a_1 + a_1^\dagger a_{-1} - a_0^\dagger a_0\right) \tag{11.25}$$

Table 11.3. Oscillator reduced matrix elements [224]

$$\langle NL \| T_0^{00} \| NL \rangle = -\tfrac{1}{\sqrt{3}} \sqrt{2L+1}\, N$$

$$\langle N+1, L+1 \| T_1^{10} \| NL \rangle = -\sqrt{(N+L+3)(L+1)}$$

$$\langle N+1, L-1 \| T_1^{10} \| NL \rangle = \sqrt{(N-L+2)L}$$

$$\langle NL \| T_1^{11} \| NL \rangle = \sqrt{L(L+1)(2L+1)}$$

$$\langle NL \| T_2^{11} \| NL \rangle = -(2N+3) \sqrt{\frac{L(L+1)(2L+1)}{(2L-1)(2L+3)}}$$

$$\langle N, L+2 \| T_2^{11} \| NL \rangle = -\sqrt{\frac{6(L+1)(L+2)(N-L)(N+L+3)}{(2L+3)}}$$

Other SU(3) reduced matrix elements may be obtained from

$$\langle NL \| T^{\lambda\mu} \| N'L' \rangle = (-1)^{L-L'} \langle N'L' \| (T^{\lambda\mu})^\dagger \| NL \rangle$$

is the most general SO(3) and SU(3) scalar. From Eq. (10.4), the Hamiltonian is

$$H = -a_1^\dagger a_{-1} - a_{-1}^\dagger a_1 + a_0^\dagger a_0 + \frac{3}{2} = -\sqrt{3}\left(T_{00}^{00} - \frac{\sqrt{3}}{2}\right), \tag{11.26}$$

and the scalar (11.25) may be expressed in terms of the Hamiltonian,

$$T_{00}^{00} = \frac{-1}{\sqrt{3}}\left(H - \frac{3}{2}\right). \tag{11.27}$$

As shown in Problem 11.12, this result may be used to construct the reduced matrix elements for the oscillator creation and annihilation operators,

$$\begin{aligned} \langle N+1, L+1 \| a^\dagger \| NL \rangle &= -\sqrt{(L+1)(N+L+3)}, \\ \langle N-1, L-1 \| a \| NL \rangle &= \sqrt{(N+L+1)L}. \end{aligned} \tag{11.28}$$

These reduced matrix elements for $a^\dagger$ and a may be used to construct more complicated SU(3) reduced matrix elements. Some important ones are summarized in Table 11.3.

11.9.3 Electromagnetic Transitions in the SD Shell

As was discussed in Ch. 10, there is experimental evidence for collective rotational bands in light atomic nuclei having valence particles that fill the $\hbar\omega = 2$ harmonic oscillator shell (the SD shell). This evidence takes the form of bands of states connected by strong electromagnetic transitions that have approximate $J(J+1)$ energy spectra, where J is the angular momentum of the state. Figure 11.6 gives an example of such a collective band in ^{20}Ne. For nuclei containing even numbers of neutrons and protons the lowest energy rotational bands correspond to angular momentum sequences $J = 0, 2, 4, \ldots$, and the strong

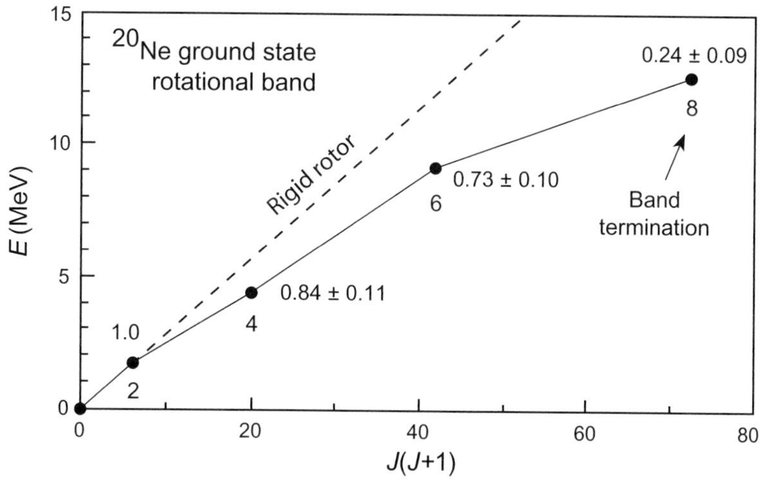

Fig. 11.6 Data indicating termination of the ground state rotational band in ^{20}Ne. Integers below points indicate angular momentum J; numbers near points represent ratios of electromagnetic transition strength for $J \longrightarrow J-2$ to that expected for a rigid collective rotor.

electromagnetic transitions correspond to emission of radiation with electric quadrupole ($E2$) character and connecting states J and $J-2$. Within the SU(3) oscillator symmetry the natural operators responsible for these $E2$ transitions are the quadrupole generators T^{11}_{2q} defined in Table 11.2. They can change the angular momentum by two units but they are generators and can cause transitions only between states in the same SU(3) representations [from Ch. 10, members of a rotational band are naturally assigned to the same SU(3) irrep]. Therefore, the reduced $E2$ transition rates $R(J+2 \to J)$ are expected to be of the form

$$R(J+2 \to J) \propto \langle N, J+2 \parallel T^{11}_{2} \parallel N, J\rangle^2 \simeq \frac{(J+1)(J+2)(N+J+3)(N-J)}{2J+3}, \tag{11.29}$$

where Table 11.3 was used. We notice three characteristic predictions of SU(3) symmetry.

Rotational Spectrum: The spectrum is of the form $J(J+1)$ [see Fig. 10.4 and Eq. (10.35)], which is characteristic of a collective rotational band.

Collectively Enhanced Transition Rates: Electromagnetic transitions are enhanced (are of greater than single-particle strength) for low-lying states.

1. For fixed J, electromagnetic rates vary as $\simeq N^2$.
2. For low-lying SD shell states, irreps are $(\lambda, \mu) \simeq (2N, 0)$ for nuclei filling the first half of the shell and $(\lambda, \mu) \simeq (0, 2N)$ for nuclei filling the second half of the shell. These states have the largest expectation values of the second-order Casimir operator.

Thus, for low-lying states larger particle number n implies larger average oscillator number N and the transition rate increases roughly quadratically with valence particle number.

Band Terminations: The factor $N - J$ in Eq. (11.29) implies an *angular momentum cutoff*, ultimately because finite angular momentum is available from the single-particle states of the valence space. For example, ^{20}Ne has two protons and two neutrons in the SD shell. The ground-band irrep is $(8, 0)$ and the maximum angular momentum from orbital motion results from placing each particle in a d orbital ($L = 2$), and aligning them to give $4 \times 2\hbar = 8\hbar$ of angular momentum. The SU(3) ground state band then terminates at angular momentum $8\hbar$ because $N = 8$ for this case. There is evidence for such band terminations. For example, Fig. 11.6 shows data for ^{20}Ne. We see that the ground band is not observed beyond $J = 8\hbar$, the rotational spectrum near band termination is beginning to be distorted, and the $E2$ transition strength is increasingly suppressed as the band termination is approached.

11.10 Lie Algebras and Many-Body Systems

We conclude from this chapter and Ch. 10 that the Elliott SU(3) model gives a reasonable description of both the spectrum and the transition rates for low-energy states in light nuclei, as a consequence of the group theory of SU(3) and its angular momentum subgroups, independent of details. This model is limited to light nuclei for which a harmonic oscillator potential is reasonable, but it gives a concrete example suggesting that Lie groups and Lie algebras may have relevance for the description of complex many-body systems. We will explore that possibility extensively in later chapters.

Background and Further Reading

The methods of this chapter are discussed in Lichtenberg [141], Wybourne [224], Harvey [106], Gasiorowicz [66], and Vergados [202].

Problems

11.1 Construct the wavefunctions and Clebsch–Gordan coefficients for the irrep **10** in the SU(3) Clebsch–Gordan series $\mathbf{6} \otimes \mathbf{3} = \mathbf{10} \oplus \mathbf{8}$. *Hint*: See Fig. 11.4. ***

11.2 Construct the wavefunctions and Clebsch–Gordan coefficients for the SU(3) irrep **8** in $\mathbf{6} \otimes \mathbf{3} = \mathbf{10} \oplus \mathbf{8}$. *Hint*: The upper right state of the octet in the following figure (indicated by the arrow) has the same weight as $\psi_2^{(10)}$.

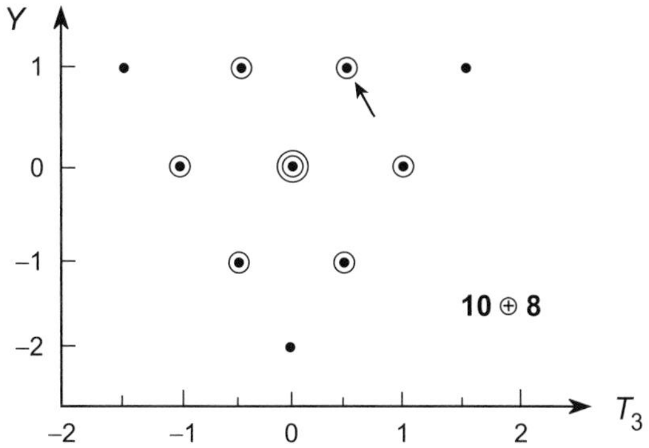

Construct a wavefunction with this weight that is orthogonal to $\psi_2^{(10)}$. Then proceed as in Problem 11.1. ***

11.3 Construct the wavefunctions and Clebsch–Gordan coefficients for the product states of $\mathbf{3} \otimes \overline{\mathbf{3}} = \mathbf{8} \oplus \mathbf{1}$ illustrated graphically in Fig. 11.2. *Hint*: The six states on the boundary of the product diagram in the left part of Fig. 11.2 clearly belong to the **8** and can be made in only one way each from the basis vectors (u_1, u_2, u_3) of the **3** and (v_1, v_2, v_3) of the $\overline{\mathbf{3}}$. Construct the remaining states starting from these and requiring that the three center states be mutually orthogonal and normalized, with two belonging to the **8** and one to the **1**. ***

11.4 Calculate the non-vanishing matrix elements of the six SU(3) raising and lowering operators $U_\pm$, $V_\pm$, and $T_\pm$ between states of the octet representation *Hint*: The required wavefunctions are given in Fig. 11.3.

11.5 Calculate the non-vanishing matrix elements of the six SU(3) raising and lowering operators $U_\pm$, $V_\pm$, and $T_\pm$ between states of the decuplet representation arising in $\mathbf{6} \otimes \mathbf{3} = \mathbf{10} \oplus \mathbf{8}$. *Hint*: The required wavefunctions are given in Fig. 11.4.

11.6 Show that in the Elliott SU(3) model in Ch. 10, the operator

$$T_{11}^{11} = -\frac{1}{\sqrt{2}}(L_x + iL_y)$$

is a vector operator [it transforms as the SU(3) adjoint representation]. *Hint*: Part of the problem is already worked in Example 11.3. ***

11.7 Beginning with Eq. (11.16), prove that

$$D_8 = \frac{1}{\sqrt{3}}\left(T^2 - \frac{1}{3}F^2 - \frac{1}{4}Y^2\right),$$

where we have defined

$$F^2 \equiv \sum_i F_i F_i \qquad T^2 \equiv F_1^2 + F_2^2 + F_3^2 \qquad Y \equiv \frac{2}{\sqrt{3}} F_8.$$

Show that this leads to Eq. (11.21) with the Gell-Mann, Okubo assumptions.

11.8 Calculate the isoscalar factors for the irrep **8** that is contained in the SU(3) direct product $\mathbf{3} \otimes \bar{\mathbf{3}} = \mathbf{8} \oplus \mathbf{1}$. ***

11.9 Determine the Clebsch–Gordan coefficients for the SU(4) product $\mathbf{4} \otimes \mathbf{4} = \mathbf{10} \oplus \mathbf{6}$.

11.10 Show that the operators a_q and $a_p^\dagger$ defined in Eq. (11.23) obey the commutators given in Eq. (11.24). ***

11.11 Find the commutators of the spherical operators $a_{0,\pm1}^\dagger$ defined in Eq. (11.23) with the angular momentum operators. Thus show that they transform as rank-1 spherical tensors under SO(3). *Hint*: See Eqs. (6.56)–(6.58).

11.12 Use Eq. (11.27) to construct the reduced matrix elements given in Eq. (11.28). *Hint*: Use $\langle H \rangle = \langle NLM | \, H \, | NLM \rangle = N + \frac{3}{2}$ and the SU(3) Wigner–Eckart theorem. ***

11.13 Show that the Gell-Mann, Okubo mass hypothesis of Section 11.8 leads to the mass formula (11.22) for the baryon SU(3) flavor representation $(3, 0)$.

12 Introduction to Non-Compact Groups

Most of the groups dealt with to this point have been compact, meaning loosely that their parameter spaces have finite volume because they are closed and bounded. We have mentioned some non-compact groups such as the translation group and the Lorentz group, but have not dwelled on them. This chapter and the next three take a more systematic look at non-compact groups. As we shall see, compact and non-compact groups share many properties but non-compact groups have certain features that are very different from those of compact groups. These can have significant implications for both the mathematical analysis and the interpretation of such groups in physical applications.

12.1 Review of the Compact Group SU(n)

Let us introduce a simple non-compact group by first recalling some basic features of the *compact group* SU(n). For an element U in a matrix representation of SU(n), $U^\dagger U = 1$. Consider the action of U on an n-dimensional vector ξ,

$$\xi' = U\xi = U\begin{pmatrix} \xi_1 \\ \xi_2 \\ \vdots \\ \xi_n \end{pmatrix}, \tag{12.1}$$

where ξ_i are complex components of the vector. But

$$(\xi')^\dagger \xi' = (U\xi)^\dagger U\xi = \xi^\dagger U^\dagger U\xi = \xi^\dagger \xi \tag{12.2}$$

and the *norm* of the vector $|\xi|$ is preserved under SU(n) transformations:

$$|\xi| \equiv \sqrt{|\xi_1|^2 + |\xi_2|^2 + \cdots + |\xi_n|^2} = |\xi'|. \tag{12.3}$$

Let us write this condition in the form

$$|\xi| = \sqrt{\xi^\dagger g \xi}, \tag{12.4}$$

where the *metric tensor* g is an $n \times n$ matrix. Obviously, for this case g is just the $n \times n$ unit matrix $\mathrm{I}^{(\mathrm{n})}$,

$$g = \mathrm{I}^{(\mathrm{n})} \equiv \mathrm{Diag}(1, 1, 1, \ldots), \tag{12.5}$$

and (12.4) seems just a pedantic way of writing (12.3). However, the form of Eq. (12.4) suggests a generalization of a preserved norm through insertion of a non-trivial metric tensor g.

12.2 The Non-Compact Group SU(l, m)

Now let us define a new Lie group that we shall label SU(l, m) with $l + m = n$, such that the elements of the new group preserve the "norm"

$$|\xi| \equiv \sqrt{\xi^\dagger g\, \xi} = \sqrt{|\xi_1|^2 + |\xi_2|^2 + \cdots + |\xi_l|^2 - |\xi_{l+1}|^2 - |\xi_{l+2}|^2 - \cdots - |\xi_n|^2}, \qquad (12.6)$$

where the metric tensor g may be written as an $(l + m) \times (l + m)$ diagonal matrix

$$g = \begin{pmatrix} \mathrm{I}^{(l)} & 0 \\ 0 & -\mathrm{I}^{(m)} \end{pmatrix}, \qquad (12.7)$$

with $\mathrm{I}^{(k)}$ representing a $k \times k$ unit matrix. The argument (l, m) of SU(l, m) then indicates that the metric has l diagonal entries of positive sign and m of negative sign.

12.2.1 Signature of the Metric

If the matrix representing the metric tensor is diagonalized and the number of positive and negative eigenvalues counted, it is said to be (1) *positive definite* if all signs are positive, (2) *negative definite* if all signs are negative, and (3) *indefinite* if some signs are positive and some are negative. Thus, the metric (12.7) is indefinite, while the metric (12.5) is positive definite. The *signature* of a metric is an indication of the number of positive and negative entries when diagonalized. In the present example the signature can be denoted by (l, m).

Example 12.1 For SU(1, 1) the explicit group transformations are of the form

$$\begin{pmatrix} \xi_1' \\ \xi_2' \end{pmatrix} = U \begin{pmatrix} \xi_1 \\ \xi_2 \end{pmatrix} \qquad |\xi_1'|^2 - |\xi_2'|^2 = |\xi_1|^2 - |\xi_2|^2,$$

with the restriction $\det U = 1$ [the "S" designation in SU(l, m)]. Since

$$(\xi')^\dagger g \xi' = \xi^\dagger U^\dagger g U \xi = \xi^\dagger g \xi,$$

we have that

$$U^\dagger g U = g \qquad g = \begin{pmatrix} 1 & 0 \\ 0 & -1 \end{pmatrix} = \sigma_3,$$

where σ_3 is given in Eq. (3.11). Therefore, the SU(1, 1) matrices satisfy $U^\dagger \sigma_3 U = \sigma_3$, which we recognize as a generalization of the usual unitarity condition $U^\dagger U = 1$.

Thus the metric for SU(1, 1) is indefinite, with signature (1, 1).

12.2.2 Parameter Space for SU(1, 1)

Let us now parameterize a general element of SU(1, 1) in the form

$$U = \begin{pmatrix} a & b \\ c & d \end{pmatrix} \qquad U^\dagger = \begin{pmatrix} a^* & c^* \\ b^* & d^* \end{pmatrix},$$

where a, b, c, and d are complex numbers. Then the requirement $U^\dagger g U = g$ implies that

$$\begin{pmatrix} a^* & c^* \\ b^* & d^* \end{pmatrix} \begin{pmatrix} 1 & 0 \\ 0 & -1 \end{pmatrix} \begin{pmatrix} a & b \\ c & d \end{pmatrix} = \begin{pmatrix} 1 & 0 \\ 0 & -1 \end{pmatrix},$$

and upon multiplying out the matrices on the left side,

$$\begin{pmatrix} a^*a - c^*c & a^*b - c^*d \\ b^*a - d^*c & b^*b - d^*d \end{pmatrix} = \begin{pmatrix} 1 & 0 \\ 0 & -1 \end{pmatrix}.$$

This matrix equation and the constraint $\det U = 1$ impose the conditions

$$a^*a - c^*c = 1 \qquad a^*b - c^*d = 0 \qquad b^*b - d^*d = -1 \qquad ad - cb = 1,$$

which require that $b^* = c$ and $a = d^*$. Hence the most general form of U for SU(1, 1) is

$$U = \begin{pmatrix} a & b \\ b^* & a^* \end{pmatrix}.$$

Writing the complex numbers $a = x_1 + ix_2$ and $b = x_3 + ix_4$ in terms of the four real quantities x_i permits an arbitrary element U of SU(1, 1) to be parameterized as

$$\begin{aligned} U &= \begin{pmatrix} x_1 + ix_2 & x_3 + ix_4 \\ x_3 - ix_4 & x_1 - ix_2 \end{pmatrix} \\ &= x_1 \begin{pmatrix} 1 & 0 \\ 0 & 1 \end{pmatrix} + ix_2 \begin{pmatrix} 1 & 0 \\ 0 & -1 \end{pmatrix} + x_3 \begin{pmatrix} 0 & 1 \\ 1 & 0 \end{pmatrix} - x_4 \begin{pmatrix} 0 & -i \\ i & 0 \end{pmatrix} \\ &= x_1\sigma_0 + x_3\sigma_1 - x_4\sigma_2 + ix_2\sigma_3, \end{aligned}$$

where the σ_i are the unit matrix and the 2×2 Pauli matrices defined in Eq. (3.11). Then the condition $\det U = 1$ requires that $x_1^2 + x_2^2 - x_3^2 - x_4^2 = 1$. This is the equation of an hyperboloid and defines an *unbounded manifold,* implying that the parameter space of SU(1, 1) is *not compact.* This result may be compared with a similar analysis for SU(2), where the parameter space was found to be that of a sphere S^3 and thus compact [see Eq. (6.77)]. The crucial difference between the manifolds of SU(2) and SU(1, 1) is seen to lie in the difference between the positive definite metric (12.5) and the indefinite metric (12.7).

12.3 The Non-Compact Group SO(*l*, *m*)

The preceding discussion may be generalized to other groups. For example, recall from Section 6.1 that SO(n) leaves invariant the length of a vector $\boldsymbol{x} = (x_1, x_2, \ldots, x_n)$ in

n-dimensional euclidean space, $|\boldsymbol{x}| = \sqrt{x^{\mathrm{T}} g x} = \sqrt{x^{\mathrm{T}} x}$, where $g = \mathrm{I}^{(\mathrm{n})}$ is the $n \times n$ unit matrix. Let us now define a group SO(l, m) such that its transformations instead leave invariant

$$|\boldsymbol{x}| = \sqrt{x^{\mathrm{T}} g x} \qquad g = \begin{pmatrix} \mathrm{I}^{(\mathrm{l})} & 0 \\ 0 & -\mathrm{I}^{(\mathrm{m})} \end{pmatrix}, \tag{12.8}$$

with a restriction that the transformation matrices have unit determinant. This group will be non-compact because of the indefinite metric in (12.8). For example, we shall explore in Ch. 13 the non-compact group of Lorentz transformations that leave the quadratic form $t^2 - x^2 - y^2 - z^2$ invariant. In this notation the Lorentz group is SO(3, 1), with the argument indicating that the metric diagonal has three entries of one sign and one with opposite sign. (Strictly one could distinguish the groups SO(3, 1) or SO(1, 3), depending on the metric signature in use.)

12.4 Euclidean Groups

The group of symmetry transformations in n-dimensional euclidean space $\mathbb{R}^n$ is termed the *euclidean group* E_n. This group is composed of the continuous linear transformations that preserve the length of vectors, which are of two general types: (1) uniform translations, and (2) rotations. Because translations are unbounded, the groups E_n are non-compact. The euclidean groups are of interest for several reasons. First, they are associated with both classical and quantum dynamics in euclidean space, as elaborated in Box 12.1. Second, they illustrate many features common to non-compact groups. Finally, they introduce terminology and mathematical techniques that will be important in analysis of the Lorentz and Poincaré groups that underlie the full four-dimensional spacetime of special relativity.

12.4.1 The Euclidean Group E_3 for 3D Space

In 3D euclidean space the momentum and angular momentum operators are ($\hbar = 1$ units)

$$P_j = -i \frac{\partial}{\partial x_j} \qquad L_j = -i\, \epsilon_{jkm}\, x_k \frac{\partial}{\partial x_m}, \tag{12.9}$$

where ϵ_{jkm} is the antisymmetric rank-3 tensor with properties given in Table 3.2. By explicit commutation these six operators generate the E_3 Lie algebra (Problem 12.1),

$$[\, L_i, L_j \,] = i\epsilon_{ijk} L_k \qquad [\, P_i, L_j \,] = i\epsilon_{ijk} P_k \qquad [\, P_i, P_j \,] = 0. \tag{12.10}$$

An analysis of E_3 may be found in Tung [199]. Here we restrict consideration to the subgroup E_2 of euclidean motions in a plane.

12.4.2 The Euclidean Group E_2 for 2D Space

The euclidean group E_2 preserves the length of all vectors in the $\mathbb{R}^2$ plane. The most general transformation may be expressed as $x_i' = R_{ij} x_j + b_i$ (sum on repeated indices), where the first term corresponds to a rotation implemented by the two-dimensional orthogonal matrix

Box 12.1 **Dynamics in Euclidean Spaces**

In classical and quantum physics the dynamical content of a system is embodied in its Hamiltonian, which is a sum of kinetic and potential energy contributions.

Kinetic Energy

The classical kinetic energy is of the form

$$T = \frac{1}{2}\sum_i m_i v_i^2 = \frac{1}{2}\sum_i m_i \left(\frac{d\boldsymbol{x}_i}{dt}\right)^2,$$

where $\boldsymbol{x}_i$ is the coordinate, m_i is the mass, and v_i is the velocity of particle i. It is invariant under all euclidean transformations because the differential $d\boldsymbol{x}_i$ is the difference of two coordinates and so is not changed by translations, while $\boldsymbol{v}_i^2$ is obviously invariant under rotation. A similar conclusion follows for the quantized version of T, by correspondence principle arguments.

Potential Energy

The classical potential energy V is a function of the coordinates but the assumed homogeneity of space implies that V should be independent of coordinate origin; thus an acceptable potential energy can depend only on $\boldsymbol{x}_{ij} = \boldsymbol{x}_i - \boldsymbol{x}_j$. Likewise, isotropy of space implies that the potential energy cannot depend on orientation, so $\boldsymbol{x}_{ij}$ can enter the potential energy only in rotationally invariant combinations. For two-particle systems, this implies that the potential must be *central*, depending only on the magnitude of the coordinate separation: $V \to V(r)$, with $r = |\boldsymbol{x}_1 - \boldsymbol{x}_2|$. Again, the correspondence principle implies similar conclusions for the quantized theory.

Euclidean Symmetry and Dynamics

Therefore, symmetries of the euclidean group are connected intimately with the dynamics of classical and quantum systems. As discussed in Section 2.1, these symmetries lead directly to conservation laws like those of linear momentum and angular momentum in the corresponding dynamical systems.

R and the second term represents a translation by a distance b^i in the direction i. For $\mathbb{R}^2$ the explicit forms of these transformations represent the generalization of the 2D rotations introduced in Section 6.2.1 to include uniform translations in the plane [199],

$$x_1' = x_1 \cos\phi - x_2 \sin\phi + b_1 \qquad x_2' = x_1 \sin\phi + x_2 \cos\phi + b_2.$$

The corresponding group multiplication law for elements $g(\boldsymbol{b}, \phi)$ is

$$g(\boldsymbol{b}_2, \phi_2)g(\boldsymbol{b}_1, \phi_1) = g(\boldsymbol{b}_3, \phi_3) \qquad \boldsymbol{b}_3 \equiv R(\phi_2)\,\boldsymbol{b}_1 + \boldsymbol{b}_2 \qquad \phi_3 \equiv \phi_1 + \phi_2, \tag{12.11}$$

where $\boldsymbol{b}_n$ is a two-dimensional translation vector. The multiplication law (12.11) is characteristic of *semidirect product groups*.

12.4.3 Semidirect Product Groups

Letting $\otimes_s$ denote the semidirect product, we will show below that E_2 may be written as

$$E_2 = T_2 \otimes_s U(1), \tag{12.12}$$

where T_2 is the two-element group of translations in the plane that forms an abelian invariant subgroup of E_2, and $U(1) \sim SO(2)$ is the group of 2D rotations. The semidirect product nature of such groups is the basis for a standard way of constructing their representation theory called the *method of induced representations,* to be considered below. Other semidirect product groups of physical interest include the semidirect product of translations and rotations in three dimensions, E_3, and the Poincaré group of Ch. 15, which is the semidirect product of the Lorentz group and the group of 4D spacetime translations. The group multiplication law (12.11) implies that a general group element $g(\boldsymbol{b}, \phi)$ of E_2 may be factored, for if we denote a pure rotation by $R(\phi)$ and a pure translation by $T(\boldsymbol{b})$,

$$\begin{aligned} g(\boldsymbol{b}, \phi)R^{-1}(\phi) &= g(\boldsymbol{b}, \phi)g(\boldsymbol{0}, -\phi) \\ &= g(R(\phi)\cdot \boldsymbol{0} + \boldsymbol{b}, \phi - \phi) \\ &= g(\boldsymbol{b}, 0) = T(\boldsymbol{b}), \end{aligned}$$

where $R^{-1}(\phi) = R(-\phi)$ was used in line one and Eq. (12.11) was used in line two. Therefore, multiplying from the right by $R(\phi)$,

$$g(\boldsymbol{b}, \phi) = T(\boldsymbol{b})R(\phi), \tag{12.13}$$

and a 2D euclidean transformation factors into a product of a translation and a rotation.

12.4.4 Algebraic Properties of E_2

From the commutation relations (12.10) for E_3, the E_2 algebra may be written as

$$[\,J, P_k\,] = i\epsilon_{km}P_m \qquad [\,P_1, P_2\,] = 0 \qquad (k = 1, 2), \tag{12.14}$$

where J is the angular momentum generator and ϵ_{km} is the antisymmetric tensor with components $\epsilon_{11} = \epsilon_{22} = 0$ and $\epsilon_{12} = -\epsilon_{21} = 1$, and satisfying $\epsilon_{ij}\epsilon_{jk} = \delta_{ik}$. As you are asked to show in Problem 12.2, $P^2 \equiv P_1^2 + P_2^2$ commutes with all generators and thus acts as a *Casimir operator* for E_2. Let us next introduce the operators $P_\pm$ through

$$P_\pm \equiv P_1 \pm iP_2. \tag{12.15}$$

From (12.14), these have the commutation properties

$$[\,J, P_\pm\,] = \pm P_\pm, \tag{12.16}$$

and if $J\,|m\rangle = m\,|m\rangle$, then $J\left(P_\pm\,|m\rangle\right) = (m \pm 1)\left(P_\pm\,|m\rangle\right)$, as shown in Problem 12.3. Thus $P_\pm$ acts as a raising and lowering operator for eigenvalues of J.

These results bring to mind the similar role of $J_\pm$ for SU(2) described in Section 3.3.3, with one big difference [125]. For SU(2) the spectrum generated by repeated application of $J_\pm$ is constrained by

$$\langle J_3^2 \rangle \leq \langle J^2 \rangle = \langle J_1^2 + J_2^2 + J_3^2 \rangle,$$

because J_1, J_2, and J_3 are components of a *single vector*. For E_2 there is *no relation* between J and the components of P, so application of P_+ generates an *infinite tower of J eigenvalues* that differ by integers. But the individual eigenvalues can be integer or half-integer (as expected for physical eigenvalues) only if allowed physical states are zero eigenvalues of the Casimir $P^2 = P_1^2 + P_2^2$, so it is necessary to require that $P_1 |m\rangle = P_2 |m\rangle = 0$. Thus, any physical representations associated with E_2 will have zero eigenvalues of P^2, and will be characterized by a single quantum number J taking on integer or half-integer values and having the dimensions of angular momentum. We shall return to this discussion of the representation space of E_2 in considering the Poincaré group in Ch. 15. There it will be seen that the zero eigenvalues of the E_2 Casimir operator P^2 are associated with *massless particles,* and that the single eigenvalue J characterizing the physical states of those particles will be interpreted as the *helicity* (projection of angular momentum in the direction of motion) for the massless particle.

12.4.5 Invariant Subgroup of Translations

The commutation relations (12.14) mean that P_k transforms like the components of a vector operator (compare discussion in Section 6.4 for 3D rotations),

$$e^{-iJ\phi} P_k e^{iJ\phi} = P_m\, R(\phi)_{mk}, \tag{12.17}$$

where J is an angular momentum generator and $R(\phi)$ is a rotation matrix. It follows that

$$e^{-iJ\phi} \boldsymbol{P} \cdot \boldsymbol{b}\; e^{iJ\phi} = e^{-iJ\phi} P_k e^{iJ\phi} b_k = P_m R(\phi)_{mk} b_k \equiv \boldsymbol{P} \cdot \boldsymbol{b}',$$

where $b'_m \equiv R(\phi)_{mk} b_k = R(\phi)\boldsymbol{b}$. Since P_1 and P_2 commute with each other, the most general translation can be written $T(\boldsymbol{b}) = e^{-iP_1 b_1} e^{-iP_2 b_2} = e^{-i\boldsymbol{P}\cdot\boldsymbol{b}}$, which implies that

$$\begin{aligned} R(\phi)T(\boldsymbol{b})R(\phi)^{-1} &= e^{-iJ\phi} T(\boldsymbol{b}) e^{iJ\phi} \\ &= e^{-iJ\phi} e^{-i\boldsymbol{P}\cdot\boldsymbol{b}} e^{iJ\phi} \\ &= e^{-i\boldsymbol{P}\cdot(R(\phi)\boldsymbol{b})} \\ &= T(R(\phi)\boldsymbol{b}). \end{aligned} \tag{12.18}$$

That is, *rotating a translation is equivalent to translating a rotation.* As shown in Problem 12.4, the preceding relations mean that the translations $T(\boldsymbol{a})$ form an *abelian invariant subgroup* of E_2, because translations conjugated with arbitrary group elements give translations (see Sections 2.12 and 2.14):

$$g(\boldsymbol{b}, \phi)T(\boldsymbol{a})g(\boldsymbol{b}, \phi)^{-1} = T(R(\phi)\boldsymbol{a}). \tag{12.19}$$

Since E_2 has an abelian invariant subgroup it is not simple or semisimple. Thus, it cannot be written as a direct product of simple groups. In fact, as noted above, E_2 is the *semidirect product* of the translation subgroup and the group of rotations in the plane. Finally, E_2 is not compact because the translation parameters are unbounded.

These observations suggest that the representation theory of E_2 will exhibit new features not found in our discussion of compact groups. Indeed, one such feature was encountered already in the infinite tower of J eigenstates discussed in Section 12.4.4. We will not pursue the detailed representation theory of E_2, since it is of limited intrinsic significance for our purposes. However, let us develop briefly one aspect of E_2 representation theory: the *method of induced representations*, which will be essential to our analysis of the Poincaré group in Ch. 15. We introduce the method in the present context because its application to E_2 is simpler than for the full Poincaré group, thus affording a pedagogical introduction to the method that will make its later application to the Poincaré group more transparent.

12.5 Method of Induced Representations for E_2

The method of induced representations was introduced by Frobenius for finite groups, but our primary interest here traces to original work by Wigner on the representation theory of the Poincaré group that has had a powerful influence on the development of both modern physics and the mathematical discipline of group representation theory. This approach and its generalizations have become a standard tool for the analysis of representations for continuous groups having an abelian invariant subgroup. In essence, the method generates all members of a representation for the full group by the action of group generators starting from a representative vector defined in the invariant subspace; this works because a vector defined in an invariant space may be reoriented by subsequent group operations but its length is invariant because it is associated with the eigenvalue of a Casimir operator that is the same for all members of the irrep. More extensive discussions may be found in Ch. 15 and Refs. [19, 199]; here we summarize only the main features.

12.5.1 Generating the Representation

The method of induced representations consists of the following steps [199].

(1) *Identify the invariant subgroup and choose a representative vector in the subspace.* For E_2 the invariant subgroup is the translation group T_2, with generators P_1 and P_2. These may be viewed as components of a vector operator $\boldsymbol{P} = (P_1, P_2)$ having eigenvalues $\boldsymbol{p} = (p_1, p_2)$. Within this subspace, we select a *standard vector* $\boldsymbol{p}_0$. For $E_2 \supset T_2$,

$$\boldsymbol{p}_0 = (p, 0) \qquad \text{with} \quad \begin{cases} P_1 \left|\boldsymbol{p}_0\right\rangle = p\left|\boldsymbol{p}_0\right\rangle \\ P_2 \left|\boldsymbol{p}_0\right\rangle = 0 \\ P^2 \left|\boldsymbol{p}_0\right\rangle = p^2 \left|\boldsymbol{p}_0\right\rangle \end{cases}$$

is an appropriate choice.

(2) *Identify the factor group G/H.* Elements of the factor group are cosets of E_2 with respect to the subgroup $H = T_2$ (with left and right cosets equal since H is invariant):

$$E_2/T_2 = \{T \cdot g(\boldsymbol{b}, \phi)\}.$$

But from Eq. (12.13) a general element of E_2 may be written $g(\boldsymbol{b}, \phi) = T(\boldsymbol{b})R(\phi)$ and

$$E_2/T_2 = \{T \cdot T(\boldsymbol{b})R(\phi)\} = \{TR(\phi)\},$$

since the product of two translations is a translation. Thus, E_2/T_2 is labeled by a single continuous parameter ϕ and must be isomorphic to the one-parameter group of rotations, $E_2/T_2 \sim SO(2) \sim U(1)$, with a single generator J.

(3) *Find the dimensionality of the subspace corresponding to the standard vector.* This step turns on whether there are operators in the factor group that leave the standard vector invariant or, equivalently, on whether there are generators in the factor group that commute with the generator of the standard vector. All elements in the factor group that leave the standard vector invariant form a subgroup called the *little group*. For E_2 the factor group is isomorphic to SO(2), which has a single generator J that does not commute with P_1 since $[J, P_1] = i\epsilon_{1k}P_k = iP_2$. Hence, the little group for E_2 is null. More generally, the little group will be non-trivial. For example, the little group for E_3 is SO(2) and we shall find in Ch. 15 that for the Poincaré group the little group is SO(3) for massive particles and E_2 for massless particles.

(4) *Generate the full irreducible invariant space.* This may be accomplished by operating on the standard vector with the group generators that do not commute with the generator of the standard vector $\boldsymbol{P}$ to produce new eigenvalues of $\boldsymbol{P}$. For the present example this is $R(\phi) = e^{-iJ\phi}$ and we must examine $R(\phi)\,|\boldsymbol{p}_0\rangle$. As shown in Problem 12.5, the momentum content of $R(\phi)\,|\boldsymbol{p}_0\rangle$ is given by

$$P_k R(\phi)\,|\boldsymbol{p}_0\rangle = R(\phi)\,|\boldsymbol{p}_0\rangle\, p_k. \tag{12.20}$$

Thus, $P_k\,|\boldsymbol{p}\rangle = p_k\,|\boldsymbol{p}\rangle$, where $|\boldsymbol{p}\rangle \equiv R(\phi)\,|\boldsymbol{p}_0\rangle$, and the set $|\boldsymbol{p}\rangle$ is closed under all group operations since

$$T(\boldsymbol{b})\,|\boldsymbol{p}\rangle = e^{-i\boldsymbol{P}\cdot\boldsymbol{b}}\,|\boldsymbol{p}\rangle \qquad R(\theta)\,|\boldsymbol{p}\rangle = R(\theta)R(\phi)\,|\boldsymbol{p}_0\rangle = R(\theta+\phi)\,|\boldsymbol{p}_0\rangle \equiv |\boldsymbol{p}'\rangle\,.$$

Therefore, $\{|\boldsymbol{p}\rangle\}$ is the *basis of a vector space that is invariant under* E_2, and it is *irreducible* because it was generated by group operations starting from a minimal invariant subspace. These irreps are *unitary* because the operators are hermitian, but they are of *infinite dimension* by virtue of the continuous label $\boldsymbol{p}$. This example illustrates a fundamental difference between compact and non-compact groups that is discussed in Box 12.2.

Box 12.2 **Representations for Compact and Non-Compact Groups**

Section 12.5.1 introduces a distinction of fundamental importance between representations of compact and non-compact groups. For compact (or finite) groups it is always possible to choose a unitary representation (see Section 3.5). For non-compact groups we no longer have this luxury.

> Irreducible representations of a non-compact group can be chosen finite, or they can be chosen unitary, but they cannot be both at the same time.

This leads to complications in applications of non-compact groups because one is forced to give up either unitarity or finite dimensionality in the analysis.

(5) *Normalize the basis vectors.* In this example there is a one to one correspondence between basis vectors and the subgroup SO(2), so it is natural to normalize using the invariant measure of SO(2). From Eq. (6.21), $dg_{\mathrm{SO}(2)} = \frac{1}{2\pi} d\phi$ and the relation

$$\langle \boldsymbol{p}' | \, \boldsymbol{p} \rangle = \left\langle p', \phi' \middle| \, p, \phi \right\rangle = 2\pi\delta(\phi' - \phi) \tag{12.21}$$

expresses the orthonormality of the basis vectors.

12.5.2 Significance of the Abelian Invariant Subgroup

We conclude this short introduction to the method of induced representations by noting that the presence of the abelian invariant subgroup was essential.

1. The basis can be labeled by $\boldsymbol{p}$ because T_2 is abelian.
2. The invariant subgroup property was used in generating all $|\boldsymbol{p}\rangle$ from the standard vector.

Hence, it is expected that similar methods may be employed for other groups with abelian invariant subgroups. We shall use this technique again for the Poincaré group in Ch. 15.

Background and Further Reading

Tung [199] gives extensive discussions for a variety of non-compact groups.

Problems

12.1 Show that in 3D the translation operators P_j and rotation operators L_j given by Eq. (12.9) generate the non-abelian Lie group E_3, with the commutators (12.10). ***

12.2 Show that $P^2 \equiv P_1^2 + P_2^2$ is a Casimir operator for the euclidean group E_2.

12.3 Show that $P_\pm$ defined in Eq. (12.15) for E_2 obey the commutation relation (12.16), and that if $J\,|m\rangle = m\,|m\rangle$, then $J\left(P_\pm\,|m\rangle\right) = (m \pm 1)\left(P_\pm\,|m\rangle\right)$. Thus prove that $P_\pm$ act as raising and lowering operators for eigenvalues of J. *Hint*: This is analogous (but not identical) to a problem for SU(2) that is discussed in Section 3.3.3. ***

12.4 Prove that the translations form an abelian invariant subgroup of the euclidean group E_2 by deriving the result (12.19).

12.5 Show that (12.20) gives the momentum content of $R(\phi)\,|\boldsymbol{p}_0\rangle$. *Hint*: Eq. (12.17).

12.6 SO(2, 1) is the analog in two spatial dimensions of the Lorentz group SO(3, 1) described in Ch. 13. Its generators (X_1, X_2, X_3) obey the Lie algebra $[\,X_i, X_j\,] = c_{ij}^k X_k$ with

$$[\,X_1, X_2\,] = X_3 \qquad [\,X_2, X_3\,] = -X_1 \qquad [\,X_3, X_1\,] = X_2.$$

Use the metric tensor computed from Eq. (7.19) to show that SO(2, 1) is semisimple and non-compact. *Hint*: Compare Problem 7.6 for SO(3).

12.7 Show that the E_2 algebra (12.14) can be rewritten in the form (7.18) as

$$[X_1, X_2] = X_3 \qquad [X_1, X_3] = -X_2 \qquad [X_2, X_3] = 0,$$

through the mappings $P_1 \to iX_2$, $P_2 \to iX_3$, and $J \to iX_1$. Compute the metric tensor (Cartan–Killing form) using Eq. (7.19) to show that E_2 is not semisimple. *Hint*: See Section 7.2.2 and Problems 7.6 and 12.6. ***

12.8 Using $g(\boldsymbol{b}, \phi) = T(\boldsymbol{b})R(\phi)$ from Eq. (12.13), show that $R(\phi)T(\boldsymbol{b})R(\phi)^{-1} = T(R(\phi)\boldsymbol{b})$ from Eq. (12.18) implies the multiplication rule (12.11) for the group E_2. *Hint*: Evaluate $g(\boldsymbol{a}, \theta)g(\boldsymbol{b}, \phi)$.

13 The Lorentz Group

Non-compact groups were introduced in Ch. 12. The most important non-compact group in physics is SO(3, 1), because it is isomorphic to the group of Lorentz transformations that underlie special relativity and relativistic quantum field theory. We now investigate the Lorentz group as a non-compact group of physical interest, and as the basis for understanding spacetime symmetries and (when extended to the Poincaré group) the meaning of spin and mass for elementary particles. This chapter discusses basic properties of the Lorentz group, Ch. 14 discusses Lorentz covariance for wave equations and fields, and Ch. 15 extends the Lorentz group to include translations, leading to the Poincaré group.

13.1 Spacetime Tensors

It is useful to introduce *spacetime tensors* and a formalism that exhibits Lorentz invariance in an obvious manner. We neglect gravity so that special relativity is valid and physics can be formulated in inertial frames (coordinate systems in which Newton's first law holds). The corresponding spatially flat manifold is termed *Minkowski spacetime.*

13.1.1 A Covariant Notation

Let us introduce some conventions that will make it more apparent whether a given equation is consistent with Lorentz invariance, utilizing the notation summarized in Box 13.1.

Vectors, Dual Vectors, and Metrics: In an inertial frame, let us introduce cartesian coordinates with unit vectors e_0, e_1, e_2, and e_3 pointing in the t, x, y, and z directions, respectively. Spacetime vectors will be called *4-vectors* and an arbitrary 4-vector A can be expanded as

$$A = \sum_{\mu} A^{\mu} e_{\mu} \equiv A^{\mu} e_{\mu} = A^0 e_0 + A^1 e_1 + A^2 e_2 + A^3 e_3,$$

where we use the *Einstein summation convention*: any index repeated once as a lower and once as an upper index in a term implies a summation on that index.[1] The upper-index

[1] For manifolds with positive definite metrics (Section 12.2.1) and orthogonal coordinate systems, the placement of indices in upper or lower positions is not crucial, but it will be for this and other chapters that use the indefinite metric of spacetime because then upper and lower indices on tensors generally are not equivalent. Thus for 4D spacetime the summation convention will *always* involve one upper and one lower repeated index.

Box 13.1 Notation and Conventions for Spacetime Tensors

In the spacetime tensor notation introduced here we adopt a convention where

1. Greek letters ($\mu, \nu, \ldots$) denote indices that can range 0–3 (either timelike or spacelike coordinates),
2. Roman letters ($i, j, \ldots$) denote indices that range only 1–3 (spacelike coordinates only).

Thus for x^μ the index can take the values 0, 1, 2, 3 but for x^i the index can take only the values 1, 2, 3. In our notation,

1. normal math font will be used for 4-vectors,
2. bold math font will be used for the usual 3-vectors.

As is conventional, a notation such as x^μ can stand for either a particular component of a 4-vector or the full 4-vector, depending on the context.

quantities $A^\mu = (A^0, A^1, A^2, A^3)$ are the *contravariant components* of the 4-vector and we will call A^μ a *vector*. Then the scalar product of 4-vectors A and B may be expressed as

$$A \cdot B = B \cdot A = (A^\mu e_\mu) \cdot (B^\nu e_\nu) = e_\mu \cdot e_\nu A^\mu B^\nu \equiv \eta_{\mu\nu} A^\mu B^\nu, \tag{13.1}$$

where the *metric tensor* components $\eta_{\mu\nu}$ are defined by the scalar products of basis vectors,

$$\eta_{\mu\nu} \equiv e_\mu \cdot e_\nu. \tag{13.2}$$

Assuming coordinates $(ct, x, y, z) \equiv x^\mu = (x^0, x^1, x^2, x^3)$, the metric tensor can be written

$$\eta_{\mu\nu} = \begin{pmatrix} 1 & 0 & 0 & 0 \\ 0 & -1 & 0 & 0 \\ 0 & 0 & -1 & 0 \\ 0 & 0 & 0 & -1 \end{pmatrix} \equiv \text{diag}\,(1, -1, -1, -1), \tag{13.3}$$

where we adopt a $(+ \; - \; - -)$ sign-pattern convention, corresponding to positive timelike (ct) and negative spacelike (x, y, z) coefficients.[2] The metric tensor obeys the relations

$$\eta_{\mu\nu} = \eta_{\nu\mu} = \eta^{\mu\nu} \qquad \eta_{\mu\lambda}\eta^{\lambda\nu} = \eta_\mu^\nu = \delta_\mu^\nu, \tag{13.4}$$

where δ_μ^ν is the 4×4 unit matrix (*Kronecker delta*). The metric $\eta_{\mu\nu}$ determines the geometry because the infinitesimal distance along a curve is given by the *line element*

$$ds^2 = c^2 d\tau^2 = \eta_{\mu\nu} dx^\mu dx^\nu = c^2 dt^2 - dx^2 - dy^2 - dz^2, \tag{13.5}$$

[2] In the curved spacetime of general relativity the metric tensor is typically denoted $g_{\mu\nu}$, but for the special case of flat Minkowski space it is conventional to reserve the symbol $\eta_{\mu\nu}$ for the metric components. A metric sign pattern $(- \; + \; + +)$ instead of $(+ \; - \; - -)$ is also in common use. This is purely a matter of choice; what is crucial is that the diagonalized spacetime metric has the $(3, 1)$ signature discussed in Sections 12.2.1 and 12.3, implying an indefinite metric with opposite signs for timelike and spacelike components in the metric. Most of the unusual features of special relativity, and many of those for general relativity, derive from this indefinite nature of the spacetime metric.

where τ is the *proper time* (the time measured by a clock at rest in an inertial frame). A lower-index quantity A_μ may be obtained by *contraction of A^μ with the metric tensor* (setting an upper and lower index equal and doing the implied sum),

$$A_\mu = \eta_{\mu\nu} A^\nu = (A_0, A_1, A_2, A_3) = (A^0, -A^1, -A^2, -A^3). \tag{13.6}$$

We shall term $A_\mu = (A_0, A_1, A_2, A_3)$ the components of a *dual vector.*[3] Then

$$A \cdot B = A^\mu A_\mu = A^0 B^0 - A^1 B^1 - A^2 B^2 - A^3 B^3, \tag{13.7}$$

defines the 4-vector scalar product.

Higher-Rank Tensors: We may generalize to tensor components carrying any number of upper or lower indices, with the tensor *rank* equal to the number of indices. For example, the metric $\eta_{\mu\nu}$ and the Kronecker delta δ^ν_μ are rank-2 tensors. It is conventional to use the adjectives *contravariant* to indicate upper indices, *covariant* to indicate lower indices, and *mixed* if there are both upper and lower indices. Any index may be raised or lowered by contraction with the metric tensor. For example, $F_{\mu\nu} = \eta_{\sigma\nu} F_\mu{}^\sigma$ lowers an index on the mixed rank-2 tensor $F_\mu{}^\sigma$ and converts it to the covariant rank-2 tensor $F_{\mu\nu}$.

Derivative Notation: We will often use the following compact notation for derivatives

$$\begin{aligned}
\partial^\mu &\equiv \frac{\partial}{\partial x_\mu} = (\partial^0, \partial^1, \partial^2, \partial^3) = \left(\frac{\partial}{\partial x^0}, -\boldsymbol{\nabla}\right), \\
\partial_\mu &\equiv \frac{\partial}{\partial x^\mu} = (\partial_0, \partial_1, \partial_2, \partial_3) = \left(\frac{\partial}{\partial x^0}, \boldsymbol{\nabla}\right), \\
\boldsymbol{\nabla} &\equiv \left(\frac{\partial}{\partial x^1}, \frac{\partial}{\partial x^2}, \frac{\partial}{\partial x^3}\right) = (\partial_1, \partial_2, \partial_3) = (-\partial^1, -\partial^2, -\partial^3).
\end{aligned} \tag{13.8}$$

For example, $\partial^1 = \partial/\partial x_1$ and $\boldsymbol{\nabla}$ is the usual 3-divergence.

13.1.2 Tensor Transformation Laws

One practical way to define tensors is in terms of how they transform under a change of coordinate system $x \to x'$. The transformation laws for some low-rank tensors are

$$\begin{aligned}
\phi'(x') &= \phi(x) && \text{(scalar)}, \\
A'^\mu(x') &= \frac{\partial x'^\mu}{\partial x^\nu} A^\nu(x) && \text{(vector)}, \\
A'_\mu(x') &= \frac{\partial x^\nu}{\partial x'^\mu} A_\nu(x) && \text{(dual vector)}, \\
T'_{\mu\nu} &= \frac{\partial x^\alpha}{\partial x'^\mu} \frac{\partial x^\beta}{\partial x'^\nu} T_{\alpha\beta} && \text{(covariant rank-2 tensor)}, \\
T'^\nu{}_\mu &= \frac{\partial x^\alpha}{\partial x'^\mu} \frac{\partial x'^\nu}{\partial x^\beta} T^\beta{}_\alpha && \text{(mixed rank-2 tensor)}, \\
T'^{\mu\nu} &= \frac{\partial x'^\mu}{\partial x^\alpha} \frac{\partial x'^\nu}{\partial x^\beta} T^{\alpha\beta} && \text{(contravariant rank-2 tensor)},
\end{aligned} \tag{13.9}$$

[3] Some authors call A^μ a *contravariant vector* and A_μ a *covariant vector*, a *co-vector*, or a *1-form*.

Box 13.2 **Spacetime Tensors and Tensor Fields**

A *tensor field* corresponds to a tensor of particular type defined at every point of spacetime. Lorentz tensors are a special case where the spacetime is flat Minkowski space, with the metric (13.3).[a] We have introduced tensors in Eqs. (13.9) and (13.10) through the transformation properties of their components when they are expressed in some basis. This is of great practical use, since in real problems it is often simplest to work with the components of tensors expressed in a basis rather than with the tensors themselves. However, this obscures considerable mathematical beauty and elegance associated with tensor properties being independent of expression in any particular basis. Mathematicians prefer to define tensors *geometrically* (independent of expression in a particular basis) in terms of linear maps to the real numbers. The two approaches embody different tradeoffs between utility and elegance, but lead to the same physical results. A more extensive introduction to spacetime tensors in possibly curved spacetime may be found in Ref. [88].

[a] The more general tensors in Eq. (13.9) are valid for curved spacetime and form the mathematical basis for the theory of general relativity. Here we will restrict to problems where gravity (spacetime curvature) is not important and the simpler Lorentz tensors of Eq. (13.3) suffice. Also, we will often use "tensor" loosely as shorthand for a tensor field defined at every point of the manifold.

where unprimed coordinates refer to the original coordinate system, primed coordinates refer to the transformed coordinate system, and all partial derivatives depend on the spacetime coordinates and are understood to be evaluated at a specific spacetime point labeled by x in one coordinate system and by x' in the other.[4] Generalizations for higher-rank tensors are straightforward. The definitions (13.9) assume the general case that derivatives may depend on the spacetime coordinates. In the special instance of Minkowski space the *derivatives are independent of the spacetime coordinates* and Eq. (13.9) reduces to

$$\begin{aligned}
\phi' &= \phi && \text{(scalar)},\\
A'^{\mu} &= \Lambda^{\mu}{}_{\nu} A^{\nu} && \text{(vector)},\\
A'_{\mu} &= \Lambda_{\mu}{}^{\nu} A_{\nu} && \text{(dual vector)},\\
T'_{\mu\nu} &= \Lambda_{\mu}{}^{\gamma} \Lambda^{\delta}{}_{\nu} T_{\gamma\delta} && \text{(covariant rank-2 tensor)},\\
T'^{\mu}{}_{\nu} &= \Lambda^{\mu}{}_{\gamma} \Lambda^{\delta}{}_{\nu} T^{\gamma}{}_{\delta} && \text{(mixed rank-2 tensor)},\\
T'^{\mu\nu} &= \Lambda^{\mu}{}_{\gamma} \Lambda_{\delta}{}^{\nu} T^{\gamma\delta} && \text{(contravariant rank-2 tensor)},
\end{aligned} \tag{13.10}$$

where the matrices Λ involve constant derivatives (having the same value at all points).

[4] A transformation where the point is unchanged but it is relabeled in a new coordinate system is termed a *passive transformation*. The practice is common in physics but we are being sloppy mathematically by referring to objects carrying indices and obeying particular transformation laws as tensors. The quantities appearing in Eq. (13.9) are in reality *tensor components expressed in a particular basis*. Tensors are *geometrical objects* and their properties are independent of expression in a particular basis, as explained further in Box 13.2.

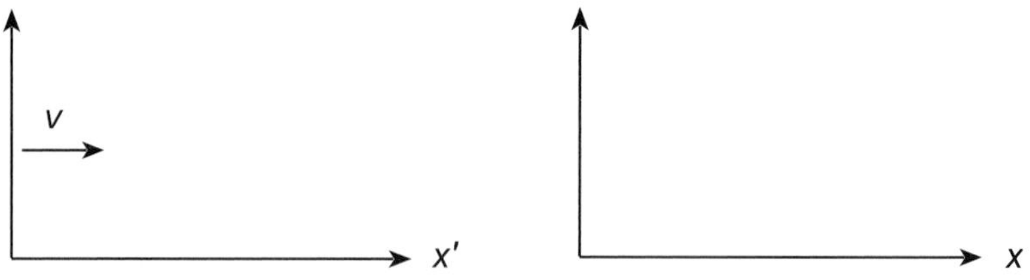

Fig. 13.1 Two inertial systems that differ by a relative velocity v along the x-axis.

13.2 Lorentz Transformations

Consider two inertial systems that move with a constant relative velocity v along the x-axis, as illustrated in Fig. 13.1. From the special theory of relativity, the *boost transformations* that connect these inertial frames are

$$x' = \frac{x + vt}{\sqrt{1 - v^2/c^2}} \qquad y' = y \qquad z' = z \qquad t' = \frac{t + vx/c^2}{\sqrt{1 - v^2/c^2}}. \tag{13.11}$$

To elucidate the group structure associated with these transformations it is desirable to change to a new set of variables.

13.2.1 Lorentz Boosts as Minkowski Rotations

Introducing the variables β and γ through

$$\gamma \equiv \frac{1}{\sqrt{1 - v^2/c^2}} \qquad \beta \equiv \frac{v}{c}, \tag{13.12}$$

and invoking the covariant notation introduced in Section 13.1.1, the Lorentz boost transformations (13.11) take the form

$$x'^0 = \gamma(x^0 + \beta x^1) \qquad x'^1 = \gamma(\beta x^0 + x^1) \qquad x'^2 = x^2 \qquad x'^3 = x^3. \tag{13.13}$$

Now observe that $\cosh^2\alpha - \sinh^2\alpha = 1$ and that

$$\gamma^2 - (\beta\gamma)^2 = \frac{1}{1 - v^2/c^2} - \frac{v^2/c^2}{1 - v^2/c^2} = 1,$$

suggesting the parameterization

$$\gamma = \cosh\xi \qquad \gamma\beta = \sinh\xi, \tag{13.14}$$

with ξ related to the velocity v through

$$\tanh\xi = \frac{\sinh\xi}{\cosh\xi} = \frac{\gamma\beta}{\gamma} = \beta = \frac{v}{c}. \tag{13.15}$$

In matrix form the x-axis Lorentz boost may then be expressed as $x' = B_x x$, where B_x is the 4×4 *boost matrix* in the x direction. Explicitly, $x' = B_x x$ is

$$\begin{pmatrix} x'^0 \\ x'^1 \\ x'^2 \\ x'^3 \end{pmatrix} = \begin{pmatrix} \cosh \xi & \sinh \xi & 0 & 0 \\ \sinh \xi & \cosh \xi & 0 & 0 \\ 0 & 0 & 1 & 0 \\ 0 & 0 & 0 & 1 \end{pmatrix} \begin{pmatrix} x^0 \\ x^1 \\ x^2 \\ x^3 \end{pmatrix}. \tag{13.16}$$

Similar boost matrices may be constructed for the y and z directions. With this parameterization boosts look like 4D rotations. However, ξ is defined in terms of hyperbolic functions and is not a rotation angle in the usual sense. For example, Eq. (13.16) may be viewed as implementing a rotation through an imaginary angle that mixes space and time coordinates.

13.2.2 Generators of Boosts and Rotations

The elements $U(\alpha_1, \alpha_2, \ldots, \alpha_N)$ of an N-parameter Lie group were written in Eq. (3.2) as

$$U(\alpha_1, \alpha_2, \ldots, \alpha_N) = e^{i\alpha_a X_a} = e^{i\alpha_1 X_1 + i\alpha_2 X_2 + \cdots + i\alpha_N X_N}, \tag{13.17}$$

indicating that the infinitesimal group generators X_k are related to the group elements U by differentiation with respect to the group parameters α_k,

$$X_k = \frac{1}{i} \frac{\partial U}{\partial \alpha_k}\bigg|_{\alpha_k = 0}.$$

Thus, from Eq. (13.16) a *Lorentz boost generator* K_1 in the x direction may be defined by

$$K_1 \equiv K_x = \frac{1}{i} \frac{\partial B_x}{\partial \xi}\bigg|_{\xi=0} = \frac{1}{i} \begin{pmatrix} \sinh \xi & \cosh \xi & 0 & 0 \\ \cosh \xi & \sinh \xi & 0 & 0 \\ 0 & 0 & 0 & 0 \\ 0 & 0 & 0 & 0 \end{pmatrix}_{\xi=0} = \begin{pmatrix} 0 & -i & 0 & 0 \\ -i & 0 & 0 & 0 \\ 0 & 0 & 0 & 0 \\ 0 & 0 & 0 & 0 \end{pmatrix}.$$

Carrying out the same procedure for the y and z axes gives for the boost generators

$$K_x = \begin{pmatrix} 0 & -i & 0 & 0 \\ -i & 0 & 0 & 0 \\ 0 & 0 & 0 & 0 \\ 0 & 0 & 0 & 0 \end{pmatrix} \qquad K_y = \begin{pmatrix} 0 & 0 & -i & 0 \\ 0 & 0 & 0 & 0 \\ -i & 0 & 0 & 0 \\ 0 & 0 & 0 & 0 \end{pmatrix},$$
$$K_z = \begin{pmatrix} 0 & 0 & 0 & -i \\ 0 & 0 & 0 & 0 \\ 0 & 0 & 0 & 0 \\ -i & 0 & 0 & 0 \end{pmatrix}. \tag{13.18}$$

The boost generators are not hermitian since $K \neq K^\dagger$; they are *antihermitian*, $K = -K^\dagger$. Explicit calculation shows that the commutator of two different boost generators is proportional to a 3-space angular momentum operator. Thus the boost generators K_i do not close under commutation and closure of a Lie algebra requires boosts to be supplemented by additional operators. The obvious choice is to add the set of (3-space) rotation operators to the boosts. In our 4-vector notation the last three components of the 4-vector correspond to the three spatial coordinates. Thus, the generators of 3D spatial rotations may be specified in this spacetime basis by appending to the usual 3D rotation matrix generators an extra null row at the top and an extra null column on the left:

$$J_x = \begin{pmatrix} 0 & 0 & 0 & 0 \\ 0 & 0 & 0 & 0 \\ 0 & 0 & 0 & -i \\ 0 & 0 & i & 0 \end{pmatrix} \quad J_y = \begin{pmatrix} 0 & 0 & 0 & 0 \\ 0 & 0 & 0 & i \\ 0 & 0 & 0 & 0 \\ 0 & -i & 0 & 0 \end{pmatrix} \quad J_z = \begin{pmatrix} 0 & 0 & 0 & 0 \\ 0 & 0 & -i & 0 \\ 0 & i & 0 & 0 \\ 0 & 0 & 0 & 0 \end{pmatrix}. \tag{13.19}$$

These matrices implement rotational transformations in the spatial coordinates only.

13.2.3 Commutation Algebra for the Lorentz Group

As may be verified by explicit calculation using the matrix representations (13.18) and (13.19) given above, the set of operators $(K_x, K_y, K_z, J_x, J_y, J_z)$ closes under commutation,

$$[\,J_i, J_j\,] = i\epsilon_{ijk} J_k \qquad [\,J_i, K_j\,] = i\epsilon_{ijk} K_k \qquad [\,K_i, K_j\,] = -i\epsilon_{ijk} J_k, \tag{13.20}$$

where the completely antisymmetric Levi-Civita symbols ϵ_{ijk} are defined in Table 3.2.

Example 13.1 From the algebra in Eq. (13.20),

$$[\,J_x, K_y\,] = i\epsilon_{xyz} K_z = iK_z \qquad [\,K_x, K_y\,] = -i\epsilon_{xyz} J_z = -iJ_z \qquad [\,J_x, K_x\,] = -i\epsilon_{xxz} J_z = 0,$$

where we have used that ϵ_{ijk} is +1 if the indices are cyclic, −1 if the indices are anticyclic, and zero if any two indices are the same.

Thus, the set of rotations plus Lorentz boosts illustrated in Table 13.1 forms a six-parameter Lie group, with a generator algebra given by Eq. (13.20). This Lie group is isomorphic to the group SO(3, 1) introduced in Section 12.3, and is termed the *proper Lorentz group*.

Table 13.1. Proper Lorentz transformations

Physical interpretation	Generators	Parameters
Three boosts	K_x, K_y, K_z	Three velocities: v_x, v_y, v_z
Three rotations	J_x, J_y, J_z	Three angles: $\theta_x, \theta_y, \theta_z$

13.3 Classification of Lorentz Transformations

The *proper Lorentz transformations* of the preceding section are continuous. More generally, the full Lorentz group comprises all boosts, rotations, and inversions that leave invariant the scalar product $x_\mu x^\mu$. The general form of a Lorentz transformation is

$$x'^\mu = \Lambda^\mu{}_\nu x^\nu, \tag{13.21}$$

where Λ is a 4×4 real matrix with components that satisfy $\Lambda^\nu{}_\lambda \Lambda^\lambda{}_\mu = \delta^\nu{}_\mu$. The invariant interval may be expressed as

$$x^\mu x_\mu = x^\mu \eta_{\mu\nu} x^\nu = x^{\mathrm{T}} \eta x, \tag{13.22}$$

with the matrix η defined in Eq. (13.3) and x^{T} denoting the transpose of the matrix x (interchange rows and columns). After a Lorentz transformation (13.21) on Eq. (13.22),

$$(x')^{\mathrm{T}} \eta x' = x^{\mathrm{T}} (\Lambda^{\mathrm{T}} \eta \Lambda) x.$$

Thus, invariance of the spacetime interval requires the matrix condition

$$\Lambda^{\mathrm{T}} \eta \Lambda = \eta, \tag{13.23}$$

or equivalently

$$\Lambda_{\mu\rho} \eta_{\lambda\nu} \Lambda^{\mu\lambda} = \Lambda_{\mu\rho} \Lambda^\mu{}_\nu = \eta_{\rho\nu}, \tag{13.24}$$

when written out in terms of matrix components.

13.3.1 The Four Pieces of the Full Lorentz Group

The conditions (13.23) and that $\det g \neq 0$ imply that $(\det \Lambda)^2 = 1$, and thus that $\det \Lambda = \pm 1$. Furthermore, in general either $\Lambda^0_0 \geq 1$ or $\Lambda^0_0 \leq -1$.

There are *four categories of Lorentz transformations,* illustrated in Table 13.2. Each is associated with a disconnected piece of the full Lorentz group.

This reminds us of O(3), which has two disconnected pieces: the part containing the identity that corresponds to continuous SO(3) rotations, and a second part associated with discrete space reflections (Section 6.1). However, the relationship of the disconnected pieces is more complex here. That the sign of the determinant fails to sort out the options, as it would for an orthogonal transformation, traces to the indefinite metric (13.3). The Lorentz transformations with $\det \Lambda = +1$ are termed *proper*, while those with $\Lambda^0_0 \geq 1$ are termed *orthochronous*. Furthermore, the proper Lorentz group is commonly termed *homogeneous*.

The Poincaré group that we shall discuss in Ch. 15 is obtained from the homogeneous Lorentz group by appending the generators of spacetime translations; it is sometimes termed the *inhomogeneous Lorentz group*. The term "Lorentz group" will be taken here to

Table 13.2. Categories of Lorentz transformations

I	$\det \Lambda = +1$	$\Lambda^0_0 \geq +1$	Contains the identity
II	$\det \Lambda = -1$	$\Lambda^0_0 \geq +1$	Contains space inversion
III	$\det \Lambda = -1$	$\Lambda^0_0 \leq -1$	Contains time inversion
IV	$\det \Lambda = +1$	$\Lambda^0_0 \leq -1$	Contains spacetime inversion

mean the proper orthochronous Lorentz group, and we will use the term "Poincaré group" for what is sometimes called the inhomogeneous Lorentz group.

13.3.2 Improper Lorentz Transformations

The proper ($\det \Lambda = +1$), orthochronous ($\Lambda^0_0 \geq 1$) transformations corresponding to category I of Table 13.2 are connected to the identity (which has unit determinant) by continuous variation of the parameters. Space inversion[5] (the parity operation) on the 4-vector basis is implemented by a matrix π and time inversion by a matrix τ, with

$$\pi = \begin{pmatrix} 1 & 0 & 0 & 0 \\ 0 & -1 & 0 & 0 \\ 0 & 0 & -1 & 0 \\ 0 & 0 & 0 & -1 \end{pmatrix} = \eta_{\mu\nu} \qquad \tau = \begin{pmatrix} -1 & 0 & 0 & 0 \\ 0 & 1 & 0 & 0 \\ 0 & 0 & 1 & 0 \\ 0 & 0 & 0 & 1 \end{pmatrix} = -\eta_{\mu\nu} \tag{13.25}$$

in the basis we are employing. For space inversion $\det \pi = -1$ and $\Lambda^0_0 = +1$, which is in category II, while for time inversion $\det \tau = -1$ and $\Lambda^0_0 = -1$ which is in category III. Combined spacetime inversion is then implemented by the matrix $\tau\pi$, which has $\det \pi = +1$ and $\Lambda^0_0 = -1$, and is in category IV of Table 13.2.

13.3.3 Lightcone Classification of Minkowski Vectors

The line element of Eq. (13.5) is of conic form and defines a *lightcone* that separates spacetime into distinct regions, as illustrated in Fig. 13.2 for two space and one time dimension.

1. Points with $ds^2 > 0$ and $x^0 > 0$ lie within the *future lightcone*. They can be reached from the origin by a signal propagating at less than the speed of light.
2. Points having $ds^2 > 0$ and $x^0 < 0$ lie in the *past lightcone*. A signal from those points can reach the origin traveling at less than the speed of light.
3. Points lying outside the lightcone have $ds^2 < 0$. They cannot be reached from the origin by signals traveling at the speed of light or less.

[5] It is popular to describe parity as mirror reflection. In 3D space parity reflects in all three spatial axes, which is not what a normal flat mirror does. However, parity is equivalent to mirror reflection in an axis and then a rotation by π about that axis. If space is isotropic, rotational invariance means that no physical outcome can depend on the final rotation, so in that sense mirror reflection and parity are equivalent operations in 3D space.

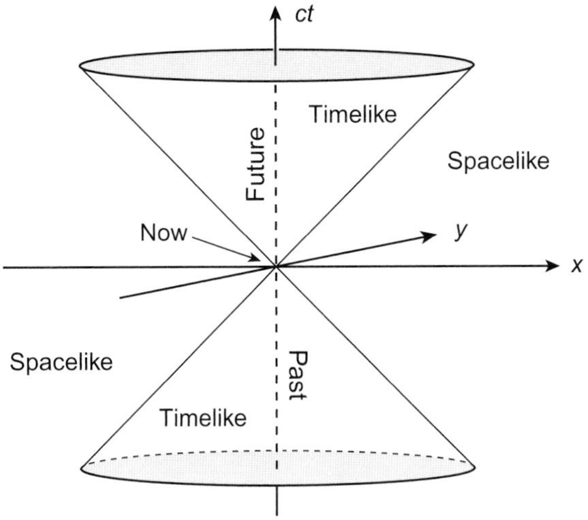

Fig. 13.2 The lightcone for two spacelike and one timelike Minkowski dimensions.

4. Points lying on the lightcone can be reached from the origin or from each other by signals traveling at exactly the speed of light.

For any point with $ds^2 < 0$, a Lorentz transformation exists that can transform the coordinate vector into one $(0, \boldsymbol{x}')$ having pure spacelike components. Therefore, points outside the lightcone are said to have a *spacelike* separation from the origin. Conversely, for any point lying within either the future or past lightcone there exists a Lorentz transformation that can bring the coordinate vector to the form $(t', \mathbf{0})$. Thus, points inside the lightcone are said to have a *timelike* separation from the origin. Points lying on the lightcone are said to have a *lightlike* or *null* separation.

13.4 Properties of the Lorentz Group

The proper Lorentz group has some important properties that may be deduced without difficulty.

1. The proper Lorentz group is non-compact. In the boost (13.16) the behavior of the parameter $\xi = \tanh^{-1}(v/c)$ is as illustrated in Fig. 13.3, and ξ has no bound as $v \to c$.
2. The rotations in three spatial dimensions form a subgroup of the proper Lorentz group since they close on themselves under commutation, $[\, J_i, J_j\,] = i\epsilon_{ijk} J_k$.
3. The boosts alone do not close under commutation and therefore do not form a subgroup.
4. The boosts transform as vectors under spatial rotations, by virtue of Eq. (6.59) and the commutators $[\, J_i, K_j\,]$ given in Eq. (13.20).

Box 13.3 discusses some physical consequences that follow from the failure of the set of Lorentz boosts to close under commutation.

Box 13.3 **Thomas Precession and Spin–Orbit Coupling**

The commutation relation $[K_x, K_y] = -iJ_z$ given in Eq. (13.20) for Lorentz boost generators implies that the commutator of two infinitesimal boosts parameterized by the "angles" $\delta\eta$ and $\delta\epsilon$ is

$$\begin{aligned} e^{iK_x\delta\eta}e^{iK_y\delta\epsilon}e^{-iK_x\delta\eta}e^{-iK_y\delta\epsilon} &= 1 - [K_x, K_y]\delta\eta\delta\epsilon + \cdots \\ &= 1 + iJ_z\Delta\theta_z + \cdots \\ &\sim e^{iJ_z\theta_z}, \end{aligned}$$

where $\Delta\theta_z \equiv \delta\eta\delta\epsilon$. *Commutation of two boosts gives a rotation!* This is why Lorentz boosts alone cannot form a Lie group. This is also the origin of *Thomas precession*, and ultimately is the reason that the spin–orbit term of the relativistic Dirac equation is a factor of two smaller than the corresponding spin–orbit term for non-relativistic quantum mechanics. These more technical issues are beyond the scope of this presentation but the interested reader may consult Itzykson and Zuber [120] or Jackson [122] for further discussion of these points.

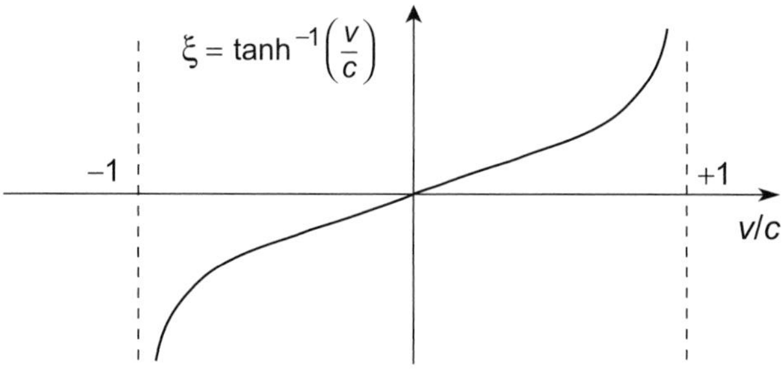

Fig. 13.3 Dependence of the Lorentz boost parameter ξ on the velocity v/c. The value of ξ is unbounded as $v/c \longrightarrow \pm 1$, implying that the Lorentz group is non-compact.

13.5 The Lorentz Group and SL(2,C)

The proper Lorentz group has six real parameters. Recall the matrix group $SL(n, C)$ introduced in Section 2.9.1 that corresponds to the group of $n \times n$ matrices of unit determinant with complex entries. An $n\times n$ complex matrix has $2n^2$ real parameters and the requirement $\det A = 1$ imposes one complex constraint, implying that $SL(2, C)$ has six real parameters, just as for the Lorentz group. This is no coincidence: as we shall now demonstrate, there is a homomorphism between $SL(2, C)$ and the proper Lorentz group $SO(3, 1)$ that may be displayed by introducing an appropriate mapping between 4-vectors and 2×2 matrices.

13.5.1 A Mapping between 4-Vectors and Matrices

In a generalization of Problem 6.17, let us associate each point x^μ in Minkowski spacetime with a 2×2 hermitian matrix X through the mapping

$$x^\mu \to X \equiv x^\mu \sigma_\mu = \eta^{\mu\nu} x_\nu \sigma_\mu, \tag{13.26}$$

with the definitions

$$\sigma^\mu \equiv (\sigma^0, \sigma^1, \sigma^2, \sigma^3) \qquad \sigma_\mu \equiv \eta_{\mu\nu}\sigma^\nu = (\sigma^0, -\sigma^1, -\sigma^2, -\sigma^3),$$
$$x^\mu \equiv (x^0, x^1, x^2, x^3) \qquad x_\mu \equiv \eta_{\mu\nu}x^\nu = (x^0, -x^1, -x^2, -x^3),$$

where the σ^i are Pauli matrices, σ^0 is the unit 2×2 matrix, and the metric tensor $\eta_{\mu\nu}$ is given by Eq. (13.3). Explicitly,

$$X = \begin{pmatrix} x_0 - x_3 & -x_1 + ix_2 \\ -x_1 - ix_2 & x_0 + x_3 \end{pmatrix}, \tag{13.27}$$

as shown in Problem 13.6. Taking the determinant of (13.27) gives

$$\det X = x_0^2 - x_1^2 - x_2^2 - x_3^2 \equiv x^2, \tag{13.28}$$

and preservation of $x^2 \equiv x^\mu x_\mu = \eta^{\mu\nu} x_\nu x_\mu$ is equivalent to preservation of $\det X$. Now a Lorentz transformation on x^μ will take the form $x' = \Lambda x$ of Eq. (13.21) and we can associate a new 2×2 hermitian matrix X' with the transformed point x' through $X' = x'^\mu \sigma_\mu$. But there must be a 2×2 matrix $A \equiv A(\Lambda)$ that can transform X into X',

$$X' = A(\Lambda) X A^\dagger(\Lambda), \tag{13.29}$$

where X' is hermitian (Problem 13.6). From Eq. (13.29), $\det X' = (\det A)(\det X)(\det A^\dagger)$. Now assume A to be an element of SL(2, C), so that $(\det A)(\det A^\dagger) = 1$ and $\det X' = \det X$. Thus, $x^2 = x^\mu x_\mu = \det X$ is invariant under transformations of SL(2, C) matrices and, since preservation of $x^\mu x_\mu$ is characteristic of the Lorentz group SO(3, 1), the groups SL(2, C) and SO(3, 1) are homomorphic.

13.5.2 The Universal Covering Group of SO(3,1)

The homomorphism between SL(2, C) and SO(3, 1) is 2:1 because the considerations of Section 13.5.1 specify A only up to a factor of ± 1. For example, if Eq. (13.29) is valid for A it is valid also for $-A$. Thus, for each proper Lorentz transformation matrix $\Lambda \in$ SO(3, 1) there are *two* SL(2, C) matrices $\pm A(\Lambda)$. In fact, SL(2, C) is the *universal covering group* for SO(3, 1) and the relationship between the non-compact group SO(3, 1) and its covering group SL(2, C) is similar to that between the compact group SO(3) and its covering group SU(2) that was discussed in Section 6.9.

13.6 Spinors and Lorentz Transformations

The two-component representation vectors of the Lorentz group associated with the SL(2, C) homomorphism may be termed *spinors* to distinguish them from three- or four-component vectors. We may again consider the analogy between the SU(2) $\to$ SO(3) and

SL(2, C) → SO(3, 1) homomorphisms described in Section 13.5.2. In each case, we can say the following.

1. The homomorphism is 2:1, with two elements of the covering group [SU(2) or SL(2, C)] mapped to a single element of the "orthogonal group" [SO(3) or SO(3, 1)].
2. The 2×2 unit-determinant (special) matrices of the covering group provide the lowest-dimensional representations of the corresponding "orthogonal group."
3. All finite-dimensional representations can be built from the direct products of spinor representations for the covering group .

For SU(2) the fundamental representations are just the two-dimensional Pauli spinors encountered in Section 3.3.1. For SL(2, C) the corresponding two-dimensional objects will be termed *Lorentz spinors,* to distinguish them from Pauli spinors. We shall sometimes just use the term "spinor" if the type of spinor is clear from the context.

13.6.1 SU(2) × SU(2) Representations of the Lorentz Group

The generators of a Lie algebra span a linear vector space, so any independent linear combination of generators also constitutes a valid set of generators. Let us introduce a new basis for the Lorentz generators by using Eqs. (13.18) and (13.19) to define

$$A_k \equiv \frac{1}{2}(J_k + iK_k) \qquad B_k \equiv \frac{1}{2}(J_k - iK_k) \qquad (k = 1,2,3). \tag{13.30}$$

From the commutation relations given in Eq. (13.20), we find that (Problem 13.3)

$$[\,A_i, A_j\,] = i\epsilon_{ijk}A_k \qquad [\,B_i, B_j\,] = i\epsilon_{ijk}B_k \qquad [\,A_i, B_j\,] = 0. \tag{13.31}$$

Thus, the Lie algebra of the Lorentz group may be written as the direct product of two SU(2) algebras, $\text{SU}(2)_A \times \text{SU}(2)_B$, and states transforming under the Lorentz group may be labeled by two "angular momenta," (J_A, J_B), one associated with the operators A_k and one with the operators B_k.[6] Apart from the trivial $(0, 0)$ representation, the lowest-dimensional representations are $(\frac{1}{2}, 0)$ or $(0, \frac{1}{2})$, with two degrees of freedom each.

13.6.2 Two Inequivalent Spinor Representations

As shown in Problem 13.4, the Lorentz commutation relations are satisfied by the choice

$$K_i = \pm\frac{i}{2}\sigma_i \qquad J_i = \frac{1}{2}\sigma_i, \tag{13.32}$$

where the σ_i are the Pauli matrices of Section 3.3.1. This corresponds to a special case of the irreps of SU(2) × SU(2) where one of the "angular momenta" is zero and the other is $\frac{1}{2}$, and implies the existence of two basic types of Lorentz spinors [174].

[6] Note that the Lie group SU(2) × SU(2) is *compact* but the Lorentz group SO(3, 1) is *non-compact.* They have the same Lie algebra, but they have different group elements. This means that properties of the representations such as unitarity may differ between the two groups.

1. *Type I Spinor* ψ_{R} : $K_i = -\frac{i}{2}\sigma_i$ and $J_i = \frac{1}{2}\sigma_i$. Then

$$A_i = \frac{1}{2}(J_i + iK_i) = \frac{1}{2}\sigma_i \qquad B_i = \frac{1}{2}(J_i - iK_i) = 0 \qquad (J_A, J_B) \to (J_A, 0) = \left(\frac{1}{2}, 0\right).$$

2. *Type II Spinor* ψ_{L} : $K_i = \frac{i}{2}\sigma_i$ and $J_i = \frac{1}{2}\sigma_i$. Then

$$A_i = \frac{1}{2}(J_i + iK_i) = 0 \qquad B_i = \frac{1}{2}(J_i - iK_i) = \frac{1}{2}\sigma_i \qquad (J_A, J_B) \to (0, J_B) = \left(0, \frac{1}{2}\right).$$

(The subscripts R and L stand for "right-handed" and "left-handed" respectively; their meaning will be explained after we introduce the concept of chirality in Section 14.5.) Thus, Lorentz spinors and Pauli spinors differ fundamentally.

Weyl Representations: Under rotations, two-component Pauli spinors transform as spin-$\frac{1}{2}$ irreps of SU(2), but under Lorentz transformations *two inequivalent two-component spinors* transform as different SU(2) × SU(2) irreps:

- *Type I*, corresponding to the irrep $(\frac{1}{2}, 0)$, and
- *Type II*, corresponding to the irrep $(0, \frac{1}{2})$.

These inequivalent spinor irreps are called *Weyl representations.*

In terms of the six Lorentz generators $\{\boldsymbol{K}, \boldsymbol{J}\}$ with $\boldsymbol{K} = (K_x, K_y, K_z)$ and $\boldsymbol{J} = (J_x, J_y, J_z)$, an element of the Lorentz group may be written [see (13.17)]

$$U = e^{i(\boldsymbol{J}\cdot\boldsymbol{\theta} + \boldsymbol{K}\cdot\boldsymbol{\xi})}, \tag{13.33}$$

with the parameters $\boldsymbol{\theta} = (\theta_1, \theta_2, \theta_3)$ associated with rotations and the parameters $\boldsymbol{\xi} = (\xi_1, \xi_2, \xi_3)$ associated with boosts. For the spinors ψ_{R} and ψ_{L} the generators are given by Eq. (13.32) and a proper Lorentz transformation on the two inequivalent spinors takes the form

$$\psi_{\mathrm{R}} = \exp\left[\frac{i}{2}\sigma \cdot (\theta - i\xi)\right] \psi_{\mathrm{R}}(0) \qquad \psi_{\mathrm{L}} = \exp\left[\frac{i}{2}\sigma \cdot (\theta + i\xi)\right] \psi_{\mathrm{L}}(0). \tag{13.34}$$

We must now understand why one type of Pauli spinor suffices for non-relativistic quantum mechanics but two inequivalent sets of Lorentz spinors are required in relativistic quantum mechanics. As will be shown in Ch. 14, the Dirac equation of relativistic quantum mechanics is in fact a relationship between these two inequivalent spinors. However, before we can understand that it is necessary to extend the discussion of the proper Lorentz group to include the improper transformation of space inversion (parity symmetry).

13.7 Space Inversion for the Lorentz Group

The discrete operations of space inversion (parity) and time reversal for the full Lorentz group are implemented by the operators π and τ in Eq. (13.25). Although these might seem rather similar in form, the discussion of time reversal involves new features (the use

Box 13.4 **Parity and the Weak Interactions**

Once it was believed that parity should be conserved in all physical processes, just like charge or energy or angular momentum. Nature was of a different mind.

Decay of Neutral Kaons

In the 1950s high-energy physicists puzzled over particles called neutral kaons that decayed to products of different parity. These data could be explained if the decays did not conserve parity, but few thought that was likely because of strong experimental evidence supporting parity conservation. But that two different elementary particles, one decaying to positive parity and one to negative parity products, should have exactly the same mass and quantum numbers seemed a most suspicious coincidence. T. D. Lee and C. N. Yang looked systematically at data and concluded that conservation of parity was indeed well supported for strong and electromagnetic interactions, but that the evidence was sketchy for weak interactions.

The Fall of Parity

In 1956 Lee and Yang advanced the bold hypothesis that the weak interactions did not conserve parity [140]. Their surmise was quickly confirmed when C. S. Wu and collaborators found evidence for parity non-conservation in the β-decay of ^{60}Co [219]. Subsequent experiments confirmed not only that parity was violated by weak interactions, but that it was *violated maximally.* Thus parity is not universally conserved. Electromagnetic and strong interactions are invariant under parity, but weak interactions are not, which has large implications for elementary particle physics.

of *antilinear operators*) that make it considerably different from that of parity. Discussion of time-reversal properties will be deferred until later. In this section we take up the issue of how parity enters the Lorentz group. This is of more than academic interest because *weak interactions do not conserve parity* (Box 13.4).

13.7.1 Action of Parity on Generators and Representations

To understand further the relationship between the Type I and Type II spinors introduced in Section 13.6, consider the action of parity on the generators and wavefunctions. From Eq. (13.25), the parity operator π inverts space but not time; thus the velocity and the boost generators change sign under parity, $\boldsymbol{v} \to -\boldsymbol{v}$ and $\boldsymbol{K} \to -\boldsymbol{K}$, but the angular momentum operators commute with parity and are unaffected, $\boldsymbol{J} \to \boldsymbol{J}$. Therefore (Problem 13.5),

$$\pi J_i \pi^{-1} = J_i \qquad \pi K_i \pi^{-1} = -K_i. \tag{13.35}$$

Now consider the effect of space inversion on the finite-dimensional (non-unitary) irreps of the Lorentz group introduced in Section 13.6. From Eq. (13.30), the generators A_k and B_k, and corresponding Casimir operators A^2 and B^2 of the $\mathrm{SU}(2)_A \times \mathrm{SU}(2)_B$ symmetry are

$$A_k = \frac{1}{2}(J_k + iK_k) \qquad B_k = \frac{1}{2}(J_k - iK_k) \qquad A^2 = A^k A_k \qquad B^2 = B^k B_k.$$

From Eq. (13.35), these transform as

$$\pi A_k \pi^{-1} = B_k \qquad \pi B_k \pi^{-1} = A_k \qquad \pi A^2 \pi^{-1} = B^2 \qquad \pi B^2 \pi^{-1} = A^2 \tag{13.36}$$

under the parity operation.

The generators A_k and B_k of proper Lorentz transformations, and the corresponding Casimir operators A^2 and B^2, are *interchanged by parity.* Hence eigenvalues of A^2 and B^2 are no longer separately conserved under parity transformations.

Let us investigate the behavior of the parity transformed basis vectors for an irreducible $\mathrm{SU}(2)_A \times \mathrm{SU}(2)_B$ representation (a, b) under the action of the proper Lorentz group generators. If the basis vectors are denoted by $|m_a m_b\rangle$ with

$$\begin{aligned} A^2 |m_a m_b\rangle &= a(a+1) |m_a m_b\rangle & B^2 |m_a m_b\rangle &= b(b+1) |m_a m_b\rangle, \\ A_3 |m_a m_b\rangle &= m_a |m_a m_b\rangle & B_3 |m_a m_b\rangle &= m_b |m_a m_b\rangle, \\ m_a &= \{-a, -a+1, \ldots, +a\} & m_b &= \{-b, -b+1, \ldots, +b\}, \end{aligned}$$

then from Eq. (13.36)

$$\begin{aligned} A_3 \pi |m_a m_b\rangle &= \pi B_3 |m_a m_b\rangle = \pi m_b |m_a m_b\rangle, \\ B_3 \pi |m_a m_b\rangle &= \pi A_3 |m_a m_b\rangle = \pi m_a |m_a m_b\rangle, \\ A^2 \pi |m_a m_b\rangle &= \pi B^2 |m_a m_b\rangle = \pi b(b+1) |m_a m_b\rangle, \\ B^2 \pi |m_a m_b\rangle &= \pi A^2 |m_a m_b\rangle = \pi a(a+1) |m_a m_b\rangle. \end{aligned} \tag{13.37}$$

Thus, for a basis vector $|m_a m_b\rangle$ of the representation (a, b), the state $\pi |m_a m_b\rangle$ transforms like a basis vector for a (b, a) representation of the proper Lorentz group.

13.7.2 General and Self-Conjugate Representations

For the special case that $a = b$ for the $\mathrm{SU}(2)_A \times \mathrm{SU}(2)_B$ Lorentz representation (a, b), one says that the representation is *self-conjugate*. Then we can have basis vectors transforming under parity like the original representation space of the proper Lorentz group. In general, the action of the parity operation π on such a self-conjugate representation can introduce a phase $\eta = \pm 1$.

Example 13.2 The scalar $(0, 0)$ representation of the Lorentz group is self-conjugate. Therefore, it may be characterized by a phase $\eta = \pm 1$ termed the *intrinsic parity.* For example, all quarks and all leptons are assigned an intrinsic parity of $+1$. The parity of a many-particle system is the product of parities for the individual particles and the Dirac equation (Ch. 14) requires the intrinsic parity of antiparticles to be opposite to that of the corresponding particles. Thus, since baryons have a qqq quark structure and mesons have a $q\bar{q}$ quark structure, baryons in the representations of Fig. 9.1 have positive intrinsic parity and mesons have negative intrinsic parity.

For $a \neq b$ the representations are said to be *general representations*. For these cases the parity transformed vectors $\{\pi |m_a m_b\rangle\}$ cannot be in the original space of (a, b) since they form a new representation space (b, a) of the *proper* Lorentz group [see Eq. (13.37)]. The minimal invariant subspace for the proper Lorentz group is now a direct sum of the two subspaces: $(a, b) \oplus (b, a)$. The vectors of the two subspaces are related by the definition $|m_b m_a\rangle \equiv \pi |m_a m_b\rangle$. Parity eigenstates may be defined by the linear combinations $\left|\psi_\pm\right\rangle = |m_a m_b\rangle \pm |m_b m_a\rangle$, such that the action of the parity operator is

$$\pi \left|\psi_\pm\right\rangle = |m_b m_a\rangle \pm |m_a m_b\rangle = \pm \left|\psi_\pm\right\rangle. \tag{13.38}$$

However, $\left|\psi_\pm\right\rangle$ does not behave in a simple way under the A and B operators of the original space of proper Lorentz transformations. The Type I and Type II spinors $(\frac{1}{2}, 0)$ and $(0, \frac{1}{2})$, respectively, of the preceding discussion provide examples of such general representations that are not parity invariant. The present considerations suggest that the minimal subspace of the proper Lorentz group for these spinors is the direct sum $(\frac{1}{2}, 0) \oplus (0, \frac{1}{2})$. In summary, there are two classes of finite-dimensional irreps of the Lorentz group extended by parity.

1. $(2u + 1) \times (2u + 1)$ dimensional *self-conjugate representations* ($u = 0, \frac{1}{2}, 1, \ldots; \eta = \pm 1$). These behave as $(a, b) = (u, u)$ irreps under *proper* Lorentz transformations and $\pi |m_a m_b\rangle = \eta |m_a m_b\rangle$, with $m_a, m_b = -u, -u + 1, \ldots, +u$.
2. *General representations* with distinct parameters (a, b). The action of parity produces new representations with respect to the proper Lorentz group and the minimal invariant subspace under parity corresponds to a direct sum $(a, b) \oplus (b, a)$.

For the general representations parity is well defined for irreducible representations of the full Lorentz group, but it is well defined only for certain reducible representations of the subgroup of proper Lorentz transformation. The direct sum $(a, b) \oplus (b, a)$ is an irrep of the Lorentz group extended by parity, but it is only a reducible representation of the subgroup of continuous transformations that leave the scalar product invariant.

13.8 Parity and 4-Spinors

As shown in Section 13.7.1, under a parity transformation the generators are interchanged

$$A_k = \frac{1}{2}(J_k + iK_k) \overset{\pi}{\longleftrightarrow} \frac{1}{2}(J_k - iK_k) = B_k,$$

as are the Type I and Type II spinors $(a, b) \overset{\pi}{\longleftrightarrow} (b, a)$, because they are not self-conjugate representations. Therefore, conservation of parity in a relativistic wave equation requires spinor representations that are symmetric in the a and b quantum numbers. Let us introduce a 4-component column vector called a *Dirac 4-spinor*,

$$\psi = \begin{pmatrix} \psi_{\mathrm{R}} \\ \psi_{\mathrm{L}} \end{pmatrix} \equiv \begin{pmatrix} \psi\left(\frac{1}{2}, 0\right) \\ \psi\left(0, \frac{1}{2}\right) \end{pmatrix}, \tag{13.39}$$

with ψ_{R} a two-component $(\frac{1}{2}, 0)$ spinor and ψ_{L} a two-component $(0, \frac{1}{2})$ spinor. The spinor ψ corresponds to a *reducible representation* $(\frac{1}{2}, 0) \oplus (0, \frac{1}{2})$ of the proper Lorentz group that is termed the *Dirac representation*. It is a particular case of the general reducible representation $(a, b) \oplus (b, a)$ introduced in the preceding section. We shall have much more to say about this when we consider the *Dirac equation* in Section 14.2.

13.9 Higher-Dimensional Lorentz Representations

From Section 13.6, all finite-dimensional representations of the Lorentz group may be constructed by compounding direct products of fundamental spinor representations.

Example 13.3 Using the Clebsch–Gordan series for the independent SU(2) factors of the $\mathrm{SU}(2)_A \times \mathrm{SU}(2)_B$ Lorentz representations, we have in terms of Young diagrams,

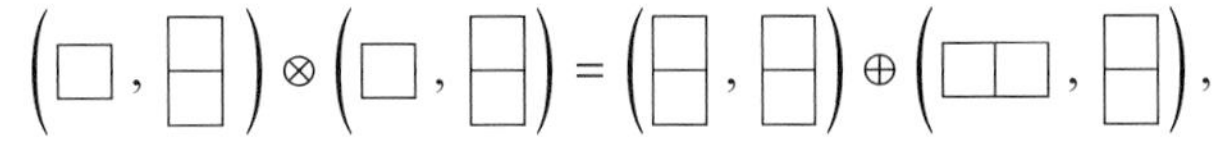

or in terms of the (J_A, J_B) representation labeling, $(\frac{1}{2}, 0) \otimes (\frac{1}{2}, 0) = (0, 0) \oplus (1, 0)$. The first term on the right side corresponds to a self-conjugate scalar representation $(0, 0)$, while the second representation $(1, 0)$ corresponds to an antisymmetric rank-2 tensor.

Example 13.4 By analogy with Example 13.3, we may construct $(0, \frac{1}{2}) \otimes (0, \frac{1}{2}) = (0, 0) \oplus (0, 1)$, which corresponds again to the scalar representation $(0, 0)$ plus a representation $(0, 1)$ corresponding to a rank-2 tensor.

Example 13.5 The electromagnetic field strength tensor $F_{\mu\nu}$ that will be introduced in Section 14.1.3 transforms as the six-dimensional reducible $(1, 0) \oplus (0, 1)$ representation, with the reducible representation with respect to the proper Lorentz group following from parity conservation of the electromagnetic field.

Example 13.6 The defining representation of the Lorentz group is $(\frac{1}{2}, \frac{1}{2})$, which is four-dimensional and transforms like a 4-vector. We may construct a general rank-2 Lorentz tensor by taking the product of two such 4-vectors. In terms of Young diagrams,

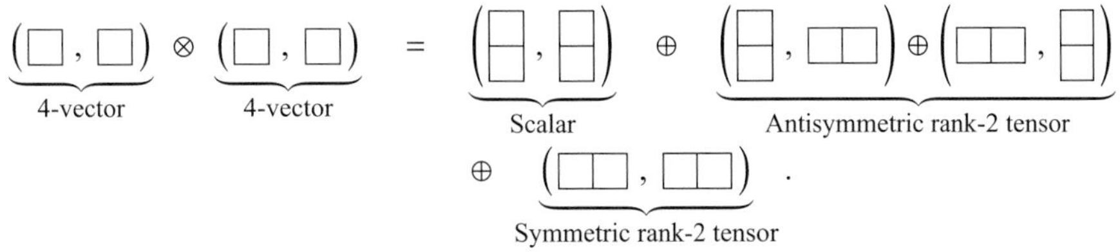

This direct product may also be expressed using (J_A, J_B) representation-labeling notation,

$$\underbrace{(\tfrac{1}{2}, \tfrac{1}{2})}_{4} \otimes \underbrace{(\tfrac{1}{2}, \tfrac{1}{2})}_{4} = \underbrace{(0, 0)}_{1} \oplus \overbrace{\underbrace{(0, 1)}_{3} \oplus \underbrace{(1, 0)}_{3}}^{6} \oplus \underbrace{(1, 1)}_{9},$$

where the dimensionality is indicated for each representation.

The tensor product in Example 13.6 gives a one-dimensional scalar, a $3 + 3 = 6$-dimensional reducible representation corresponding to an antisymmetric rank-2 tensor, and a 9-dimensional representation $(1, 1)$ corresponding to a symmetric rank-2 tensor. Some physical examples are the following.

1. The scalar Klein–Gordon field transforms as a one-dimensional $(0, 0)$ irrep.
2. As noted in Example 13.5, the electromagnetic field tensor $F_{\mu\nu}$ transforms as a rank-2 antisymmetric tensor with six independent components.
3. The stress–energy tensor $T^{\mu\nu}$ of special relativity transforms as a symmetric rank-2 tensor with nine components.

Example 13.6 may be viewed as the organization of the 16 independent components of a general rank-2 tensor into (1) one trace (the scalar representation), (2) six independent antisymmetric combinations of the $(1, 0) \oplus (0, 1)$ representation, and (3) nine symmetric combinations independent of the trace that correspond to the $(1, 1)$ representation.

13.10 Non-Unitarity of Representations

The $\mathrm{SU}(2)_A \times \mathrm{SU}(2)_B$ Lorentz representations considered above are of *finite dimensionality* $(2J_A + 1) \times (2J_B + 1)$, but they are *not unitary*.

Example 13.7 The matrix corresponding to $U = \exp[\frac{i}{2}\sigma \cdot (\theta + i\,\xi)]$ is not unitary since

$$U^{-1} = \exp[-\tfrac{i}{2}\sigma \cdot (\theta + i\,\xi)] \qquad U^{\dagger} = \exp[-\tfrac{i}{2}\sigma \cdot (\theta - i\,\xi)]$$

and $U^{\dagger} \neq U^{-1}$, ultimately because the generators of rotations may be chosen hermitian $(J = J^{\dagger})$, but then the generators of boosts are necessarily anti-hermitian $(K = -K^{\dagger})$.

This example illustrates once more the point of Box 12.2 that irreducible representations of non-compact groups can be finite or they can be unitary, but not both simultaneously.

13.11 Meaning of Non-Unitary Representations

The results of the preceding section raise an important issue with respect to physical interpretation of Lorentz representations. In quantum theory observables are associated with eigenvalues of hermitian operators, implying that the corresponding representation matrices should be unitary. Then, how are non-unitary Lorentz representations related to physical observables? The short answer is that this relationship follows from the complementarity of describing quantum systems in terms of particles (which are observable) or fields (which are not observable). To elaborate on this we must

1. explore the connections among Lorentz representations, quantum wave equations, and equations of motion for fields in relativistic quantum field theory, and
2. introduce the Poincaré group by appending spacetime translations to the proper Lorentz group.

In Ch. 14 we shall establish a connection between Lorentz spinor representations and the Dirac equation of relativistic quantum mechanics, demonstrating that the information content of the free-particle Dirac equation is equivalent to a knowledge of spinor representation theory for the Lorentz group. Then, in Ch. 15 the full Poincaré group will be introduced, which finally will allow the qualitative discussion of this section to be made more rigorous.

Background and Further Reading

The Lorentz group is discussed in Jones [125], Ryder [174], Sternberg [185], and Tung [199]. A more comprehensive discussion of tensors in both flat and curved 4D spacetime may be found in Guidry [88].

Problems

13.1 Verify that the Lorentz transformation (13.16) leaves invariant the squared Minkowski line element (13.5).

13.2 The indefinite metric of Minkowski space endows it with properties that seem strange to our (euclidean-influenced) intuition. For example, show that if a Minkowski vector is lightlike, it must be orthogonal to itself. *Hint*: The scalar product vanishes for orthogonal vectors. ***

13.3 Derive the commutation relations (13.31) using Eq. (13.20). ***

13.4 Show that the Lorentz group commutation relations (13.20) are satisfied by the choices $K_i = \pm\frac{i}{2}\sigma_i$ and $J_i = \frac{1}{2}\sigma_i$, where the σ_i are Pauli matrices.

13.5 Show that the action of a parity transformation on a Lorentz boost $B_i(\xi)$ is given by $\pi B_i(\xi)\pi^{-1} = B_i(-\xi)$. Use this result to prove that the action of a parity transformation on a boost generator K_i is given by $\pi K_i \pi^{-1} = -K_i$.

13.6 Show that Eq. (13.27) follows from Eq. (13.26). Prove that X' obtained by the transformation in Eq. (13.29) is hermitian if X is hermitian. ***

13.7 Show that the mapping $X = x^\mu \sigma_\mu$ given in Eq. (13.27) can be inverted to give $x^\mu = \frac{1}{2}\mathrm{Tr}\,(X\sigma^\mu)$. *Hint*: Multiply $X = x^\mu \sigma_\mu$ by σ^ν, take the trace of both sides, and use the properties in Table 3.1 for Pauli matrices. ***

13.8 Prove that $\Lambda_\mu{}^\nu = \eta_{\mu\alpha}\eta^{\nu\beta}\Lambda^\alpha{}_\beta$ is the inverse of the Lorentz transformation $\Lambda^\mu{}_\nu$.

14 Lorentz-Covariant Fields

Perhaps the most important application of the Lorentz group is to relativistic quantum field theory, where wave equations are interpreted as defining the motion of a classical field. When the field equations are quantized, the resulting theory provides a powerful description of physical reality in which the quantum fields interact through terms in the Lagrangian densities, and the field quanta appear as physical particles or antiparticles. Because such theories are (special) relativistic, the symmetries of the Lorentz group are central to their use. In this chapter we discuss two representative examples of such field theories and their associated wave equations: (1) the classical electromagnetic field, which produces photons upon being quantized, and (2) the classical (that is, first-quantized) Dirac wave equation, which produces fermions and antifermions upon second quantization. Our emphasis in this chapter will be on the Lorentz covariance properties at the classical or first-quantized level. Discussion of second-quantization for such fields and the formalism of relativistic quantum field theory will be deferred until subsequent chapters. However, we shall take up briefly in this chapter the group-theoretical relationship between particles and fields, with a more complete discussion to follow in Ch. 15 and subsequent material.

14.1 Lorentz Covariance of Maxwell's Equations

The Maxwell equations governing classical electromagnetic theory may be written in free space with Heaviside–Lorentz, $c = 1$ natural units (see Appendix B) as

$$\nabla \cdot \boldsymbol{E} = \rho \qquad \text{(Gauss' law)}, \tag{14.1a}$$

$$\frac{\partial \boldsymbol{B}}{\partial t} + \nabla \times \boldsymbol{E} = 0 \qquad \text{(Faraday's law)}, \tag{14.1b}$$

$$\nabla \cdot \boldsymbol{B} = 0 \qquad \text{(No magnetic charges)}, \tag{14.1c}$$

$$\nabla \times \boldsymbol{B} - \frac{\partial \boldsymbol{E}}{\partial t} = \boldsymbol{j} \qquad \text{(Ampère's law, as modified by Maxwell)}, \tag{14.1d}$$

where $\boldsymbol{E}$ is the electric field, $\boldsymbol{B}$ is the magnetic field, ρ is the charge density, and $\boldsymbol{j}$ is the current vector, with the density and current vector satisfying the continuity equation

$$\frac{\partial \rho}{\partial t} + \nabla \cdot \boldsymbol{j} = 0 \tag{14.2}$$

that ensures conservation of charge. The Maxwell equations (14.1) are consistent with special relativity (*covariant* with respect to Lorentz transformations). However, in the form (14.1) this is not manifest (cannot be seen at a glance) because these equations are formulated in terms of 3-vectors and separate derivatives with respect to space and time.[1] It will prove useful to reformulate the Maxwell equations so that they are manifestly covariant with respect to Lorentz transformations.

14.1.1 Scalar and Vector Potentials

The electric and magnetic fields appearing in the Maxwell equations may be replaced by a *3-vector potential* $\boldsymbol{A}$ and a *scalar potential* ϕ, through the definitions

$$\boldsymbol{B} \equiv \boldsymbol{\nabla} \times \boldsymbol{A} \qquad \boldsymbol{E} \equiv -\boldsymbol{\nabla}\phi - \frac{\partial \boldsymbol{A}}{\partial t}. \tag{14.3}$$

Then, because of the vector identities $\boldsymbol{\nabla} \cdot (\boldsymbol{\nabla} \times \boldsymbol{B}) = 0$ and $\boldsymbol{\nabla} \times \boldsymbol{\nabla}\phi = 0$, the second Maxwell equation (14.1b) and the third Maxwell equation (14.1c) are satisfied identically. Upon employing the vector identity $\boldsymbol{\nabla} \times (\boldsymbol{\nabla} \times \boldsymbol{A}) = \boldsymbol{\nabla}(\boldsymbol{\nabla} \cdot \boldsymbol{A}) - \boldsymbol{\nabla}^2 \boldsymbol{A}$, the remaining two Maxwell equations become the coupled second-order equations

$$\boldsymbol{\nabla}^2 \phi + \frac{\partial}{\partial t}(\boldsymbol{\nabla} \cdot \boldsymbol{A}) = -\rho, \tag{14.4a}$$

$$\boldsymbol{\nabla}^2 \boldsymbol{A} - \frac{\partial^2 \boldsymbol{A}}{\partial t^2} - \boldsymbol{\nabla}\left(\boldsymbol{\nabla} \cdot \boldsymbol{A} + \frac{\partial \phi}{\partial t}\right) = -\boldsymbol{j}. \tag{14.4b}$$

Now a fundamental symmetry of electromagnetism termed *gauge invariance* may be used to decouple these equations.

14.1.2 Gauge Transformations

The preceding definitions of the potentials $\boldsymbol{A}$ and ϕ are not unique. Because of the identity $\boldsymbol{\nabla} \times \boldsymbol{\nabla}\phi = 0$, the simultaneous transformations

$$\boldsymbol{A} \to \boldsymbol{A} + \boldsymbol{\nabla}\chi \qquad \phi \to \phi - \frac{\partial \chi}{\partial t} \tag{14.5}$$

for an arbitrary scalar function χ leave the $\boldsymbol{E}$ and $\boldsymbol{B}$ fields unchanged; thus, the Maxwell equations are invariant under (14.5), which is termed a classical *gauge transformation*. This invariance under gauge transformations may be used to decouple Eqs. (14.4).

Lorenz Gauge: For example, if we choose a set of potentials $(\boldsymbol{A}, \phi)$ that satisfy

$$\boldsymbol{\nabla} \cdot \boldsymbol{A} + \frac{\partial \phi}{\partial t} = 0, \tag{14.6}$$

then Eqs. (14.4) decouple to yield the independent equations

$$\boldsymbol{\nabla}^2 \phi - \frac{\partial^2 \phi}{\partial t^2} = -\rho \qquad \boldsymbol{\nabla}^2 \boldsymbol{A} - \frac{\partial^2 \boldsymbol{A}}{\partial t^2} = -\boldsymbol{j}. \tag{14.7}$$

[1] In special relativity space and time must enter on an equal footing. The use of 3-vectors instead of 4-vectors, and of separate derivatives for space and time, obscures whether this condition is satisfied. Covariance of the Maxwell equations with respect to Lorentz transformations means that their validity is unchanged by the transformation.

A constraint like (14.6) is termed a *gauge-fixing condition*, and imposing such a constraint is termed *fixing the gauge*. The gauge choice implied by Eq. (14.6) is called the *Lorenz gauge*.[2]

Coulomb Gauge: Another common gauge choice is the *Coulomb gauge* (also called the *radiation gauge*), with a gauge-fixing condition $\nabla \cdot \boldsymbol{A} = 0$ that leads to the equations

$$\nabla^2 \phi = -\rho \qquad \nabla^2 \boldsymbol{A} - \frac{\partial^2 \boldsymbol{A}}{\partial^2 t} = \nabla \frac{\partial \phi}{\partial t} - \boldsymbol{j}. \tag{14.8}$$

Covariant Gauge Conditions: Upon introducing the 4-vector potential A^μ, the 4-current j^μ, and the *d'Alembertian operator* $\Box$,

$$A^\mu \equiv (\phi, \boldsymbol{A}) = (A^0, \boldsymbol{A}) \qquad j^\mu \equiv (\rho, \boldsymbol{j}) \qquad \Box \equiv \partial_\mu \partial^\mu, \tag{14.9}$$

a gauge transformation takes the form

$$A^\mu \rightarrow A^\mu - \partial^\mu \chi \equiv A'^\mu, \tag{14.10}$$

and the preceding examples of gauge-fixing constraints become

$$\partial_\mu A^\mu = 0 \qquad \text{(Lorenz gauge)}, \tag{14.11}$$

$$\nabla \cdot \boldsymbol{A} = 0 \qquad \text{(Coulomb gauge)}. \tag{14.12}$$

This notation makes clear that the gauge condition (14.11) is a covariant constraint but the gauge condition (14.12) is not, because it involves only three of the components of a 4-vector: $A^\mu = (A^0, \boldsymbol{A})$. The operator $\Box$ is a Lorentz invariant, since from Eq. (13.10)

$$\Box' = \partial'_\mu \partial'^\mu = \Lambda^\nu{}_\mu \Lambda_\lambda{}^\mu \partial_\nu \partial^\lambda = \partial_\mu \partial^\mu = \Box.$$

Thus, the Lorenz-gauge wave equations (14.7) and the continuity equation (14.2) may be expressed as

$$\Box A^\mu = j^\mu \qquad \partial_\mu j^\mu = 0, \tag{14.13}$$

which are manifestly covariant because they are formulated entirely in terms of objects that are Lorentz tensors. The covariance of the Maxwell wave equation (14.13) in the Lorenz gauge and the gauge invariance of electromagnetism suggests that the theory is generally Lorentz covariant. However, this covariance may not be transparent in a choice of gauge other than Lorenz gauge.

14.1.3 Manifestly Covariant Form of the Maxwell Equations

A manifestly covariant form of the Maxwell equations may be constructed by using Eqs. (14.3) to express components of the electric and magnetic fields in terms of derivatives of the potential components. For example, $B^2 = \partial^1 A^3 - \partial^3 A^1 = F^{13} = -F^{31}$. Proceeding in

[2] The Lorenz gauge is named in honor of *Ludvig Lorenz.* It is a Lorentz-invariant condition (named in honor of *Hendrik Lorentz*). Thus there is ample opportunity for confusion and the Lorenz gauge is often called the Lorentz gauge in the literature.

this manner, we find that the six independent components of $\boldsymbol{E}$ and $\boldsymbol{B}$ are elements of an antisymmetric rank-2 *electromagnetic field tensor* (Problem 14.19)

$$F^{\mu\nu} = -F^{\nu\mu} = \partial^\mu A^\nu - \partial^\nu A^\mu, \tag{14.14}$$

which may be written as the matrix[3]

$$F^{\mu\nu} = \begin{pmatrix} 0 & -E^1 & -E^2 & -E^3 \\ E^1 & 0 & -B^3 & B^2 \\ E^2 & B^3 & 0 & -B^1 \\ E^3 & -B^2 & B^1 & 0 \end{pmatrix}. \tag{14.15}$$

Now let us introduce the completely antisymmetric rank-4 tensor $\epsilon_{\alpha\beta\gamma\delta}$, which has the value $+1$ for $\alpha\beta\gamma\delta = 0123$ and cyclic permutations, -1 for odd permutations, and zero if any two indices are equal, and satisfies $\epsilon_{\alpha\beta\gamma\delta} = -\epsilon^{\alpha\beta\gamma\delta}$. If we then define

$$\mathscr{F}^{\mu\nu} \equiv \frac{1}{2}\epsilon^{\mu\nu\gamma\delta}F_{\gamma\delta} = \begin{pmatrix} 0 & -B^1 & -B^2 & -B^3 \\ B^1 & 0 & E^3 & -E^2 \\ B^2 & -E^3 & 0 & E^1 \\ B^3 & E^2 & -E^1 & 0 \end{pmatrix}, \tag{14.16}$$

where $\mathscr{F}^{\mu\nu}$ is the *dual field tensor*, the Maxwell equations (14.1a) and (14.1d) become

$$\partial_\mu F^{\mu\nu} = j^\nu, \tag{14.17}$$

and the Maxwell equations (14.1b) and (14.1c) become

$$\partial_\mu \mathscr{F}^{\mu\nu} = 0. \tag{14.18}$$

The Maxwell equations (14.17) and (14.18) are now formulated exclusively in terms of Lorentz tensors, so they are manifestly covariant. From Eqs. (14.17) and (14.14), the 4-vector potential A^μ is required to obey (Problem 14.10)

$$\Box A^\mu - \partial^\mu(\partial_\nu A^\nu) = j^\mu. \tag{14.19}$$

For the *specific case of Lorenz gauge*, $\partial_\nu A^\nu = 0$ and this reduces to $\Box A^\nu = j^\nu$, which is Eq. (14.13), but Eq. (14.19) is valid generally in any gauge.

14.2 The Dirac Equation

We shall now demonstrate that the representation theory of the Lorentz group that was developed in Ch. 13 may be used to derive the free-particle Dirac equation of relativistic quantum mechanics. This implies that Lorentz group representation theory and the free-particle Dirac equation have equivalent physical content.

[3] The electric field $\boldsymbol{E}$ and magnetic field $\boldsymbol{B}$ are each *vectors* under SO(3) transformations in 3D, but in 4D Minkowski space their components together form the rank-2 tensor (14.14) under Lorentz transformations. The rank-2 Lorentz tensor $F^{\mu\nu}$ is also invariant under gauge transformations (see Ch. 16 and Problem 16.7).

14.2.1 Lorentz-Boosted Spinors

Consider a pure Lorentz boost ($\boldsymbol{\theta} = 0$) for the spinor ψ_{R} in Eq. (13.34),

$$\psi_{\mathrm{R}} = \exp\left(-\frac{i}{2}\boldsymbol{\sigma}\cdot(i\boldsymbol{\xi})\right)\psi_{\mathrm{R}}(0) = \left(\cosh\frac{\xi}{2} + \boldsymbol{\sigma}\cdot\boldsymbol{n}\,\sinh\frac{\xi}{2}\right)\psi_{\mathrm{R}}(0), \tag{14.20}$$

where $\boldsymbol{n}$ is a unit vector in the direction of the boost and the second form follows from an identity proved in Problem 14.1. Let the boosted spinor $\psi_{\mathrm{R}}(\boldsymbol{p})$ refer to a particle moving with a momentum $\boldsymbol{p}$ and let the original (unboosted) spinor $\psi_{\mathrm{R}}(0)$ refer to the same particle at rest. As shown in Problem 14.2, the corresponding boost transformation may be written

$$\psi_{\mathrm{R}}(\boldsymbol{p}) = \frac{E + m + \boldsymbol{\sigma}\cdot\boldsymbol{p}}{\sqrt{2m(E+m)}}\psi_{\mathrm{R}}(0), \tag{14.21}$$

where the energy E and the mass m of the particle are related through the Einstein equation $E^2 = \boldsymbol{p}^2 + m^2$ (in $c = 1$ units). By a similar procedure

$$\psi_{\mathrm{L}}(\boldsymbol{p}) = \frac{E + m - \boldsymbol{\sigma}\cdot\boldsymbol{p}}{\sqrt{2m(E+m)}}\psi_{\mathrm{L}}(0). \tag{14.22}$$

We shall justify below that ψ_{R} and ψ_{L} define different helicities (spin components in the direction of motion) for spin-$\frac{1}{2}$ particles. For a particle at rest the two helicities are indistinguishable and $\psi_{\mathrm{R}}(0) = \psi_{\mathrm{L}}(0)$. As shown in Problem 14.5, this may be used in conjunction with Eqs. (14.21) and (14.22) to establish a relationship between $\psi_{\mathrm{R}}(\boldsymbol{p})$ and $\psi_{\mathrm{L}}(\boldsymbol{p})$,

$$\psi_{\mathrm{R}}(\boldsymbol{p}) = \frac{E + \boldsymbol{\sigma}\cdot\boldsymbol{p}}{m}\psi_{\mathrm{L}}(\boldsymbol{p}) \qquad \psi_{\mathrm{L}}(\boldsymbol{p}) = \frac{E - \boldsymbol{\sigma}\cdot\boldsymbol{p}}{m}\psi_{\mathrm{R}}(\boldsymbol{p}). \tag{14.23}$$

Upon introducing a momentum 4-vector,

$$p^\mu = \begin{pmatrix} E \\ \boldsymbol{p} \end{pmatrix} = \begin{pmatrix} E \\ p^1 \\ p^2 \\ p^3 \end{pmatrix} \equiv \begin{pmatrix} p^0 \\ p^1 \\ p^2 \\ p^3 \end{pmatrix} \qquad p_\mu = \eta_{\mu\nu}p^\nu, \tag{14.24}$$

these equations may be written in the matrix form

$$\begin{pmatrix} -m & p_0 + \boldsymbol{\sigma}\cdot\boldsymbol{p} \\ p_0 - \boldsymbol{\sigma}\cdot\boldsymbol{p} & -m \end{pmatrix}\begin{pmatrix} \psi_{\mathrm{R}} \\ \psi_{\mathrm{L}} \end{pmatrix} = 0. \tag{14.25}$$

It may not yet be obvious but Eq. (14.25) is the Dirac equation of relativistic quantum mechanics. This will become clear if we rewrite it in a more covariant notation.

14.2.2 A Lorentz-Covariant Notation

Define a Dirac 4-spinor [see Eq. (13.39)] in terms of two-component spinors $\psi_{\mathrm{R}}(\boldsymbol{p})$ and $\psi_{\mathrm{L}}(\boldsymbol{p})$,

$$\psi(p) \equiv \begin{pmatrix} \psi_{\mathrm{R}}(\boldsymbol{p}) \\ \psi_{\mathrm{L}}(\boldsymbol{p}) \end{pmatrix}, \tag{14.26}$$

and introduce 4×4 matrices γ^μ satisfying the anticommutator

$$\{ \gamma_\mu, \gamma_\nu \} \equiv \gamma_\mu \gamma_\nu + \gamma_\nu \gamma_\mu = 2\eta_{\mu\nu} \tag{14.27}$$

that are termed the (Dirac) *γ-matrices.* A specific realization that is useful in the present context is the *Weyl representation* (also termed the *chiral representation*) of the γ-matrices,

$$\gamma^0 \equiv \begin{pmatrix} 0 & 1 \\ 1 & 0 \end{pmatrix} \qquad \gamma^i \equiv \begin{pmatrix} 0 & -\sigma^i \\ \sigma^i & 0 \end{pmatrix}, \tag{14.28}$$

where $i = 1, 2, 3$, the $\sigma^i = \sigma_i$ are the Pauli matrices given in Eq. (3.11), and a compact notation is used with each entry being itself a 2×2 matrix. For example,

$$\gamma^0 = \begin{pmatrix} 0 & 1 \\ 1 & 0 \end{pmatrix} = \left(\begin{array}{cc|cc} 0 & 0 & 1 & 0 \\ 0 & 0 & 0 & 1 \\ \hline 1 & 0 & 0 & 0 \\ 0 & 1 & 0 & 0 \end{array}\right),$$

$$\gamma^2 = \begin{pmatrix} 0 & -\sigma^2 \\ \sigma^2 & 0 \end{pmatrix} = \left(\begin{array}{cc|cc} 0 & 0 & 0 & i \\ 0 & 0 & -i & 0 \\ \hline 0 & -i & 0 & 0 \\ i & 0 & 0 & 0 \end{array}\right),$$

where the horizontal and vertical lines indicate the 2×2 submatrices. Let us also introduce another 4×4 matrix γ_5 through the definition

$$\gamma_5 = \gamma^5 \equiv i\gamma^0\gamma^1\gamma^2\gamma^3. \tag{14.29}$$

It is common to call γ_5 the *chirality operator*, for reasons to be explained in Section 14.5. In the Weyl representation (14.28), the explicit form of γ_5 is

$$\gamma_5 = \begin{pmatrix} 1 & 0 \\ 0 & -1 \end{pmatrix}. \tag{14.30}$$

Important properties of the γ-matrices follow from the definition (14.27) and are summarized in Table 14.1. Now from Eq. (14.24) we have $\gamma^\mu p_\mu = \gamma^0 p_0 + \gamma^i p_i$ and Eq. (14.25) may be expressed as (Problem 14.3)

$$(\gamma^\mu p_\mu - m)\psi(p) = (i\gamma^\mu \partial_\mu - m)\psi(p) = 0, \tag{14.31}$$

where we have made the standard quantum operator substitution $p_\mu \to i\partial/\partial x^\mu = i\partial_\mu$. This is the *Dirac equation* in standard covariant notation, with $\hbar = c = 1$ units.

14.3 Dirac Bilinear Covariants

Manifestly Lorentz-covariant theories must be formulated in terms of Lorentz tensors but Dirac spinors are not tensors. However, *physical observables* involve forms bilinear in the Dirac spinors and these *bilinear covariants* do transform as Lorentz tensors, as we shall now discuss.

Table 14.1. Properties of the Dirac γ-matrices

$$\{\alpha_i, \alpha_j\} = 2\delta_{ij} \qquad \{\alpha_i, \beta\} = 0 \qquad \alpha_i^2 = \beta^2 = 1$$

$$\gamma_0 = \gamma^0 \equiv \beta \qquad \gamma_i \equiv \gamma^0 \alpha_i \qquad \gamma_5 = \gamma^5 \equiv i\gamma^0\gamma^1\gamma^2\gamma^3 \qquad \sigma^{\mu\nu} \equiv \tfrac{i}{2}[\gamma^\mu, \gamma^\nu]$$

$$\{\gamma^\mu, \gamma^\nu\} = 2\eta^{\mu\nu} \qquad \gamma_0\gamma_\mu^\dagger\gamma_0 = \gamma_\mu \qquad \{\gamma_5, \gamma^\mu\} = 0 \qquad (\gamma^i)^2 = -1$$

$$(\gamma^0)^2 = 1 \qquad (\gamma^5)^2 = 1 \qquad (\gamma_0)^\dagger = \gamma_0 \qquad (\gamma^i)^\dagger = -\gamma^i \qquad (\gamma^5)^\dagger = \gamma^5$$

Pauli–Dirac representation (4 × 4 matrices)

$$\gamma^0 = \begin{pmatrix} 1 & 0 \\ 0 & -1 \end{pmatrix} \quad \gamma^i = \begin{pmatrix} 0 & \sigma_i \\ -\sigma_i & 0 \end{pmatrix} \quad \gamma^5 = \begin{pmatrix} 0 & 1 \\ 1 & 0 \end{pmatrix} \quad \alpha_i = \begin{pmatrix} 0 & \sigma_i \\ \sigma_i & 0 \end{pmatrix}$$

$$\beta = \begin{pmatrix} 1 & 0 \\ 0 & -1 \end{pmatrix} \qquad \sigma^{ij} = \epsilon_{ijk} \begin{pmatrix} \sigma_k & 0 \\ 0 & \sigma_k \end{pmatrix} \qquad \sigma^{0k} = i \begin{pmatrix} 0 & \sigma_k \\ \sigma_k & 0 \end{pmatrix}$$

Weyl (chiral) representation (4 × 4 matrices)

$$\gamma^0 = \begin{pmatrix} 0 & 1 \\ 1 & 0 \end{pmatrix} \qquad \gamma^i = \begin{pmatrix} 0 & -\sigma_i \\ \sigma_i & 0 \end{pmatrix} \qquad \gamma^5 = \begin{pmatrix} 1 & 0 \\ 0 & -1 \end{pmatrix}$$

Majorana representation (4 × 4 matrices)

$$\gamma^0 = \begin{pmatrix} 0 & \sigma_2 \\ \sigma_2 & 0 \end{pmatrix} \quad \gamma^1 = \begin{pmatrix} i\sigma_1 & 0 \\ 0 & i\sigma_1 \end{pmatrix} \quad \gamma^2 = \begin{pmatrix} 0 & \sigma_2 \\ -\sigma_2 & 0 \end{pmatrix}$$

$$\gamma^3 = \begin{pmatrix} -i\sigma_3 & 0 \\ 0 & i\sigma^3 \end{pmatrix} \qquad \gamma^5 = \begin{pmatrix} \sigma_2 & 0 \\ 0 & \sigma_2 \end{pmatrix}$$

Standard Pauli matrices (2 × 2 matrices)

$$\sigma_1 = \begin{pmatrix} 0 & 1 \\ 1 & 0 \end{pmatrix} \qquad \sigma_2 = \begin{pmatrix} 0 & -i \\ i & 0 \end{pmatrix} \qquad \sigma_3 = \begin{pmatrix} 1 & 0 \\ 0 & -1 \end{pmatrix}$$

14.3.1 Covariance of the Dirac Equation

Let us examine the covariance of the Dirac equation (14.31) expressed in the form

$$\left(i\gamma^\mu \frac{\partial}{\partial x^\mu} - m\right)\psi(x) = 0. \tag{14.32}$$

Under a Lorentz transformation to a new (primed) coordinate system the principle of covariance requires this equation to retain the same form:[4]

$$\left(i\gamma^\mu \frac{\partial}{\partial x'^\mu} - m\right)\psi'(x') = 0,$$

[4] The matrices γ and γ' are equivalent up to a unitary transformation, so we drop primes on the γ-matrices.

where $\psi(x)$ and $\psi'(x')$ describe the same physical state and x and x' are related through a Lorentz transformation (13.21). The transformation between ψ and ψ' is assumed to be implemented by a matrix S, and to be linear,

$$\psi'(x') = S\psi(x), \tag{14.33}$$

and invertible. In terms of these transformations, the original Dirac equation is

$$\left(iS\gamma^\mu S^{-1}\frac{\partial}{\partial x^\mu} - m\right)\psi'(x') = 0,$$

and after employing a Lorentz transformation on the derivatives,

$$\left(iS\gamma^\mu S^{-1}\Lambda^\nu{}_\mu\frac{\partial}{\partial x'^\nu} - m\right)\psi'(x') = 0.$$

Comparing this result with Eq. (14.32), form invariance requires that

$$S\gamma^\mu S^{-1}\Lambda^\nu{}_\mu = \gamma^\nu, \tag{14.34}$$

which defines the 4×4 matrix S that transforms the wavefunction according to (14.30).[5]

14.3.2 Transformation Properties of Bilinear Products

There are five independent bilinear forms in ψ and $\psi^\dagger$ that have definite Lorentz tensor properties. Let us look at some examples.

Example 14.1 Consider $\overline{\psi}\psi$, where the *adjoint spinor* is defined by $\overline{\psi}(x) \equiv \psi^\dagger\gamma^0$, with the Lorentz transformation behavior

$$\overline{\psi'}(x') = \overline{\psi}(x)S^{-1}. \tag{14.35}$$

From (14.30) and (14.35), $\overline{\psi'}\psi' = \overline{\psi}S^{-1}S\psi = \overline{\psi}\psi$, which is the transformation law for a Lorentz scalar [see Eq. (13.10)]. Under parity, for which we may take $S = S^{-1} = \gamma^0$, this quantity is also unchanged, so it transforms as a *true scalar.*

Example 14.2 Consider the combination $\overline{\psi}\gamma^5\psi$ under proper Lorentz transformations:

$$\overline{\psi'}\gamma^5\psi' = \overline{\psi}S^{-1}\gamma^5 S\psi = \overline{\psi}\gamma^5\psi,$$

where the last step follows from a general result that γ^5 always commutes with S for proper Lorentz transformations. Now consider the action of the parity operator γ^0,

$$\overline{\psi}\gamma^0\gamma^5\gamma^0\psi = -\overline{\psi}\gamma^5\gamma^0\gamma^0\psi = -\overline{\psi}\gamma^5\psi,$$

[5] The 4×4 matrices S and γ^μ shuffle the spinor indices (of which there are four). The 4×4 matrix Λ shuffles the spacetime indices (of which there are also four). However, these correspond to matrix operations in *completely different spaces,* so one should always be aware that in an equation such as this there are both (explicit or implicit) spacetime and spinor indices, and they are associated with independent degrees of freedom.

Table 14.2. Dirac bilinear covariants

Bilinear form	Transforms as Lorentz	Components
$\overline{\psi}\psi$	Scalar	1
$\overline{\psi}\gamma^5\psi$	Pseudoscalar	1
$\overline{\psi}\gamma^\mu\psi$	Vector	4
$\overline{\psi}\gamma^\mu\gamma^5\psi$	Axial vector	4
$\overline{\psi}\sigma^{\mu\nu}\psi$	Antisymmetric tensor	6

where we have used the properties $(\gamma^0)^2 = 1$ and $\{\gamma^5, \gamma^\mu\} = 0$ from Table 14.1. Thus, $\overline{\psi}\gamma^5\psi$ transforms as a scalar under proper Lorentz transformations but changes sign under a parity transformation; it is said to transform as a *pseudoscalar*.

Example 14.3 The *Dirac current* $j^\mu = \overline{\psi}\gamma^\mu\psi$ is important in many field-theory applications. As shown in Problem 14.13, for proper Lorentz transformations $j'^\mu = \Lambda^\mu{}_\nu j^\nu$, which from Eq. (13.10) is the transformation law for a 4-vector. Furthermore, under a parity transformation, $j^0 \to j^0$ and $j^k \to -j^k$. Thus $\overline{\psi}\gamma^\mu\psi$ is a true 4-vector.

Example 14.4 As you are asked to show in Problem 14.7, the bilinear form $\overline{\psi}\sigma^{\mu\nu}\psi$, with $\sigma^{\mu\nu} \equiv \frac{i}{2}[\gamma^\mu, \gamma^\nu]$, transforms as an antisymmetric rank-2 tensor.

The complete set of bilinear covariants is summarized in Table 14.2. As discussed in Box 14.1, the 16 components of these bilinear products satisfy an anticommutator algebra called a *Clifford algebra,* long known to mathematicians in the study of *quaternions.*

14.4 Weyl Equations and Massless Fermions

Let us return to Eq. (14.25); doing the matrix multiplication gives the coupled equations

$$m\psi_R = (p_0 + \boldsymbol{\sigma}\cdot\boldsymbol{p})\psi_L \qquad m\psi_L = (p_0 - \boldsymbol{\sigma}\cdot\boldsymbol{p})\psi_R. \tag{14.36}$$

These equations decouple in the limit $m \to 0$ to yield the *Weyl equations,*

$$(p_0 + \boldsymbol{\sigma}\cdot\boldsymbol{p})\psi_L = 0 \qquad (p_0 - \boldsymbol{\sigma}\cdot\boldsymbol{p})\psi_R = 0, \tag{14.37}$$

and the corresponding wavefunctions are termed *Weyl spinors.* For relativistic particles $E^2 - \boldsymbol{p}^2 = m^2$ and for massless particles $E = p_0 = |\boldsymbol{p}|$, so

$$\boldsymbol{\sigma}\cdot\boldsymbol{p} = \boldsymbol{\sigma}\cdot|\boldsymbol{p}|\,\hat{\boldsymbol{p}} = p_0(\boldsymbol{\sigma}\cdot\hat{\boldsymbol{p}})$$

where $\hat{\boldsymbol{p}}$ is a unit momentum vector and the Weyl equations take the decoupled form

$$(\boldsymbol{\sigma}\cdot\hat{\boldsymbol{p}})\psi_L = -\psi_L \qquad (\boldsymbol{\sigma}\cdot\hat{\boldsymbol{p}})\psi_R = \psi_R. \tag{14.38}$$

Box 14.1 **Quaternions and Clifford Algebras**

The total number of components for the bilinear covariants in the last column of Table 14.2 is 16, which is the number of entries in a 4×4 matrix. Thus, the products of γ matrices contained in the bilinear covariants form a basis for the expansion of arbitrary 4×4 matrices. The algebra obeyed by these products is known as a *Clifford algebra*, which was studied by mathematicians within the theory of quaternions long before the Dirac equation was introduced.

Quaternions: Quaternions are a generalization of complex numbers, $a + b\mathbf{i}$ (where $\mathbf{i} = \sqrt{-1}$), to

$$a + b\mathbf{i} + c\mathbf{j} + d\mathbf{k},$$

where a, b, c, and d are real numbers and $\mathbf{i}, \mathbf{j}$, and $\mathbf{k}$ may be interpreted as unit vectors along the three spatial axes. The multiplication formula for quaternions is

$$\mathbf{i}^2 = \mathbf{j}^2 = \mathbf{k}^2 = \mathbf{ijk} = -1.$$

They were discovered by Irish mathematician and physicist William Rowan Hamilton in 1843, when he realized how the idea that complex numbers can be represented by points in a 2D plane could be generalized to define the kind of number that points in a 3D space represent.[a] Quaternions do not commute under multiplication, which means that some standard formulas of linear algebra that are valid for real and complex numbers (which do commute) are no longer valid for quaternions.

Quaternions fell out of favor in physics with the rise of modern forms of vector analysis in the late 1800s, which could describe the same problems to which quaternions had been applied with simpler notation and greater transparency. They enjoyed a revival in the late twentieth century, particularly in animated computer graphics where they were found to have some speed advantages for rotation of 3D objects over standard formulations using rotation matrices parameterized by Euler angles.

Clifford Algebras: Clifford algebras have a broader and more abstract definition in mathematics, but in the physics applications discussed in this chapter a Clifford algebra is assumed be be defined by the anticommutation relation

$$\{\gamma^\mu, \gamma^\nu\} = 2\eta^{\mu\nu} \qquad (\mu, \nu = 0, 1, 2, 3),$$

among Dirac γ-matrices that is given in Eq. (14.27) and Table 14.1, and is associated specifically with the $(3, 1)$ indefinite metric of 4D Minkowski spacetime.

[a] Hamilton was appointed to a chair professorship at Trinity College, Dublin, while he was still an undergraduate student. He considered himself to be a mathematician but made many important contributions to physics in the areas of mechanics and optics. For example, his reformulation of Newtonian mechanics is now taught in all physics departments as *Hamiltonian mechanics*, and as one legacy of that work the *Hamiltonian operator* of quantum mechanics bears his name. In mathematics Hamilton is best known for quaternions.

Thus, for *massless fermions* the Type I spinors ψ_{R} and the Type II spinors ψ_{L} defined in Section 13.6.2 obey separate relativistic wave equations. Since by the discussion in Section 13.7 a parity-conserving state must be a superposition of ψ_{R} and ψ_{L}, we have the following result.

Parity is not conserved by the Weyl equations for massless fermions because they propagate ψ_R and ψ_L independently.

Failure to conserve parity was initially a motivation for rejecting the Weyl equations, but the discovery that weak interactions break parity symmetry brought them to the fore as a possible description of the (nearly) massless neutrinos participating in such interactions.

14.5 Chiral Invariance

Consider the chiral operator γ^5 introduced in Eq. (14.29) . To appreciate its significance, let us turn to a more detailed examination of spin in Lorentz-invariant wave equations.

14.5.1 Helicity States for Fermions

We begin by introducing a 4×4 matrix operator

$$S \equiv \begin{pmatrix} \sigma & 0 \\ 0 & \sigma \end{pmatrix}, \tag{14.39}$$

where $\boldsymbol{\sigma} \equiv (\sigma_1, \sigma_2, \sigma_3)$ denotes a vector of 2×2 Pauli matrices. A *helicity operator* h may then be introduced through

$$h \equiv \frac{1}{2} \boldsymbol{S} \cdot \hat{\boldsymbol{p}} = \frac{1}{2} \begin{pmatrix} \sigma & 0 \\ 0 & \sigma \end{pmatrix} \cdot \hat{\boldsymbol{p}} = \frac{1}{2} \begin{pmatrix} \boldsymbol{\sigma} \cdot \hat{\boldsymbol{p}} & 0 \\ 0 & \boldsymbol{\sigma} \cdot \hat{\boldsymbol{p}} \end{pmatrix}, \tag{14.40}$$

where $\hat{\boldsymbol{p}} \equiv \boldsymbol{p} / |\boldsymbol{p}|$, and by explicit multiplication,

$$\boldsymbol{\sigma} \cdot \boldsymbol{p} = \begin{pmatrix} p_3 & p_1 - i p_2 \\ p_1 + i p_2 & -p_3 \end{pmatrix}. \tag{14.41}$$

Choosing the z-axis parallel to the momentum, the free-particle Dirac spinors $u^{(\alpha)}$ are eigenstates of this operator with[6]

$$h u^{(\alpha)}(\boldsymbol{p}) = \left(\frac{1}{2} \boldsymbol{S} \cdot \hat{\boldsymbol{p}} \right) u^{(\alpha)}(\boldsymbol{p}) = \lambda_\alpha u^{(\alpha)}(\boldsymbol{p}), \tag{14.42}$$

where $\alpha = (1, 2)$ and the allowed values of λ_α for spin-$\frac{1}{2}$ are $\lambda_1 = +\frac{1}{2}$ and $\lambda_2 = -\frac{1}{2}$. The *helicity quantum number* λ may be interpreted physically as the projection of the spin on the direction of motion. It is a constant of motion because it commutes with the Dirac Hamiltonian, as will be shown below. It is a 3-vector scalar product, so it is rotationally invariant, but it is not Lorentz invariant (for finite-mass particles) because it involves only

[6] In this section we assume positive-energy solutions of the Dirac equation, denoted by $u^{(\alpha)}$, with α labeling the helicity. There also are two negative-energy solutions, denoted by $v^{(\alpha)}$, which will be interpreted later as antiparticle spinors. Each of these spinors is a 4-component column vector but the spinor component indices will be suppressed in most contexts.

the 3-momentum $\boldsymbol{p}$ and not the full 4-momentum. This failure of Lorentz invariance can be understood intuitively.

> Because helicity is the component of spin in the direction of motion, it can be reversed for a massive particle by boosting to a new Lorentz frame with speed greater than the particle but still less than c. Thus helicity is not generally conserved in a Lorentz transformation for particles with finite mass.

However, if $m \to 0$ the helicity becomes Lorentz invariant since then no boost to a faster frame is possible because massless particles move at light velocity in all frames.

14.5.2 Dirac Equation in Pauli–Dirac Representation

Another useful form for the Dirac equation results from a new representation of the γ-matrices termed the *Pauli–Dirac representation.*[7] For this representation (see Table 14.1),

$$\begin{aligned} \gamma^0 &= \beta = \begin{pmatrix} 1 & 0 \\ 0 & -1 \end{pmatrix} & \alpha &\equiv \begin{pmatrix} 0 & \sigma \\ \sigma & 0 \end{pmatrix}, \\ \gamma^i &= \beta\alpha^i = \begin{pmatrix} 0 & \sigma_i \\ -\sigma_i & 0 \end{pmatrix} & \gamma^5 &= \begin{pmatrix} 0 & 1 \\ 1 & 0 \end{pmatrix}, \end{aligned} \tag{14.43}$$

which serves to define the new matrices α_i and β in this representation. In terms of these matrices the Dirac equation (14.31) may be written as

$$i\frac{\partial\psi}{\partial t}(\boldsymbol{x},t) = (\boldsymbol{\alpha}\cdot\boldsymbol{p} + \beta m)\psi(\boldsymbol{x},t) \equiv H\psi(\boldsymbol{x},t), \tag{14.44}$$

where $\boldsymbol{\alpha}\cdot\boldsymbol{p} = -i\,\boldsymbol{\alpha}\cdot\boldsymbol{\nabla} = -i\alpha^k\partial_k$. This is termed the "Hamiltonian form" of the Dirac equation because Eq. (14.44) looks like a non-relativistic Schrödinger equation with a Hamiltonian operator H. The single-particle spectrum that results is illustrated in Fig. 14.1(a).

Such an interpretation is legitimate only for *weak, slowly varying fields,* because the Dirac equation has properties like negative-energy solutions and a spectrum with no lower bound that have no counterpart in non-relativistic quantum mechanics. In the electronic ground state the negative-energy states are assumed all filled and the positive-energy states are assumed empty, with an energy gap of $2mc^2$ between filled and empty states. If an energy of at least $2mc^2$ is supplied in the ground state, a particle in a negative-energy state can be promoted to a positive-energy state where it is interpreted as an electron and the hole left behind is interpreted as a positron, giving an electron–positron pair; Fig. 14.1(b) illustrates. For strong fields many pairs can be created as in Fig. 14.1(b) and the single-particle picture fails. Then the Dirac equation must be viewed as describing a relativistic field that creates and annihilates fermions and antifermions. This is not a book about

[7] Different representations of the γ-matrices such as the Weyl, Pauli–Dirac, or Majorana representations displayed in Table 14.1 correspond to choice of different bases for the γ-matrices. Thus representations A and B for the γ^μ are related by a basis transformation $\gamma^\mu_A = U\gamma^\mu_B U^\dagger$, where U is a unitary 4×4 matrix.

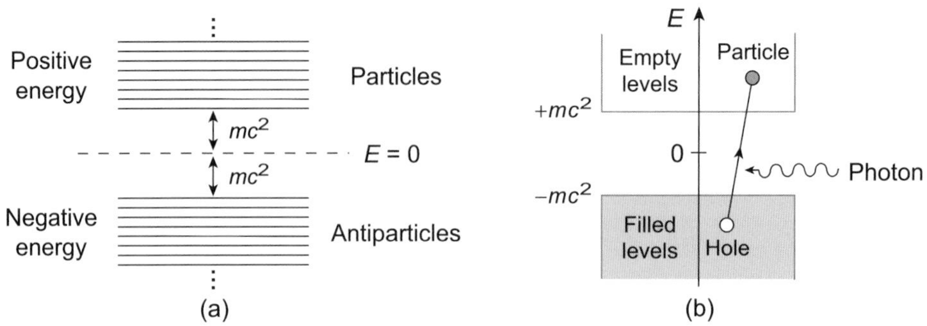

Fig. 14.1 (a) Positive-energy and negative-energy states in the spectrum that results from the Dirac equation (14.44). (b) Particle–antiparticle pair creation. The hole in the negative-energy sea acts as an antiparticle. A minimum energy $E = 2mc^2$ is required to create the pair.

field theory but later chapters will introduce sufficient relativistic quantum field theory to describe such a system (see also Appendix A).

14.5.3 Helicity and Chirality for Dirac Fermions

Let us investigate now the commutation properties of the helicity and chirality operators with respect to the Dirac Hamiltonian. From Problem 14.15, the helicity operator $h = \frac{1}{2}\boldsymbol{S} \cdot \hat{\boldsymbol{p}}$ defined in Eq. (14.40) commutes with the Dirac Hamiltonian, $[\,h, H\,] = 0$, so its eigenvalues can be used to label the spinor eigenstates. Now let us consider commutation of the chirality operator γ^5 with the Dirac Hamiltonian. Utilizing Eq. (14.44) and $\alpha^i = \gamma_0\gamma^i$,

$$H = (\boldsymbol{\alpha} \cdot \boldsymbol{p} + \beta m) = -i\alpha^i p_i + \beta m = -i\gamma_0\gamma^i p_i + \gamma_0 m. \tag{14.45}$$

As shown in Problem 14.12, chirality commutes with this Hamiltonian *only in the limit of vanishing fermion mass.* Thus for *massless* Dirac fermions chirality and helicity are interchangeable labels for the "handedness" of spin-$\frac{1}{2}$ particles, as summarized in Box 14.2. However, as shown in Problem 14.15, this is no longer true for massive fermions.

For free *massive* fermions chirality is Lorentz invariant but it is not conserved, while helicity is conserved but it is not Lorentz invariant.

Thus neither helicity nor chirality characterizes uniquely a massive fermion described by the free-particle Dirac equation because neither is a constant of motion that has the same value in all Lorentz frames.

14.5.4 Projection Operators for Chiral Fermions

From Box 14.2 and Eq. (14.42), for *massless* Dirac particle spinors u and antiparticle (negative-energy) spinors v,

Box 14.2 Relationship between Chirality and Helicity

In the limit of vanishing mass a direct relationship may be established between chirality and helicity. From Eq. (14.31), the massless Dirac equation is

$$\gamma^\mu p_\mu \psi(p) = (\gamma^0 p_0 - \boldsymbol{\gamma} \cdot \boldsymbol{p})\psi(p) = 0,$$

which we multiply by $\gamma^5\gamma^0$ and use $\gamma^0\gamma^0 = 1$ and $\alpha^i = \gamma^0\gamma^i$ to write as

$$(\gamma^0 p_0 - \boldsymbol{\gamma} \cdot \boldsymbol{p})\psi = (\gamma^5 p_0 - \gamma^5\gamma^0\gamma^i p^i)\psi = (\gamma^5 p_0 - \gamma^5\alpha^i p^i)\psi = 0,$$

and insert explicit matrices in the Pauli–Dirac representation to obtain

$$\left[\gamma^5 p_0 - \begin{pmatrix} 0 & 1 \\ 1 & 0 \end{pmatrix}\begin{pmatrix} 0 & \boldsymbol{\sigma} \cdot \boldsymbol{p} \\ \boldsymbol{\sigma} \cdot \boldsymbol{p} & 0 \end{pmatrix}\right]\psi = 0,$$

which is equivalent to

$$|\boldsymbol{p}|\begin{pmatrix} \boldsymbol{\sigma} \cdot \hat{\boldsymbol{p}} & 0 \\ 0 & \boldsymbol{\sigma} \cdot \hat{\boldsymbol{p}} \end{pmatrix}\psi = \gamma^5 p_0 \psi,$$

where $\hat{\boldsymbol{p}} = \boldsymbol{p}/|\boldsymbol{p}|$. But $E = p_0 = \pm\sqrt{\boldsymbol{p}^2 + m^2}$ and for $m \to 0$,

$$E = \pm\sqrt{\boldsymbol{p}^2} = \pm\,|\boldsymbol{p}| = p_0.$$

Therefore, for the positive-energy and negative-energy solutions of the massless Dirac equation, respectively,

$$(\boldsymbol{S} \cdot \hat{\boldsymbol{p}})\psi = \gamma^5\psi \text{ (positive energy)} \qquad (\boldsymbol{S} \cdot \hat{\boldsymbol{p}})\psi = -\gamma^5\psi \text{ (negative energy)}.$$

The eigenvalues of $\boldsymbol{S} \cdot \hat{\boldsymbol{p}}$ and γ^5 are equal for positive-energy particle solutions of the massless Dirac equation, and of equal magnitude but opposite sign for the corresponding negative-energy or antiparticle solutions. Thus chirality and helicity are interchangeable labels *only for massless fermions*. For massive fermions, neither helicity nor chirality is simultaneously conserved and independent of Lorentz frame.

$$\gamma_5 u_R = u_R \qquad \gamma_5 u_L = -u_L \qquad \gamma_5 v_R = -v_R \qquad \gamma_5 v_L = v_L, \tag{14.46}$$

where right-handed massless particles (R) are associated with helicity $+\frac{1}{2}$, left-handed massless particles (L) with helicity $-\frac{1}{2}$, right-handed massless antiparticles with helicity $-\frac{1}{2}$, and left-handed massless antiparticles with helicity $+\frac{1}{2}$. Now let us introduce new operators $P_\pm$ through

$$P_\pm \equiv \frac{1 \pm \gamma_5}{2}. \tag{14.47}$$

As demonstrated in Problem 14.9, these have the properties

$$P_\pm^2 = P_\pm \qquad P_+ + P_- = 1 \qquad P_- P_+ = P_+ P_- = 0 \qquad P_\pm \gamma^\mu = \gamma^\mu P_\pm. \tag{14.48}$$

The first three relations imply that the $P_\pm$ are *projection operators,* with P_+ projecting right-handed and P_- projecting left-handed chiral components for particles (and the opposite for antiparticles). Wavefunctions may be decomposed into left and right chiral components:

$$u = u_L + u_R = P_- u + P_+ u = \left(\frac{1-\gamma_5}{2}\right) u + \left(\frac{1+\gamma_5}{2}\right) u. \tag{14.49}$$

This chiral decomposition is always possible, but only for vanishing mass can the u_L and u_R components be identified with the helicities $-\frac{1}{2}$ and $+\frac{1}{2}$, respectively.

14.5.5 Interactions and Chiral Symmetry

For free massless fermions chirality is a good quantum number, so it is of interest to know whether fermion interaction terms conserve chiral symmetry. Such terms typically involve the bilinear covariants listed in Table 14.2. Decomposing the wavefunction into left-handed and right-handed components gives

$$\psi = R + L \equiv u_R + u_L = P_+\psi + P_-\psi. \tag{14.50}$$

This implies for the corresponding adjoint wavefunction $\overline{\psi} \equiv \psi^\dagger \gamma_0$,

$$\frac{1}{2}\overline{\psi}(1+\gamma_5) = \bar{u}_L \qquad \frac{1}{2}\overline{\psi}(1-\gamma_5) = \bar{u}_R \tag{14.51}$$

where $\bar{u} \equiv u^\dagger \gamma_0$ (see Problem 14.8), and we may write

$$\overline{\psi} = \bar{R} + \bar{L} = \overline{\psi} P_- + \overline{\psi} P_+. \tag{14.52}$$

Therefore, from (14.50) and (14.52),

$$\overline{\psi}\psi = \bar{R}L + \bar{L}R, \tag{14.53}$$

where the properties (14.48) have been used (see Problem 14.11). Thus, terms of the form $\overline{\psi}\psi$, which typically enter relativistic quantum field theory as mass terms for fermions or as the interaction of fermion fields with scalar fields, break chiral symmetry by mixing left-handed and right-handed components. However, from Problem 14.11

$$\overline{\psi}\gamma^\mu\psi = \bar{R}\gamma^\mu R + \bar{L}\gamma^\mu L \qquad \overline{\psi}\gamma^\mu\gamma^5\psi = \bar{R}\gamma^\mu\gamma^5 R + \bar{L}\gamma^\mu\gamma^5 L, \tag{14.54}$$

so vector (γ^μ) and axial vector ($\gamma^\mu\gamma^5$) interactions conserve chiral symmetry.

14.6 The Majorana Equation

The *Majorana equation* is a variation on the Dirac equation that can be written

$$i\gamma^\mu \partial_\mu \psi - m\psi^c = 0. \tag{14.55}$$

It differs from the Dirac equation (14.32) in the presence of ψ^c, which is the *charge conjugate* of ψ. In the *Majorana representation* of the γ-matrices given in Table 14.1, $\psi^c \equiv i\psi^*$, where ψ^* is the complex conjugate of ψ. In relativistic quantum field theory ψ^c is associated with the antiparticle of the particle that corresponds to ψ. Because both ψ and ψ^c appear in Eq. (14.55), a field described by the Majorana equation *cannot be charged*: the antiparticle of a charged particle has the opposite charge and coupling of the Majorana equation for charged particles to the electromagnetic field would violate conservation of electrical charge.[8]

The Majorana equation is viable *only for neutral particles that carry no electrical or other gauge charge,* which allows for two classes of solutions: a neutral particle field and its corresponding neutral antiparticle field.

If a spin-$\frac{1}{2}$ particle carries no conserved charge of any kind, it may be viewed as its own antiparticle and $\psi = \psi^c$. Such particles are said to be *Majorana spinors.* These stringent conditions raise the question of whether the Majorana equation is a suitable description of any physical particles, or whether it is just a mathematical curiosity.

14.6.1 Dirac and Majorana Masses

All particles of the Standard Model displayed in Fig. 19.1 carry one or more gauge charges, except for neutrinos and antineutrinos. Neutrinos were thought originally to be massless, parity-conserving particles described by the Dirac equation (14.32) with $m = 0$. Then when parity violation of the weak interactions was discovered in the late 1950s, the parity non-conserving Weyl equations (14.37) – valid only for *massless particles* – gained favor as a description of neutrinos. More recently, observation of neutrino flavor oscillations indicates that neutrinos have a tiny but non-zero mass and the Weyl equations are only approximately valid for neutrinos. Thus we may ask whether neutrinos are more properly described by the Dirac equation or the Majorana equation. The mass m appearing in Eq. (14.55) is called the *Majorana mass,* to distinguish it from the *Dirac mass* of Eq. (14.32).

Standard Model particles other than neutrinos cannot have Majorana masses but *neutrinos could have Dirac or Majorana masses*; present data are inconclusive.

One fundamental distinction between Dirac and Majorana neutrinos is that Dirac processes conserve lepton number but Majorana processes do not.[9] Section 14.6.2 describes an experiment that has the potential to settle the issue of whether neutrinos are Dirac fermions described by Eq. (14.32), or Majorana fermions described by Eq. (14.55).

[8] This statement applies also to generalizations of electrical charge. Broadly, *a charge is a generator of a continuous symmetry,* since Noether's theorem (Section 16.2.1) implies a corresponding conserved current representing flow of the charge that can be integrated over all space to define the total charge. Examples are the *gauge charges* associated with the Standard Model such as the SU(3) color charge or the SU(2) weak isospin charge; see the discussion in Chs. 16 and 19.

[9] Lepton number is $L \equiv n - \bar{n}$, where n is the number of leptons and $\bar{n}$ is the number of antileptons. Lepton number is conserved within generations of the Standard Model of Ch. 19, so Majorana neutrinos would entail physics beyond the Standard Model.

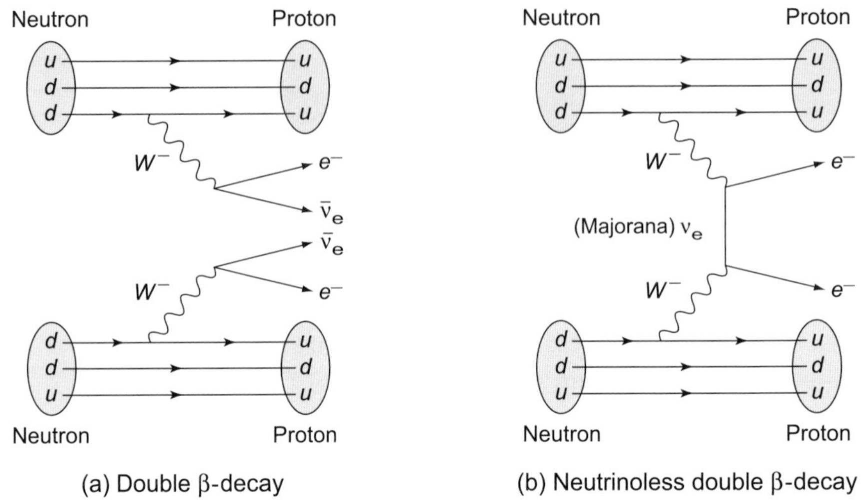

Fig. 14.2 (a) Normal double β-decay and (b) neutrinoless double β-decay in terms of Feynman diagrams (see Box 14.3) [89]. Here neutron β^- decay is viewed at the quark level, with a down quark (d) converted into an up quark (u), transforming a neutron (*udd* quark content) into a proton (*uud* quark content). Reproduced with permission from Cambridge University Press: *Stars and Stellar Processes*, M. Guidry (2019).

14.6.2 Neutrinoless Double β-Decay

In rare instances certain nuclei can undergo *double β-decay*, where two neutrons are converted to two protons by weak interactions. Figure 14.2 illustrates in terms of the Feynman diagrams described in Box 14.3. In the normal double β-decay process of Fig. 14.2(a), two neutrons are converted to two protons, two e^-, and two $\bar{\nu}_e$; it is a very unlikely process but it has been observed in a few cases. If neutrinos are Majorana particles (which are their own antiparticles), then the *neutrinoless double β-decay* illustrated in Fig. 14.2(b) is possible. This process has not been observed yet but if it occurs it would show that neutrinos are Majorana particles, which would imply physics beyond the Standard Model because neutrinoless double β-decay does not conserve lepton number (Problem 14.16).

The existence of Majorana neutrinos would permit an interesting conjecture that is described in Box 14.4 and explored in Problem 14.14. The *seesaw model* gives possible (phenomenological) reasons for several otherwise puzzling aspects of the weak interactions.

1. Observed neutrinos are left-handed and have tiny but non-zero masses. *Reason*: these are characteristics of the lowest-mass seesaw eigenstate X_-.
2. No right-handed neutrinos are observed. *Reason*: the high-mass seesaw eigenstate X_+ requires much more energy to produce than is available in accelerators or cosmic rays.
3. Weak interactions violate parity maximally. *Reason*: at low energy, right-handed neutrinos are too massive to participate at all, breaking parity to the greatest extent possible.

Box 14.3 **Feynman Diagrams**

Intuitive pictorial representations of interaction matrix elements (probability amplitudes) called *Feynman diagrams* are extremely useful in quantum field theory.

> A Feynman diagram implies a corresponding quantum matrix element and a matrix element implies a corresponding Feynman diagram.

Examples of Feynman diagrams for some weak interaction matrix elements:

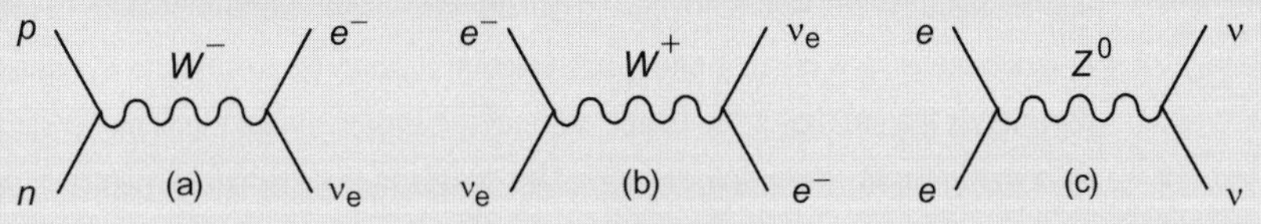

Feynman diagrams are evaluated quantitatively by a highly technical formalism, but they can be interpreted qualitatively using only a few simple rules.

1. Solid lines denote (fermion) matter fields.
2. Wiggly or coiled lines denote gauge bosons.
3. A point where two or more lines meet (a *vertex*) represents an interaction.
4. Open-ended lines like ν_e are *external lines* denoting real detectable particles.
5. Lines with no open ends as for $W^{\pm}$ are *internal lines*; they represent *virtual particles* that are not observable because of the uncertainty principle.
6. These rules illustrate concisely that exchange of virtual gauge bosons mediates the forces between fermionic matter fields.

For example, reading from the bottom, diagram (a) above represents an interaction in which a neutron n exchanges a virtual W^- vector boson with an electron neutrino ν_e, which converts the neutron n to a proton p and the neutrino ν_e to an electron e^-. We will seldom need to evaluate the actual matrix elements but various phenomena discussed in this book will be described qualitatively using Feynman diagrams.

4. Neutrino masses are much smaller than quark masses. *Reason*: the mass scale for observed neutrinos has been driven down from its natural value by the seesaw mechanism.

Thus the seesaw mechanism suggests plausible explanations of various weak interaction puzzles, though no direct experimental evidence supports it yet.

14.7 Summary: Possible Spinor Types

Non-relativistic spin-$\frac{1}{2}$ particles are described by SO(3) Pauli spinors, but there are three possible $SL(2,C)$ spinors for spin-$\frac{1}{2}$ particles in Lorentz-invariant quantum field theories.

Box 14.4

Neutrino Masses and the Seesaw Conjecture

An explanation for why parity symmetry is broken because only left-handed neutrinos and right-handed antineutrinos enter the weak interactions is possible if neutrinos are Majorana particles that are their own antiparticles.

Right-Handed and Left-Handed Neutrinos

This explanation begins by speculating that there actually is a right-handed component of the neutrino (and a left-handed component of the antineutrino) but that it is very massive relative to the tiny mass of the left-handed neutrino and so has no influence in our low-energy world. This leads to maximal violation of parity in weak interactions because parity conservation requires equal mixtures of right-handed and left-handed particles. It follows that if the temperature of the Universe were much higher (as in the early GUTs period discussed in Ch. 34), the average energy of left-handed and right-handed neutrinos could be comparable since total energy would dwarf restmass energy and they would enter the weak interactions on equal terms, thus restoring parity symmetry.

The Seesaw Mechanism

By arguments from quantum field theory, the *mass matrix* $\mathscr{M}$ describing the interactions of a very heavy Majorana neutrino (which violates lepton number conservation) and a much lighter Dirac neutrino is conjectured to take the form

$$\mathscr{M} = \begin{pmatrix} 0 & m \\ m & M \end{pmatrix},$$

where $m \sim 100$ GeV is the electroweak symmetry breaking scale (Ch. 19) and $M \sim 10^{15}$ GeV is the Grand Unified (GUTs) symmetry breaking scale (Ch. 34). The eigenvalues and eigenstates resulting from the interaction are obtained by diagonalizing $\mathscr{M}$ (Problem 14.14), which gives a high-mass eigenstate,

$$\lambda_+ \simeq M \simeq 10^{15}\,\text{GeV} \qquad X_+ = \frac{m}{M}\nu_{\text{L}} + \nu_{\text{R}} \simeq \nu_{\text{R}},$$

and a low-mass eigenstate,

$$\lambda_- \simeq \frac{-m^2}{M} \simeq -10^{-2}\,\text{eV} \qquad X_- = \nu_{\text{L}} - \frac{m}{M}\nu_{\text{R}} \simeq \nu_{\text{L}},$$

where ν_{L} denotes the wavefunction of a pure left-handed neutrino and ν_{R} the wavefunction of a pure right-handed neutrino. This has the following interpretation.

1. The eigenvector X_- associated with λ_- is the left-handed neutrino of the Standard Model, but with a tiny mass.
2. The eigenvector X_+ associated with λ_+ is a right-handed neutrino that is presumably much too massive to observe at currently available energies.

Increasing M in these formulas drives λ_+ up and λ_- down, so this is called the *seesaw mechanism.* As discussed in Section 14.6.2, it has the potential to explain a number of otherwise puzzling features of the weak interactions.

1. *Dirac spinors* describe massive particles having two degrees of freedom with particles and antiparticles being distinct, implying four spinor components. They conserve parity (P) and charge conjugation (C) symmetries.
2. *Weyl spinors* are two-component spinors that describe massless left-handed particles and massless right-handed antiparticles. The components are related by CP symmetry.
3. *Majorana spinors* are two-component spinors corresponding to massive spin-$\frac{1}{2}$ particles that are their own antiparticles.

Neutrinos have small but finite mass, so they could be Dirac or Majorana spinors. The critical issue is whether they are their own antiparticles, which is difficult to assess because they carry no charge that would distinguish particle from antiparticle. Neutrinoless double β-decay [Fig. 14.2(b)] is possible for Majorana but not for Dirac neutrinos. This process has not been observed but experiments may not yet be sensitive enough to see it. Thus, as of this writing in 2021 it is not known whether neutrinos are Dirac or Majorana particles.

14.8 Spinor Symmetry in the Weak Interactions

Weak interactions involve Lorentz spinors for the charged leptons and quarks, and the uncharged neutrinos. It is useful to summarize what we know about the symmetries of those spinors, since this has a significant influence on weak interaction phenomenology.

14.8.1 The Left Hand of the Neutrino

Clear experimental evidence indicates that only left-handed neutrinos $\nu_{\rm L}$ and right-handed antineutrinos $\bar{\nu}_{\rm R}$ participate in the weak interactions. Observation of neutrino flavor oscillations has given proof that at least some neutrino flavors carry a tiny mass, with an implied small mixing of left- and right-handed components. However, the masses of neutrinos are very small and in many applications they may be assumed massless. The fundamental reason for this is not understood but the consequences are clear: in the weak interactions the Universe is not ambidextrous. From our discussions of the Dirac and Weyl spinors, we expect that this will imply a violation of parity symmetry P and charge conjugation symmetry C in the weak interactions, as discussed further below.

14.8.2 Violation of Parity P

The charged-current weak interactions (mediated by the charged vector bosons $W^\pm$ to be introduced in Ch. 19) are found to be approximated well by the "vector minus axial vector" or $V - A$ form, with the bilinear covariants $V \sim \overline{\psi}\gamma^\mu\psi$ and $A \sim \overline{\psi}\gamma^\mu\gamma^5\psi$ given in Table 14.2. In the simplest model, weak interactions are described by the product of two 4-currents J^μ at a point (see Box 19.1). Schematically, the interaction is of the form

$$\mathscr{L} \propto J^\mu J_\mu^\dagger \sim (V - A)(V - A)^\dagger = VV^\dagger + AA^\dagger - AV^\dagger - VA^\dagger.$$

But vectors and axial vectors transform oppositely under parity P, so the $AV^\dagger$ and $VA^\dagger$ terms change sign under P and the weak interactions do not conserve parity. In more detail, leptonic $V - A$ weak interaction matrix elements are of the form

$$M = \langle \mathrm{e} | \, l^\mu \, | \nu_\mathrm{e} \rangle \simeq \frac{1}{2} \bar{u}_\mathrm{e} \gamma^\mu (1 - \gamma_5) u_\nu ,$$

where l^μ represents the leptonic charge-changing weak current. But utilizing Eq. (14.48),

$$M = \bar{u}_\mathrm{e} \gamma^\mu P_- u_\nu = \bar{u}_\mathrm{e} \gamma^\mu P_- P_- u_\nu = \bar{u}_\mathrm{e} P_+ \gamma^\mu P_- u_\nu = \bar{u}_\mathrm{e}^\mathrm{L} \gamma^\mu u_\nu^\mathrm{L}$$

and participation is limited to left-handed particles (or right-handed antiparticles). Thus weak interactions violate parity *maximally*, since a parity-conserving state must be an equal mixture of right- and left-handed components.

14.8.3 C, CP, and T Symmetries

Charge conjugation C corresponds to replacement of a particle with its antiparticle. The action of C on a left-handed neutrino gives a left-handed *antineutrino*. But only right-handed antineutrinos are observed to participate in the weak interactions, so C is also violated maximally. However, the combined operation CP (invert the coordinate system and exchange particle for antiparticle) converts a left-handed neutrino into a right-handed antineutrino, so CP is conserved by the $V - A$ approximation (though not exactly by more realistic weak interactions). Likewise, time reversal T inverts both momentum and spin, so it does not affect helicity and T is conserved by $V - A$ interactions. For interactions more realistic than the $V - A$ approximation the weak interactions violate CP and T by very small amounts (for reasons not understood), but the combined symmetry CPT is expected to be conserved by all interactions on fundamental grounds and no violation of CPT has been found.

14.8.4 A More Complete Picture

A fully realistic discussion of weak interactions requires features not considered here. The present example of a leptonic charged current is simpler than the corresponding hadronic current because strong interactions renormalize ("dress") the weak interactions by modifying the environment in which they occur.[10] Furthermore hadronic and leptonic weak currents require additional empirical parameters called *mixing angles* that are associated with a mismatch between the weak eigenstates and the mass eigenstates.[11] In addition, there are weak neutral currents as well as charged ones, and they are not of pure $V - A$ form. Introductions to hadronic charged weak currents, neutral weak currents, and quark mixing angles are given in Ref. [85], and to neutrino mixing angles in Refs. [89, 90]. Our goal

[10] But the renormalization is smaller than expected, due to symmetries protecting the weak current; see Ch. 33.

[11] Neutrinos and quarks produced in weak interactions are *created in weak eigenstates* but *propagate in mass eigenstates*. In the simplest of universes these eigenstates would be equivalent but it seems that we do not live in that universe! Mixing angles account for this phenomenologically but we have little fundamental understanding of why our universe opts for a mismatch between weak eigenstates and mass eigenstates.

here has been to show the influence of a variety of symmetries on the weak interactions, and this streamlined discussion is adequate for that purpose.

Background and Further Reading

The Lorentz properties of the Maxwell and Dirac equations are discussed in Guidry [85], Jackson [122], Ryder [174], and Tung [199]. An overview of the relationship among Dirac, Weyl, and Majorana spinors is given in Pal [159]. Neutrino masses and flavor oscillations are discussed in Refs. [89, 90].

Problems

14.1 Prove the useful identity employed in Eq. (14.20) that $e^{iL\phi} = \cosh\phi + iL\sinh\phi$. *Hint*: Expand the exponential in a power series and compare the odd and even terms to the power series expansions of $\cosh\phi$ and $\sinh\phi$.

14.2 Prove that the boosted right-handed spinor $\psi_{\rm R}(\boldsymbol{p})$ is related to the corresponding rest spinor by Eq. (14.21). *Hint*: Begin from Eq. (14.20) and use Eq. (13.14), hyperbolic trigonometry identities, and that $\gamma = E/m$ for a relativistic particle with energy given by $E^2 = p^2 + m^2$. ***

14.3 Use the γ-matrices in the Weyl representation to show that the Dirac equation (14.31) is equivalent to Eq. (14.25). ***

14.4 Prove the identity $(\sigma \cdot \boldsymbol{p})^2 = \mathrm{I}^{(2)} p^2$, where $\boldsymbol{\sigma} = (\sigma_1, \sigma_2, \sigma_3)$ are the Pauli matrices, $\boldsymbol{p}$ is the 3-momentum, $p = |\boldsymbol{p}|$, and $\mathrm{I}^{(2)}$ is the unit 2×2 matrix.

14.5 Show that the relations (14.23) follow from Eqs. (14.21) and (14.22). *Hint*: Use the identity in Problem 14.4 and $E^2 = p^2 + m^2$ to simplify the final expression. ***

14.6 Verify that the *Pauli–Dirac representation* (14.43) satisfies Eq. (14.27). ***

14.7 Show that the quantity $\overline{\psi}\sigma^{\mu\nu}\psi$ of Table 14.2 transforms as an antisymmetric rank-2 Lorentz tensor. *Hint*: Use $\sigma^{\mu\nu} \equiv \frac{i}{2}[\gamma^\mu, \gamma^\nu]$, Eq. (14.34), and $S^{-1}S = SS^{-1} = 1$. ***

14.8 Verify the projection characteristics implied by equations (14.51).

14.9 Verify the results of Eq. (14.48) for the properties of the chiral projection operators.

14.10 Starting from Eq. (14.17), show that the equation for the 4-vector potential A^μ takes the covariant form $\Box A^\mu = j^\mu$ of Eq. (14.13) in Lorenz gauge.

14.11 Beginning with Eq. (14.50), prove Eq. (14.52), and derive the chiral decompositions given for $\overline{\psi}\psi$, $\overline{\psi}\gamma^\mu\psi$, and $\overline{\psi}\gamma^\mu\gamma^5\psi$ interactions in Eqs. (14.53) and (14.54). ***

14.12 Using the Dirac Hamiltonian $H = -i\gamma_0\gamma^i p_i + \gamma_0 m$ given in (14.45), show that the chirality operator γ_5 commutes with this Hamiltonian only if the mass $m \to 0$.

14.13 Show that the Dirac current $j^\mu = \overline{\psi}\gamma^\mu\psi$ of Example 14.3 transforms as a true vector under Lorentz transformations.

14.14 (a) Prove that the eigenvalues for a general 2×2 matrix A are given by $\lambda_\pm = \frac{1}{2}T \pm (\frac{1}{4}T^2 - D)^{1/2}$, where $T \equiv \mathrm{Tr}\, A$ and $D \equiv \det A$. (b) Find the eigenvectors X corresponding to the eigenvalues found in part (a), assuming a two-component basis

$$\chi_1 \equiv \begin{pmatrix} 1 \\ 0 \end{pmatrix} \qquad \chi_2 \equiv \begin{pmatrix} 0 \\ 1 \end{pmatrix} \qquad X = a\chi_1 + b\chi_2.$$

Hint: Solve the linear system $AX = \lambda X$ using the eigenvalues λ found in part (a). (c) Find the eigenvalues of the mass matrix $\mathscr{M}$ given in Box 14.4. Find the corresponding eigenvectors and argue that they can be interpreted approximately as a left-handed neutrino but with a small but non-zero mass, and a right-handed neutrino with mass much too large to be detected in present experiments. ***

14.15 Operators giving quantum numbers to label states should be conserved (they should commute with the Hamiltonian) and should not depend on choice of reference frame (they should be Lorentz invariant). Prove that for *massive fermions* described by the free-particle Dirac equation chirality is Lorentz invariant but is not conserved, while helicity is conserved but is not Lorentz invariant. ***

14.16 Show that in Fig. 14.2, normal double β-decay conserves lepton number but neutrinoless double β-decay does not. *Hint*: Define a lepton number by $L \equiv n_\ell - n_{\bar{\ell}}$, where n_ℓ is the total number of leptons and $n_{\bar{\ell}}$ is the total number of antileptons.

14.17 Show that a gauge transformation of the 4-vector potential A^μ leaves the Maxwell field tensor $F^{\mu\nu}$ invariant.

14.18 Show that the Maxwell equations written in the manifestly covariant form (14.17) and (14.18) are equivalent to the standard form given by Eqs. (14.1a)–(14.1d).

14.19 Prove that Eq. (14.3) implies the Maxwell field tensor of Eqs. (14.14) and (14.15).

14.20 In Section 6.2.5 we emphasized the utility of an integration measure that respects the symmetry for continuous groups. Consider a Lorentz boost along the x-axis of the 4-momentum vector $(p_0, p_1, p_2, p_3) = (E, p_1, p_2, p_3)$ analogous to that of Eq. (13.16). The relationship between momentum volume elements in the two frames is given generally by $d^3p' = J d^3p$, where $J \equiv \det(\partial p'/\partial p)$ is the Jacobian determinant of the transformation. Evaluate the Jacobian of the transformation between the unboosted (unprimed) and boosted (primed) coordinate systems to show that $d^3p'/E' = d^3p/E$, which suggests defining an invariant volume element in momentum space, $\mathscr{D}^3p \equiv \text{constant} \times d^3p/E$. ***

14.21 It is common to use the normalization $u^{(\alpha)\dagger}(\boldsymbol{p})u^{(\beta)}(\boldsymbol{p}) = (E/m)\delta_{\alpha\beta}$ for a massive free Dirac spinor $u^{(\alpha)}$, where E is energy, m is restmass, and α and β are spinor indices. Show that this spinor normalization is covariant with respect to a change of Lorentz frame. *Hint*: The free fermion will propagate in a plane-wave state. In a particular Lorentz frame, what is the probability of finding the fermion in a box of volume V? How does that probability change under a Lorentz transformation? ***

14.22 In the Pauli–Dirac representation of Table 14.1, a suitable charge conjugation operator is $C = i\gamma^2\gamma^0$. Show that C is given explicitly by the matrix

$$C = \begin{pmatrix} 0 & -i\sigma_2 \\ -i\sigma_2 & 0 \end{pmatrix},$$

in this representation. Verify by matrix multiplication the matrices for γ^5 shown in Table 14.1 for the Pauli–Dirac and Weyl (chiral) representations. *Hint*: A useful identity for Pauli matrices is $\sigma_1\sigma_2\sigma_3 = i\,\mathrm{I}^{(2)}$, where $\mathrm{I}^{(2)}$ is the 2×2 unit matrix.

14.23 Denote right-handed and left-handed Dirac positive-energy (particle) spinors by u_{R} and u_{L}, respectively, and denote right-handed and left-handed negative-energy (antiparticle) spinors by v_{R} and v_{L}, respectively, where R corresponds to helicity $\lambda_{\mathrm{R}} = +\frac{1}{2}$ and L corresponds to helicity $\lambda_{\mathrm{L}} = -\frac{1}{2}$. Use Box 14.2 and Eq. (14.42) to verify Eq. (14.46) for massless fermions.

15 Poincaré Invariance

The Lorentz group described in Ch. 13 encompasses two sets of generators that are of obvious importance for physical problems: boosts between inertial frames and spatial rotations within an inertial frame. However, it does not include another class of generators that may be expected to be significant for those same problems: spacetime translations. This leads us to consider the 10-parameter *Poincaré* or *inhomogeneous Lorentz group*, which is obtained by appending the set of four continuous spacetime translations to the set of six continuous Lorentz transformations that we have considered previously. The resulting group is non-compact since the Lorentz group is non-compact and the translation parameters are unbounded. As we shall see, the structure of the Poincaré group is tied intimately to our understanding of the spin and mass of elementary particles, and the relationship between representations of the Poincaré and Lorentz groups is the basis for our understanding of the relationship between particles and fields in relativistic quantum field theory.

15.1 The Poincaré Multiplication Rule

The most general Poincaré transformation corresponds to a Lorentz transformation plus a spacetime translation,

$$x'_\mu = \Lambda_\mu{}^\nu x_\nu + b_\mu, \tag{15.1}$$

where b_μ is a constant 4-vector, independent of x_ν. Equation (15.1) may be written in matrix notation as $x' = \Lambda x + b$. As shown in Problem 15.1, application of two successive Poincaré transformations leads to the multiplication rule for the Poincaré group,

$$g(b'\Lambda')g(b,\Lambda) = g(\Lambda' b + b', \Lambda'\Lambda), \tag{15.2}$$

where Λ parameterizes a Lorentz transformation and b parameterizes a spacetime translation. As noted previously in conjunction with the euclidean group E_2 [compare Eq. (12.11)], this multiplication law is characteristic of a *semidirect product group.* The Poincaré group obeys the multiplication rule (15.2) because it is a semidirect product of the group of four-dimensional spacetime translations T_4 and the Lorentz group SO(3, 1),

$$\text{Poincaré group} = T_4 \otimes_s SO(3,1), \tag{15.3}$$

where $\otimes_s$ indicates the semidirect product. Conceptually, the Poincaré group has a structure similar to the euclidean groups E_n of Section 12.4, but with the euclidean spatial manifold replaced by the Minkowski spacetime manifold (spatial translations replaced by spacetime translations and spatial rotations replaced by proper Lorentz transformations).

As shown in Problems 15.2 and 15.3, a general Poincaré transformation may be factored into a product of a proper Lorentz transformation Λ and a translation $T(b)$,

$$g(b, \Lambda) = T(b)\Lambda, \tag{15.4}$$

a proper Lorentz transformation of a translation is a translation,

$$\Lambda T(b)\Lambda^{-1} = T(\Lambda b), \tag{15.5}$$

and the action of a general group element on a translation is again a translation,

$$g(b, \Lambda)T(a)g(b, \Lambda)^{-1} = T(\Lambda a). \tag{15.6}$$

Thus, the translations conjugated with all group elements return translations and they form an *(abelian) invariant subgroup* of the full Poincaré group (see Section 2.12).

15.2 Generators of Poincaré Transformations

We shall explore the properties of the Poincaré group by examining the infinitesimal transformations directly.

15.2.1 Proper Lorentz Transformations

Consider first an infinitesimal Lorentz transformation, which takes the form

$$x_\mu = \Lambda_{\mu\nu} x^\nu = (\eta_{\mu\nu} - \epsilon_{\mu\nu})x^\nu = x_\mu - \epsilon_\mu{}^\nu x_\nu. \tag{15.7}$$

As shown in Problem 15.4, the infinitesimal $\epsilon_{\mu\nu}$ is an antisymmetric rank-2 Lorentz tensor, which has six independent components (the 16 components of a general rank-2 tensor are reduced to six by the antisymmetry condition). Physically, the three components ϵ_{ij} correspond to rotations and the three components ϵ_{0k} correspond to boosts. Let us now introduce

$$L_{\mu\nu} \equiv i(x_\mu \partial_\nu - x_\nu \partial_\mu). \tag{15.8}$$

As shown in Problem 15.5, for infinitesimal $\epsilon_{\mu\nu}$

$$e^{-\frac{1}{2}i\epsilon_{\mu\nu}L^{\mu\nu}} x_\rho \simeq x_\rho - \epsilon_\rho{}^\mu x_\mu. \tag{15.9}$$

This is equivalent to Eq. (15.7), implying that Eq. (15.8) defines generators of proper Lorentz transformations. The explicit rotation and boost operators $(K_x, K_y, K_z, J_x, J_y, J_z)$ employed for the Lorentz group in Section 13.2 are recovered if we make the identifications

$$J_i = \frac{1}{2}\epsilon_{ijk}L_{jk} \qquad K_i = L_{0i}, \tag{15.10}$$

which obey the commutation relations (13.20). That takes care of the six proper Lorentz transformations of the Poincaré group. Now let us consider the four translations.

15.2.2 Four-Dimensional Spacetime Translations

In infinitesimal form, translations may be written as

$$x'_\mu = x_\mu - \epsilon_\mu. \tag{15.11}$$

Now introduce the Lorentz 4-momentum operator P^μ,

$$P^\mu = i\partial^\mu = i(\partial^0, \partial^1, \partial^2, \partial^3) \qquad P_\mu = i\partial_\mu = i(\partial^0, -\partial^1, -\partial^2, -\partial^3), \tag{15.12}$$

where $P_\mu = \eta_{\mu\nu} P^\nu$. As shown in Problem 15.6,

$$e^{i\epsilon_\nu P^\nu} x_\mu \simeq (1 + i\epsilon_\nu P^\nu) x_\mu = x_\mu - \epsilon_\mu. \tag{15.13}$$

This is equivalent to (15.11), so the P_μ are generators of translations.

15.2.3 Commutators for Poincaré Generators

The commutation relations of the Poincaré generators are [125]

$$\begin{gathered}[P_\mu, P_\nu] = 0 \qquad [P_\mu, L_{\rho\sigma}] = i(\eta_{\mu\rho} P_\sigma - \eta_{\mu\sigma} P_\rho), \\ [L_{\mu\nu}, L_{\rho\sigma}] = -i(\eta_{\mu\rho} L_{\nu\sigma} - \eta_{\mu\sigma} L_{\nu\rho} + \eta_{\nu\sigma} L_{\mu\rho} - \eta_{\nu\rho} L_{\mu\sigma}).\end{gathered} \tag{15.14}$$

The commutation relations between P_μ and $L_{\rho\sigma}$ may be rewritten in terms of the $\boldsymbol{J}$ and $\boldsymbol{K}$ operators (15.10),

$$[P_0, J_i] = 0 \qquad [P_i, J_j] = i\epsilon_{ijk} P_k \qquad [P_0, K_i] = iP_i \qquad [P_i, K_j] = iP_0\, \delta_{ij}. \tag{15.15}$$

The first two commutators in (15.15) indicate that P_0 is a scalar and P_i a vector under rotations. The second two imply that a boost along the ith axis influences P_0 and P_i, but not the other two components. The Poincaré generators define the full symmetry of Minkowski spacetime, as discussed in Box 15.1.

15.3 Representation Theory of the Poincaré Group

Let us now consider the representation theory of the Poincaré group for cases of physical interest. The group is not compact (the Lorentz transformations are not closed and the translations are unbounded), nor is it semisimple (the translations form an abelian invariant subgroup), so our approach will differ from that for compact groups.

15.3.1 Casimir Operators for the Poincaré Group

The Poincaré group has two Casimir operators. The first is just $P^2 = P^\mu P_\mu$, since from the previous commutation relations P^2 is Lorentz and translationally invariant:

$$[P^2, L_{\mu\nu}] = 0 \qquad [P^2, P_\mu] = 0. \tag{15.16}$$

Box 15.1

Isometries of Minkowski Space

A metric is said to have a symmetry under a transformation if it is unchanged by that transformation. Such a symmetry of the metric is termed an *isometry*.

> *Example:* Consider the Minkowski metric $\eta_{\mu\nu} = \text{diag}\,(1, -1, -1, -1)$. It has constant entries so it is unaltered by changing a spacetime coordinate. For example, $\eta_{\mu\nu}$ is (trivially) invariant under $x^0 \longrightarrow x^0 +$ constant, so this defines an isometry of the metric.

Killing Vectors

If an isometry exists for a metric, a unit vector may be defined pointing in a direction in which the metric is constant that is termed a *Killing vector.*[a] In the preceding example the unit vector corresponding to the isometry under $x^0 \longrightarrow x^0 +$ constant is $K_0^\mu \equiv (1, 0, 0, 0)$. Since $x^0 = ct$, the Killing vector K_0^μ implies time independence of the metric. We are usually interested in cases where a Killing vector may be defined at every point of the manifold, which we term a *Killing (vector) field.*

Isometry Groups

The set of one to one, distance-preserving maps of a metric space to itself forms a group under application of two successive maps (composition) that is termed the *isometry group* of that space. The following are examples.

1. The isometry group of the 2-sphere S^2 is the group O(3) of 3D rotations and reflections, because a 2-sphere transforms into itself under the actions of the rotations and reflections corresponding to the O(3) group generators.
2. The isometry group of Minkowski spacetime is the *Poincaré group*. Thus, the ten generators of the Poincaré group represent the full symmetry of special relativity, or equivalently of Minkowski space.

From a group-theoretical perspective, Killing fields are the infinitesimal generators of isometries for the metric.

Spacetime Conservation Laws

The Poincaré symmetry of Minkowski space implies ten continuous transformations associated with the generators: four translations, three rotations, and three boosts. Noether's theorem (Section 16.2.1) implies that there are ten conserved quantities associated with the Poincaré isometries. These correspond to conservation of energy (one time-translation generator), conservation of linear momentum (three spatial-translation generators), conservation of angular momentum (three rotational generators), and conservation of the center of mass (three boost generators).

[a] Killing vectors are named for the German mathematician Wilhelm Killing (1847–1923). Sophus Lie (1842–1899) and Killing provided much of the initial foundation for Lie algebras and Lie groups.

Table 15.1. Classes of vectors in Minkowski space [185]

Specification of vector†	Characterization	Standard vector $\bar{p}^\mu$
$p^2 > 0, p_0 > 0$	Positive timelike	$(p, 0, 0, 0)$
$p^2 > 0, p_0 < 0$	Negative timelike	$(-p, 0, 0, 0)$
$p^2 < 0$	Spacelike	$(0, 0, 0, p)$
$p^2 = 0, p_0 > 0$	Positive lightlike	$(p, 0, 0, p)$
$p^2 = 0, p_0 < 0$	Negative lightlike	$(-p, 0, 0, p)$
$p^2 = 0, p_0 = 0$	Null	$(0, 0, 0, 0)$

†Assuming our metric signature pattern $\{+ - - -\}$.

The associated physical invariant is the particle restmass, since if $P^2 |p\rangle = p^2 |p\rangle$, by the Einstein energy–momentum relation

$$p^2 = p^\mu p_\mu = p_0^2 - \boldsymbol{p}^2 = E^2 - \boldsymbol{p}^2 = m^2.$$

To exhibit the second Casimir operator let us introduce the *Pauli–Lubanski pseudovector,*

$$W_\mu \equiv -\frac{1}{2}\,\epsilon_{\mu\nu\rho\sigma} L^{\nu\rho} P^\sigma, \tag{15.17}$$

where $\epsilon_{\mu\nu\rho\sigma}$ is the completely antisymmetric rank-4 tensor.[1] Then $W^2 = W^\mu W_\mu$ is both translationally and Lorentz invariant,

$$[\,W^2, P_\mu\,] = 0 \qquad [\,W^2, L_{\rho\sigma}\,] = 0, \tag{15.18}$$

so it serves as a second Casimir operator. As will be seen, W^2 is associated with the spin of the particle but the nature of that association is nuanced.

15.3.2 Classification of Poincaré States

Mathematically there are six classes of states, with not all having obvious physical relevance. These are listed in Table 15.1, along with a *standard vector* of the corresponding type (see the discussion of standard vectors in Section 12.5.1 and the lightcone classification in Fig. 13.2). For each class an arbitrary vector can be converted into the standard vector by an appropriate Lorentz transformation but a Lorentz transformation cannot connect standard vectors that lie in different classes. The subsequent discussion will be concerned with the two classes of most obvious physical import in Table 15.1.

1. *Massive particles:* $p^2 = m^2 > 0$ and $p_0 > 0$ (positive timelike),
2. *Massless particles:* $p^2 = 0$ and $p_0 > 0$ (positive lightlike).

[1] With $\epsilon_{0123} \equiv +1$ and $\epsilon_{\alpha\beta\gamma\delta}$ equal to +1 if it is related to ϵ_{0123} by an even permutation of indices, equal to −1 if it is related to ϵ_{0123} by an odd permutation of indices, and equal to zero if any two indices are equivalent. For example, $\epsilon_{0132} = -1$ and $\epsilon_{0121} = 0$.

Among the other classes, we note that some ($p_0 < 0$) are probably not physical, that spacelike momenta can occur for virtual particles but not for physical ones because they would violate causality, and that the null category presumably corresponds to the physical vacuum (ground state) of a relativistic quantum field theory.

15.3.3 Method of Induced Representations

Let us now construct the unitary representations of the Poincaré group, which will necessarily be of infinite dimensionality by virtue of the non-compact group manifold (Box 12.2). The standard method to accomplish this uses the *method of induced representations* that was introduced in conjunction with the euclidean group E_2 in Section 12.5. By this method all members of a representation are generated by actions of the group starting from a representative vector in an invariant subspace. Recall the essential steps.

1. Identify the invariant subgroup and choose a *standard vector* in this space.
2. Identify the factor group associated with the invariant subgroup.
3. Determine whether there are any group generators in the factor group that leave the standard vector invariant (that commute with the generator of the standard vector). The set of generators satisfying this condition defines the *little group*.
4. Generate the full irreducible space by operating on the standard vector with those group generators that do not commute with the generator of the standard vector.
5. Normalize as appropriate.

The four-dimensional translations were identified as an invariant subgroup of the Poincaré group in Section 15.1. Thus, let us apply the method of induced representations to the Poincaré group, beginning with the invariant translation subgroup. A key feature of the method of induced representations is the nature of the little group of step 3 above.

> The Lie algebra of the little group is generated by the components of the Pauli–Lubanski pseudovector W_μ within the standard vector subspace. For every unitary irrep of the little group an *induced representation* of the full Poincaré group results from successive application of proper Lorentz transformations [199].

The little group and the corresponding representation structure differ fundamentally for massive and massless particles. We analyze the massive case first.

15.4 Massive Representations of the Poincaré Group

The factor group of the Poincaré group with respect to the abelian invariant subgroup of translations T_4 is the group of proper Lorentz transformations. A massive particle is positive timelike, and from Table 15.1 a standard vector is $\bar{p}_\mu = (p, 0, 0, 0) = (m, 0, 0, 0)$. Physically this corresponds to a state at rest with restmass m, and thus energy $E = mc^2$.

15.4.1 Quantum Numbers for Massive States

Multiplying Eq. (15.17) from the right by P^μ and doing the implied summations, the Pauli–Lubanski pseudovector must satisfy

$$W_\mu P^\mu = -\frac{1}{2}\epsilon_{\mu\nu\rho\sigma} L^{\nu\rho} P^\sigma P^\mu = 0, \tag{15.19}$$

as may be verified by inspection: $\epsilon_{\mu\nu\rho\sigma}$ is antisymmetric under exchange of the summation indices σ and μ but the product $P^\sigma P^\mu$ is symmetric under exchange of σ and μ, and from basic tensor analysis the contraction of a tensor symmetric on two indices with a tensor antisymmetric in those indices is identically zero. Therefore, since $\bar{p}^\mu$ has only a timelike component, Eq. (15.19) requires that W_μ have only spacelike components, $W_\mu = (0, \boldsymbol{W}) = (0, W_1, W_2, W_3)$, with

$$W_i = -\frac{1}{2}\,\epsilon_{i\nu\rho\sigma}\, L^{\nu\rho} P^\sigma. \tag{15.20}$$

In the restframe only $\bar{p}^\sigma = \bar{p}^0 = m$ is non-zero for $\bar{p}^\sigma$ and $W_i = -\frac{1}{2}\,\epsilon_{ijk}\, L^{jk} m$. Then from Eq. (15.10), $W_\mu = (0, \boldsymbol{w})$ with $w_i = mJ_i$, and for massive particles

$$W^2 = W_\mu W^\mu = W_0^2 - \boldsymbol{W}^2 = -m^2 J^2. \tag{15.21}$$

This result was derived in the restframe but it is true in all inertial frames since the Casimir W^2 commutes with boost transformations between inertial frames. Therefore, massive states can be labeled by the restmass m and the spin $s(s+1)$, which is the eigenvalue of J^2 in the restframe of the massive particle.

Additional quantum number labels must correspond to operators that commute among themselves and with the mass and spin operators already found. From Eq. (15.14), the components p^μ commute among themselves so p^μ is a constant of motion. But $p^\mu = (E, \boldsymbol{p})$ and since $E = (m^2 + \boldsymbol{p}^2)^{1/2}$ and m^2 is already given, we need only specify $\boldsymbol{p}$ to fix p^μ. Fixing an angular momentum component is not allowed because $\boldsymbol{p}$ is a label and from Eq. (15.15), p_i does not commute with J_i. However, the *helicity operator* $\boldsymbol{J} \cdot \hat{\boldsymbol{p}}$ (spin component in the direction of motion) commutes with $\boldsymbol{p}$ and the *helicity quantum number* λ can be used as a label. Summarizing, eigenstates for a massive particle may be labeled $|m, s, \boldsymbol{p}, \lambda\rangle$, with

$$\begin{aligned} P^2\,|m, s, \boldsymbol{p}, \lambda\rangle &= m^2\,|m, s, \boldsymbol{p}, \lambda\rangle \qquad & W^2\,|m, s, \boldsymbol{p}, \lambda\rangle &= -m^2 s(s+1)\,|m, s, \boldsymbol{p}, \lambda\rangle, \\ P_\mu\,|m, s, \boldsymbol{p}, \lambda\rangle &= p_\mu\,|m, s, \boldsymbol{p}, \lambda\rangle \qquad & \boldsymbol{J} \cdot \hat{\boldsymbol{p}}\,|m, s, \boldsymbol{p}, \lambda\rangle &= \lambda\,|m, s, \boldsymbol{p}, \lambda\rangle, \end{aligned} \tag{15.22}$$

where $\hat{\boldsymbol{p}} = \boldsymbol{p}/|\boldsymbol{p}|$ and s may be interpreted in terms of the eigenvalue of angular momentum in the restframe: $J^2\,|s\rangle = s(s+1)\,|s\rangle$. The Poincaré group acts on the vector space (15.22), which is of infinite dimensionality because the momentum label $\boldsymbol{p}$ is continuous.

15.4.2 Action of the Poincaré Group on Massive States

How do the states of Eq. (15.22) behave under Poincaré transformations [125]? For translations $T(b)$ the result is immediate: the basis states are explicit eigenstates of the generator P_μ and under translations $x \to x - b$ the states just get multiplied by a phase,

$$T(b)\,|m,s,\boldsymbol{p},\lambda\rangle = e^{-ib^\mu p_\mu}\,|m,s,\boldsymbol{p},\lambda\rangle\,. \tag{15.23}$$

Now consider proper Lorentz transformations (boosts and rotations). Note that $|m,s,\boldsymbol{p},\lambda\rangle$ can be obtained from a restframe state $|m,s,\boldsymbol{0},s_3=\lambda\rangle$ by a Lorentz transformation:

$$|m,s,\boldsymbol{p},\lambda\rangle = \Lambda_p\,|m,s,\boldsymbol{0},\lambda=s_3\rangle\,. \tag{15.24}$$

(In the restframe the helicity is indeterminate; it can be replaced by any component of the restframe spin, typically s_3.) Now consider a second Lorentz transformation from p to p',

$$\begin{aligned}\Lambda_{p'\leftarrow p}\,|m,s,\boldsymbol{p},\lambda\rangle &= \Lambda_{p'\leftarrow p}\Lambda_p\,|m,s,\boldsymbol{0},s_3\rangle\\ &= \Lambda_{p'}(\Lambda_{p'}^{-1}\Lambda_{p'\leftarrow p}\Lambda_p)\,|m,s,\boldsymbol{0},s_3\rangle\\ &\equiv \Lambda_{p'}R_{\rm W}\,|m,s,\boldsymbol{0},s_3\rangle\,.\end{aligned}$$

But $R_{\rm W}\equiv\Lambda_{p'}^{-1}\Lambda_{p'\leftarrow p}\Lambda_p$ takes the rest frame momentum $\bar p=(m,\boldsymbol{0})$ to p, then from p to p', and then from p' back to the restframe value of $(m,\boldsymbol{0})$. Since $R_{\rm W}$ leaves the restframe 4-momentum invariant, it is a (complicated) *rotation*. But this is familiar territory from Ch. 6: restframe states transform under rotations according to the SU(2) irrep $D^{(s)}$. Therefore

$$\begin{aligned}|m,s,\boldsymbol{p}',\lambda\rangle \equiv \Lambda_{p'\leftarrow p}\,|m,s,\boldsymbol{p},\lambda\rangle &= \Lambda_{p'}\sum_{\lambda'} D^s_{\lambda\lambda'}(R_{\rm W})\,|m,s,\boldsymbol{0},\lambda'\rangle\\ &= \sum_{\lambda'} D^s_{\lambda\lambda'}(R_{\rm W})\,|m,s,\boldsymbol{p}',\lambda'\rangle,\end{aligned} \tag{15.25}$$

where the $D^s_{\lambda\lambda'}(R_{\rm W})$ are the usual unitary rotation matrices (Section 6.3).

> For *massive particles* the subset of transformations that leave the restframe vector $\bar p_\mu$ invariant corresponds to rotations, and the little group is SU(2) [or SO(3)].

In the particular case of the restframe for massive particles, the Pauli–Lubanski pseudovector is associated with SU(2) angular momentum in the usual manner. Thus the previous assertion that the little group is generated by the components of W^μ in the subspace of the representative vector is seen to hold for the timelike case.

15.4.3 Summary: Representations for Massive States

We have just obtained a result of fundamental importance, so let us pause and take stock. By starting from the invariant subspace of translations for massive particles a representation space for the full Poincaré group was generated that behaves under proper Lorentz transformations as Eq. (15.25), and under translations as Eq. (15.23). This representation space has the following properties.

1. Equations (15.23) and (15.25) imply that the full space $\{|m,s,\boldsymbol{p},\lambda\rangle\}$ is invariant under Poincaré transformations because the translations and proper Lorentz transformations give back linear combinations of the original space.

2. This space is *irreducible* because the basis vectors have been generated from a single vector by actions of the generators and group elements; thus there is no smaller subspace from which one could have implemented this procedure.
3. The space can be generated in an unambiguous way by starting from $|m, s, \boldsymbol{p}, \lambda = s_3\rangle$, creating other values of λ using the $j_\pm$ operators of the SU(2) algebra as in Section 6.3.9, and then using proper Lorentz transformations to produce the other states.
4. The representations are *unitary* because the generators are hermitian operators.
5. The representations are of *infinite dimensionality,* since $\boldsymbol{p}$ takes a continuum of values.

Let us now apply the same methods to the other case of direct physical interest, the massless particles having lightlike Lorentz behavior.

15.5 Massless Representations

For lightlike particles the restmass is zero and the particle moves at light velocity in all frames. Thus, the device of going to the restframe that simplified the discussion of Section 15.4 is not available and we may expect qualitative differences in the representation theory of massless particles relative to massive ones. We shall find as a consequence that the meaning of "spin" differs substantially between the massive and massless cases.

15.5.1 The Standard Lightlike Vector

By preceding considerations we expect that the relevant little group in the massless particle case is again generated by the components of the Pauli–Lubanski pseudovector (15.17), but now in the subspace of a representative positive lightlike vector from Table 15.1. For the massless (lightlike) case, the lightcone condition for a particle traveling on the z-axis is that $p_\mu = (p, 0, 0, p)$. Assuming the massless particle to have energy $E = \hbar\omega = \omega$ in natural units, we may take as a standard momentum vector

$$\bar{p}_\mu = (\omega, 0, 0, \omega). \tag{15.26}$$

Then utilizing the identifications

$$L_{\mu\nu} = -L_{\nu\mu} \qquad J_i = \tfrac{1}{2}\epsilon_{ijk}L_{jk} \qquad K_i = L_{0i}, \tag{15.27}$$

the components of the Pauli–Lubanski pseudovector W^μ may be constructed explicitly for the standard lightlike vector,

$$W_\mu = (W_0, W_1, W_2, W_3) = (\omega J_3,\ \omega(J_1 + K_2),\ \omega(J_2 - K_1),\ \omega J_3), \tag{15.28}$$

as shown in the solution of Problem 15.9.

15.5.2 Lie Algebra of the Little Group

According to our earlier discussion the components (15.28) generate the Lie algebra of the appropriate little group for lightlike Poincaré representations. Let us evaluate the commutators to deduce the structure of the little group. From the solution of Problem 15.10,

$$[W_1, W_2] = 0 \qquad [W_1, W_0] = [W_1, W_3] = -i\omega W_2,$$
$$[W_2, W_0] = [W_2, W_3] = i\omega W_1. \tag{15.29}$$

But this is equivalent to the euclidean algebra E_2 displayed in Eq. (12.14) if the identifications $\omega W_i = P_i$, and $W_3 = W_0 = \omega J$ are made.

We conclude that for massless particles the little group for Poincaré transformations is *not the rotation group SU(2),* as it was for the massive case, but instead is the group E_2 of euclidean motion in two dimensions.

Problem 15.12 verifies that E_2 is the minimal group leaving the standard vector (15.26) invariant, confirming that E_2 is the little group for massless particles.

15.5.3 Quantum Numbers for Massless States

Using the notation for E_2 generators, the most general Pauli–Lubanski pseudovector (15.28) for massless particles is $W_\mu = \omega(J, P_1, P_2, J)$. Hence,

$$W^2 = W_\mu W^\mu = -\omega^2(P_1^2 + P_2^2). \tag{15.30}$$

But it was shown in Section 12.4.4 that $P^2 = P_1^2 + P_2^2$ must have eigenvalue zero for the physically realized massless case. Therefore, for massless particles,

$$W^2 \left|p\right\rangle = 0 \qquad P^2 \left|p\right\rangle = 0 \qquad W_\mu P^\mu \left|p\right\rangle = 0 \tag{15.31}$$

[see Eq. (15.19)]. It follows that P and W are lightlike and orthogonal, and hence must be proportional: $(W^\mu - \lambda P^\mu)\left|p\right\rangle = 0$ (see Problem 13.2 in this regard). States are characterized by the constant of proportionality λ, which is just the helicity. Thus, for the lightlike states the subspace of the standard vector is one-dimensional and the basis vector $\left|\boldsymbol{p}\lambda\right\rangle$ may be labeled by the momentum $\boldsymbol{p}$ and the helicity λ. Generally,

$$P^\mu \left|\boldsymbol{p}\lambda\right\rangle = p^\mu \left|\boldsymbol{p}\lambda\right\rangle \qquad J_3 \left|\boldsymbol{p}\lambda\right\rangle = \lambda \left|\boldsymbol{p}\lambda\right\rangle \qquad W_i \left|\boldsymbol{p}\lambda\right\rangle = 0, \tag{15.32}$$

where the standard vector is $(\omega, 0, 0, \omega)$ and helicity takes the values $\lambda = 0, \pm\frac{1}{2}, \pm 1, \pm\frac{3}{2}, \ldots$ Notice the difference between this sequence of allowed helicity values for massless particles and the corresponding $2s + 1$ permitted states of polarization for massive particles of spin s discussed in Section 15.4.2. The full unitary representations may be produced by successive Lorentz transformations on the standard vector.

Box 15.2 It Depends on What You Mean by Spin

It is common to speak loosely of the spin of a photon and the spin of an electron as if they were two instances of the same thing, but we see from the present discussion that the nature of the corresponding quantum number is quite different in the two cases. For massive particles there are $2s + 1$ states of polarization available to an SU(2) spin vector; for massless particles only two states of polarization are available because s is no longer an SU(2) quantum number but instead is the SO(2) helicity quantum number. Consider some specific examples.

1. A massive spin-1 particle has three states of polarization.
2. The spin-1 photon has only two states of polarization because it is *massless*. It has helicity states $\lambda = \pm 1$ corresponding to left and right circular polarization, but it lacks the $\lambda = 0$ state that a massive spin-1 particle would possess.
3. Gravitational waves have two states of polarization because the graviton mediating the interaction is associated with a *massless* spin-2 field.

The missing states of polarization for massless particles relative to their massive counterparts have deep implications for the physical particle spectrum and for technical issues like renormalizability, as we shall discuss in Chs. 18 and 19.

15.6 Mass and Spin for Poincaré Representations

To summarize, the irreps of the Poincaré group are characterized by parameters m and s, where m is the restmass and the meaning of s depends on the mass of the particle.

1. For *massive particles,* $m > 0$ and s may be interpreted as the intrinsic spin of the particle in its restframe. States are SU(2) eigenstates and spin can take the values $s = 0, \frac{1}{2}, 1, \frac{3}{2}, \ldots$, with each value of s having the usual $2s + 1$ polarization states.
2. For *massless particles,* $m = 0$ and s is not an SU(2) quantum number; instead, it corresponds to an *SO(2) quantum number* λ (helicity), which takes allowed values $s = \lambda = 0, \pm\frac{1}{2}, \pm 1, \pm\frac{3}{2}, \ldots$, with only *two states of polarization* for a given s instead of $2s + 1$.

Thus, as discussed further in Box 15.2, the meaning of spin differs essentially between a massive particle like an electron and a massless particle like a photon.

15.7 Lorentz and Poincaré Representations

Let us conclude this chapter with a discussion of the relationship between representations of the Lorentz group and representations of the Poincaré group that was first broached in Section 13.11. Recall the following.

1. The Lorentz and Poincaré groups are non-compact, so *their representations cannot be finite and unitary at the same time.*
2. Particles associated with the Poincaré representations are produced by fields having a finite number of Lorentz degrees of freedom, but each such particle takes a continuum of 4-momentum values.

Physical states of definite mass, spin, and other quantum numbers arise as unitary representations of the Poincaré group. However, the *relativistic wave equations* that describe quantum fields have components that transform as finite-dimensional (but non-unitary) irreps of the Lorentz group. Thus, the dichotomy between the Lorentz and Poincaré representations reflects the general dichotomy between the descriptions of nature in terms of quantum fields and in terms of particles that are the quanta of those fields. Wavefunctions and fields are not observables; thus, they may be described by *non-unitary representations* (of the Lorentz group). The corresponding particle states are physical and are most conveniently described by *unitary representations* (of the Poincaré group).

15.7.1 Operators for Relativistic Quantum Fields

A *relativistic wave function* respects special relativity. It is a set of n functions of spacetime $\psi^\mu(x)$ transforming under proper Lorentz transformations as [199]

$$\psi'^\mu(x) = D(\Lambda)^\mu{}_\nu \psi^\nu(\Lambda^{-1}x), \tag{15.33}$$

where the matrix $D(\Lambda)^\mu{}_\nu$ is a finite-dimensional, $n\times n$ representation of the proper Lorentz group and the arguments on the right side reflect the action of a Lorentz transformation on functions defined in Hilbert space (see Example 2.8). These wavefunctions describe particles subject to a relativistic wave equation like the Dirac equation (14.31).

However, it is often preferable to formulate (relativistic or non-relativistic) problems in terms of fields rather than particles, because fields provide a natural way to formulate the many-body problem in which interactions can be handled more conveniently. In relativistic quantum field theory the *field operators* $\Psi(x)$ transform under proper Lorentz transformations as

$$U(\Lambda)\Psi^\mu(x)U(\Lambda^{-1}) = D(\Lambda^{-1})^\mu{}_\nu \Psi^\nu(\Lambda x), \tag{15.34}$$

where $U(\Lambda)$ is the operator representing the Lorentz transformation Λ in the Hilbert space of $\Psi^{(\alpha)}$. The field operators are required to obey wave equations that are analogs of the corresponding single-particle wave equations.[2] The field operator $\Psi^\mu(x)$ and the wavefunction $\psi^\mu(x)$ describe the same physics, and are governed by wave equations of the same form, but those wave equations govern *complex-valued functions* in the particle case and *operators* in the field case.

[2] More precisely, it is common to derive properties of field operators by starting with an appropriate Lagrangian density and either a variational condition deriving from this Lagrangian density, or a sum over paths weighted by an action constructed from the Lagrangian density (for example, see Ch. 16). However, this is a formal device in the present context because the content of the appropriate Lagrangian density is fixed precisely by the requirement that it yield the right wave equation. At any rate, our primary concern here is not to derive quantum field theory from first principles but to analyze quantum field theory for symmetry implications.

15.7.2 Wave Equations for Quantum Fields

Let us summarize a few important wave equations entering into formulations of relativistic quantum field theory that will be useful in later chapters.

1. The *Klein–Gordon equation* for scalar (single-component) fields $\Phi(x)$
$$(\Box + m^2)\Phi(x) = 0 \qquad \Box \equiv \partial_\mu \partial^\mu, \tag{15.35}$$
which is quadratic in ∂_μ and carries no Lorentz index.
2. The *Dirac equation* for 4-component Lorentz spinor fields $\Psi(x)$
$$(-i\gamma^\mu \partial_\mu + m)\Psi(x) = 0, \tag{15.36}$$
which is linear in ∂^μ.
3. The *Maxwell wave equation* for massless vector fields $A^\mu(x)$
$$\Box A^\mu(x) = j^\mu \qquad \text{(Lorenz gauge)}, \tag{15.37}$$
which is quadratic in ∂^μ.
4. The *massive vector* or *Proca field* $A^\mu(x)$, with an equation
$$(\Box + m^2)A^\mu(x) = 0 \tag{15.38}$$
that is quadratic in ∂^μ. As we shall now discuss, the connection between fields and particles may be exhibited explicitly through a Fourier expansion of the field operators.

15.7.3 Plane-Wave Expansion of the Fields

Let us use a Dirac field as an example. It may be expanded within a 3D box of volume Ω as
$$\Psi(x) = \frac{1}{\sqrt{\Omega}} \sum_p \sum_\lambda \sqrt{\frac{m}{E}} (a_{p\lambda} u_{p\lambda} e^{-ip\cdot x} + b^\dagger_{p\lambda} v_{p\lambda} e^{ip\cdot x}), \tag{15.39}$$
where m is the fermion mass, $p^\mu = (E, \boldsymbol{p})$, the helicity index is λ, the scalar product is $p \cdot x = p^\mu x_\mu$, the particle spinors are denoted by $u_{p\lambda}$ and the antiparticle spinors by $v_{p\lambda}$, and the field $\Psi(x)$ is an operator. These sums become Fourier integrals as the volume Ω of the quantization box tends to infinity according to the standard prescription[3]
$$\frac{1}{\Omega}\sum_p \underset{\Omega\to\infty}{\longrightarrow} \frac{1}{(2\pi)^3}\int d^3p. \tag{15.40}$$
Thus, a generic plane-wave expansion in the infinite-volume limit can be expressed as
$$\Psi^{(\mu)}(x) = \sum_\lambda \int \mathscr{D}p \, (a(p,\lambda)u^{(\mu)}(p,\lambda)e^{ip\cdot x} + b^\dagger(p,\lambda)v^{(\mu)}(p,\lambda)e^{-ip\cdot x}), \tag{15.41}$$

[3] Only $\Omega \to \infty$ gives full Poincaré symmetry, because translational invariance is broken by a box of finite size.

where μ is a Lorentz index, $\mathscr{D}p$ is a Lorentz-invariant integration measure (see Problems 14.20 and 14.21), the first term of the integrand is associated with particles, and the second is associated with antiparticles.

15.7.4 The Relationship of Fields and Particles

The left side of Eq. (15.41) is a Hilbert space operator, with the corresponding operator value on the right side carried by $a(p, \lambda)$ and $b^\dagger(p, \lambda)$. By defining an operator $\Pi(x)$ canonically conjugate to $\Psi^{(\mu)}(x)$ in the sense of Hamiltonian dynamics [thus $\Psi^{(\mu)}(x)$ will act as a "coordinate" Q and $\Pi(x)$ will act as a "momentum" P], expanding it in a Fourier series, and imposing a continuum analog of the canonical quantization condition for discrete mechanics, $[P, Q] = -i\hbar$, we may determine the algebra obeyed by $a(p, \lambda)$ and $b^\dagger(p, \lambda)$. Such manipulations show that these operators and their hermitian conjugates create and annihilate particles and antiparticles, in a relativistic generalization of the second quantization formalism of Appendix A. This is the detailed connection between fields and particles, and between Lorentz and Poincaré representations, that we have pursued beginning in Ch. 13 [199].

The field operator $\Psi^{(\mu)}(x)$ on the left side of Eq. (15.41) transforms as a *finite-dimensional but non-unitary representation of the Lorentz group* labeled by Lorentz index μ. The quantities $a(p, \lambda)$ and $b^\dagger(p, \lambda)$ on the right side of Eq. (15.41) are particle or antiparticle creation and annihilation operators that transform as *infinite-dimensional but unitary representations of the Poincaré group.*

These creation and annihilation operators will be labeled by momentum p, mass m, spin s (massive particles) or helicity λ (massless particles), and quantum numbers describing any internal degrees of freedom. The connection between the field and particle points of view is provided by the plane-wave functions $u^{(\mu)}(p, \lambda)e^{ip\cdot x}$ and $v^{(\mu)}(p, \lambda)e^{-ip\cdot x}$, which serve as expansion coefficients carrying both Lorentz (μ) and Poincaré (p and λ) indices.

15.7.5 Symmetry and the Wave Equation

For *free particles* the wave equation contains no information beyond that of representation theory. However, the wave equation provides a natural way to introduce *interactions between fields* and thus to explore dynamical processes that depend on the nature of the interactions and are not specified by the static group representation theory of the non-interacting states. Extension of symmetry considerations to include interactions must consider symmetries that are *dynamical*: they have something to say about the Hamiltonian or Lagrangian of the interacting theory. Symmetries with dynamical implications were introduced in Section 3.6, and will be addressed at length in Chs. 16, 18–20, and 31–32.

Background and Further Reading

The classic paper on unitary representations of the Poincaré group is Wigner [210]. Textbook discussions are given in Jones [125], Ryder [174], Sternberg [185], and Tung [199]. The discussion of the relationship between Lorentz and Poincaré representations has been strongly influenced by the presentation in Tung [199].

Problems

15.1 By applying two successive Poincaré transformations (15.1), show that the Poincaré multiplication rule is given by Eq. (15.2). ***

15.2 Prove that the transformations of the Poincaré group factor into a product $g(b,\Lambda) = T(b)\Lambda$ of a translation $T(b)$ and a Lorentz transformation Λ. *Hint*: Insert a pure translation $g(b',\Lambda') = g(b',I)$ (where I is the unit matrix) and a pure Lorentz transformation $g(b,\Lambda) = g(0,\Lambda)$ into the Poincaré multiplication law (15.2).

15.3 (a) Show that in the Poincaré group the Lorentz transformation Λ of a translation $T(b)$ is a translation: $\Lambda T(b)\Lambda^{-1} = T(\Lambda b)$. (b) Use the results of part (a) to demonstrate that translations in Minkowski space form an invariant subgroup of the Poincaré group. *Hint*: Set $g(b',\Lambda') = g(b',I)$ (where I is the unit matrix) and $g(b,\Lambda) = g(0,\Lambda)$ in the Poincaré multiplication law (15.2), and use Eq. (15.4). ***

15.4 Prove that the infinitesimal rank-2 tensor $\epsilon_{\mu\nu}$ introduced in Eq. (15.7) is antisymmetric in its indices, $\epsilon_{\mu\nu} = -\epsilon_{\nu\mu}$, by requiring that Eq. (15.7) be consistent with Eq. (13.24) to first order in $\epsilon_{\mu\nu}$.

15.5 Prove the result in Eq. (15.9) that the $L_{\mu\nu}$ defined in Eq. (15.8) are generators of proper Lorentz transformations. *Hint*: Utilize $\partial^\alpha x_\rho = \eta^\alpha_\rho = \delta^\alpha_\rho$ and the antisymmetry of $\epsilon_{\mu\nu}$ under exchange of indices. ***

15.6 Prove the result in Eq. (15.13) that the P_μ defined in Eq. (15.12) are generators of translations. *Hint*: Remember that $\partial^\alpha x_\rho = \eta^\alpha_\rho = \delta^\alpha_\rho$.

15.7 Verify $[\,P^2, P_\mu\,] = 0$ [Eq. (15.16)] and $[\,P_\mu, L_{\rho\sigma}\,] = i(\eta_{\mu\rho}P_\sigma - \eta_{\mu\sigma}P_\rho)$ [Eq. (15.14)].

15.8 Use Eq. (15.10) to prove that $J_1 = L_{23}$, $J_2 = L_{31}$, and $J_3 = L_{12}$.

15.9 Prove Eq. (15.28) for lightlike particles. *Hint*: Use Eq. (15.26) for the standard vector and take note of the Minkowski metric, so $p_\mu = \eta_{\mu\nu}p^\nu$ and $L^{\mu\nu} = \eta^{\mu\lambda}L_{\lambda\sigma}\eta^{\sigma\nu}$. You also will need Eq. (15.27) and results of Problem 15.8. ***

15.10 (a) Verify the commutation relations in Eq. (15.29) for the components of the Pauli–Lubanski pseudovector W^μ in the case of massless particles. *Hint*: Use the commutation relations in Eq. (13.20). (b) Show that the Lie algebra (15.29) is isomorphic to the algebra for the euclidean group E_2 that was given in Section 12.4.4.

15.11 Prove Eq. (15.30) for massless Poincaré particles.

15.12 Demonstrate that the identification of E_2 as the little group for lightlike particles in Section 15.5.2 is valid by showing explicitly that E_2 is the minimal group that leaves the standard lightlike momentum vector (15.26) invariant. *Hint*: Take for the little group a set of E_2 generators with the commutators $[L_1, L_2] = 0$, $[L_1, J_3] = -iL_2$, and $[L_2, J_3] = iL_1$, and note that for the standard vector $|\bar{p}_\mu\rangle = \omega(1, 0, 0, 1)$,

$$P_0 |\bar{p}_\mu\rangle = \omega |\bar{p}_\mu\rangle \qquad P_1 |\bar{p}_\mu\rangle = 0 \qquad P_2 |\bar{p}_\mu\rangle = 0 \qquad P_3 |\bar{p}_\mu\rangle = \omega |\bar{p}_\mu\rangle .$$

Use this to show that the generators of E_2 commute with all components of the standard vector. ***

16 Gauge Invariance

In this chapter we introduce the symmetry of *gauge invariance* and its physical implications. Gauge symmetries occur in many different contexts but the best known concern elementary particles described in terms of relativistic quantum field theories. Our primary interest will be in the importance of symmetries, not in the detailed formalism of such theories. However, appreciating the (profound) symmetry implications of gauge theories requires understanding the basics of how quantum field theories are formulated and used to construct physical descriptions. Therefore, we begin with a quick introduction to the essential elements of such theories. Considerable foundation for this has already been laid in the discussion of Lorentz and Poincaré covariant fields in Chs. 13–15, and the presentation in this chapter will assume familiarity with the material in those chapters.

16.1 Relativistic Quantum Field Theory

By relativistic quantum field theory we mean a quantum theory that respects *special relativity* (is Lorentz covariant), not general relativity, since gravity is negligible in practical applications of quantum mechanics. Let us summarize, in broad overview, the essential ingredients of Lorentz-covariant quantum field theories.

16.1.1 Quantization of Classical Fields

The standard approach to quantum field theory is to define fields that obey equations of motion appropriate to the classical limit of the problem at hand, and then to quantize those fields. Two methods are commonly used for field quantization.

1. *Canonical quantization,* where classical coordinates and momenta are converted into quantum field operators by requiring particular commutation relations for boson fields and anticommutation relations for fermion fields.
2. *Path integral quantization,* where quantum transition amplitudes are represented by an infinite sum over paths, weighted by an exponentiated factor of the classical action for each path (a *functional integral*).

We will not deal much with field quantization itself but will use the results of these methods extensively. The classical action S is defined by

$$S = \int d^4x\, \mathscr{L} = \int_{t_1}^{t_2} L\, dt, \tag{16.1}$$

where $\mathscr{L} = \mathscr{L}(x)$ is the Lagrangian defined at a Minkowski spacetime point (the *Lagrangian density*) and the total Lagrangian L is related to the Lagrangian density by

$$L = \int d^3x\, \mathscr{L}. \tag{16.2}$$

Formally, the action S is the central quantity for field quantization. In the canonical method, S determines the equations of motion for the classical fields that are to be quantized through a variational procedure. In the path integral method, each path in the functional integral is weighted by a factor $\exp(iS)$. Therefore, the starting point for developing a quantum field theory is to construct the appropriate Lagrangian density, which then leads to the action through Eq. (16.1).

16.1.2 Symmetries of the Classical Action

Symmetry enters our discussion through constraints on the Lagrangian density and thus on the action. At the most general level the Lagrangian density will be assumed local, depending only on the fields and their first derivatives at each spacetime point, and the action will be assumed real and Poincaré invariant, ensuring conservation of probability, energy, and momentum. In addition to such spacetime restrictions, we may choose to impose further constraints with respect to *internal symmetries,* such as gauge invariance.

16.1.3 Lagrangian Densities for Free Fields

In this section the Lagrangian densities appropriate to particular field theories that will be of interest will be summarized. For our purposes these may be viewed simply as prescriptions tailored to lead to the correct final results.

Example 16.1 The Lagrangian density given in Eq. (16.10) below for the Dirac field leads, through the Euler–Lagrange equations of motion (which follow by requiring that the action be an extremum under parameter variation), to the Dirac wave equation (15.36) as the classical or "first-quantized" equation of motion for the field.[1] See Section 15.7.2 for a summary of the wave equations that will be relevant to our discussion.

The Lagrangian densities to be quoted will be appropriate for free (that is, non-interacting) fields. Interesting physical theories involve interactions among fields and typically will have Lagrangian densities that may be expressed as

[1] "Classical" is used here in its usual sense in a second-quantization formalism: the classical wave equation is the appropriate single-particle wave equation (Schrödinger, Dirac, Klein–Gordon, ...); the "second quantization" of this equation then interprets the wave equation in terms of field operators that create and annihilate the quanta of the field. As described in Appendix A, second quantization adds no new physics but the resulting formalism makes it much easier to deal with the many-body problem in quantum mechanics.

$$\mathscr{L} = \mathscr{L}_0 + \mathscr{L}_{\text{int}}, \tag{16.3}$$

where $\mathscr{L}_0$ is the free-field Lagrangian density and $\mathscr{L}_{\text{int}}$ is an interaction term, with a form depending on the physics of the problem. The complete solution of such a problem is generally difficult. In some (certainly not all) cases of physical interest the problem can be *solved using perturbation theory.*

1. The effect of the free-field Lagrangian density $\mathscr{L}_0$ is determined exactly or by some reliable approximation, and
2. the influence of the interaction term $\mathscr{L}_{\text{int}}$ is incorporated as an expansion in powers of an appropriate small quantity (*perturbation series*).

Terms of the perturbation series may be enumerated through *Feynman diagrams* (Box 14.3), and systematic *Feynman rules* may be constructed that relate diagrams and matrix elements. We shall skate lightly over details, referring the reader to more specialized treatments of quantum field theory (for example, Refs. [41, 85, 174]). However, the *symmetries* exhibited by both free and interacting fields and their physical implications will be of great interest to us.

Real Scalar or Pseudoscalar Fields: A real spin-0 field $\phi(x)$ (of positive or negative intrinsic parity; see Section 13.7) corresponds to a free Lagrangian density

$$\mathscr{L}_0 = \frac{1}{2}(\partial^\mu \phi)(\partial_\mu \phi) - \frac{1}{2}m^2\phi^2 \qquad \text{(real scalar field)}, \tag{16.4}$$

where m may be interpreted as a mass,[2] and the partial derivative operators are defined in Eq. (13.8). Introducing the d'Alembertian operator $\Box \equiv \partial_\mu \partial^\mu$, this Lagrangian density leads, by the Euler–Lagrange equations (16.14), to the Klein–Gordon equation (15.35),

$$(\Box + m^2)\,\phi(x) = 0, \tag{16.5}$$

for an uncharged scalar ($J^P = 0^+$) or pseudoscalar ($J^P = 0^-$) field.

Complex Scalar or Pseudoscalar Fields: A complex scalar (or pseudoscalar) field $\phi(x)$ may be expressed in terms of two real scalar fields $\phi_1(x)$ and $\phi_2(x)$ having the same mass ($m_1 = m_2 \equiv m$),

$$\phi = \frac{1}{\sqrt{2}}\left(\phi_1 + i\phi_2\right) \qquad \phi^\dagger = \frac{1}{\sqrt{2}}\left(\phi_1 - i\phi_2\right), \tag{16.6}$$

where the corresponding free Lagrangian density is

$$\mathscr{L}_0 = (\partial^\mu \phi)^\dagger(\partial_\mu \phi) - m^2\phi^\dagger\phi \qquad \text{(complex scalar field)}. \tag{16.7}$$

Separate variations of ϕ and $\phi^\dagger$ then lead to two independent Klein–Gordon equations,

$$(\Box + m^2)\phi(x) = 0 \qquad (\Box + m^2)\phi^\dagger(x) = 0. \tag{16.8}$$

Charged scalar or pseudoscalar particles are associated with complex scalar fields.

[2] In a Lagrangian density mass terms for a field ψ will generally be of the quadratic form $\frac{1}{2}m_\psi^2\psi^2$, where m_ψ is the mass parameter, except for Dirac fields where they take the form $-\overline{\psi}m_\psi\psi$, with $\overline{\psi}$ defined in Eq. (16.9).

Dirac Spinor Field: For a Dirac field $\psi(x)$ and a conjugate (adjoint spinor) field $\overline{\psi}(x)$ defined by

$$\overline{\psi}(x) \equiv \psi^\dagger(x)\,\gamma^0, \tag{16.9}$$

where the matrix γ_0 is defined through Eq. (14.27), the free-field Lagrangian density for a Dirac spinor field may be written as

$$\mathscr{L}_0 = \overline{\psi}(i\partial\!\!\!/ - m)\psi \qquad \text{(Dirac spinor field)}, \tag{16.10}$$

where m is the mass and we introduce *Feynman slash notation*: a slash through a quantity indicates that it is a 4-vector contracted completely with the γ-matrices (14.27), $\partial\!\!\!/ \equiv \gamma^\mu \partial_\mu$. The Lagrangian density (16.10) then yields the Dirac equation (15.36),

$$(i\partial\!\!\!/ - m)\psi = 0, \tag{16.11}$$

as the equation of motion for a spinor field appropriate for describing fermions.

Massless Vector Field: In terms of the rank-2 *electromagnetic field tensor* $F^{\mu\nu} = \partial^\mu A^\nu - \partial^\nu A^\mu$ introduced in Eq. (14.14) through the 4-vector potential A^μ defined in Eq. (14.9), the free-field Lagrangian density appropriate for the massless vector (photon) field is

$$\mathscr{L}_0 = -\frac{1}{4}F_{\mu\nu}F^{\mu\nu} \qquad \text{(massless vector field)}. \tag{16.12}$$

This Lagrangian density describes massless vector particles like the photon.

Massive Vector Field: For a vector field of finite mass m, the appropriate free-field Lagrangian density is obtained from Eq. (16.12) by adding a Lorentz-invariant mass term,

$$\mathscr{L}_0 = -\frac{1}{4}F_{\mu\nu}F^{\mu\nu} + \frac{1}{2}m^2 A^\mu A_\mu \qquad \text{(massive vector field)}. \tag{16.13}$$

This implies an equation of motion $(\Box + m^2)A^\mu = 0$ for a massive vector field.

16.1.4 Euler–Lagrange Field Equations

For a scalar field ϕ the Lagrangian density $\mathscr{L}$ satisfies the *Euler–Lagrange equation,*

$$\partial_\mu \frac{\partial \mathscr{L}}{\partial(\partial_\mu \phi)} - \frac{\partial \mathscr{L}}{\partial \phi} = 0, \tag{16.14}$$

which follows from the requirement of *Hamilton's principle* that the classical action be an extremum under variation of the classical path with fixed endpoints (see Problem 16.8). Lagrangian densities for other fields obey similar Euler–Lagrange equations with the appropriate field substituted for ϕ. Solution of (16.14) provides a systematic way to determine the equation of motion obeyed by a field described by some Lagrangian density $\mathscr{L}$.

16.2 Conserved Currents and Charges

Observables in quantum field theory are often formulated in terms of *currents* that generalize electrical current, and *charges* (defined as integrals of such currents) that generalize electrical charge. Many symmetries associated with field theories may be expressed concisely through conservation of such currents and charges.

16.2.1 Noether's Theorem

The conservation laws for field theories that are formulated in terms of Lagrangian densities are often expressed in terms of *Noether's theorem.*

> ***Noether's Theorem:*** For every continuous symmetry of a field theory Lagrangian there is a corresponding conserved quantity.

These conserved quantities may involve spacetime or internal symmetries and are termed *Noether charges* or more generally *Noether tensors.* Problem 16.3 illustrates by deriving conservation of 4-momentum from invariance of a field under spacetime translations.

Conserved Isospin Current: As an example of Noether's theorem for an internal rather than spacetime symmetry, let us consider conservation of isospin for neutrons and protons [166]. Define the neutron–proton *isospinor* by

$$\psi = \begin{pmatrix} p \\ n \end{pmatrix}, \tag{16.15}$$

where p and n denote Dirac spinor wavefunctions for the proton and neutron, respectively.[3] Because of the assumed isospin symmetry, the neutron and proton have the same mass and the Lagrangian density for non-interacting fields is a generalization of (16.10):

$$\mathscr{L}_0 = \bar{p}(i\not{\partial} - m)p + \bar{n}(i\not{\partial} - m)n = \overline{\psi}(i\not{\partial} - m)\psi. \tag{16.16}$$

Isospin rotations are generated by transformations like

$$\psi(x) \to e^{\frac{i}{2}\tau\cdot\alpha}\psi(x) = e^{\frac{i}{2}\tau^a\alpha_a}\psi(x), \tag{16.17}$$

which is in infinitesimal form

$$\psi(x) \simeq \psi(x) + \frac{i}{2}\tau^a\alpha_a\psi(x) \equiv \psi + \delta\psi, \tag{16.18}$$

[3] The isospin symmetry in this example is phenomenological rather than fundamental, but it is useful as a simple illustration. Do not confuse the two-dimensional isospinor space with the Dirac spinor space. Each of the two isospin components of ψ in Eq. (16.15) is a four-component Dirac spinor.

where τ^a are Pauli matrices and the parameters α_a are assumed constant, *independent of the spacetime coordinates.* The corresponding symmetry is termed *global isospin invariance.* As shown in Problem 16.14, invariance under (16.18) implies an isospin current

$$J_\mu^a = \frac{1}{2}\overline{\psi}\gamma^\mu \tau^a \psi \tag{16.19}$$

that is conserved because it has vanishing 4-divergence, $\partial^\mu J_\mu^a = 0$ (see Section 16.2.2).

General Internal Currents: For an internal symmetry described by a set of matrix generators T^a and a Lagrangian density $\mathscr{L}$, the conserved Noether currents are

$$J_\mu^a = -i \frac{\partial \mathscr{L}}{\partial(\partial^\mu \phi_i)} T_{ij}^a \phi_j, \tag{16.20}$$

where μ is a Lorentz index, a labels an internal-symmetry generator, i and j are matrix indices, and the fields ϕ are transformed by the matrices T^a. In Eq. (16.19), $T^a = \frac{1}{2}\tau^a$.

16.2.2 Conserved Charges

Given a 4-current $J^\mu = (J^0, J^1, J^2, J^3)$, an associated charge $Q(t)$ may be defined by

$$Q(t) \equiv \int d^3x\, J^0(x). \tag{16.21}$$

For example, the electromagnetic 4-current $j = (\rho, \boldsymbol{j})$ implies an electrical charge

$$Q_e = \int d^3x\, \rho(x) = \int d^3x\, j^0. \tag{16.22}$$

The charge operator satisfies the *Heisenberg equation of motion* for quantum operators,

$$\dot{Q} \equiv \frac{dQ(t)}{dt} = i[\,H, Q(t)\,], \tag{16.23}$$

where H is the Hamiltonian. As shown in Problem 16.13, if J^μ has vanishing 4-divergence,

$$\partial_\mu J^\mu \equiv \partial_0 J^0 + \boldsymbol{\nabla} \cdot \boldsymbol{J} = \partial_0 J^0 + \partial_k J^k = 0, \tag{16.24}$$

then $[\,H, Q\,] = 0$ and Q is a constant of motion.

Conserved Charges: If a 4-current $J^\mu(x)$ satisfies $\partial_\mu J^\mu = 0$, the charge Q defined by the spatial integral of $J^0(x)$ is conserved.

This establishes a formal link between conserved Noether currents in a Lagrangian formulation (signified by vanishing 4-divergence of the current) and conserved charges as constants of motion in a Hamiltonian formulation (signified by the charge commuting with the Hamiltonian).

Example 16.2 The electromagnetic 4-current $j^\mu = (\rho, \boldsymbol{j})$ has $\partial_\mu j^\mu = 0$, implying that the electrical charge Q_e of Eq. (16.22) is conserved.

Example 16.2 is likely well known to readers. However, the preceding derivation implies that *any charge* defined formally through Eq. (16.21) is conserved (is a constant of motion) if the 4-divergence of the associated current J^μ vanishes.

16.2.3 Symmetries for Interacting Fields

In a quantized, interacting field theory the conserved currents and charges depend on the symmetries of both the free-field Lagrangian density $\mathscr{L}_0$ and the interaction density $\mathscr{L}_{\text{int}}$ appearing in Eq. (16.3). For example, consider the Lagrangian density for two real scalar fields, ϕ_1 and ϕ_2, of equivalent mass m and interacting through

$$\mathscr{L}_{\text{int}} = \frac{1}{4}\lambda(\phi_1^2 + \phi_2^2)^2, \tag{16.25}$$

where λ is a coupling constant. From Eqs. (16.3) and (16.4), the Lagrangian density is then

$$\mathscr{L} = \mathscr{L}_0 + \mathscr{L}_{\text{int}} = \frac{1}{2}(\partial^\mu\phi_1)(\partial_\mu\phi_1) + \frac{1}{2}(\partial^\mu\phi_2)(\partial_\mu\phi_2) - \frac{1}{2}m^2(\phi_1^2 + \phi_2^2). \tag{16.26}$$

Expressing the two scalar fields in terms of a column vector,

$$\phi \equiv \begin{pmatrix}\phi_1\\ \phi_2\end{pmatrix}, \tag{16.27}$$

it may be verified (Problem 16.5) that the Lagrangian density (16.26) is invariant under SO(2) rotations, with a generator L and a conserved Noether current J_μ given by

$$L \equiv \begin{pmatrix}0 & i\\ -i & 0\end{pmatrix} \qquad J_\mu = -i\frac{\partial\mathscr{L}}{\partial(\partial^\mu\phi_i)}L_{ij}\phi_j = (\partial_\mu\phi_1)\phi_2 - (\partial_\mu\phi_2)\phi_1. \tag{16.28}$$

Alternatively, the two real fields may be expressed as a complex scalar field $\phi(x)$ with

$$\phi = \frac{1}{\sqrt{2}}(\phi_1 + i\phi_2) \qquad \phi^\dagger = \frac{1}{\sqrt{2}}(\phi_1 - i\phi_2). \tag{16.29}$$

The corresponding Lagrangian density is

$$\mathscr{L} = (\partial_\mu\phi)^\dagger(\partial^\mu\phi) - m^2(\phi^\dagger\phi) - \lambda(\phi^\dagger\phi)^2, \tag{16.30}$$

which is invariant under U(1) phase rotations $\phi \to e^{i\alpha}\phi$ and $\phi^\dagger \to \phi^\dagger e^{-i\alpha}$. The conserved Noether current associated with the U(1) invariance is (Problem 16.6)

$$J_\mu = -i\frac{\partial\mathscr{L}}{\partial(\partial^\mu\phi)}(1)\phi - i\frac{\partial\mathscr{L}}{\partial(\partial_\mu\phi^\dagger)}(-1)\phi^\dagger = i\left[(\partial_\mu\phi)\phi^\dagger - (\partial_\mu\phi^\dagger)\phi\right], \tag{16.31}$$

which is equivalent to the current in Eq. (16.28).[4]

[4] This alternative description of the invariance as an SO(2) symmetry for two real scalar fields, or a U(1) symmetry for a complex scalar field, parallels the discussion in Section 6.2.2.

16.2.4 Partially Conserved Currents

In the example just discussed the full Lagrangian density (free-field plus interaction terms) was invariant under SO(2) or U(1) transformations. If either $\mathscr{L}_0$ or $\mathscr{L}_{\text{int}}$ breaks the symmetry the corresponding currents and charges will not be conserved. However, in many instances the Lagrangian density may be written in a form where the major portion respects a symmetry and a small portion that may be treated as a perturbation does not. Typically, $\mathscr{L}_0$ might be invariant and $\mathscr{L}_{\text{int}}$ might break the symmetry weakly. In such cases it may be useful to associate a *partially conserved current or charge* with the system. If $\partial_\mu j^\mu = \epsilon \simeq 0$, we obtain $[H, Q] \simeq \mathscr{O}(\epsilon)$ for the associated charge Q, implying that it is an *approximate constant of motion.* For example, our modern theory of the weak interactions owes much to attempts to understand partially conserved weak interaction currents (in particular, the *partially conserved axial vector current* introduced in Section 33.1.2).

16.3 Gauge Invariance in Quantum Mechanics

Gauge invariance and its implications for the classical electromagnetic field and the Maxwell equations were introduced in Section 14.1. Let us now extend that discussion and consider gauge invariance for quantum mechanics in the presence of an electromagnetic field [7]. We note first that a particle of charge q moving with 3-velocity $\boldsymbol{v}$ in an electromagnetic field experiences a *Lorentz force*

$$\boldsymbol{F} = q(\boldsymbol{E} + \boldsymbol{v} \times \boldsymbol{B}), \tag{16.32}$$

if the classical Hamiltonian function is

$$H = \frac{1}{2m}(\boldsymbol{p} - q\boldsymbol{A})^2 + q\phi, \tag{16.33}$$

where $\boldsymbol{E}$ is the electric field, $\boldsymbol{B}$ is the magnetic field, $\boldsymbol{p}$ is the 3-momentum, $\boldsymbol{A}$ is the 3-vector potential, and ϕ is the scalar potential. Quantizing this Hamiltonian gives

$$\left[\frac{1}{2m}(-i\boldsymbol{\nabla} - q\boldsymbol{A})^2 + q\phi\right]\psi(\boldsymbol{x}, t) = i\,\frac{\partial\psi(\boldsymbol{x}, t)}{\partial t}, \tag{16.34}$$

for the Schrödinger equation. Comparison with the free-particle Schrödinger equation

$$-\frac{1}{2m}\boldsymbol{\nabla}^2\psi(\boldsymbol{x}, t) = i\,\frac{\partial\psi(\boldsymbol{x}, t)}{\partial t}$$

suggests that a Schrödinger equation accounting for the effect of an electromagnetic field on a particle of charge q can be constructed from the free-particle Schrödinger equation through the operator substitutions

$$\boldsymbol{\nabla} \to \boldsymbol{D} \equiv \boldsymbol{\nabla} - iq\boldsymbol{A} \qquad \frac{\partial}{\partial t} \to D^0 \equiv \frac{\partial}{\partial t} + iq\phi. \tag{16.35}$$

This can be expressed in covariant notation as

$$\partial^\mu \to D^\mu \equiv \partial^\mu + iqA^\mu, \tag{16.36}$$

in terms of the 4-vectors $A^\mu = (\phi, \boldsymbol{A})$ and $D^\mu = (D^0, \boldsymbol{D})$. The replacement (16.36) is termed *minimal substitution* and the modified derivative D^μ is termed the *covariant derivative.* Covariant derivatives are central to the understanding of gauge invariance.[5] The Schrödinger equation in this notation is

$$\frac{1}{2m}(-i\boldsymbol{D})^2\psi(\boldsymbol{x},t) = iD^0\psi(\boldsymbol{x},t). \tag{16.37}$$

It is not difficult to show (for example, see Ref. [7]) that the Schrödinger equation with the minimal substitution is invariant under a local gauge transformation

$$\boldsymbol{A} \to \boldsymbol{A} + \boldsymbol{\nabla}\chi(\boldsymbol{x},t) \qquad \phi \to \phi - \frac{\partial}{\partial t}\chi(\boldsymbol{x},t), \tag{16.38}$$

provided that the wavefunction is simultaneously transformed as

$$\psi(\boldsymbol{x},t) \to e^{iq\chi(\boldsymbol{x},t)}\psi(\boldsymbol{x},t), \tag{16.39}$$

where we note explicitly that this is a *local transformation* (it can vary from point to point) because the scalar $\chi(\boldsymbol{x},t)$ is a function of the spacetime coordinates. Similar arguments may be applied to Lorentz-covariant wave equations, as illustrated in Example 16.3.

Example 16.3 Minimal substitution in the Klein–Gordon equation (15.35) and in the Dirac equation (15.36) give

$$(D^\mu D_\mu + m^2)\psi(x) = 0 \qquad (i\gamma_\mu D^\mu - m)\psi(x) = 0,$$

respectively, and both equations are invariant under a transformation

$$A^\mu \to A^\mu - \partial^\mu\chi(x) \qquad \psi \to e^{iq\chi(x)}\psi. \tag{16.40}$$

These are again *local transformations* because $\chi(x)$ is a function of the coordinates.

Such considerations imply a simple gauge-invariant prescription for including an electromagnetic field in quantum mechanics.

> ***Minimal Substitution:*** In a free-particle wave equation, replace all derivatives with covariant derivatives $\partial^\mu \to D^\mu \equiv \partial^\mu + iqA^\mu$. The resulting wave equation will be *invariant under a local gauge transformation* $A^\mu \to A^\mu - \partial^\mu\chi(x)$ and $\psi \to e^{iq\chi(x)}\psi$, where q is charge, A^μ is the 4-vector potential, and $\chi(x)$ is a *local* scalar field.

This is called the *minimal substitution* because it, and not a more complicated coupling prescription, appears to be adequate for implementing gauge-invariant coupling of charged particles to the electromagnetic field.

[5] A mathematically analogous covariant derivative is important for general relativity. There covariant derivatives ensure that spacetime tensors remain tensors under differentiation, and define a prescription for parallel transport of tensors in possibly curved spacetime. This relationship is discussed more extensively in Ch. 26.

16.4 Gauge Invariance and the Photon Mass

The Lagrangian density for a *massless* vector field is given by Eq. (16.12). Since $F^{\mu\nu}F_{\mu\nu}$ is gauge invariant, the massless vector field (photon field) is gauge invariant. A *massive* vector field has a Lagrangian density given by Eq. (16.13). The first term for (16.13) is gauge invariant but the second (mass) term is not, since under the gauge transformation $A^\mu \to A^\mu - \partial^\mu \chi$,

$$A^\mu A_\mu \to (A^\mu - \partial^\mu \chi)(A_\mu - \partial_\mu \chi) \neq A^\mu A_\mu.$$

Therefore, we reach a conclusion that will have far-reaching implications for gauge symmetry and its influence on modern physics.

> Gauge invariance of the electromagnetic field is tied directly both to charge conservation and to masslessness of the vector boson (the photon) that mediates the electromagnetic interaction. If the photon had a finite mass the corresponding vector field would not be gauge invariant.

That local gauge symmetry implies massless gauge bosons will be crucial in our discussion of non-abelian gauge theories in Section 16.6 and the Standard Model in Ch. 19.

16.5 Quantum Electrodynamics

The Lorentz-invariant equation of motion for a free charged-fermion field is the Dirac equation, which follows by the Euler–Lagrange equation (16.14) from Eq. (16.10).

16.5.1 Global U(1) Gauge Invariance

Equation (16.10) is invariant under a *global* U(1) phase transformation (Problem 16.4) $\psi(x) \to e^{i\alpha}\psi(x)$ and $\overline{\psi}(x) \to \overline{\psi}(x)e^{-i\alpha}$, where α is independent of the spacetime coordinates x. Then the Dirac current, $j^\mu(x) = \overline{\psi}(x)\gamma^\mu\psi(x)$, is conserved, $\partial_\mu j^\mu = 0$, and the electrical charge,

$$Q = \int d^3x\, j^0(x) = \int d^3x\, \psi^\dagger(x)\psi(x), \tag{16.41}$$

is also conserved, $[\, H, Q\,] = 0$: global U(1) symmetry implies global charge conservation.

16.5.2 Local U(1) Gauge Invariance

The free Dirac Lagrangian density is not invariant under *local* U(1) transformations

$$\psi(x) \to e^{i\alpha(x)}\psi(x) \qquad \overline{\psi}(x) \to \overline{\psi}(x)e^{-i\alpha(x)} \tag{16.42}$$

because of the derivative terms in Eq. (16.10). It can be made locally gauge invariant, but only if an additional field is introduced giving terms that cancel the non-invariant terms arising from the derivatives of Eq. (16.10) acting on Eq. (16.42). In fact, we have already seen how to convert the global gauge invariance of the Dirac equation to a local gauge invariance: replace all derivatives with covariant derivatives according to the minimal substitution (16.36). Then $\partial^\mu \to \partial^\mu + iqA^\mu \equiv D^\mu$, where q is the particle charge and A^μ is the 4-vector potential. Under this substitution the Dirac Lagrangian density becomes

$$\begin{aligned}\mathscr{L}(x) &= i\overline{\psi}(x)\gamma^\mu[\partial_\mu + iqA_\mu(x)]\psi(x) - m\overline{\psi}(x)\psi(x)\\ &= \overline{\psi}(x)(i\not{D} - m)\psi(x),\end{aligned} \tag{16.43}$$

which is invariant under the *local gauge transformation*

$$\psi(x) \to e^{i\alpha(x)}\psi(x) \qquad \overline{\psi}(x) \to \overline{\psi}(x)e^{-i\alpha(x)} \qquad A_\mu(x) \to A_\mu(x) - \frac{1}{q}\partial_\mu\alpha(x),$$

where $\alpha(x)$ is related to $\chi(x)$ appearing in Eq. (16.40) through $\alpha(x) \equiv q\chi(x)$. Comparing Eq. (16.43) with the free Lagrangian density (16.10) and with Eq. (16.3),

$$\mathscr{L}_{\text{int}} = -q\overline{\psi}(x)\gamma^\mu A_\mu(x)\psi(x) \equiv -q\overline{\psi}\not{A}\,\psi \tag{16.44}$$

may be identified as the interaction of the fermion and photon fields: upon quantization, A_μ will create and annihilate photons while $\overline{\psi}$ and ψ will create and annihilate fermions.

16.5.3 Gauging the U(1) Symmetry

The preceding construction of the locally gauge-invariant Lagrangian density associated with the quantum electrodynamics of charged fermions suggests three important principles.

1. Imposing local gauge invariance requires introducing massless vector fields through the vector potential A^μ that are termed *gauge bosons.* Photons are the gauge bosons associated with local gauge invariance for charged particle fields. The symmetry in this case is an abelian U(1) invariance and the photon is termed an *abelian gauge boson.*
2. Local gauge invariance has a non-trivial dynamical content. As was seen in Eq. (16.44),
   - imposing local gauge invariance *specifies the form of the interaction,* and
   - requires that the interaction strength be *proportional to the electrical charge.*
3. Global gauge invariance implies a global (integral over all space) conservation law. Local gauge invariance is more restrictive, requiring that charges be *conserved locally.* Destroying charge at a point while creating an equivalent charge at another point is permitted by a global conservation law, but not by a local one.

Extending a global symmetry to a local one is termed *gauging the symmetry.* Generally, this procedure is useful only if the global symmetry is *exact,* ultimately because renormalizability of the corresponding theories (Box 16.1) can be ensured only if the symmetry before gauging is exact. Global charge conservation is thought to be an exact symmetry and gauging it leads to quantum electrodynamics, which is renormalizable. On the other hand, we do not expect gauging a global symmetry like flavor SU(3) to be useful because it is only an approximate symmetry.

Box 16.1 **Renormalizability of Quantum Field Theories**

Relativistic quantum field theories typically involve integrals that can diverge, leading to (unacceptable) infinite values for matrix elements. A systematic and highly technical process called *renormalization* has been developed to remove these infinities from matrix elements corresponding to observables. It is believed that valid quantum field theories must be *renormalizable* (to all orders in a perturbation series for the interactions) by such a procedure. We shall mostly eschew these technical issues, except that it will be assumed (explicitly or implicitly) that valid symmetries for quantum fields must be consistent with renormalizability of those theories.

16.6 Yang–Mills Fields

We have seen that extension of global charge symmetry for the Dirac equation to a local symmetry through minimal substitution leads to quantum electrodynamics (QED). The local gauge symmetry for QED involves U(1) gauge rotations that commute, so it is abelian. Yang and Mills [226] extended the idea of local gauge invariance to allow for non-abelian gauge groups. Initially it was far from clear whether such theories had any physical application, but non-abelian gauge symmetry is now viewed as the cornerstone of elementary particle physics and non-abelian gauge fields are commonly termed *Yang–Mills fields*.

16.6.1 Non-Abelian Gauge Invariance

Let us describe how to construct fields with non-abelian local gauge symmetry by generalizing the procedure employed for abelian gauge invariance [85]. We begin by assuming a set of N non-abelian group generators T_i obeying a Lie algebra

$$[T_j, T_k] = i f_{jkl} T_l \tag{16.45}$$

that operate on a set of n fields

$$\Psi = \begin{pmatrix} \Psi_1 \\ \Psi_2 \\ \vdots \\ \Psi_n \end{pmatrix}$$

transforming as

$$\Psi(x) \to \Psi'(x) = e^{-i\tau_i \theta_i(x)} \Psi(x) \equiv U(\boldsymbol{\theta}) \Psi(x), \tag{16.46}$$

where the N operators τ_j are $n \times n$ matrix representations of the generators and the N parameters $\theta_j(x)$ depend on the spacetime coordinates. From Eq. (16.46),

$$\partial_\mu \Psi(x) \to U(\boldsymbol{\theta}) \partial_\mu \Psi(x) + \left(\partial_\mu U(\boldsymbol{\theta})\right) \Psi(x), \tag{16.47}$$

for the transformation of the field gradients. As for the abelian case, it is the derivatives that will cause problems in attempting to impose local gauge invariance.

16.6.2 Covariant Derivatives

Using the earlier abelian example as guidance, we surmise that the unwanted derivative terms arising from a local gauge transformation may be canceled by new vector fields entering through a generalized form of the gauge-covariant derivative. Introduce one new vector field $A^j_\mu(x)$ for each of the N group generators and define [compare Eq. (16.36)]

$$D_\mu \Psi(x) \equiv \left(\partial_\mu + igA_\mu(x)\right)\Psi(x), \tag{16.48}$$

where g is a gauge coupling constant and

$$\begin{aligned} \boldsymbol{A}_\mu(x) &\equiv \left(A^1_\mu(x), A^2_\mu(x), \dots, A^N_\mu(x)\right), \\ A_\mu(x) &\equiv \boldsymbol{\tau}\cdot\boldsymbol{A}_\mu(x) = \tau_i A^i_\mu(x) \qquad \boldsymbol{\tau} \equiv (\tau_1, \tau_2, \dots, \tau_N). \end{aligned} \tag{16.49}$$

The following example illustrates.

Example 16.4 Consider an SU(2) Yang–Mills field. The generators are $\tau_i = \frac{1}{2}\sigma_i$, where the σ_i are defined in Eq. (3.11). Then

$$\begin{aligned} A^{\mathrm{SU(2)}}_\mu &= \frac{1}{2}\sigma_i A^i_\mu = \frac{1}{2}\left(\sigma_1 A^1_\mu + \sigma_2 A^2_\mu + \sigma_3 A^3_\mu\right) \\ &= \frac{1}{2}\begin{pmatrix} 0 & A^1_\mu \\ A^1_\mu & 0 \end{pmatrix} + \frac{1}{2}\begin{pmatrix} 0 & -iA^2_\mu \\ iA^2_\mu & 0 \end{pmatrix} + \frac{1}{2}\begin{pmatrix} A^3_\mu & 0 \\ 0 & -A^3_\mu \end{pmatrix} \\ &= \frac{1}{2}\begin{pmatrix} A^3_\mu & A^1_\mu - iA^2_\mu \\ A^1_\mu + iA^2_\mu & -A^3_\mu \end{pmatrix}, \end{aligned}$$

for an SU(2) non-abelian gauge field in the basis (3.11).

Local gauge invariance requires that the covariant derivative D_μ transform as

$$D_\mu \Psi(x) \to D'_\mu \Psi'(x) = U(\boldsymbol{\theta}) D_\mu \Psi(x), \tag{16.50}$$

where D_μ carries a Lorentz index μ but also is an $n \times n$ matrix operating on the n-component internal field $\Psi(x)$ (often termed the *charge space*). Some algebra applied to Eqs. (16.46), (16.50), and (16.48) shows that the matrix vector potential $A_\mu(x)$ must transform as

$$A'_\mu = UA_\mu U^{-1} + \frac{i}{g}(\partial_\mu U)U^{-1} = UA_\mu U^{-1} - \frac{i}{g}U\partial_\mu U^{-1}, \tag{16.51}$$

to ensure local gauge invariance. For an infinitesimal transformation, $U \simeq 1 - \tau_k \theta^k$, and further algebra on Eq. (16.51) indicates that individual vector fields A^j_μ must transform as

$$A'^j_\mu(x) = A^j_\mu(x) + \frac{1}{g}\partial_\mu \theta^j(x) + f_{jkl}\theta^k A^l_\mu(x), \tag{16.52}$$

as shown in Problem 16.12. Comparing this expression with Eq. (16.40) for an abelian gauge field indicates that the third term is new. The dependence on the structure constant f_{jkl} is clear evidence of its non-abelian origin. The rank-2 field tensor $F_{\mu\nu}$ of electromagnetism is replaced in a non-abelian theory by a generalized rank-2 field tensor

$$\begin{aligned} F_{\mu\nu} &\equiv F^j_{\mu\nu}\tau_j = \partial_\mu A_\nu - \partial_\nu A_\mu + ig[\,A_\mu, A_\nu\,], \\ F^j_{\mu\nu} &= \partial_\mu A^j_\nu - \partial_\nu A^j_\mu - g f_{jkl} A^k_\mu A^l_\nu. \end{aligned} \tag{16.53}$$

In an abelian theory the field tensor $F_{\mu\nu}$ is gauge invariant so $F^{\mu\nu}F_{\mu\nu}$ appearing in the Lagrangian density (16.12) is gauge invariant. In a non-abelian theory, $F_{\mu\nu}$ is not gauge invariant, transforming instead as the adjoint representation of the gauge group; however, the contraction $\boldsymbol{F}_{\mu\nu} \cdot \boldsymbol{F}^{\mu\nu} = F^j_{\mu\nu}F^{\mu\nu}_j$ is gauge invariant. Just as for an abelian gauge theory, no mass terms are permitted for the fields A^j_μ. They would be proportional to $\boldsymbol{A}_\mu \cdot \boldsymbol{A}^\mu$, which breaks gauge invariance [see Eq. (16.13) and Section 16.4].

16.6.3 Non-Abelian Generalization of QED

We conclude that introducing non-abelian vector fields to convert a global gauge invariance to a local one leads to a non-abelian generalization of QED in which N gauge bosons (one for each gauge-group generator) mediate interactions among particles carrying a non-abelian "charge" that generalizes electrical charge. A schematic form for a Lagrangian density that couples Yang–Mills fields to matter fields through a local gauge interaction is

$$\mathscr{L} = -\frac{1}{4}\boldsymbol{F}_{\mu\nu} \cdot \boldsymbol{F}^{\mu\nu} + \mathscr{L}_{\rm m} + \mathscr{L}_{\rm int}(\Psi^j, D_\mu \Psi^j). \tag{16.54}$$

The factor of $\frac{1}{4}$ is conventional and the field tensors are given by

$$\boldsymbol{F}_{\mu\nu} \cdot \boldsymbol{F}^{\mu\nu} = F^j_{\mu\nu}F^{\mu\nu}_j = 2\mathrm{Tr}\, F_{\mu\nu}F^{\mu\nu} \qquad \boldsymbol{F}_{\mu\nu} \equiv (F^1_{\mu\nu}, F^2_{\mu\nu}, \ldots, F^N_{\mu\nu}), \tag{16.55}$$

where we have assumed the non-abelian gauge group to be compact and have chosen to normalize its generators in accord with Eq. (8.5). The term $\mathscr{L}_{\rm m}$ is the locally gauge-invariant Lagrangian density of the free matter fields and the term $\mathscr{L}_{\rm int}$ is the locally gauge-invariant coupling between the non-abelian vector fields and the matter fields.

Just as for QED, the form of the coupling is *determined completely by local gauge invariance* through the minimal substitution. The coupling term in Eq. (16.54) is a function of the *covariant derivatives* D_μ and the matter fields Ψ^j.

Unlike for QED, the non-abelian algebra implies a *universal gauge coupling strength.*

16.6.4 Properties of Non-Abelian Gauge Fields

From Eqs. (16.45)–(16.55), we may deduce the following general properties of non-abelian fields with local gauge symmetry (Yang–Mills fields).

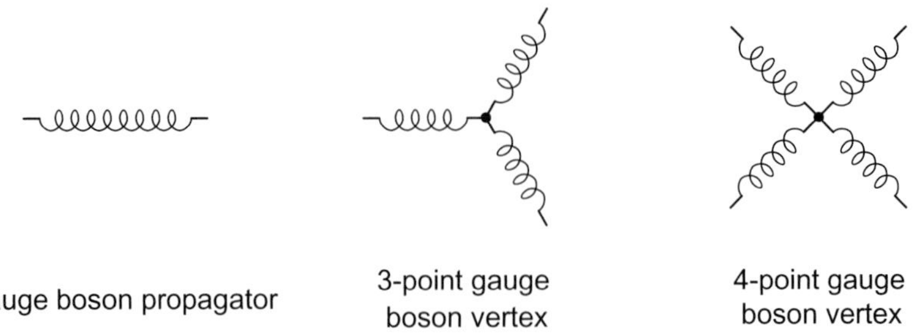

Fig. 16.1 Feynman diagrams (see Box 14.3) for a gauge boson propagator and self-interactions of gauge bosons for a pure non-abelian gauge theory without matter. Coils for the gauge propagators denote explicitly that these are non-abelian gauge bosons.

(1) Physical applications of gauge groups require that the non-abelian symmetry before gauging be exact. Therefore, we expect conserved currents and charges associated with the non-abelian symmetry that generalize the conserved electrical current and charge in QED.

(2) As for QED, the local gauge symmetry prescribes the form of the interaction $\mathscr{L}_{\text{int}}$ between the gauge fields and matter fields that carry the generalized gauge charge.

(3) Even without matter fields the Lagrangian density contains self-coupling of the gauge fields through the first term of Eq. (16.54). Substitution of Eqs. (16.53) into Eq. (16.54) gives terms that are trilinear and quadrilinear in the gauge fields, implying Feynman diagrams of the form shown in Fig. 16.1 for a pure Yang–Mills field.

(4) Self-coupling of Yang–Mills gauge bosons means that non-abelian gauge fields are *intrinsically nonlinear*: Yang–Mills gauge bosons carry the charge of the Yang–Mills field. Conversely, photons do not carry the U(1) gauge charge and QED is linear.[6]

(5) The number of gauge bosons is equal to the number of group generators, so non-abelian gauge fields transform as the *adjoint representation* of the gauge group (see Section 3.2.2).

(6) If the gauge group is not a direct product of simple groups (see Section 2.15), there is only *one* gauge coupling constant g. Thus, non-abelian gauge fields interact with themselves and with matter fields in a manner prescribed by symmetry and gauge couplings have a *universal strength* set by g. This differs from theories without local gauge symmetry where the coupling strength is not constrained by symmetry, and from QED where *abelian* local gauge invariance requires that photon coupling to matter be proportional to the electrical charge but different particles (for example, electrons and positrons) can have different charges that are not related by the abelian gauge symmetry.

Example 16.5 For the SU(3) color symmetry QCD described in Section 19.2 a *single strength* dictates the coupling of the eight gauge bosons (*gluons*) among themselves and to all matter fields undergoing strong interactions.

---

[6] Non-linearities can occur in electromagnetism because of higher-order interactions between photons and matter fields. However, there are no photon–photon coupling terms in QED analogous to those in Fig. 16.1.

(7) If the gauge group G factors into k direct products, $G = G_1 \times G_2 \times \cdots \times G_k$, there are k independent gauge coupling constants. These can be related to each other through symmetry only if the gauge group can be embedded in a larger group that does not factor.

Example 16.6 The gauge group of the Standard Electroweak Model described in Section 19.1 is SU(2) × U(1), implying *two independent gauge coupling constants.* Their ratio can be specified empirically through the *Weinberg angle* of Eq. (19.25), or if SU(2) × U(1) is embedded in a larger group the commutators of the larger group can fix the relationship of the two coupling constants in the subgroup symmetry. In the grand unified theories introduced in Ch. 34, the SU(2) × U(1) gauge group can be embedded in a larger gauge group, which may then determine *by symmetry* the relationship of the two SU(2) × U(1) gauge coupling constants.

---

(8) We will not prove it, but because of the local gauge invariance Yang–Mills fields are perturbatively renormalizable. (However, see the discussion of anomalies in Box 34.1.)

(9) Abelian and non-abelian local gauge fields have elegant geometrical interpretations in terms of parallel transport of vectors. This formulation has much in common mathematically with general relativity, and is discussed further in Ch. 26.

(10) Yang–Mills fields, just as for abelian gauge fields, must be *identically massless* because normal mass terms would spoil the gauge invariance and endanger renormalizability.

This last point would appear to relegate Yang–Mills theories to the realm of mathematical curiosities because few actual particles are massless. However, we shall find that this conclusion is premature when the *Higgs mechanism* is discussed in Chs. 17 and 18.

Background and Further Reading

An introduction to gauge fields for non-specialists is given in Ref. [85]. Introductions assuming some background in elementary particle physics include Aitchison and Hey [7], Cheng and Li [41], Halzen and Martin [103], Quigg [166], Ryder [174], and Zee [227].

Problems

16.1 The Euler–Lagrange equation (16.14) is used in a field theory context in this chapter, but it is applicable to a broad range of problems. Show that inserting a Lagrangian $L(x, \dot{x}) = \frac{1}{2}m\dot{x}^2 - V(x)$ in Eq. (16.14) leads to Newton's second law of motion.

16.2 Show that for the Lagrangian density (16.7) of a complex scalar field, the field equation (16.14) reduces to the two Klein–Gordon equations given by Eq. (16.8).

16.3 Poincaré invariance requires that the action of a scalar field be unchanged under an infinitesimal spacetime translation $x_\mu \to x'_\mu = x_\mu + a_\mu$. Show that this requirement implies the conservation law

$$\partial_\mu \Theta^{\mu\nu} = 0 \qquad \Theta^{\mu\nu} \equiv \frac{\partial \mathscr{L}}{\partial(\partial_\mu \phi)} \partial^\nu \phi - \eta^{\mu\nu} \mathscr{L},$$

where $\Theta^{\mu\nu}$ is a conserved Noether tensor, $\eta^{\mu\nu}$ is the metric tensor, and $\mathscr{L}$ is the Lagrangian density for the scalar field. Show that the components Θ^{00} and Θ^{0k} with $k = 1, 2, 3$ may be interpreted as the energy and momentum densities, respectively, and that generally $\partial_\mu \Theta^{\mu\nu} = 0$ is a statement of 4-momentum conservation. ***

16.4 Show that Eq. (16.10) is invariant under *global* U(1) rotations $\psi(x) \to e^{i\alpha}\psi(x)$, where α is assumed to be independent of the spacetime coordinate x. Find the corresponding conserved current. *Hint*: See Eq. (16.20). ***

16.5 Verify that the Lagrangian density Eq. (16.26) is invariant under SO(2) rotations. Show that the SO(2) generator L and the conserved Noether current J_μ associated with this symmetry are given by Eq. (16.28). ***

16.6 Show that the Lagrangian density (16.30) is invariant under U(1) phase rotations, find the corresponding conserved Noether current, and show that the conserved current is equivalent to that of Eq. (16.28). ***

16.7 The electromagnetic field tensor $F^{\mu\nu}$ given in Eq. (14.14) is Lorentz covariant since it is a rank-2 Lorentz tensor. Show that it is also invariant under the local gauge transformation given by Eq. (16.40).

16.8 Consider spacetime paths between fixed endpoints A and B. For a Lagrangian function $L(x^\mu(\sigma), \dot{x}^\mu(\sigma))$, where σ parameterizes the position on the path and $\dot{x}^\mu \equiv dx^\mu/d\sigma$, define an integral over a path

$$S = \int_A^B L(x^\mu(\sigma), \dot{x}^\mu(\sigma))\, d\sigma.$$

Show that for an arbitrary small variation in the path $x^\mu(\sigma) \to x^\mu(\sigma) + \delta x^\mu(\sigma)$, the corresponding variation in the value of the integral is

$$\delta S \equiv \int_A^B \delta L\, d\sigma = \int_A^B \left(\frac{\partial L}{\partial \dot{x}^\mu(\sigma)} \delta \dot{x}^\mu(\sigma) + \frac{\partial L}{\partial x^\mu(\sigma)} \delta x^\mu(\sigma) \right) d\sigma.$$

Integrate this by parts and use that the variation vanishes at the endpoints (by definition) to show that this leads to the Euler–Lagrange equations (16.14). Thus show that the variational condition $\delta S = 0$ (Hamilton's principle) is equivalent to satisfaction of the Euler–Lagrange equations. ***

16.9 Construct the matrix generator $A_\mu(x) = \tau_i A^i_\mu(x)$ of Eq. (16.49) for an SU(3) Yang–Mills field assuming the representation (8.2). *Hint*: See Example 16.4.

16.10 Evaluate the commutator $[D_\mu, D_\nu]$, where D_μ is the covariant derivative defined in Eq. (16.36). Show that for the U(1) electromagnetic field

$$(iq)^{-1}[D_\mu, D_\nu] = \partial_\mu \partial_\nu - \partial_\nu \partial_\mu = F_{\mu\nu},$$

where q is the gauge coupling strength (gauge charge), A_μ is the vector field, and $F_{\mu\nu}$ is the field strength tensor of Eq. (14.14). Conversely, show that for Yang–Mills gauge fields this commutator suggests the form (16.53) for $F_{\mu\nu}$. *Hint*: Operate with the commutator of the covariant derivatives on an arbitrary gauge field Ψ.

16.11 For an abelian gauge field the field strength tensor $F_{\mu\nu}$ of Eq. (14.14) is gauge invariant (see Problem 14.17). For a non-abelian gauge field the complete contraction $\boldsymbol{F}_{\mu\nu} \cdot \boldsymbol{F}^{\mu\nu} = F^i_{\mu\nu} F^{\mu\nu}_i$ appearing in the Lagrangian density is gauge invariant, but $F^i_{\mu\nu}$ is not. Show that instead $F^i_{\mu\nu}$ transforms as the adjoint representation of the gauge group (see Section 3.2.2). *Hint*: Evaluate $U(\theta)\tau_l F^l_{\mu\nu} U(\theta)^{-1}$, where

$$U(\theta) = \exp(-i\boldsymbol{\tau} \cdot \boldsymbol{\theta}) \simeq 1 - i\tau_k \theta^k$$

implements the gauge transformation. Assume that the gauge generators obey the commutator $[\tau_i, \tau_j] = if_{ijk}\tau_k$, where the f_{ijk} are structure constants for the group, and are normalized according to $\mathrm{Tr}\,(\tau_i, \tau_j) = 2\delta_{ij}$. See also Problem 16.12.

16.12 Prove that Eq. (16.52) follows from Eq. (16.51). *Hint*: Take $A_\mu(x) = \tau_j A^j_\mu(x)$ and

$$U(\boldsymbol{\theta}) \simeq 1 - i\tau_k \theta^k \qquad [\tau_i, \tau_j] = if_{ijk}\tau_k \qquad \mathrm{Tr}\,(\tau_i, \tau_j) = 2\delta_{ij}$$

for the gauge transformation and its generators τ_i. ***

16.13 Show that if Eq. (16.24) is true, the charge Q of Eq. (16.21) is conserved. *Hint*: Use Eq. (16.23) and the *divergence theorem*, which says that for a 2D surface S bounding a volume V,

$$\int_V d^3x\, \boldsymbol{\nabla} \cdot \boldsymbol{J} = \int_S \boldsymbol{J} \cdot \boldsymbol{n}\, ds,$$

where $\boldsymbol{n}$ is an outward normal to S and ds is a surface differential element. ***

16.14 Show, beginning from Eqs. (16.15)–(16.18) for a global isospin symmetry, that there is a global Noether current

$$J^a_\mu \equiv -\frac{i}{2}\frac{\partial \mathscr{L}_0}{\partial(\partial_\mu \psi)}\tau^a \psi = \frac{1}{2}\bar{\psi}\gamma^\mu \tau^a \psi,$$

which is conserved because $\partial^\mu J^a_\mu = 0$. ***

PART II

BROKEN SYMMETRY

17 Spontaneous Symmetry Breaking

This is a book about symmetry, but it is at the same time a book about broken symmetry. In modern usage, symmetry breaking takes on two distinct meanings: a broken symmetry may be broken well and truly, or a broken symmetry may actually be conserved but may appear to be broken unless one looks very deeply at relationships in the system. This latter case should more properly be termed hidden rather than broken symmetry, but it is standard to say that a hidden symmetry is *broken spontaneously.* This chapter introduces the concept of spontaneous symmetry breaking. It builds on the introduction to relativistic quantum fields given in Section 16.1 and assumes a basic familiarity with that material.

17.1 Modes of Symmetry Breaking

At least five modes of symmetry breaking (or symmetry hiding) may be identified in modern physical theories. The following list gives an overview.

1. The *Wigner mode* or *explicit symmetry breaking.* This occurs through terms included in the Lagrangian or Hamiltonian that violate a symmetry explicitly.
2. The *Nambu–Goldstone mode* or *spontaneous symmetry breaking.* In this mode the states of the theory do not have the same symmetry as the Lagrangian or Hamiltonian, because a global symmetry for them appears to be broken by the physical states of the system.
3. The *Higgs mode.* The Higgs mode is the Nambu–Goldstone mode, but with the proviso that the symmetry broken spontaneously is a local gauge symmetry, not a global one.
4. *Dynamical symmetry.* As suggested in Section 3.6, a dynamical symmetry is actually a particular pattern of symmetry breaking that occurs when the Hamiltonian or Lagrangian can be written as a polynomial in the invariants of a subgroup chain.
5. *Anomalous symmetry breaking.* Local gauge symmetries respected by classical fields are sometimes broken by the act of quantizing the fields. This is termed an *anomaly.*

In this chapter we discuss the first two cases, the Higgs mode and dynamical symmetry will be described in Chs. 18 and 20, respectively, and anomalies will be defined in Box 34.1.

17.2 Explicit Symmetry Breaking

Explicit symmetry breaking, which is often termed the Wigner mode of symmetry realization, is illustrated in Fig. 17.1. It is characterized by breaking of the degeneracy

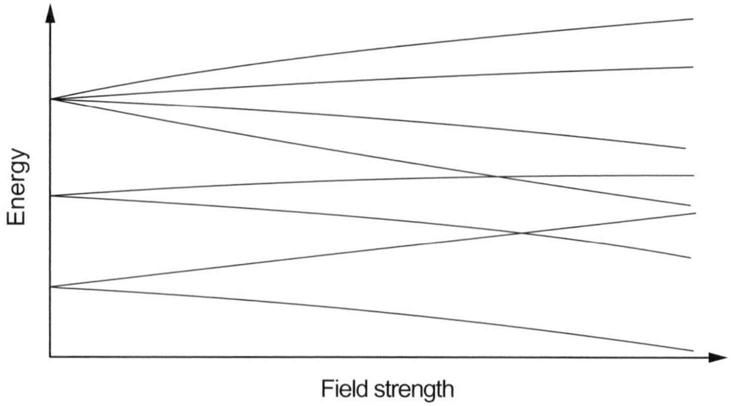

Fig. 17.1 Wigner symmetry mode (explicit symmetry breaking).

in multiplet structure by explicit symmetry breaking terms in the Hamiltonian. The prototype example is the breaking of rotational SU(2) symmetry by an explicit term in the Hamiltonian corresponding to a magnetic field along a chosen axis (*Zeeman effect*). Because the corresponding term favors one axis over the others, the rotational symmetry exemplified by degenerate angular momentum multiplets is split by the field according to the magnetic substate quantum number, and the amount of this splitting increases with the strength of the symmetry breaking term (the magnetic field in this example).

In symmetry language the original states are symmetric under angular momentum SU(2), with the states labeled by an angular momentum quantum number J. The states are $2J + 1$ degenerate, corresponding to the possible magnetic substates M for the states J. Addition of a term to the Hamiltonian representing a uniform magnetic field along the z-axis breaks the SU(2) symmetry (invariance under all rotations) down to a U(1) symmetry (invariance only under rotations about a single axis). The irreps of SU(2) are $2J+1$ degenerate but the irreps of U(1) are one-dimensional, so each SU(2) multiplet splits into $2J + 1$ states that each respond differently to increasing the strength of the magnetic field, giving the characteristic pattern illustrated in Fig. 17.1.

17.3 The Vacuum and Hidden Symmetry

Of more interest in many applications are the possibilities of symmetry breaking in the Nambu–Goldstone and Higgs modes. The essential difference among the Wigner, Nambu–Goldstone, and Higgs modes of symmetry realization concerns the structure of the vacuum. In quantum field theory states are built by operating on the vacuum state with particle creation operators. Then the nature of the physical states depends on both the symmetry of the Lagrangian or Hamiltonian *and* the symmetry of the vacuum, and these need not be the same. Specifically, if the vacuum state is denoted by $|0\rangle$ the symmetry is implemented as follows.

1. The *Wigner mode* if the Lagrangian is invariant under a set of symmetry transformations U and the vacuum $|0\rangle$ is also invariant: $U|0\rangle = |0\rangle$.
2. The *Nambu–Goldstone mode* if the Lagrangian is invariant under a global symmetry but the vacuum is not: $U|0\rangle \neq |0\rangle$, where U implements a *global symmetry*.
3. The *Higgs mode* if the vacuum is not invariant and the symmetry of the Lagrangian is a local gauge symmetry: $U|0\rangle \neq |0\rangle$, where U implements a *local symmetry*.

As you are asked to demonstrate in Problem 17.1, the condition $U|0\rangle \neq |0\rangle$ is equivalent to a statement that at least one group generator fails to give zero when applied to the vacuum state (it "fails to annihilate the vacuum"). Let us investigate the case of spontaneously broken symmetry where the Lagrangian has a global symmetry but the vacuum state does not.

17.4 Spontaneously Broken Discrete Symmetry

The idea of spontaneously broken symmetry can be introduced using a simple model corresponding to the Lagrangian density

$$\mathscr{L} = \frac{1}{2}(\partial_\mu \phi)(\partial^\mu \phi) - \frac{1}{2}\mu^2\phi^2 - \frac{1}{4}\lambda\phi^4, \tag{17.1}$$

where we require $\lambda > 0$, which ensures that the energy has a lower bound, and the parameter μ^2 will be discussed further below. This Lagrangian density is appropriate for a real (uncharged) scalar field ϕ interacting through the ϕ^4 term. Clearly the Lagrangian density (17.1) is invariant under the discrete transformation $\phi \to -\phi$. Let us consider the particle spectrum expected for such a Lagrangian density. Two qualitatively different cases may be distinguished, depending on the sign of the parameter μ^2. Case (a), with $\mu^2 > 0$, is illustrated in Fig. 17.2(a) and case (b), with $\mu^2 < 0$, is illustrated in Fig. 17.2(b).

17.4.1 Symmetry in the Wigner Mode

Case (a) corresponds to the Wigner symmetry mode. From Fig. 17.2(a), the vacuum expectation value of the classical field is $\langle\phi\rangle_0 \equiv \langle 0|\phi|0\rangle = 0$. In the absence of

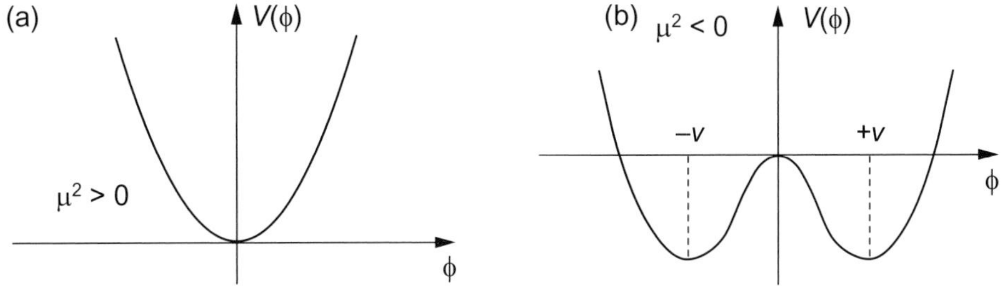

Fig. 17.2 Effective potentials for Eq. (17.1): (a) symmetric, (b) symmetry broken spontaneously.

interactions the particle spectrum associated with this Lagrangian density may be obtained by expanding to second order in a Taylor series about the classical vacuum state $\phi = 0$, yielding[1]

$$\mathscr{L} \simeq \frac{1}{2}(\partial_\mu \phi)(\partial^\mu \phi) - \frac{1}{2}\mu^2 \phi^2. \tag{17.2}$$

Comparison with Eq. (16.4) suggests that (17.2) is the Lagrangian density of a free scalar field of mass μ. For $\mu^2 > 0$, small quantized oscillations of the field around its classical vacuum may be interpreted as particles of mass μ.

17.4.2 Spontaneously Broken Symmetry

Case (b) is rather different. If $\mu^2 < 0$ the effective potential shown in Fig. 17.2(b) now has *two degenerate minima* (classical vacuum states) at

$$\langle \phi \rangle_0 = \pm\sqrt{-\mu^2/\lambda} \equiv \pm v. \tag{17.3}$$

Figure 17.2(b) is an example of spontaneously broken or hidden symmetry. To examine the particle spectrum we must again expand to quadratic order. The origin cannot be chosen as the point of expansion as in case (a) because it is unstable, nor can μ^2 be interpreted as a mass because it is negative. Instead, we must expand around the state of *lowest energy* (the classical vacuum). But there are *two degenerate classical vacuum states* at $\phi = \pm v$. They are equivalent because they are related by the symmetry $\phi \to -\phi$, but one must be chosen arbitrarily to proceed. Let us agree to choose the minimum at $+v$ as the expansion point.

Now the Lagrangian (17.1) is invariant under reflection $\phi \to -\phi$ but the vacuum state $\langle \phi \rangle_0 = +v$ is not. The symmetry has been *broken spontaneously* by the choice of vacuum state.

Furthermore, excited states built by expanding around this chosen vacuum will also break the symmetry because their foundation (the classical vacuum) does so. To facilitate expansion around the vacuum state $\phi = +v$ we may introduce

$$\xi(x) \equiv \phi(x) - \langle \phi \rangle_0 = \phi(x) - v. \tag{17.4}$$

Now the vacuum state corresponds to $\langle \xi \rangle = 0$ and Eq. (17.1) becomes

$$\mathscr{L} = \frac{1}{2}(\partial_\mu \xi)(\partial^\mu \xi) - \lambda v^2 \xi^2 - \lambda v \xi^3 - \frac{1}{4}\lambda \xi^4, \tag{17.5}$$

where constant terms have been ignored. This has no obvious reflection symmetry; it has been hidden by the choice of vacuum state and the parameterization. For small oscillations around the classical vacuum, quadratic expansion yields

$$\mathscr{L} \simeq \frac{1}{2}(\partial_\mu \xi)(\partial^\mu \xi) - \lambda v^2 \xi^2. \tag{17.6}$$

[1] Recall that constant coefficients of terms in a Lagrangian density that are quadratic in a field variable may be interpreted as the mass or square of the mass for particles of that field; see examples in Section 16.1.3.

This describes a free scalar field ξ of mass $m_\xi = (2\lambda v^2)^{1/2} = (-2\mu^2)^{1/2}$, upon comparing with Eq. (16.4) and using Eq. (17.3). The mass m_ξ is real and positive since μ^2 is negative, and differs from the mass found for case (a).

17.4.3 Summary of Spontaneously Broken Discrete Symmetry

The preceding example is simple, involving a discrete rather than continuous symmetry. Nevertheless, it exhibits many characteristic features of a spontaneously broken symmetry.

1. For $\mu^2 < 0$, there is a non-zero expectation value of a classical field in the ground state.
2. For $\mu^2 < 0$, the classical vacuum is degenerate, with the choice of one of the degenerate vacua as the ground state arbitrary since all vacuum states are related by symmetry.
3. The qualitative differences between cases (a) and (b) suggest fundamentally different phases of the theory for positive μ^2 relative to negative μ^2, as is confirmed by different masses for the free fields in the two cases. The transition from symmetric vacuum in case (a) to degenerate vacuum in case (b) is a *phase transition* that occurs as a parameter (μ^2 in this case) is varied continuously.
4. The chosen vacuum does not have the same symmetry as the original Lagrangian density.
5. The original symmetry of the Lagrangian density is hidden by expansion around the chosen broken symmetry vacuum. The symmetry is still there, because the two degenerate vacua in Fig. 17.2(b) are related by the symmetry operation $\phi \to -\phi$, but it is not manifest in the Lagrangian density (17.5).
6. Once the theory develops degenerate vacua ($\mu^2 < 0$), the origin is no longer a minimum but is rather an unstable maximum. Therefore, an infinitesimal fluctuation is sufficient to drive the system away from the symmetric point ($\phi = 0$) and into one of the broken symmetry minima. That is the origin of the term "spontaneous symmetry breaking"; even if it is initialized in the classical symmetric state, the system can break the symmetry spontaneously in response to an infinitesimal fluctuation.
7. The mass of particles with and without spontaneous symmetry breaking [μ in the former case and $(-2\mu^2)^{1/2}$ in the latter case] can differ substantially. We say that the mass $(-2\mu^2)^{1/2}$ has been *acquired spontaneously* in case (b).

This discrete symmetry example exemplifies many properties of spontaneously broken symmetry but some important features appear only when the symmetry that is broken is continuous. Therefore, let us now consider breaking a *continuous symmetry* spontaneously.

17.5 Spontaneously Broken Continuous Symmetry

Consider a complex scalar field ϕ and a Lagrangian density[2]

$$\mathscr{L} = (\partial_\mu \phi)^\dagger (\partial^\mu \phi) - \mu^2 \phi^\dagger \phi - \lambda (\phi^\dagger \phi)^2, \tag{17.7}$$

[2] Uncharged particles of zero spin are described by real scalar fields and charged particles of zero spin are described by complex scalar fields in relativistic quantum field theory (see Section 16.1.3). Thus the Lagrangian density (17.7) is appropriate for charged, spin-0 particles.

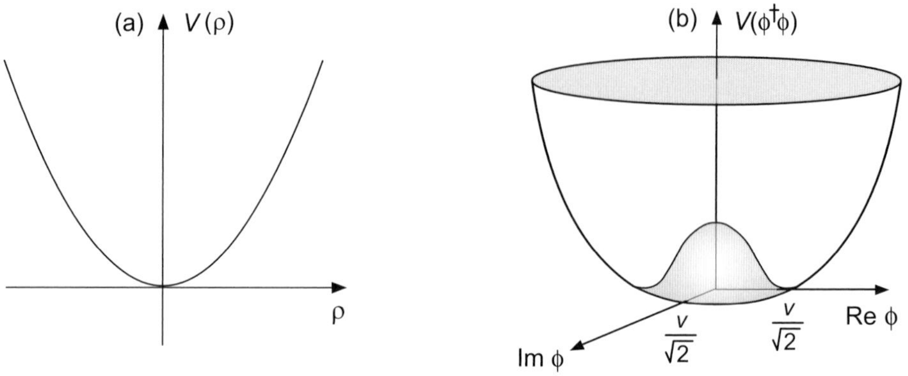

Fig. 17.3 Potentials for the complex scalar field Lagrangian density (17.7). (a) Manifest symmetry for $\mu^2 > 0$, with $\rho \equiv \phi^\dagger \phi$. (b) Spontaneously broken symmetry for $\mu^2 < 0$.

where we assume that $\lambda > 0$. This $\mathscr{L}$ is invariant under a global U(1) transformation

$$\phi(x) \rightarrow \phi'(x) = e^{i\theta}\phi(x), \tag{17.8}$$

with θ independent of x. We may identify a potential

$$V(\rho) = \mu^2\rho + \lambda\rho^2 \qquad \rho \equiv \phi^\dagger\phi, \tag{17.9}$$

and may distinguish two qualitatively different cases, depending on the sign of μ^2.

17.5.1 Symmetric Classical Vacuum

For $\mu^2 > 0$ the classical potential is symmetric about a minimum at $\rho = 0$, as illustrated in Fig. 17.3(a). This corresponds to implementation of the symmetry in Wigner mode and is analogous to the case in Fig. 17.2(a). Particles correspond to quadratic oscillations of the field about the symmetric minimum and μ^2 determines the mass for the free field.

17.5.2 Hidden Continuous Symmetry

For the case $\mu^2 < 0$, the minima of the potential correspond to a circle of radius

$$|\phi| = \sqrt{-\mu^2/2\lambda} \equiv v/\sqrt{2} \tag{17.10}$$

in the complex ϕ plane, as illustrated in Fig. 17.3(b). This is spontaneously broken symmetry similar to the previous discrete symmetry example, but now there are *infinitely many* degenerate classical ground states corresponding to different positions around the ring of minima (17.10), with each classical vacuum state related to all others by the global phase transformation (17.8). As before, a classical ground state must be chosen. Let us select the minimum point lying on the real ϕ-axis as the classical vacuum, $\mathrm{Re}(\phi) = v/\sqrt{2}$, and expand around it to investigate the particle spectrum. This expansion may be parameterized in terms of

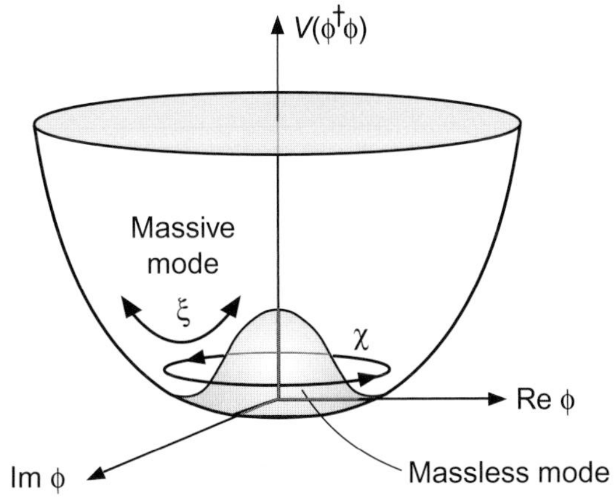

Fig. 17.4 Spontaneously broken global symmetry for a complex scalar field with Lagrangian density (17.12). The field ξ is massive but the field χ corresponds to a massless Goldstone boson.

$$\phi(x) = \frac{1}{\sqrt{2}}\left[v + \xi(x) + i\chi(x)\right], \tag{17.11}$$

which, when substituted into Eq. (17.7), gives

$$\mathscr{L} = \frac{1}{2}(\partial_\mu \xi)^2 + \frac{1}{2}(\partial_\mu \chi)^2 - \lambda v^2 \xi^2 - \lambda v \xi(\xi^2 + \chi^2) - \frac{1}{4}\lambda(\xi^2 + \chi^2)^2, \tag{17.12}$$

where constant terms have been dropped. This Lagrangian density looks like that for two fields, ξ and χ, plus some interaction terms. There is no term in $\mathscr{L}$ proportional to χ^2 so the χ field is massless, but the field ξ has a mass

$$m_\xi = \sqrt{2\lambda v^2} = \sqrt{-2\mu^2} \qquad (\mu^2 < 0), \tag{17.13}$$

which was acquired spontaneously. The interpretation of these fields and their masses is suggested in Fig. 17.4. The massive field ξ corresponds to "radial" oscillations in the potential. Because of the curvature in the radial direction there is a restoring force against the oscillations of the ξ field and it acquires an effective mass given by Eq. (17.13). On the other hand, the χ mode corresponds to circular motion in the flat valley of the potential, for which there is no restoring force and the corresponding field χ has vanishing mass.[3]

17.5.3 The Goldstone Theorem

The appearance of the massless scalar field χ as a consequence of the spontaneously broken symmetry in Fig. 17.4 is not an isolated example. Systematic analysis of spontaneous symmetry breaking for continuous global symmetries leads to a celebrated result called the *Goldstone theorem* [26, 80, 81, 153, 154, 155].

[3] Do not be confused by the abstract nature of Fig. 17.4, which is plotted in field space, not spacetime. The complex scalar field $\phi(x)$ is a function of the spacetime coordinates x, but the potential $V(\phi^\dagger\phi)$ in Fig. 17.4 is plotted in a 2D space with axes corresponding to the real and imaginary parts of the field $\phi(x)$.

Goldstone Theorem: If a continuous global symmetry is broken spontaneously, for each broken generator a massless *Goldstone boson* appears in the spectrum.

Goldstone Bosons: In the preceding example the broken continuous symmetry is a global U(1) invariance under the transformation (17.8). Since U(1) has a single generator, one massless Goldstone boson field appears that corresponds to motion induced by the generator that was broken (circular motion in the valley of the potential in Fig. 17.4). The Goldstone boson is said to carry the quantum numbers of the broken symmetry generator. The symmetry breaking can involve more than one broken group generator and more than one Goldstone boson. The following example illustrates for multiple scalar fields.

Example 17.1 The Lagrangian density for n real scalar fields interacting through a ϕ^4 interaction is a generalization of Eq. (16.4),

$$\mathscr{L} = \frac{1}{2}(\partial_\alpha \phi_i)(\partial^\alpha \phi_i) - \frac{1}{2}\mu^2 \phi_i \phi_i - \frac{1}{4}(\phi_i \phi_i)^2, \tag{17.14}$$

where α is a Lorentz index and i labels the scalar fields (summation on repeated indices α and i) [1]. This $\mathscr{L}$ is invariant under SO(n) rotations. If the symmetry is broken spontaneously by choosing $\mu^2 < 0$ in Eq. (17.14), generalization of Eq. (17.10) implies an infinity of vacuum states satisfying $\phi_i \phi_i = -\mu^2/\lambda \equiv v$. The fields entering Eq. (17.14) may be viewed as the components ϕ_i of a vector ϕ, with the SO(n) symmetry implying that the Lagrangian density preserves the length of the vector but not its direction. This is the generalization to a higher-dimensional space of the circle of minima found in Eq. (17.11). In this notation the vacuum state may be chosen to be

$$\langle \phi \rangle_0 \equiv \begin{pmatrix} \phi_1 \\ \phi_2 \\ \vdots \\ \phi_n \end{pmatrix}_{\text{vac}} = \begin{pmatrix} 0 \\ 0 \\ \vdots \\ -\mu^2/\lambda \end{pmatrix}. \tag{17.15}$$

All other possible vacuum states (an infinity of them) are related to this choice by continuous SO(n) rotations because of the original symmetry of the Lagrangian density.

Counting Broken Generators: In the example of Fig. 17.4 the U(1) symmetry had no continuous subgroups but for SO(n) in Example 17.1 the vacuum state remains invariant under the subgroup SO($n-1$) that does not mix the nth field in the vector (17.15) with the others. From Appendix D the group SO(n) has $\frac{1}{2}n(n-1)$ generators so the subgroup SO($n-1$) has $\frac{1}{2}(n-1)(n-2)$ generators and the number broken spontaneously is

$$\text{Number of broken generators} = \frac{1}{2}n(n-1) - \frac{1}{2}(n-1)(n-2) = n-1.$$

Carrying out an analysis generalizing the earlier U(1) example then indicates that only one of the original n fields acquires a mass spontaneously and $n-1$ massless scalar particles (Goldstone bosons) are generated by the spontaneous breaking of $n-1$ group generators through the choice $\mu^2 < 0$ in Eq. (17.14).

17.5.4 The Stability Subgroup

Consider spontaneous breaking of a continuous symmetry described by a group G. The largest subgroup that leaves the vacuum invariant is termed the *stability subgroup* or the *little group,* commonly denoted by H.

> We say that G has been broken spontaneously down to H, and the number of Goldstone bosons that result is equal to the difference between the number of generators for the full group G and that for the stability subgroup H.

Examples 17.2 and 17.3 illustrate the consequences of breaking spontaneously a full group G to a stability subgroup H.

Example 17.2 As shown in Problem 17.3, a Lorentz scalar field with three components ϕ_i $(i = 1, 2, 3)$ and a Lagrangian density of the form (17.14) is invariant under SO(3) rotations. The ground state after spontaneous symmetry breaking no longer has SO(3) symmetry but it is still invariant under the SO(2) subgroup of rotations about a single axis. The symmetry has been broken spontaneously from $G = \text{SO}(3)$, with three generators, to $H = \text{SO}(2)$, with one generator, and the spectrum corresponds to $3 - 1 = 2$ massless scalar fields and $3 - 2 = 1$ massive scalar field.

Example 17.3 The fields corresponding to Example 17.2 may be constructed explicitly by introducing a new field $\chi(x)$ through $\phi_3(x) = \chi(x) + v$. After rewriting the Lagrangian density (17.14) for $i = 1, 2, 3$ in terms of the fields ϕ_1, ϕ_2, and χ, we find upon comparing with the Lagrangian density for free scalar fields that the spectrum consists of

1. *a massive scalar* χ, with mass $m_\chi = (2\lambda v^2)^{1/2}$ and
2. *two Goldstone bosons* with $m_{\phi_1} = m_{\phi_2} = 0$.

The interpretation of these fields corresponds to a higher-dimensional generalization of Fig. 17.4. The massive field corresponds to "radial" oscillations with a non-zero restoring force and the massless Goldstone bosons correspond to motion in flat valleys (the motion associated with the generators that were broken), which experiences no restoring force.

The symmetries considered in this chapter were global ones that did not depend on the spacetime coordinates. In Ch. 18 we shall find that the already quite interesting results of global spontaneous symmetry breaking take on new and unexpected properties if the symmetry that is broken is a *local gauge symmetry.*

Background and Further Reading

For an introduction to spontaneously broken symmetry similar in spirit to the treatment in this chapter but related more systematically to relativistic quantum field theory, see Ref. [85]. For more extensive discussions, primarily from an elementary particle physics perspective, see Abers and Lee [1], Aitchison and Hey [6, 7], Coleman [43], and Quigg [166]. For discussions of spontaneously broken symmetry in non-relativistic applications see Aitchison and Hey [7], and Ring and Schuck [171].

Problems

17.1 Prove that if a symmetry is broken spontaneously, at least one generator of the symmetry group gives a non-zero value when applied to the vacuum state. ***

17.2 Use Eq. (17.1) to find the minima of Fig. 17.2. Confirm Eq. (17.5) for $\mu^2 < 0$ and show that the corresponding mass of the ξ field is given by $m_\xi = (2\lambda v^2)^{1/2} = (-2\mu^2)^{1/2}$ after the symmetry is broken spontaneously.

17.3 Assume the Lagrangian density (17.14) with $i = 1, 2, 3$ for an isovector Lorentz scalar field to be invariant under a global internal SO(3) symmetry. Show that for $\mu^2 < 1$ the symmetry is broken spontaneously from SO(3) to SO(2). What is the resulting particle spectrum? ***

18 The Higgs Mechanism

In Ch. 16 we found that the quanta of gauge fields (gauge bosons) must be *identically massless* to preserve gauge invariance. The non-abelian gauge invariance described in Section 16.6 is an attractive principle for theories of fundamental interactions because it represents the generalization of a highly successful theory, quantum electrodynamics (QED), that is renormalizable. However, the required masslessness of the gauge bosons is a huge stumbling block to application of non-abelian gauge theories to, say, the weak interactions. Phenomenologically, the weak interactions are of very short range. By uncertainty principle arguments a short-range interaction in quantum field theory must be mediated by a massive virtual exchange particle. Therefore, any attempt to generalize QED to include "non-abelian photons" that might mediate the weak interactions seems doomed, since massless particles are associated with fields of long range, not short range.

A way forward might be provided by spontaneous symmetry breaking, for we saw in Ch. 17 that particles can acquire mass by this mechanism and it might be hoped that some form of spontaneous symmetry breaking would allow massless gauge bosons to be given an effective mass, while still preserving gauge invariance. However, the Goldstone theorem seems to undermine this hope because spontaneously breaking a continuous *global* symmetry implies the appearance of new massless particles, for which there is little evidence in nature. But there is a possible way out! Derivation of the Goldstone theorem rests on assuming that field theories have "normal" properties such as locality, Lorentz invariance, and positive definite norm on the Hilbert space.[1] Careful analysis of *local* gauge fields indicates that they do not simultaneously obey all of these "normal" conditions, thus possibly evading the physically undesirable massless Goldstone bosons! This can be illustrated by considering the quantization of the abelian local gauge field in QED.

18.1 Photons and the Higgs Loophole

Free photons have two transverse states of polarization but Lorentz covariance requires that a photon correspond to a 4-vector field with four states of polarization (a timelike and longitudinal polarization, in addition to the two transverse polarizations). A common

[1] Locality means that particles are influenced directly only by nearby particles (no "action at a distance"). Lorentz invariance means symmetry under the Lorentz transformations. Positive definite norm means that quantum state vectors have "lengths" that are never negative, as would be expected of physical particles.

approach to quantizing the electromagnetic field is to work in the radiation or Coulomb gauge (see Section 14.1.2) implied by the gauge condition $\nabla \cdot \boldsymbol{A} = 0$, which identifies clearly the physical transverse photon states but is not manifestly Lorentz invariant since ∇ and $\boldsymbol{A}$ are not 4-vectors. Conversely, if one works in the Lorenz gauge by imposing the covariant gauge condition $\partial_\mu A^\mu = 0$, it is found that the timelike and longitudinal polarizations for A^μ must give exactly canceling contributions in physical matrix elements for free photons. Specifically, the indefinite Minkowski metric $\eta_{\mu\nu}$ requires that the timelike component A^0 of the 4-vector potential must have *negative norm.* For massless fields, the A^0 contribution then exactly cancels contributions from the longitudinal spacelike components $\boldsymbol{p} \cdot \boldsymbol{A}$, where $\boldsymbol{p}$ is the 3-momentum and $\boldsymbol{A}$ the 3-vector potential, leaving two physical transverse components, $\boldsymbol{p} \times \boldsymbol{A}$. These considerations raise the question of whether the Goldstone theorem applies in a theory having local gauge invariance. The answer is remarkable.

> In the presence of a *local* gauge invariance there is a conspiracy between the massless scalar fields required by the Goldstone theorem and the massless vector fields required by local gauge invariance that can (1) eliminate the Goldstone bosons and (2) bestow effective mass on the gauge bosons.

This is called the *Higgs mechanism,* and the way in which the Higgs mechanism allows the Goldstone theorem to be circumvented is termed the *Higgs loophole* [12, 58, 100, 109, 110]. The next section presents a simple Lorentz-covariant model showing the Higgs mechanism in action, and following sections show that many aspects of the Higgs mechanism can be understood intuitively by analogy with the properties of an ordinary superconductor.

18.2 The Abelian Higgs Model

A simple example of the Higgs mechanism results from generalizing the spontaneously broken global U(1) invariance of Section 17.5 to spontaneously broken *local* U(1) invariance in the presence of an electromagnetic field. This is called the *abelian Higgs model.* We describe it guided by Ref. [85].

18.2.1 Lagrangian Density

The Lagrangian density for the abelian Higgs model is

$$\mathscr{L} = (D_\mu \phi)^\dagger (D^\mu \phi) - \mu^2 \phi^\dagger \phi - \lambda (\phi^\dagger \phi)^2 - \frac{1}{4} F_{\mu\nu} F^{\mu\nu}, \tag{18.1}$$

where the covariant derivative D^μ and gauge-invariant field tensor $F^{\mu\nu}$ are

$$D^\mu = \partial^\mu + iqA^\mu \qquad F^{\mu\nu} = \partial^\mu A^\nu - \partial^\nu A^\mu, \tag{18.2}$$

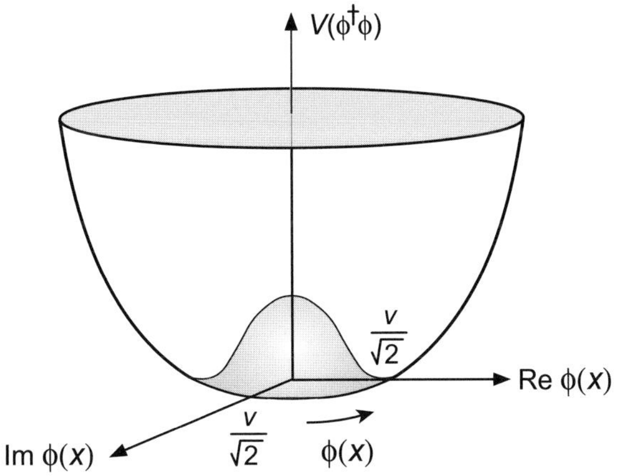

Fig. 18.1 Potential corresponding to the Lagrangian density (18.1) for the abelian Higgs model with $\mu^2 < 0$. The ground state (18.6) is infinitely degenerate, corresponding to different values of $\phi(x)$ in the circular minimum of the potential.

the parameter λ is positive, μ^2 is a parameter to be discussed further below, and the complex scalar field ϕ may be expressed in terms of two real fields, ϕ_1 and ϕ_2, as[2]

$$\phi = \frac{1}{\sqrt{2}}(\phi_1 + i\phi_2) \qquad \phi^\dagger = \frac{1}{\sqrt{2}}(\phi_1 - i\phi_2). \tag{18.3}$$

With no spontaneous symmetry breaking, $\mathscr{L}$ describes the usual electrodynamics of a scalar field carrying charge q. Equation (18.1) is invariant under global U(1) phase rotations on the fields and also invariant under the local gauge transformations

$$\phi(x) \to e^{iq\alpha(x)}\phi(x) \qquad A_\mu(x) \to A_\mu(x) - \partial_\mu \alpha(x), \tag{18.4}$$

since it employs the locally gauge-invariant covariant derivative D^μ and field tensor $F^{\mu\nu}$.

18.2.2 Symmetry Breaking

As for the example in Section 17.5 with global U(1) symmetry, two qualitatively different results may be obtained for a field theory based on the Lagrangian density (18.1), depending on the choice of sign for the parameter μ^2. For $\mu^2 > 0$ the potential has a unique minimum at $\phi = \phi^\dagger = 0$, the vacuum is not degenerate, symmetry is realized in the Wigner mode, and the spectrum consists of a massless photon associated with A^μ and two scalar fields, ϕ and $\phi^\dagger$, having a common mass μ. Of more interest is the case $\mu^2 < 0$, which corresponds to a spontaneously broken local gauge symmetry. As displayed in Fig. 18.1, the degenerate vacua occur for

$$|\phi|^2 = -\frac{\mu^2}{2\lambda} \equiv \frac{v^2}{2}, \tag{18.5}$$

[2] Recall from Section 16.1.3 that spin-0 charged particles are described by complex scalar fields. Thus Eq. (18.1) describes charged spin-0 particles coupled to electromagnetism in a Lorentz-covariant way.

where v is real and positive, implying an infinitely degenerate set of vacuum states

$$\langle\phi\rangle_0 = \frac{v}{\sqrt{2}}\, e^{i\alpha(x)}, \tag{18.6}$$

with each equivalent vacuum state labeled by a value of $\alpha(x)$, corresponding to different positions in the circular valley of Fig. 18.1. To proceed we break the symmetry spontaneously by choosing $\alpha(x) = 0$, giving a vacuum state $\langle\phi\rangle_0 = v/\sqrt{2}$ lying at the minimum on the real axis of Fig. 18.1, and expand in polar coordinates,

$$\phi(x) = \frac{1}{\sqrt{2}}\left[v + \eta(x)\right] e^{i\xi(x)/v} \simeq \frac{1}{\sqrt{2}}\left[v + \eta(x) + i\xi(x)\right]. \tag{18.7}$$

Inserting this expansion in Eq. (18.3), keeping only low-order terms, and invoking the local gauge transformation

$$\phi(x) \to e^{-i\xi(x)/v}\phi(x) = \frac{v + \eta(x)}{\sqrt{2}} \qquad A_\mu(x) \to A_\mu(x) + \frac{1}{qv}\partial_\mu \xi(x), \tag{18.8}$$

gives the Lagrangian density (see Problem 18.1)

$$\mathscr{L} = \frac{1}{2}(\partial_\mu \eta)(\partial^\mu \eta) + \mu^2\eta^2 + \frac{1}{2}q^2v^2 A_\mu A^\mu - \frac{1}{4}F_{\mu\nu}F^{\mu\nu}. \tag{18.9}$$

The spectrum associated with this Lagrangian density differs fundamentally from that found when the symmetry is not broken spontaneously, and also differs fundamentally from the analogous spontaneous breaking of a *global U(1) symmetry* discussed in Section 17.5. Inspecting the quadratic terms, we see that (18.9) corresponds to a *massive vector field* A_μ with a mass $m = qv$, and a *massive scalar* η with mass $m = (-2\mu^2)^{1/2}$.

The spontaneous breaking of a U(1) *local gauge symmetry* has given the photon a mass, without violating gauge invariance, and there is a massive scalar but there are *no massless Goldstone bosons!*

The mathematics of the abelian Higgs model has led to quite intriguing and unexpected results. Let us see if we can understand the physics of what we have found.

18.2.3 Understanding the Higgs Mechanism

We begin by counting degrees of freedom (always an advisable check!). Before breaking the symmetry spontaneously there were four degrees of freedom:

- two associated with (real and imaginary parts of) the complex scalar field and
- two associated with the transverse states of polarization for a massless vector field.

After spontaneous symmetry breaking there are still four degrees of freedom, but in a rather different arrangement:

- one degree of freedom associated with one real scalar field and
- *three* degrees of freedom associated with a *massive* vector field.[3]

[3] Recall (Box 15.2) that massless vector fields have two polarization states but massive vector fields have three.

Detailed examination reveals that the parameter ξ appearing in Eq. (18.7), which would have become a massless Goldstone field for a spontaneously broken *global* symmetry, has effectively been absorbed by the photon field through the gauge transformation (18.8), becoming a third state of polarization for the photon and thereby rendering it *massive*. Thus, for spontaneously broken local U(1) gauge symmetry the abelian Higgs model predicts

- *no Goldstone bosons*,
- a *massive photon*, and
- a *massive scalar* η that is termed a *Higgs boson.*

This means of symmetry realization is termed the *Higgs mode.* The Wigner mode was characterized by explicit breaking of degenerate multiplets and the Nambu–Goldstone mode by the appearance of one massless Goldstone boson for each generator broken spontaneously. The signature of the Higgs mode is the acquisition of mass by gauge bosons at the expense of would-be Goldstone bosons that vanish from the theory, and the appearance of massive scalars (Higgs bosons).

18.3 Vacuum Screening Currents

The abelian Higgs model was formulated in Lorentz-covariant terms. However, there is nothing intrinsically relativistic about the Higgs mechanism and there are many aspects of spontaneous symmetry breaking in the Higgs mode that can be understood in terms of familiar concepts from non-relativistic problems in atomic and condensed matter physics. These concepts will involve the idea of *vacuum screening currents*; to understand their importance, we must first revisit the relationship between gauge invariance and mass.

18.3.1 Gauge Invariance and Mass

Of central importance to understanding the Higgs mechanism is an appreciation of the exact relationship between massive and massless vector fields, and how this is related to (local) gauge invariance [7]. From Eq. (14.19) the electromagnetic current j^ν satisfies

$$\Box A^\nu - \partial^\nu(\partial_\mu A^\mu) = j^\nu, \tag{18.10}$$

where A^ν is the 4-vector potential. On the other hand, the corresponding wave equation for a massive vector field A^ν is

$$(\Box + m^2)A^\nu - \partial^\nu(\partial_\mu A^\mu) = j^\nu, \tag{18.11}$$

where m is the mass. Taking the 4-divergence of both sides of Eq. (18.11) gives the constraint $\partial_\mu A^\mu = 0$ (Problem 18.2), and Eq. (18.11) for the massive vector field becomes

$$(\Box + m^2)\, A^\nu = \begin{cases} j^\nu & \text{(interacting field),} \\ 0 & \text{(free field).} \end{cases} \tag{18.12}$$

As discussed in Section 16.4, under a local gauge transformation $A^\mu \to A'^\mu = A^\mu - \partial^\mu \chi$, we see that Eq. (18.10) is invariant but Eq. (18.11) is not, precisely because of the mass term $m^2 A^\nu$. Therefore, *photons must be identically massless because of gauge invariance,* and this generalizes to non-abelian gauge invariance where one finds that the *gauge bosons associated with Yang–Mills fields must also be identically massless.*

18.3.2 Screening Currents and Effective Mass

An object's mass may be determined either gravitationally (weigh it) or inertially (push it). By the weak equivalence principle of general relativity the inertial and gravitational masses are equivalent, so we lose no generality by thinking of mass in inertial terms. But it is well known that objects can acquire an effective mass (resistance to changes in the state of motion) because of interaction with a medium. Think of the contrast between pushing a metal ball through the air and pushing it through a container of honey, for example. Furthermore, modern quantum field theory has taught us that the vacuum is not a bare stage but is itself a medium with properties that can be as complex as any physical medium. Therefore, let us look more carefully at the issue of mass and gauge invariance to see whether interactions with the vacuum can impart an effective mass to gauge bosons, thereby providing some intuitive understanding of the Higgs mechanism.

First notice that for the *massless* vector field interacting through a current j^μ in Eq. (18.10), if the current takes the special form that it is proportional to the 4-vector potential,

$$j^\nu = -\mu^2 A^\nu, \tag{18.13}$$

then from (18.10) and (18.13) for the photon field

$$(\Box + \mu^2)A^\nu - \partial^\nu(\partial_\mu A^\mu) = 0. \tag{18.14}$$

But this is exactly Eq. (18.11) with $j^\nu = 0$, which is the equation obeyed by a *free, massive* vector field of mass μ.

> An *interacting, massless* vector field, with an interaction specified by a current proportional to the 4-vector potential as in Eq. (18.13), has the same equation of motion as a *free, massive* vector field of mass μ.

This is of more than academic interest because screening currents associated with atomic diamagnetism and the Meissner effect in superconductivity have precisely the form (18.13), as we shall now discuss, guided by the presentation in Aitchison and Hey [7].

18.3.3 Atomic Screening Currents

For a free, non-relativistic particle of charge q the 3-current is

$$\boldsymbol{j} = \frac{q}{2m}\left[\psi^*(-i\boldsymbol{D}\psi) + \psi(i\boldsymbol{D}\psi^*\right], \tag{18.15}$$

where the use of the covariant derivative in place of the normal derivative

$$\boldsymbol{\nabla} \to \boldsymbol{D} \equiv \boldsymbol{\nabla} - iq\boldsymbol{A}, \tag{18.16}$$

(the *minimal substitution*, see Section 16.3), ensures gauge invariance. Expanding (18.15) using (18.16) and parameterizing the complex wavefunction ψ in terms of a real modulus and phase, $\psi = |\psi| e^{i\theta}$, permits the 3-current (18.15) to be written as

$$\boldsymbol{j} = \frac{q}{m} |\psi|^2 (\boldsymbol{\nabla}\theta - q\boldsymbol{A}). \tag{18.17}$$

This current is invariant under the gauge transformation

$$\boldsymbol{A} \to \boldsymbol{A} + \boldsymbol{\nabla}\chi \equiv \boldsymbol{A}' \qquad \theta \to \theta + q\chi \equiv \theta', \tag{18.18}$$

for an arbitrary scalar function $\chi = \chi(x)$. As shown in Problem 18.3, the general description of atomic screening currents requires simultaneous solution of the equations

$$\nabla^2 \boldsymbol{A} = -\boldsymbol{j} \qquad -\frac{1}{2m}(\boldsymbol{\nabla} - q\boldsymbol{A})^2\psi = E\psi, \tag{18.19}$$

where the first expression follows from Maxwell's equations for the static limit in Coulomb gauge with the current (18.17), and the second from the Schrödinger equation with minimal substitution to ensure gauge-invariant coupling, with ψ the electron wavefunction and m the electron mass. We may gain considerable insight into solutions of these equations in a weak field approximation that assumes ψ to not be appreciably disturbed by $\boldsymbol{A}$.

The first term ($\sim \boldsymbol{\nabla}\theta$) in Eq. (18.17) is associated with paramagnetism and does not concern us here, but the second term is of potential interest because it is responsible for *diamagnetic screening currents.*[4] Denoting this term by $\boldsymbol{j}_{\rm s}$,

$$\boldsymbol{j}_{\rm s} \equiv \frac{-q^2}{m} |\psi|^2 \boldsymbol{A}, \tag{18.20}$$

we observe it to be of the form $\boldsymbol{j} \propto \boldsymbol{A}$ suggested by Eq. (18.13). Therefore, by the arguments of Section 18.3.2 this current will be expected to give an effective mass to photons interacting with electrons in the system described by Eqs. (18.19). Equation (18.20) and Maxwell's equations imply a differential equation (Problem 18.4)

$$\nabla^2 \boldsymbol{B} = \frac{q^2}{m} |\psi|^2 \boldsymbol{B}. \tag{18.21}$$

Comparing with a static massive vector field $\boldsymbol{V}$, which would obey an equation of the form $\nabla^2 \boldsymbol{V} = \mu^2 \boldsymbol{V}$, we see that the effective photon mass generated by the screening current $\boldsymbol{j}_{\rm s}$ is

$$\mu = \sqrt{\frac{q^2}{m} |\psi|^2}, \tag{18.22}$$

[4] The responses of paramagnets and diamagnets are caused by the nature of the screening currents induced in the material by an external magnetic field. In paramagnetic material the screening currents produce a magnetic field sympathetic to the external field and thus paramagnetic material is attracted by a magnet. Diamagnetic material produces a screening current that opposes the external field, so it is repelled by a magnet.

and the solution of Eq. (18.21) in one dimension is of the form $B_x = B_0 e^{-\mu x}$, corresponding to a characteristic screening length

$$\lambda \sim \frac{1}{\mu} = \sqrt{\frac{m}{q^2 |\psi|^2}}, \tag{18.23}$$

for exponential decay of the magnetic field as it penetrates the superconductor.

Example 18.1 Assume a number density of one conduction electron per atom and approximate $|\psi|^2 \sim (a_0)^{-3}$, where a_0 is the Bohr radius. Then the screening length λ is several hundred angstroms in typical atomic systems, much larger than atomic dimensions.

We conclude that diamagnetic screening is small in normal matter. But in a superconductor the coherent wavefunction can be spatially extended and largely undisturbed by the magnetic field. As we now discuss, diamagnetic screening can then have a dramatic effect.

18.3.4 The Meissner Effect and Massive Photons

If a superconductor above the superconducting transition temperature T_c is placed in a magnetic field and the temperature lowered, below T_c surface electrical currents flow without resistance and the magnetic field is expelled from the superconductor, except for a thin surface layer where the field penetrates with exponentially decaying strength. This *Meissner effect* is illustrated in Fig. 18.2(a,b) and the characteristic length scale over which the field falls off near the surface (*London penetration depth*) is illustrated in Fig. 18.2(c).

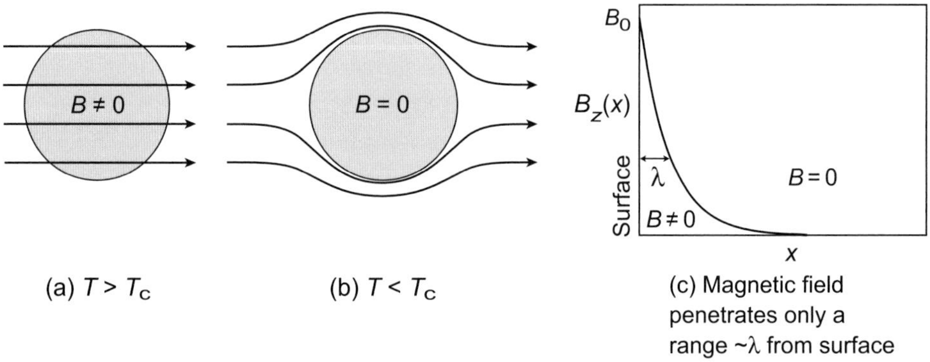

Fig. 18.2 Meissner effect for a superconductor. (a) Sample in a weak magnetic field but with the temperature T above the critical temperature T_c. The sample is not superconducting and is penetrated by the magnetic field. (b) Sample cooled below T_c so that it is superconducting. The magnetic field is expelled from the superconductor, except in a thin layer at the surface. (c) The magnetic field penetrates the superconductor with exponentially decaying strength over a range characterized by the London penetration depth λ of Eq. (18.23).

The Superconducting Wavefunction: In a superconductor the charge carriers are paired electrons called *Cooper pairs* (Box 32.2). Below the superconducting transition temperature the number density n_c of Cooper pairs is a large fraction of the total number density of conducting electrons n and $n_c \simeq \frac{1}{2}n$ for $T \ll T_c$ (each Cooper pair has two electrons).

Example 18.2 The macroscopic pair wavefunction $\psi(x)$ of the superconductor may be approximated by

$$\psi(x) \simeq \left|\psi\right| e^{i\chi(x)} \simeq n_c^{1/2} e^{i\chi(x)}.$$

Well below the superconducting transition temperature T_c the pair density is $\left|\psi\right|^2 = n_c \simeq \frac{1}{2}n \propto a_0^{-3}$, where we have assumed one conducting electron per atomic site as in Example 18.1. Inserting reasonable numbers, a typical Cooper pair density is $n_c \sim 10^{22}\ \mathrm{cm}^{-3}$. As for Example 18.1, this implies a penetration depth of several hundred angstroms, but now this is *much smaller than the characteristic length scale of the superconductor,* which has a collective wavefunction with long-range coherence.

Example 18.2 shows explicitly that screening currents can exclude a magnetic field from a superconductor in the manner illustrated in Fig. 18.2.

Screening Currents and the Effective Photon Mass: Magnetic interactions result from exchange of virtual photons. Exponential decay of a force with distance indicates exchange of a *massive* virtual particle mediating the interaction, with the distance to fall by $1/e$ given by $\lambda \sim 1/\mu$, where μ is the effective mass of the exchange particle.

> In the Meissner effect photons acquire an effective mass within a superconductor, which limits the magnetic field to a finite range. This may be viewed as a nonrelativistic version of the Higgs mechanism

From the estimate of a few hundred angstroms for λ obtained in Example 18.2, the effective mass of the photon propagating within the superconducting vacuum is $mc^2 \sim 5\,\mathrm{eV}$. This exercise suggests that the Higgs mechanism of elementary particle physics may be construed as a kind of "superconductivity" viewed from *within* a "superconducting medium" that corresponds to the entire Universe [212].

Diamagnetic Currents and the Lenz Law: The Meissner effect is made plausible by the empirical *Lenz law* of classical electromagnetism: the current induced in a conductor by applying a magnetic field is in a direction such that the magnetic field created by the induced current opposes the applied field. Thus, a magnetic field applied to mobile charged particles accelerates the particles and the resulting screening currents create magnetic fields that oppose the applied field. Screening is decisive in the Meissner effect because the superconducting wavefunction has a large spatial extent relative to the screening length, and because the screening currents encounter no resistance so they are persistent.

18.3.5 Gauge Invariance and Longitudinal Polarization

The preceding discussion of diamagnetic screening currents in atomic systems has helped to paint a physical picture of the Higgs mechanism but it has not provided details of (1) how the photon acquires a third state of polarization, and (2) how exactly the Goldstone bosons were vanquished. In this section we shall interpret the Higgs mechanism for a superconductor in a manner that elucidates these issues.

Cooper Pair Screening Currents: The electrons in a superconductor are fermions but the Cooper pairs responsible for normal superconductivity have their spin angular momentum coupled to zero and behave approximately as charged scalar bosons at low energy. Thus, a complex scalar field, as embodied in Eq. (18.1), is a reasonable approximate description. The electrical current associated with a charged scalar field is

$$j^\mu \equiv iq[\phi^* \partial^\mu \phi - (\partial^\mu \phi)^* \phi], \tag{18.24}$$

where the partial derivatives are presumed to act only within the parentheses in expressions like $(\partial^\mu \phi)^* \phi$. Introducing gauge invariance for this current by the minimal substitution $\partial^\mu \to D^\mu \equiv \partial^\mu + iqA^\mu$ gives for the charged scalar current

$$j^\mu = iq[\phi^* \partial^\mu \phi - (\partial^\mu \phi^*)\phi] - 2q^2 A^\mu |\phi|^2 . \tag{18.25}$$

Therefore, in the superconductor ground state (18.6) the vacuum screening current $\langle j^\mu \rangle_0$ of Cooper pairs is

$$\begin{aligned}\langle j^\mu \rangle_0 &= iq\left(\frac{v^2}{2} e^{-i\alpha(x)} \partial^\mu e^{i\alpha(x)} - \frac{v^2}{2}(\partial^\mu e^{-i\alpha(x)}) e^{i\alpha(x)}\right) - 2q^2 A^\mu \left(\frac{v^2}{2}\right) \\ &= -q^2 v^2 \left(\frac{1}{q}\partial^\mu \alpha + A^\mu\right).\end{aligned} \tag{18.26}$$

Inserting this current in Eq. (18.10) gives

$$\Box A^\nu - \partial^\nu (\partial_\mu A^\mu) = -q^2 v^2 \left(\frac{1}{q}\partial^\nu \alpha + A^\nu\right). \tag{18.27}$$

Now, if we define the gauge transformation [see the second equation in (18.8)]

$$A'^\nu = A^\nu - \partial^\nu \chi = A^\nu + \frac{1}{q}\partial^\nu \alpha, \tag{18.28}$$

Eq. (18.27) may be expressed as

$$\Box A'^\nu - \partial^\nu (\partial_\mu A'^\mu) = -q^2 v^2 A'^\nu, \tag{18.29}$$

which becomes, upon defining an effective mass $\mu \equiv qv$ and dropping primes,

$$(\Box + \mu^2) A^\nu - \partial^\nu (\partial_\mu A^\mu) = 0. \tag{18.30}$$

Taking the 4-divergence of both sides gives $\partial_\mu A^\mu = 0$ (see Problem 18.2); thus,

$$(\Box + \mu^2) A^\mu = 0. \tag{18.31}$$

This is the wave equation (18.12) for a free *massive* vector field, but it is here obeyed by a vector field that, by construction, is a gauge-invariant electromagnetic 4-vector potential that would be massless in the absence of the Higgs mechanism.

Goldstones and Third States of Polarization: The preceding derivation shows explicitly the origin of the third state of polarization for the effectively massive photon in the superconductor, and the fate of the Goldstone boson that would have been expected from breaking the global continuous phase symmetry spontaneously. It is clear from Eq. (18.28) that the additional polarization state for A^μ arises from using gauge invariance to absorb the phase α of the superconductor vacuum state into the 4-vector potential. From discussion of the Nambu–Goldstone mode in Ch. 17, this degree of freedom would have become a Goldstone boson (corresponding to motion of the χ field in Fig. 17.4 with zero restoring force, implying that it is a massless mode), were it not for the Higgs mechanism. Instead, because of the local gauge invariance the would-be Goldstone degree of freedom has been absorbed by the photon, converting it into a massive vector field with three states of polarization. As a result, α does not appear explicitly in the final expression (18.31), which describes a *massive* photon, the condensate of Cooper pairs that plays the role of a massive Higgs field, and no massless Goldstone fields.

Further, if we pursued more advanced issues we would find that the Higgs mechanism does not compromise renormalizability of the gauge theory, and that the original gauge invariance is no longer manifest but can still be discerned through specific relations among parameters of the theory. The gauge symmetry has not actually been broken, it has been hidden. The abelian Higgs mechanism generalizes to non-abelian local gauge symmetries (Yang–Mills fields) that were introduced in Ch. 16 and will be discussed further in Ch. 19. Therefore, we come to a remarkable conclusion.

> In abelian and non-abelian local gauge theories spontaneous symmetry breaking can endow gauge bosons with mass, without producing troublesome massless Goldstone bosons and without compromising renormalizability.

Spontaneous breaking of a local symmetry also implies the emergence of new particles: massive scalars (Higgs bosons) are expected if spontaneously broken local gauge symmetry is realized in nature. Let us address this final issue.

18.4 The Higgs Boson

We have shown that the Higgs mechanism can be interpreted in terms of concepts from non-relativistic field theory. However, the most important use of the Higgs mechanism is for relativistic quantum field theory applied to elementary particle physics, where it is postulated that the vacuum state of the Universe itself is a kind of relativistic superconductor in which there are non-zero scalar fields that break local gauge symmetries spontaneously in the Higgs mode. In Ch. 19 we shall find that the requisite Higgs fields are central to all modern understanding of elementary particles. In particular, a Higgs field is presently the only known way to give mass consistently to either gauge bosons or matter (fermion) fields. Thus, in our present understanding *all mass in the Universe for non-interacting particles is a consequence of Lorentz-invariant coupling to Higgs fields.*[5]

For a number of decades it has been clear that the vacuum state of the Universe must contain Higgs fields, or something that mimics their properties. As will be seen in Ch. 19, the entire Standard Model of elementary particle physics is predicated on their existence and the many successful precision tests of the Standard Model would be inexplicable if there were no Higgs fields. However, it remained an open question as to whether Higgs fields are fundamental (meaning that the Higgs boson is a true elementary particle, detectable as a narrow resonance), or whether Higgs fields are a consequence of many-body interactions among fields at a more fundamental level that produce "condensates" behaving like effective scalar fields. (See the discussion in Section 18.3, where the Cooper pair condensate of electrons acted as an effective Higgs field for a superconductor.) This question was answered in 2012 when the Large Hadron Collider reported the robust detection of a 125.1 GeV resonance having the properties of the long-sought Higgs boson.

Background and Further Reading

The Higgs mechanism was first discussed by Anderson [12], Englert and Brout [58], Guralnik, Hagen, and Kibble [100], and Higgs [109, 110]. Modern textbook discussions may be found in Cheng and Li [41], Guidry [85], Ryder [174], and Quigg [166]. The discussion of screening currents as non-relativistic analogs of the Higgs mechanism was strongly influenced by the presentation in Aitchison and Hey [7].

Problems

18.1 Show that the expansion (18.7) substituted into Eq. (18.1) gives a Lagrangian density for which there appear to be five degrees of freedom: one from a massive scalar η, one from a massive scalar ξ, and three from a massive photon A_μ. Argue that this is unphysical because the original Lagrangian density had only four degrees of freedom. Show that the gauge transformation (18.8) eliminates the explicit ξ degree of freedom and that the resulting Lagrangian density is Eq. (18.9), with four degrees of freedom: one from a massive Higgs boson and three from a massive photon. ***

18.2 Prove that taking the 4-divergence of both sides of Eq. (18.11) leads to the constraint $\partial_\mu A^\mu = 0$, so that Eq. (18.11) reduces to Eq. (18.12). ***

[5] This statement applies specifically to *non-interacting particles.* Most of the effective mass/energy in the actual Universe is a result of interactions. This statement also entails a caveat that the nature of (tiny but non-zero) neutrino masses remains unclear as of this writing in 2021 (see the discussion in Section 14.6).

18.3 Show that the current $\boldsymbol{j}$ in Eq. (18.17) is invariant under the gauge transformation $\boldsymbol{A} \to \boldsymbol{A} + \boldsymbol{\nabla}\chi \equiv \boldsymbol{A}'$ and $\theta \to \theta + q\chi \equiv \theta'$, where $\chi = \chi(x)$ is an arbitrary scalar function. Assume the static limit (no time dependence) and work in the Coulomb gauge (14.12). Show that a general solution of the atomic screening-current problem requires solution of the coupled equations

$$\nabla^2 \boldsymbol{A} = -\boldsymbol{j} \qquad -\frac{1}{2m}(\boldsymbol{\nabla} - q\boldsymbol{A})^2\psi = E\psi,$$

where the first equation describes the electromagnetic field and the second describes the motion of the electrons.

18.4 Show the validity of Eq. (18.21) by taking the curl ($\boldsymbol{\nabla}\times$) of Eq. (18.20) and using Ampere's law (14.1d) assuming static fields. *Hint*: Use $\boldsymbol{B} = \boldsymbol{\nabla} \times \boldsymbol{A}$ from Eq. (14.3), the identity $\boldsymbol{\nabla} \times (\boldsymbol{\nabla} \times \boldsymbol{B}) = \boldsymbol{\nabla}(\boldsymbol{\nabla} \cdot \boldsymbol{B}) - \nabla^2 \boldsymbol{B}$, and the Maxwell equation $\boldsymbol{\nabla} \cdot \boldsymbol{B} = 0$.

19 The Standard Model

Chapters 16–18 introduced the tools required to develop a theory of the weak interactions based on local gauge symmetry. In this chapter we formulate that theory using Yang–Mills fields and the Higgs mechanism to break local gauge symmetry spontaneously. As a bonus, we will find that this framework can partially unify the weak and electromagnetic interactions in a *Standard Electroweak Model.* We will then extend the idea of local gauge invariance to implement a theory of the strong interactions called *quantum chromodynamics (QCD)* that employs a local SU(3) symmetry built on a quark and gluon degree of freedom called *color.* This combined local gauge theory of the electromagnetic, weak, and strong interactions is termed the *Standard Model* of elementary particle physics.

19.1 The Standard Electroweak Model

Let us first consider a local gauge theory of the electromagnetic and weak interaction sector. We begin by summarizing the rich phenomenology of the weak interactions.

19.1.1 Guidance from Data

The low-energy properties of the weak interactions summarized in Box 19.1 suggest that they are mediated by exchange of virtual spin-1 bosons. Observation of charged weak currents requires two of these to be charged ($W^{\pm}$); observation of neutral weak currents requires a third neutral vector boson, Z^0. The universality of the weak interactions argues for a theory based on Yang–Mills fields but their short range implies that $W^{\pm}$ and Z^0 must be massive, which would break gauge invariance and compromise renormalizability if we simply added mass terms to the Lagrangian density. However, we have seen in Ch. 18 that massive gauge bosons can be introduced without spoiling renormalizability by breaking local gauge symmetry spontaneously in the Higgs mode. A general recipe for constructing such a theory may be given.

1. Choose a gauge group and assign fermions to representations consistent with low-energy phenomenology and with renormalizability.
2. Introduce scalar fields to give masses eventually to (and only to) those gauge bosons that are massive, and to all massive fermions by the Higgs mechanism.

Box 19.1

Phenomenology of the Weak Interactions

By the 1950s some consensus had been reached that weak interactions were described by the *Fermi current–current* Lagrangian density

$$\mathscr{L}_{\mathrm{F}} = -\frac{G_{\mathrm{F}}}{\sqrt{2}} J^{\mu}(x) J^{\dagger}_{\mu}(x),$$

where the Fermi coupling constant is $G_{\mathrm{F}} = 1.166 \times 10^{-5}\ \mathrm{GeV}^{-2}$ and the current is

$$J^{\mu}(x) = l^{\mu}(x) + h^{\mu}(x),$$

with l^{μ} the leptonic and h^{μ} the hadronic contributions.

Current–Current Interactions

The Lagrangian density $\mathscr{L}_{\mathrm{F}}$ implies three kinds of interactions: (1) leptonic–leptonic, (2) leptonic–hadronic, and (3) hadronic–hadronic. We shall illustrate using leptonic–leptonic interactions. Because only left-handed neutrinos participate in the weak interactions, the leptonic current is of *vector minus axial vector* $(V - A)$ form (see Table 14.2). Restricting to the leptons of generation I in Fig. 19.1 for illustration,

$$l^{\mu}(x) = \bar{e}(x)\gamma^{\mu}(1 - \gamma_5)\nu_{\mathrm{e}}(x),$$

and a representative matrix element of the leptonic current is

$$\langle e|\, l^{\mu}\, |\nu_{\mathrm{e}}\rangle \sim \bar{u}_{\mathrm{e}}\gamma^{\mu}(1 - \gamma_5)u_{\nu},$$

where e and ν_{e} are fields and $\bar{u}_{\mathrm{e}}$ and ν_{e} are spinors for electrons and neutrinos.

Intermediate Vector Bosons

The Fermi theory works at low energy but the point-like nature of the interaction leads to unacceptable violation of unitarity (conservation of probability) at very high energy. Initial attempts to fix this introduced exchange particles called *intermediate vector bosons* to mediate the interaction, by analogy with electromagnetism being mediated by exchange of virtual photons. Unlike electromagnetic interactions, weak interactions can transfer charge. Thus, some intermediate vector bosons must be charged and a minimal description of *weak charged currents* corresponds to adding vector bosons coupled to the current–current Lagrangian density by hand:

$$\mathscr{L} = g_{\mathrm{w}}[J^{\mu}(x)W^{+}_{\mu}(x) + J^{\mu\dagger}(x)W^{-}_{\mu}(x)],$$

where $W^{\pm}_{\mu}(x)$ denotes fields of the charged vector bosons $W^{\pm}$ and g_{w} is a coupling strength [compare Eq. (16.44) for QED]. But unlike photons, which are massless and can mediate long-range interactions, intermediate vector bosons must be quite massive because weak interactions have very short range. Thus the Fermi theory with massive exchange bosons added by hand cannot satisfy gauge invariance as massless photons do (Section 16.4), and ultimately cannot be renormalized.

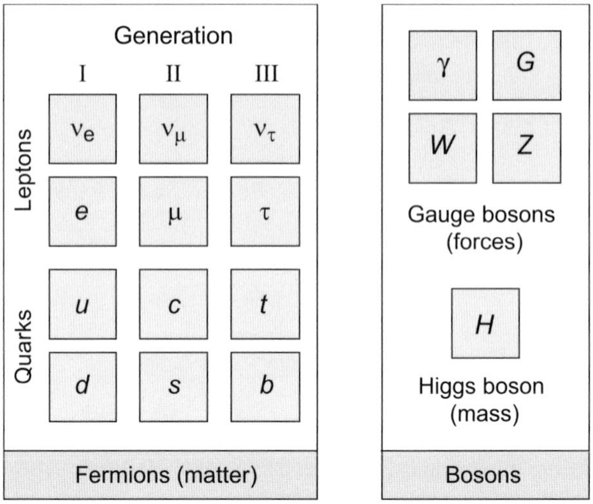

Fig. 19.1 Elementary particles of the Standard Model in the quark and lepton sectors for each generation, and the gauge and Higgs bosons. Photons are labeled by γ and gluons by G. Elementary particles of half-integer spin that do not undergo strong interactions are called *leptons*; electrons and electron neutrinos are examples. Particles made from quarks, antiquarks, and gluons (and thus that undergo strong interactions) are called *hadrons*; pions (pi mesons) and protons are examples. A subset of hadrons corresponding to more massive particles containing three quarks are called *baryons*; protons and neutrons are examples. The different types of neutrinos (ν_e, ν_μ, ...) and the different types of *quarks* ($u, d, s, \ldots$) are called *flavors*. For simplicity the largely parallel classification of antiparticles has been omitted. Reproduced with permission from Cambridge University Press: *Stars and Stellar Processes*, M. Guidry (2019).

3. Choose parameters to break the gauge symmetries spontaneously in Higgs mode, consistent with phenomenology.

Let us now implement this recipe to construct a local gauge theory of weak interactions.

19.1.2 The Gauge Group

The first choice to be made is that of the weak gauge group. The elementary particles of the Standard Model are summarized in Fig. 19.1. A clue is provided by the observations that only left-handed neutrinos participate in the weak interactions and that the weak charged currents induce transitions within generations (also called families) labeled by I, II, and III, but not across generational lines. Each leptonic generation consists of a charged lepton and its associated neutrino (e^- and ν_e for leptonic generation I), suggesting that left-handed matter fields should transform as 2D irreps (doublets) under some gauge group. The simplest possibility is the fundamental representation of SU(2),[1] with the leptonic assignments

[1] This symmetry is termed *weak isospin* and denoted $SU(2)_W$. It has a mathematical structure analogous to, but should not be confused physically with, strong interaction isospin. There will also be similar assignments for antiparticles, but for simplicity we will concentrate on the behavior of particles in this discussion.

Table 19.1. Weak isospin and weak hypercharge

Particle	t	t_3	y	Q
ν_e, ν_μ	$\frac{1}{2}$	$\frac{1}{2}$	-1	0
e_L, μ_L	$\frac{1}{2}$	$-\frac{1}{2}$	-1	-1
e_R, μ_R	0	0	-2	-1

$$t = \tfrac{1}{2} \text{ doublets:} \quad \begin{Bmatrix} t_3 = +1/2 \\ t_3 = -1/2 \end{Bmatrix} \quad \left\{ \begin{pmatrix} \nu_e \\ e^- \end{pmatrix}_L \begin{pmatrix} \nu_\mu \\ \mu^- \end{pmatrix}_L \begin{pmatrix} \nu_\tau \\ \tau^- \end{pmatrix}_L \right\}, \tag{19.1}$$

where t denotes the weak isospin quantum number and t_3 its third component, and the subscript L indicates that only left-handed fermions participate in the charged-current weak interactions. Since right-handed electrons do not enter the charged-current interactions, it may be assumed that they transform as 1D irreducible representations of SU(2) (singlets)

$$t = t_3 = 0 \text{ singlets:} \quad \left\{ (e^-)_R \quad (\mu^-)_R \quad (\tau^-)_R \right\}. \tag{19.2}$$

The gauge group $SU(2)_w$ has three generators, so three gauge boson fields transforming under the adjoint representation are required. The members of the weak isospin doublets in Eq. (19.1) differ in charge by one unit. Linear combinations of generators allow us to step through a multiplet and reach all members, so one gauge boson (W^+) must raise the charge by one unit and one (W^-) must lower the charge by one unit. In addition, the existence of neutral weak currents requires the third gauge boson (Z^0) to be uncharged.

Experiments indicate that weak neutral currents do not have a pure $V - A$ structure. That, and the prospect of unifying the weak interactions with the electromagnetic interactions, argues for expanding the group structure by adding a second neutral gauge boson. Then by taking appropriate orthogonal linear combinations we may hope to get a Z^0 gauge boson and a photon γ, with their currents in accord with weak and electromagnetic phenomenology. This suggests enlarging to the *electroweak gauge group* $SU(2)_w \times U(1)_y$, where $SU(2)_w$ is the weak isospin symmetry introduced above and the abelian $U(1)_y$ factor is associated with an additional quantum number y called the *weak hypercharge,* which is related to the electrical charge Q through

$$Q = t_3 + \frac{y}{2}, \tag{19.3}$$

where t_3 is the third component of weak isospin. A comparison of Eqs. (19.1)–(19.3) then gives the quantum number assignments in Table 19.1 for the first two generations.

19.1.3 Electroweak Lagrangian Density

The gauge group $SU(2)_w \times U(1)_y$ requires four gauge bosons, which we will label $\boldsymbol{b}_\mu = (b^1_\mu, b^2_\mu, b^3_\mu)$ for $SU(2)_w$ and a_μ for $U(1)_y$. The Lagrangian density is

$$\mathscr{L} = \mathscr{L}_g + \mathscr{L}_f + \mathscr{L}_s + \mathscr{L}_{f\text{-}s}, \tag{19.4}$$

with $\mathscr{L}_{\rm g}$ representing gauge fields, $\mathscr{L}_{\rm f}$ representing fermion fields and their coupling with gauge fields, $\mathscr{L}_{\rm s}$ representing scalar fields, and $\mathscr{L}_{\rm f\text{-}s}$ representing the fermion–scalar couplings. The gauge field part of Eq. (19.4) is

$$\mathscr{L}_{\rm g} = -\frac{1}{4}F^j_{\mu\nu}F^{\mu\nu}_j - \frac{1}{4}f_{\mu\nu}f^{\mu\nu}, \tag{19.5}$$

where the abelian field tensor $f_{\mu\nu}$ [see Eq. (14.14)] and non-abelian field tensor $F^j_{\mu\nu}$ [see Eq. (16.53)] are given by

$$f_{\mu\nu} = \partial_\mu a_\nu - \partial_\nu a_\mu \qquad F^j_{\mu\nu} = \partial_\mu b^j_\nu - \partial_\nu b^j_\mu - g\,\epsilon_{jkl} b^k_\mu b^l_\nu. \tag{19.6}$$

The contribution of the fermion matter fields is

$$\mathscr{L}_{\rm f} = \bar{R}\, i\,\gamma^\mu \left(\partial_\mu + \frac{ig'}{2}a_\mu y\right) R + \bar{L}\, i\,\gamma^\mu \left(\partial_\mu + \frac{ig'}{2}a_\mu y + \frac{ig}{2}\boldsymbol{\tau}\cdot\mathbf{b}_\mu\right) L, \tag{19.7}$$

with right-handed and left-handed components of the generation I lepton fields given by

$$R \equiv e_{\rm R} = \frac{1+\gamma_5}{2}e \qquad L \equiv \begin{pmatrix}\nu\\ e\end{pmatrix}_{\rm L} = \frac{1-\gamma_5}{2}\begin{pmatrix}\nu\\ e\end{pmatrix}, \tag{19.8}$$

where e denotes the electron field and ν the electron neutrino field, with the chiral projectors $\frac{1}{2}(1 \pm \gamma_5)$ defined in Section 14.5.4. Terms for lepton generations II and III have the same structure. *Two coupling constants* are required by the direct product group structure.

1. The coupling constant associated with the weak isospin $\mathrm{SU(2)_w}$ is g.
2. The coupling constant associated with weak hypercharge $\mathrm{U(1)_y}$ is $\frac{1}{2}g'$.

No mass terms can appear in the Lagrangian density because explicit masses for either lepton or gauge fields would break gauge invariance; all masses for free fields must come through the Higgs mechanism described in Ch. 18 (see Problem 19.2).

Phenomenology demands that the electron and the three intermediate vector bosons $(W^\pm, Z^0)$ must acquire masses spontaneously, while a fourth gauge boson (the photon γ) and the neutrino remain massless. This will require more Higgs fields than for the prototype U(1) abelian Higgs model of Section 18.2. A convenient choice is to introduce a $t = \frac{1}{2}$, $y_\phi = 1$ complex SU(2) doublet of Lorentz scalar fields (see Problem 19.1)

$$\phi \equiv \begin{pmatrix}\phi^+\\ \phi^0\end{pmatrix} = \frac{1}{\sqrt{2}}\begin{pmatrix}\phi_1 + i\phi_2\\ \phi_3 + i\phi_4\end{pmatrix}, \tag{19.9}$$

where ϕ^+ is charged and ϕ^0 is uncharged. This contributes a term

$$\mathscr{L}_{\rm s} = (D^\mu\phi)^\dagger(D_\mu\phi) - V(\phi^\dagger\phi), \tag{19.10}$$

to the Lagrangian density, with a covariant derivative defined by

$$D_\mu = \partial_\mu + \frac{ig'}{2}a_\mu y + \frac{ig}{2}\boldsymbol{\tau}\cdot\boldsymbol{b}_\mu, \tag{19.11}$$

and a potential

$$V(\phi^\dagger\phi) = \mu^2\phi^\dagger\phi + \lambda(\phi^\dagger\phi)^2, \tag{19.12}$$

with $\lambda > 0$. To give the electron mass without jeopardizing gauge symmetry the scalar field must be coupled to the fermions and the symmetry broken spontaneously using the potential (19.12). A typical choice for the fermion–scalar interaction is the gauge-invariant and Lorentz-invariant *Yukawa coupling* (Problem 19.2)

$$\mathscr{L}_{\text{f-s}} = -G_e[\bar{R}(\phi^\dagger L) + (\bar{L}\phi)R], \tag{19.13}$$

where G_e is independent of g and g'.

19.1.4 The Electroweak Higgs Mechanism

Proceeding as in earlier simple models, we may break the local gauge symmetry *spontaneously* by setting $\mu^2 < 0$ in Eq. (19.12) and choosing

$$\langle\phi\rangle_0 \equiv \langle 0|\,\phi\,|0\rangle = \begin{pmatrix} 0 \\ v/\sqrt{2} \end{pmatrix} \qquad v = \sqrt{\frac{-\mu^2}{\lambda}} \tag{19.14}$$

as the vacuum expectation value of the scalar field. It is then natural to write the generators of the gauge group in a 2×2 matrix representation. The $SU(2)_w$ generators can be taken as $t_i \equiv \frac{1}{2}\tau_i$ with the τ_i corresponding to the Pauli matrices of Eq. (3.11), and the single generator y of the $U(1)_y$ symmetry can be taken as the 2×2 unit matrix:

$$\tau_1 = \begin{pmatrix} 0 & 1 \\ 1 & 0 \end{pmatrix} \qquad \tau_2 = \begin{pmatrix} 0 & -i \\ i & 0 \end{pmatrix} \qquad \tau_3 = \begin{pmatrix} 1 & 0 \\ 0 & -1 \end{pmatrix} \qquad y = \begin{pmatrix} 1 & 0 \\ 0 & 1 \end{pmatrix}. \tag{19.15}$$

From the generators τ_3 and y it is convenient to construct two new generators

$$K \equiv \frac{\tau_3 - y}{2} = \begin{pmatrix} 0 & 0 \\ 0 & -1 \end{pmatrix} \qquad Q \equiv \frac{\tau_3 + y}{2} = \begin{pmatrix} 1 & 0 \\ 0 & 0 \end{pmatrix}, \tag{19.16}$$

and to choose as the generators of the gauge symmetry the set (τ_1, τ_2, K, Q) with

$$\tau_1 = \begin{pmatrix} 0 & 1 \\ 1 & 0 \end{pmatrix} \qquad \tau_2 = \begin{pmatrix} 0 & -i \\ i & 0 \end{pmatrix} \qquad K = \begin{pmatrix} 0 & 0 \\ 0 & -1 \end{pmatrix} \qquad Q = \begin{pmatrix} 1 & 0 \\ 0 & 0 \end{pmatrix}, \tag{19.17}$$

where Q may be interpreted as the electrical charge [see Eq. (19.3)]. By the Higgs mechanism, each gauge-symmetry generator (19.17) that fails to annihilate the vacuum will give mass spontaneously to a gauge boson. By explicit matrix multiplication (Problem 19.4),

$$\tau_1\langle\phi\rangle_0 \neq 0 \qquad \tau_2\langle\phi\rangle_0 \neq 0 \qquad K\langle\phi\rangle_0 \neq 0 \qquad Q\langle\phi\rangle_0 = 0.$$

Therefore, by virtue of spontaneous symmetry breaking in the Higgs mode, we obtain the following results.

> The symmetry is broken spontaneously to $SU(2)_w \times U(1)_y \supset U(1)_{QED}$, the vacuum remains invariant under $U(1)_{QED}$ generated by Q, three gauge bosons acquire mass, the photon remains massless, and electrical charge is conserved.

19.1.5 Particle Spectrum

By analogy with Section 18.2, the particle spectrum may be obtained by expanding around the classical vacuum. To do so it is convenient to parameterize the scalar field ϕ as [1]

$$\phi(x) = \exp\left(i\,\frac{\boldsymbol{\xi}\cdot\boldsymbol{\tau}}{2v}\right)\begin{pmatrix} 0 \\ \dfrac{v+\eta(x)}{\sqrt{2}} \end{pmatrix} \equiv U^{-1}(\boldsymbol{\xi})\begin{pmatrix} 0 \\ \dfrac{v+\eta}{\sqrt{2}} \end{pmatrix} \tag{19.18}$$

[compare Eq. (18.7)], with the four real components ϕ_i of (19.9) traded for a scalar field $\eta(x)$ and the three components of a field $\boldsymbol{\xi}(x) = (\xi_1, \xi_2, \xi_3)$.

Transformation to Unitary Gauge: The particle spectrum becomes particularly clear after a local gauge transformation that is the analog of Eq. (18.8),[2]

$$\begin{gathered} \phi \to \phi' = U(\boldsymbol{\xi})\phi = \frac{1}{\sqrt{2}}\begin{pmatrix} 0 \\ v+\eta \end{pmatrix} \\ \boldsymbol{\tau}\cdot\boldsymbol{b}_\mu \to \boldsymbol{\tau}\cdot\boldsymbol{b}'_\mu \qquad a_\mu \to a_\mu \qquad R \to R \qquad L \to L' = U(\boldsymbol{\xi})L, \end{gathered} \tag{19.19}$$

where $\boldsymbol{\tau}\cdot\mathbf{b}'_\mu$ is given by Eqs. (16.51) and (16.49). Substituting in Eq. (19.13) and dropping primes, the fermion–scalar interaction may be written (Problem 19.2)

$$\mathscr{L}_{\text{f-s}} = -G_{\text{e}}\,\frac{v+\eta}{\sqrt{2}}\,(\bar{e}_{\text{R}}e_{\text{L}} + \bar{e}_{\text{L}}e_{\text{R}}) = -\frac{G_{\text{e}}v}{\sqrt{2}}\,\bar{e}e - \frac{G_{\text{e}}}{\sqrt{2}}\,\eta\bar{e}e. \tag{19.20}$$

Comparing with the Lagrangian density (16.10) for a free Dirac field, the first term in the last expression indicates that the electron has acquired a mass $m_{\text{e}} = G_{\text{e}}v/\sqrt{2}$ spontaneously.[3] However, the electron neutrino will remain massless because it *has only left-handed components* and will not couple through Eq. (19.13).

This is how the Standard Model builds maximal parity violation into the weak interactions: by fiat only left-handed neutrinos (and right-handed antineutrinos) participate. As a corollary, Standard Model neutrinos are *identically massless.*

Substitution of Eq. (19.19) into Eq. (19.10) gives

$$\begin{aligned} \mathscr{L}_{\text{s}} &= \frac{1}{2}(\partial^\mu\eta)(\partial_\mu\eta) - \frac{1}{2}m_\eta^2\eta^2 + \frac{1}{2}m_{\text{W}}^2\left(|W_\mu^+|^2 + |W_\mu^-|^2\right) \\ &\quad + \frac{1}{2}m_{\text{Z}}^2|Z_\mu^0|^2 + \text{interaction terms}, \end{aligned} \tag{19.21}$$

[2] This new gauge is called *unitary gauge.* The physical particle spectrum of a local gauge theory with spontaneous symmetry breaking is exhibited particularly clearly in unitary gauge. Since we are building a gauge-invariant theory, a change of gauge should not affect its validity.

[3] The two terms on the right side of Eq. (19.20) have similar form. However, in the first term the coefficient $G_{\text{e}}v/\sqrt{2}$ is a constant multiplying a factor quadratic in the fields, so this is a mass term. But in the second term η is a scalar field, so this is not a mass term but instead represents coupling of the electron and scalar fields.

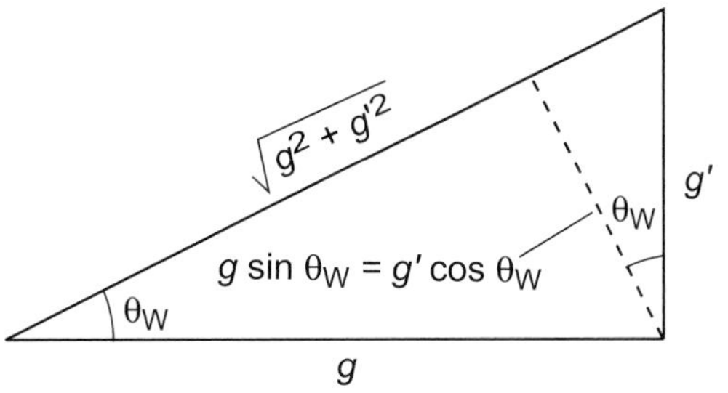

Fig. 19.2 Geometrical relationship of the Weinberg angle θ_W and the gauge coupling constants g and g' in the Standard Electroweak Model.

where we have defined two *charged gauge boson fields*

$$W_\mu^+ \equiv \frac{1}{\sqrt{2}}(b_\mu^1 - ib_\mu^2) \qquad W_\mu^- \equiv \frac{1}{\sqrt{2}}(b_\mu^1 + ib_\mu^2), \tag{19.22}$$

and two *neutral gauge boson fields*

$$Z_\mu^0 \equiv \frac{-g'a_\mu + gb_\mu^3}{\sqrt{g^2+g'^2}} \qquad A_\mu \equiv \frac{ga_\mu + g'b_\mu^3}{\sqrt{g^2+g'^2}}, \tag{19.23}$$

and where the mass terms in Eq. (19.21) are

$$m_\eta \equiv \sqrt{-2\mu^2} \qquad m_W = \frac{gv}{2} \qquad m_Z = m_W\sqrt{1+(g'/g)^2}. \tag{19.24}$$

Comparing Eq. (19.21) with the Lagrangian densities for free fields given in Section 16.1.3, we may identify (1) a scalar field η with mass m_η, (2) two charged gauge boson fields $W_\mu^\pm$, with mass m_W, (3) a massive neutral gauge boson field Z_μ^0 with mass m_Z, and (4) a massless gauge boson field A_μ. Extensive comparison of calculated interaction couplings and other properties with data indicates that these fields may be identified physically with

- the massive *Higgs boson* η,
- the three massive *intermediate vector bosons* $W^\pm$ and Z^0, and
- the *photon* A, which remains massless because of the residual U(1) charge symmetry.

These correspond to the particles in the boson sector of Table 19.1. All masses were acquired by the Higgs mechanism, so we may expect that the theory is renormalizable.

The Weinberg Angle: Because the electroweak gauge symmetry $SU(2)_w \times U(1)_y$ is a direct product of two groups there are two coupling constants, g and g', and their relationship is not specified by the theory. It is customary to parameterize the relationship between the weak isospin $SU(2)_w$ gauge coupling g and the weak hypercharge $U(1)_y$ gauge coupling g' in terms of the empirical *Weinberg angle* θ_W, which is defined by

$$\tan\theta_W = \frac{g'}{g}, \tag{19.25}$$

and illustrated in Fig. 19.2. Then Eq. (19.23) can be written (Problem 19.3)

$$Z_\mu^0 = -a_\mu \sin\theta_W + b_\mu^3\cos\theta_W \qquad A_\mu = a_\mu\cos\theta_W + b_\mu^3\sin\theta_W. \tag{19.26}$$

The Weinberg angle can be determined in various experiments and depends on the mass scale.[4] Its value evaluated at the restmass of the Z^0 gauge boson is [192]

$$\sin^2 \theta_W = 0.23122 \pm 0.00017. \tag{19.27}$$

The Weinberg angle could be specified by a more comprehensive theory based on a larger gauge group that contained $SU(2)_w \times U(1)_y$ as a subgroup. We will return to this issue when grand unified theories are discussed in Ch. 34. The Standard Electroweak Model outlined above is found to be in precise agreement with a remarkably broad range of data.

19.2 Quantum Chromodynamics

We now wish to extend local gauge symmetry to incorporate the strong interactions, thereby giving a (partially) unified picture of the electromagnetic, weak, and strong interactions.

19.2.1 A Color Gauge Theory

As suggested in Problem 9.11, there is a potential conflict with the Pauli principle for some baryonic states in the SU(3) flavor **10** representation displayed in Fig. 9.1(d). If we assume an additional degree of freedom (let us call it "color") associated with an SU(3) symmetry of the quarks that is *independent of the flavor SU(3) symmetry,* and the quarks transform as the fundamental representation under this color SU(3) symmetry, then the Pauli principle can be satisfied if the physical particles of the baryon **10** transform as a **10** with respect to flavor SU(3) but as a **1** with respect to color SU(3). The gauge theory of the strong interactions is termed *quantum chromodynamics (QCD)*, and is based on two premises.

1. The strong interactions are described by a *local gauge theory* (Yang–Mills fields).
2. The gauged symmetry is an *exact* SU(3) symmetry based on a hypothesized *color degree of freedom* carried by quarks and gluons.

We will denote this symmetry $SU(3)_c$, which should not be confused with the approximate global flavor symmetry $SU(3)_f$ discussed in earlier chapters: flavor and color are two completely different properties of strongly interacting particles. The adjoint representation of SU(3) is of dimension $N^2 - 1 = 8$, so we expect eight gauge bosons that we call *gluons.*

Universality of Color Interactions: It is assumed that the gluons see color but not flavor. Thus each flavor of quark experiences the same strong interactions, except for effects associated with differences in quark masses.

[4] The value of θ_W is a function of momentum transfer q because in the Standard Model the gauge couplings are predicted to change in specific ways with momentum (in the jargon these are called *running coupling constants* since they "run" with momentum; see Section 19.2.4). It is customary to report θ_W at momentum transfer $q = 91.2\,\text{GeV}/c$, corresponding to the restmass $m_z c^2 = 91.2\,\text{GeV}$ of the Z^0 gauge boson.

Gluon Masses: Gluons are the gauge quanta of a Yang–Mills field so they necessarily are *massless,* which would imply long-range interactions that are not observed for the strong force. This problem has two possible solutions.

1. Mimic the electroweak gauge theory and give the gluons mass through spontaneous symmetry breaking in the Higgs mode.
2. Hypothesize that the non-linear nature of the color gauge fields leads to a spatial confinement of the gluons and quarks, so that they are never observed as free particles.

There are strong hints from non-perturbative calculations that local $SU(3)_c$ is indeed confining. This, coupled with the absence of experimental candidates for free quarks or for free massive gluons and other phenomenology of the strong interactions, strongly favors color confinement.

Color Singlet States: One way to impose color confinement is to demand that all observable states be *color singlets* [must transform as the **1** representation of $SU(3)_c$], meaning that observable states carry no net color. If quarks q and antiquarks $\bar{q}$ form the **3** and $\bar{\mathbf{3}}$ representations, respectively, of color SU(3), the representations arising from the simplest quark and antiquark combinations of two or three quarks and antiquarks are

$$\begin{aligned} q\bar{q} &= \mathbf{3} \otimes \bar{\mathbf{3}} = \mathbf{1} \oplus \mathbf{8} \qquad qq = \mathbf{3} \otimes \mathbf{3} = \mathbf{6} \oplus \bar{\mathbf{3}}, \\ qq\bar{q} &= \mathbf{3} \otimes \mathbf{3} \otimes \bar{\mathbf{3}} = \mathbf{3} \oplus \bar{\mathbf{6}} \oplus \mathbf{3} \oplus \mathbf{15}, \\ qqq &= \mathbf{3} \otimes \mathbf{3} \otimes \mathbf{3} = \mathbf{1} \oplus \mathbf{8} \oplus \mathbf{8} \oplus \mathbf{10}, \end{aligned} \tag{19.28}$$

and quarks in the combinations qqq and $q\bar{q}$ can form color singlet states, corresponding to hadrons of finite mass. We can express this more formally as the requirement that all hadrons correspond to color singlets with wavefunctions of the form

$$\Psi^{ij}_{\text{meson}} \simeq \delta_{\alpha\beta}\, \bar{q}_i^{\alpha} q_j^{\beta} \qquad \Psi^{ijk}_{\text{baryon}} \simeq \epsilon_{\alpha\beta\gamma}\, q_i^{\alpha} q_j^{\beta} q_k^{\gamma},$$

where α, β, γ are color indices and i, j, k are flavor/spin indices, and implied sums over repeated color indices ensure that these are color singlet states. Then the confinement of color in hadrons and the observation of the quark combinations qqq and $q\bar{q}$ predominantly in physical hadrons is (in a sense) explained: *no observable state can have a net color.*[5]

19.2.2 The QCD Lagrangian Density

The Lagrangian density for QCD is taken to be

$$\mathscr{L}_{\text{QCD}} = \sum_f \overline{\psi}_f \left(i\not{D} - m_f \right) \psi_f - \frac{1}{4} F^j_{\mu\nu} F^{\mu\nu}_j, \tag{19.29}$$

[5] This does not address color confinement at a fundamental level. We have simply issued an edict that color is confined by restricting physical states to color SU(3) singlet representations. The proof that color is confined in QCD is a difficult strongly interacting problem that can be solved only by large-scale numerical *lattice gauge simulations.* Such calculations generally support the hypothesis that color is confined by non-linear interactions of the gluon fields.

where the covariant derivative is

$$D_\mu = \partial_\mu + \frac{i}{2} g \lambda_\ell A_\mu^\ell \qquad \not{D} \equiv \gamma^\mu D_\mu, \tag{19.30}$$

λ_ℓ denotes the SU(3) matrices of Eq. (8.2), A_μ^ℓ represents the μth spacetime component of the ℓth gauge boson vector potential with $\ell = 1, 2, 3, \ldots, 8$, flavor is labeled by f, and g is the $SU(3)_c$ gauge coupling strength. From Eq. (16.53) the gluon field tensor is

$$F_{\mu\nu}^j = \partial_\mu A_\nu^j - \partial_\nu A_\mu^j - g f_{jk\ell} A_\mu^k A_\nu^\ell, \tag{19.31}$$

where the SU(3) structure constants $f_{jk\ell}$ were given in Eq. (8.4). The quarks are color triplets, with a composite spinor for each flavor f,

$$\psi_f = \begin{pmatrix} q_{\text{red}} \\ q_{\text{blue}} \\ q_{\text{green}} \end{pmatrix}_f \equiv \begin{pmatrix} q_1 \\ q_2 \\ q_3 \end{pmatrix}_f \equiv \begin{pmatrix} R \\ B \\ G \end{pmatrix}_f . \tag{19.32}$$

Gluon coupling to all quark flavors in Eq. (19.29) is characterized by a single gauge coupling constant g, so flavor symmetry is broken only through inequivalent masses m_f.

19.2.3 Symmetries of the QCD Lagrangian Density

The QCD Lagrangian density (19.29) has several important symmetries, in addition to invariance under the discrete symmetries P, C, and T (but see Box 25.2).

1. The QCD Lagrangian density is invariant under global phase rotations $\psi_f \to e^{i\theta_f} \psi_f$, implying conservation of the number of quarks of flavor f.
2. It has approximate flavor symmetry if differences in quark masses m_f are ignored so that $\mathscr{L}_{\text{QCD}}$ is invariant under $\psi_j = U_{jk}\psi_k$, with U_{jk} a unitary matrix acting on flavor indices. Then the light quarks u and d lead to good SU(2) isospin symmetry, while including the somewhat heavier strange quark s yields approximate $SU(3)_f$ flavor symmetry.
3. If quark masses are ignored, $\mathscr{L}_{\text{QCD}}$ may be separated into terms involving only left-handed quarks q_L and terms involving only right-handed quarks q_R, using the chiral projection operators of Section 14.5.4:

$$q_L \equiv \frac{1}{2}(1 - \gamma_5)q \qquad q_R \equiv \frac{1}{2}(1 + \gamma_5)q.$$

   For the three lightest flavors (u, d, s), this implies a symmetry of the Lagrangian density under the direct product of global SU(3) flavor rotations on left-handed and right-handed fields. This $SU(3)_L \times SU(3)_R$ symmetry is called *chiral invariance*.[6]

These flavor and chiral symmetries are broken (explicitly) by the small but not zero quark masses, as discussed further in Box 19.2.

[6] Restricting to only u and d quarks leads to the chiral $SU(2)_L \times SU(2)_R$ symmetry described in Section 33.1.1.

Box 19.2

Current and Constituent Masses for Quarks

For a three-flavor quark model, explicit breaking of chiral symmetry in the QCD Lagrangian density (19.29) comes from the mass terms for up (u), down (d), and strange (s) quarks (using shorthand like $\overline{\psi}_u m_u \psi_u \equiv m_u \bar{u}u$),

$$\mathscr{L}' = m_u \bar{u}u + m_d \bar{d}d + m_s \bar{s}s,$$

where comparison with data indicates that

$$m_u \simeq 4\ \text{MeV} \qquad m_d \simeq 3\ \text{MeV} \qquad m_s \simeq 100\ \text{MeV}.$$

These masses appear directly in the QCD Lagrangian density and are called the *current masses* of the quarks. They are considerably smaller than the *constituent masses* of the quarks

$$M_u \simeq 300\ \text{MeV} \qquad M_d \simeq 300\ \text{MeV} \qquad M_s \simeq 500\ \text{MeV},$$

which are effective masses for quarks in composite hadrons. The constituent masses are larger than the current masses because they include the effect of interactions in addition to the "bare" current mass. These masses explain the approximate phenomenological symmetries of the strong interactions.

1. The approximate chiral symmetry of the strong interactions is a consequence of the current masses of the three lightest quarks being smaller than the characteristic strong interaction energy scale of several hundred MeV.
2. Approximate isospin symmetry results from u and d quarks having constituent masses that are almost equal.
3. Approximate SU(3) flavor symmetry is a result of the u, d, and s quark constituent masses being relatively similar.

Isospin SU(2) and flavor SU(3) are of utility as phenomenological symmetries. However, we have limited fundamental insight into the constituent masses responsible for them because much of the effective mass is generated by interactions that are non-perturbative and thus difficult to calculate and understand.

19.2.4 Asymptotic Freedom and Confinement

The effective coupling $\alpha(Q^2)$ for gauge interactions depends on $Q^2 \equiv -q^2$, where q^2 is the squared momentum transfer. It may be evaluated in perturbation theory by summing the contributions of relevant Feynman diagrams. The results of this complex task for quantum electrodynamics (QED) and quantum chromodynamics (QCD) are [41, 166, 174]

$$\alpha_{\text{QED}}(Q^2) = \frac{\alpha(m^2)}{1 - \dfrac{\alpha(m^2)}{3\pi} \ln\left(\dfrac{Q^2}{m^2}\right)} \qquad \alpha_{\text{QCD}}(Q^2) = \frac{12\pi}{(33 - 2N_{\text{f}}) \ln\left(\dfrac{Q^2}{\Lambda^2}\right)}, \tag{19.33}$$

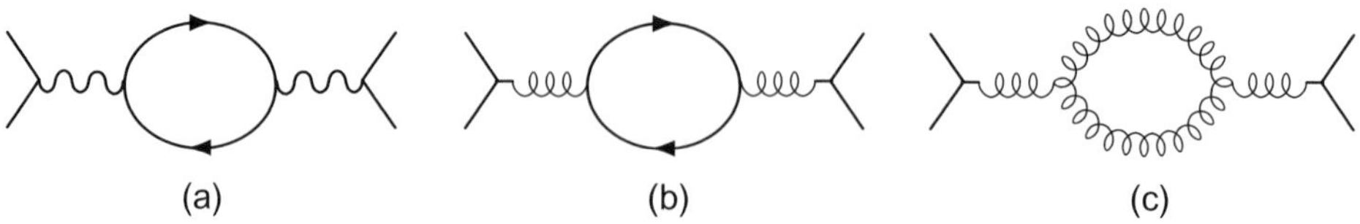

Fig. 19.3 Feynman diagrams (see Box 14.3) for vacuum polarization in abelian and non-abelian theories. Wiggly lines indicate abelian gauge bosons (photons), coiled lines indicate non-abelian gauge bosons (gluons), and solid lines indicate particles and antiparticles. The physical meaning of diagrams (a), (b), and (c) is explained in Box 19.3.

where m is the particle restmass, the momentum transfer is assumed to be much larger than the restmass, $Q^2 \gg m^2$, the QED coupling at mass scale m is $\alpha(m^2)$, the number of quark flavors energetically accessible at Q^2 is N_f, and Λ is the empirical *QCD scale parameter,* which is determined experimentally to be $\Lambda \sim 200$ MeV. Thus we see that the gauge coupling strengths vary with momentum transfer ("running coupling constants"). Comparison of QED and QCD couplings in Eq. (19.33) suggests quite different behavior: the number of known quark flavors is much less than $33/2$, so *the QCD running coupling tends to zero as* $Q^2 \to \infty$. Since by uncertainty principle arguments *large momentum transfer implies short-range interactions,* QCD interactions become *weaker* at smaller separations.

Asymptotic Freedom: This property is called *asymptotic freedom,* since the QCD interaction strength approaches zero asymptotically at large momentum transfer.

The asymptotically free behavior of QCD differs from that of QED, where interaction *increases* with momentum transfer. This difference is a consequence of gauge symmetry, as explained in Box 19.3 and Fig. 19.3. Conversely, Eq. (19.33) indicates that for increasing separation between interacting particles the strength of QCD *increases.* This is only suggestive, since Eqs. (19.33) were obtained in perturbation theory and are not valid for large interaction strength. However, non-perturbative QCD calculations and data indicate that the interaction strength indeed continues to grow with increased distance, which leads to *color confinement.*

Color Confinement: QCD coupling increases at low momentum transfer, which prevents the propagation of states that are not $SU(3)_c$ singlets. This *color confinement* precludes the appearance of free quarks or gluons in experiments.

As already noted, color confinement is thought to be a consequence of the strong non-linear gluon interactions expected in QCD at low momentum transfer.

19.2.5 Exotic Hadrons and Glueballs

Only qqq, $\bar{q}\bar{q}\bar{q}$, and $\bar{q}q$ color-singlet states are seen in typical experiments but QCD allows configurations with more than three quarks or antiquarks if they can couple to color singlet

Box 19.3

Vacuum Polarization and Local Gauge Invariance

Asymptotic freedom for QCD is a consequence of the momentum dependence of local gauge couplings arising from *vacuum polarization* processes in which gauge bosons create virtual excitations of the vacuum, as illustrated in Fig. 19.3. The difference between the momentum dependence of QED coupling and QCD coupling in Eq. (19.33) lies in the types of vacuum polarization processes that can occur, which ultimately traces to fundamental differences between the abelian U(1) gauge symmetry of QED and the non-abelian $SU(3)_c$ gauge symmetry of QCD.

Vacuum Polarization in Abelian Gauge Theories

The Feynman diagram in Fig. 19.3(a) is a typical abelian QED vacuum polarization process where (reading the diagram from left to right) interaction of two particles produces a photon (wiggly line), the photon produces a virtual particle–antiparticle excitation of the vacuum (the loop, with a particle indicated by a right arrowhead and an antiparticle by a left arrowhead), and then the virtual particle–antiparticle pair annihilates to produce another photon, which can then interact with other particles. Such virtual interactions with the vacuum modify the effective coupling in a momentum-dependent way.

Vacuum Polarization in Non-Abelian Gauge Theories

In a non-abelian gauge theory a similar vacuum polarization process as for QED occurs, with abelian photons replaced by non-abelian gauge bosons (gluons for QCD); see Fig. 19.3(b). However, for non-abelian gauge theories like QCD there is another possibility that is illustrated in Fig. 19.3(c): a particle interaction creates a gluon and the gluon produces a gluon loop corresponding to two virtual gluons, which annihilate to produce a gluon. This type of diagram can occur only for a non-abelian gauge symmetry like $SU(3)_c$ because *only in non-abelian theories can gauge bosons couple to themselves,* as required in Fig. 19.3(c).

Counting Colors

The first term in the factor $(33-2N_f)$ for the QCD coupling in Eq. (19.33) comes from evaluation of gluon loops as in Fig. 19.3(c), while the second term comes from evaluation of fermion loops, as in Fig. 19.3(b). The absence of the first term in QED is responsible for its opposite momentum dependence relative to QCD. For a universe such as ours with only a few quark flavors apparent at low energies, $(33 - 2N_f)$ is positive and $\alpha_{QCD}(Q^2)$ *decreases with increasing momentum transfer,* leading to asymptotic freedom for the strong interactions. The ultimate reason is related to dominance of gluon loops over quark loops in vacuum polarization, which depends on group theory: there are only three quark colors [fundamental representation of $SU(3)_c$], but there are eight gluons [adjoint representation of $SU(3)_c$].

states. For example, you are asked to show in Problem 19.5 that $qq\bar{q}\bar{q}$ has a color singlet in its decomposition and so is a valid multi-quark state. Such states are called *exotic hadrons.* Even more unusually, since QCD is a non-abelian theory the gauge bosons (gluons) self-interact and could bind into a multi-gluon state, either alone or in the presence of quarks. Bound states of pure gluons are called *glueballs*; bound states of gluons and quarks are called *hybrids*. Quarks transform as a **3** and gluons as an **8** under $SU(3)_c$, so the simplest combinations of gluons (G), quarks (q), and antiquarks ($\bar{q}$) yielding color singlets are

$$\begin{aligned} &(GG)_1 : (\mathbf{8} \otimes \mathbf{8})_1 \qquad &(GGG)_1 : (\mathbf{8} \otimes \mathbf{8} \otimes \mathbf{8})_1 \\ &(Gq\bar{q})_1 : [\mathbf{8} \otimes (\mathbf{3} \otimes \bar{\mathbf{3}})_8]_1 \qquad &(Gqqq)_1 : [\mathbf{8} \otimes (\mathbf{3} \otimes \mathbf{3} \otimes \mathbf{3})_8]_1, \end{aligned} \tag{19.34}$$

where the subscripts **1** and **8** denote SU(3) coupling to singlet and octet representations, respectively, and a notation like $[\mathbf{8} \otimes (\mathbf{3} \otimes \bar{\mathbf{3}})_8]_1$ means that the **3** and $\bar{\mathbf{3}}$ are first coupled to an **8**, and then that **8** is coupled to another **8** to give a **1**. A few such states have been reported experimentally but as of 2021 the vast majority of observed states are traditional mesonic and baryonic particles.

19.3 The Gauge Theory of Fundamental Interactions

In summary, the modern theory of the electromagnetic, weak, and strong interactions (the Standard Model) is a Yang–Mills theory based on the local gauge group

$$G = \mathrm{SU}(2)_{\mathrm{w}} \times \mathrm{U}(1)_y \times \mathrm{SU}(3)_{\mathrm{c}}, \tag{19.35}$$

where w denotes weak isospin, y denotes weak hypercharge, and c indicates the color degree of freedom. The electroweak sector is partially unified in the $\mathrm{SU}(2)_{\mathrm{w}} \times \mathrm{U}(1)_y$ gauge symmetry. The unification is incomplete because the direct product structure of the electroweak gauge group means that there are two independent gauge coupling constants for electroweak interactions, with the relationship between them specified by the empirical Weinberg angle θ_{W} defined in Eq. (19.25). In contrast the strong interactions are described by the $\mathrm{SU}(3)_{\mathrm{c}}$ gauge group, which is semisimple and has a single gauge coupling constant. Thus, in the Standard Model gauge group (19.35) there are three independent gauge coupling constants because of the overall direct product structure. In Ch. 34 we shall address the possibility that this gauge structure might be further unified in a larger gauge group that is not a direct product and thus could be characterized by a single universal gauge coupling strength (*grand unification*).

Background and Further Reading

Pedagogical introductions to the Standard Model may be found in Aitchison and Hey [7], Cheng and Li [41], Guidry [85], Halzen and Martin [103], Quigg [166], and Ryder [174].

Problems

19.1 Verify the weak isospin and weak hypercharge quantum number assignments in Table 19.1, given the charges Q in the last column. Verify the weak isospin and weak hypercharge assignments for the complex scalar doublet in Eq. (19.9).

19.2 Demonstrate that an explicit Dirac mass term in the electroweak Lagrangian of the form $m\bar{\psi}\psi$ as in Eq. (16.10) would violate gauge symmetry. Show that a mass introduced by breaking the gauge symmetry spontaneously using Eq. (19.13) with $\mu^2 < 0$ is gauge invariant, and obtain Eq. (19.20) from Eq. (19.13). ***

19.3 Show that the Weinberg angle θ_W is related to the coupling strengths g and g' by

$$\sqrt{g^2 + g'^2} = \frac{g}{\cos\theta_W} = \frac{g'}{\sin\theta_W},$$

for the electroweak model, and that Eq. (19.23) is equivalent to Eq. (19.26). ***

19.4 Prove that the generators (τ_1, τ_2, K) of Eq. (19.17) for the local $SU(2)_w \times U(1)_y$ standard electroweak symmetry annihilate the vacuum state but that the charge generator Q does not.

19.5 One possible exotic QCD hadronic structure is $qq\bar{q}\bar{q}$. Assuming the quarks and antiquarks to transform according to the fundamental and conjugate representations of an SU(3) color symmetry, respectively, find the color SU(3) irreps corresponding to $qq\bar{q}\bar{q}$ and show that SU(3) color singlets occur in this product.

19.6 Assume the basis vectors of the fundamental representation for color SU(3) to correspond to the "colors" r, g, and b. Write the properly symmetrized wavefunction corresponding to a singlet color SU(3) state. Show that if a gluon G transforms as the adjoint $SU(3)_c$ representation, both GG and GGG have a $\mathbf{1}$ in their SU(3) irrep content; thus, particles of this composition (glueballs) might exist in nature. ***

19.7 Demonstrate that for colors r, g, and b, color SU(3) two-quark states are of the form $qq = \mathbf{3} \otimes \mathbf{3} = \mathbf{6} \oplus \bar{\mathbf{3}}$, with

$$\bar{\mathbf{3}}\,(\text{antisymmetric}) = (rg - gr, rb - br, bg - gb),$$
$$\mathbf{6}\,(\text{symmetric}) = (rg + gr, rb + br, gb + bg, rr, gg, bb).$$

Show that this implies that the non-color part of any two-quark wavefunction in a baryon must be symmetric. ***

19.8 Verify the color SU(3) representations for combinations of three or fewer quarks and antiquarks given in Eq. (19.28).

19.9 Prove that Eq. (19.34) gives the simplest multi-gluon and gluon–quark states that contain an SU(3) color singlet in the decomposition.

20 Dynamical Symmetry

Historically the primary function of groups and algebras in physics was to impose conservation laws that have nothing directly to do with dynamics. For example, the entire machinery of angular momentum coupling and recoupling discussed in Chs. 6 and 30 is only about systematic angular momentum conservation in quantum states. Other than requiring angular momentum to be a good quantum number in all processes, it places no constraints on dynamics. However, as suggested in Section 3.6, it has been realized increasingly that Lie algebras can also determine the *dynamics* of physical systems. Such applications are much more powerful than those that merely impose conservation laws and selection rules because they go beyond constraining what is allowed to prediction of *what actually happens.* Two broad theoretical categories exploit the dynamical implications of Lie algebras: (1) *local gauge theories* and (2) *dynamical symmetries.* Local gauge fields were described in Chs. 16 and 19. In this chapter we introduce the formalism of dynamical symmetries and demonstrate their power by applying them to emergent states of graphene in a magnetic field, while Chs. 23, 31, and 32 will describe applications in various other contexts.

20.1 The Microscopic Dynamical Symmetry Method

The dynamical symmetry method applied to many-body systems uses generators of groups and subgroups to construct a microscopic theory imposing a truncation of the full Hilbert space to a tractable subspace; Fig. 20.1(a) illustrates. Such a drastic truncation is justified if it leads to correct matrix elements for physical observables, as illustrated in Fig. 20.1(b). Even though a dynamical symmetry theory and some other theory may use very different methodologies, they both must produce matrix elements of observables as physical output. Thus valid comparisons are through matrix elements; wavefunctions and operators separately are not observables and are relevant only in that they may be helpful pedagogically. We now give a formal statement of the dynamical symmetry method applied microscopically to a strongly correlated system, beginning with the following assertion [218].

> Strongly correlated states in fermion or boson many-body quantum systems imply a corresponding dynamical symmetry described by a Lie algebra obeyed by the second-quantized operators representing the emergent modes (Box 20.1).

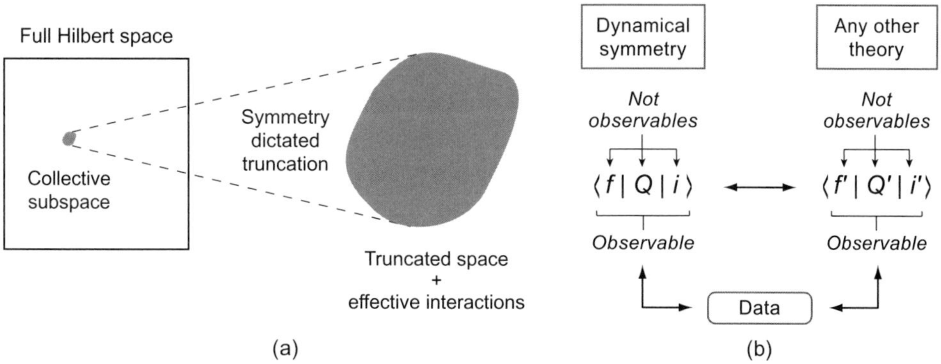

Fig. 20.1 (a) Emergent-symmetry truncation of the full Hilbert space to a collective subspace using principles of fermion dynamical symmetry. (b) Comparison of matrix elements among different theories and data. Wavefunctions and operators are *not observables*. Only *matrix elements* are related directly to experimental data and serve as valid comparison criteria.

Box 20.1 Emergent Modes in Many-Body Systems

Many topics in this book deal with *emergent states,* which are strongly correlated modes that are not present in the weakly interacting system and that *emerge* spontaneously (without external intervention) because of many-body correlations. Thus they are closely associated with spontaneous symmetry breaking. Such states are also described as *collective* (common in nuclear physics) or are said to be states with *long-range order* (common in condensed matter physics). Examples include collective rotational bands in heavy nuclei and superconductivity/superfluidity observed across many disciplines.

Emergent states are characterized by strongly collective effects, suggesting the correlated motion of many particles, and typically exhibit large values of particular matrix elements indicative of that correlation. Typically they are not connected adiabatically (not linked by a continuous series of infinitesimal changes in configuration) to states of the weakly interacting system because they are separated from weakly correlated states by phase transitions. Thus, emergent states will be prominent in the discussion of quantum phases and quantum phase transitions in Ch. 23. Comprehensive descriptions of emergent states using fermion dynamical symmetries are given in this chapter and in Chs. 31 and 32.

This conjecture is supported by a large amount of evidence from various fields of many-body physics, as we shall document.

20.1.1 Solution Algorithm

Assuming the preceding conjecture, we may find microscopic solutions for emergent states in quantum many-body systems by the following dynamical symmetry algorithm.

1. Identify a set of emergent degrees of freedom thought to be physically relevant for the problem at hand, guided by phenomenology and theory.

2. Close a commutation algebra on the second-quantized operators creating, annihilating, and counting these modes. This Lie algebra is termed the *highest symmetry*, and may be specified completely in terms of microscopic generators for emergent physical modes.

3. Identify a *collective subspace* of the full Hilbert space by requiring that matrix elements of the operators found in the preceding step do not cause transitions out of the collective subspace. This dramatic reduction of the space is termed *symmetry dictated truncation.* Collective states in this subspace are of low energy but their wavefunction components are *selected by symmetry, not energy,* and will often contain strongly correlated low-energy and high-energy components of a basis appropriate for the weakly interacting system.

4. Identify subalgebra chains of the highest symmetry ending in algebras for relevant conservation laws, like for charge and spin. Associated with these Lie algebras will be corresponding *Lie groups.* Each subalgebra chain or corresponding subgroup chain defines a *dynamical symmetry* of the highest symmetry. Multiple dynamical symmetries may be associated with a given highest symmetry, each defining a quantum phase of the system, and transitions between dynamical symmetries will be quantum phase transitions (Ch. 23).

5. Construct Hamiltonians that are polynomials in the Casimir invariants (*dynamical symmetry Hamiltonians*) for subgroup chains. Each chain defines a wavefunction basis labeled by eigenvalues of chain invariants (Casimirs and elements of the Cartan subalgebras), and a Hamiltonian that is diagonal in that basis because it is constructed from invariants. Thus, the Schrödinger equation can be solved analytically for each chain, by construction.

6. Examine the physical content of each dynamical symmetry by calculating relevant matrix elements. This is possible because of the eigenvalues and eigenvectors obtained in step 4, and because consistency requires that transition operators be related to group generators; otherwise transitions would mix irreducible multiplets and break the symmetry.

7. Construct the most general Hamiltonian in the model space as a linear combination of terms in the Hamiltonians for each symmetry group chain. Invariant operators of different subgroup chains do not generally commute, so an invariant for one chain may be a source of symmetry breaking for another. Thus the competition between different dynamical symmetries and the corresponding quantum phase transitions may be studied.

8. More ambitious formulations can be solved by (a) perturbation theory around the symmetric solutions (which corresponds to perturbation theory around a non-perturbative vacuum), (b) numerical diagonalization of symmetry breaking terms, or (c) coherent state (see Ch. 21) or other approximations to the full Hamiltonian described above in step 7. Applications in various fields are given in Refs. [24, 93, 98, 116, 117, 118, 218, 223].

20.1.2 Validity and Utility of the Approach

The *only approximation* in the microscopic dynamical symmetry approach is the Hilbert space truncation. If all degrees of freedom are incorporated the resulting theory is microscopic and exact. Practically, only select degrees of freedom can be accommodated and the effect of the excluded space must be incorporated through renormalized (effective) interactions operating in the truncated space. It follows that the utility of this approach depends on (1) making a wise choice for the relevant emergent degrees of freedom and corresponding symmetries, and (2) having sufficient phenomenological or theoretical information to specify the effective interactions. The validity of the resulting formalism then stands on whether predicted matrix elements agree with corresponding physical observables, once a small set of effective interaction parameters has been fixed by comparison with the global data set.

20.1.3 Spontaneously Broken Symmetry and Dynamical Symmetry

Spontaneous symmetry breaking (Ch. 17) is closely related to dynamical symmetry because spontaneously broken symmetry may be viewed as the restriction of a vector freely rotating in a multidimensional configuration space to precession about a limited number of axes. That is, spontaneously broken symmetry corresponds to selecting dynamically a preferred direction in an otherwise isotropic configuration space, so we may expect a close relationship among dynamical symmetries, spontaneously broken symmetries, and the appearance of collective condensates and long-range order. Later we shall use the method of generalized coherent states (see Ch. 21) to demonstrate explicitly the connection between dynamical symmetry and spontaneous symmetry breaking at a mean-field level.

20.1.4 Kinematics and Dynamics

The approach discussed above has much in common with the local gauge theories discussed in Ch. 16 (though methodologies are different), in that the symmetries in question are *internal,* associated not with the structure of spacetime but with properties of a Hamiltonian and wavefunction defined at each spacetime point. Just as gauge theories represent internal symmetries at local spacetime points and imply dynamics, the symmetries introduced here are internal symmetries that entail dynamics. That is, gauge field theories are to Poincaré invariance (Ch. 15) as the dynamical symmetries to be discussed here are to space group classifications of lattice electronic states (Ch. 5). Spacetime classifications constrain only kinematics but the new symmetries introduced here, just as for local gauge theories, do much more: they imply *dynamics.* We note in addition that the idea emerging in later discussion of dynamical symmetries based on non-abelian symmetries being richer with fewer free parameters than abelian counterparts because of constraints imposed by commutators is also a characteristic ingredient of local gauge field theories.

20.2 Monolayer Graphene in a Strong Magnetic Field

Later chapters will give various examples of applying the microscopic dynamical symmetry method outlined in Section 20.1.1 to strongly correlated fermionic systems. Here we illustrate the method by studying electronic states for monolayer graphene in a strong magnetic field, guided by the presentations in Refs. [87, 221, 223].

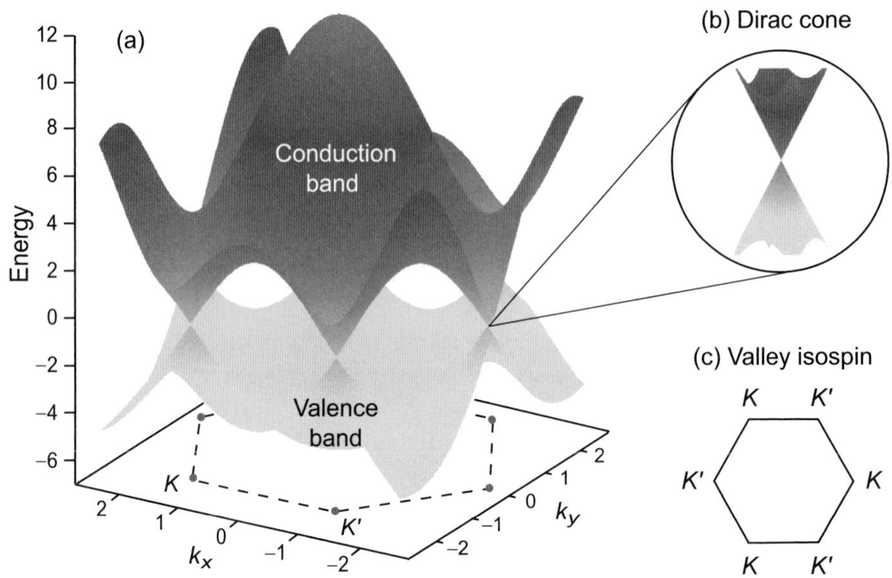

Fig. 20.2 (a) Electronic dispersion [energy versus momentum (k_x, k_y)] of graphene calculated in a tight-binding model with no magnetic field. Two inequivalent points in the Brillouin zone (points not connected by reciprocal lattice vectors; see Section 5.3) are labeled K and K'. (b) Near these *K-points* the dispersion is *linear*, leading to six *Dirac cones*. The Fermi surface for undoped graphene lies at $E = 0$ in this diagram. (c) The six minima in the conduction band at the Dirac cones are called *valleys*, which are labeled by K or K'. The two possible valley labels K or K' for an electron are termed *valley isospin*.

20.2.1 Electronic Dispersion in Monolayer Graphene

Comprehensive reviews of graphene physics may be found in Refs. [78, 157]. Here we recall only select features relevant to the present discussion. In real space undoped graphene has a bipartite 2D honeycomb lattice structure that corresponds to two interlocking triangular sublattices, labeled A and B.[1] The two-fold degree of freedom specifying whether an electron is on the A or B sublattice is termed the *sublattice pseudospin*. The electronic dispersion of graphene is illustrated in Fig. 20.2(a). Inequivalent points in the Brillouin zone are labeled K and K'. Near these points the dispersion is approximately linear, leading to the *Dirac cones* shown in the expanded view of Fig. 20.2(b). For undoped graphene the Fermi surface lies at the apex of the cones, where the level density vanishes and the effective electronic mass tends to zero.[2] Thus low-energy electrons

[1] Recall the real-space direct lattice, the momentum-space reciprocal lattice, and the Brillouin zone discussed in Ch. 5. A lattice consisting of two interlocking sublattices is called *bipartite*. By undoped we mean graphene in its pure state, without added electron donors or receptors.

[2] Graphene is an example of a *semimetal* (see Box 5.1). The valence and conduction bands just touch at the Fermi surface for the discrete Dirac cones in Fig. 20.2.

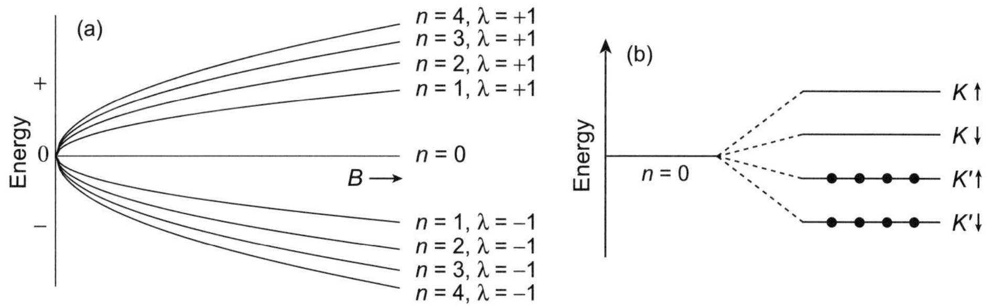

Fig. 20.3 (a) Energy for massless Dirac electrons as a function of magnetic field strength B. (b) One configuration for occupation of the $n = 0$ Landau level in monolayer graphene, where the spin of the electron is indicated by an up or down arrow and the valley quantum number is labeled by *K* or *K′*. Splitting and occupation are schematic only [221, 223]. Reproduced with permission from *Scientific Reports*: The Ground State of Monolayer Graphene in a Strong Magnetic Field, L.-A. Wu and M. W. Guidry, **6**, 22423 (2016).

are governed approximately by a Dirac equation for massless electrons in which the part of the speed of light is played by the Fermi velocity (velocity of electrons at the Fermi surface).

Whether a conduction electron is in one of the six locations for the Dirac cones is termed the *valley degree of freedom,* labeled by the two distinct values K and K' in Fig. 20.2(c). The 2D electronic spin degree of freedom and the 2D electronic valley (K) degree of freedom are most elegantly expressed in terms of independent spin and valley isospin symmetries.[3] Spin may be described by a vector of Pauli matrices $\boldsymbol{\sigma} = (\sigma_x, \sigma_y, \sigma_z)$, with the standard representation in terms of 2×2 matrices (3.11) obeying the SU(2) Lie algebra (3.12). The σ_i operate on a (Pauli) spinor basis of spin-up and spin-down electrons denoted by $|\uparrow\rangle$ and $|\downarrow\rangle$, respectively. Analogous valley isospin equations result if we define SU(2) Pauli matrix representations for the valley isospin operators $\boldsymbol{\tau} = (\tau_x, \tau_y, \tau_z)$ that operate on the valley isospinor basis with components $|K\rangle$ and $|K'\rangle$.

20.2.2 Landau Levels for Massless Dirac Electrons

Monolayer graphene in a strong magnetic field is of particular interest because it exhibits *quantum Hall effects* such as those described in Ch. 28 for 2D electron gases in semiconductor heterostructures, but with new features associated with modification of the 2D electron gas by the structure of graphene that are summarized in Box 20.2. Thus we are motivated to study massless Dirac electrons in a strong magnetic field.

Landau Levels for Graphene: Figure 20.3(a) illustrates the energy for massless Dirac electrons in a magnetic field of strength B, obtained by solving the Dirac equation with a vector potential describing the magnetic field. States are labeled by a principal quantum

[3] For brevity "valley isospin" is sometimes termed simply "isospin." Of course valley isospin has nothing to do with elementary particle isospin. It is just suggestive terminology indicating that there are two possible valley states, which can be described concisely by an SU(2) symmetry if they are degenerate.

Box 20.2 Quantum Hall Effects for Graphene in a Magnetic Field

As discussed in Ch. 28, a quantum Hall effect is signaled by a plateau in the Hall conductance having quantized values $\sigma_{xy} = \nu e^2/h$, where the *filling factor* ν is defined in Eq. (28.12). Such plateaus indicate formation of an *incompressible quantum liquid,* with a ground state separated from excited states by an *energy gap.*

Integer Quantum Hall States for Graphene

The graphene integer quantum Hall effect (IQHE) is similar to the integer quantum Hall effect in conventional 2D semiconductor heterostructures with the following exceptions.

1. In addition to the two-fold spin degeneracy (neglecting Zeeman splitting), there is a two-fold valley degeneracy associated with distinct K and K' points in the Brillouin zone. Thus filling factors change in *steps of four* between plateaus in the Hall resistance.
2. For graphene the filling factor ν defined in Eq. (28.12) vanishes for half filling of the lattice (ground state), since the electron density n_e tends to zero there. Thus no integer quantum Hall effect is expected in graphene for $\nu = 0$.

Integer quantum Hall effects have been observed in graphene for filling factors $\nu = \pm 2, \pm 6, \pm 10, \ldots$, implying Hall resistance quantization at filling factors

$$\nu = \frac{hn_e}{eB} = 4\left(n + \frac{1}{2}\right) = 4n + 2.$$

This sequence is quite different from the IQHE sequence described in Ch. 28 for a 2D electron gas, but likely results from modification of the same underlying physics by the four-fold spin–valley graphene degeneracies.

Electron–Electron Correlations in Graphene

A fractional quantum Hall effect (FQHE) requires strong electron–electron correlations, which are enhanced by high degeneracy. The Dirac cone dispersion with the Fermi level located at the apex of the cones in Fig. 20.2 implies that low-energy excitations occur in regions of reduced electron density, disfavoring correlations. But by applying a strong perpendicular magnetic field the resulting Landau quantization leads to bunching of levels into regions of locally high degeneracy that are more favorable for the development of strong correlations.

Fractional Quantum Hall States for Graphene

Landau levels (LL) become strongly correlated when low-energy excitations involve only transitions between states within the same LL. This limit has two important physical implications. (1) The four-fold spin–valley degeneracy leads to SU(4) quantum Hall ferromagnetic states, discussed in Section 20.2.3. (2) The strong correlations can produce incompressible states at partial LL filling that are reminiscent of FQHE states in semiconductor devices. Experiments at higher magnetic field strengths have observed graphene FQHE states at filling factors such as $0, \pm 1, \pm 4$.

number n and a quantum number λ indicating particle-like (+) and hole-like (−) states. Each Landau level labeled by n has a high orbital degeneracy Ω_k, and an additional four-fold degeneracy associated with spin and valley isospin. In Fig. 20.3(b) one configuration for occupying the $n = 0$ level is exhibited. The ground state would be a superposition of such configurations and in the limit of Coulomb-only interactions the four levels shown labeled by valley K or K' and spin up or down would be degenerate. This suggests an SU(4) symmetry associated with the valley isospin and spin degrees of freedom, since the fundamental representation of SU(4) consists of four degenerate states.

Degeneracies and Level Filling: The single-particle states within a single Landau level may be labeled by the quantum numbers (n, m_k), where n indicates the Landau level and m_k distinguishes degenerate states within the Landau level. In the absence of spin and valley degrees of freedom the states (n, m_k) of the Landau level hold a maximum of $2\Omega_k$ electrons, where from solution of the Dirac equation in a magnetic field

$$2\Omega_k = \frac{BS}{(h/e)}, \tag{20.1}$$

with B the strength of the magnetic field, S the area of the two-dimensional sample, and $h/e = 4.136 \times 10^{-15}$ Wb the magnetic flux quantum. But graphene has four additional internal degrees of freedom associated with the $|\text{spin}\rangle \otimes |\text{isospin}\rangle$ space, so there are four copies of each Landau level and the total electron degeneracy 2Ω is given by

$$2\Omega = 4(2\Omega_k) = \frac{4BS}{(h/e)}. \tag{20.2}$$

The *fractional occupation* f of the single Landau level may be defined as[4]

$$f \equiv \frac{n}{2\Omega} = \frac{N}{\Omega}, \tag{20.3}$$

where n is the number of electrons and $N = \frac{1}{2}n$ is the number of electron pairs. For half filling of the $n = 0$ Landau level located at the Fermi surface (ground state of undoped graphene) the electron number n_{gs} is then

$$n_{\text{gs}} = \Omega = \frac{2BS}{(h/e)}. \tag{20.4}$$

These degeneracies and occupation numbers are just the standard results for relativistic Landau levels in a 2D electron gas subject to a strong perpendicular magnetic field, but modified by the graphene valley degree of freedom.

Effective Hamiltonian: Let us consider a single $n = 0$ Landau level, with a Hamiltonian

$$H = H_{\text{C}} + H_{\text{v}} + H_{\text{Z}}, \tag{20.5}$$

where H_{C} is the (valley independent) Coulomb interaction, H_{v} is the valley isospin and spin interaction, H_{Z} is the magnetic Zeeman interaction, and the z direction for the spin space

[4] The Landau level fractional occupation f should not be confused with the quantum Hall filling factor ν defined in Eq. (28.12). For monolayer graphene the two are related by $\nu = 4(f - \frac{1}{2})$ [223].

is assumed aligned with the magnetic field [130, 220]. The short-range, valley-dependent interaction may be written

$$H_{\rm v} = \frac{1}{2} \sum_{i \neq j} \left[g_z \tau_z^i \tau_z^j + g_\perp (\tau_x^i \tau_x^j + \tau_y^i \tau_y^j) \right] \delta(\boldsymbol{r}_i - \boldsymbol{r}_j), \tag{20.6}$$

in terms of the parameters g_z and $g_\perp$. For realistic situations we may expect the Coulomb interaction to dominate, with $H_{\rm v}$ and $H_{\rm Z}$ acting as relatively small perturbations.

20.2.3 SU(4) Quantum Hall Ferromagnetism

Letting $\alpha = (x, y, z)$, $\beta = (x, y)$, and using m_k to label Landau states, the set of 15 operators

$$\mathcal{S}_\alpha = \sum_{m_k} \sum_{\tau\sigma\sigma'} \langle \sigma' | \, \sigma_\alpha \, | \sigma \rangle \, c^\dagger_{\tau\sigma' m_k} c_{\tau\sigma m_k} \tag{20.7a}$$

$$T_\alpha = \sum_{m_k} \sum_{\sigma\tau\tau'} \langle \tau' | \, \tau_\alpha \, | \tau \rangle \, c^\dagger_{\tau'\sigma m_k} c_{\tau\sigma m_k} \tag{20.7b}$$

$$N_\alpha = \frac{1}{2} \sum_{m_k} \sum_{\sigma\sigma'\tau} \langle \tau | \, \tau_z \, | \tau \rangle \langle \sigma' | \, \sigma_\alpha \, | \sigma \rangle \, c^\dagger_{\tau\sigma' m_k} c_{\tau\sigma m_k} \tag{20.7c}$$

$$\Pi_{\alpha\beta} = \frac{1}{2} \sum_{m_k} \sum_{\sigma\sigma'\tau\tau'} \langle \tau' | \, \tau_\beta \, | \tau \rangle \langle \sigma' | \, \sigma_\alpha \, | \sigma \rangle \, c^\dagger_{\tau'\sigma' m_k} c_{\tau\sigma m_k} \tag{20.7d}$$

closes an SU(4) Lie algebra that *commutes with* $H_{\rm C}$ [130, 220]. Thus, if $H_{\rm v}$ and $H_{\rm Z}$ are small compared with $H_{\rm C}$, the Hamiltonian (20.5) has an approximate SU(4) invariance that becomes exact if $H_{\rm v} = H_{\rm Z} = 0$. In Eqs. (20.7) the generator $\mathcal{S}$ represents the total electronic spin, $\boldsymbol{T}$ represents the total valley isospin, in the $n = 0$ Landau level $\boldsymbol{N}$ is a Néel vector measuring the difference in spins on the A and B sublattices, and $\Pi_{\alpha\beta}$ couples the spin and valley isospin. Four explicit symmetry breaking patterns may be identified, depending on the relative values of the valley parameters g_z and $g_\perp$, as shown in Fig. 20.4 [130, 220].

Example 20.1 From the leftmost subgroup chain in Fig. 20.4, for arbitrary non-zero values of g_z and $g_\perp$ the SU(4) symmetry is broken to $\mathrm{SU(4)} \supset \mathrm{SU(2)_s} \times \mathrm{U(1)_v} \supset \mathrm{U(1)_s} \times \mathrm{U(1)_v}$, where $\mathrm{SU(2)_s}$ is associated with global conservation of spin and $\mathrm{U(1)_s}$ with conservation of its z component, and $\mathrm{U(1)_v}$ is associated with conservation of the T_z component of the valley isospin. In the absence of Zeeman splitting spin is conserved but only the z component of the valley isospin is conserved. The full Hamiltonian (20.5) conserves only the z components of spin and valley isospin, corresponding to the subgroup $\mathrm{U(1)_s} \times \mathrm{U(1)_v}$.

The examples in Fig. 20.4 correspond to breaking the symmetry through explicit terms added to the Hamiltonian (see Section 17.2). This can describe states that differ only slightly from the symmetric SU(4) state but cannot account for highly collective states observed in this system. These represent *spontaneous* (not explicit) breaking of the SU(4) symmetry for the weakly perturbed system, which creates fundamentally new states that cannot be deformed continuously into states of the weakly perturbed system.

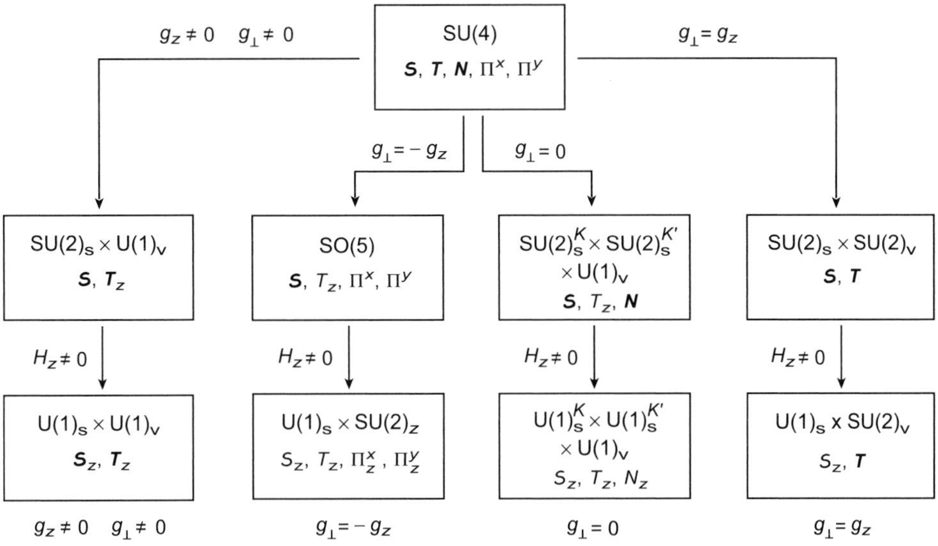

Fig. 20.4 Explicit symmetry breaking pattern for SU(4) quantum Hall ferromagnetism generated by the operators in Eq. (20.7) [130, 220]. Reproduced with permission from *Scientific Reports*: The Ground State of Monolayer Graphene in a Strong Magnetic Field, L.-A. Wu and M. W. Guidry, **6**, 22423 (2016).

(a) Quantum numbers

Valley	τ	σ	m_i	a
K	+	↑	+3/2	1
K	+	↓	+1/2	2
K'	−	↑	−1/2	3
K'	−	↓	−3/2	4

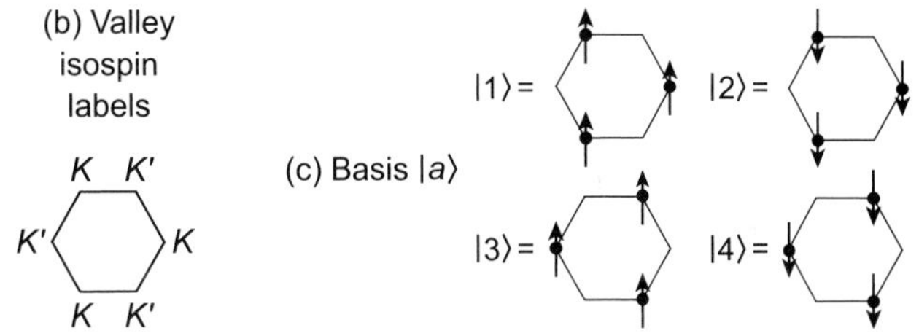

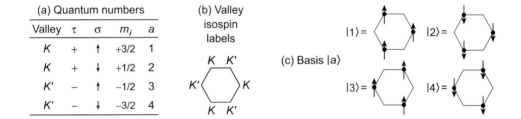

Fig. 20.5 (a) Valley isospin (τ) and spin (σ) quantum numbers. (b) The two valley isospin labels K and K'. (c) The four corresponding basis vectors.

Such emergent states can be studied numerically but we now demonstrate a dynamical symmetry solution that can recover such emergent states *analytically.* This is simpler, easier to visualize, avoids errors associated with use of small bases in many numerical simulations, and leads to new emergent modes that have not been discussed previously. We begin by enlarging the group structure through inclusion of spin–isospin pair (particle–particle and hole–hole) operators, in addition to the particle–hole operators of Eq. (20.7).

20.2.4 Fermion Dynamical Symmetries for Graphene

Spin and valley isospin quantum number assignments for a useful basis are illustrated in Fig. 20.5. The table in Fig. 20.5(a) also displays a unique mapping of these four states to a label m_i that takes the four possible projection quantum numbers $\left\{\pm\frac{1}{2},\pm\frac{3}{2}\right\}$ of a fictitious angular momentum $i = \frac{3}{2}$; the motivation for this mapping will become apparent later.

Example 20.2 Consider the basis state labeled $|2\rangle$ in Fig. 20.5(c). The electron occupies valleys labeled by valley isospin K ($\tau = +$) [see Fig. 20.5(b)] with spin down ($\sigma = \downarrow$); thus this basis state corresponds to the second row of the table in Fig. 20.5(a).

We shall use both the label a and the label m_i to distinguish the basis states in Fig. 20.5(a).

SO(8) Generators: Let us introduce an operator $A^\dagger_{ab}$ that creates a pair of electrons, one in the $a = (\tau_1, \sigma_1)$ level and one in the $b = (\tau_2, \sigma_2)$ level, with the total m_k labeling different $n = 0$ Landau states coupled to zero term by term,

$$A^\dagger_{ab} = \sum_{m_k} c^\dagger_{am_k} c^\dagger_{b-m_k}, \tag{20.8}$$

and its hermitian conjugate $A_{ab} = (A^\dagger_{ab})^\dagger$, which annihilates a corresponding electron pair. Each index a or b ranges over four values, implying 16 components in Eq. (20.8). However, the antisymmetry requirement for fermionic wavefunctions reduces this to six independent operators $A^\dagger$, with six independent hermitian conjugates A. Introduce also the 16 particle–hole operators B_{ab} through

$$B_{ab} = \sum_{m_k} c^\dagger_{am_k} c_{bm_k} - \frac{1}{4}\delta_{ab}\Omega, \tag{20.9}$$

where δ_{ab} is the Kronecker delta and Ω is the total degeneracy of the single Landau level. The commutators for the 28 operators A, $A^\dagger$, and B are found to be [39]

$$[A_{ab}, A^\dagger_{cd}] = -B_{db}\delta_{ac} - B_{ca}\delta_{bd} + B_{cb}\delta_{ad} + B_{da}\delta_{bc} \qquad [B_{ab}, B_{cd}] = \delta_{bc}B_{ad} - \delta_{ad}B_{cb}$$
$$[B_{ab}, A^\dagger_{cd}] = \delta_{bc}A^\dagger_{ad} + \delta_{bd}A^\dagger_{ca} \qquad [B_{ab}, A_{cd}] = -\delta_{ac}A_{bd} - \delta_{ad}A_{cb}, \tag{20.10}$$

which is isomorphic to an SO(8) Lie algebra.

A More Physical Basis: We wish to exploit dynamical symmetries of the SO(8) algebra (20.10), but to simplify interpretation it is useful to transform to a new basis. First note that the SO(8) particle–hole operators (20.9) can be replaced by the operators of Eq. (20.7) through a comparison of their definitions, as illustrated in the following example.

Example 20.3 Consider the spin operator $\mathcal{S}_y$. From Eq. (20.7a),

$$\begin{aligned}\mathcal{S}_y &= \sum_{m_k}\sum_{\tau\sigma\sigma'} \langle\sigma'|\,\sigma_y\,|\sigma\rangle\, c^\dagger_{\tau\sigma' m_k} c_{\tau\sigma m_k} \\ &= \sum_{m_k}\left(-ic^\dagger_{+\uparrow m_k}c_{+\downarrow m_k} + ic^\dagger_{+\downarrow m_k}c_{+\uparrow m_k} - ic^\dagger_{-\uparrow m_k}c_{-\downarrow m_k} + ic^\dagger_{-\downarrow m_k}c_{-\uparrow m_k}\right) \\ &= -iB_{12} + iB_{21} - iB_{34} + iB_{43},\end{aligned}$$

where the representation (3.11) for $\sigma_2 = \sigma_y$ was employed and equivalences between the indices a and (τ, σ) in the table of Fig. 20.5(a) were used to map to indices for B_{ab}.

By a similar procedure the other 14 operators of Eq. (20.7) may be expressed in terms of the B_{ab} operators, as shown in Problem 20.2 and Ref. [223].

Next, because spin is conserved if Zeeman splitting is ignored, and valley isospin is approximately conserved for low-lying states, it is useful to rewrite the pairing operators in a form coupled to states of good spin and isospin. An electron pair creation operator coupled to good spin and good valley isospin may be defined by (see Section 6.3)

$$A^{\dagger ST}_{M_S M_T} \equiv \sum_{m_1 m_k} \sum_{n_1 n_2} \left\langle \tfrac{1}{2}\, m_1\, \tfrac{1}{2}\, m_2 \right| S\, M_S \rangle \left\langle \tfrac{1}{2}\, n_1\, \tfrac{1}{2}\, n_2 \right| T\, M_T \rangle\, c^{\dagger}_{m_1 n_1 m_k} c^{\dagger}_{m_2 n_2 - m_k}, \tag{20.11}$$

where S is the total spin of the pair with M_S its projection, T is the total valley isospin of the pair with M_T its projection, the SU(2) Clebsch–Gordan coefficient $\langle \frac{1}{2}\, m_1\, \frac{1}{2}\, m_2 |\, S\, M_S\rangle$ couples the spins to good total spin $|SM_S\rangle$, and the SU(2) Clebsch–Gordan coefficient $\langle \frac{1}{2}\, n_1\, \frac{1}{2}\, n_2 |\, T\, M_T\rangle$ couples the valley isospins to good total isospin $|TM_T\rangle$. Antisymmetry restricts the fermion pair wavefunction to $(S = 1, T = 0)$ or $(S = 0, T = 1)$. A convenient set satisfying these conditions consists of the six coupled pairing operators,

$$\begin{gathered} S^{\dagger} = \frac{1}{\sqrt{2}}\left(A^{\dagger}_{14} - A^{\dagger}_{23}\right) \qquad D^{\dagger}_{0} = \frac{1}{\sqrt{2}}\left(A^{\dagger}_{14} + A^{\dagger}_{23}\right) \\ D^{\dagger}_{1} = A^{\dagger}_{13} \qquad D^{\dagger}_{-1} = A^{\dagger}_{24} \qquad D^{\dagger}_{2} = A^{\dagger}_{12} \qquad D^{\dagger}_{-2} = A^{\dagger}_{34} \end{gathered} \tag{20.12}$$

and the six corresponding hermitian conjugates $S = (S^{\dagger})^{\dagger}$ and $D_{\mu} = (D^{\dagger}_{\mu})^{\dagger}$ (see Problem 20.4). Since Eq. (20.12) and the SU(4) generators (20.7) expressed in terms of the operators B_{ab} (as in Example 20.3 and Problem 20.2) represent independent linear combinations of the generators of SO(8) defined in Eqs. (20.8) and (20.9), the 28 operators

$$G_{\mathrm{SO(8)}} = \{\mathcal{S}_{\alpha}, T_{\alpha}, N_{\alpha}, \Pi_{\alpha x}, \Pi_{\alpha y}, S_0, S, S^{\dagger}, D_{\mu}, D^{\dagger}_{\mu}\} \tag{20.13}$$

are also generators of SO(8) that will now be used to explore dynamical symmetry for monolayer graphene in a magnetic field.

SO(8) Pair States in Graphene: The configurations resulting from application of the creation operators to the pair vacuum are illustrated in Fig. 20.6, as are the configurations generated by the linear combinations

$$|Q_{\pm}\rangle = Q^{\dagger}_{\pm}\,|0\rangle \equiv \frac{1}{2}(S^{\dagger} \pm D^{\dagger}_{0})\,|0\rangle = \frac{1}{2}(\,|S\rangle \pm |D_0\rangle), \tag{20.14}$$

which will be useful later. The physical meaning of the states in Fig. 20.6 may be elucidated by constructing the corresponding electronic configurations, as in the following example.

Example 20.4 Consider $D^{\dagger}_{2}$ in Fig. 20.6. From Eqs. (20.12) and (20.8)

$$D^{\dagger}_{2} = \frac{1}{\sqrt{2}} A^{\dagger 10}_{10} = A^{\dagger}_{12} = \sum_{m_k} c^{\dagger}_{1 m_k} c^{\dagger}_{2 - m_k} = \sum_{m_k} c^{\dagger}_{K\uparrow m_k} c^{\dagger}_{K\downarrow - m_k},$$

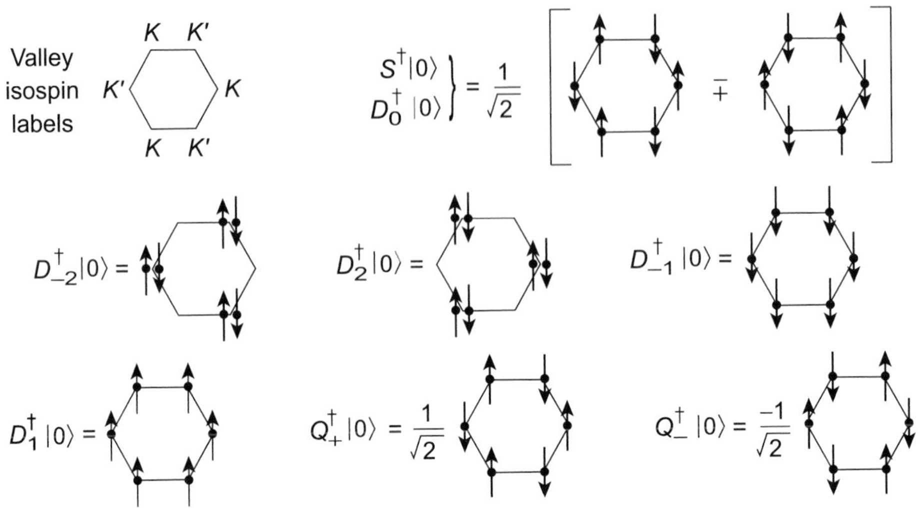

Fig. 20.6 Configurations resulting from the pair creation operators operating on the vacuum $|0\rangle$. Location of the dots (K or K' site) indicates valley isospin; arrows indicate spin polarization.

where the correspondence between the index $a = 1, 2, 3, 4$ and the valley (K or K') and spin ($\uparrow\downarrow$) labels in Fig. 20.5(a) was used. Thus $D_2^\dagger|0\rangle$ creates a state with a spin-up and a spin-down electron on each equivalent site K. Going around the hexagon the net spin is zero on each site but the charge changes from 0 to -2 between adjacent sites, which is called a *charge density wave*. Likewise $D_{-2}^\dagger$ generates a charge density wave on the K' sites.

It is convenient to characterize these states in terms of order parameters defined in the basis (20.13) (see Problem 20.6):

$$\langle \mathcal{S}_z \rangle = \langle \hat{n}_1 \rangle - \langle \hat{n}_2 \rangle + \langle \hat{n}_3 \rangle - \langle \hat{n}_4 \rangle \qquad \langle T_z \rangle = \langle \hat{n}_1 \rangle + \langle \hat{n}_2 \rangle - \langle \hat{n}_3 \rangle - \langle \hat{n}_4 \rangle$$
$$\langle N_z \rangle = \langle \hat{n}_1 \rangle - \langle \hat{n}_2 \rangle - \langle \hat{n}_3 \rangle + \langle \hat{n}_4 \rangle, \tag{20.15}$$

where $\hat{n}_i$ is the number operator counting particles in basis state $|i\rangle$ and the expectation value is taken with respect to the collective wavefunction.

1. The net spin is measured by $\langle \mathcal{S}_z \rangle$, which characterizes *ferromagnetic order.*
2. The difference in charge between the K and K' sites is measured by $\langle T_z \rangle$, which characterizes *charge density wave order.*
3. The difference in spins between the K and K' sites is measured by $\langle N_z \rangle$, which characterizes *antiferromagnetic* (also termed Néel or spin density wave) order.

The order parameters for the configurations in Fig. 20.6 are worked out in Problem 20.3.

Example 20.5 In Example 20.4 we concluded that $D_2^\dagger |0\rangle$ is a component of a charge density wave. The solution of Problem 20.3 supports this interpretation since it gives $\langle T_z \rangle = 2$ and $\langle S_z \rangle = \langle N_z \rangle = 0$ for this state, which is consistent with a state having charge density wave order, but no ferromagnetic or antiferromagnetic order.

From the order parameters we find the following physical interpretation of the operators creating the pair states in Fig. 20.6.

1. The operators $D_{\pm 2}^\dagger$ applied to the pair vacuum generate *charge density waves.*
2. The operators $D_{\pm 1}^\dagger$ applied to the pair vacuum distribute electrons equally on all sites with all spins aligned. These are components of a *ferromagnetic state.*
3. The operators $S^\dagger$, $D_0^\dagger$, and the linear combinations $Q_\pm^\dagger$ involve configurations with a single spin on each site but with spin direction alternating between adjacent sites. This is termed an *antiferromagnetic spin wave.*

The SO(8) Collective Subspace: A $2N$-particle state with no broken pairs is produced by acting N times on the pair vacuum with SO(8) pair creation operators [98, 218],

$$|\mathrm{SO(8)}\rangle = (S^\dagger)^{N_S}(D^\dagger)^{N_D}\,|0\rangle\,, \tag{20.16}$$

where $S^\dagger$ and $D^\dagger$ are defined in Eq. (20.12) and Fig. 20.6, N_S is the number of S pairs, N_D is the number of D pairs, and $N = N_S + D$.[5] The portion of the full Hilbert space spanned by the states (20.16) is the *collective subspace* of Fig. 20.1(a). Following the procedure described in Section 20.1.1, the SO(8) symmetry can be used to construct effective Hamiltonians that are diagonal in this space, and the SO(8) generators do not couple this subspace to the rest of the Hilbert space. Since the pair basis described in Fig. 20.6 exhibits finite charge density wave, ferromagnetic, and antiferromagnetic expectation values, the pair condensate (20.16) can produce a rich variety of strongly correlated states corresponding to spontaneously broken symmetries. Let us examine some of those solutions.

20.2.5 Graphene SO(8) Dynamical Symmetries

Graphene SO(8) subgroup chains that end in the group $\mathrm{SU(2)}_\sigma \times \mathrm{U(1)}_\mathrm{c}$ imposing spin and charge conservation are shown in Fig. 20.7,with Zeeman splitting ignored. If Zeeman splitting is included it will influence directly only the spin sector and break $\mathrm{SU(2)}_\sigma$ to $\mathrm{U(1)}_\sigma$ generated by S_z. Seven distinct subgroup chains are displayed. Each defines an emergent mode and corresponding quantum phase for which matrix elements and therefore observables such as energies, order parameters, and transition rates may be calculated analytically. The following examples describe several of these subgroup chains.

[5] More generally, the wavefunction can contain u broken pairs. For simplicity we consider only the lowest-energy states here, which will be assumed to correspond to $u = 0$ since those are the most collective states.

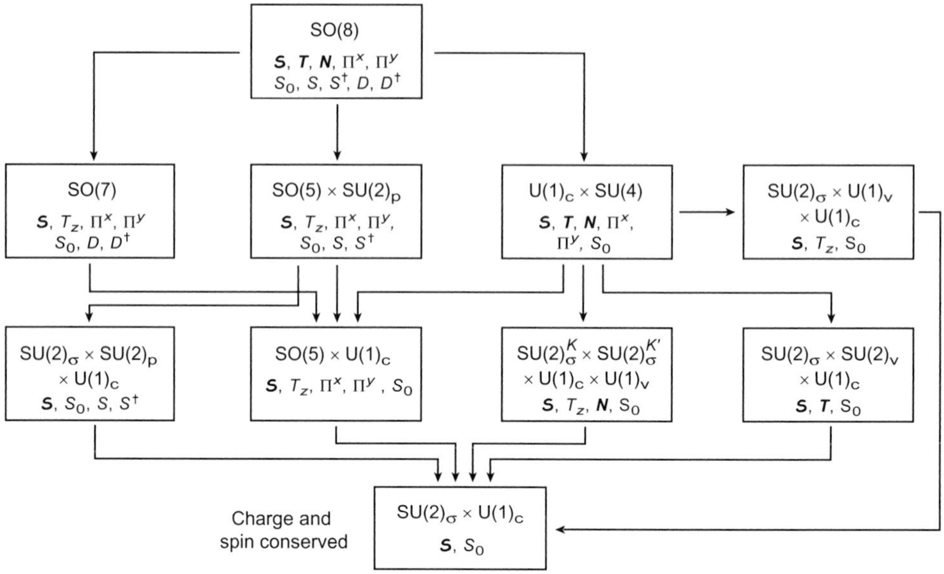

Fig. 20.7 SO(8) dynamical symmetry chains with group generators for graphene in a magnetic field [87, 221, 223]. The subgroup chains define seven distinct dynamical symmetries that correspond to different emergent states compatible with the SO(8) highest symmetry. Reproduced with permission from *Scientific Reports*: The Ground State of Monolayer Graphene in a Strong Magnetic Field, L.-A. Wu and M. W. Guidry, **6**, 22423 (2016).

Example 20.6 The generators $(S, S^\dagger, S_0)$ close an $\mathrm{SU}(2)_\mathrm{p}$ algebra and the generators $\{\mathcal{S}_\alpha, \Pi_{\alpha x}, \Pi_{\alpha y}, T_z\}$ close an SO(5) algebra and commute with the $\mathrm{SU}(2)_\mathrm{p}$ generators. Furthermore, the components of the total spin $\mathcal{S}_\alpha$ generate an $\mathrm{SU}(2)_\sigma$ subgroup of SO(5) and S_0 generates a $\mathrm{U}(1)_\mathrm{c}$ subgroup of $\mathrm{SU}(2)_\mathrm{p}$. Thus one emergent state corresponds to the dynamical symmetry subgroup chain

$$\mathrm{SO}(8) \supset \mathrm{SO}(5) \times \mathrm{SU}(2)_\mathrm{p} \supset \mathrm{SU}(2)_\sigma \times \mathrm{SU}(2)_\mathrm{p} \supset \mathrm{SU}(2)_\sigma \times \mathrm{U}(1)_\mathrm{c},$$

where the final group $\mathrm{SU}(2)_\sigma \times \mathrm{U}(1)_\mathrm{c}$ imposes conservation of spin and charge. Alternatively, SO(5) may be broken according to the dynamical symmetry pattern

$$\mathrm{SO}(8) \supset \mathrm{SO}(5) \times \mathrm{SU}(2)_\mathrm{p} \supset \mathrm{SO}(5) \times \mathrm{U}(1)_\mathrm{c} \supset \mathrm{SU}(2)_\sigma \times \mathrm{U}(1)_\mathrm{c},$$

which also conserves spin and charge but corresponds to a different emergent state.

Example 20.7 A $\mathrm{U}(4) \supset \mathrm{U}(1)_\mathrm{c} \times \mathrm{SU}(4)$ subgroup of SO(8) may be obtained by removing the 12 pairing operators from the SO(8) generator set. The $\mathrm{U}(1)_\mathrm{c}$ subgroup is generated by the particle number (charge) and the SU(4) subgroup is generated by the 15 remaining operators defined in Eq. (20.7). There are then several options for subgroup chains.

(a) The subset $\{S_\alpha, \Pi_{\alpha x}, \Pi_{\alpha y}, T_z\}$ defines generators of the SO(5) symmetry discussed in Example 20.6 and so forms an SO(5) subgroup of SU(4). Hence

$$\mathrm{SO(8)} \supset \mathrm{U(1)_c} \times \mathrm{SU(4)} \supset \mathrm{SO(5)} \times \mathrm{U(1)_c} \supset \mathrm{SU(2)}_\sigma \times \mathrm{U(1)_c}$$

is one possible SU(4) subgroup chain.

(b) If there is little inter-valley scattering the spin within each valley is separately conserved, implying $\mathrm{SU(2)}_\sigma^K \times \mathrm{SU(2)}_\sigma^{K'}$ symmetry. Thus a second SU(4) subgroup chain is

$$\mathrm{SO(8)} \supset \mathrm{U(1)_c} \times \mathrm{SU(4)} \supset \mathrm{SU(2)}_\sigma^K \times \mathrm{SU(2)}_\sigma^{K'} \times \mathrm{U(1)_c} \times \mathrm{U(1)_v} \supset \mathrm{SU(2)}_\sigma \times \mathrm{U(1)_c},$$

where $\mathrm{U(1)_v}$ is generated by T_z.

(c) A third possible SU(4) subgroup chain is

$$\mathrm{SO(8)} \supset \mathrm{U(1)_c} \times \mathrm{SU(4)} \supset \mathrm{SU(2)}_\sigma \times \mathrm{SU(2)_v} \times \mathrm{U(1)_c} \supset \mathrm{SU(2)}_\sigma \times \mathrm{U(1)_c},$$

which corresponds to simultaneous conservation of both spin and valley isospin. These three dynamical symmetry chains correspond to different emergent states having the same SO(8) highest symmetry but different values of effective interaction parameters.

Example 20.8 The 21 operators $\{\mathcal{S}_\alpha, \Pi_{\alpha x}, \Pi_{\alpha y}, T_z, S_0, D_\mu^\dagger, D_\mu\}$ close an SO(7) subalgebra of SO(8) and the subset $\{\mathcal{S}_\alpha, \Pi_{\alpha x}, \Pi_{\alpha y}, T_z, S_0\}$ closes an $\mathrm{SO(5)} \times \mathrm{U(1)_c}$ subalgebra of SO(7), implying a third subgroup chain

$$\mathrm{SO(8)} \supset \mathrm{SO(7)} \supset \mathrm{SO(5)} \times \mathrm{U(1)_c} \supset \mathrm{SU(2)}_\sigma \times \mathrm{U(1)_c}.$$

This chain is of special interest because it defines a *critical dynamical symmetry* that represents an entire phase exhibiting critical behavior, as discussed in Box 20.3 below.

Emergent and Perturbative States: Figure 20.7 bears a superficial resemblance to Fig. 20.4 but the implied physics differs fundamentally. In Fig. 20.4 the symmetry is broken *explicitly* by terms in the Hamiltonian. The resulting states are perturbations of a state corresponding to a spin–isospin multiplet that would be degenerate [exact SU(4) symmetry] if $H_\mathrm{v} = H_Z = 0$ in Eq. (20.5). Conversely, in the dynamical symmetry chains of Fig. 20.7 the symmetry has been broken *spontaneously*. These states are emergent and represent different quantum phases of the theory that cannot be related perturbatively to each other or to the original weakly interacting states.

20.2.6 Generalized Coherent States for Graphene

The dynamical symmetry limits displayed in Fig. 20.7 result from particular choices of the coupling parameters appearing in the Hamiltonian. They have exact analytical solutions for physical matrix elements. For *arbitrary values of coupling parameters* solutions will correspond to superpositions of the different symmetry-limit solutions that do not have exact analytical forms. In this more general case solutions could be obtained numerically since the collective subspace is highly truncated. However, there is a powerful alternative: the *generalized coherent state approximation* described in Ch. 21, which permits analytical solutions for arbitrary coupling-parameter values.

The simplest way to visualize the nature of the collective states implied by the group chains of Fig. 20.7 is to examine their total energy surfaces in coherent state approximation.

We illustrate by restricting to the three SO(8) subgroup chains of Fig. 20.7 that contain the $SO(5) \times U(1)_c$ subgroup. Thus the corresponding coherent state solutions will represent a superposition of the symmetry-limit solutions for the

$$\begin{gathered} SO(8) \supset SO(5) \times SU(2)_p \supset SO(5) \times U(1)_c \\ SO(8) \supset U(1)_c \times SU(4) \supset SO(5) \times U(1)_c \qquad SO(8) \supset SO(7) \supset SO(5) \times U(1)_c \end{gathered} \tag{20.17}$$

chains of Fig. 20.7, which will be termed the SO(5) × SU(2), SU(4), and SO(7) symmetries, respectively. The ground state energy may be determined through the variational condition $\delta \langle \eta | H | \eta \rangle = 0$, where $|\eta\rangle$ is the coherent state and H is the SO(8) Hamiltonian.

Another Useful Basis: For evaluation of coherent states it is useful to introduce a new basis for the particle–hole operators given by[6]

$$P_\mu^r = \sum_{m_j m_l} (-1)^{\frac{3}{2}+m_l} \left\langle \tfrac{3}{2}\, m_j\, \tfrac{3}{2}\, m_l \right| r\; \mu \Big\rangle B_{m_j - m_l}, \tag{20.18}$$

with the definition

$$B_{m_j - m_l} \equiv \sum_{m_k} c^\dagger_{m_j m_k} c_{-m_l m_k} - \frac{1}{4} \delta_{m_j - m_l} \Omega, \tag{20.19}$$

where m_j and m_l take the values of the fictitious angular momentum projection m_i in Fig. 20.5(a), providing a labeling equivalent to that of a and b in B_{ab}, with m_j or m_l values $\left\{\frac{3}{2}, \frac{1}{2}, -\frac{1}{2}, -\frac{3}{2}\right\}$ mapping to a or b values $\{1, 2, 3, 4\}$, respectively.

Example 20.9 From Fig. 20.5(a), we see that $B_{ab} = B_{12}$ and $B_{m_j m_l} = B_{3/2,1/2}$ label the same quantity, which is defined through either Eq. (20.9) or Eq. (20.19).

The index r in Eq. (20.18) can take values $r = 0, 1, 2, 3$, with $2r + 1$ projections μ for each possibility, giving a total of 16 operators P_μ^r. By inserting the explicit values of the Clebsch–Gordan coefficients in Eq. (20.18) the 16 independent P_μ^r may be evaluated in terms of the B_{ab}. These are worked out in Problem 20.8 and summarized in Ref. [223].

Example 20.10 From the solution of Problem 20.8,

$$\begin{aligned} P_0^0 &= \sum_{m_j m_l} (-1)^{\frac{3}{2}+m_l} \left\langle \tfrac{3}{2}\, m_j\, \tfrac{3}{2}\, m_l \right| 0\, 0 \rangle B_{m_j - m_l} \\ &= \frac{1}{2}(B_{-3/2,-3/2} + B_{-1/2,-1/2} + B_{1/2,1/2} + B_{3/2,3/2}) \\ &= \frac{1}{2}(B_{44} + B_{33} + B_{22} + B_{11}), \end{aligned}$$

where the mapping between the labels m_i in line 2 and a in line 3 is given in Fig. 20.5(a), and B_{ab} is defined in Eq. (20.9).

---

6 This basis is convenient because the Ginocchio SO(8) model of nuclear structure physics discussed in Section 31.1 was formulated using this basis. Thus, much of the group theory required for applying SO(8) symmetry to graphene has already been worked out in terms of this basis, most notably in Ref. [230].

It is convenient to introduce a number operator n_i through

$$n_i \equiv B_{ii} = \sum_{m_k} c^\dagger_{im_k} c_{im_k} - \frac{\Omega}{4}, \qquad (20.20)$$

where 2Ω is the degeneracy of the space for particles contributing to SO(8) symmetry. Introducing a total particle number operator $n \equiv n_1 + n_2 + n_3 + n_4$, we may also define

$$S_0 \equiv \frac{n - \Omega}{2} = P^0_0, \qquad (20.21)$$

where the last equality is proved in Problem 20.9. Physically, S_0 is half the particle number measured from half filling ($n = \Omega$). Thus, we shall now work in an SO(8) generator basis

$$G'_{\mathrm{SO(8)}} = \{P^1, P^2, P^3, S_0, S, S^\dagger, D_\mu, D^\dagger_\mu\}, \qquad (20.22)$$

where the particle–hole operators in this basis and those in (20.13) can be related by expressing both in terms of the B_{ab} defined in Eq. (20.9), as illustrated in Problem 20.7. The explicit SO(8) commutation algebra for the generators (20.22) is given in Ref. [223].

Coherent State Energy Surfaces: The Hamiltonian may now be written

$$H = H_0 + G_0 S^\dagger S + G_2 D^\dagger \cdot D + \sum_{r=1,2,3} b_r P^r \cdot P^r, \qquad (20.23)$$

where the coefficients G_0, G_2, and b_r are effective interaction strengths in the truncated collective subspace, $D^\dagger \cdot D \equiv \sum_\mu D^\dagger_\mu D_\mu$ and $P^r \cdot P^r \equiv \sum_\mu P^r_\mu P^r_\mu$, and we assume H_0 to be constant in first approximation. The coherent state is a general SO(8) solution that is not restricted to the dynamical symmetry limits but it is highly instructive to evaluate the coherent state energy surface for each dynamical symmetry limit. For the chains (20.17) the total energy surface is given by [230]

$$E_{\mathrm{g}}(n,\beta) = \langle H \rangle = N_{\mathrm{g}} \left[A_{\mathrm{g}}\beta^4 + B_{\mathrm{g}}(n)\beta^2 + C_{\mathrm{g}}(n) + D_{\mathrm{g}}(n,\beta) \right], \qquad (20.24)$$

where n is the particle number, β is the single order parameter characterizing these states (which measures antiferromagnetic order and indicates the mixture of S and D_μ pairs in the ground state),[7] and the parameters N_{g}, A_{g}, $B_{\mathrm{g}}(n)$, $C_{\mathrm{g}}(n)$, and $D_{\mathrm{g}}(n,\beta)$, which depend on the group g, are tabulated in Ref. [223]. The ground states in the coherent state approximation at fixed $n/2\Omega$ will be given by those values of $\beta \equiv \beta_0$ that correspond to minima of the energy surface $E(n,\beta)$. Evaluation of these constraints for Eq. (20.24) indicates that [230]

$$\beta_0^{\mathrm{SU(2)\times SO(5)}} = \beta_0^{\mathrm{SO(7)}} = 0 \qquad \beta_0^{\mathrm{SU(4)}} = \pm\sqrt{n/4\Omega}. \qquad (20.25)$$

Energy surfaces for these SO(8) symmetry limits are shown in Fig. 20.8(a–c).

[7] As shown in Problem 20.10, the AF order parameter $\langle N_z \rangle$ is maximized at the values of β that correspond to minimum total energy.

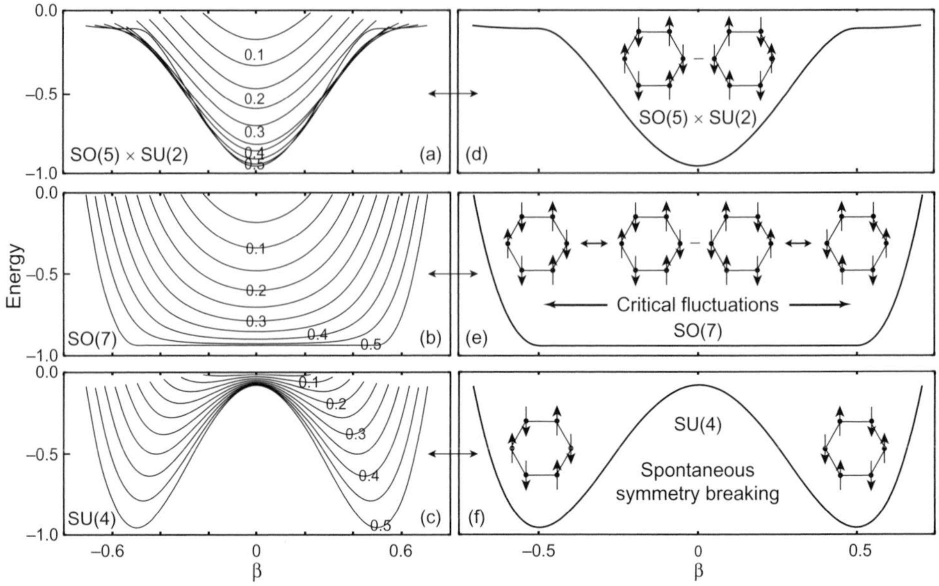

Fig. 20.8 Coherent state energy surfaces for monolayer graphene in a magnetic field [221]. (a)–(c) Energy surfaces versus the AF order parameter β for the three dynamical symmetry limits of Fig. 20.7 displayed in Eq. (20.17). Curves labeled by fractional occupation $f = n/2\Omega$ defined in Eq. (20.3). (d)–(f) Ground state ($f = 0.5$) energy surfaces corresponding to (a)–(c). The inset diagrams in (d)–(f) indicate schematically the corresponding ground state wavefunctions in terms of the configurations in Fig. 20.6. Reproduced with permission from *Scientific Reports*: The Ground State of Monolayer Graphene in a Strong Magnetic Field, L.-A. Wu and M. W. Guidry, **6**, 22423 (2016).

20.2.7 Physical Interpretation of the Energy Surfaces

The physical meaning of the energy surfaces displayed in Fig. 20.8(a–c) is illustrated in Fig. 20.8(d–f), and may be understood from the following considerations. The coherent state wavefunction corresponding to $N = n/2$ pairs conserves particle number only on average, so it is a superposition of terms having different pair numbers p [223, 230],

$$
\begin{aligned}
|\mathrm{SO(5)} \times \mathrm{SU(2)}\rangle &= \sum_p C_p \left(S^\dagger\right)^p |0\rangle \simeq \left(S^\dagger\right)^N |0\rangle , \\
|\mathrm{SU(4)}\rangle = \sum_p C_p \left(S^\dagger \pm D_0^\dagger\right)^p |0\rangle &= 2\sum_p C_p \left(Q_\pm^\dagger\right)^p |0\rangle \simeq \left(Q_\pm^\dagger\right)^N |0\rangle ,
\end{aligned}
\tag{20.26}
$$

where the coefficients C_p are given in Ref. [230], the pair operators $S^\dagger$ and $Q_\pm^\dagger$ are defined in Eqs. (20.12) and (20.14) (see also Fig. 20.6), and the final approximations are justified because the fluctuation in particle number is small (Problem 20.5) and thus dominated by terms in the sum with $p \sim N$. From Eq. (20.26) and the results of Problem 20.3, we have the following.

1. The SO(5) × SU(2) state in Fig. 20.8(d) is a coherent superposition of S pairs, with vanishing order parameters $\langle \mathcal{S}_z \rangle$, $\langle T_z \rangle$, and $\langle N_z \rangle$ or $\langle \beta \rangle$.
2. The SU(4) state in Fig. 20.8(f) is a coherent superposition of Q_- or Q_+ pairs that has vanishing ferromagnetic order $\langle \mathcal{S}_z \rangle$ and charge density wave order $\langle T_z \rangle$, but finite AF order parameters $\langle N_z \rangle$ or $\langle \beta \rangle$. The degenerate energy minima imply spontaneous breaking of the symmetry if one of them is chosen as the physical ground state.[8]
3. The SO(7) state in Fig. 20.8(e) is a superposition of S and $Q_\pm$ pairs that looks like SO(5) × SU(2) states for $\beta \sim 0$ and SU(4) AF states for $\beta \sim \pm\beta_0$:

$$\left|\beta = 0\right\rangle \sim (S^\dagger)^N \left|0\right\rangle \quad \left|\beta = \beta_0\right\rangle \sim (Q_+)^N \left|0\right\rangle \quad \left|\beta = -\beta_0\right\rangle \sim (Q_-)^N \left|0\right\rangle, \tag{20.27}$$

where $\beta = \beta_0 = \pm(n/4\Omega)^{1/2}$ corresponds to a minimum of the energy in the SU(4) limit.[9] In the SO(7) ground state these configurations are *nearly degenerate in energy.*

Thus the SO(5) × SU(2) and SU(4) dynamical symmetries may be distinguished by the order parameters $\langle N_z \rangle$ or $\langle \beta \rangle$, which vanish in the SO(5) × SU(2) state and are finite in the SU(4) state. The SO(7) dynamical symmetry is characterized by *maximal fluctuations in* $\langle \beta \rangle$. It is an example of a *critical dynamical symmetry,* discussed further in Box 20.3.

20.2.8 Quantum Phase Transitions in Graphene

The transitions between different dynamical symmetry chains that inherit from a given highest symmetry correspond to quantum phase transitions, which can be studied by varying control parameters in the coherent state approximation. In Ref. [223] the approximate SO(8) coherent state Hamiltonian

$$H = G_0 S^\dagger S + b_2 P^2 \cdot P^2 \tag{20.28}$$

was used to study transitions among the quantum phases defined by the dynamical symmetries (20.17). The approximate Casimir expectation values associated with dominant symmetries of the subgroup chains are [223]

$$\langle C_{\mathrm{SO(5)\times SU(2)}} \rangle \sim \langle S^\dagger S \rangle \qquad \langle C_{\mathrm{SU(4)}} \rangle \sim \langle P^2 \cdot P^2 \rangle \qquad \langle C_{\mathrm{SO(7)}} \rangle \sim \langle S^\dagger S \rangle + \langle P^2 \cdot P^2 \rangle.$$

Defining a control parameter $q \equiv b_2/G_0$, the Hamiltonian (20.28) can be rewritten,

$$H = G_0 (S^\dagger S + q P^2 \cdot P^2) \tag{20.29}$$

and q tunes the Hamiltonian (20.29) between SU(2) × SO(5) and SU(4) quantum phases, via an intermediate SO(7) quantum phase displaying quantum critical behavior.

1. If $q \ll 1$ the ground state energy surface is approximated by Fig. 20.8(d), with a minimum at $\beta = 0$, no antiferromagnetic order, and SU(2) × SO(5) dynamical symmetry.

[8] Figures 20.8 are slices along a particular axis in order-parameter space. The full diagrams are multidimensional. For example, the full version of Fig. 20.8(f) is similar to a multidimensional version of Fig. 17.4.

[9] For the undoped ground state of graphene $n = 2\Omega$ and $\beta_0 = (n/4\Omega)^{1/2} = \pm\frac{1}{2}$. Thus the fluctuations in β suggested by the ground state energy surface in Fig. 20.8(e) are *the largest possible,* since they represent excursions over the full range of β permitted by SO(8) symmetry for a Landau level containing n electrons.

Box 20.3 Critical Dynamical Symmetries

The SO(8) $\supset$ SO(7) dynamical symmetry chain discussed in Section 20.2.7 is a specific example of a general phenomenon called *critical dynamical symmetry* [229].

Universality of Critical Dynamical Symmetry

Critical dynamical symmetry has been shown to occur with strikingly similar features in nuclear structure physics [218, 229, 230], high-temperature superconductors [93, 98, 222], and monolayer graphene in magnetic fields [87, 221, 223]. Figure 20.10 illustrates this surprising universality. The

- SO(8) $\supset$ SO(7) dynamical symmetry in Fig. 20.10(a),
- SU(4) $\supset$ SO(5) dynamical symmetry in Fig. 20.10(b), and
- SO(8) $\supset$ SO(7) dynamical symmetry in Fig. 20.10(c)

all exhibit a common set of properties associated with critical dynamical symmetry.

1. The energy surface is extremely flat as a function of an order parameter, implying classically an infinity of nearly degenerate states that can have extremely different wavefunctions and values of the order parameter.
2. They exhibit interpolation properties as in Eq. (20.27), where fluctuations in an order parameter provide a doorway between two other phases of the theory.
3. In cases studied thus far, the doorway provided by critical dynamical symmetry typically connects a phase with spontaneously broken symmetry to one where the symmetry has been restored (see Fig. 20.9).
4. The flat energy surfaces may produce systems having strong *complexity* (extreme sensitivity to initial conditions) and fluctuations [97, 98].
5. The critical dynamical symmetry may be viewed as a generalization of a quantum critical point to an entire *quantum critical phase.*

As a consequence, it has been proposed that critical dynamical symmetry may be a fundamental organizing principle for quantum criticality and quantum phase transitions in complex systems exhibiting more than one emergent mode [98].

Degeneracy and a Better Wavefunction

The flat energy surfaces for critical dynamical symmetries imply the existence of many nearly degenerate classical ground states distinguished by different values of an order parameter and possibly very different wavefunction components. As discussed in Ch. 22, this suggests that a better ground state with lower energy could be found by using the *generator coordinate method* of Section 22.2 to obtain a new ground state that is a superposition of the nearly degenerate quantum critical states.

2. If $q \gg 1$ the ground state energy surface is approximated by Fig. 20.8(f), with spontaneously broken symmetry corresponding to a set of degenerate energy minima at $|\beta| \neq 0$, SU(4) dynamical symmetry, and antiferromagnetic order.
3. If $q \sim 1$, the ground state energy surface is approximated by Fig. 20.8(e), with SO(7) critical dynamical symmetry and with large fluctuations in the AF order parameter β.

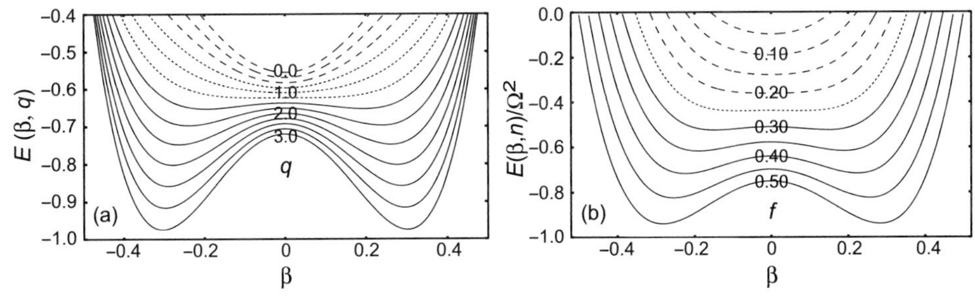

Fig. 20.9 Total energy surfaces illustrating SO(8) quantum phase transitions for graphene in a magnetic field. Dashed curves correspond to $\sim$SO(5) $\times$ SU(2) symmetry, solid curves to $\sim$SU(4) symmetry, and dotted curves to SO(7) critical dynamical symmetry mediating the transition from SO(5) $\times$ SU(2) to SU(4). (a) Particle number fixed and coupling strength ratio q as control parameter. The figure shows coherent state energy surfaces as a function of $q \equiv G_0/b_2$ for a fractional occupation $f = n/2\Omega = 0.5$. (b) Coupling strength fixed and particle number as the control parameter. The figure shows coherent state energy surfaces for different filling fractions $f = n/2\Omega$ at fixed $q = 2.5$.

Figure 20.9(a) illustrates SO(8) quantum phase transitions controlled by varying the coupling strength q. The dotted curves near $q \sim 1$ correspond to $\sim$SO(7) symmetry mediating the quantum phase transition from SO(5) $\times$ SU(2) to SU(4). Quantum phase transitions also may be implemented by varying particle number (filling fraction) at constant q. This is illustrated in Fig. 20.9(b), with dotted curves near $n/2\Omega \sim 0.25$ mediating the transition from SO(5) $\times$ SU(2) to SU(4). Section 23.7 gives a more general discussion of the close relationship between dynamical symmetry and quantum phase transitions.

20.3 Universality of Emergent States

Figure 20.10 illustrates the remarkable *universality of emergent dynamical symmetries* for a broad range of disciplines. The systems compared there differ fundamentally at the microscopic level and yet exhibit a clear universality manifested through a similar Lie group structure for the generators of symmetries responsible for emergent states.

> Dynamical symmetries encode collective macroscopic similarities of emergent states; effective interactions parameterize smoothly the differences that follow from underlying microscopic structure for those macroscopic states.

A striking aspect of this universality is the systematic occurrence of *critical dynamical symmetries,* as discussed in Box 20.3. This universality extends beyond qualitative similarities: dynamical symmetries provide microscopic descriptions of phenomena within given subfields. Thus Fig. 20.10(a,b) provides a unified picture of collective states in heavy

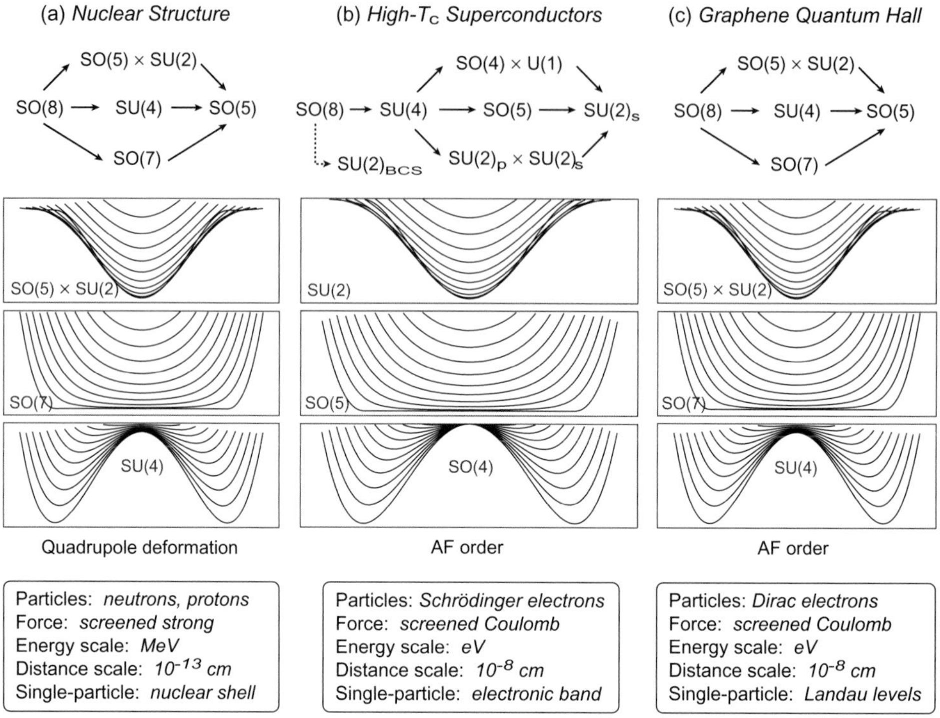

Fig. 20.10 Similarity in the dynamical symmetry chains and the ground coherent state energy surfaces for dynamical symmetry in (a) nuclear structure [218], (b) high-temperature superconductivity [93, 222], and (c) monolayer graphene in a strong magnetic field [221]. The plot contours show total energy as a function of an appropriate order parameter, with each curve corresponding to a different value of a particle number parameter.

atomic nuclei and the antiferromagnetic and superconducting states of a high-temperature superconductor, but at the same time, Figs. 31.6, 31.7, 31.8, and 32.8(a) demonstrate computation of observables in quantitative agreement with data within each subfield.

20.3.1 Topological and Algebraic Constraints

The universality of dynamical symmetries for collective modes across multiple fields having completely different underlying microscopic physics that is suggested by Fig. 20.10 is perhaps surprising. We might think that many possible Lie algebras could describe these modes, which raises the issue of uniqueness. How does a dynamical symmetry description pick out a unique set of Lie algebras for problems within a field, or even *across completely disparate fields* that leads to the universality of a Fig. 20.10? The ultimate reason lies in topological properties of the Lie group manifolds and in the restrictiveness of Lie algebras, which leads to a *"quantization" of candidate Lie groups* such that only a few relatively small groups are viable for description of emergent states.

1. *Topological constraints:* Emergent states are typically *bound states*, which restricts to the subset of *compact Lie groups*. This is a topological constraint (on the group manifold), since compactness is a topological invariant; see Section 24.3.1.
2. *Algebraic constraints:* Compact groups have the further restriction that emergent physical degrees of freedom be represented by generators that are functions of fermion creation and annihilation operators, and only some sets of generators constructed from acceptable fermion operators close a Lie algebra under commutation. This algebraic constraint further restricts the candidate Lie groups to a subset of the compact Lie groups.
3. *Dimensionality constraints:* Emergent states are characterized by *only a few effective degrees of freedom* (often expressed as order parameters), implying that their emergent dynamical behavior is described by groups with a relatively small number of generators.

As a consequence of these restrictions, only groups having certain discrete numbers of generators are candidates for dynamical symmetries in realistic physical systems.

These ideas are illustrated further in Fig. 20.11, where we use the Lie group classification in Appendix D to enumerate all compact Lie groups having fewer than 50 generators. Of

(a) Compact Lie groups

Group	Dimension	Cases	Generators
SO(2n + 1)	$2n^2 + n$	SO(3)	3
		SO(5)	10
		SO(7)	21
		SO(9)	36
SO(2n)	$2n^2 - n$	SO(2)	1
		SO(4)	6
		SO(6)	15
		SO(8)	28
		SO(10)	45
SU(n)	$n^2 - 1$	SU(2)	3
		SU(3)	8
		SU(4)	15
		SU(5)	25
		SU(6)	35
		SU(7)	48
Sp(2n)	$2n^2 + 1$	Sp(2)	3
		Sp(4)	10
		Sp(6)	21
		Sp(8)	36

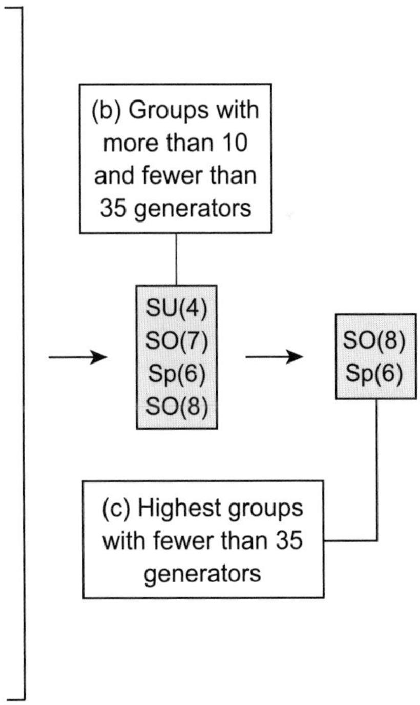

Fig. 20.11 (a) Number of generators (dimension) for compact Lie groups with fewer than 50 generators. (b) Only four compact groups have more than 10 and fewer than 35 generators, and also include pair generators. (c) Of these SU(4) ~ SO(6) and SO(7) commonly appear as subgroups of SO(8), leaving SO(8) and Sp(6) as unique highest groups.

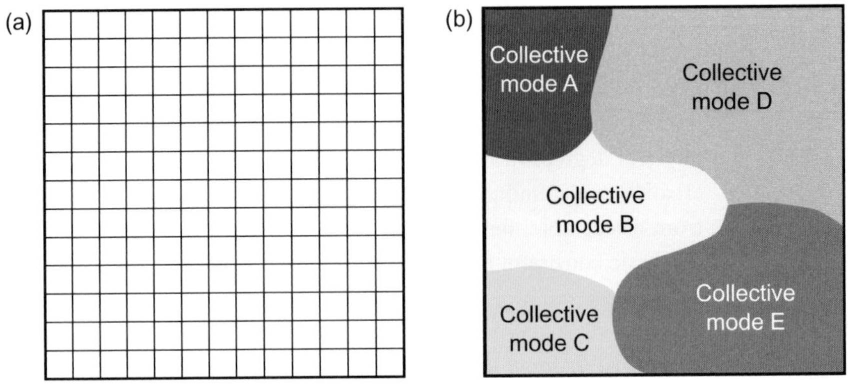

Fig. 20.12 Fitting competing but different collective modes into a collective subspace as in (b) is much more constrained than fitting similar uncorrelated modes into the same subspace as in (a).

these compact groups, only four have more than 10 and fewer than 35 generators, and also include pair (two-particle or two-hole) generators: SU(4), SO(7), Sp(6), and SO(8).[10] Finally, we note that SU(4) $\sim$ SO(6) and SO(7) commonly appear as subgroups of SO(8), leaving SO(8) and Sp(6) as the candidate highest groups having fewer than 35 generators (with many of the smaller compact groups appearing as subgroups of these highest symmetries). Figure 20.12 illustrates in a geometrical analogy the constraint imposed by fitting five collective modes (the "puzzle pieces" A–E) into the same highest symmetry (the container box). In Fig. 20.12(a) the small squares fit together in many ways to fill the box. Conversely, in Fig. 20.12(b) there is *only one way* to arrange pieces to exactly fill the box. Any other shape for the pieces, any other spatial relationship between them, the absence of any pieces, or the presence of any additional pieces, will prevent filling the box exactly.

Example 20.11 In the theory of non-abelian superconductivity described in Section 32.3, the minimal symmetry accommodating antiferromagnetism (AF), superconductivity based on singlet (spin-0) pairing (SC), and conserved spin and charge is SU(4). However, AF, SC, spin, and charge operators alone do not close under commutation of generators. If triplet (spin-1) pairing operators are added, the full set of 15 operators defined in Eqs. (32.2) exactly closes SU(4). The set of AF, singlet pairing, triplet pairing, spin, and charge operators defined in Eqs. (32.2) is the analog of the set of collective modes A–E in Fig. 20.12(b), and the commutator algebra obeyed by these SU(4) generators is the algebraic analog of the geometrical relationship among "puzzle pieces" required to exactly tile the container.

---

[10] The restriction to compact groups accommodating pair operators as generators is motivated physically since efficient description of emergent states often requires such generators. A major reason is that pair operators allow a natural definition of a pair-condensate collective subspace, as in Eq. (20.16). The cases displayed in Fig. 20.10 all include pairing operators among their generators and a collective subspace of the form (20.16).

The other cases in Fig. 20.10 exhibit a similar minimal higher symmetry that can just accommodate physical operators characteristic of the emergent modes, as in Example 20.11.

20.3.2 Analogy with General Relativity

The view of Hilbert space constrained by dynamical symmetries advocated here has an abstract similarity to general relativity. Einstein realized that the universality of gravity (independence of gravity from the nature of the object experiencing a gravitational force; for example, gravitational acceleration does not depend on mass or any other intrinsic property) meant that gravity must be a consequence of spacetime manifold structure, not of interactions between specific objects in spacetime. In like manner, the universality observed for emergent modes in many-body systems across many disciplines derives from the structure of a similar Hilbert space submanifold selected by dynamical symmetries [91].

20.3.3 Analogy with Renormalization Group Flow

For those with some knowledge of the *renormalization group,*[11] there also is an analogy of the present discussion with renormalization group flow because dynamical symmetries distinguish "relevant operators" characterizing the collective subspace from "irrelevant operators" that enter only parametrically into the collective behavior. The "flow" is in the dimensionality of the generator space; as it is decreased from the full Hilbert space toward the collective subspace by taking successive subgroups of the full Hilbert space symmetry, the influence of irrelevant operators is neglected or absorbed into smoothly varying parameters, leaving only relevant operators to define explicitly the collective subspace. Universality is implied because differences between problems are represented by irrelevant operators but the relevant operators define collective subspaces having common algebraic structures across problems. In this view the *flow to fixed points* for the renormalization group is analogous to the dynamical symmetry chains discussed here.

Background and Further Reading

Additional applications of dynamical symmetries to physical problems are given in Chs. 31 and 32, and the deep connections between quantum phase transitions and dynamical symmetries are elaborated in Section 23.7. Extensive reviews of dynamical symmetries applied in both fermionic and bosonic many-body systems may be found in Guidry, Sun, Wu, and Wu [98], Iachello and Arima [116], and Wu, Feng, and Guidry [218].

[11] The renormalization group (RG) is a formalism that describes how the properties of a physical system evolve under change of scale. The RG is a rather technical subject but those aspects relevant to the present discussion are simple and are explained concisely in the Wikipedia article *Renormalization group.* The "renormalization group" is not actually a group; it is instead a *semigroup,* which is a set with an associated binary operation. Thus it differs from a group in that it need not have inverses or an identity. The renormalization group is a semigroup and not a group because an RG transformation often does not have a unique inverse.

Problems

20.1 Consider the isospin subgroup chain, $U(2) \supset U(1)_B \times SU(2) \supset U(1)_B \times U(1)_{T_3}$, where subscripts distinguish the U(1) group generated by baryon number B from the U(1) group generated by the third component of isospin, T_3. Write a Hamiltonian consisting of a linear combination of the Casimir invariants for this chain and show that the resulting spectrum can be obtained analytically as

$$E(N_B, T, Q) = \left(a - \frac{c}{2}\right) N_{\rm B} + bT(T+1) + cQ,$$

where a, b, and c are constants, and $Q \equiv T_3 + \frac{1}{2}B$.

20.2 Write the SU(4) generators of Eq. (20.7) as linear combinations of the operators B_{ab} defined in Eq. (20.9). *Hint*: See Example 20.3. ***

20.3 Evaluate the order parameters $\langle S_z \rangle$ (ferromagnetic order), $\langle T_z \rangle$ (charge density wave order), and $\langle N_z \rangle$ (antiferromagnetic order) for the configurations in Fig. 20.6. ***

20.4 Evaluate Eq. (20.11) for the six possible pair creation operators corresponding to $S = 1, T = 0$ and $S = 0, T = 1$ in terms of the uncoupled pair operators (20.8). *Hint*: Clebsch–Gordan coefficients may be found in the solution of Problem 6.5 and Table C.3; the mapping $a = (\tau, \sigma)$ between the spin (σ) and isospin (τ) labels in Eq. (20.11), and the single label a used in Eq. (20.8), is given in Fig. 20.5(a). ***

20.5 The generalized coherent state approximation is a (sophisticated) form of mean-field theory, which means that it does not conserve particle number exactly. For the SO(8) coherent state applied to graphene in Section 20.2.6, the fractional uncertainty in electron number Δn is given by [230]

$$(\Delta n)^2 = \langle \hat{n}^2 \rangle - \langle \hat{n} \rangle = 2n - \frac{n^2}{\Omega} + 16\Omega\beta_0^4 - 8n\beta_0^2,$$

where β_0 is the value of β at the minimum energy. Show that in the SO(5) × SU(2) and SU(4) limits, respectively,

$$\left[\frac{\Delta n}{n}\right]_{\rm SO(5)\times SU(2)} = \sqrt{\frac{1-f}{f\Omega}} \qquad \left[\frac{\Delta n}{n}\right]_{\rm SU(4)} = \sqrt{\frac{1-2f}{f\Omega}},$$

where $f = n/2\Omega$ is the fractional occupation of the Landau level.

20.6 Verify the equations for the order parameters in Eq. (20.15). *Hint*: Use the results of Problem 20.2 to write S_z, T_z, and N_z in terms of the B_{ab}, the definition (20.9), and that the number operator $\hat{n}_i$ counting particles in state $|i\rangle$ is given by Eq. (20.20). ***

20.7 Show that P_0^2 defined in Eq. (20.18) and N_z defined in Eq. (20.7c) are equivalent. Thus the expectation value of either serves as an antiferromagnetic order parameter. *Hint*: Both can be expressed in terms of the operators B_{ab} given in Eq. (20.9).

20.8 Express the 16 generators P^r_μ of Eq. (20.18) as linear combinations of the operators B_{ab} defined in Eq. (20.9). ***

20.9 Show that for graphene in a magnetic field, P^0_0 defined in Eq. (20.18) satisfies Eq. (20.21). Thus, P^0_0, S_0, and n defined in Eq. (20.20) can serve as number operators.

20.10 The AF order parameter N_z is related to the coherent state order parameter β by $\langle N_z \rangle = 2\Omega |b_2| (f - \beta^2)^{1/2} \beta$, where b_2 is a coupling strength and $f = 2\Omega$ is the fractional occupation. Prove that for the graphene SU(4) coherent state the maximum of N_z occurs for the β value that minimizes total energy. Show that $\langle N_z \rangle_{\max} = \Omega |b_2| f$, so that the maximum value of the AF order parameter occurs for half filling of the $n = 0$ Landau level. *Hint*: The SO(8) model is particle–hole symmetric, so f counts electrons up to half filling and holes (absence of electrons) after half filling.

21 Generalized Coherent States

The idea of coherent states originated with Schrödinger in 1926 [181], but the modern applications that concern us date from seminal work by Glauber [76, 77] in quantum optics and its subsequent extension to generalized coherent states by Gilmore [70, 71, 72] and Perelomov [161]. In this chapter we give an introduction to these methods, which are powerful because they permit dynamical symmetry solutions for arbitrary coupling strengths, and because they define a connection between dynamical symmetries and the Hartree–Fock–Bogoliubov and Ginzburg–Landau methods that are staples of more traditional approaches to the physics of emergent broken symmetry states. Actual implementation is rather technical. This chapter concentrates on an overview. In Chs. 20, 31, and 32 we will give results obtained using coherent state approximation, but even there we will largely relegate details to the references.

21.1 Glauber Coherent States

Glauber's original work dealt with coherent states of the electromagnetic field, which is a boson field that may be modeled by a set of harmonic oscillators. In terms of the oscillator creation operator $a^\dagger$ and the corresponding annihilation operator a (see Section 10.1), Glauber demonstrated that there are three equivalent definitions of coherent states [77].

Definition 1 Coherent states $|\alpha\rangle$ are eigenstates of the harmonic oscillator annihilation operator a, with $a\,|\alpha\rangle = \alpha\,|\alpha\rangle$, where α is a complex number.

Definition 2 Introducing a *displacement operator* $D(\alpha)$ by

$$D(\alpha) \equiv e^{\alpha a^\dagger - \alpha^* a}, \tag{21.1}$$

a coherent state $|\alpha\rangle$ may be defined through the action of this operator,

$$|\alpha\rangle = D(\alpha)\,|0\rangle\,, \tag{21.2}$$

where $|0\rangle$ is the oscillator ground (vacuum) state.

Definition 3 Defining coordinate and momentum operators (q, p) using Eq. (10.2)

$$q = \frac{1}{\sqrt{2}}\left(a + a^\dagger\right) \qquad p = \frac{-i}{\sqrt{2}}\left(a - a^\dagger\right), \tag{21.3}$$

coherent states exhibit a *minimum uncertainty relationship*

$$(\Delta p)^2(\Delta q)^2 = \left(\frac{1}{2}\right)^2. \tag{21.4}$$

This definition motivated Schrödinger's study of coherent states for the harmonic oscillator [181]. It is not unique since more than one state can satisfy Eq. (21.3), with differing trade-offs between Δp and Δq (which is the origin of *squeezed states* in quantum optics).

The coherent states resulting from these definitions are extremely useful because they behave in many respects as classical fields. It is of interest to extend these ideas to fields other than photon fields, but the preceding definitions depend on the properties of harmonic oscillators and not all fields can be described as collections of harmonic oscillators. In particular, matter fields are fermion fields and the creation and annihilation operators of a fermion field do not obey boson commutators. Furthermore, even if the fields admit an approximate oscillator description, one often finds that oscillator wavefunctions are too schematic to describe the quantitative features of strongly correlated, many-body systems. Let us now discuss an alternative formulation of electromagnetic coherent states that suggests a way to generalize coherent states to particles other than photons.

21.2 Symmetry and Coherent Electromagnetic States

The Hamiltonian operator and the associated Hilbert space determine the quantum dynamical properties of a system. Let us begin our reformulation of Glauber coherent state theory by introducing a Hamiltonian for the interaction of photons with an atomic system. In this and following material we follow the review by Zhang, Feng, and Gilmore [231].

21.2.1 Quantum Optics Hamiltonian

A simple Hamiltonian for interaction of an electromagnetic field with a set of atoms is

$$H = \sum_k \hbar\omega_k\, a_k^\dagger a_k + \sum_\alpha \epsilon\sigma_0^{(\alpha)} + \sum_{k,\alpha} \gamma_{k\alpha}\left(\frac{\sigma_+^{(\alpha)}}{\sqrt{N}}\,a_k + \frac{\sigma_-^{(\alpha)}}{\sqrt{N}}\,a_k^\dagger\right), \tag{21.5}$$

where $\hbar\omega_k\, a_k^\dagger a_k$ is the energy of the field mode k, and the coupling strength between the atomic and electromagnetic systems is specified by a state-dependent $\gamma_{k\alpha}$, which we shall approximate by a constant γ_k. Each of the N atoms labeled by the index α is assumed to be a two-state system, so its dynamical variables can be specified by SU(2) "spin" operators, $(\sigma_\pm, \sigma_0)$. Regarding the atomic system as a classical source, the σ operators become complex numbers and the Hamiltonian may be written

$$H = H_0 + H_{\rm int} = \sum_k \hbar n_k\omega_k + \sum_k \left(\lambda_k(t)a_k^\dagger + \lambda_k^*(t)a_k\right), \tag{21.6}$$

where $n_k \equiv a_k^\dagger a_k$ is the photon number operator and a constant term has been neglected. This is the semiclassical Hamiltonian of a quantized harmonic oscillator system (the photon field) in a classical external field provided by the atoms. The term $H_0 = \hbar n_k \omega_k$ is the energy of the free electromagnetic field and the term H_{int} approximates the interaction of this field with the atoms. We now use the semiclassical Hamiltonian (21.6) and the associated Hilbert space to construct an alternative algorithm for generating coherent states of the electromagnetic field.

21.2.2 Symmetry of the Hamiltonian

Let us drop the index k and treat the modes of the field independently. The Hamiltonian (21.6) is expressed in terms of four operators, $\{\hat{n}, a^\dagger, a, I\}$, where $\hat{n} \equiv a^\dagger a$ and I is the unit operator. Using the oscillator commutation relations $[\, a, a^\dagger \,] = 1$, it is easily verified that

$$\begin{array}{lll} [\, \hat{n}, a^\dagger \,] = a^\dagger & [\, \hat{n}, I \,] = 0 & [\, \hat{n}, a \,] = -a, \\ [\, a^\dagger, I \,] = 0 & [\, a, a^\dagger \,] = I & [\, a, I \,] = 0. \end{array} \tag{21.7}$$

Thus the operators $\{\hat{n}, a^\dagger, a, I\}$ are closed under commutation and generate a Lie algebra. The corresponding Lie group is termed the *Heisenberg–Weyl group* H_4; its continuous group elements g are obtained by exponentiating the generators. We shall term this a dynamical symmetry, since it is associated with an algebraic and group structure for the Hamiltonian.

21.2.3 Hilbert Space

The Hilbert space associated with the group H_4 is spanned by the number eigenstates $\{|n\rangle = |0\rangle, |1\rangle, |2\rangle, \dots\}$, with

$$\hat{n}\,|n\rangle = n\,|n\rangle \qquad |n\rangle = \frac{(a^\dagger)^n}{\sqrt{n!}}\,|0\rangle\,. \tag{21.8}$$

The energies of the unperturbed states are given by $H_0\,|n\rangle = \hbar\omega\, a^\dagger a\,|n\rangle = \hbar\omega\hat{n}\,|n\rangle$, so the unperturbed ground state is the field vacuum state $|0\rangle$. The ground state of the oscillator corresponds to $n = 0$ (no quanta). We shall refer to this state $|0\rangle$ as an *extremal state*, since from the properties of the oscillator operator algebra no lower-energy state can exist.

21.2.4 Stability Subgroup

A *stability subgroup* of H_4 is generated by a subset of generators closed under commutation and leaves the extremal state invariant (up to a possible phase). The subalgebra of H_4 spanned by the generators $\{\hat{n}, I\}$ corresponds to a subgroup with elements

$$h = e^{i(\theta\hat{n}+\phi I)}, \tag{21.9}$$

where θ and ϕ are angular variables conjugate to the operators $\hat{n}$ and I, respectively. Operating on the extremal state with h then gives

$$h\,|0\rangle = e^{i(\theta\hat{n}+\phi I)}\,|0\rangle = |0\rangle\, e^{i\phi}, \tag{21.10}$$

where the last step follows from Problem 3.11(c) and $\hat{n}\,|0\rangle = 0\,|0\rangle$. Thus, h leaves the extremal state $|0\rangle$ invariant (up to a phase), and since $\hat{n}$ and I commute with each other h is a U(1) × U(1) subgroup of H_4 corresponding to two independent U(1) phase rotations. Usually the phase ϕ will be irrelevant because we shall be interested in expectation values.

21.2.5 Coset Space

The final ingredient needed to construct the coherent state is the *coset space* of H_4 with respect to the stability subgroup U(1) × U(1). Recall from Section 2.14.2 that if H is an invariant subgroup of G the cosets of G with respect to H form a group G/H that is termed the factor group, and $G/H = H_4/U(1) \times U(1)$ is the set of elements Ω that provide a unique decomposition $g = \Omega h$ of every $g \in H_4$. A typical representative of $H_4/U(1) \times U(1)$ is

$$\Omega(\alpha) = e^{(\alpha a^\dagger - \alpha^* a)}, \tag{21.11}$$

with α an arbitrary complex parameter. Note that $\Omega(\alpha)$ is just the displacement operator D of Eq. (21.1).

21.2.6 The Coherent State

From the properties exhibited in Eqs. (21.9)–(21.11), the group element $g \in H_4$ acting on the oscillator ground state $|0\rangle$ can be factored in the form

$$g\,|0\rangle = \Omega(\alpha) h\,|0\rangle = \Omega(\alpha)\,|0\rangle\, e^{i\phi} \equiv |\alpha\rangle\, e^{i\phi}, \tag{21.12}$$

where $h \in U(1) \times U(1)$ and $\Omega(\alpha) \in H_4/U(1) \times U(1)$. But this is the coherent state definition (21.2), if we identify $|\alpha\rangle$ with the state produced by the actions of the coset elements $\Omega(\alpha)$ on the extremal state $|\Phi_0\rangle$; that is, $|\alpha\rangle \equiv \Omega(\alpha)\,|0\rangle$.

> We have obtained Glauber coherent states for the electromagnetic field by using a dynamical symmetry of the Hamiltonian and well-studied properties of Lie algebras and Lie groups, with no reference to bosonic fields or oscillators.

This suggests a way to obtain coherent states for any fermionic or bosonic system described dynamically by a Lie algebra.This method of *generalized coherent states* is illustrated schematically in Fig. 21.1 and described further in Section 21.3.

21.3 Construction of Generalized Coherent States

Still following Ref. [231], consider a general quantum dynamical system with a Hamiltonian operator H whose matrix elements define the energy and a set of transition operators $\{A\}$ whose matrix elements define the dynamical properties. We shall assume that the Hamiltonian and transition matrix elements are defined by a complete set of operators $\{T_i\} \equiv \mathbb{T}$, with $[T_i, T_j] \in \mathbb{T}$. Thus the set $\mathbb{T}$ spans a Lie algebra and the operators $T_i \in \mathbb{T}$

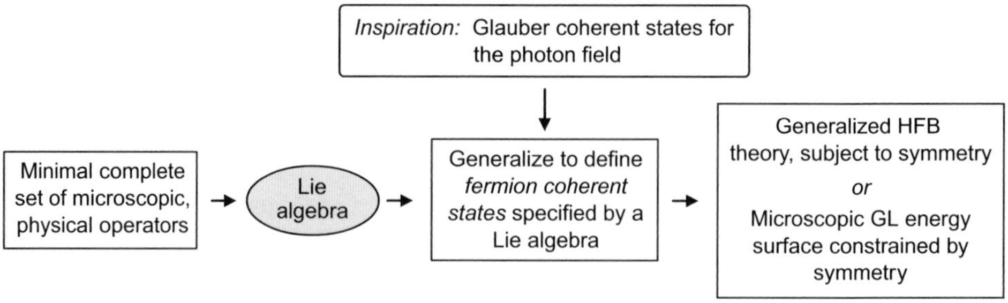

Fig. 21.1 Method of generalized coherent states. Coherent state calculations may be viewed as equivalent to a symmetry-constrained Hartree–Fock–Bogoliubov (HFB) approximation [231], or as a microscopic Ginzburg–Landau (GL) theory. Thus generalized coherent states are related to various conventional theoretical methods for emergent states.

are generators of the algebra. Since the generators define a basis for a linear vector space, any new independent linear combination of these generators will also correspond to the generators of the same algebra.

Hamiltonian: To simplify we restrict to Hamiltonians linear or quadratic in the T_i,

$$H = \sum_i c_i T_i + \sum_{ij} c_{ij} T_i T_j. \tag{21.13}$$

For example, a typical n-particle Hamiltonian may take the second-quantized form

$$H = \sum_{ij} k_{ij} a_i^\dagger a_j + \sum_{ijlm} V_{ijlm} a_i^\dagger a_j^\dagger a_m a_l \qquad (i, j, l, m = 1, \ldots, n), \tag{21.14}$$

which implies physically that interactions are restricted to two-body or less.[1] As in the electromagnetic field example we now require (1) a dynamical group, (2) a corresponding Hilbert space, and (3) identification of an extremal state in the Hilbert space.

Dynamical Symmetry: The dynamical symmetry is defined through a Lie algebra

$$[T_i, T_j] = \sum_k C_{ij}^k T_k, \tag{21.15}$$

where C_{ij}^k are the structure constants of the algebra. For the cases of most interest the algebra is semisimple and it is convenient to transform the generators $\{T_i\}$ into the standard Cartan–Weyl basis $\{H_i, E_\alpha, E_{-\alpha}\}$ of Section 7.2, with the commutation relations

$$\begin{aligned} [H_i, H_j] &= 0 \qquad & [H_i, E_\alpha] &= \alpha_i E_\alpha, \\ [E_\alpha, E_{-\alpha}] &= \alpha^i H_i \qquad & [E_\alpha, E_\beta] &= N_{\alpha;\beta} E_{\alpha+\beta}, \end{aligned} \tag{21.16}$$

as in Eq. (7.17). The group associated with this algebraic structure will be labeled G.

[1] For a Hamiltonian of the form (21.14), the operator set $\{a_i^\dagger a_j, a_i^\dagger a_j^\dagger, a_i a_j\}$ generates a closed algebra that is SO($2n$) for fermions and Sp($2n$) for bosons [72, 224].

Hilbert Space: The Hilbert space for a given Hamiltonian carries an irreducible representation Γ^Λ of the group G.

Reference State: The reference state $|\Phi_0\rangle$ is a state in the Hilbert space that can be normalized to unity. In principle it can be arbitrary; in practice the nature and thus utility of the coherent states depends on this choice. It usually makes sense to choose as a reference state an unperturbed ground state. If G is semisimple, it is normal to select for $|\Phi_0\rangle$ the highest-weight or lowest-weight state of the representation Γ^Λ.

Maximum Stability Subgroup: The maximum stability subgroup H is a subgroup of the full dynamical group G that leaves the reference state invariant,

$$h\,|\Phi_0\rangle = |\Phi_0\rangle\, e^{i\phi(h)} \qquad h \in H, \tag{21.17}$$

up to a phase $e^{i\phi(h)}$ that is usually irrelevant.

The Coset Space G/H: Each dynamical group element $g \in G$ has a unique decomposition into a product of one element h in H and one element Ω in the coset G/H,

$$g = \Omega h \qquad (g \in G,\ \ h \in H,\ \ \Omega \in G/H). \tag{21.18}$$

Thus, each choice of extremal state $|\Phi_0\rangle$ implies a unique coset space.

Coherent States: The coherent state $|\Lambda, \Omega\rangle$ is then defined by

$$|\Lambda, \Omega\rangle \equiv \Omega\,|\Phi_0\rangle, \tag{21.19}$$

where the preceding definitions ensure that the action of an arbitrary group element $g \in G$ on the reference state is

$$g\,|\Phi_0\rangle = \Omega h\,|\Phi_0\rangle = \Omega\,|\Phi_0\rangle\, e^{i\phi(h)} = |\Lambda, \Omega\rangle\, e^{i\phi(h)}, \tag{21.20}$$

and where the definition (21.19) ensures that the generalized coherent states preserve the algebraic and topological properties of the coset space G/H.

21.4 Atoms Interacting with Classical Radiation

Let us illustrate the generalized coherent state method outlined in Section 21.3 by considering the Hamiltonian (21.5) again, but now in the limit where the atomic system is quantized in terms of two-level atoms but the electromagnetic field is described classically [231]. The Hamiltonian is

$$H = \sum_\alpha \left(\Delta E \sigma_0^{(\alpha)} + \gamma(t)\sigma_+^{(\alpha)} + \gamma^*(t)\sigma_-^{(\alpha)} \right), \tag{21.21}$$

where a constant has been discarded.

Dynamical Symmetry: The quantum-mechanical operators $(\sigma_0^{(\alpha)}, \sigma_+^{(\alpha)}, \sigma_-^{(\alpha)})$ appearing in Eq. (21.21) are Pauli matrices and generate an SU(2) algebra. Assuming the atoms to be independent, we may define many-atom operators

$$J_{\pm} \equiv \sum_{\alpha} \sigma_{\pm}^{(\alpha)} \qquad J_0 \equiv \sum_{\alpha} \sigma_0^{(\alpha)}, \tag{21.22}$$

and these too will obey an SU(2) angular momentum algebra

$$[\, J_0, J_{\pm}\,] = \pm J_{\pm} \qquad [\, J_+, J_-\,] = 2J_0. \tag{21.23}$$

Therefore, the Hamiltonian may be expressed as

$$H = H_0 + H_{\text{int}} = \Delta E J_0 + \gamma(t) J_+ + \gamma^*(t) J_-, \tag{21.24}$$

with dynamical group SU(2).

Hilbert Space and Reference State: The Hilbert space of SU(2) consists of the states $|jm\rangle$ with $m = -j, -j+1, \ldots, j-1, +j$, where

$$J^2 |jm\rangle = j(j+1) |jm\rangle \qquad J_0 |jm\rangle = m |jm\rangle, \tag{21.25}$$

and where all states $|jm\rangle$ can be constructed from an extremal state $|j - j\rangle$ through $j + m$ applications of the raising operator J_+,

$$|jm\rangle = \sqrt{\begin{pmatrix} 2j \\ j+m \end{pmatrix}} \frac{(j_+)^{j+m}}{(j+m)!} |j - j\rangle. \tag{21.26}$$

The eigenvalues of H_0 are $H_0 |jm\rangle = \Delta E J_0 |jm\rangle = (\Delta E)m |jm\rangle$, so the ground state is the extremal state $|j - j\rangle$.

Maximum Stability Subgroup: Selecting this extremal state as the reference state, the actions of the group generators on the extremal state are given by

$$\begin{gathered} J_+ |j,-j\rangle \propto |j,-j+1\rangle \neq |j,-j\rangle \qquad J_- |j,-j\rangle = 0 \neq |j,-j\rangle \\ J_0 |j,-j\rangle = -j\, |j,-j\rangle. \end{gathered}$$

Therefore, only J_0 leaves the extremal state invariant and it defines a single-generator U(1) subgroup that is the maximum stability subgroup,

$$h\, |j,-j\rangle = |j,-j\rangle\, e^{i\phi} \qquad (h \in \text{U}(1)), \tag{21.27}$$

with explicit elements $h = e^{i\alpha J_0}$, where $\alpha \equiv -\phi/j$.
Coset Space: The factor group is SU(2)/U(1). For g an element of SU(2), we obtain a unique coset decomposition $g = \Omega h$ by taking

$$\Omega(\xi) = e^{\xi J_+ - \xi^* J_-}, \tag{21.28}$$

where ξ is a complex number. Then $g = g\Omega = e^{\xi J_+ - \xi^* J_- + i\alpha J_0} \in \text{SU}(2)$.

Coherent States: From Eq. (21.28), the atomic coherent states associated with the Hamiltonian (21.21) may be defined as

$$|j, \xi\rangle \equiv \Omega(\xi)\, |j,-j\rangle = e^{\xi J_+ - \xi^* J_-}\, |j,-j\rangle, \tag{21.29}$$

so that for $g \in$ SU(2), $h \in$ U(1), and $\Omega(\xi) \in$ SU(2)/U(1),

$$g\,|j,-j\rangle = \Omega(\xi)h\,|j,-j\rangle = \Omega(\xi)\,|j,-j\rangle\, e^{i\phi} \equiv |j,\xi\rangle\, e^{i\phi}, \tag{21.30}$$

gives the action of a group element g on the extremal state.

Geometry of the Coset Space: Because of the correspondence between the coset space and the coherent states, *the geometry of the coherent-state space is determined uniquely by the geometry of the coset space.* As shown in Problem 21.2,

$$\Omega(\xi) = e^{\xi J_+ - \xi^* J_-} = \exp\begin{pmatrix} 0 & \xi \\ -\xi^* & 0 \end{pmatrix} = \begin{pmatrix} \cos|\xi| & \dfrac{\xi}{|\xi|}\sin|\xi| \\ -\dfrac{\xi^*}{|\xi|}\sin|\xi| & \cos|\xi| \end{pmatrix}, \tag{21.31}$$

where the last step follows from expanding the exponential. Since $\Omega(\xi)$ is an element of SU(2) we have $\det\Omega(\xi) = 1$, which when applied to (21.31) gives $\cos^2|\xi| + \sin^2|\xi| = 1$. Thus, the coset space corresponds to the unit 2-sphere S^2, which can be parameterized as

$$\xi = \frac{1}{2}\theta e^{-i\phi} \qquad (0 \le \theta \le \pi;\ 0 \le \phi \le 2\pi). \tag{21.32}$$

Since S^2 is compact, the topology of our atomic coherent-state manifold is compact.

21.5 Fermion Coherent States

Generalized coherent states for fermions are of particular interest because fundamental matter fields are always fermionic. The most important new ingredient for fermionic coherent states relative to bosonic ones is the antisymmetry of the wavefunction (Pauli principle), which has varied physical implications.

Example 21.1 Many features of fermion coherent states are exhibited by the simple case of a single fermion obeying the anticommutation relations $\{a, a^\dagger\} = 1$ and $\{a, a\} = \{a^\dagger, a^\dagger\} = 0$, where $a^\dagger$ creates a fermion, a annihilates a fermion, and $\{a, b\} \equiv ab + ba$. Problems 21.3–21.7 will guide you through finding the coherent state for this case.

Comprehensive results for generalized coherent states in many-fermion systems are given in Chs. 20, 31, and 32. Perhaps the most far-reaching outcome of generalized coherent states applied to fermionic systems is that it permits a systematic introduction of quasiparticles like those commonly encountered in condensed matter and nuclear physics, but subject to powerful symmetry constraints. This permits application of variational principles with states strongly restricted by symmetry. Such *symmetry-constrained variational methods* permit construction of realistic theories for correlated fermions that are also tractable because of the simplification enabled by the symmetry.

Background and Further Reading

For applications of coherent states in quantum optics and extensions to arbitrary fermion or boson fields see Gilmore [70, 71, 72], Glauber [76, 77], Perelomov [161], Sudarshan [186], and Zhang, Feng, and Gilmore [231]. Much of our discussion in this chapter has been influenced by the presentation in Ref. [231]. Fermionic applications of generalized coherent states are described in Chs. 20, 31, and 32 of the present text.

Problems

21.1 Verify the commutation relations for the H_4 algebra displayed in Eq. (21.7).

21.2 For the coherent state of atoms in Section 21.4, prove that

$$\Omega(\xi) = \exp\begin{pmatrix} 0 & \xi \\ -\xi^* & 0 \end{pmatrix},$$

starting from Eq. (21.29) and that $J_\pm \equiv J_1 \pm iJ_2$.

21.3 For Example 21.1, show that the fermion operator set $\{a,\ a^\dagger,\ a^\dagger a - \frac{1}{2}\}$ obeys

$$[a^\dagger, a] = 2(a^\dagger a - \tfrac{1}{2}) \qquad [a^\dagger a - \tfrac{1}{2}, a] = -a \qquad [a^\dagger a - \tfrac{1}{2}, a^\dagger] = a^\dagger,$$

and that this is equivalent to the SU(2) Lie algebra of Eq. (3.18).

21.4 Construct the Hilbert space corresponding to the Lie algebra for a single fermion in Problem 21.3. *Hint*: Remember that for a fermion the Pauli principle must be obeyed, which greatly restricts allowed states in the Hilbert space. ***

21.5 For the single-fermion example worked out in Problems 21.3 and 21.4, take as a reference state the minimal weight SU(2) state $|\frac{1}{2} - \frac{1}{2}\rangle$ corresponding to the fermion vacuum. Find the stability subgroup and the coset space. ***

21.6 Show that the coset representative is

$$\Omega(\xi) = e^{\xi J_+ - \xi^* J_-} = e^{\xi a^\dagger - \xi^* a},$$

for the generalized coherent state approximation corresponding to the single-fermion example worked out in Problems 21.3, 21.4, and 21.5. ***

21.7 Show that the coherent state corresponding to the single-fermion problem worked out in Problems 21.3 through 21.6 is

$$\left|\tfrac{1}{2}\,\xi\right\rangle = \left|\tfrac{1}{2}\,\theta\,\phi\right\rangle = e^{-i\phi}\sin\left(\tfrac{\theta}{2}\right)\left|\tfrac{1}{2}\tfrac{1}{2}\right\rangle + \cos\left(\tfrac{\theta}{2}\right)\left|\tfrac{1}{2} - \tfrac{1}{2}\right\rangle,$$

where θ and ϕ are angular variables parameterizing a sphere S^2. *Hint*: The coset representative $\Omega(\xi)$ is the same as for the example worked out in Section 21.4. ***

22 Restoring Symmetry by Projection

The low-energy spectrum of a quantum many-body system often is described concisely in terms of collective rotations of some equilibrium configuration and elementary excitations representing low-amplitude collective fluctuations about that equilibrium configuration. For example, in molecular physics the low-energy excitations may often be approximated as collective rotations of the molecule and vibrations of its bond lengths and angles. In nuclear physics also, the low-lying modes often resemble collective rotations or vibrations of the nuclear shape characterizing the equilibrium density distribution. For example, letting J denote angular momentum and π parity, low-lying $J^\pi = 0^+$ and 2^+ excitations in nuclei with even number of neutrons and protons often correspond to collective excitations called β- and γ-vibrational modes [27, 28]. The collective rotations and shape oscillations occurring in nuclei typically have equilibrium non-spherical quadrupole shapes. Such states have been pivotal in understanding nuclear structure but they break rotational invariance. In this chapter we address methods that allow the best of both worlds, permitting the use of broken symmetry states for ease of calculation and physical interpretation, while at the same time allowing recovery of the symmetry that is broken through projection of a symmetry-conserving state from an integral over broken symmetry states.

22.1 Rotational Symmetry in Atomic Nuclei

As shown in Ch. 17, for *spontaneous symmetry breaking* the states of a theory do not have the same symmetry as the Lagrangian or Hamiltonian of that theory. Collective states in atomic nuclei provide an example. Nuclear structure calculations often find (approximate) mean-field solutions having wavefunctions that imply *deformed* shapes for the density distribution and for which angular momentum is not a good quantum number. However, angular momentum conservation corresponds to rotational symmetry, which is one of the few universally conserved symmetries in nature: an isolated nucleus *must* conserve angular momentum, as described by a rotational invariant (spherically symmetric) Hamiltonian.

> Nuclear deformation is an emergent consequence (see Box 20.1) of spontaneous symmetry breaking in an *approximate solution* of the nuclear many-body problem, but the *exact solution* must conserve rotational symmetry.

Let us consider an approximate nuclear state corresponding to a prolate quadrupole deformation of the density, as in Fig. 22.1. Define an *intrinsic or body-fixed coordinate*

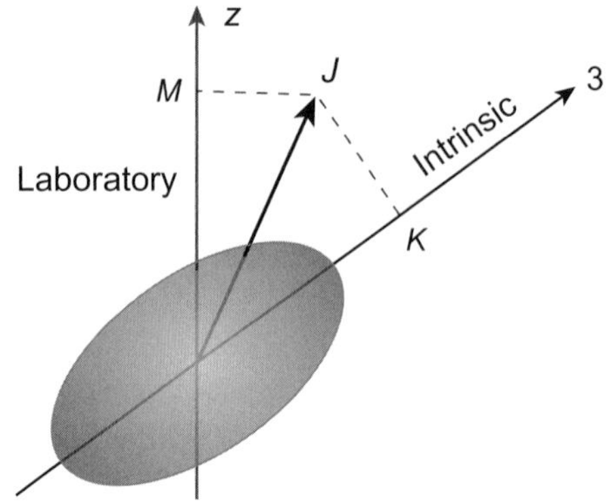

Fig. 22.1 Intrinsic and laboratory frames for a deformed nucleus of angular momentum J. The projection of J is M on the laboratory z-axis and K on the intrinsic 3-axis.

system with the 3-axis oriented along the long axis of the quadrupole. This deformed shape represents an *intrinsic state* of the deformed nucleus, and if the intrinsic state rotates the intrinsic coordinate system is assumed to rotate with it. Conversely, experiments will reference a fixed *laboratory coordinate system,* with the z-axis oriented as illustrated in Fig. 22.1.

The intrinsic states are spatially deformed, so they break rotational symmetry and do not conserve angular momentum. The nucleus is shown with a fixed orientation relative to the laboratory coordinate system but any orientation will have the same energy, so the lowest-energy deformed state is infinitely degenerate in the orientation of the intrinsic axis relative to the laboratory axis. This is reminiscent of the infinite angular degeneracy for the ground state of the complex scalar field illustrated in Fig. 17.4, and is in fact a non-relativistic analog of the spontaneously broken symmetry discussed in Section 17.5 for a complex scalar field. The intrinsic state is a useful theoretical construction but it breaks rotational invariance. States measured in experiments are observed in the laboratory reference frame, and they must be rotationally invariant and conserve angular momentum. This example motivates us to consider general methods to relate "intrinsic states," which break some fundamental symmetry, to "laboratory-frame states" that conserve these symmetries.

22.2 The Method of Generator Coordinates

To describe the relationship of symmetry breaking intrinsic states to symmetry conserving physical states, we introduce an approach called the *Generator Coordinate Method (GCM)*.

22.2.1 Generator Coordinates and Generating Functions

Generator coordinate methods start from a general prescription for a trial wavefunction

$$|\Psi\rangle = \int da f(a)\,|\Phi(a)\rangle\,. \tag{22.1}$$

This is a linear superposition of continuous functions, $|\Phi(a)\rangle$, termed *generating functions,* which are labeled by an unlimited number of (real or complex) parameters $\{a\}$ called *generator coordinates*. In Eq. (22.1), $|\Phi(a)\rangle$ can be any well-behaved wavefunction of a many-body Hilbert space. We do not need to display generating functions explicitly because they are not important for the current discussion. What is important here are the generator coordinates themselves $\{a\} = \{a_1, a_2, \ldots\}$.[1] One can always include a large set of generating functions $|\Phi(a)\rangle$ in the integration of Eq. (22.1) to better express the physics, but such a calculation becomes complicated very quickly. Practical applications must limit the number of generating functions by choosing only those general enough to describe the physics but simple enough to make the problem treatable; this requires physical intuition.

22.2.2 The Hill–Wheeler Equation

The *variational principle* of quantum mechanics that is reviewed in Box 22.1 asserts that for a set of trial wavefunctions the ones that best describe a system are those that minimize the energy E, which requires that the variation of the energy vanish, $\delta E = 0$. The key step in using the method of generator coordinates is to determine the weight function $f(a)$ in Eq. (22.1) by applying the variational principle in the form

$$\delta E = \delta \frac{\langle \Psi | \hat{H} | \Psi \rangle}{\langle \Psi | \Psi \rangle} = 0, \tag{22.2}$$

where $\hat{H}$ is the Hamiltonian operator. Putting Eq. (22.1) into Eq. (22.2) and performing a variational calculation with respect to $f(a)$ gives an integral equation

$$\int da' \langle \Phi(a) | \hat{H} | \Phi(a') \rangle f(a') = E \int da' \langle \Phi(a) | \Phi(a') \rangle f(a'), \tag{22.3}$$

called the *Hill–Wheeler equation* [111], which may be written formally as

$$\hat{H} f = E \hat{N} f, \tag{22.4}$$

with the overlap matrix elements for the Hamiltonian $H(a, a')$ and the norm $N(a, a')$,

$$H(a, a') = \langle \Phi(a) | \hat{H} | \Phi(a') \rangle \qquad N(a, a') = \langle \Phi(a) | \Phi(a') \rangle, \tag{22.5}$$

as integral kernels. Note that the norm matrix is not an identity matrix as in a normal eigenvalue equation because the basis formed by $|\Phi(a)\rangle$ is generally *not orthonormal.* Thus mathematical care must be exercised because assumptions employed in the usual formulation of quantum mechanics in an orthonormal basis are not always valid in the present discussion.

Finding solutions of Eq. (22.4) is easy formally: first invert the norm matrix N and then diagonalize $N^{-1}H$, giving a non-hermitian eigenvalue problem (with the restriction that N has no zero eigenvalues). In practice, solving the Hill–Wheeler equation to determine the weight function $f(a)$ can be a non-trivial numerical procedure, with the degree of complexity depending on the nature of the problem. There are particular examples where group representations may be used to great advantage in solving the Hill–Wheeler

[1] Note that the name "generator coordinates" is somewhat misleading because these are parameters in $|\Phi(a)\rangle$, not really coordinates in the usual sense. But it is the standard terminology, so we use it here.

Box 22.1 The Variational Principle

If we wish to calculate the ground state energy E_g for a quantum system described by a Hamiltonian $\hat{H}$ but cannot solve the Schrödinger equation $\hat{H}\,|\Psi\rangle = E\,|\Psi\rangle$ exactly, we can apply the *variational principle*, which is based on restricting $|\Psi\rangle$ to a set of mathematically simple *trial functions*. If the true wavefunction is not in the set the variational solution is *only an approximation*, but it can be a very good one.

The Variational Approximation

We start with eigenfunctions $\{|\psi_n\rangle\}$ of $\hat{H}$ that form a complete set, expressing $|\Psi\rangle$ as a linear combination of the orthonormalized set $\{|\psi_n\rangle\}$,

$$|\Psi\rangle = \sum_i c_i\,|\psi_i\rangle \qquad \hat{H}\,|\psi_i\rangle = E_i\,|\psi_i\rangle \qquad \langle\psi_i|\,\psi_j\rangle = \delta_{ij}.$$

The normalization condition for $|\Psi\rangle$ yields

$$1 = \langle\Psi|\,\Psi\rangle = \Big\langle\sum_i c_i\psi_i|\sum_j c_j\psi_j\Big\rangle = \sum_i\sum_j c_i^* c_j\,\langle\psi_i|\,\psi_j\rangle = \sum_i |c_i|^2,$$

and the expectation value of $\hat{H}$ can be calculated as

$$\langle\Psi|\,\hat{H}\,|\Psi\rangle = \Big\langle\sum_i c_i\psi_i|\hat{H}\sum_j c_j\psi_j\Big\rangle = \sum_i\sum_j c_i^* E_j c_j\,\langle\psi_i|\,\psi_j\rangle = \sum_i E_i|c_i|^2.$$

The true ground state energy corresponds to the smallest eigenvalue, $E_g \le E_i$, so

$$\langle\Psi|\,\hat{H}\,|\Psi\rangle = \sum_i E_i|c_i|^2 \ge E_g\sum_i |c_i|^2 = E_g,$$

and by the variational principle E_g is the lower bound of variational expectation values and the best $|\Psi\rangle$ gives an energy closest to E_g. A wisely chosen trial function can give a variational result that approximates the true ground state quite well.

Minimizing the Energy with Physical Constraints

Finding the smallest expectation value requires minimizing the energy functional,

$$\delta E(\Psi) = \delta\,\langle\Psi|\,\hat{H}\,|\Psi\rangle = 0,$$

with respect to variational parameters in the trial function. The u_k and v_k in the BCS trial wavefunction (22.24) are examples of variational parameters. The minimization is often subjected to a set of constraints $f_i(x_1, x_2, \ldots) = 0$. For example, in the BCS calculation a Lagrange multiplier term is added to the original energy functional to impose the condition that the solution have the correct average particle number. This is accomplished by adding constraint terms $\lambda_i f_i$,

$$G(x_1, x_2, \ldots, \lambda_1, \lambda_2, \ldots) = G_0(x_1, x_2, \ldots) + \lambda_1 f_1 + \lambda_2 f_2 + \cdots,$$

with $G_0(x_1, x_2, \ldots)$ the original energy functional. Equating all derivatives to zero,

$$\frac{\partial G}{\partial x_i} = 0 \qquad \frac{\partial G}{\partial \lambda_i} = 0,$$

then leads to a set of differential equations that must be solved numerically.

equation, not only by simplifying the numerical procedure, but also by giving physical meaning to the numerical results. Angular momentum projection is one such example.

22.3 Angular Momentum Projection

As discussed in Section 22.1, broken rotational symmetry implies a failure to conserve angular momentum. *Angular momentum projection* is a procedure to restore the broken rotational symmetry for approximate wavefunctions that do not conserve angular momentum. It may be viewed as a special application of the generator coordinate method to many-body systems executing the quantum equivalent of rotational motion. The generator coordinates in Eq. (22.1) are now chosen as the Euler angles $\Omega = (\alpha, \beta, \gamma)$ described in Section 6.3.2 and the generating functions are wavefunctions describing the quantum rotor oriented in a particular spatial direction Ω, which are called *intrinsic rotational states*.

22.3.1 The Rotation Operator and its Representations

Elements of the 3D rotation group SO(3) may be specified by the group parameter $\Omega = (\alpha, \beta, \gamma)$, which represents a set of Euler angles ($\alpha, \gamma = [0, 2\pi]$ and $\beta = [0, \pi]$) illustrated in Fig. 6.4 and discussed in Section 6.3.2. The explicit form of the rotation operator is given in Eq. (6.30) as

$$D(\Omega) = e^{-\imath\alpha J_z} e^{-\imath\beta J_y} e^{-\imath\gamma J_z}, \tag{22.6}$$

where J_i ($i = x, y, z$) are the angular momentum operators and $\Omega = (\alpha, \beta, \gamma)$. Its (unitary) representation is

$$\left\langle \mu JM \right| D(\Omega) \left| \nu JM' \right\rangle = \delta_{\mu\nu} D^J_{MM'}(\Omega)^*, \tag{22.7}$$

where * denotes complex conjugation of the Wigner D-matrix given in Eq. (6.31a). Equation (22.7) is defined for rotation within a single coordinate system. Now we consider rotation by Ω *between* the laboratory frame and the body-fixed intrinsic frame. A generalization for the representation of a rotation of Ω between the two coordinate systems is accomplished by using $D^J_{MK}(\Omega)$, in which a subindex K replaces M' in Eq. (22.7):

$$\left\langle \mu JM \right| D(\Omega) \left| \nu JK \right\rangle = \delta_{\mu\nu} D^J_{MK}(\Omega)^*, \tag{22.8}$$

where K denotes the third component of angular momentum in the body-fixed frame illustrated in Fig. 22.1. For justification, see Box 22.2.

For a state $\left|\mu JM\right\rangle$ having angular momentum quantum numbers J and M, the parameter μ designates any additional quantum numbers that are required to specify the quantum state uniquely, so that the closure condition

$$\sum_{\mu JM} \left|\mu JM\right\rangle \left\langle \mu JM\right| = 1 \tag{22.9}$$

Box 22.2 Rotation between Laboratory and Body-Fixed Frames

In Section 6.3.2 we discussed 3D rotation by an Euler angle Ω (see Fig. 6.4) for rotations of angular momentum in a single coordinate system. Now we extend the picture to rotation by Ω between the laboratory system and the body-fixed system illustrated in Fig. 22.1. We shall summarize the results here, with the technical details to be found in Edmonds [51]. Let us use $(L_i, i = x, y, z)$ to denote the three angular momentum components in the laboratory system and $(I_\mu, \mu = 1, 2, 3)$ for the corresponding components in the body-fixed system. It is convenient to write the angular momentum vector as a spherical tensor of rank 1 (see Section 6.5):

$$L_{-1} = \frac{1}{\sqrt{2}}(L_x - iL_y) \qquad L_0 = L_z \qquad L_{+1} = -\frac{1}{\sqrt{2}}(L_x + iL_y).$$

The spherical components of the angular momentum with respect to the body-fixed frame are

$$I_\mu = \sum_i D^1_{i\,\mu}(\Omega) L_i,$$

which relates the components of the two coordinate frames. After working out the commutation relations between the angular momentum components of the two coordinate frames, we can determine the tensor properties of $D^J_{MK}(\Omega)$ under rotations around the axis in the laboratory and in the body-fixed frame. Specifically,

$$[L_i, D^{J^*}_{MK}] = \sum_{M'} D^J_{M'K} \langle JM'|\, J_i \,|JM\rangle \qquad [I_\mu, D^{J^*}_{MK}] = \sum_{K'} D^J_{MK'} \langle JK|\, J_\mu \,|JK'\rangle.$$

Comparing with Eq. (6.56), the function $D^{J^*}_{MK}$ behaves as a spherical tensor of rank J. The normalized wavefunctions

$$|JMK\rangle = \sqrt{\frac{2J+1}{8\pi^2}}\, D^J_{MK}(\Omega)$$

are simultaneous eigenfunctions of J^2, $L_z (\equiv J_z)$, and I_3,

$$J^2 \,|JMK\rangle = J(J+1)\,|JMK\rangle \quad L_z \,|JMK\rangle = M\,|JMK\rangle \quad I_3 \,|JMK\rangle = K\,|JMK\rangle,$$

and the $|JMK\rangle$ form a complete and orthogonal set of spatial functions depending on the Euler angles Ω.

holds. We need not know details of the state $\left|\mu JM\right\rangle$, except that it belongs to a complete set of orthonormal vectors in a Hilbert space in which the rotation operator (22.6) acts. The D-functions in Eq. (22.8) form a complete set of functions in the parameter space of Ω (see Box 22.2). Applying $\sum_{\mu JM} \left|\mu JM\right\rangle$ from the left to both sides of Eq. (22.8) and considering Eq. (22.9) gives

$$D(\Omega)\,|\nu JK\rangle = \sum_M |\nu JM\rangle\, D^J_{MK}(\Omega)^*, \tag{22.10}$$

which is the multiplet relation between the states belonging to the representation labeled by angular momentum J.

22.3.2 The Angular Momentum Projection Operator

Now let us suppose that $|\Phi\rangle$ in Eq. (22.1) is a deformed state obtained as a solution of a variational calculation for a many-particle system. Thus $|\Phi\rangle$ is an intrinsic state that violates rotational invariance and is not an eigenstate of angular momentum. A rotationally invariant Hamiltonian obeys

$$D^{\dagger}(\Omega)\hat{H}D(\Omega) = \hat{H} \tag{22.11}$$

and the following identity for the energy expectation value holds

$$\frac{\langle\Phi|\,\hat{H}\,|\Phi\rangle}{\langle\Phi|\,\Phi\rangle} = \frac{\langle\Phi|\,D^{\dagger}(\Omega)\hat{H}D(\Omega)\,|\Phi\rangle}{\langle\Phi|\,D^{\dagger}(\Omega)D(\Omega)\,|\Phi\rangle}. \tag{22.12}$$

This means that the energy expectation value remains the same when the state $|\Phi\rangle$ is rotated spatially by an angle Ω. In other words, all rotated states, $|\Phi(\Omega)\rangle \equiv D(\Omega)\,|\Phi\rangle$, having different orientations Ω, are degenerate. As $|\Phi(\Omega)\rangle$ represents a state with definite spatial orientation Ω, it is linearly independent of any other rotated state with $|\Phi(\Omega')\rangle$ and the most general wavefunction corresponds to a superposition of all the rotated states [160]

$$|\Psi\rangle = \int d\Omega\, F(\Omega)\,|\Phi(\Omega)\rangle = \int d\Omega\, F(\Omega)D(\Omega)\,|\Phi\rangle\,, \tag{22.13}$$

where $F(\Omega)$ is a weight function specified further below. We recognize immediately that the form of Eq. (22.13) is the same as that of (22.1), and Eq. (22.13) is just a special case of the generator coordinate method with the Euler angles Ω acting as generator coordinates.

The weight function $F(\Omega)$ appearing in Eq. (22.13) may be evaluated by taking advantage of the group representations associated with rotations and making use of the completeness of the D-functions. We first expand the weight function $F(\Omega)$ as

$$F(\Omega) = \sum_{JMK} \frac{2J+1}{8\pi^2}\, F^{J}_{MK} D^{J}_{MK}(\Omega), \tag{22.14}$$

and then insert it into Eq. (22.13) to obtain

$$|\Psi\rangle = \sum_{JMK} F^{J}_{MK}\hat{P}^{J}_{MK}\,|\Phi\rangle\,, \tag{22.15}$$

where $\hat{P}^{J}_{MK}$ is defined as

$$\hat{P}^{J}_{MK} = \frac{2J+1}{8\pi^2}\int d\Omega\, D^{J}_{MK}(\Omega)\hat{D}(\Omega) \tag{22.16}$$

and is termed the *angular momentum projection operator*. The coefficients F^{J}_{MK} in Eq. (22.15) play the part of variational parameters in place of the weight function $F(\Omega)$ in Eq. (22.13). Now $|\Psi\rangle$ in Eq. (22.15) is expressed as a linear combination of a set of states created by the operator $\hat{P}^{J}_{MK}$ acting on the deformed state $|\Phi\rangle$. Thus, $\hat{P}^{J}_{MK}\,|\Phi\rangle$ in Eq. (22.15) is the (angular momentum) projected state. Using Eqs. (22.10) and (22.16), and the orthogonality of the D-functions given in Eq. (6.38), one obtains the relation

$$\hat{P}^{J}_{MK}\,|\nu J'K'\rangle = \delta_{JJ'}\delta_{KK'}\,|\nu JM\rangle\,. \tag{22.17}$$

From Eqs. (22.9) and (22.17), one can obtain its spectral representation and the "sum rule"

$$\hat{P}^J_{MK} = \sum_\nu |\nu JM\rangle\langle\nu JK| \qquad \sum_{JM} \hat{P}^J_{MM} = 1. \tag{22.18}$$

Using the spectral representation, one can easily derive the properties for the angular momentum projection operator

$$\left(\hat{P}^J_{MK}\right)^\dagger = \hat{P}^J_{KM} \qquad \hat{P}^J_{KM}\hat{P}^{J'}_{M'K'} = \delta_{JJ'}\delta_{MM'}\hat{P}^J_{KK'}. \tag{22.19}$$

If we carry out a variational procedure with the trial wavefunction of Eq. (22.15), it is easy to show that the summation over J and M drops out due to Eqs. (22.19) and (22.11); therefore, the state has sharp values of J and M, and it is sufficient to carry out the variational calculation with

$$|\Psi\rangle = \sum_K F^J_K \hat{P}^J_{MK}|\Phi\rangle, \tag{22.20}$$

without the summation over J and M. This means that $|\Psi\rangle$ becomes an eigenstate of angular momentum. The rotational symmetry violated in the deformed state $|\Phi\rangle$ is thus recovered in the new state $|\Psi\rangle$, as discussed further in Box 22.3.

22.3.3 Solving the Eigenvalue Equation

From the results of the preceding discussions, we can write explicitly the eigenvalue equation in the (angular momentum) projected basis by putting $\hat{P}^J_{MK}|\Phi\rangle$ from Eq. (22.15) into Eq. (22.5) to replace $\Phi(a)$:

$$H = \left\langle\Phi\left|\hat{P}^{J\dagger}_{MK}\hat{H}\hat{P}^{J'}_{M'K'}\right|\Phi\right\rangle \qquad N = \left\langle\Phi\left|\hat{P}^{J\dagger}_{MK}\hat{P}^{J'}_{M'K'}\right|\Phi\right\rangle. \tag{22.21}$$

The Hamiltonian is spherically symmetric so it behaves like a scalar and commutes with the projector,

$$\hat{P}^{J\dagger}_{MK}\hat{H}\hat{P}^{J'}_{M'K'} = \hat{H}\hat{P}^{J\dagger}_{MK}\hat{P}^{J'}_{M'K'}.$$

Considering the properties of the angular momentum projector given in Eq. (22.19), we obtain a GCM equation of the Hill–Wheeler type [111]. This is an eigenvalue equation with the normalization condition written in a non-orthogonal basis:

$$\sum_{K'}\left(H^J_{KK'} - EN^J_{KK'}\right)F^J_{K'} = 0 \qquad \sum_{KK'} F^J_K N^J_{KK'} F^J_{K'} = 1, \tag{22.22}$$

where the Hamiltonian matrix and norm matrix are defined as

$$H^J_{KK'} = \left\langle\Phi\left|\hat{H}\hat{P}^J_{KK'}\right|\Phi\right\rangle \qquad N^J_{KK'} = \left\langle\Phi\left|\hat{P}^J_{KK'}\right|\Phi\right\rangle. \tag{22.23}$$

An eigenvalue equation is usually solved by diagonalizing the Hamiltonian matrix. In introductory quantum mechanics we normally work with basis states that are orthogonal, but for the present case the calculation is more involved because the matrix in the projection theory is represented in a non-orthogonal basis. This means that the norm matrix must be diagonalized first and then the Hamiltonian matrix transformed into the representation in which the norm matrix is diagonal. We shall omit the technical details, which may be found in Ref. [187].

Box 22.3 The Projected State $|\Psi\rangle$ Is an Eigenstate of Angular Momentum

A rotated state $|\Phi(\Omega)\rangle = D(\Omega)\,|\Phi\rangle$ corresponds to a particular spatial orientation, Ω. The wavefunction $|\Psi\rangle$ in Eq. (22.13) is a superposition of all these rotated states. Equation (22.13) indicates that $|\Psi\rangle$ lies in the space spanned by all $|\Phi(\Omega)\rangle$. We can thus define a projection operator for this space:

$$P_\Phi = \int d\Omega\,|\Phi(\Omega)\rangle\,\langle\Phi(\Omega)|\,.$$

It is easy to prove (see Problem 22.1) that P_Φ and the rotation operator $D(\Omega)$ commute: $D(\Omega)P_\Phi = P_\Phi D(\Omega)$. On the other hand, a rotationally invariant Hamiltonian $\hat{H}$ commutes with $D(\Omega)$ and $P_\Phi \hat{H} P_\Phi$ also commutes with $D(\Omega)$. Thus $P_\Phi \hat{H} P_\Phi$ and J^2 have a simultaneous eigenfunction, which is temporarily denoted as $|\psi\rangle$,

$$P_\Phi \hat{H} P_\Phi|\psi\rangle = E|\psi\rangle \qquad J^2|\psi\rangle = J(J+1)|\psi\rangle.$$

Remember that the projection operator P_Φ defines a linear transformation P_Φ that maps a vector space to itself such that $P_\Phi^2 = P_\Phi$. Thus $P_\Phi|\psi\rangle = |\psi\rangle$ and $|\psi\rangle$ lies completely in the space spanned by $\Phi(\Omega)$. The energy minimization expression is

$$\delta\frac{\langle\psi|P_\Phi\,\hat{H}P_\Phi|\psi\rangle}{\langle\psi|\psi\rangle} = \delta\frac{\langle\psi|\hat{H}|\psi\rangle}{\langle\psi|\psi\rangle},$$

but $|\psi\rangle$ is an eigenstate of $P_\Phi \hat{H} P_\Phi$ so this variation is zero. We conclude the following.

1. The state $|\psi\rangle$ minimizes the energy.
2. The state $|\psi\rangle$ is an eigenstate of angular momentum.
3. The state $|\psi\rangle$ lies in the space spanned by $|\Phi(\Omega)\rangle$.

Therefore, it should be possible to find an appropriate $F(\Omega)$ in Eq. (22.13) so that $|\Psi\rangle = |\psi\rangle$ and is rotationally invariant.

22.4 Particle Number Projection

In microscopic systems of identical fermions the spin-$\frac{1}{2}$ particles can under certain conditions form pairs with total spin zero called *Cooper pairs* (see Box 32.2) [44]. The Bardeen–Cooper–Schrieffer or *BCS theory* [18] describes the phenomenon of superconductivity as caused by a condensation of Cooper pairs into a collective coherent state with long-range pairing order.[2] It gives an *approximation* for the quantum many-body

[2] Superconductivity is characterized by the ability of charge to flow without resistance and the expulsion of magnetic fields from the superconductor (the *Meissner effect*). A similar phenomenon corresponding to a current of uncharged particles that flows without resistance is called *superfluidity*. The original BCS theory postulated a specific mechanism for binding of electron Cooper pairs by phonon (vibrational) interactions in a solid. We adopt a broader view that the BCS theory corresponds to wavefunctions of the BCS form, irrespective of the specific type of fermion or of the specific mechanism binding them into Cooper pairs.

state corresponding to the superconductor known as the *BCS state,* which is a specific example in a non-relativistic context of the *spontaneous symmetry breaking* described in Ch. 17. The symmetry that is broken corresponds to a *global gauge invariance* associated with phase rotations of the particle wavefunctions. The broken symmetry then implies that particle number, which is a variable conjugate to the gauge rotation angle in a quantized theory, is not conserved.

Just as the breaking of rotational symmetry implies that angular momentum is not conserved, breaking of (global) gauge symmetry implies a failure to conserve quantities such as electrical or more general charges related to particle number. If the effective number of fermions participating in a state with spontaneously broken gauge symmetry is large, the implied fluctuation in particle number has minimal effect. However, for fermionic systems with a relatively small number of particles, failure to conserve particle number in intrinsic states can cause large unphysical effects. As we now discuss, an effective way to deal with this is to project a state of good particle number from the broken symmetry state.

22.4.1 Violation of Particle Number in BCS Theory

The *BCS wavefunction* is given by [18],

$$|\Psi_{\rm BCS}\rangle = \prod_{k>0}(u_k + v_k c^\dagger_{k\uparrow} c^\dagger_{\bar{k}\downarrow})\,|0\rangle\,, \tag{22.24}$$

where k is a momentum label, the arrows indicate up or down spin projections, and u_k and v_k represent variational parameters that are determined by energy minimization. The product in Eq. (22.24) runs over only half the configuration space, as indicated by $k > 0$. For each state with $k > 0$, there exists a conjugate state $\bar{k} < 0$ that couples to the $k > 0$ state to form a Cooper pair. If the Hamiltonian is invariant under time reversal, the conjugate state can be chosen as the time-reversed state. Example 22.1 illustrates.

Example 22.1 In a spherical single-particle basis, $|k\rangle = |jm\rangle$ and $\left|\bar{k}\right\rangle = |j\,-m\rangle$ for $m > 0$, where j is angular momentum and m is its projection. The m and $-m$ states correspond to spin-up and spin-down particles, which couple to a (Cooper) pair with total spin zero.

The factors u_k and v_k are the probability amplitudes that a Cooper pair is or is not occupied, respectively, which are constrained by $|\,u_k\,|^2 + |\,v_k\,|^2 = 1$. To carry out a variational calculation, one first constructs a variational Hamiltonian $\hat{H}' = \hat{H} - \lambda\hat{N}$ and then solves $\delta\,\langle\Psi_{\rm BCS}|\,\hat{H}'\,|\Psi_{\rm BCS}\rangle = 0$ to find the energy minimum. The Lagrange multiplier λ is fixed by the condition

$$\langle\Psi_{\rm BCS}|\,\hat{N}\,|\Psi_{\rm BCS}\rangle = 2\sum_{k>0} v_k^2 = N, \tag{22.25}$$

which implies that (see Problem 22.2)

$$\lambda = \frac{dE}{dN}, \tag{22.26}$$

where E is the energy and N is the particle number. Thus λ, which is called the *chemical potential,* represents the change in energy associated with a change in particle number. In reactions a system tends to move from higher to lower chemical potential. At zero temperature λ is also termed the *Fermi energy* ϵ_F, with a corresponding *Fermi momentum* $p_f = \hbar k_F$. At a temperature of absolute zero, ϵ_F is the energy of the highest occupied single-fermion level. Let us introduce a pair creation operator

$$A^\dagger \equiv \sum_{k>0} \frac{v_k}{u_k} c^\dagger_{k\uparrow} c^\dagger_{\bar{k}\downarrow} \tag{22.27}$$

and rewrite Eq. (22.24) as

$$|\Psi_{BCS}\rangle \propto \exp\left(A^\dagger\right)|0\rangle = \sum_{i=0}^{\infty} \frac{1}{i!}\left(A^\dagger\right)^i |0\rangle, \tag{22.28}$$

implying that $|\Psi_{BCS}\rangle$ is a *superposition of different numbers of pairs* and thus does not have a definite particle number. This approximation can cause serious error if the particle number is small.

Example 22.2 Nuclear structure physics typically deals explicitly with only tens to hundreds of valence nucleons. A BCS wavefunction for a nucleus (Z, N) of proton number Z and neutron number N can be constrained to have the correct *average* values of Z and N, but it will have non-zero components that belong to *different nuclei* with a range of proton and neutron numbers around the average Z and average N. This is illustrated schematically in Fig. 22.2, and the fluctuation in particle number for a BCS wavefunction is discussed more quantitatively in Box 22.4.

The physical consequences of particle number fluctuations depend on what quantities are being calculated in BCS approximation. For the nuclear structure problem illustrated in Example 22.2 the error is often not very consequential for matrix elements of ground state properties, but can be significant for matrix elements involving excited states of nuclei or the transfer of particles between nuclei in reactions.

22.4.2 Bogoliubov Quasiparticles

A *quasiparticle* is a “dressed” particle formed from a bare particle by absorbing correlations from a field. A “bare” particle in a strongly correlated field then behaves as if it were a different non-interacting particle in free space. The quasiparticle concept is important since it is one of the few known systematic ways of simplifying a quantum-mechanical many-body problem.[3] In addition to the quasielectrons that we discuss in this subsection,

[3] However, the simple quasiparticle approach outlined here is not useful for *all* systems. In many strongly correlated materials, not all field effects can be dressed into a bare particle and the elementary excitations are so far from being independent that it is not even useful as a starting point to treat them as independent. This is often true for *emergent states* (Box 20.1), which are typically separated from the weakly interacting states by a phase transition (see Box 32.1 and Section 29.1.1). This is particularly important when there are *competing emergent states,* such as for deformed nuclei with strong pairing fields or high-temperature superconductivity

Box 22.4 Particle Number Fluctuation in the BCS Wavefunction

The BCS approximation gives a wavefunction $|\Psi_{\rm BCS}\rangle$ that is not an eigenstate of the particle number operator because the approximation breaks U(1) gauge symmetry. Thus, we may expect that the BCS theory is well grounded only if $|\Psi_{\rm BCS}\rangle$ is localized around the actual particle number N. To estimate the *particle number fluctuation* about N, one can calculate the *mean square deviation* of the particle number. As shown in Problem 22.5, this is given by

$$(\Delta N)^2 = \langle \Psi_{\rm BCS}|\,\hat{N}^2\,|\Psi_{\rm BCS}\rangle - N^2$$

for the BCS wavefunction. In the second-quantized representation (see Appendix A), the particle number operator is

$$\hat{N} = \sum_{k>0} (c_k^\dagger c_k + c_{-k}^\dagger c_{-k}).$$

By using the BCS wavefunction (22.24) and the constraint condition (22.25), one obtains from the first equation above

$$(\Delta N)^2 = 4\sum_{k>0} (u_k v_k)^2.$$

The sum extends over all states but only those with energies close to the Fermi energy $\varepsilon_{\rm F}$ contribute significantly, as illustrated in the following figure.

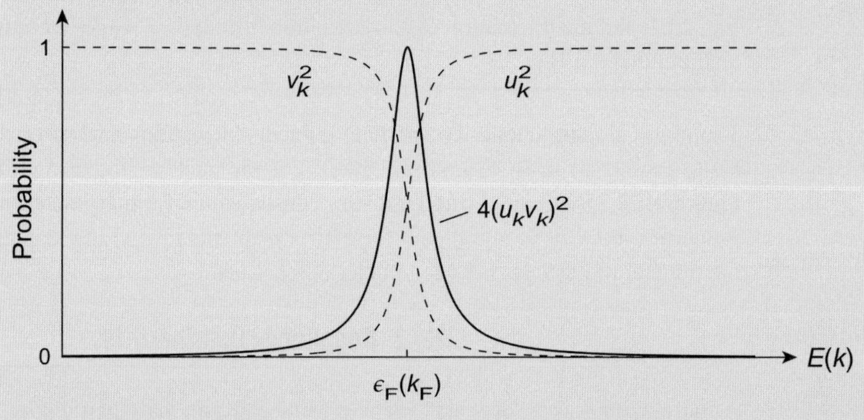

The value of $(\Delta N)^2$ depends on the specifics of the system because the single-particle distribution and interactions determine the values of u_k and v_k in the variational calculation. For example, consider a *pure pairing force* Hamiltonian,

$$\hat{H} = \sum_{k>0} \epsilon_k (c_k^\dagger c_k + c_{-k}^\dagger c_{-k}) - G \sum_{kk'>0} c_k^\dagger c_{-k}^\dagger c_{-k'} c_{k'},$$

where the $\{\epsilon_k\}$ are single-particle energies and G is the pairing strength. Both terms influence u_k and v_k; thus they both affect the BCS particle number fluctuation.

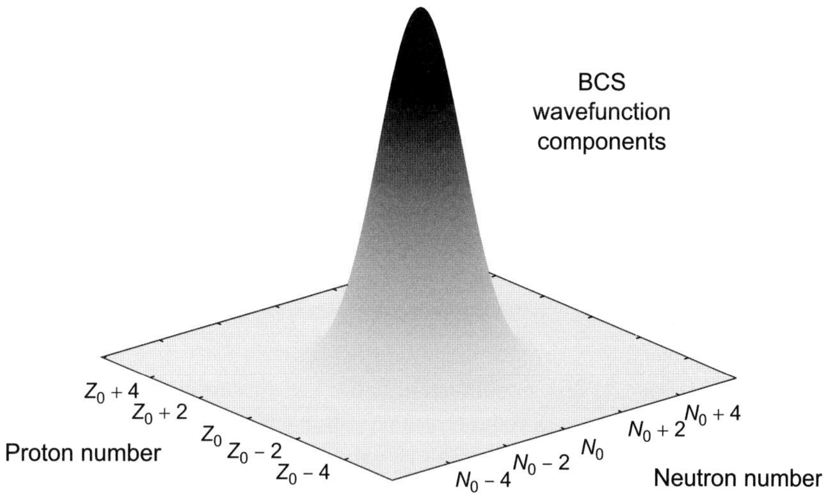

Fig. 22.2 Components of different particle number in a BCS wavefunction for a nucleus with proton number Z_0 and neutron number N_0. The wavefunction has the correct average (Z_0, N_0), but has components corresponding to nuclei with $Z_0 \pm 2, Z_0 \pm 4, \ldots$, and $N_0 \pm 2, N_0 \pm 4, \ldots$, unless wavefunctions of good proton number and good neutron number are projected.

other examples of quasiparticles include phonons (particles derived from the vibrations of atoms in a solid), plasmons (particles derived from plasma oscillations), and quasiprotons and quasineutrons (protons and neutrons in strong nuclear pairing fields).

The BCS wavefunction may be viewed as a product state of new types of fermions, called *Bogoliubov quasiparticles*. If we label the time-reversed state of $|k\rangle$ as $|-k\rangle$, the quasiparticles (created and annihilated by $\alpha_k^\dagger$ and α_k, respectively) and the "bare" particles (created and annihilated by $c_k^\dagger$ and c_k, respectively) are related by the *Bogoliubov–Valatin transformations* (which we will term *Bogoliubov transformations* for brevity),

$$\begin{aligned} \alpha_k^\dagger &= u_k c_k^\dagger - v_k c_{-k} & \alpha_{-k}^\dagger &= u_k c_{-k}^\dagger + v_k c_k, \\ \alpha_k &= u_k c_k - v_k c_{-k}^\dagger & \alpha_{-k} &= u_k c_{-k} + v_k c_k^\dagger, \end{aligned} \tag{22.29}$$

where u_k and v_k are real and defined only for positive k, with $u_k^2 + v_k^2 = 1$. Crucially, (22.29) preserves the fermion algebraic structure and the quasiparticles are also fermions, obeying

$$\{\alpha_k, \alpha_{k'}\} = \{\alpha_k^\dagger, \alpha_{k'}^\dagger\} = 0, \qquad \{\alpha_k, \alpha_{k'}^\dagger\} = \delta_{kk'} \tag{22.30}$$

(see Problem 22.4), which can be used with Eq. (22.29) to express the BCS wavefunction (22.24) as a product of quasiparticle creation operators acting on the vacuum,

$$|\Psi_{\rm BCS}\rangle \propto \prod_k \alpha_k^\dagger \, |0\rangle . \tag{22.31}$$

where strong pairing fields, antiferromagnetism, and Coulomb repulsion compete to determine the structure of highly collective states. As we shall demonstrate in Chs. 31 and 32, in that case *dynamical symmetries* based on compact Lie algebras may provide a powerful method to deal with the competing emergent states.

From Eqs. (22.29), a Bogoliubov quasiparticle is a linear combination of a bare particle and a bare hole. Let us simplify by assuming zero temperature, so that λ corresponds to the Fermi energy ϵ_F. When a quasiparticle lies far above the Fermi energy ϵ_F (so that u_k^2 is large and v_k^2 small) it is approximately a particle, while far below ϵ_F (so that u_k^2 is small and v_k^2 large) it is approximately a hole. Close to ϵ_F it is part particle and part hole.

> Through a Bogoliubov transformation (22.29) the complicated state (22.24) of pairwise *interacting particles* has been replaced by a simple product wavefunction (22.31) of *non-interacting quasiparticles*.

The price of this simplification is the loss of particle number conservation (spontaneous breaking of gauge symmetry) that follows from mixing particle creation and annihilation operators in Eq. (22.29). But as we saw in Ch. 17, symmetries broken spontaneously are not truly broken; they are just hidden. This suggests that it should be possible to recover the broken symmetry. We shall now discuss how to do so by using projection techniques.

22.4.3 The Particle Number Projection Operator

As discussed in Section 3.4.3, for U(N) internal symmetries the physical meaning of the U(1) factor in the homomorphism U(N) $\to$ U(1) $\times$ SU(N) is that some quantity related to a particle number is conserved. As a simple but illustrative example, we discuss the U(1) factor for the U(2) $\supset$ U(1) $\times$ SU(2) case in Box 22.5 where the particle number operator is represented in terms of gauge angles, which is analogous to representation of the angular momentum operator J_z in the real space in terms of Euler angles that was discussed in Section 6.3.1. Thus particle number projection has much in common with angular momentum projection but is simpler, since we work in a 1D gauge space associated with U(1) instead of the 3D Euler angle space associated with SO(3).

To implement particle number projection let us again start from the GCM trial wavefunction in Eq. (22.1). Our interest lies in the gauge group connected with particle number so the generator coordinate is the gauge group phase ϕ. From Box 22.5 the number operator $\frac{1}{2}\hat{N}$ has the representation $i\partial/\partial\phi$, with eigenfunctions

$$f(\phi) = \frac{1}{2\pi}\sum_n g_n e^{-in\phi}, \tag{22.32}$$

where the g_n are constants and the integer n reflects the periodicity in ϕ. Proceeding as for the treatment of angular momentum projection, we insert (22.32) into (22.1) to give

$$|\Psi\rangle = \sum_n g_n \hat{P}^{2n} |\Phi\rangle, \tag{22.33}$$

where $|\Phi\rangle$ is a particle number violating state such as one described by a BCS wavefunction. The *particle number projection operator* for a state of good particle number N is

$$\hat{P}^N = \frac{1}{2\pi}\int_0^{2\pi} e^{-i\phi(N-\hat{N})}d\phi, \tag{22.34}$$

Box 22.5 Particle Number Is Related to Gauge Angle

In Problems 31.1 through 31.6 we will work with a single-j shell model, where *quasispin operators* S_i close under the SU(2) commutator algebra. The effective Hamiltonian of the single-j model is

$$\hat{H} = -GS_+S_- = -G(\boldsymbol{S} \cdot \boldsymbol{S} - S_z^2 + S_z),$$

where G is a coupling constant. The quasispin operators act in a three-dimensional quasispin space and $\boldsymbol{S}^2$ and S_z commute with the single-j Hamiltonian. Therefore one can classify the eigenstates according to their *seniority* s (see Problem 31.4) and the particle number N,

$$s = \Omega - 2S \qquad \hat{N} = 2S_z + \Omega,$$

where $\Omega = \frac{1}{2}(2j + 1)$ denotes half the shell degeneracy. We see that the particle number operator $\hat{N}$ is related to S_z and acts as an infinitesimal U(1) generator of rotations around the z-axis in the abstract *quasispin space*. Such rotations represent *global gauge transformations* with the rotational angle ϕ regarded as the *gauge angle*. We can thus express the particle number operator $\hat{N}$ as

$$\frac{\hat{N}}{2} = -\frac{1}{i}\frac{\partial}{\partial\phi} + \text{constant},$$

in terms of the gauge angle ϕ.

where only the term $2n = N$ has been retained in the summation (22.33). Comparing the particle number projector Eq. (22.34) with the angular momentum projector (22.16) we can say the following.

- The Euler angles Ω have been replaced by the gauge angle ϕ.
- The rotation operator $\hat{D}(\Omega)$ has been replaced by the particle number operator $\hat{N}(\phi)$.
- The rotational matrix elements $D^J_{MK}(\Omega)$ have been replaced by $\sim e^{-i\phi N}$.

The projected wavefunction with conserved particle number is then

$$|\Psi\rangle = \hat{P}^N |\Phi\rangle = \mathcal{N} \int_0^{2\pi} d\phi e^{-i\phi(N-\hat{N})} |\Phi\rangle, \tag{22.35}$$

where $\mathcal{N}$ is a normalization. This wavefunction may now be used to calculate physical observables, as in Example 22.3.

Example 22.3 The energy of a particle number projected state can be expressed as an integral over the gauge angle ϕ of the Hamiltonian matrix elements between states labeled by gauge angles. Formally,

$$E^N = \frac{\langle\Psi| \hat{H} |\Psi\rangle}{\langle\Psi| \Psi\rangle} = \frac{\langle\Phi| \hat{H}\hat{P}^N |\Phi\rangle}{\langle\Phi| \hat{P}^N |\Phi\rangle}.$$

Then the following integrations,

$$\langle\Phi| \left\{ \begin{matrix} \hat{H} \\ 1 \end{matrix} \right\} \hat{P}^N |\Phi\rangle = \frac{1}{2\pi} \int_0^{2\pi} d\phi e^{-i\phi N} \langle\Phi| \left\{ \begin{matrix} \hat{H} \\ 1 \end{matrix} \right\} e^{i\phi\hat{N}} |\Phi\rangle,$$

may be used to calculate the matrix elements required in the expression for E^N.

Thus, projection of a state of good particle number from an approximate many-body state that breaks global U(1) gauge invariance spontaneously is analogous to projection of a state of good angular momentum from an approximate many-body state that breaks rotational SO(3) invariance spontaneously, with the Euler angle generator coordinate Ω for rotations replaced by the gauge angle generator coordinate ϕ for particle number.

22.5 Parity Projection

Parity transformations were introduced in Box 6.1 and Section 13.7. They involve a flip in the sign of spatial coordinates that changes a right-handed coordinate system into a left-handed one or vice versa. As discussed in Box 13.4, the strong and electromagnetic interactions conserve parity but the weak interactions violate parity symmetry with abandon.

22.5.1 The Parity Transformation

The group of reflections through a point is composed of the reflection operation $U_{\mathscr{R}}$ and the identity operation I, and is isomorphic to the group Z_2. In the transformation induced by $\mathscr{R}$ the polar coordinate $\boldsymbol{r}$ and momentum $\boldsymbol{p}$ change sign, while the axial vectors $\boldsymbol{r} \times \boldsymbol{p}$ and spin $\boldsymbol{s}$ are invariant, and since $\boldsymbol{r}$ and $\boldsymbol{p}$ change sign simultaneously the transformation conserves the commutation relations of the orbital angular momentum $\boldsymbol{L}$. In summary,

$$U_{\mathscr{R}}\, \boldsymbol{r}\, U_{\mathscr{R}}^\dagger = -\boldsymbol{r} \qquad U_{\mathscr{R}}\, \boldsymbol{p}\, U_{\mathscr{R}}^\dagger = -\boldsymbol{p} \qquad U_{\mathscr{R}}\, \boldsymbol{L}\, U_{\mathscr{R}}^\dagger = \boldsymbol{L} \qquad U_{\mathscr{R}}\, \boldsymbol{s}\, U_{\mathscr{R}}^\dagger = \boldsymbol{s}.$$

The reflection operator acts generally on a spatial wavefunction as $U_{\mathscr{R}}|\psi(\boldsymbol{r})\rangle = |\psi(-\boldsymbol{r})\rangle$. It is obvious that $U_{\mathscr{R}}^2 = I$ because two applications of the parity transformation restores the coordinate system to its original state. Spatial inversion is an example of an improper rotation because it is a discrete rather than continuous transformation (see Box 6.1); it is a generator of O(3) but not of the subgroup O(3) $\supset$ SO(3).

Let us further examine consequences of the reflection operation on quantum states. We can introduce the parity operator Π, which acts on states containing spatial degrees of freedom. We discuss the 1D case to start with, but the conclusions are valid for 3D. The definition of the parity operator means that $\Pi\,|x\rangle = |-x\rangle$, which implies several general properties for Π (Problem 22.6).

1. The parity operator is its own inverse, $\Pi^{-1} = \Pi$.
2. The parity operator is hermitian, $\Pi^{\dagger} = \Pi$.
3. Because of 1 and 2, the parity operator is unitary, $\Pi^{\dagger} = \Pi^{-1}$.

To solve the eigenvalue equation $\Pi\,|\pi\rangle = \pi\,|\pi\rangle$ we apply Π on both sides, which yields $\Pi^2\,|\pi\rangle = \pi^2\,|\pi\rangle$. But $\Pi^2 = 1$, which means $|\pi\rangle = \pi^2\,|\pi\rangle$ and thus that $\pi^2 = 1$. Therefore, we find two eigenvalues $\pi = \pm 1$ for parity corresponding to the eigenfunctions $|+\rangle$ and $|-\rangle$, respectively. Under a parity transformation $|+\rangle$ is unchanged and is labeled *even parity*, while $|-\rangle$ acquires a negative sign and is labeled *odd parity*. Thus, if the Hamiltonian commutes with Π the Hilbert space decomposes into eigenspaces of parity, with $\pi = \pm 1$.

22.5.2 Breaking Parity Spontaneously

In some quantum systems that are dominated by electromagnetic or strong interactions (which should be invariant under the parity transformations) *spontaneous breaking of the parity symmetry* may occur. In Section 22.1 examples were shown of (approximate) mean-field solutions that imply *deformed* shapes for the density distribution, which break rotational symmetry. Now we present examples in which reflection symmetry also can be violated in mean-field solutions. To illustrate *shapes* in quantum systems, we may describe a surface in terms of a multipole expansion in spherical harmonics,

$$R(\Omega) = c(\alpha)\left[1 + \sum_{\lambda=2}^{\infty}\sum_{\mu=-\lambda}^{+\lambda} \alpha_{\lambda\mu} Y^*_{\lambda\mu}(\Omega)\right]. \tag{22.36}$$

In the above expression $c(\alpha)$ has the meaning of the radius of a sphere (corresponding to the $\lambda = 0$ term of the multipole expansion). It is typically determined from a volume conservation condition that results if the quantum fluid is assumed to be incompressible.

The $\lambda = 1$ term in Eq. (22.36) describes a translation of the whole system. Thus the *deformed* shapes are illustrated starting from the multipolarity $\lambda = 2$. The quadrupole deformation with $\lambda = 2$ was illustrated in Fig. 22.1 when the *intrinsic state* of the deformed nucleus was introduced. The next term in Eq. (22.36) is $\lambda = 3$, corresponding to an *octupole* deformation. Box 22.6 illustrates parity classification for angular momentum states. As $\lambda = 3$ is an odd integer, by parity selection rules[4] the two states involved in octupole matrix elements must have *opposite* parities. Thus a non-vanishing $\lambda = 3$ term in Eq. (22.36) implies a mixing of states with opposite parities and motion in a potential with octupole deformation does not conserve parity.

Example 22.4 It has been proposed that a non-vanishing α_{30} in Eq. (22.36) leads to pear-shaped intrinsic states for some radium isotopes [5], and that non-vanishing α_{32} leads to potato-shaped intrinsic states for some superheavy nuclei [40].

[4] The solution of Problem 6.14 requires that spherical harmonic matrix elements vanish unless the sum of the spherical harmonic multipolarity and the orbital angular momenta of the two wavefunctions is an even integer. From Box 22.6 the parity of a spherical harmonic is $(-1)^l$, so this is a parity selection rule.

Box 22.6 Parity for Spherical Harmonics

A spherically symmetric Hamiltonian commutes with the parity operator Π and each energy level has a definite parity. The angular eigenfunctions are *spherical harmonics* $Y_{lm}(\theta, \phi)$, which are $(2l + 1)$-dimensional representations of SO(3) (see Ch. 6):

$$Y_{lm}(\theta, \phi) \sim e^{im\phi} P_l^m(\cos\theta),$$

where P_l^m is the *associated Legendre function*. The spherical harmonics can be classified by the parity quantum number $\pi = (-1)^l$, so the parity of $Y_{lm}(\theta, \phi)$ is

1. even if l is even, and
2. odd if l is odd.

The reason is that replacing $\boldsymbol{r}$ by $-\boldsymbol{r}$ in spherical coordinates means that

$$\theta \to \pi - \theta \qquad \phi \to \phi + \pi.$$

The change in θ modifies $\cos\theta$ to $-\cos\theta$, which leaves $P_l(\cos\theta)$ unchanged for even l but changes its sign for odd l, so that the sign varies as $(-1)^l$. This is not modified by the change in ϕ because the factor $e^{im\phi}$ in the spherical harmonic gives a sign $(-1)^{|m|}$ under parity, but this is compensated exactly by a sign of $P_l^{|m|}(\cos\theta)$ that alternates with integer changes in $|m|$.

22.5.3 The Parity Projection Operator

For intrinsic states that violate parity, one can construct the *parity projector* [52]

$$\hat{P}^{\pi} = \frac{1}{2}(1 + \pi\Pi), \tag{22.37}$$

where $\pi = \pm 1$ are eigenvalues of Π, and apply the projector to the intrinsic states. $\hat{P}^{\pi}$ has the same properties as Π; in particular it obviously satisfies $(\hat{P}^{\pi})^{\dagger} = \hat{P}^{\pi}$, and $(\hat{P}^{\pi})^2 = \hat{P}^{\pi}$. It is easy to see that when the operator $\hat{P}^{\pi}$ is applied to an arbitrary state $|\psi\rangle$ that is not an eigenstate of parity, it produces eigenstates of parity with the eigenvalues indicated in the projector. For example, in a one-dimensional case with $\left|\psi(x)\right\rangle$ being an arbitrary wave function but not necessarily an eigenfunction of parity,

$$\hat{P}^{+}|\psi(x)\rangle = \frac{1}{2}(|\psi(x)\rangle + |\psi(-x)\rangle) \qquad \hat{P}^{-}|\psi(x)\rangle = \frac{1}{2}(|\psi(x)\rangle - |\psi(-x)\rangle).$$

Now suppose $|\Phi\rangle$ to be a parity-violating state from a mean-field calculation. It is simple to apply the parity projection operator to obtain projected wavefunction with good parity

$$|\Psi^{\pi}\rangle = \mathcal{N}\hat{P}^{\pi}|\Phi\rangle, \tag{22.38}$$

with $\mathcal{N}$ a normalization constant. The energy level with good parity can be evaluated as

$$E^{\pi} = \langle\Psi^{\pi}|H|\Psi^{\pi}\rangle = \frac{\langle\Phi|\hat{P}^{\pi}H\hat{P}^{\pi}|\Phi\rangle}{\langle\Phi|\hat{P}^{\pi}\hat{P}^{\pi}|\Phi\rangle} = \frac{\langle\Phi|H\hat{P}^{\pi}|\Phi\rangle}{\langle\Phi|\hat{P}^{\pi}|\Phi\rangle}, \tag{22.39}$$

where in the last two steps we have applied $(\hat{P}^{\pi})^2 = \hat{P}^{\pi}$ and assumed that parity projection commutes with the Hamiltonian. Projection of parity is relatively easy because it works with a discrete group of only two elements. In contrast, both angular momentum and particle number projections involve infinite numbers of elements in a continuous group, which may require numerical integration in the Euler or gauge spaces, respectively.

22.6 Spin and Momentum Projection for Electrons

Condensed matter physics usually involves many electrons and the dimension of the Hilbert space increases exponentially with the number of particles. Representing a quantum state for such a large many-electron system through the coefficients in a mathematically simple basis is highly inefficient. However, not all quantum states in the Hilbert space are equally important. In many condensed matter problems the interaction between particles tends to be local (for example, between nearest or next-nearest neighbors). The *Hubbard model* [114] exploits this by assuming that at sufficiently low temperatures all electrons remain in the lowest Bloch band (Section 5.5) and long-range interactions can be ignored.

22.6.1 Hartree–Fock Approximation for the Hubbard Model

Even for the greatly simplified Hubbard model Hamiltonian, an exact solution of the 2D problem for arbitrary lattice size is still not known. For small lattices, one can resort to exact diagonalization; for larger systems, further approximations must be introduced. Here we show that a projection method that is better known in nuclear structure physics can be applied also to many-electron systems. Let us consider a Hamiltonian [172]

$$H = -t \sum_{\boldsymbol{j},\sigma} \left[c^{\dagger}_{\boldsymbol{j}+\boldsymbol{x}\sigma} c_{\boldsymbol{j}\sigma} + c^{\dagger}_{\boldsymbol{j}+\boldsymbol{y}\sigma} c_{\boldsymbol{j}\sigma} + \text{h. c.} \right] + U \sum_{\boldsymbol{j}} c^{\dagger}_{\boldsymbol{j}\uparrow} c^{\dagger}_{\boldsymbol{j}\downarrow} c_{\boldsymbol{j}\downarrow} c_{\boldsymbol{j}\uparrow}, \tag{22.40}$$

which is a one-band version of the 2D Hubbard Hamiltonian in which

1. the first term (with $t > 0$) represents hopping between nearest-neighbor sites, with unit hopping vectors $\boldsymbol{x} = (1, 0)$ and $\boldsymbol{y} = (0, 1)$, and
2. the second term (with $U > 0$) is a repulsive onsite interaction.

The operator $c^{\dagger}_{\boldsymbol{j}\sigma}$ ($c_{\boldsymbol{j}\sigma}$) in (22.40) creates (destroys) a particle at the lattice site $\boldsymbol{j} = (j_x, j_y)$ with spin projection $\sigma = +\frac{1}{2}$ (denoted by $\uparrow$) or $\sigma = -\frac{1}{2}$ (denoted by $\downarrow$). The total number of sites is $N_{\text{s}} = N_x \times N_y$, with N_x and N_y the number of sites along the x- and y-axes, respectively. It is convenient to work in the momentum space instead of the coordinate space by applying a 2D Fourier transformation

$$c^{\dagger}_{\alpha\sigma} = \frac{1}{\sqrt{N_{\text{s}}}} \sum_{\boldsymbol{j}} e^{-i\boldsymbol{k}_{\alpha}\cdot\boldsymbol{j}}\, c^{\dagger}_{\boldsymbol{j}\sigma} \tag{22.41}$$

to obtain new operators with the momentum index α

$$\boldsymbol{k}_{\alpha} = (k_{\alpha_x}, k_{\alpha_y}) = \left(\frac{2\pi\alpha_x}{N_x}, \frac{2\pi\alpha_y}{N_y} \right), \tag{22.42}$$

and rewrite the Hamiltonian (22.40) using the momentum operators (22.41). It is not necessary to show explicitly the new form of the Hamiltonian, but we note that the indices for the operators are changed from (j_x, j_y) to (α_x, α_y) and the total number of sites remains N_{s}.

In the Hartree–Fock (HF) approximation the ground state of an N-electron system is represented by a Slater determinant (see Appendix A)

$$|\mathbb{D}\rangle = \prod_{a=1}^{N} b_a^\dagger \, |0\rangle \,, \tag{22.43}$$

in which the lowest N single-electron states are occupied and the new *HF-quasiparticle operators* are related to the operators of Eq. (22.41) by a unitary transformation

$$b_a^\dagger = \sum_{\alpha\sigma} Q^*_{\alpha\sigma,a} \, c^\dagger_{\alpha\sigma}. \tag{22.44}$$

The transformation matrix Q in Eq. (22.44) is a $2N_{\text{s}} \times 2N_{\text{s}}$ matrix with quasiparticle index $a \equiv (a_x, a_y, \sigma_a)$ that specifies the quasiparticles and associated quasiparticle vacuum state in Eq. (22.43), thus defining the Slater determinant. This transformation changes the particle representation into a quasiparticle representation. The matrix Q is not necessarily known at this stage; it will be determined later by variational calculations.

The transformation (22.44) generally mixes all linear momentum and spin projection states defined in Eq. (22.41), thus breaking translational and rotational (in spin space) invariance, so it is an example of *spontaneous symmetry breaking.* By analogy with the discussion in Section 22.1, one may expect that by (deliberately) breaking symmetries of the original Hamiltonian the states of the symmetry-violating system may approximate the same physics that would be difficult or impossible to calculate for the representation using the original single-electron basis [187]. Likewise, by analogy with the discussion in Sections 22.3 and 22.4, we may expect that the broken symmetries can be restored by projection of good quantum numbers from the broken symmetry solutions, as we we shall now discuss.

22.6.2 Spin and Momentum Projection in the Hubbard Model

To obtain states with a good spin quantum number S we may apply the spin projection operator

$$\hat{P}^S_{\Sigma\Sigma'} = \frac{2S+1}{8\pi^2} \int d\Omega \, D^{S*}_{\Sigma\Sigma'}(\Omega) \hat{D}_S(\Omega), \tag{22.45}$$

where $\hat{D}_S(\Omega) = e^{-\imath\alpha S_z} e^{-\imath\beta S_y} e^{-\imath\gamma S_z}$ is the rotation operator in spin space, $D^{S*}_{\Sigma\Sigma'}(\Omega)$ is the Wigner D-function (Section 6.3.2), and $\Omega = (\alpha, \beta, \gamma)$ are Euler angles. The index Σ is introduced for spin projection in symmetry-violating systems. The projector in Eq. (22.45) has the same form as the angular momentum projection operator defined in Eq. (22.16). To recover states with good linear momenta k_{ξ_x} and k_{ξ_y} and restore translational symmetry we introduce the momentum projector [172]

$$\hat{P}(\xi) = \frac{1}{N_{\text{s}}} \sum_{\boldsymbol{j}} e^{i(j_x + j_y)\hat{P}_{\text{T}}} \, e^{-i\boldsymbol{k}_\xi \cdot \boldsymbol{j}}, \tag{22.46}$$

where $\hat{P}_{\text{T}} = \sum_{\alpha\sigma} (k_{\alpha_x} + k_{\alpha_y}) c^\dagger_{\alpha\sigma} c_{\alpha\sigma}$ is the generator of the lattice translation. Note that this operator is not the usual linear momentum operator. Rather, it is associated with the quasi-momentum resulting from translational invariance of the lattice. Applying these projectors to the symmetry-violating function in Eq. (22.43), the projected wavefunction is

$$|\Psi^\Theta(\mathbb{D}, \Sigma)\rangle = \sum_{\Sigma'=-S}^{S} f^\Theta_{\Sigma'} \, \hat{P}^\Theta_{\Sigma\Sigma'} \, |\mathbb{D}\rangle \qquad \hat{P}^\Theta_{\Sigma'} \equiv \hat{P}^S_{\Sigma\Sigma'} \, \hat{P}(\xi), \tag{22.47}$$

where the $f^\Theta_{\Sigma'}$ are variational parameters, $\Theta \equiv (S, \xi_x, \xi_y)$ denotes the set of good spin and momentum quantum numbers, and the composite projector $\hat{P}^\Theta_{\Sigma\Sigma'}$ combines the spin and momentum projections.

By using the wavefunction of Eq. (22.47) to sandwich the Hamiltonian, the energy of the symmetry-restored state can be expressed as

$$E^\Theta = \frac{f^{\Theta\dagger} \mathscr{H}^\Theta f^\Theta}{f^{\Theta\dagger} \mathscr{N}^\Theta f^\Theta}, \tag{22.48}$$

where the Hamiltonian and norm matrix elements are

$$\mathscr{H}^\Theta_{\Sigma\Sigma'} = \langle \mathbb{D} | H \hat{P}^\Theta_{\Sigma\Sigma'} | \mathbb{D} \rangle \qquad \mathscr{N}^\Theta_{\Sigma\Sigma'} = \langle \mathbb{D} | \hat{P}^\Theta_{\Sigma\Sigma'} | \mathbb{D} \rangle. \tag{22.49}$$

This may be viewed as a more general form of Eq. (22.21). The ground state energy is obtained by minimizing Eq. (22.48) with respect to the coefficients f^Θ and the HF transformation $\mathbb{D}$, all of which are generally complex quantities. The actual number of minimization variables depends on the numbers N (electrons) and $2N_{\text{s}} - N$ (electron holes). The computational footprint depends also on the numerical grid in the spin projection (22.45), which is a 3D integration. Thus, efficient numerical algorithms and significant computing power are required to apply this method.

Background and Further Reading

An overview of many concepts discussed in this chapter may be found in Ring and Schuck [171]. Angular momentum projection is reviewed in Hara and Sun [105]. Basic projection methods are often discussed in standard quantum-mechanical texts such as Griffiths [83].

Problems

22.1 Prove that the projection operator

$$P_\Phi = \int d\Omega \, |\Phi(\Omega)\rangle \langle\Phi(\Omega)|$$

defined in Box 22.3 commutes with the rotation operator $D(\Omega)$. *Hint*: Use the property $D^\dagger(\Omega) = D(-\Omega)$ given in Eq. (6.35a).

22.2 Show that a variational calculation with a BCS wavefunction as the variational state,

$$\delta \langle \Psi_{\rm BCS}| \hat{H} - \lambda |\Psi_{\rm BCS}\rangle \equiv \langle \Psi_{\rm BCS}| \hat{H}' |\Psi_{\rm BCS}\rangle = 0,$$

where $\hat{H}$ is the Hamiltonian and λ is the variational parameter, leads to $\lambda = dE/dN$, where E is the energy and N is the average particle number. ***

22.3 Show that the transformation (22.29) can be inverted to give

$$c_k = u_k \alpha_k + v_k \alpha^\dagger_{-k} \qquad c_{-k} = u_k \alpha_{-k} - v_k \alpha^\dagger_k$$

for the bare fermion operators $\{c, c^\dagger\}$ in terms of the quasiparticle operators $\{\alpha, \alpha^\dagger\}$.

22.4 Demonstrate that the Bogoliubov quasiparticle creation and annihilation operators obey the anticommutators $\{ \alpha_k, \alpha_{k'} \} = \{ \alpha^\dagger_k, \alpha^\dagger_{k'} \} = 0$ and $\{ \alpha_k, \alpha^\dagger_{k'} \} = \delta_{kk'}$ given in Eq. (22.30), if the bare fermion operators satisfy the

$$\{ c_i, c^\dagger_j \} = \delta_{ij} \qquad \{ c^\dagger_i, c^\dagger_j \} = \{ c_i, c_j \} = 0$$

fermionic anticommutator algebra. ***

22.5 Show that the mean square deviation of the particle number from the actual particle number for a BCS wavefunction is given by

$$(\Delta N)^2 = \langle \Psi_{\rm BCS}| \hat{N}^2 |\Psi_{\rm BCS}\rangle - N^2,$$

where $\hat{N}$ is the particle number operator and $N = \langle \hat{N} \rangle$ is the average particle number. *Hint*: In quantum mechanics the spread in a distribution of measurements with respect to the average or *expectation value* is given by the *standard deviation* σ, with

$$\sigma = \sqrt{\langle (\Delta \hat{O})^2 \rangle} \qquad \Delta \hat{O} \equiv \hat{O} - \langle \hat{O} \rangle.$$

Use this to prove the general result that $\sigma^2 = \langle \hat{O}^2 \rangle - O^2$, and then specialize to particle number in the BCS approximation.

22.6 Show that the parity operator Π is its own inverse and is hermitian, so it is unitary.

22.7 Prove that the parity operator Π does not commute with the position operator $\hat{x}$ but it does commute with $\hat{x}^2$. ***

22.8 Discuss the parity selection rule for electric dipole transitions of a single-electron atom (the *Laporte rule*). *Hint*: The electric dipole matrix element is $\langle n'l'm'|\, \boldsymbol{r}\, |nlm\rangle$, where $\boldsymbol{r}$ is the coordinate operator, n is a principal quantum number, and (l, m) are angular momentum quantum numbers. ***

23 Quantum Phase Transitions

Phase transitions are germane to our discussion of symmetry and broken symmetry because they often are characterized by a change in the symmetry properties of a system. For example, a ferromagnet corresponds classically to a large set of atomic spins all aligned approximately in the same direction, which establishes a macroscopic state having a preferred spatial direction that breaks rotational invariance. But if the ferromagnet is heated above a certain temperature it undergoes a phase transition to a state in which the spins point in random directions, which restores the rotational invariance (no preferred direction). Thus the phase transition represents a qualitative change of state between a disordered high-temperature phase having a symmetry and an ordered low-temperature phase in which that symmetry is broken (usually spontaneously; see Ch. 17).

23.1 Classical and Quantum Phases

The spins of a ferromagnet are described microscopically by quantum mechanics but the ordered magnetic state is macroscopic and may be understood in classical terms. The corresponding phase transition may be termed a *classical phase transition.* But phase transitions may occur in a purely quantum context that correspond to a qualitative change in matrix elements of a microscopic wavefunction. These are called *quantum phase transitions.* In this chapter we introduce the basics of quantum phase transitions and their contrast with classical phase transitions, and show that quantum phase transitions have a natural affinity with the *dynamical symmetries* that were described in Section 3.6 and Ch. 20.

23.1.1 Thermal and Quantum Fluctuations

Classical phase transitions are typically induced by varying an external control parameter and generally lead to qualitative changes in a physical system. At finite temperatures, thermal fluctuations can be the agents of the phase transition. For example, a crystal undergoes increasing thermal fluctuations as the temperature is raised to the melting point. These fluctuations are classical in nature for a macroscopic system and melt the crystal by disrupting its crystalline order. Quantum systems undergo thermal fluctuations too but there is another source of fluctuations. Even at zero temperature there exist quantum fluctuations because of the uncertainty principle. They can destroy order and cause phase transitions, but they are purely quantum in nature and are distinct from temperature induced fluctuations since they remain finite in the limit $T \to 0$. The corresponding phase transitions are typically controlled by a non-thermal external parameter like a magnetic field strength. Phase transitions induced by fluctuations having a quantum origin are termed *quantum phase transitions*. As discussed further below, under appropriate conditions the corresponding transition point for the quantum phase transition may be a *quantum critical point*.

23.1.2 Quantum Critical Behavior

Quantum critical points are of special interest because they can lead to *quantum critical behavior*, which is associated with a peculiar excitation spectrum of the quantum critical ground state. Because of quantum critical behavior a quantum critical point can spread its influence through regions of a phase diagram far removed from the actual critical point. In addition, the quantum fluctuations that destroy long-range order even at zero temperature can lead to novel effects not found in ordinary classical phase transitions.

23.2 Classification of Phase Transitions

Phase transitions may be classified as *first-order* or *second-order* (sometimes termed *continuous*). The distinction between the two is illustrated in Fig. 23.1. In first-order

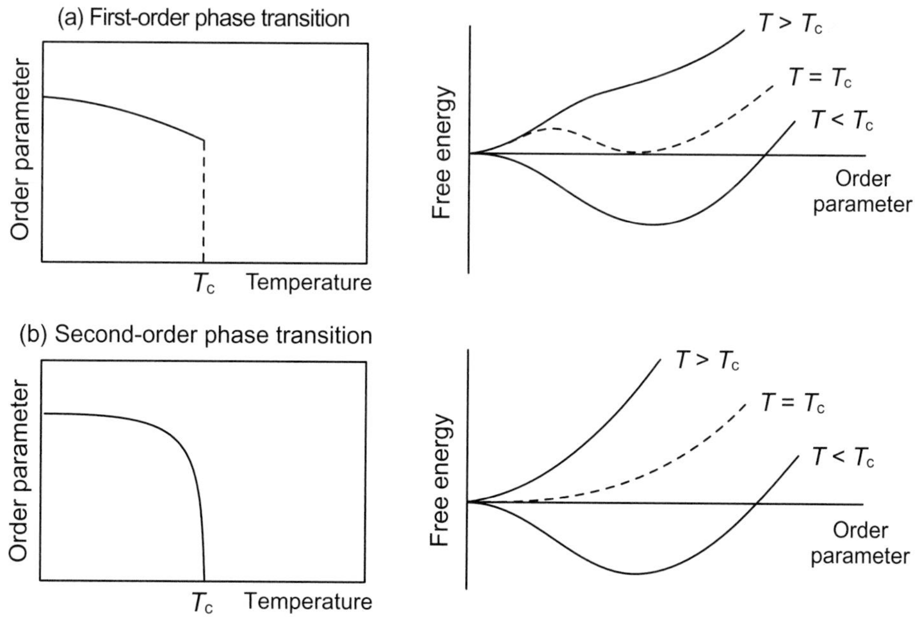

Fig. 23.1 (a) First-order and (b) second-order phase transitions.

transitions the two phases coexist at the transition point while in second-order phase transitions they do not. A a first-order transition may be characterized as one that involves a *latent heat* released as the system cools through an infinitesimally small temperature range around T_c. The release of latent heat indicates a dramatic restructuring at the transition point. In contrast, there is no latent heat associated with a second-order transition and the restructuring that leads from one phase to the other is a continuous process.

Example 23.1 The transition between ice and water at 0° C is a first-order phase transition. The ferromagnetic transition of iron where the magnetic moment vanishes above the *Curie point* at 1043 K is a second-order phase transition.

As we shall now discuss, second-order phase transitions are of particular interest because they can lead to *critical behavior.*

23.3 Classical Second-Order Phase Transitions

The ferromagnetic phase transition of Example 23.1 occurs because thermal fluctuations destroy the ordering of magnetic moments that characterize iron at lower temperatures. A natural *order parameter* distinguishing these phases is the magnetization m. In the ferromagnetic transition the magnetization vanishes continuously as the transition temperature is approached from below. Then the *ordered* phase below the Curie point has a finite value of the order parameter, while in the *disordered* phase above the Curie point the order parameter vanishes. The point of continuous second-order phase transition where the order parameter vanishes is termed the *critical point.*

23.3.1 Critical Exponents

The thermodynamical average of the order parameter is finite in the ordered phase and zero in the disordered phase, but it has non-zero fluctuations and very near the critical temperature the spatial correlations of the order parameter become long range. If a quantity t is defined that is a dimensionless measure of the distance from the critical temperature T_c,

$$t \equiv \frac{|T - T_c|}{T_c}, \tag{23.1}$$

then the correlation length ξ typically diverges as a power of t,[1]

$$\xi \propto |t|^{-\nu}, \tag{23.2}$$

[1] Physically the correlation length measures the largest distance over which order is correlated. For example, in a 2D spin system the correlation length is related to the characteristic size of the largest patches containing spins that all point in the same direction. Note that in Eq. (23.1) we are assuming that $T_C \neq 0$.

Table 23.1. Some critical exponents for magnets [144, 203]

Quantity	Exponent	Power law	Conditions
Specific heat	α	$C \propto \|t\|^{-\alpha}$	$t \to 0,\ B = 0$
Order parameter	β	$m \propto (-t)^{\beta}$	$t \to 0,^{\dagger}\ B = 0$
Susceptibility	γ	$\chi \propto \|t\|^{-\gamma}$	$t \to 0,\ B = 0$
Correlation length	ν	$\xi \propto \|t\|^{-\nu}$	$t \to 0,\ B = 0$
Correlation function	η	$G(r) \propto \|r\|^{-d+2-\eta}$	$t = 0,\ B = 0$
Dynamic	z	$\tau_c \propto \xi^z$	$t \to 0,\ B = 0$

†From below.

where ν is an example of a (correlation length) *critical exponent*. As the critical point is approached, fluctuations of the order parameter that are long range in time also will occur. Near the critical point the correlation time τ_c (a measure of the characteristic equilibration time for the system) diverges as

$$\tau_c \propto \xi^z \propto |t|^{-\nu z}, \tag{23.3}$$

where z is a (dynamical) critical exponent. These ideas lead to a far-reaching conjecture for critical systems.

> ***Scaling Hypothesis:*** Near a critical point the only characteristic length scale is ξ and the only characteristic time scale is τ_c. Very near the critical point the divergences in (23.2) and (23.3) ensure that fluctuations occur on all length and time scales, and the system becomes *scale invariant* because it is no longer sensitive to any length or time scales set by the microscopic physics of the problem.

As a result, physical observables depend on external parameters through power laws and the critical exponents characterize completely the critical behavior near the transition. Critical exponents for the magnetization critical point are summarized in Table 23.1.

23.3.2 Universality

The behavior encapsulated in the scaling hypothesis implies *universality in critical phenomena.* Because fluctuations destroy all length or time scales near the critical point except those set by the correlation lengths and correlation times, the detailed microscopic structure of the sample becomes irrelevant and one finds that the critical exponents are the same for physically very different systems that have the same *universality class.*

> ***Universality Classes:*** The universality class for critical behavior is determined only by (1) dimensionality n of the order parameter, (2) symmetry G of any local couplings between variables, and (3) dimensionality d of the space.

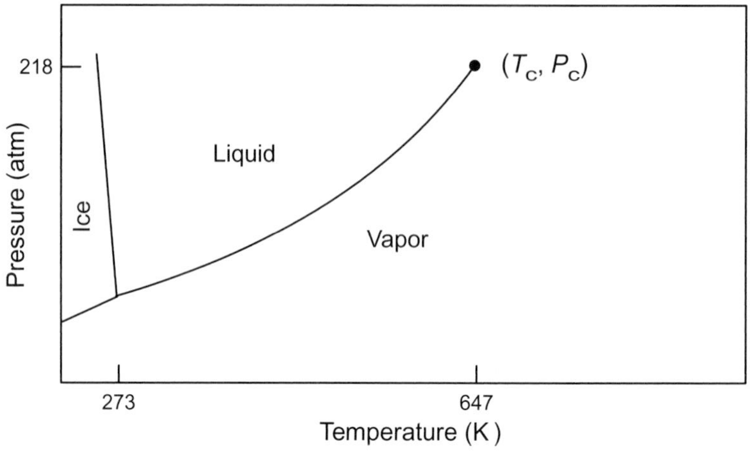

Fig. 23.2 Gas–liquid critical point of water. Beyond this critical point there is no clear distinction between vapor and liquid; only a single undifferentiated mixture of gas and liquid exists.

The following two examples illustrate that a single universality class can accommodate physical systems that are very different microscopically.

Example 23.2 Water has a second-order phase transition with a critical point at $T_c = 647$ K and $P_c = 218$ atm, as illustrated in Fig. 23.2. A suitable order parameter is the density difference between liquid and vapor phases, which is a scalar. Thus in a 3D volume the order parameter dimensionality is $n = 1$ and the space dimensionality is $d = 3$.

Example 23.3 Consider a 3D Ising model of a ferromagnet (a 3D lattice with an up or down spin on each site). A natural order parameter is the difference in the number of up and down spins on the lattice. This difference is a scalar and for the 3D Ising model $n = 1$ and $d = 3$.

Examples 23.2 and 23.3 indicate that the gas–liquid critical point for water and the magnetization critical point for a 3D Ising model belong to the same universality class. Indeed, these systems are observed to exhibit scaling behavior near these critical points with similar critical exponents. Water and the 3D Ising spin lattice are very different microscopically, but they are in the same universality class and exhibit the same critical behavior. The physical reason was given above: near a critical point fluctuations erase knowledge of small-scale microphysics, leaving only the scaling properties to characterize the system.

23.4 Continuous Quantum Phase Transitions

Quantum effects can influence continuous phase transitions for two distinct reasons.

1. The ordered phase may have a quantum origin (for example, superconductivity).
2. At low temperature quantum fluctuations may influence critical behavior.

Let us now ask what happens to continuous classical phase transitions and their critical properties as the temperature is lowered toward absolute zero so that quantum fluctuations assume increased importance [175, 203]. Two energy scales are relevant. The first is the classical thermal energy $\sim kT$. The second is the quantum energy characterizing long-wavelength fluctuations of the order parameter near the critical point, which may be estimated as

$$\hbar\omega_c \propto |t|^{\nu z}. \tag{23.4}$$

Quantum effects will be less important for critical behavior if $\hbar\omega_c < kT_c$, or equivalently

$$|t| \lesssim T_c^{1/\nu z}. \tag{23.5}$$

This condition is satisfied for T asymptotically close to T_c and quantum effects may contribute, but classical thermal fluctuations dominate the macroscopic scales controlling critical behavior. At the other extreme, for $T = 0$ the fluctuations are entirely quantum in origin. Understanding the interplay of classical and quantum fluctuations near the critical point requires distinguishing systems having long-range order only at $T = 0$ from systems where order may also exist at finite T. Let us consider these two possibilities in turn [203].

23.4.1 Order Only at Zero Temperature

In terms of a tuning parameter x (for example, a concentration variable, applied pressure, or an external field), the quantum critical phase diagram is illustrated in Fig. 23.3(a) for the case where there is no order at finite temperature. The (quantum) ordered phase exists only at $T = 0$ up to a critical value x_c of the control parameter, which defines the position of the quantum critical point. In the left-hand portion of Fig. 23.3(a) the system is thermally disordered for $T > 0$; in the right-hand portion of Fig. 23.3(a) quantum fluctuations (controlled by x) are larger and the disorder for all values of T is quantum in origin, so this region is labeled "Quantum disordered." The most interesting feature in Fig. 23.3(a) is the

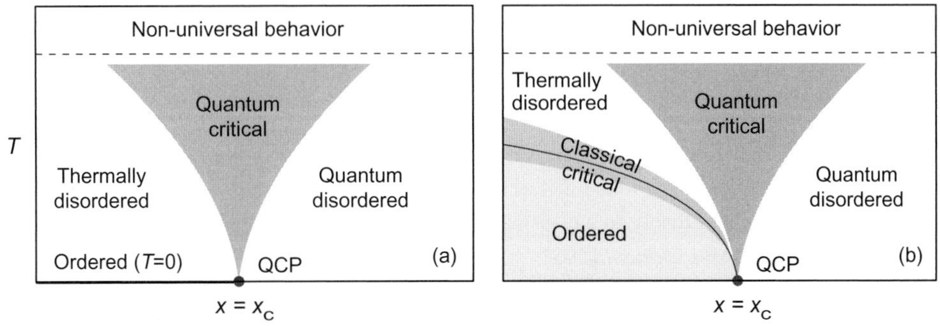

Fig. 23.3 Schematic quantum critical phase diagrams for temperature T and control parameter x, with a single quantum critical point (QCP) at $x = x_c$ [203]. (a) Case where the ordered phase exists only for $T = 0$. (b) Case where classical order exists at finite temperature.

dark gray region fanning out from the quantum critical point with increasing temperature that is labeled "Quantum critical." The curves marking its boundary are defined by

$$kT > \hbar\omega_c \propto |x - x_c|^{\nu z}. \tag{23.6}$$

At finite temperature this region receives significant contributions from both quantum and thermal fluctuations. The boundaries of the quantum critical region mark crossovers between types of behavior, not true phase transitions. Above the dashed horizontal lines, kT exceeds the characteristic coupling energy and universality is lost.

23.4.2 Order Also at Finite Temperature

The case where order may also exist at finite temperature is illustrated in Fig. 23.3(b). The quantum disordered, quantum critical, and non-universal behavior regions of this diagram are similar to the corresponding ones in Fig. 23.3(a). The basic difference is that now a line of classical second-order phase transitions extends from finite temperature, terminating in the quantum critical point at $T = 0$. Below this line the system is ordered while above it the system is disordered by thermal fluctuations. In the region marked "Classical critical" we may expect classical critical behavior except at the $T = 0$ quantum critical point.

23.5 Quantum to Classical Crossover

Given a Hamiltonian $H = H_{\text{kin}} + H_{\text{pot}}$ that is a sum of kinetic and potential energies, thermodynamical properties may be derived from the *partition function* Z,

$$Z \equiv \text{Tr}\, e^{-H/kT}, \tag{23.7}$$

where Tr denotes the matrix trace. Classically kinetic and potential energies commute and the partition function factors as $Z = Z_{\text{kin}} \times Z_{\text{pot}}$. Therefore, for classical systems statics and dynamics decouple and classical phase transitions can be studied using effective time-independent methods in d space dimensions. But quantum mechanically the kinetic and potential energies do not generally commute (see Problem 23.3), so the partition function cannot be factored into a product of kinetic and potential energy contributions. Hence, in quantum phase transitions the order parameters must be expressed in terms of fields with both space and time dependence, and statics and dynamics are *always coupled.* As we now demonstrate, these differences imply that a quantum phase transition in d dimensions is related closely to a classical phase transition in $d + z$ dimensions, where z is the *dynamical critical exponent* [203].

23.5.1 The Classical–Quantum Mapping

The quantum evolution operator for propagation through a real time t is $\exp(iHt/\hbar)$. If time is continued to imaginary values by the replacement $t \to -i\tau$ where τ is real,

$$e^{\frac{i}{\hbar}Ht} \to e^{H\tau/\hbar} = e^{-H/kT}, \tag{23.8}$$

provided that we identify $\tau = -\hbar/kT$.

The partition function of classical statistical mechanics is related to the propagator of quantum mechanics through rotation of the time axis in the complex plane from the real to imaginary direction.

Analytical continuation to imaginary time is called a *Wick rotation* and the resulting problem is said to be formulated in *euclidean space* (see also Box 25.1). Since $\tau \propto T^{-1}$, at zero temperature the euclidean time has infinite extent and behaves as if it were an extra spatial dimension. From Eq. (23.3), time scales like a length to the z power, which suggests the following.

A quantum phase transition in d real spatial dimensions is related to a classical phase transition in $d + z$ dimensions, where often $z = 1$ but it can be fractional.

We may then expect that quantum phase transitions often belong to a universality class for the corresponding classical problem formulated in one higher spatial dimension. Near the transition the behavior is governed by the relationship between τ and the correlation time τ_c. The crossover from quantum to classical behavior will occur when $\tau_c > \tau$, which implies that $|t|^{-\nu z} > (kT)^{-1}$ and therefore is equivalent to a condition $|t|^{\nu z} < kT$. While $\tau_c < \tau$ (suggesting that quantum effects are important), the system behaves effectively as if it were $(d + z)$-dimensional but once $\tau_c > \tau$ (suggesting classical critical behavior) the system realizes that it is only d-dimensional. On the other hand, if the critical point is approached by lowering the temperature at $x = x_c$ both τ and τ_c diverge. Then the condition $\tau_c > \tau$ is not generally fulfilled, quantum effects are always important, and the system behaves as if it had dimension $d + z$.

23.5.2 Optimal Dimensionality

General arguments suggest that two spatial dimensions are optimal for the observation of quantum critical points in strongly correlated electron systems [176]. In 1D, quantum fluctuations of the order parameters are so strong that they tend to suppress long-range order. In 3D, correlated electron systems tend to form good Fermi liquids (Box 32.1) or ordered states with order parameters having weak fluctuations that do not lead to unusual quasiparticle behavior. Two dimensions strikes a balance between order and fluctuations that permits quantum critical points with non-trivial universality properties to be identified.

23.5.3 Quantum versus Classical Phase Transitions

Given the preceding discussion suggesting that quantum phase transitions in some number of dimensions can be associated with classical phase transitions in a different number of dimensions, one may well ask why quantum phase transitions need be considered as separate entities. In fact, we cannot simply appropriate results indiscriminately from the classical analog to describe a quantum phase transition, for several important reasons [175, 203].

1. The quantum–classical mapping gives imaginary-time quantum correlation functions. To relate these theoretical quantities to real-time dynamics requires analytically continuing to real time. This continuation may be poorly defined and approximations that work in imaginary time may fail to translate into valid approximations in real time.
2. A fundamental new timescale characterizes dynamics near a quantum critical point: the *phase coherence time,* which is the timescale for the quantum system to lose phase memory. This can lead to quantum interference effects in local measurements that have no analog for a classical critical point.
3. Some quantum critical systems have no simple classical analog. For example, topological effects deriving from Berry phases (see Section 28.4) may lead to quantum critical points with properties that do not appear in any classical theory.

Thus the quantum–classical correspondence is illuminating but quantum phase transitions must be studied in their own right to obtain a full picture of quantum critical transitions.

23.6 Example: Ising Spins in a Transverse Field

Let us now illustrate some of the preceding ideas by examining a real system that exhibits quantum critical behavior. The magnetic properties of $LiHoF_4$ at low temperature provide an example of a quantum phase transition that may be tuned using the strength of a magnetic field and that can be approximated by a simple mathematical model. This compound has only the spins of the Ho atoms as magnetic degrees of freedom at low temperature, and these prefer to point up or down with respect to a particular crystal direction. The observed magnetic properties of $LiHoF_4$ as a function of temperature T and the strength of an external magnetic field H_t applied perpendicular to the preferred spin direction are summarized in the phase diagram of Fig. 23.4.

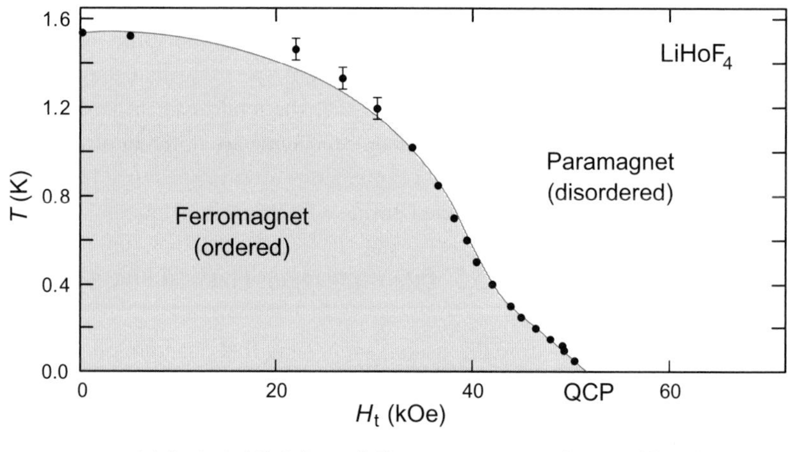

Fig. 23.4 Low-temperature magnetic behavior in $LiHoF_4$ [25, 203]. The curve represents a theoretical fit to the data points. A quantum critical point is labeled QCP.

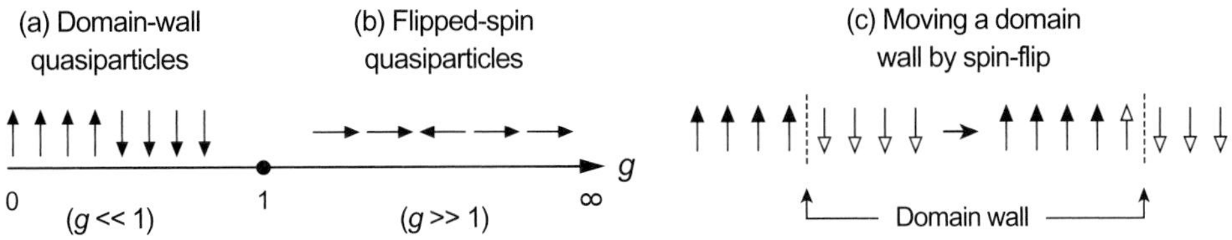

Fig. 23.5 Quasiparticles in a 1D Ising model with transverse magnetic field for large and small values of the tuning parameter g. (a) Domain-wall quasiparticles for $g \ll 1$. (b) Flipped-spin quasiparticles for $g \gg 1$. (c) Moving the domain wall in (a) by flipping a spin.

23.6.1 Hamiltonian

A 1D chain of Ising spins interacting with a transverse magnetic field has been suggested as a simple model for $LiHoF_4$ [163, 175, 176, 203]. The corresponding Hamiltonian is

$$H = -J \sum_j (g\sigma_j^x + \sigma_j^z \sigma_{j+1}^z), \tag{23.9}$$

where the $\sigma_j^{x,z}$ are Pauli matrices, J and g are positive, and j labels sites in a 1D chain where the ions reside. Each site j hosts two possible spin states, $|\uparrow\rangle_j$ and $|\downarrow\rangle_j$, corresponding to eigenstates of σ_j^x with eigenvalues $+1$ and -1, respectively. The two terms in Eq. (23.9) do not commute (Problem 23.4), and represent competing physical tendencies.

1. The first term mixes $|\uparrow\rangle_j$ and $|\downarrow\rangle_j$, and favors spin alignment in the x direction.
2. The second term favors parallel spin alignment on neighboring sites (ferromagnetism).

This model exhibits many basic features of quantum phase transitions, and accounts for the general features of Fig. 23.4.[2] Let us analyze it in a quasiparticle approach, where we (1) identify the ground state and (2) attempt to associate the low-energy excited states with quasiparticle excitations of this ground state.

23.6.2 Ground States and Quasiparticle States for $g \to 0$

The Hamiltonian (23.9) has simple ground states for large and small g. If $g \ll 1$, the second term dominates and the ground state has either all spins up or all spins down.[3] Low-energy excitations of this ground state are *domain-wall quasiparticles,* where all spins to the left of a particular point (the *domain wall*) are up and all spins to the right are down (or vice–versa), as illustrated in Fig. 23.5(a).These are excited states because they have less ferromagnetic stabilization energy from the second term of Eq. (23.9)] than the ground state (all spins pointing in the same direction) for $g \ll 1$. If $g = 0$ such states are stationary

[2] A 2D Ising model gives critical behavior similar to that of a 1D model. In 3D the behavior is not quantum critical. This illustrates the earlier comments that quantum criticality is a low-dimensional effect.

[3] This corresponds to spontaneously broken Z_2 symmetry, since the two ground state choices are equivalent and related by symmetry. It is similar in spirit to the spontaneous breaking of reflection symmetry for a scalar field in Section 17.4.2 and parity symmetry for intrinsic nuclear states in Section 22.5.2.

because they are eigenstates of Eq. (23.9). For small $g \neq 0$ the domain walls become mobile because perturbation from the first term in Eq. (23.9) can flip a spin at the domain wall; this effectively moves the location of the domain wall by one site, as illustrated in Fig. 23.5(c).

23.6.3 Ground States and Quasiparticle States for $g \to \infty$

In the opposite limit $g \to \infty$ the wavefunctions become eigenstates of the σ_j^x operator. These states are linear superpositions of the up and down eigenstates and the ground state has

$$|{\rightarrow}\rangle_j = \frac{1}{\sqrt{2}}\left(\,|{\uparrow}\rangle_j + |{\downarrow}\rangle_j\,\right), \tag{23.10}$$

(with eigenvalue +1) for all sites, which may be interpreted as a state in which all spins in the chain point to the right. The low-lying quasiparticle excitations result from flipping a single spin to point to the left, corresponding to the state on that site

$$|{\leftarrow}\rangle_j = \frac{1}{\sqrt{2}}\left(\,|{\uparrow}\rangle_j - |{\downarrow}\rangle_j\,\right), \tag{23.11}$$

with eigenvalue −1. The corresponding state for the full chain has a higher energy than the ground state because the product Jg is negative, which disfavors left-pointing spins. The quasiparticle excitations in this large-g limit are illustrated schematically in Fig. 23.5(b). For $g = \infty$ these states are stationary but they develop dynamics because of the perturbation from the second term in the Hamiltonian if g is large but not infinite.

In the limits of large g and small g a quasiparticle description appears to be valid for our model, but the quasiparticles corresponding to the $g \ll 1$ limit differ fundamentally from the quasiparticles corresponding to the $g \gg 1$ limit. This suggests interesting possibilities for the transition between these two limits at intermediate g, which we shall now explore.

23.6.4 Competing Ground States

The spin system described by the Hamiltonian (23.9) has eigenstates of the operators σ_j^x for $g \gg 1$ and of σ_j^z for $g \ll 1$. Because $[\,\sigma_j^x, \sigma_j^z\,] \neq 0$, neither of these ground states can be realized over the entire range of g, implying an essential tension between them at intermediate g. This suggests the possibility of a quantum critical point at intermediate g where the zero-temperature system undergoes a quantum phase transition between the competing quantum ground states expected in the $g \gg 1$ and $g \ll 1$ limits. In fact, this model has an exact solution and a quantum critical point at $g = g_c = 1$ [162]. The essentially quantum nature of this transition at $g = 1$ can be brought into clear focus by considering the non-commuting operators σ_j^x and σ_j^z, and their eigenvalues.

1. In the $g \gg 1$ limit the system has sharp values of σ_j^x but the uncertainty in σ_j^z is large since the eigenstates of σ_j^x are linear combinations of σ_j^z eigenstates.
2. In the $g \ll 1$ limit the uncertainty in values of σ_j^x is large but σ_j^z has sharp values.

As g is tuned through the quantum critical point, uncertainties in the z and x components of spin vary continuously but their product remains non-zero because the spin components are eigenvalues of non-commuting operators. This indicates clearly that the $T = 0$ phase transition at $g = 1$ is driven by fluctuations of purely quantum origin.

Quantum phase transitions are enabled by a Hamiltonian having non-commuting terms representing distinct ordering tendencies for the system, with qualitatively different quantum ground states and quasiparticle states for each.

In the quantum phase transition the tuning parameter g then represents the fundamental tension between these quantum ground states and their conflicted ordering agendas, and the quantum phase transition is driven by fluctuations in that ordering.

23.6.5 The Quantum Critical Region

At $g = g_c$ the description of the system is not simple in terms of either the quasiparticles valid for $g \to \infty$ or those valid for $g \to 0$. However, $g = g_c$ is a quantum critical point that exhibits *scale invariance.* This means that near $g = g_c$ it is not possible to determine the distance between well-separated spins from the ground state wavefunction because the correlation between spins at the critical point is sufficiently long range that nothing changes if we increase the length scale over which the spins are observed. At finite temperature new features enter as temperature fluctuations modify the purely quantum aspects of the problem. The quantum phase transition takes place at zero temperature but remnants of it appear in the finite-temperature phase diagram as long as time constraints set by the natural timescale $\hbar/kT$ are satisfied. In the Ising model example, the zero-momentum dynamic response function at low temperature is found to be [176]

$$\chi(\omega) \equiv \frac{i}{\hbar} \sum_k \int_0^\infty \left\langle [\,\sigma_j^z(t), \sigma_k^z(0)\,] \right\rangle e^{i\omega t}\, dt \sim T^{-7/4} \Phi\left(\frac{\hbar\omega}{kT}\right), \tag{23.12}$$

where Φ is a universal function depending only on T.

23.6.6 Phase Diagram

The preceding considerations permit the phase diagram of Fig. 23.6 to be sketched. In the lightly shaded regions outside the diagonal lines quasiparticle behavior is exhibited, with domain-wall quasiparticles for $g < g_c$ and flipped-spin quasiparticles for $g > g_c$. The V-shaped quantum critical region in darker gray is characterized by the response (23.12). The diagonal lines bounding it are set by the requirement $\hbar\omega > kT$. The horizontal dashed line corresponds to a temperature where kT is large enough that properties are no longer determined solely by scaling. This diagram should be compared with Fig. 23.3(a). We see in a simple model that the QCP at $(T = 0, g = g_c)$ influences a large region of the phase diagram at finite temperature (the dark V-shaped area in Fig. 23.6). For $T > 0$ quantum criticality extends over a broad range of g values around g_c. In the quantum critical

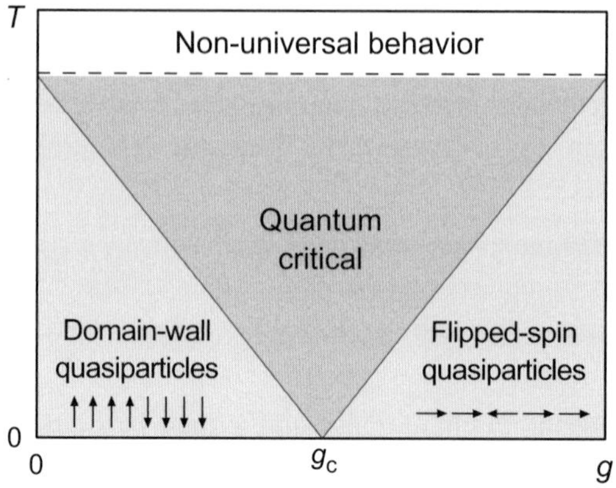

Fig. 23.6 Quantum critical behavior in a one-dimensional Ising model with a transverse magnetic field.

region the system displays unique properties like the response (23.12) that are determined by quantum critical fluctuations and cannot be described by quasiparticles characterizing regions of the phase diagram dominated by one or the other forms of competing order.

23.7 Dynamical Symmetry and Quantum Phases

The preceding discussion has reviewed the standard picture of quantum phase transitions. Now we wish to show that quantum phase transitions can also be understood as a natural consequence of dynamical symmetries for non-abelian symmetry groups. This different perspective leads to new insights for transitions between quantum phases, and to a new view of universality in quantum phase transitions across diverse physical systems.

23.7.1 Quantum Phases in Superconductors

From the discussion in Section 20.1, the microscopic dynamical symmetry method leads to subgroup chains of the highest symmetry, each representing a unique quantum ground state. Thus fermion dynamical symmetries describe quantum phases that are related by quantum phase transitions. Let us illustrate using the SU(4) model of high-T_c superconductors (SC) described in Section 32.3. Dynamical symmetry subgroup chains and their physical interpretation based on calculated matrix elements are described in Eq. (32.6), Table 32.1, and Figs. 32.5 and 32.6. These are summarized in Fig. 23.7, where four quantum phases relevant to a unified understanding of conventional and unconventional superconductivity are displayed.

1. Conventional SC corresponds to quantum phase I, defined by the dynamical symmetry subgroup chain $SO(8) \supset \cdots \supset SU(2)_{BCS}$, where ellipses indicate possible intervening groups. This phase is favored when onsite Coulomb repulsion is weak, so that onsite

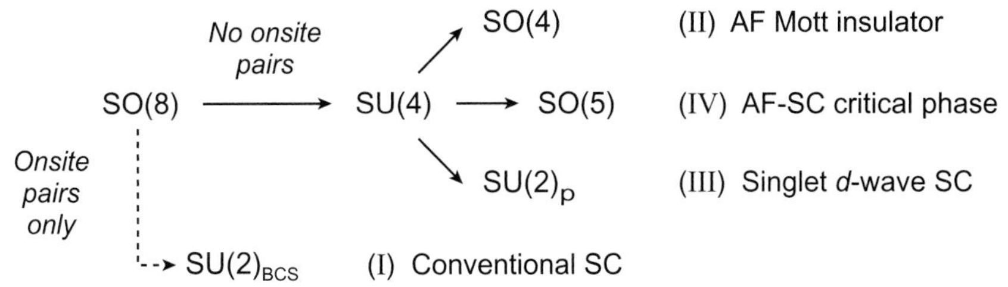

Fig. 23.7 Schematic overview of dynamical symmetries for superconductors discussed in Section 32.3. Quantum phase I describes conventional superconductivity and quantum phases II–IV describe unconventional superconductivity.

pairs are low in energy and there is little competition from non-pairing emergent modes. Phase I typically is characterized by the simplest pairing order: singlet *s*-wave.

2. Antiferromagnetic (AF) Mott insulators correspond to quantum phase II, defined by SO(8) ⊃ SU(4) ⊃ SO(4). This is favored when Coulomb repulsion suppresses double site occupancy and favors bondwise pairs [SO(8) ⊃ SU(4) subgroup], and AF correlations favor antiferromagnetic states [SU(4) ⊃ SO(4) subgroup]. It is characterized by large AF order, Mott insulator transport properties, and vanishing pairing order.
3. Singlet *d*-wave SC corresponds to quantum phase III, defined by SO(8) ⊃ SU(4) ⊃ $SU(2)_p$. It is favored when Coulomb repulsion favors bondwise pairs [SO(8) ⊃ SU(4) subgroup] and pairing correlations favor SC states [SU(4) ⊃ $SU(2)_p$ subgroup]. It is characterized by vanishing AF order and large *d*-wave singlet pairing order.
4. Quantum phase IV corresponds to SO(8) ⊃ SU(4) ⊃ SO(5). It is implicated when Coulomb repulsion favors bondwise over onsite pairs [SO(8) ⊃ SU(4) subgroup] and doping favors contribution of both AF and SC correlations [SU(4) ⊃ SO(5) subgroup]. This phase exhibits remarkable universality associated with *fluctuations in Hilbert space,* since the SO(5) wavefunction at low doping is a rich superposition of SC and AF states with zero expectation value for AF and SC order but large fluctuations in both.

Each quantum phase I–IV issues from the same highest symmetry, SO(8) or SO(8) ⊃ SU(4), which implies that they are *related to each other though constraints imposed by the highest symmetry.* This suggests that dynamical symmetries offer an important tool to understand quantum phase transitions from a comprehensive and unified perspective.

23.7.2 Unique Perspective of Dynamical Symmetries

The dynamical symmetry picture of quantum phases and quantum phase transitions offers a unique perspective and some potential advantages over traditional understanding.

Exact Quantum Solutions: Dynamical symmetries arise as *exact quantum solutions* within a symmetry truncated Hilbert space. Thus they provide a unified microscopic description of quantum phases and their physical interpretation. This is similar in spirit to the Ising model discussed in Section 23.6, but dynamical symmetries can provide solutions for models that are much richer and more complex than that of magnetism in a 1D spin chain.

Physical Pictures with a Microscopic Basis: Generalized coherent state wavefunctions (Ch. 21) provide a physical interpretation of quantum phases and also a direct microscopic connection to Ginzburg–Landau energy surfaces and Hartree–Fock–Bogoliubov variational wavefunctions for interpreting quantum phases and their fluctuations. An example for graphene states in a strong magnetic field is given in Fig. 20.8.

Microscopic Relationship of Quantum Phases: Dynamical symmetries arising from the same highest symmetry define quantum phases that are related by the highest symmetry. Thus they provide microscopic predictions for the properties of quantum phase transitions. This is illustrated in Examples 23.4 and 23.5 below.

A More Abstract Universality: Quantum phases arising from dynamical symmetry lead to critical points and universality classes as do traditional theories, but they suggest a higher abstraction through similar dynamical Lie group structure across diverse systems. Striking examples of Lie group universality are given in Figs. 20.10 and 32.12. This universality arises because systems that are fundamentally different microscopically can exhibit emergent states corresponding to similar symmetry dictated truncations of their Hilbert spaces.

23.7.3 Quantum Phases and Insights from Symmetry

It was suggested above that relationships between quantum phases implied by dynamical symmetry lead to new insights and new predictions for phase transitions. Here we discuss two representative cases: Examples 23.4 and 23.5 provide plausible and self-consistent explanations for two enduring mysteries in the physics of high-temperature superconductors.

Example 23.4 The SU(4) model of Section 32.3 describes the transition from the AF Mott insulator normal state to a superconducting state as a doping-driven quantum phase transition from the SO(4) AF Mott quantum phase to the SU(2) SC quantum phase (Fig. 32.10). This transition can occur *spontaneously* for infinitesimal doping through quantum fluctuations in the order parameter with SO(5) symmetry. Thus SU(4) symmetry provides a natural explanation for why cuprate AF Mott insulator ground states become superconducting with only a tiny amount of electron hole doping: the AF Mott insulator state is fundamentally unstable against condensing d-wave pairs with hole doping because the AF Mott insulator derives from the same highest symmetry as the superconducting state [96, 98].

Example 23.5 The SU(4) model described in Section 32.3 suggests a generic reason for unusually high transition temperatures for cuprate superconductors that is illustrated in Fig. 32.11 [98, 99]. The parent AF Mott insulator state and the d-wave SC state are both generated by subsets of SU(4) generators [SU(4) $\supset$ SO(4) and SU(4) $\supset$ SU(2) subgroups, respectively]. Thus the AF Mott state can be rotated collectively into the SC state in the group space with minimal entropy generation, which can occur with minimal cooling.

It is important to understand the essential role of dynamical symmetry in the preceding examples. In Example 23.4 it is clear from data that there is a quantum phase transition between the cuprate normal ground state and superconducting state. However, only the insight that this quantum phase transition is between dynamical symmetries deriving from the same higher SU(4) symmetry leads to the prediction that the normal AF Mott insulator ground state is unstable against producing the SC state because (in essence) it already contains the superconductor hidden inside it. This is non-intuitive from the traditional point of view, where AF Mott insulators and superconductors appear to be fundamentally different quantum phases with little obvious similarity. The SU(4) analysis shows that Example 23.4 is in fact just the venerable Cooper instability described in Box 32.2, but modified by polarization of the Fermi sea caused by strong AF correlation and Coulomb repulsion. Thus, the SU(4) picture of quantum phase transitions in cuprate superconductors provides a natural generalization of the Cooper instability to doped AF Mott insulators.

In Example 23.5 the SC transition is predicted to be a low-entropy (and therefore unusually high T_c) quantum phase transition *only* because the two quantum phases are intimately related through the higher SU(4) symmetry, so one can be rotated into the other at minimal entropy cost. If the two phases were not related by this symmetry one would generally expect this transition to be *high-entropy* (implying low T_c), because to make the SC state requires first deconstructing the AF state and then reconstructing the SC state from the pieces. The insight from SU(4) dynamical symmetry is that *the SC state has already been constructed* and is masquerading as an AF Mott insulator; we just need to point it in the right direction in the SU(4) group space (for example, with a small amount of hole doping).

Example 23.6 Consider phases for the ground state of monolayer graphene in a strong magnetic field shown in Fig. 20.7. The dynamical symmetries imply seven emergent quantum phases corresponding to distinct subgroup chains deriving from SO(8). Some have been predicted before but some, like those involving SO(7), appear to be new.

Thus, in Example 23.6 the fermion dynamical symmetry method accounts for phases that have been proposed before, but also predicts new quantum phases and associated quantum phase transitions that have not yet been discussed in standard graphene physics.

Background and Further Reading

Ma [144] is a classic introduction to critical phenomena. Quantum phase transitions are reviewed in Phillips [163], Sachdev [175, 176], and Vojta [203]. Examples of fermion dynamical symmetries that imply quantum phase transitions are given in Chs. 20, 31, and 32 of this book. The review [98] deals extensively with dynamical symmetry quantum phase transitions for high-temperature superconductors.

Problems

23.1 This problem and Problem 23.2 give a qualitative feeling for the difference between a phase dominated by classical thermal effects and one dominated by quantum effects by considering the corresponding velocities and pressures. Assume a gas of electrons, each having mass m. Show that a "thermal velocity" $v_T \sim (kT/m)^{1/2}$ may be associated with thermal motion at a temperature T for classical electrons and a "quantum velocity" $v_Q \sim \hbar n^{1/3}/m$ may be associated with the uncertainty principle, assuming a number density n for the Fermi–Dirac gas particles. Compare the behavior of the thermal and quantum velocities with density and temperature. *Hint*: Average spacing between particles may be estimated as $\Delta x \sim n^{1/3}$, and it is convenient to drop constant factors of order one.

23.2 Use the classical and thermal velocities derived in Problem 23.1 to estimate the ratio of the corresponding "quantum pressure" and classical "thermal pressure." *Hint*: Assume thermal electrons to obey an ideal gas equation of state and quantum electrons to obey the Fermi–Dirac equation of state $P \sim (\hbar^2/m)n^{5/3}$, where m is the mass and n the number density of electrons. ***

23.3 Assume a 2D lattice with islands of superconductivity surrounded by regions of insulating behavior. Within each SC island patch i, assume a set of electron Cooper pairs having the same wavefunction phase that can be approximated as bosons. A simple Hamiltonian describing this system is given by the *Bose–Hubbard model* [163]

$$H = H_{\text{pot}} + H_{\text{kin}} = E_{\text{C}} \left(\sum_i (\hat{n} - n_0)^2 - \frac{t}{E_{\text{C}}} \sum_{\langle ij \rangle} (b_i^\dagger b_j + b_j^\dagger b_i) \right),$$

where $b_i^\dagger$ creates a Cooper pair on patch i and b_i annihilates it, n_0 is the average Cooper pair density, E_{C} is the energy cost to convert an isolated Cooper pair into charge carriers in a conductor, t is the energy associated with a Cooper pair hopping between adjacent patches i and j, and $\langle ij \rangle$ indicates that the summation is only over patches i and j that are nearest neighbors. Describe physically the competing ground state tendencies of this Hamiltonian as t/E_{C} is varied. Prove that the kinetic and potential energies do not commute, suggesting that there should be a critical point at a quantum phase transition between SC and insulating states as t/E_{C} is varied. ***

23.4 Show that the two terms in the Ising model Hamiltonian (23.9) do not commute and thus represent competing, incompatible tendencies in the corresponding system.

PART III

TOPOLOGY AND GEOMETRY

24 Topology, Manifolds, and Metrics

Various physical problems formulated in euclidean or Minkowski space, or more abstractly in a quantum-mechanical Hilbert space, have properties that are not determined by local symmetries and depend on the global nature of the manifold for the theory. To understand such properties and their increasingly important role in modern physics, we must consider more formally the subjects of topology and topological spaces, differentiable manifolds, and metrics and metric spaces. Loosely, the first deals with continuity, the second with smoothness, and the third with measurement of distance. Let us now give a more detailed description of each of these, beginning with topology.

24.1 Basic Concepts of Topology

Geometry is largely *quantitative*: distances, angles, curvature, ... Topology is largely *qualitative*; it is "what is left of geometry when measurement is removed" [16]. A famous example is shown in Fig. 24.1: for the topologist a doughnut and a coffee cup are one and the same because they are characterized by a topological invariant called the *genus,* which is intuitively the number of "holes" or "handles" for a surface. Figure 24.2 illustrates. Loosely, topology studies continuous deformations that do not change the most basic properties of objects such as the number of "holes," as in the continuous deformation of a doughnut into a coffee cup.

24.1.1 Discrete Categories Distinguished Qualitatively

Topology often leads to classifications into discrete categories with a qualitative (global) distinction; Fig. 24.3 gives an example.The number of times a length of string with fixed endpoints is wound around the cylinder takes on discrete integer values called the *winding number*, and configurations with different winding numbers cannot be deformed continuously into each other. This is an example of a *topological conservation law,* since the impossibility of deforming the winding number +1 configuration continuously into the winding number +2 configuration in Fig. 24.3 has nothing to do with physics. It has been decreed once and for all by the gods of mathematics; physics considerations like dynamical equations for the strings and their local symmetries have no power over the winding number because it originates in topology. This discrete nature of topology is of great import for physics. In fact, one would not be too far off the mark to assert as follows.

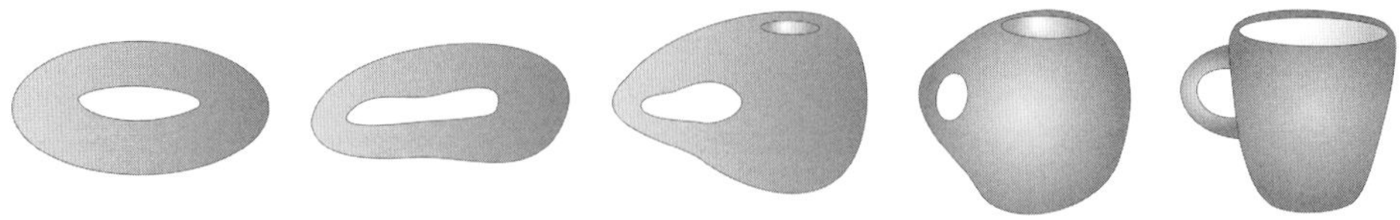

Fig. 24.1 A doughnut and a coffee cup are equivalent topologically, because each is a surface with a single hole and one can be deformed continuously into the other without tearing, puncturing, or gluing. Roughly, the number of "holes" or "handles" for a surface is its *genus*.

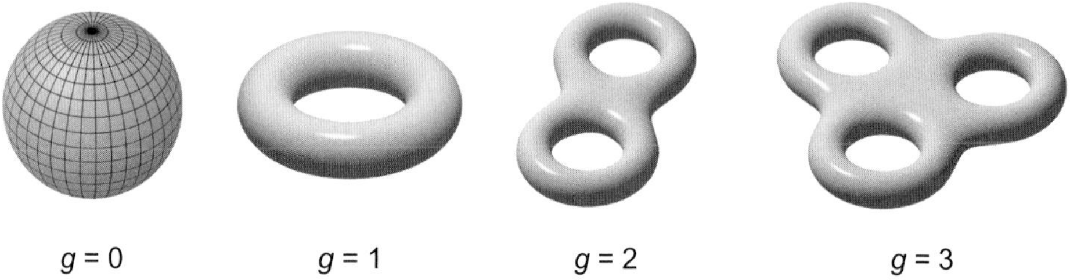

Fig. 24.2 Two-dimensional manifolds with genus g equal to 0, 1, 2, and 3. Images reproduced from https://en.wikipedia.org/wiki/Genus_(mathematics), CC BY-SA.

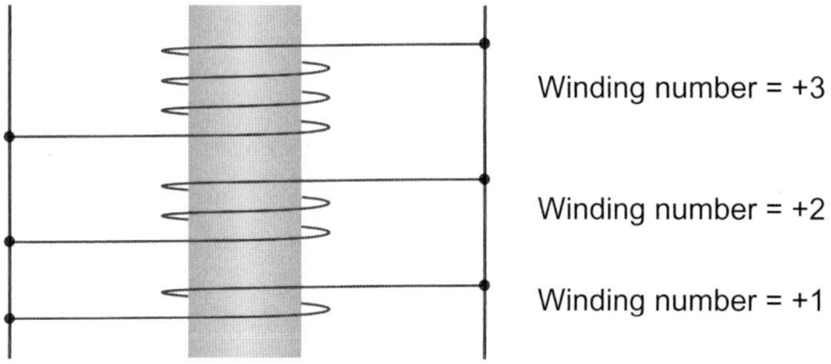

Fig. 24.3 For integers $n \neq m$ and fixed endpoints, a string wound n times around a cylinder of infinite length cannot be deformed continuously into one wound m times. These topologically distinct cases are characterized by a (positive or negative) integer *winding number*.

> Physicists are interested in topology mostly because of the rather obvious notion that *integers cannot morph continuously into other integers*.

This seeming banality has non-trivial implications because it can lead to conservation laws that largely transcend any particular physics manifestation.

24.1.2 The Nature of Topological Proofs

Topology is characterized by many concepts that are "intuitively obvious" but difficult to prove in a formal mathematical sense. Since this is not a mathematical treatise and our interest lies primarily in physical applications of rather elementary topological principles,

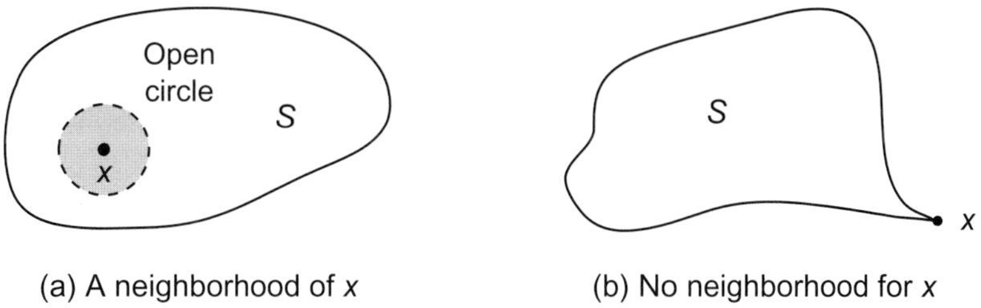

Fig. 24.4 (a) The shaded area is a neighborhood of x. (b) A point x where no neighborhood can be defined because an open ball contained entirely in S cannot be drawn around x.

some basic mathematical proofs will be demonstrated in the text and in problems but often we will appeal more to intuition and schematic mathematics ("coffee cup and doughnut" arguments) than rigorous proof in our discussion.

24.1.3 Neighborhoods

An essential concept in topology is that of *continuity* and a key aspect of continuity is the idea of a *neighborhood*. Intuitively, a neighborhood for a point x is a set of points near x that completely surround it. More formally, for a point x in a space, we have the following definition.

> A neighborhood of x is any set S containing an open solid sphere with center x, where an open solid sphere is a solid sphere without its surface points.

An open solid sphere is commonly called an *open ball* (see Box 24.1); open balls in some number of dimensions are central concepts in the explication of topology.

Example 24.1 An interval on the real number line that contains its endpoints a and b (a *closed interval*, denoted by $[a, b] \subset \mathbb{R}$), is a neighborhood of each of its points except for the endpoints a and b.

Figure 24.4 shows an example of a neighborhood of x in 2D space, and a point x for which no neighborhood exists.

24.2 Topology and Topological Spaces

Let us introduce some formal definitions that will allow us to put many of the preceding ideas on a firm mathematical footing.

Box 24.1 Open Sets and Open Balls

Sets (see Box 2.1) can be classified as *closed* or *open*. Open sets are important in topology, so we review the concept here. Generally a set can be rather abstract but physics usually assumes *metric spaces* (Section 24.5.2), for which the concept of distance between points has meaning. This simplifies the definition of open sets.

Open and Closed Intervals

An *interval* is a segment of the real number line bounded by two points. A *closed interval* is an interval that contains its boundary points, which is denoted by square brackets: the closed interval $[a, b]$ means the real numbers greater than or equal to a and less than or equal to b. An *open interval* does not contain its boundary points, which is denoted by parentheses: the open interval (a, b) means the real numbers greater than a and less than b.

Open and Closed Sets

Closed sets generalize closed intervals in that they include their boundary points. *Open sets* generalize open intervals in that they exclude their boundary points. The following figure illustrates for a filled circle in a plane.

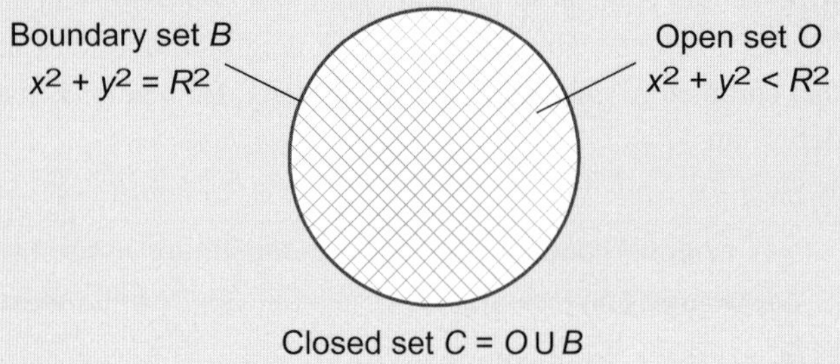

In this example the boundary set B (circle) is given by points for which $x^2 + y^2 = R^2$, the open set O is given by points in the hatched area for which $x^2 + y^2 < R^2$, and the closed set C is the union of O and B. Thus in this example the open set is the interior of the filled circle, without its boundary set. More formally, an *open set* O is a set that for each of its points x a neighborhood U can be defined that is a subset of the original set: $U \subset O$.

Open and Closed Balls

In n-dimensional euclidean space, an *open n-ball* of radius R and center at X is the set of points having distance from X that is less than R. A *closed n-ball* of radius R is the set of points having distance from X that is less than or equal to R. Thus, in euclidean n-space every ball is bounded by a hypersphere (sphere of dimension $n - 1$), and *open balls are closed balls minus their boundary*. Examples: a ball is (1) a bounded interval for $n = 1$, (2) a disk bounded by a circle for $n = 2$ (example shown above), (3) a volume bounded by a 2-sphere S^2 for $n = 3$, and so on.

24.2.1 Formal Definition of a Topology

Let M be a set and $\tau = \{S\} = \{S_1, S_2, \ldots\}$ be a collection of subsets S_i of M. A *topology* τ on the set M results if τ satisfies the following axioms.[1]

1. The collection of subsets τ contains the null set $\emptyset$ and the full set M: $\emptyset \in \tau$ and $M \in \tau$.
2. The union of elements of subcollections in τ is also in τ: if $S_i, S_j \subset \tau$, then $S_i \cup S_j \in \tau$.
3. If $S_i, S_j \subset \tau$ and $\{S\}$ is finite, then the intersection of S_i and S_j is also in τ: $S_i \cap S_j \in \tau$.

Any choice of subsets that satisfies these conditions defines a *topology for (or on) M*, and τ is said to give a topology to M. Loosely then, a topology for a set is a family of subsets closed under unions and finite intersections; Problem 24.1 gives a simple example. The set M (the *underlying set*) and its topology τ together constitute a *topological space* T,

$$T \equiv \{M, \tau\} \qquad \text{(topological space)}, \tag{24.1}$$

and the members of τ are termed the *open sets*. Typically more than one topology can be associated with a given set M. Let us give a few examples.

Example 24.2 For any set M, the *discrete topology* for $T = \{M, \tau\}$ is the topology introduced by taking for τ every subset of M. Clearly these are closed under unions and finite intersections, and satisfy the three topological axioms given above.

Example 24.3 If M is any set then $\tau = \{\emptyset, M\}$ satisfies the topological conditions. This is termed the *indiscrete (or trivial) topology.*

Example 24.4 Consider the set of real numbers $\mathbb{R}$ and all subsets $\tau = \{S\}$, with $S \subset \mathbb{R}$ so that for any point $x \in S$ there is a positive number ϵ such that the interval $\{x - \epsilon < x < x + \epsilon\}$ is contained in S. These are the usual *open sets of* the real number line (a, b) and this topology $T = \{\mathbb{R}, \tau\}$ is termed the *natural (or usual) topology* of $\mathbb{R}$.

Example 24.5 Consider n-dimensional Euclidean spaces, $\mathbb{R}^n$. For $\mathbb{R}^1$ the topology is the topology of the real number line (for example, as in Example 24.4), while $\mathbb{R}^2$ has the topology of two copies of the real number line (the 2D euclidean plane).

Box 24.2 describes a type of topological space called a *Hausdorff space.* Physical theories are almost always formulated in topological spaces that satisfy the Hausdorff condition.

24.2.2 Continuity

The concept of continuity is central to topology. Intuitively, continuous functions take nearby points to nearby points. We are now in a position to make this idea precise. A continuous function from a topological space (S, σ) (the *domain*) to a topological space (T, τ) (the *range*) is a function $F \colon S \to T$ such that $F^{-1}(U)$ (termed the *inverse image;*

[1] Our discussion assumes a familiarity with basic concepts for sets that were reviewed in Boxes 2.1 and 24.1.

Box 24.2 Hausdorff Spaces

A *Hausdorff space* H is a topological space with the property that for any two distinct points x and y of the space there exists a pair of open sets S_x and S_y such that

$$S_x \cap S_y = \emptyset \qquad x \in S_x;\ y \in S_y;\ x \neq y.$$

In words, a topological space is a Hausdorff space if there always exist neighborhoods of arbitrary distinct points x and y that do not intersect. Hausdorff spaces are sometimes called *separated spaces* because of this property. Spaces of interest in physics are normally Hausdorff. For example, n-dimensional euclidean space $\mathbb{R}^n$, and more generally any metric space (Section 24.5.2), are Hausdorff.

see Box 2.3) is an open set in (S, σ) when U is an open set in (T, τ). We use the terminology *continuous map* for such a continuous function. Each choice of topologies σ and τ for the fixed sets S and T selects a subset of functions from S to T that are continuous. This definition is basically what a physicist speaking in the vernacular means by "smooth and well-behaved".

24.2.3 Compactness

A physicist's understanding of compactness was employed often in discussion of group properties in earlier chapters. We now wish to give a more precise mathematical definition, and to show that the precise definition reduces to the earlier one for typical situations encountered in physics. First it is necessary to define the *cover* of a set. For a family of sets $S = \{S_i\}$, if $\cup S_i$ contains a set U, then S is a cover of U. If the S_i are open sets, the cover is an *open cover.* Now, if for every open covering $\{S_i\}$ of a set U with $\cup S_i \supset U$ there is a *finite subcovering* $\{S_1, S_2, \ldots, S_n\}$ with $S_1 \cup S_2 \cup S_3 \cup \cdots \cup S_n \supset U$, then the set U is *compact*. In words: compact sets can be covered by a finite number of sets of arbitrarily small size. Let us consider several examples of applying this definition to determine compactness [156].

Example 24.6 Consider the *infinite vertical strip* X displayed in Fig. 24.5(a), and choose as an open covering the set of overlapping open rectangles (dashed boxes) satisfying

$$S_\alpha = \begin{cases} |x| < a + \epsilon \\ \alpha/2 < y < \alpha/2 + 1 \end{cases}$$

where $\epsilon > 0$ and $\alpha = \ldots - 2, -1, 0, 1, 2, \ldots$

1. Thus, each S_i is open and overlaps its two immediate neighbors and $\cup S_i \equiv S_1 \cup S_2 \cup S_3 \cdots \supset X$, implying that $\{S_i\}$ is an open cover of X.
2. However, no *finite* subcovering $\{S_1, S_2, \ldots, S_n\}$ exists because the area of X is infinite but the area covered by the rectangles of $\{S_1, S_2, \ldots, S_n\}$ must be finite, so it cannot be a covering of X.

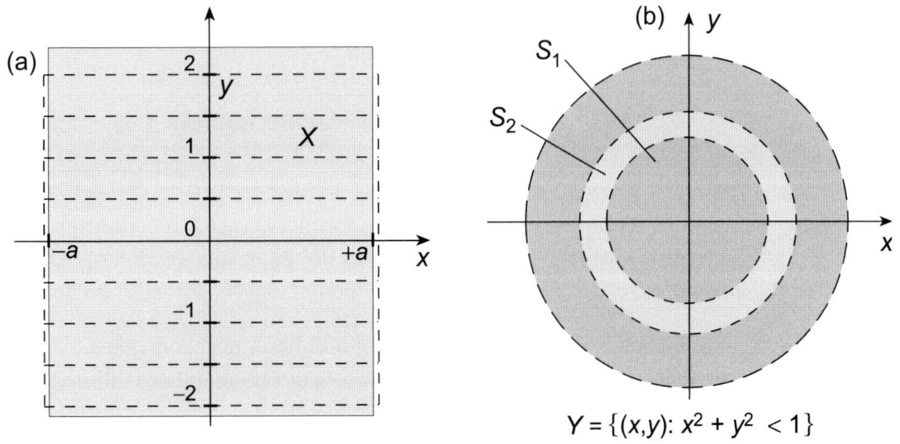

Fig. 24.5 Two spaces in $\mathbb{R}^2$ that are not compact: (a) an infinite vertical strip; (b) an open disk. Dashed boundaries indicate an open set.

We conclude that the infinite strip is non-compact. Clearly the lack of compactness is associated with the absence of bounds in the y direction for the space X.

Example 24.7 Consider the *closed disk* $Y = \{(x, y) : x^2 + y^2 \leq 1\}$. Choose as an open covering the set of concentric disks

$$S_\alpha = \left\{(x, y) : x^2 + y^2 < \left(1 - \frac{1}{1+\alpha} + \epsilon\right)\right\},$$

where $\epsilon > 0$ and small, and $\alpha = 1, 2, \ldots$

1. Then this is a covering of Y, since $\cup S_\alpha \supset Y$.
2. But also there is a *finite subcover* consisting of $\{S_1, S_2, \ldots, S_n\}$, where n is the smallest integer satisfying $n > 1/\epsilon - 1$ (that is, the first n for which the radius exceeds 1).

Thus the closed disk Y is compact.

Example 24.8 As you are asked to show in Problem 24.5, the *open disk*

$$Y = \{(x, y) : x^2 + y^2 < 1\}$$

displayed in Fig. 24.5(b) is non-compact. Comparing Problem 24.5 with Example 24.7, we see that it is the open property of the disk that is responsible for its non-compact topology.

For n-dimensional euclidean spaces $\mathbb{R}^n$, if $Y \subset \mathbb{R}^n$ then Y is compact only if it is *closed and bounded*.[2] This is the definition that we have employed for compactness in much of our earlier discussion. If Y is not a subset of $\mathbb{R}^n$, the question of compactness requires the preceding more general considerations in terms of coverings and subcoverings to test for

[2] This is called the *Heine–Borel theorem*. Notice that the non-compact space in Example 24.6 is not bounded, while the non-compact space in Example 24.8 is bounded, but it is not closed.

compactness. Compactness is significant because we shall show in Section 24.3.1 that it is a topological invariant for a space.

24.2.4 Connectedness

Intuitively, a *connected set* cannot be divided into parts that are "far apart" (the usual concept of "being in one piece"). More formally, a set Y is connected if it cannot be written in the form $Y = Y_1 \cup Y_2$, where Y_1 and Y_2 are open sets and $Y_1 \cap Y_2 = \emptyset$. The following are two examples.[3]

1. The closed interval $[a, b]$ on the real number line is connected.
2. Any discrete subset of $\mathbb{R}^n$ with more than one member is disconnected. For example, the set of rational numbers is disconnected.

It will be shown in Section 24.3.2 that connectedness is a topological invariant, and we shall explore the nature of connectedness in more detail in Section 24.4.

24.2.5 Homeomorphism

We now wish to make precise the qualitative idea that topological properties are conserved under smooth deformations. Consider topological spaces T_1 and T_2, and a map between them $f : T_1 \rightarrow T_2$. This map is a *homeomorphism* if it is continuous and has an inverse $f^{-1} : T_2 \rightarrow T_1$ that is also continuous. This permits a definition of *topological equivalence.*

> ***Topological Equivalence:*** A homeomorphism establishes a relationship between one topological space and a second topological space such that open sets in the two spaces are in one to one correspondence. If two spaces admit such a homeomorphism, they are *topologically equivalent.*

Homeomorphism is an *equivalence relation* (see Box 2.5 and Section 2.11). Using T_i to denote topological spaces and the symbol $\sim$ to denote homeomorphism, by equivalence

1. $T_1 \sim T_1$,
2. if $T_1 \sim T_2$, then $T_2 \sim T_1$,
3. if $T_1 \sim T_2$ and $T_2 \sim T_3$, then $T_1 \sim T_3$.

Sets of topological spaces can be divided into *equivalence classes* and topological spaces that are homeomorphic to each other belong to the same equivalence class. As we now discuss, these equivalence characterizations allow the definition of *topological invariants* that are *conserved under homeomorphisms.*

[3] Experience suggests that a space is connected if any two points can be joined by a continuous path, which is called *pathwise connectivity.* But this is too restrictive by the formal definition of connectivity: all pathwise connected spaces are connected, but not all connected spaces are pathwise connected. However, our interest here lies in pathwise connectivity, so for brevity we use "connected" to mean pathwise connected.

24.3 Topological Invariants

In the preceding section topological invariants were defined as quantities that do not change under a homeomorphism. We shall now prove that three fundamental properties of topological spaces, *compactness*, *connectedness*, and *dimensionality*, are topological invariants.[4]

24.3.1 Compactness Is a Topological Invariant

Suppose that X is a compact topological space, that $f : X \to Y$ is a homeomorphism from X to a second topological space Y, and that $\{S_i\}$ is an open cover of Y [156]. Then $f^{-1}(S_i)$ is an open set in X because f is continuous and, since f is invertible, $\cup f^{-1}(S_i)$ is an open cover of X. But X is compact, so there is a finite subcover $\{f^{-1}(S_1), f^{-1}(S_2), \ldots, f^{-1}(S_n)\}$. This is also a finite subcover of Y, so Y is compact. Thus, if a space is compact a homeomorphism leads to a space that is also compact and we conclude that *compactness is a topological invariant* because it is unchanged by a homeomorphism.

24.3.2 Connectedness Is a Topological Invariant

Consider a connected topological space X with a homeomorphism f to Y [156]. If Y were disconnected, $Y = Y_1 \cup Y_2$ and $Y_1 \cap Y_2 = \emptyset$, with Y_1 and Y_2 open sets. But f is continuous, so $f^{-1}(Y_1)$ and $f^{-1}(Y_2)$ are open sets in X. However, they also satisfy $f^{-1}(Y_1) \cup f^{-1}(Y_2) = X$, which would imply that X is not connected, contradicting the original hypothesis. It follows that Y is necessarily connected and *connectedness is a topological invariant.*

24.3.3 Dimensionality Is a Topological Invariant

We shall now demonstrate that the dimensionality of $\mathbb{R}^n$ is a topological invariant by showing inductively that there can be no homeomorphism between $\mathbb{R}^n$ and $\mathbb{R}^m$ if $n \neq m$ [156].

Conjectured Homeomorphism of $\mathbb{R}$ ***and*** $\mathbb{R}^2$***:*** First consider the special case of $\mathbb{R}$ and $\mathbb{R}^2$. If there were a homeomorphism $f : \mathbb{R} \to \mathbb{R}^2$, then the dimensionality of $\mathbb{R}$ and $\mathbb{R}^2$ could be changed by the homeomorphism and dimensionality would not be topologically invariant. However, we shall now show that such a homeomorphism is logically contradictory. Let $\mathbb{R}^2$ be represented by the x–y plane and $\mathbb{R}$ by its x-axis, but delete the point $(x, y) = (0, 0)$ from $\mathbb{R}^2$ (which is the point $x = 0$ in $\mathbb{R}$). Then, the following is evident from Fig. 24.6(a).

1. The space $\mathbb{R} - \{(0)\}$ is disconnected while the space $\mathbb{R}^2 - \{(0, 0)\}$ is connected.
2. But connectedness is a topological invariant, as proved in Section 24.3.2.
3. Hence the spaces $\mathbb{R} - \{(0)\}$ and $\mathbb{R}^2 - \{(0, 0)\}$ cannot be homeomorphic.

[4] Conversely, Problem 24.3 will demonstrate that the property of *boundedness* is not a topological invariant.

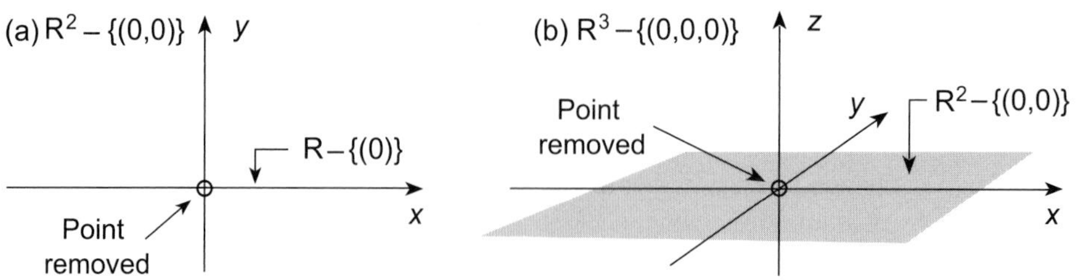

Fig. 24.6 (a) The spaces $\mathbb{R}$ (x-axis) and $\mathbb{R}^2$ (xy-axes), with a point at the origin removed. (b) The spaces $\mathbb{R}^2$ and $\mathbb{R}^3$ with a point at the origin removed.

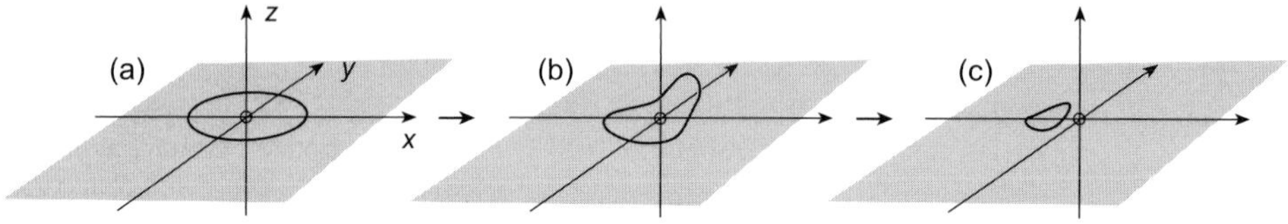

Fig. 24.7 Proof that $\mathbb{R}^3$ with the origin removed is simply connected. The circle in the x–y plane shown in (a) can be distorted out of the plane as in (b), moved away from the origin as in (c), and then shrunk to a point without encountering the origin.

Now, suppose the homeomorphism $f : \mathbb{R} \to \mathbb{R}^2$ to exist and consider its restriction to $\mathbb{R} - \{(0)\}$, which implies a homeomorphism from $\mathbb{R} - \{(0)\}$ to $\mathbb{R}^2 - \{(0,0)\}$. But we just proved that *this homeomorphism is non-existent*, so no homeomorphism can exist between $\mathbb{R}$ and $\mathbb{R}^2$ either. Now we need only generalize this reasoning to $\mathbb{R}^n$ and $\mathbb{R}^{n+1}$ for $n \geq 2$.

Conjectured Homeomorphism of $\mathbb{R}^n$ ***and*** $\mathbb{R}^{n+1}$**:** For the special case of $\mathbb{R}^2$ and $\mathbb{R}^3$, take $\mathbb{R}^2$ as the x–y plane and $\mathbb{R}^3$ as the (x, y, z) space, and delete the point at the origin, as illustrated in Fig. 24.6(b). Assume that there is a homeomorphism $f : \mathbb{R}^2 \to \mathbb{R}^3$, which would imply that the dimensionalities of $\mathbb{R}^2$ and $\mathbb{R}^3$ are not topological invariants. Restrict f to $\mathbb{R}^2 - \{(0,0)\}$ to define a homeomorphism $f : \mathbb{R}^2 - \{(0,0)\} \to \mathbb{R}^3 - f\{(0,0)\}$. Now place a circle $x^2 + y^2 = r^2$ around the origin and consider the limit $r \to 0$.

1. In $\mathbb{R}^3 - f\{(0,0)\}$, the circle may be shrunk to a point since it can be deformed out of the x–y plane to avoid $(0,0)$, as illustrated in Fig. 24.7.
2. On the other hand, in $\mathbb{R}^2 - \{(0,0)\}$ no deformation can avoid the point $(0,0)$ and the circle cannot be shrunk to a point.

Thus, continuity of the homeomorphism breaks down and it does not exist. By induction we may continue, deleting a point p from $\mathbb{R}^n$ and $\mathbb{R}^{n+1}$ to obtain the sets $\mathbb{R}^n - \{p\}$ and $\mathbb{R}^{n+1} - \{p\}$, and showing that no homeomorphism exists between $\mathbb{R}^n$ and $\mathbb{R}^{n+1}$ by surrounding p with spheres S^{n-1} of radius r and taking the limit $r \to 0$. We conclude that $\mathbb{R}^n$ and $\mathbb{R}^m$ are homeomorphic only if $m = n$, and *dimensionality is a topological invariant.*

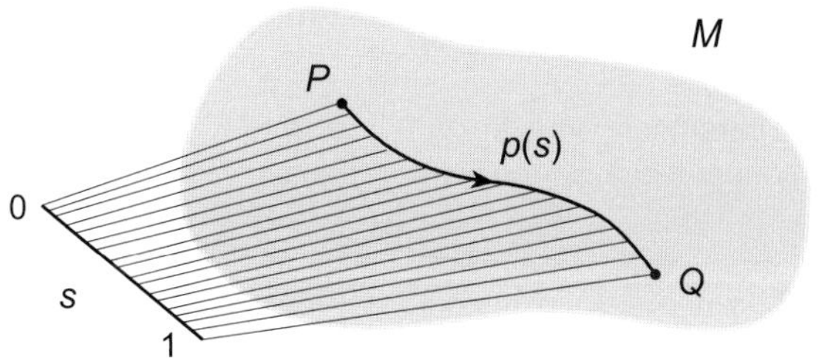

Fig. 24.8 A path from P to Q in the space M. A path $p(s)$ may be parameterized by a 1D variable s that ranges continuously from 0 to 1 as the path goes continuously from P to Q.

24.4 Homotopies

A *path* in a topological space X may be defined as a continuous function $p(s)$ of some real parameter $0 \leq s \leq 1$ that associates each value of s with a point $p(s)$ in the space, as in Fig. 24.8. Such a path connects the points P and Q if $p(0) = P$ and $p(1) = Q$, or vice versa since paths may be directed, with a direction "to" and "from" the endpoints that may be indicated graphically by an arrow. A great deal can be learned about the connectedness of a topological space by studying the possible independent paths that can be constructed in it. Of particular interest are *closed paths,* which have the same starting and ending points.

24.4.1 Homotopic Equivalence Classes

A *loop* or a *closed path* at the point S corresponds to $p(0) = p(1) \equiv p(S)$. If for two closed paths $p(S)$ and $p'(S)$ a function $h(t, S)$ exists satisfying

$$h(0, S) = p(S) \qquad h(1, S) = p'(S), \tag{24.2}$$

where t is a parameter ranging over the interval $[0, 1]$, then the paths $p(S)$ and $p'(S)$ are said to be *homotopic* (continuously deformable one into the other by the function $h(t, S)$). The space of continuous maps $C(X, Y)$ from X to Y is divided by homotopy into equivalence classes: all continuous maps $C(X, Y)$ that are homotopic belong to the same equivalence (homotopic) class.[5]

24.4.2 Homotopy Classes Are Topological Invariants

Since homeomorphism is a continuous map, the homotopy equivalence classes are unchanged under a homeomorphism of either X or Y; thus, they are *topological invariants* of the pair of spaces X and Y. If we choose a fixed space for X and allow Y to vary over all

[5] Note as an aside that (1) for groups, *conjugation* produces equivalence classes that have *group elements* as members (see Section 2.11 and Box 2.5). (2) For topologies, *homeomorphism* produces equivalence classes that have *topological spaces* as members (see Section 24.2.5). (3) For closed paths in topological spaces, *homotopy* produces equivalence classes that have *continuous maps* as members. These observations illustrate the utility and abstract nature of equivalence classes that was emphasized in Box 2.5.

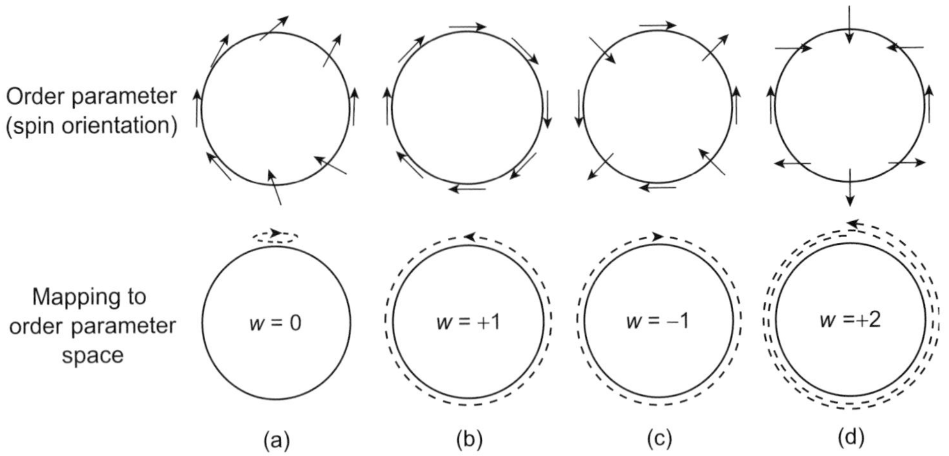

Fig. 24.9 Winding of a planar spin around a circle [194]. The spin orientation is an order parameter.

topological spaces of interest, we can then study the topological differences between two spaces Y and Y' by comparing their equivalence classes under maps with the fixed space X. The equivalence classes under homotopy of the continuous maps $C(X, Y)$ from X to Y are denoted $[X, Y]$. A common choice is $X = S^n$, where S^n is the n-dimensional unit sphere. Then the equivalence classes under homotopy $[S^n, Y]$ of $C(S^n, Y)$ cannot be deformed one into another because of the conservation of topological invariants for the pair of spaces S^n and Y. Many topological quantities of interest in physics are homotopy invariants of this sort. An example is shown in Fig. 24.9.

24.4.3 The First Homotopy Group

We shall now show that the equivalence classes of $C(S^n, Y)$ have a *group structure* [146, 174]. Because we associate a direction with a path, an inverse is defined naturally by traversing the path in the opposite direction, $p^{-1}(s) = p(1-s)$. We can also introduce a natural definition for a *product of paths*. Consider two paths $a(s)$ and $b(s)$. If the endpoint of $a(s)$ coincides with the beginning of $b(s)$, we may define the product path $c = ab$ by requiring that

$$c(s) = \begin{cases} a(2s) & 0 \le s \le \frac{1}{2} \\ b(2s-1) & \frac{1}{2} \le s \le 1 \end{cases} \tag{24.3}$$

as Fig. 24.10 illustrates. A *null path* can be defined by the product aa^{-1} (or ab^{-1} if $a \sim b$).

Let $[a]$ denote the set of paths having the same endpoints that are homotopic (continuously deformable) to the path a, and introduce a multiplication law for the space X for these *homotopy classes*,

$$[a][b] = [ab], \tag{24.4}$$

where any representative of a class may be used in the multiplication. Now consider the multiplication of the classes of closed paths having a common basepoint in a space

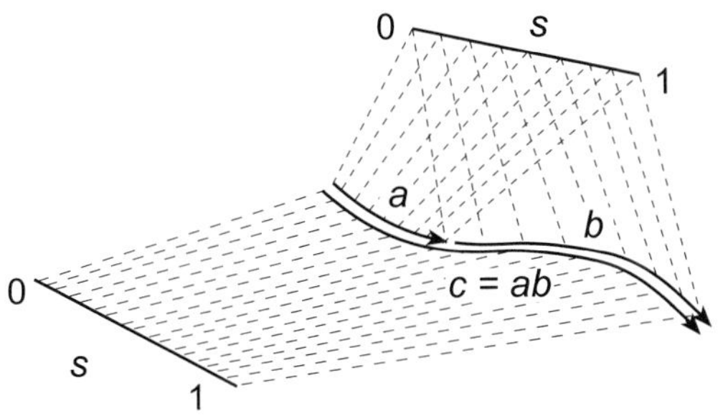

Fig. 24.10 Product of two paths, $c = ab$. The product path c is formed by connecting the end of path a and beginning of path b. All paths may be parameterized by a variable ranging from 0 to 1.

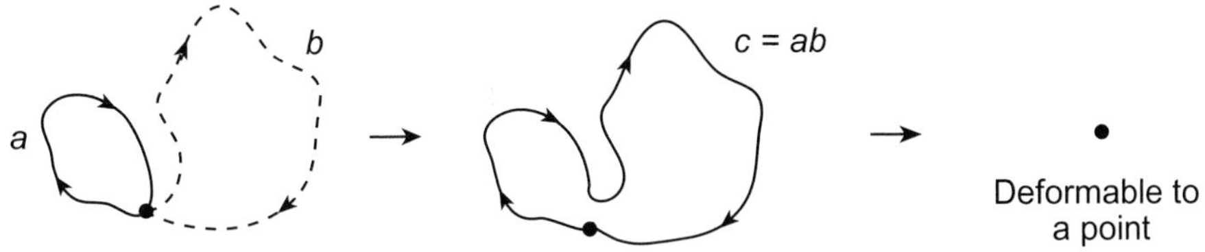

Fig. 24.11 Path multiplication in a space that is simply connected.

(denoted by a dot).[6] This defines a group called the *first homotopy group* $\pi_1(X)$ of the space X, since the group postulates of Section 2.2 are satisfied [174].

1. *Closure:* If $[a] \in \pi_1(X)$ and $[b] \in \pi_1(X)$, then $[a][b] = [ab] \in \pi_1(X)$.
2. *Associativity:* $([a][b])[c] = [a]([b][c])$, because $(ab)c \sim a(bc)$.
3. *Identity:* The class of null paths [1] constitutes an identity element since $[a][1] = [a]$.
4. *Inverse:* $[a^{-1}][a] = [1]$, so an inverse exists for every class.

The first homotopy group is also called the *fundamental group* of the corresponding space. For a simply connected space, the product of closed paths is a path that is always deformable to a point, as illustrated in Fig. 24.11. Thus, for a simply connected space $\pi_1(X) = [1]$. We may, in fact, use this as a definition.

A space is simply connected if its fundamental group consists of only the identity. Conversely, the fundamental group is no longer trivial for a space that is not simply connected.

Example 24.9 Consider the 2D euclidean plane with a hole in it, as in Fig. 24.12. Topologically, this space is the product of a line and a circle: $\mathbb{R} \times S^1$. It may be shown that $\pi_1(A \times B) = \pi_1(A) \times \pi_1(B)$. Thus, in this example $\pi_1(\mathbb{R} \times S^1) = \pi_1(S^1)$, since $\mathbb{R}$ is simply connected with a trivial first homotopy group. This corresponds to the group

[6] Attaching the loops to a common basepoint is necessary to be definite, but it can be shown that the choice of basepoint is irrelevant if the space is pathwise connected.

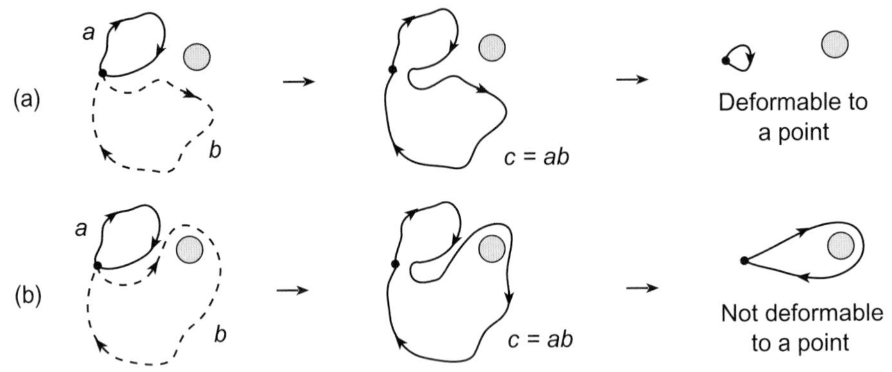

Fig. 24.12 Path multiplication in a space that is not simply connected.

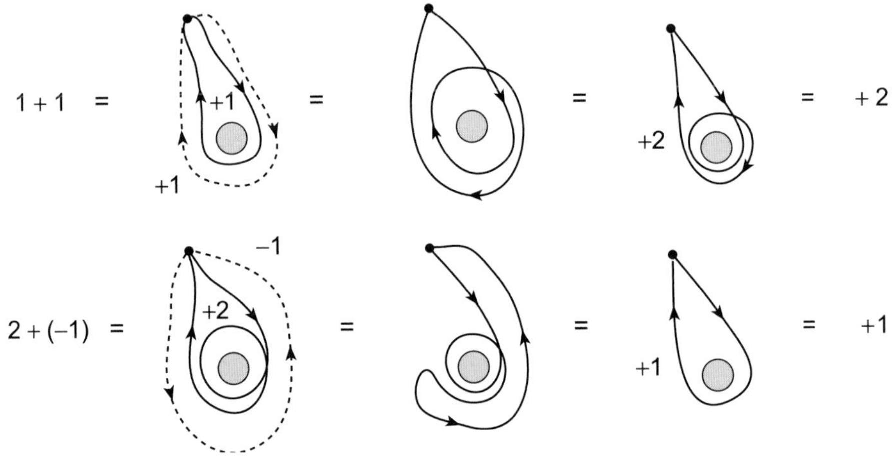

Fig. 24.13 Equivalence of $\pi_1(S^1 \times \mathbb{R}) = \pi_1(S^1)$ and the additive group of integers Z in Example 24.9.

structure associated with the mapping of circles to circles, $S^1 \to S^1$. As illustrated in Fig. 24.13, the fundamental group is in one to one correspondence with the additive group of integers Z, where (signed) integers label the number of loops around the hole for a closed path. In this example positive integers indicate clockwise and negative integers indicate counterclockwise loops.

Let us look at the mapping $S^1 \to S^1$ of Example 24.9 more carefully, as a prototype of mappings for higher-dimensional spheres [164]. The general mapping may be characterized by an angle ϕ associated with the first circle and a corresponding angle $\Lambda(\phi)$ associated with the second circle. Consider first the trivial mappings

$$\Lambda_0(\phi) = 0 \quad \text{(for all } \phi) \qquad \Lambda_0'(\phi) = \begin{cases} t\phi & 0 \le \phi < \pi \\ t(2\pi - \phi) & \pi \le \phi < 2\pi \end{cases} \tag{24.5}$$

where t is a parameter in the range $[0, 1]$. It is obvious algebraically that the mapping Λ_0' may be deformed continuously to Λ_0 by letting $t \to 0$, so these mappings are all in the

Box 24.3 Of Coffee Cups, Doughnuts, and Winding Numbers

A coffee cup and a doughnut are equivalent topologically, as illustrated in Fig. 24.1. We may wind a string around the handle of the cup and define a winding number n as the number of complete windings of the string around the handle. Everything changes smoothly in the continuous deformation between a coffee cup and a doughnut, so any change in this winding number must also be smooth.

> But winding numbers are integers and *integers cannot change smoothly into other integers.* Therefore, the winding number cannot change in the continuous, smooth deformation and it is a topological invariant [79].

Such winding numbers are conserved *topologically,* for reasons that have nothing to do with the details of the deforming transition. This simple example illustrates the essence of many conserved topological invariants in physical systems.

same homotopy class. Now consider the more general mapping

$$\Lambda_n(\phi) = n\phi, \tag{24.6}$$

where n is an integer. For fixed $n \neq 0$ this cannot be deformed continuously to Λ_0. Formally we may define a *winding number* Q by

$$Q \equiv \frac{1}{2\pi}\int_0^{2\pi} \frac{d\Lambda}{d\phi}d\phi. \tag{24.7}$$

For the mapping (24.6) this gives $Q = n$, which is a topological index giving the number of times one circle is wrapped around the other circle in $S^1 \to S^1$, with the sign indicating the direction of winding. A physical example of topological invariance for a winding number is given in Box 24.3. The general result is $\pi_1(S^1) = Z$, with Z the additive group of integers. Equation (24.7) is an overly formal way to extract the winding number for this simple example, but it can be generalized to situations where the topological invariant is less obvious.

In the examples discussed so far the first homotopy group has been abelian because the order of path multiplication did not matter. However, consider Fig. 24.14(a). The paths f and g are obviously not homotopic, and from Fig. 24.14(b), $fc \neq cf$. Thus, the fundamental group is non-abelian for this case and the order of path multiplication matters.

24.4.4 Higher Homotopy Groups

The first homotopy group is associated with mapping a space X to the circle S^1; higher homotopy groups $\pi_n(X)$ involve mapping X to higher-dimensional spheres S^n. For example, the second homotopy group $\pi_2(X)$ corresponds to equivalence classes of mappings $X \to S^2$, the third homotopy group $\pi_3(X)$ corresponds to equivalence classes of mappings $X \to S^3$, and so on. The principles are the same as for $\pi_1(X)$, but the geometry is harder to visualize. The first homotopy group can be non-abelian but all higher homotopy groups are abelian. In many cases of physical interest X may be taken as an m-dimensional

Table 24.1. Some properties of homotopy groups [2]

$$\pi_1(\mathrm{U(1)}) = \pi_1(S^1) = \mathrm{Z} \qquad \pi_1(\mathrm{U(1)} \times \mathrm{U(1)}) = \pi_1(\mathrm{U(1)}) \times \pi_1(\mathrm{U(1)}) = \mathrm{Z} \times \mathrm{Z}$$
$$\pi_1(\mathrm{SU(2)}) = \pi_1(S^3) = 1 \qquad \pi_1(\mathrm{U(2)}) = \pi_1(S^3 \times S^1) = \pi_1(S^3) \times \pi_1(S^1) = \mathrm{Z}$$
$$\pi_n(A \times B) = \pi_n(A) \times \pi_n(B) \qquad \pi_n(S^m) = 1 \quad (n < m) \qquad \pi_n(S^n) = \mathrm{Z}$$
$$\pi_n(S^1) = 1 \quad (n > 1) \qquad \pi_2(S^n) = 1 \quad (n \neq 2) \qquad \pi_3(S^n) = 1 \quad (n \neq 2,3) \qquad \pi_3(S^2) = \pi_3(S^3) = \mathrm{Z}$$
$$\pi_1(\mathrm{SO}(n)) = \mathrm{Z}_2 \quad (n > 2) \qquad \pi_1(\mathrm{U}(n)) = \mathrm{Z} \quad (n \geq 1) \qquad \pi_1(\mathrm{SU}(n)) = 1 \quad (n \geq 2)$$
$$\pi_1(\mathrm{SU}(n)/\mathrm{Z}_n) = \mathrm{Z}_n \quad (n \geq 2) \qquad \pi_2(G) = 1 \quad \text{(for any Lie group } G)$$
$$\pi_3(G) = \mathrm{Z} \quad \text{(for any } simple \text{ Lie group } G) \qquad \pi_3(\mathrm{SO(4)}) = \mathrm{Z} \times \mathrm{Z}$$
$$\pi_3(\mathrm{SO}(n)) = \mathrm{Z} \quad (n > 4) \qquad \pi_n(\mathrm{U(1)}) = 1 \quad (n > 1) \qquad \pi_n(\mathrm{U}(n)) = \begin{cases} 1 & n \text{ even} \\ \mathrm{Z} & n \text{ odd} \end{cases}$$
$$\pi_{2n}(\mathrm{U}(n)) = \mathrm{Z}_{n!} \qquad \pi_2(G/H) = \pi_1(H) \quad (G \text{ simply connected})$$

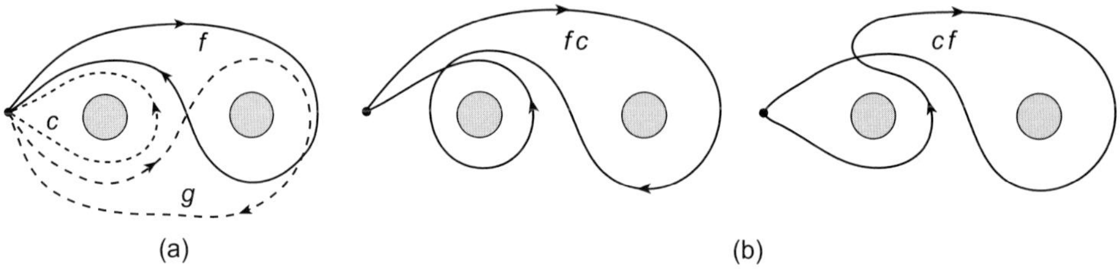

Fig. 24.14 (a) Paths f, c, and g in a 2D euclidean plane with two holes. (b) Path multiplication indicating that $fc \neq cf$, so the first homotopy group is non-abelian [149].

sphere, implying mappings of the form $S^m \to S^n$. For these cases the important topological implications will reside in a generalization of the winding number (24.7) that measures the number of times that one sphere is "wrapped around" the other sphere in the mapping $S^m \to S^n$. Some important results for homotopy groups are summarized in Table 24.1, where $\pi_n(X) = 1$ indicates that the homotopy group is trivial, consisting only of the identity.

24.5 Manifolds and Metric Spaces

Usually we take for granted that spaces of physical interest are "smooth," and that they have metrics (a concept of distance); however, topological spaces need not be smooth or possess metrics. As we now discuss, differentiable manifolds, which are topological spaces that look locally like euclidean space and satisfy certain smoothness criteria, are a subset of topological spaces, and typically in physics applications a metric can be assigned to the manifold, making it a metric space.

24.5.1 Differentiable Manifolds

Manifolds generalize familiar ideas about curves and surfaces to an arbitrary number of dimensions.

A manifold is a topological space that is *homeomorphic to n-dimensional euclidean space* $\mathbb{R}^n$ *locally.*

For example, a smooth curve is locally homeomorphic to $\mathbb{R}$ and a smooth 2D surface is locally homeomorphic to $\mathbb{R}^2$; thus, these constitute manifolds. However, the locally euclidean nature of any manifold does not guarantee that a specific manifold is *globally* homeomorphic to euclidean space. The following example illustrates.

Example 24.10 If the surface of the Earth is idealized as a sphere S^2, a region localized around a point looks like 2D euclidean space (hence flat maps over small regions of the Earth's surface have approximately the correct geometry). We conclude that S^2 is a manifold by virtue of its local properties. However, S^2 and $\mathbb{R}^2$ clearly are not equivalent globally because S^2 is compact and intrinsically curved while $\mathbb{R}^2$ is neither compact nor curved.

Thus S^2 and $\mathbb{R}^2$ are 2D manifolds that are locally, but not globally, equivalent.

Dimensionality of Manifolds: A *dimensionality n* may be associated with a manifold that is equal to the dimensionality of the euclidean space $\mathbb{R}^n$ to which it is locally homeomorphic. Thus, a curve is a one-dimensional manifold and a surface is a two-dimensional manifold. Since we have shown in Section 24.3.3 that the dimensionality of $\mathbb{R}^n$ is a topological invariant, *the dimensionality of a manifold is a topological invariant.*

Local Coordinates: Since an n-dimensional manifold is a topological space that is locally homeomorphic to $\mathbb{R}^n$, we may assign to each point in the manifold a set of n numbers that are termed the *local coordinates* of that point. If the manifold is not globally homeomorphic to $\mathbb{R}^n$, different local coordinates must be introduced in different parts of the global space and a given point may have associated with it more than one set of local coordinates. The formal definition of a manifold is then implemented through a requirement that the transition between these different sets of local coordinates be sufficiently smooth.

Formal Definition of a Manifold: Let us now give a precise formal definition of a manifold. A space U is an n-dimensional (differentiable) manifold if the following hold [156].

1. U is a topological space, as defined in Eq. (24.1).
2. U has a family of pairs $\{(U_i, \phi_i)\}$, where
   a. $\{U_i\}$ is a family of open sets with $\cup_i U_i = U$, so the U_i form an *open cover* of U,
   b. ϕ_i is a *homeomorphism* $\phi_i : U_i \to U_i'$ from U_i to an open subset U_i' of $\mathbb{R}^n$.
3. For U_i and U_j such that $U_i \cap U_j \neq \emptyset$, the map from $\phi_j(U_i \cap U_j)$ to $\phi_i(U_i \cap U_j)$ is *infinitely differentiable.*

Atlases and Charts: The family of pairs $\{(U_i, \phi_i)\}$ is termed an *atlas* and individual members of the family (U_i, ϕ_i) are called *charts*. Thus an atlas is the minimal set of local coordinate systems (charts) necessary to supply unique coordinates for all points in a manifold.

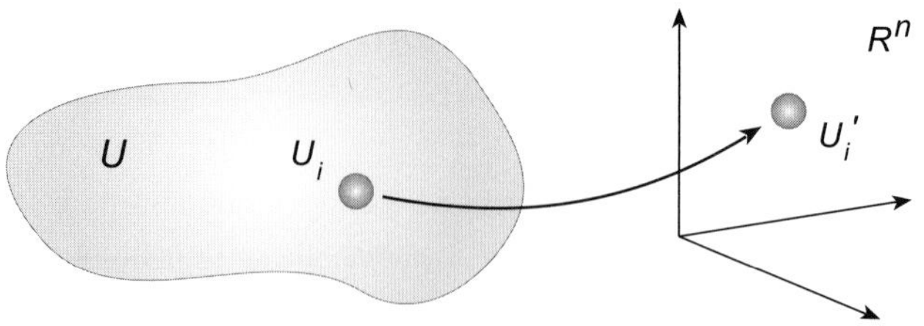

Fig. 24.15 An n-dimensional manifold U is locally homeomorphic to n-dimensional euclidean space.

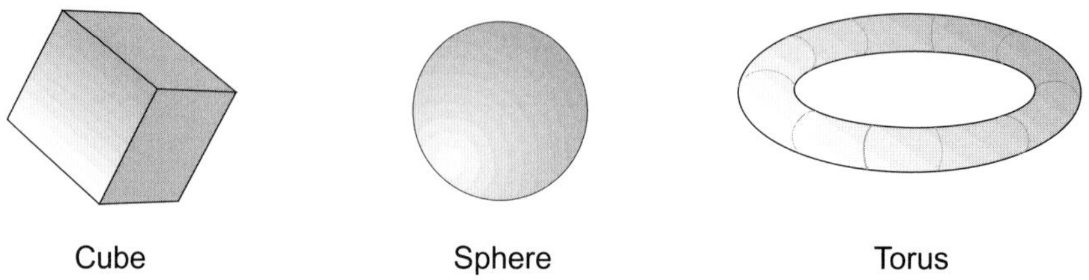

Fig. 24.16 Illustration of the relationship between topological spaces, which are characterized by continuity and connectedness, and differentiable manifolds, which are characterized by smoothness. The cube and sphere are equivalent topologically but of these only the sphere is a manifold. The sphere and torus are both manifolds, but they differ topologically.

Example 24.11 The Earth's surface cannot be covered uniquely by a single set of polar coordinates because at the poles an infinite number of longitudes map to a single point. It is well known that at least two overlapping coordinate systems (charts) are required to assign a unique coordinate to every point on the surface of the Earth.

In the preceding definition of a manifold, requirements (1) and (2) assert that U is locally euclidean; that is, that U can be covered with patches U_i that are assigned coordinates in $\mathbb{R}^n$ by the ϕ_i, as illustrated in Fig. 24.15. Within one of these patches U looks like a subset of euclidean n-space, $\mathbb{R}^n$. Generally, U need not look like $\mathbb{R}^n$ globally because that depends on how the individual patches are fitted together. Requirement (3) implies that, if two patches U_i and U_j overlap, in the overlap region $U_i \cap U_j$

1. there are two sets of $\mathbb{R}^n$ coordinates available, and
2. the transition between the two sets is smooth.

This prescription ensures that (1) a manifold is locally euclidean and (2) on any path the coordinates will vary smoothly, even if the manifold is not euclidean globally.

Continuity, Smoothness, and Connectedness: The distinction between (1) *continuity* and *connectedness,* which are the domain of topology, and (2) *smoothness* (more precisely *differentiability*), which is a property of manifolds, is illustrated in Fig. 24.16. The sphere and the cube are topologically equivalent because one can be distorted continuously into the other. However, while we would accept the sphere as a smooth two-dimensional manifold, the sharp edges and corners of the topologically equivalent cube would lead us to reject it as a candidate for a differentiable manifold. On the other hand, the sphere and the torus are both locally homeomorphic to $\mathbb{R}^2$ so we may expect that they are 2-manifolds, but clearly they differ topologically: the hole in the torus implies different connectedness for the 2-sphere and 2-torus.

24.5.2 Metric Spaces

Metric spaces were introduced earlier as manifolds having a prescription to measure distances. We now give a formal definition in accord with this intuitive idea. A metric space $(M, \mathscr{M})$ is a set M and a function $\mathscr{M} : M \times M \to \mathbb{R}$, such that for $x, y, z \in M$,

1. $\mathscr{M}(x, y) \geq 0$,
2. $\mathscr{M}(x, y) = 0$ if and only if $x = y$,
3. $\mathscr{M}(x, y) = \mathscr{M}(y, x)$,
4. $\mathscr{M}(x, z) \leq \mathscr{M}(x, y) + \mathscr{M}(y, z)$.

This space is said to have a metric $\mathscr{M}(x, y)$, and these requirements coincide with common-sense notions of a function $\mathscr{M}(x, y)$ to measure distance. A metric space is a topological space but not all topological spaces have a metric. However, spaces employed in physics almost always come equipped with a metric, which simplifies many topological proofs that would otherwise have to use abstract set properties to ascertain the "closeness" of two points in a space.

Background and Further Reading

McCarty [146], Mendelson [148], and Kosniowski [134] give introductions to topology. Frankel [63], Naber [151], Nakahara [152], Nash and Sen [156], and Pires [164] review topology and geometry for physicists. For the absolute beginner an introduction to topology that assumes little mathematics background at all may be found in Warner [206]. Concise introductions to homotopy theory for physics may be found in Ryder [174] and Appendix I of Actor [2], and a broader mathematical treatment is given in Kosniowski [134].

Problems

24.1 Consider a set $M = \{a, b, c, d, e\}$ and the collection of subsets

$$\tau \equiv \{\emptyset, M, \{a\}, \{c, d\}, \{a, c, d\}, \{b, c, d, e\}\},$$

where $\emptyset$ is the empty set. Prove that τ defines a topology on the set M. ***

24.2 Consider a set $M = \{a, b, c, d\}$ and a collection of subsets $\tau = \{\emptyset, M, \{a\}, \{b\}\}$. Show that τ is not a topology on M.

24.3 Prove that an interval without endpoints is homeomorphic to the real number line $\mathbb{R}$. Thus, boundedness is not a topological invariant. *Hint*: Take $X = (-\frac{\pi}{2}, \frac{\pi}{2})$ and $Y = \mathbb{R}$, and consider the map $f : X \to Y$ given by $f(x) = \tan x$. ***

24.4 What is the first homotopy group of a two-dimensional torus T^2? *Hint*: For the 2-torus, $T^2 = S^1 \times S^1$. ***

24.5 Using an open covering corresponding to the family of concentric open disks

$$S_\alpha = \left\{(x, y) : x^2 + y^2 < \left(1 - \frac{1}{1+\alpha}\right)\right\},$$

where $\alpha = 1, 2, \ldots$, show that the open unit disk of Fig. 24.5(b) is not compact. ***

24.6 Show that the winding number Q of Eq. (24.7) gives $Q = n$ for the mapping (24.6).

24.7 From Problems 2.9 and 2.11, the group $\mathrm{D}_2 = \{e, a, b, c\}$ has a factor (quotient) group with respect to the abelian invariant subgroup $H = \{e, a\}$,

$$\mathrm{D}_2/H = H + M = \{e, a\} + \{b, c\},$$

with a map ϕ from D_2 to D_2/H given by

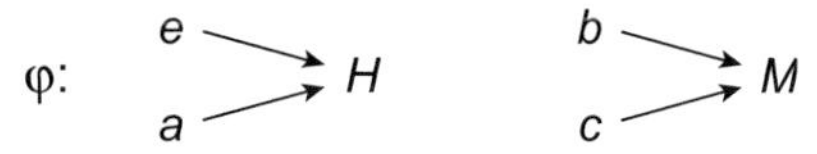

Show that if the space for D_2 is equipped with a topology defined by the open sets

$$\tau = \{\emptyset, \{e\}, \{a\}, \{b\}, \{c\}, \{e, a, b, c\}\},$$

the inverse map ϕ^{-1} implies that the quotient space has a topology also. ***

24.8 Are the following homeomorphic? (a) A closed interval and an open interval for the real numbers $\mathbb{R}$? (b) A parabola and a hyperbola? (c) A circle S^1 and $\mathbb{R}$?

24.9 Sketch the results of path multiplications (indicated by $\times$) for these examples:

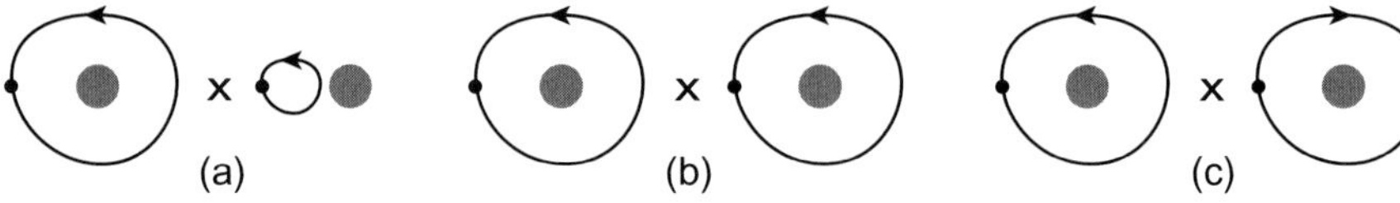

where each loop is defined in the same 2D euclidean plane with a single hole.

25 Topological Solitons

The solutions of all but the simplest wave equations exhibit the property of *dispersion,* where different wavelengths travel at different velocities. For example, the Klein–Gordon equation $(\Box + m^2)\phi(x,t) = 0$, exhibits a dispersion relation $\omega^2 = k^2 + m^2$, where ω is frequency, k is momentum, and m is mass. Then a localized wave packet formed by superposing solutions at time $t = 0$ will spread as time goes on because different wavelengths travel at different velocities $\omega(k)/k$. However, under certain conditions wave solutions may exhibit dispersionless propagation because dispersion is exactly compensated by nonlinearities. Such solutions are called *solitons*. In some cases the suppression of dispersion may occur because of explicit dynamics; we will not consider such cases. The more interesting situation is when dispersionless solutions follow from *topological constraints* rather than from dynamics. The corresponding *topological solitons* are the subject of this chapter.

25.1 Models in (1+1) Dimensions

Let us introduce the basic ideas of topological solitons by considering wave equations in one space and one time (1+1) dimension [85].

25.1.1 Equations of Motion

Solitons can occur only for wave equations that are *non-linear*. Consider a simple Lagrangian density in (1+1) dimension,

$$\mathscr{L} = \frac{1}{2}(\partial_\mu \phi)(\partial^\mu \phi) - U(\phi) = \frac{1}{2}\left(\frac{\partial \phi}{\partial t}\right)^2 - \frac{1}{2}\left(\frac{\partial \phi}{\partial x}\right)^2 - U(\phi), \tag{25.1}$$

where $x^\mu \equiv (t, x)$, ϕ is a scalar field and $U(\phi)$ characterizes the non-linearity of the field. Restricting to time-independent solutions, the equation of motion follows from Hamilton's principle [the Euler–Lagrange equation (16.14); see Problem 25.1],

$$\frac{d^2\phi}{d^2x} = \frac{dU}{d\phi}, \tag{25.2}$$

and the total energy $E(\phi)$ is the integral over all space of the energy density,

$$E(\phi) = \int_{-\infty}^{+\infty} dx\, \mathscr{E}(\phi(x)) \qquad \mathscr{E}(\phi) = \frac{1}{2}\left(\frac{d\phi}{dx}\right)^2 + U(\phi), \tag{25.3}$$

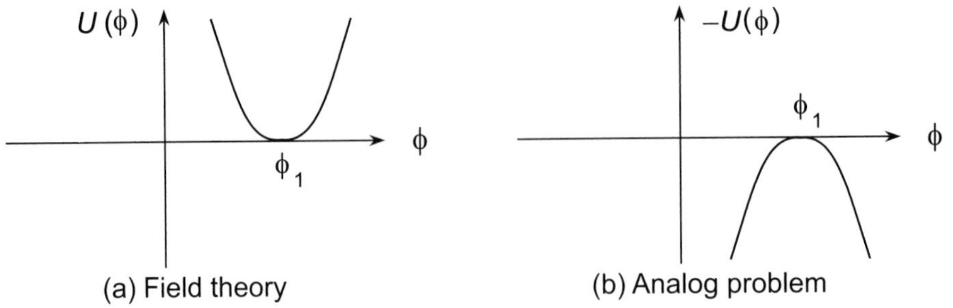

Fig. 25.1 Potentials $U(\phi)$ for a field theory and $-U(\phi)$ for the classical particle analog problem for a non-degenerate classical ground [85]. Topological solitons are impossible because the required boundary conditions cannot be satisfied for these potentials. Reproduced with permission from Wiley Interscience: *Gauge Field Theories – An Introduction with Applications,* M. Guidry (1991).

with the energy density $\mathscr{E}(\phi)$ following from a *Legendre transformation,*

$$\mathscr{E}(\phi) = \mathscr{H}(\phi) = \textstyle\sum_i p_i \dot{q}_i - \mathscr{L}(\phi), \tag{25.4}$$

where the q_i are coordinates and the p_i are the corresponding conjugate momenta (see Problem 25.2).

We will assume the potential $U(\phi)$ to have one or more minima corresponding to $U = 0$. Soliton solutions of Eq. (25.2) must have a localized energy density $\mathscr{E}(\phi(x))$ so that the integral in Eq. (25.3) converges, which is possible only for boundary conditions where $\phi(x)$ tends to a zero of $U(\phi)$ as $x \to \pm\infty$. Finding and interpreting such solutions is aided by realizing that the field theory problem of Eq. (25.2) is identical mathematically to one for the motion of a frictionless classical particle of unit mass in a potential $-U(x)$, for which Newton's second law would be $d^2\phi/dx^2 = -dU/d\phi$, with x in the field theory standing in for time in the particle problem and ϕ in the field theory problem standing in for the position coordinate in the classical analog problem.[1]

25.1.2 Vacuum States and Boundary Conditions

First assume that $U(\phi)$ has a single unique minimum, as illustrated in Fig. 25.1(a), along with the potential $-U(\phi)$ for the particle analog problem in Fig. 25.1(b). By inspection, Fig. 25.1(b) is inconsistent with topological solitons. The boundary conditions necessary for (25.3) to converge require a particle trajectory beginning and ending at $\phi = \phi_1$, which is impossible: a particle launched with a gentle push at ϕ_1 for $x = -\infty$ will never return to ϕ_1 at $x = +\infty$. Next let us consider a case where there is spontaneous symmetry breaking. In Fig. 25.2 there are three degenerate minima for the field theory problem and three degenerate maxima for the particle analog problem.The field must approach one of the zeros of $U(\phi)$ as $x \to \pm\infty$ for the energy calculated in Eq. (25.3) to be finite.

[1] Do not be confused by the abstraction we are employing here. The field $\phi = \phi(x)$ behaves as if it were a coordinate in the field theory, but $\phi(x)$ also is a function of the *actual* spacetime coordinate x.

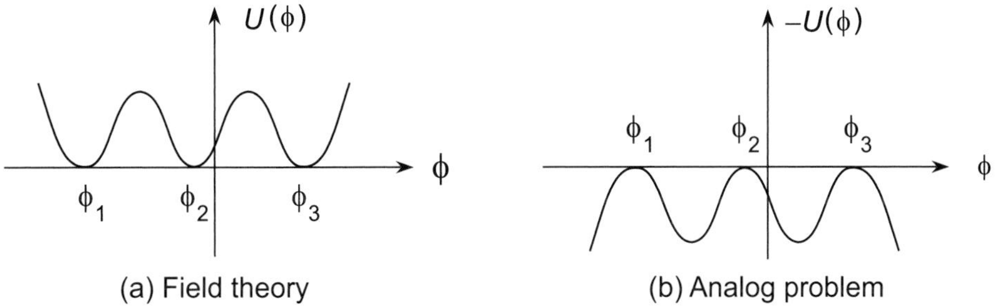

Fig. 25.2 As for Fig. 25.1, but for a triply degenerate classical ground state.

Example 25.1 In the analog problem of Fig. 25.2(b), four possibilities satisfy the finite-energy soliton boundary conditions, beginning at $x = -\infty$.

1. Start with a push to the right at ϕ_1 so that as $x \to +\infty$ the particle approaches ϕ_2.
2. Start with a push to the left at ϕ_2 so that as $x \to +\infty$ the particle approaches ϕ_1.
3. Start with a push to the right at ϕ_2 so that as $x \to +\infty$ the particle approaches ϕ_3.
4. Start with a push to the left at ϕ_3 so that as $x \to +\infty$ the particle approaches ϕ_2.

Each of these four cases gives a finite-energy solution of Eq. (25.2).

The four solutions of Example 25.1 are *topological solitons.* They are stable because the boundary conditions at infinity are topologically distinct.

We conclude that for scalar fields a necessary condition for existence of topological solitons is degeneracy of the vacuum (ground) state.

Hence, for scalar fields we expect a close relationship between topological solitons and the vacuum degeneracy associated with spontaneous symmetry breaking (see Ch. 17).

25.1.3 Topological Charges

A convenient labeling for these scalar-field topological solitons is to specify the values of ϕ that the field takes at spatial infinity. In Example 25.1 the solutions are characterized by the values of $(\phi(-\infty), \phi(+\infty))$ shown in Fig. 25.3. These solutions are in distinct topological classes because to convert one solution into another in Fig. 25.3 requires a finite change $\Delta\phi$ in the field at each point x but over an *infinite expanse of space.* The difference between vacuum field configurations at spatial infinity defines the *topological charge* Q. The stability of solitons may be attributed then to conservation of Q, which is associated with *boundary conditions,* not with dynamics. Such conserved topological charges have a fundamentally different origin than the conserved quantities following from Noether's theorem (Section 16.2.1) for continuous symmetries of the Lagrangian.

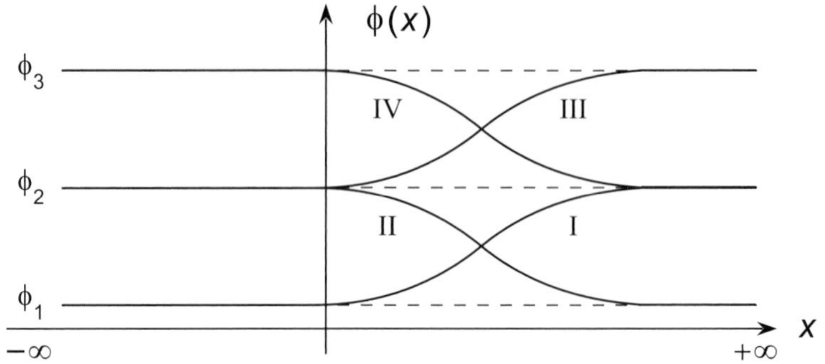

Fig. 25.3 The four soliton solutions given in Example 25.1. They may be labeled by the pair (ϕ_n, ϕ_m) that the field approaches at $x = \pm\infty$. For example, solution IV corresponds to (ϕ_3, ϕ_2).

25.1.4 Soliton Solutions in (1+1) Dimensions

As you are asked to show in Problem 25.3, the equation of motion (25.2) has a formal solution

$$x = x_0 \pm \int_{\phi(x_0)}^{\phi(x)} \frac{d\phi}{\sqrt{2U(\phi)}}, \tag{25.5}$$

where the arbitrary constant x_0 implies translational invariance. Topological soliton solutions may be obtained by choosing a potential $U(\phi)$, integrating Eq. (25.5), and solving the resulting equation for $\phi(x)$. Example 25.2 illustrates.

Example 25.2 Assume a scalar field theory with a field ϕ and potential

$$U(\phi) = \frac{1}{4}\lambda\left(\phi^2 - \frac{m^2}{\lambda}\right)^2, \tag{25.6}$$

where λ and m^2 are positive constants. As shown in Problem 25.5, the solution is

$$\phi_\pm(x) = \pm\frac{m}{\sqrt{\lambda}} \tanh\left(\frac{m}{\sqrt{2}}(x - x_0)\right), \tag{25.7}$$

where $\phi_+(x)$ is called the *kink* and $\phi_-(x)$ is called the *antikink*. The energy density $\mathscr{E}(x)$ and total energy E are found to be

$$\mathscr{E}(x) = \frac{m^4}{2\lambda}\operatorname{sech}^4\left(\frac{m}{\sqrt{2}}(x - x_0)\right) \qquad E = \frac{2\sqrt{2}}{3}\frac{m^3}{\lambda}, \tag{25.8}$$

for both the kink and antikink. The solutions and energy density are plotted in Fig. 25.4.

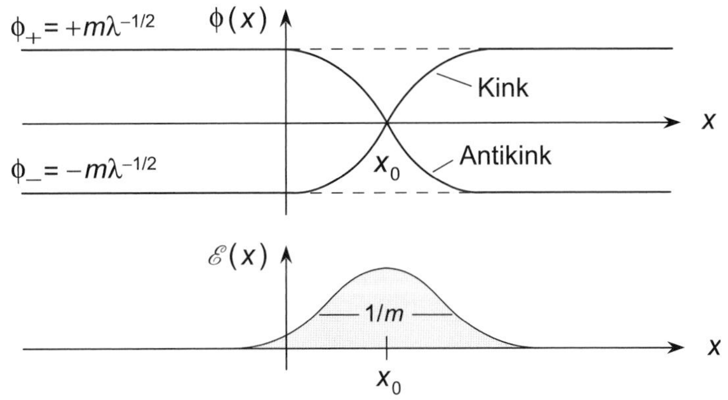

Fig. 25.4 Kink and antikink solutions $\phi_\pm(x)$, and associated energy density $\mathscr{E}(x)$, for Example 25.2 [85].

25.2 Solitons in (2+1) and (3+1) Dimensions

Encouraged by interesting results for scalar fields in (1+1) dimensions, we would like to forge ahead and consider topological solutions in (2+1) or (3+1) spacetime dimensions, but we encounter an immediate problem. For scalar fields it may be shown that no time-independent topological solutions exist in more than one spatial dimension. Thus, we have two options: (1) find multidimensional *time-dependent* soliton solutions, or (2) find multidimensional time-independent solutions by *including fields of non-zero spin.* We shall follow the second option and consider time-independent soliton solutions in (2+1) and (3+1) dimensions that include vector (gauge) fields.[2]

25.2.1 Homotopy Groups

As in (1+1) dimensions, finite-energy solutions in higher dimensions require that fields approach a zero of the potential at spatial infinity. In (3+1) dimensions with scalar and vector fields the topological classification of soliton solutions is more complicated than for scalar fields in (1+1) dimensions, but the basic question remains the same: how is spatial infinity mapped to the zeros of the potential? To answer this question for more complicated systems we will appeal to the theory of homotopy groups introduced in Section 24.4.

Let us first consider systems having both gauge and scalar fields, and assume that in the ground state the gauge fields vanish and the scalar fields are constant and correspond to a zero of the potential. We shall also assume the gauge groups to be simple, compact, and connected, and that accidental degeneracies and non-gauge internal symmetries are excluded, so that the zeros of U may be identified with the *coset space* G/H, where G

[2] To keep things simple we will not discuss fermion fields, which would require a detour into *Grassmann variables* and take us beyond the scope of our intended discussion.

is the gauge group and H is an invariant subgroup of G (see Section 2.14). Thus in the current discussion, G/H may be viewed as just eccentric shorthand for the set of zeros for the potential U [43].

The search for topological conservation laws becomes a question of how spatial infinity is mapped to zeros G/H of U. This mapping will often take the form

$$S^n \text{(ordinary space)} \longrightarrow G/H \text{ (field space)},$$

with $n = 1$ for two spatial dimensions and $n = 2$ for three spatial dimensions.

The existence of non-trivial homotopic classes permitting topological solutions depends on whether all mappings $S^n \longrightarrow G/H$ are continuously deformable into each other. For the examples in this chapter, only two results from Table 24.1 will be required:

$$\pi_n(S^n) = Z \qquad \pi_n(S^m) = 1 \quad (n < m), \tag{25.9}$$

where the homotopy group for the mapping $S^n \longrightarrow S^m$ is designated by $\pi_n(S^m)$, Z is the additive group of integers, and a one on the right side means that all mappings can be deformed into the trivial mapping. Thus for mapping spheres to spheres, if $n = m$ the mapping is non-trivial and a topological invariant can be defined that is an integer winding number specifying how many times one sphere is "wrapped around" the other in the mapping.

25.2.2 Mapping Spheres to Spheres

Let us give some examples of finding topological conservation laws in (2+1) and (3+1) dimensions. We assume for these examples a gauge-invariant Lagrangian density

$$\mathcal{L} = -\frac{1}{4}F^a_{\mu\nu}F^{\mu\nu}_a + (D_\mu\phi)^\dagger(D^\mu\phi) - U(\phi),$$

with D_μ the covariant derivative and $F^a_{\mu\nu}$ the non-abelian gauge field tensor. The potential is $U = 0$ for classical ground states and positive otherwise.

Example 25.3 Assume a dimensionality (2+1) with spatial infinity corresponding to S^1, a gauge group $U(1)$, a complex scalar field ϕ, and a potential $U = \frac{1}{2}\lambda(\phi^*\phi - a^2)^2$, where λ and a are positive. The zeros G/H of U occur for a circle defined by $\phi = ae^{i\sigma}$, and the relevant homotopy mapping is S^1(ordinary space) $\longrightarrow$ S^1(field space), which is characterized by an integer winding number n. This model is mathematically identical to the Landau–Ginzburg theory of Type II superconductors, and the solitons are 2D cross sections of the magnetic flux tubes penetrating a Type II superconductor [174].

Example 25.4 Assume that the dimensionality is (3+1), spatial infinity is S^2, the gauge group is SO(3), the scalar fields comprise an isovector, and $U = \lambda(\Phi^2 - a^2)^2/2$. The zeros of U are distributed on a sphere S^2 and the mapping is now S^2(ordinary space) $\rightarrow$ S^2(field space), which admits homotopy classes characterized by a winding number (called the *Pontryagin index* in this context) that labels the number of times one sphere is wrapped around the other. This non-trivial structure admits a topological solution called the *Polyakov–'t Hooft magnetic monopole* [174].

Additional examples in this vein are given in Coleman [43]. Note that we have not actually found the solutions for Examples 25.3 and 25.4. We have only given topological classifications of possible soliton solutions. Methods for obtaining solutions and discussion of their physical interpretation may be found in Refs. [43, 174].

25.3 Yang–Mills Fields and Instantons

The Standard Model of elementary particle physics in Ch. 19 is built on non-abelian Yang–Mills gauge fields. Can Yang–Mills fields have topological soliton solutions? The answer at first sight seems discouraging: a pure Yang–Mills theory does not have topological soliton solutions in (3+1) dimensional Minkowski space, though it can have static solitons in four *euclidean* dimensions [with a metric tensor $\eta = \text{diag}\,(+1\,,+1\,,+1\,,+1)$]. This seems not very useful for physics, since actual spacetime has a Minkowski metric $\eta = \text{diag}\,(+1\,,-1\,,-1\,,-1)$, not a euclidean metric. However, as discussed in Box 25.1,

> Quantum tunneling through a barrier may be viewed as *classical evolution in imaginary time,* which corresponds to classical motion in euclidean space.

Thus the occurrence of Yang–Mills solitons in (fictitious) euclidean space suggests that they represent a *quantum tunneling effect* in the (actual) Minkowski space, and therefore could be of physical interest.

25.3.1 Solitons in the Euclidean Yang–Mills Field

Yang–Mills instantons are finite-action solutions of the euclidean Yang–Mills equations. The steps to obtain them are similar to those for any topological soliton.

- Identify the boundary conditions for finite-action states.
- Make a homotopy classification using the boundary conditions.
- Search for explicit solutions within a given homotopic sector.

In addition to solutions being sought in euclidean rather than Minkowski space, Yang–Mills instantons differ from the solitons considered so far in that we will find that non-trivial homotopic classifications do not require spontaneous symmetry breaking.

Box 25.1 Propagation in Euclidean Space and Quantum Tunneling

Semiclassically, tunneling through quantum barriers is associated with *complex classical trajectories* that represent classical evolution in imaginary time during the barrier penetration. This may be anticipated by noting that the euclidean form of Newton's second law is given by transforming the Minkowski time to euclidean time τ through a *Wick rotation*, $t \to -i\tau$,

$$m\frac{d^2x}{dt^2} = -\frac{\partial V}{\partial x} \quad \xrightarrow[t\to -i\tau]{} \quad m\frac{d^2x}{d\tau^2} = +\frac{\partial V}{\partial x}.$$

Therefore, the replacement $t \to -i\tau$ effectively changes the sign of the potential and barriers become valleys.

A Wick rotation inverts the barrier and motion that is classically forbidden in real time becomes permitted in imaginary time.

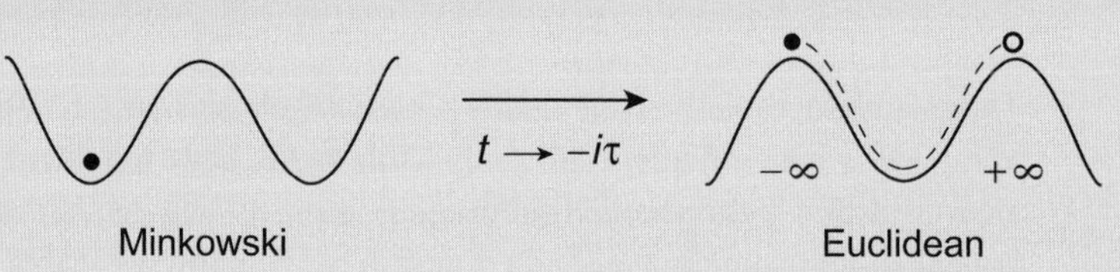

Although the motion becomes classically allowed in euclidean space it occurs with exponentially damped probability, as would be expected for a quantum barrier penetration process.

Thus classical propagation in euclidean space approximates a quantum barrier penetration process in Minkowski space.

25.3.2 Boundary Conditions

Let us introduce a rescaled vector potential B_μ and a rescaled field tensor $G_{\mu\nu}$ by

$$B_\mu \equiv \frac{g}{i}A_\mu = \frac{g}{i}\tau_a A_\mu^a \qquad G_{\mu\nu} \equiv \frac{g}{i}\tau_a F_{\mu\nu}^a,$$

with τ_a a gauge group generator. A gauge transformation on the vector potentials is [see Eq. (16.51)]

$$B_\mu \to U B_\mu U^{-1} - U\partial_\mu U^{-1}, \tag{25.10}$$

where U is a matrix representation of the gauge group. If $B_\mu = 0$, a gauge transformation yields,

$$B_\mu(x) = -U(x)\partial_\mu U^{-1}(x), \tag{25.11}$$

which is called a *pure gauge*. For a pure gauge it may be shown that $G_{\mu\nu} = 0$, so the euclidean Yang–Mills action,

$$S = -\frac{1}{2g^2}\int d^4x\,\mathrm{Tr}\,(G_{\mu\nu}G_{\mu\nu}), \tag{25.12}$$

is zero. Let us now consider finite-action solutions of the euclidean Yang–Mills equations. Parameterizing using the real euclidean coordinates (x_1, x_2, x_3, x_4), we may take the boundary of euclidean 4-space to be the 3D spherical surface S^3(space) at $r = \infty$, with

$$r = \sqrt{x_1^2 + x_2^2 + x_3^2 + x_4^2}.$$

The condition that Eq. (25.12) give finite action requires $G_{\mu\nu}$ to decrease faster than $1/r^2$ at the boundary, meaning that the vector potential must tend to a pure gauge at spatial infinity faster than $1/r$,

$$B_\mu(x) \xrightarrow[r\to\infty]{} -U\partial_\mu U^{-1} + \mathcal{O}\left(r^{-2}\right). \tag{25.13}$$

The matrix functions U define a mapping of S^3(space) to the gauge group manifold. Clearly one finite-action solution results if we take U as the identity matrix, giving $B_\mu(x) = 0$ over all space. This is the normal ground state. But Eq. (25.13) implies that finite-action solutions are also possible if the vector potential does not vanish over all space *as long as it tends to a pure gauge on the boundary.* Therefore, we are led to investigate the general mapping between S^3(space) and the manifold of the gauge group G.

25.3.3 Topological Classification of Solutions

It can be shown that any continuous mapping of S^3 into a simple Lie group G may be continuously deformed to a mapping into an SU(2) subgroup of G. Therefore, the homotopy classification of euclidean Yang–Mills theory with a simple gauge group reduces to evaluation of the mapping S^3(space) $\to$ S^3(field), since an SU(2) manifold is topologically S^3(field) (see Section 6.9.1). From Eq. (25.9) this mapping is generally non-trivial, giving a discrete infinity of homotopy classes labeled by a topological index Q that relates the sphere S^3(field) to the sphere S^3(space) at infinity. Explicitly,

$$Q = -\frac{1}{16\pi^2}\int d^4x\,\mathrm{Tr}\left(\tilde{G}_{\mu\nu}G_{\mu\nu}\right) \qquad \tilde{G}_{\mu\nu} \equiv \frac{1}{2}\epsilon_{\mu\nu\rho\sigma}G_{\rho\sigma}. \tag{25.14}$$

Then the trivial solution corresponds to $Q = 0$, the instanton solution corresponds to $Q = +1$, the anti-instanton solution corresponds to $Q = -1$, and various multiple instanton or multiple anti-instanton solutions correspond to $|Q| > 1$. We will not construct the actual solutions here but further discussion may be found in Refs. [167, 174].

25.3.4 Physical Interpretation of Instantons

Figure 25.5(a) illustrates an instanton configuration schematically. Finite action for a solution requires $G_{\mu\nu} = 0$ on the boundary of euclidean 4-space, which is satisfied if $B_\mu = 0$ on the boundary, as for the normal ground state, but is also satisfied if B_μ is pure

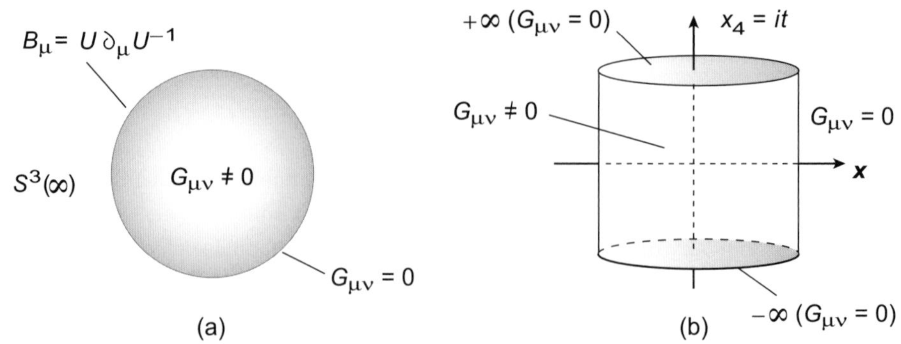

Fig. 25.5 (a) Instantons have a pure gauge configuration $B_\mu = U\partial_\mu U^{-1} \neq 0$ giving $G_{\mu\nu} = 0$ on the euclidean boundary S^3. (b) Distortion of the euclidean space boundaries in (a). The top and bottom surfaces correspond to all of 3D space $\boldsymbol{x}$ at $x_4 = it = \pm\infty$; the cylindrical surface represents all of time x_4 at spatial infinity [167, 174]. Reproduced with permission from Wiley Interscience: *Gauge Field Theories – An Introduction with Applications*, M. Guidry (1991).

Box 25.2 The θ-Vacuum and the Strong CP Problem

Instanton solutions with their implied tunneling between Minkowski vacuum states suggest that the true ground state of quantum chromodynamics (QCD) is a coherent superposition of winding number vacua [31, 84, 121],

$$|0\rangle_\theta = \sum_{n=-\infty}^{+\infty} e^{in\theta}|n\rangle,$$

where n labels degenerate vacuum states with different winding numbers and the parameter θ with $|\theta| \leq \pi$ labels independent sectors of QCD. This superposition is called the *θ-vacuum.* If we use a simple vacuum of fixed n instead of the superposition given above, an additional term is required in the effective QCD Lagrangian density that depends on θ:

$$\mathscr{L}_{\text{QCD}}^{\text{eff}} = \mathscr{L}_{\text{QCD}} + \frac{\theta}{16\pi^2}\text{Tr}\,(G_{\mu\nu}\tilde{G}_{\mu\nu}),$$

where $\mathscr{L}_{\text{QCD}}$ is the QCD Lagrangian density without instantons. Because P and CP are not symmetries of the QCD ground state unless $\theta = 0$, experimental limits on CP violation for strong interactions require $|\theta| \leq 10^{-9}$. The absence of a fundamental reason for why θ should be such a small number is called the *strong CP problem.*

gauge on S^3. Then a non-trivial mapping (an instanton) results if $G_{\mu\nu} \neq 0$ over some region of space but $G_{\mu\nu} = 0$ on the boundary S^3 because B_μ is pure gauge there.

Further insight is afforded by distorting the boundary in Fig. 25.5(a) to that of Fig. 25.5(b) [174]. The instanton is then seen to be a solution of the euclidean Yang–Mills equations in which a vacuum characterized by some topological class at $x_4 = -\infty$ evolves by propagation in imaginary time to a vacuum belonging to a different topological class at $x_4 = +\infty$. This generalizes the simpler boundary conditions illustrated in Fig. 25.3. The region between the vacuum states at $+\infty$ and $-\infty$ has positive field energy because

$G_{\mu\nu} \neq 0$ there; this represents an energy barrier and the imaginary-time propagation implies penetration of the barrier. Comparison with Box 25.1 then supports our earlier interpretation of instantons as states representing tunneling between classical vacuum states in Minkowski space. We close this chapter by noting an unresolved implication of instantons for the theory of strong interactions that is discussed in Box 25.2.

Background and Further Reading

Coleman [43], Ryder [174], and Ward [205] give readable introductions to topological solitons. Lee [139] discusses both topological and non-topological solitons. The book by Rajaraman [167] gives a comprehensive overview of solitons and instantons in quantum field theory. Various examples in this chapter are considered more extensively in Ref. [85].

Problems

25.1 Show that the equation of motion (25.2) follows from Hamilton's principle [the Euler–Lagrange equation given by Eq. (16.14)]. ***

25.2 Show that if fields are independent of time the Lagrangian density (25.1) leads to the energy density $\mathscr{E}(\phi)$ given in Eq. (25.3). *Hint*: Evaluate the Legendre transform (25.4) for time-independent fields.

25.3 Obtain the formal solution (25.5) to Eq. (25.2). *Hint*: Multiply Eq. (25.2) by $d\phi/dx$, integrate over dx, and use that $d\phi/dx$ and $U(\phi)$ both vanish at spatial infinity to give $\pm\, d\phi = \sqrt{2U(\phi)}\, dx$; then obtain Eq. (25.5) by integration of this differential equation. ***

25.4 Consider Example 25.3 in (2+1) dimensions but assume a gauge group SO(3), an isovector of scalar fields $\Phi = (\Phi_1, \Phi_2, \Phi_3)$, and a potential $U = \frac{1}{2}\lambda(\Phi^2 - a^2)^2$. Show that there is only one homotopy class and thus no topological soliton in this case.

25.5 Show that Eqs. (25.6) and (25.5) lead to the solution (25.7) of Example 25.2. ***

26 Geometry and Gauge Theories

In Ch. 16 we used the idea of a covariant derivative in formulating theories with local gauge invariance. Readers familiar with general relativity will recall that objects also called covariant derivatives play a central role in constructing a description of gravity in curved spacetime. In fact, the use of the same terminology in gauge field theories and in general relativity is not an accident. The covariant derivatives arising in these two different theories have a related geometrical origin: covariant derivatives both in general relativity and in local gauge theories originate in devising a formalism to permit the comparison of vectors located at different points of a spacetime manifold. This issue is central to the calculation of derivatives and therefore of dynamics because, by definition, the derivative of a vector requires taking the difference of vectors located at two different points within the manifold. In this chapter, we explore a geometrical interpretation of local gauge fields that has much in common with the formalism of general relativity for dealing with the calculation of derivatives in curved spacetime.

26.1 Parallel Transport

In a curved manifold the simple view of a vector as a directed line segment is not helpful because an extended straight line has no meaning. This difficulty may be resolved by separating the "directed" and "line segment" portions of the simple vector definition. Specifically, a vector is not defined in the curved n-dimensional manifold itself but rather in an n-dimensional euclidean space $\mathbb{R}^n$ attached to each point of the manifold called the *tangent space,* with basis vectors defined intrinsically by directional derivatives evaluated entirely within the manifold. The construction of these basis vectors is described in Box 26.1 and Fig. 26.1.

26.1.1 Flat and Curved Manifolds

If we wish to compare vectors defined at two different points, one of the vectors must be transported to the position of the other vector to enable the comparison. This procedure will be ill defined unless we ensure that the intrinsic properties of the vector being transported are not changed in this process; transport that fulfills this condition will be termed *parallel transport.* For a flat manifold the tangent space where vectors are defined coincides with the manifold itself and there is no conceptual difficulty in comparing vectors at two

Box 26.1

Holonomic Coordinate Systems

Introductory physics is often formulated in euclidean manifolds using cartesian coordinates. Then it is convenient to define basis vectors that (1) apply globally to the manifold, (2) can be chosen mutually orthogonal, and (3) are normalized. Such *coordinate-independent* or *anholonomic* bases simplify elementary problems but can be a poor choice for curved manifolds and/or non-cartesian coordinates. In that case it may be more useful to employ basis vectors that are *coordinate dependent* or *holonomic*. These basis vectors need not be orthogonal, and because they differ from point to point it is typically not useful to normalize them.

Coordinate Curves and Position-Dependent Basis Vectors

A *coordinate curve* x^μ is a curve in an n-dimensional manifold along which one coordinate x^μ varies while the other $n-1$ coordinates $x^\nu\,(\nu \neq \mu)$ are held constant. As shown in Fig. 26.1, n coordinate curves $x^\mu\,(\mu = 0, 1, \ldots, n)$ pass through any point P. Position-dependent basis vectors e_μ can be defined at each point by

$$e_\mu = \lim_{\delta x^\mu \to 0} \frac{\delta s}{\delta x^\mu},$$

where δs is the infinitesimal distance along the coordinate curve x^μ between the point P with coordinate x^μ and a nearby point Q with coordinate $x^\mu + \delta x^\mu$.

Directional Derivatives and the ∂_μ Basis

For a parameterized curve $x^\mu(\lambda)$ with tangent vector components $t^\mu = dx^\mu/d\lambda$,

$$t = t^\mu e_\mu = \frac{dx^\mu}{d\lambda} e_\mu,$$

the *directional derivative* of an arbitrary scalar function $f(x^\mu)$ is

$$\frac{df}{d\lambda} \equiv \lim_{\epsilon \to 0} \left[\frac{f(x^\mu(\lambda + \epsilon)) - f(x^\mu(\lambda))}{\epsilon} \right] = \frac{dx^\mu}{d\lambda}\frac{\partial f}{\partial x^\mu} = t^\mu \frac{\partial f}{\partial x^\mu},$$

and since $f(x)$ is arbitrary this implies the operator relation

$$\frac{d}{d\lambda} = \frac{dx^\mu}{d\lambda}\frac{\partial}{\partial x^\mu} = t^\mu \frac{\partial}{\partial x^\mu}.$$

Hence each component t^μ is associated with a unique directional derivative and *the derivative operators* $\partial_\mu \equiv \partial/\partial x^\mu$ *may be identified with the basis vectors* e_μ,

$$e_\mu = \frac{\partial}{\partial x^\mu} \equiv \partial_\mu,$$

which permits an arbitrary vector to be expanded as

$$V = V^\mu e_\mu = V^\mu \frac{\partial}{\partial x^\mu} = V^\mu \partial_\mu.$$

Position-dependent basis vectors ∂_μ specified in this way define a *coordinate or holonomic basis.* As shown in Problem 26.2, the metric tensor components are scalar products of the holonomic basis vectors, $g_{\mu\nu} = e_\mu \cdot e_\nu$.

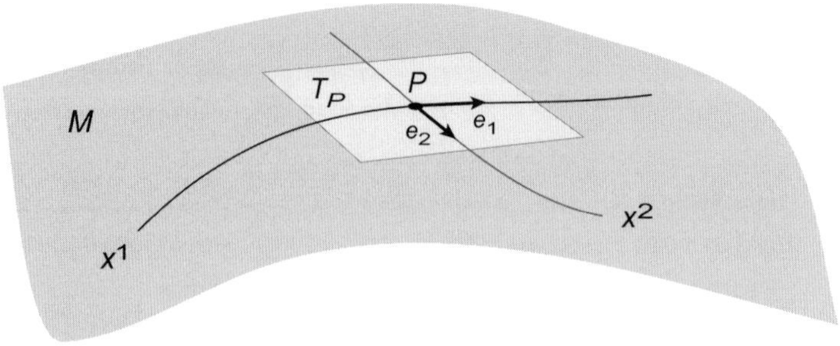

Fig. 26.1 Tangent space T_P at a point P for a curved 2D manifold M [88]. The vectors tangent to the coordinate curves at each point define a coordinate or holonomic basis, as described in Box 26.1. The basis vectors e_1 and e_2 of the tangent space are specified by directional derivatives of the coordinate curves evaluated entirely in M at the point P. Reproduced with permission from Cambridge University Press: *Modern General Relativity – Black Holes, Gravitational Waves, and Cosmology*, M. Guidry (2019).

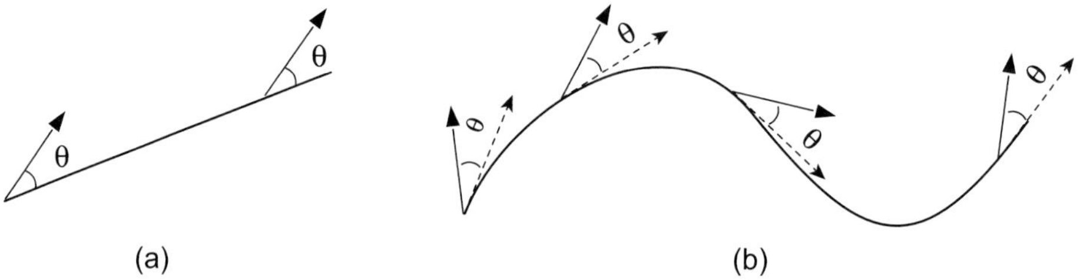

Fig. 26.2 Parallel transport of vectors in (a) flat euclidean space and (b) a curved manifold. The solid vector is being transported; the dashed vector is the local tangent to the curve in (b).

different points. We can just move one vector, keeping its orientation and length fixed with respect to a global set of coordinate axes, to the position of the other vector and compare. Figure 26.2(a) illustrates. However, for a curved manifold this prescription fails because the tangent space in which the vectors are defined also changes between two points and there is no global coordinate system common to all the tangent spaces. For a curved manifold the best that we can do is to move the vector while maintaining a constant angle θ between the transported vector and the local tangent vector to the curve, as illustrated in Fig. 26.2(b). For a flat space this prescription reduces to that in Fig. 26.2(a), but in the general curved case the net change in transporting the vector is a combination of intrinsic change in the vector and change in the (generally position-dependent) coordinate system described in Box 26.1.

It follows that a comparison of vectors at two points must separate the part of the difference that is an intrinsic change in the vectors and the part that is caused by the transformation of the coordinate system in which the vectors are defined. That this is a non-trivial consideration is suggested by Fig. 26.3, which shows that vectors transported on a curved 2D surface by the prescription in Fig. 26.2(b) generally get rotated while they are transported, with the amount of rotation depending on the transport path.

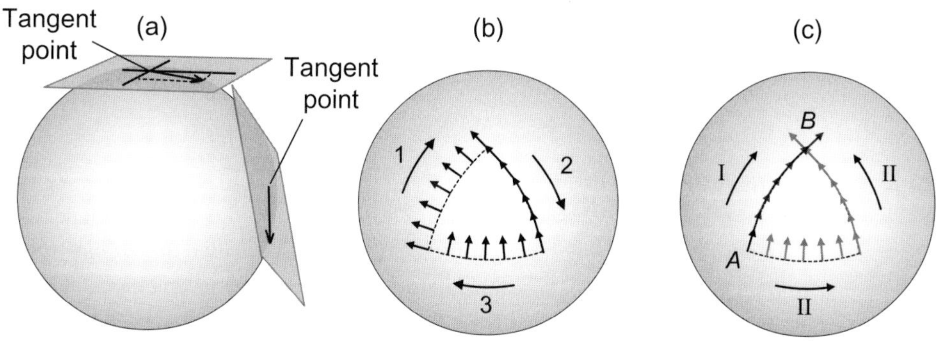

Fig. 26.3 (a) Tangent spaces and vectors for a 2-sphere S^2. The manifold is 2D so the tangent spaces are euclidean planes $\mathbb{R}^2$. (b) Parallel transport of a vector in a closed path by the prescription of Fig. 26.2(b). The vector rotates by 90° on the path $1 \longrightarrow 2 \longrightarrow 3$. (c) Parallel transport from A to B on the direct path labeled I rotates the vector by a different amount than for parallel transport from A to B on the two-segment path labeled II [88]. Reproduced with permission from Cambridge University Press: *Modern General Relativity – Black Holes, Gravitational Waves, and Cosmology*, M. Guidry (2019).

26.1.2 Connections and Covariant Derivatives

Consider a vector V expanded in a coordinate-dependent basis, $V(x) = V^\mu(x)e_\mu(x)$, and let λ parameterize a 1D path in the manifold over which the vector is assumed to be transported. Then by the usual product rule, differentiation of the 4-vector V gives two terms

$$\begin{aligned}\frac{dV}{d\lambda} &= \frac{d}{d\lambda}\left(V^\mu e_\mu\right)\\ &= \frac{dV^\mu}{d\lambda}e_\mu + \frac{de_\mu}{d\lambda}V^\mu\\ &= \frac{\partial V^\mu}{\partial x^\nu}\frac{dx^\nu}{d\lambda}e_\mu + \frac{\partial e_\mu}{\partial x^\nu}\frac{dx^\nu}{d\lambda}V^\mu.\end{aligned} \tag{26.1}$$

Using the chain rule and defining $u^\nu \equiv dx^\nu/d\lambda$ and $\partial_\nu \equiv \partial/\partial x^\nu$, this may be written

$$\frac{dV}{d\lambda} = \partial_\nu V^\mu u^\nu e_\mu + V^\mu u^\nu \partial_\nu e_\mu. \tag{26.2}$$

This equation represents the total change of the vector expressed in a position-dependent basis for an infinitesimal change in position λ, which has two contributions.

1. The first term in Eq. (26.2) represents the change in vector components for a fixed coordinate system defined by the basis vectors e_μ.
2. The second term in Eq. (26.2) represents the change in the coordinate-dependent basis vectors in moving from one point to another along the curve parameterized by λ.

In the second term $\partial_\nu e_\mu$ is itself a vector which may be expanded in the vector basis, $\partial_\nu e_\mu = \Gamma^\alpha_{\mu\nu} e_\alpha$ (where the expansion coefficient $\Gamma^\alpha_{\mu\nu}$ is assumed to be symmetric under exchange of its lower indices) to give

$$\begin{aligned}\frac{dV}{d\lambda} &= \partial_\nu V^\mu u^\nu e_\mu + \Gamma^\alpha_{\mu\nu} V^\mu u^\nu e_\alpha \\ &= \partial_\nu V^\mu u^\nu e_\mu + \Gamma^\mu_{\alpha\nu} V^\alpha u^\nu e_\mu \\ &= \left(\partial_\nu V^\mu + \Gamma^\mu_{\alpha\nu} V^\alpha\right) u^\nu e_\mu,\end{aligned}$$

where in the second step the dummy indices were switched $\mu \leftrightarrow \alpha$ in the implied summation of the second term. The change in the position-dependent basis is now parameterized through the coefficient $\Gamma^\mu_{\alpha\nu}$, which is called the *(affine) connection* in general relativity. It defines a relationship (a "connection") between tangent spaces at two points of the manifold, which provides a prescription to parallel transport a vector (move, preserving lengths and angles locally) and compare it to a vector at another point, which is defined in a different tangent space.[1] In general relativity the affine connection

1. can be constructed from the metric and its derivatives,
2. vanishes in a manifold with constant metric, and
3. can be used to construct the Riemann curvature tensor describing the local intrinsic curvature of spacetime.

The covariant derivative ∇_α for general relativity is then defined in terms of the connection coefficients by

$$\nabla_\alpha V^\mu \equiv \partial_\alpha V^\mu + \Gamma^\mu_{\alpha\nu} V^\nu. \tag{26.3}$$

The similarity with the gauge-covariant derivatives defined in Ch. 16 is then apparent upon comparison with Eqs. (16.36) and (16.48), with the vector potential A_μ in the gauge theory playing the role of the connection coefficient in general relativity. In both cases the covariant derivative is the ordinary partial derivative plus a correction term involving the connection or the vector potential that serves two related purposes. Use of the covariant derivative in place of the partial derivative

- ensures that derivatives of tensors are themselves tensors, and
- implements parallel transport according to a consistent prescription on the manifold.

We shall elaborate further on this relationship in Section 26.3.

26.1.3 Curvature and Parallel Transport

A crucial issue for a manifold is whether it exhibits curvature, and how to quantify that curvature.[2] Gauss showed that for a 2D manifold the intrinsic curvature could be specified

[1] Connection coefficients are not determined by the differential geometry of the manifold. They are an added structure that *defines* parallel transport on the manifold. In general relativity the structure of parallel transport is constrained by the assumption invoked above that the connection coefficients are symmetric in their lower indices (which ensures that vector scalar products are preserved under parallel transport). The validity of this assumption must then be verified after the fact by the agreement of the resulting theory (general relativity) with all experimental and observational tests to date.

[2] A manifold may exhibit two kinds of curvature. (1) *Intrinsic curvature* can be detected by measurements made entirely within the manifold. Gaussian curvature for a 2D manifold is an example. (2) *Extrinsic curvature* results from embedding the manifold in one of higher dimension. A famous example is that a cylinder has zero intrinsic curvature (it can be cut and rolled out into a flat sheet), but it appears curved when viewed embedded in 3D space. Our interest here is in the intrinsic curvature of a manifold.

by the *Gaussian curvature,* which can be determined from distance measurements made entirely from within the manifold. Riemann generalized the idea of Gaussian curvature to higher dimensions and in 4D spacetime the *Riemann curvature tensor* is the standard measure of spacetime curvature. As suggested by comparing Fig. 26.3 to vector transport on a flat manifold, the Gaussian curvature and more generally the Riemann curvature are related to how much a vector is rotated under parallel transport on a closed path in the manifold.

26.2 Absolute Derivatives

Absolute derivatives are similar to covariant derivatives except that covariant derivatives are defined over an entire manifold but absolute derivatives are defined only along a constrained path in the manifold. Using $D/D\lambda$ to denote the absolute derivative evaluated on a curve parameterized by λ, the absolute derivatives for vector fields V^μ and dual vector fields V_μ are given by

$$\frac{DV^\alpha}{D\lambda} = \frac{dV^\alpha}{d\lambda} + \Gamma^\alpha_{\beta\gamma} V^\beta \frac{dx^\gamma}{d\lambda} \qquad \text{(vectors)} \tag{26.4}$$

$$\frac{DV_\alpha}{D\lambda} = \frac{dV_\alpha}{d\lambda} - \Gamma^\beta_{\alpha\gamma} V_\beta \frac{dx^\gamma}{d\lambda} \qquad \text{(dual vectors)} \tag{26.5}$$

(see Problem 26.7), with similar formulas for tensors of higher rank.

The absolute derivative is of central importance for parallel transport because the condition for parallel transport along a path parameterized by λ in a curved manifold is that the absolute derivative vanish.

For example, the condition for parallel transport of a vector field V is that $DV^\mu/D\lambda = 0$ at each point of the path, which reduces to $dV^\mu/d\lambda = 0$ only if the manifold is flat.

26.3 Parallel Transport in Charge Space

The relevance of the preceding discussion to local gauge invariance is that in the local gauge theories introduced in Ch. 16, the freedom to define different phases at different spacetime points also implies that the comparison of vectors is poorly defined unless we implement consistently a notion of parallel transport in the (gauge) charge space analogous to the parallel transport discussed above. There is a direct analogy between the group of all coordinate transformations in general relativity and the group of all gauge transformations in a local gauge theory. The gauge fields define parallel transport in charge space with

- A^μ acting as the connection coefficient, and
- $F^{\mu\nu}$ acting as the Riemann curvature tensor in the general relativistic analog.

For this reason, the vector potentials A_a^μ in gauge theories are sometimes also called *connections*. The gauge connection coefficients "connect" the components of a vector at one point with the components of a vector at another point by specifying a prescription for parallel transport of one vector to the location of the other one.

A partial derivative $\partial\psi/\partial x^\mu$ is not covariant with respect to local gauge symmetry because $\psi(x)$ and $\psi(x + dx) \equiv \psi(x) + d\psi$ are *measured in different coordinate systems*: freedom to adopt different phases at different spacetime points is equivalent to having different orientations of the coordinate axes at those points.

To form a properly covariant derivative, $\psi(x+dx)$ should be compared with the value $\psi(x)$ would have if carried by parallel transport from x to $x + dx$. Therefore, we may interpret the action of the connection coefficients in general relativity, and the gauge fields in the internal charge space, as compensating for the variation of local frames at each spacetime point when attempting to compare tensor quantities defined at different points. This action is implemented automatically in each case through the systematic replacement of ordinary partial derivatives with appropriately defined covariant derivatives.

26.4 Fiber Bundles and Gauge Manifolds

An elegant geometrical description of local gauge invariance may be given in terms of *fiber bundles*. We give here a qualitative introduction [88], since the language of fiber bundles is common in more mathematical discussions of gauge field theories.

26.4.1 Tangent Spaces and Tangent Bundles

The prototype fiber bundles are the *tangent bundles* defined for spacetime manifolds. As illustrated in Fig. 26.3(a), in a curved manifold vectors at a point do not lie in the manifold but rather in a (euclidean) *tangent space* attached to the manifold. The tangent space is *defined intrinsically* because its basis vectors can be constructed from local directional derivatives evaluated entirely within the manifold (see Box 26.1). For an n-dimensional manifold the tangent spaces are n-dimensional euclidean manifolds defined at each point.[3]

Thus an n-dimensional manifold M may associate with each point P an n-dimensional euclidean vector space T_P, with a basis defined by n directional derivatives evaluated in the manifold at P.

[3] Dual vectors (see Section 13.1.1) may be defined in an analogous way in a *cotangent space* T_P^* at P. We introduce the tangent space only for illustration here, so we will not bother defining the cotangent space T_P^*.

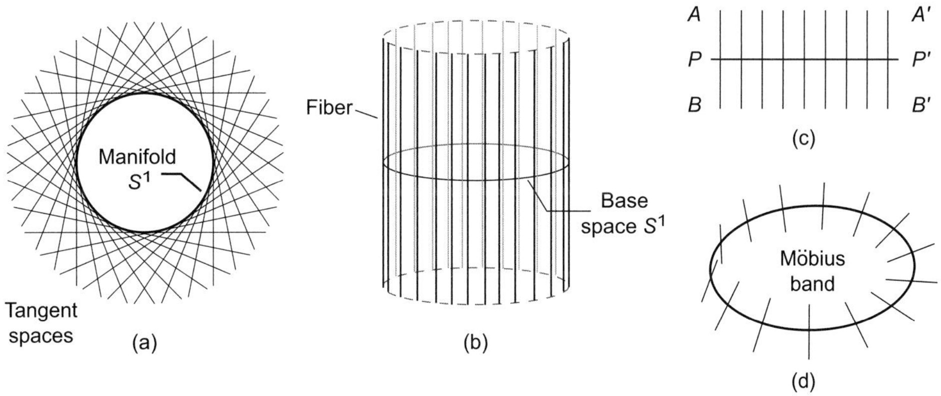

Fig. 26.4 Tangent bundle TS^1 for S^1 [88]. (a) The manifold S^1 and some of its tangent spaces (lines tangent to the circle). (b) The corresponding tangent bundle (locally and globally $R^1 \times S^1$). (c) Figure (b) cut vertically and rolled out flat. Figure (b) corresponds to identifying $A \leftrightarrow A'$ and $B \leftrightarrow B'$. (d) Non-trivial Möbius-band topology generated by identifying $A \leftrightarrow B'$ and $B \leftrightarrow A'$ in (c): $R^1 \times S^1$ locally, but not globally. Reproduced with permission from Cambridge University Press: *Modern General Relativity – Black Holes, Gravitational Waves, and Cosmology,* M. Guidry (2019).

The *tangent bundle* TM of a manifold M is itself a manifold that is the disjoint union of all the tangent spaces T_P, one at each point P of the manifold. A tangent bundle is illustrated in Fig. 26.4(a–b)for the manifold S^1. This manifold is 1D so the tangent spaces are also 1D and may be represented by straight lines drawn tangent to the circle at each point. Each tangent space is independent and the apparent overlaps of these lines in Fig. 26.4(a) are meaningless. Therefore, in Fig. 26.4(b) the lines (*fibers*) have been arrayed perpendicular to the base manifold to avoid the illusion of crossing. A location y on a fiber exists at the point $P = x$ where the fiber intersects the base space, with y (vector length) given by the distance to the intersection of the fiber with the base space.

26.4.2 Fiber Bundles

Fiber bundles are abstractions of the basic idea represented by tangent bundles for spacetime manifolds that generalize the product of two manifolds.

> A fiber bundle is a manifold E that is *locally* the cartesian product (Box 2.1) of two spaces, $E = F \times B$, where B is called the *base space* and F is called the *fiber space,* but that *globally* may have a different topological structure.

In words, a fiber bundle is a topological space that looks locally like the direct product of two topological spaces, which may be conceptualized by attaching a copy of one space to each point of the other space. If a fiber bundle is both globally and locally $F \times B$, the bundle is said to be *trivial*. The following example illustrates trivial and non-trivial bundles.

Example 26.1 The tangent bundle on S^1 that is illustrated in Fig. 26.4(a–b) is both globally and locally $\mathbb{R}^1 \times S^1$, so it is trivial. Conversely, the circular strip with a twist (Möbius band) illustrated in Fig. 26.4(d) is locally $\mathbb{R}^1 \times S^1$ but its global topology is non-trivial: for case (d) the orientation of the fiber winds through an angle π for once around the base space, so case (d) cannot be deformed continuously into case (b) and they are distinct topologically (see Section 27.4).

More advanced discussions of local gauge theories often employ fiber bundles in which the base space is Minkowski spacetime and the fibers represent the internal (gauge) degrees of freedom at each spacetime point. To go further would exceed our scope but a concise overview is given by Pires [164], and a more systematic discussion may be found in Nakahara [152].

26.5 Gauge Symmetry on a Spacetime Lattice

It is sometimes useful to consider continuous gauge fields approximated on a discrete spacetime lattice. For example, computer simulations of gauge theories are implemented on such lattices. *Path-dependent representations* of a gauge group are important within this context. We shall first examine path-dependent representations within a full continuum theory, and then consider the effect of approximating the continuum theory by a discrete spacetime lattice for quantum chromodynamics (QCD) [85].

26.5.1 Path-Dependent Gauge Representations

Suppose a complex field $\psi(x)$ coupled to a U(1) gauge field. Invariance under the displacement $x \longrightarrow x + dx$ requires that

$$dx_\mu D^\mu \psi(x) = dx_\mu (\partial^\mu + igA^\mu(x))\psi(x) = 0, \tag{26.6}$$

where from Eq. (16.36) the covariant derivative is given by $D^\mu = \partial^\mu + igA^\mu(x)$. This may be extended to non-abelian gauge fields by introducing the matrix potential of Eq. (16.49),

$$A^\mu(x) = A_i^\mu(x)\tau^i \qquad A_i^\mu(x) \equiv 2\mathrm{Tr}\,(\tau_i A^\mu), \tag{26.7}$$

where the $A_i^\mu(x)$ are vector potentials with a Lorentz index μ and internal index i, and the τ^i are matrix generators for the gauge group. For a Yang–Mills field the solution of the differential equation (26.6) over a path between x_0 and x_1 labeled by a parameter γ is

$$\psi(x_1) = P\exp\left(ig\int_\gamma A_\mu dx^\mu\right)\psi(x_0), \tag{26.8}$$

where P is a *path-ordering operator* that orders the matrices $A_\mu(x)$ in the sequence encountered on the path for each term in a power series expansion of the exponential. [For non-abelian groups, $A_\mu(x)$ and $A_\mu(x')$ do not generally commute.] Thus

$$U_\gamma(x_0, x_1) = P \exp\left(ig \int_\gamma A^i_\mu \tau_i dx^\mu\right) \tag{26.9}$$

defines a matrix associated with each path γ between spacetime points x_0 and x_1. The matrix $U_\gamma(x_0, x_1)$ is a *path-dependent representation of a gauge group element* because it is a product of group elements associated with infinitesimal segments of the path and a product of group elements necessarily is itself a group element. Physically, $U_\gamma(x_0, x_1)$ implements a transformation $\psi(x_1) = U_\gamma(x_0, x_1)\psi(x_0)$ that transports the system from spacetime points x_0 to x_1 by parallel displacement along the path labeled by γ. Under a local gauge transformation by $G(x)$, U_γ is sensitive to $G(x)$ *only at the endpoints of the path,*

$$U_\gamma(x_0, x_1) \rightarrow U'_\gamma(x_0, x_1) = G(x_1)U_\gamma(x_0, x_1)G^{-1}(x_0), \tag{26.10}$$

as proved in Problem 26.4. Example 26.2 illustrates for an abelian gauge field.

Example 26.2 For an abelian gauge group the path-ordering operator P is not needed since everything commutes and there is only one gauge field, so Eq. (26.9) reduces to

$$U(A, B) = \exp\left(ig \int_A^B A_\mu(x)\, dx^\mu\right).$$

Performing a gauge transformation $A_\mu(x) \rightarrow A_\mu(x) - \partial_\mu\phi(x)$ on $U(A, B)$ gives

$$\begin{aligned} U'(A, B) &= \exp\left(ig \int_A^B A_\mu(x)\, dx^\mu - ig \int_A^B \partial_\mu\phi(x)\, dx^\mu\right) \\ &= U(A, B) \exp\left(-ig \int_A^B d\phi\right) \\ &= e^{ig\phi(A)} U(A, B) e^{-ig\phi(B)}, \end{aligned}$$

where $\partial_\mu\phi(x)dx^\mu = (\partial\phi/\partial x^\mu)dx^\mu = d\phi$ was used. This is the transformation law for a field with opposite abelian gauge charges at endpoints A and B [compare Eq. (16.39)].

From Eq. (26.10) it may be demonstrated that for a closed path ($x_1 = x_0 \equiv x$) the trace $\mathrm{Tr}\, U_\gamma(x, x)$ is *gauge invariant* (see Problem 26.5).

26.5.2 Lattice Gauge Symmetries

Now let us consider discretizing spacetime to define a spacetime lattice. We assume cubic lattices for simplicity and illustrate with a lattice formulation of quantum chromodynamics (QCD, described in Section 19.2) called *lattice gauge theory.* A 2D set of lattice points is illustrated in Fig. 26.5(a). A gauge theory can be constructed on a lattice by the following prescription [213, 214].

Matter Fields: Quarks and antiquarks are restricted to lattice sites. Each is described by a wavefunction $|n\alpha i\sigma\rangle$, where n is a lattice vector giving the site location, α is a flavor

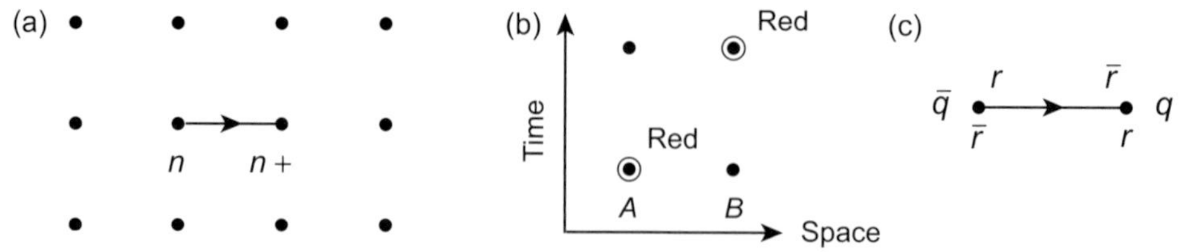

Fig. 26.5 (a) Plane section of a cubic lattice with a directed string bit from site n to site $n + \mu$, where μ is a unit lattice vector. (b) Violation of local color conservation on the lattice. A circle around a point indicates occupation by a red quark. (c) A color singlet $q\bar{q}$ lattice state.

index, i is a QCD color index, and σ is a Dirac spinor index. Multi-quark states can be formed by putting more than one quark or antiquark on the lattice. The number of quarks or antiquarks occupying a single lattice site is limited only by the Pauli principle.

Example 26.3 For three flavors u, d, and s, each with three color states and two helicity states, $3 \times 3 \times 2 = 18$ quarks could occupy a site without violating the Pauli principle.

The lattice Dirac operators that create and annihilate quarks and antiquarks are like those of the continuum theory, except that they act *only at the discrete sites of the lattice.*

Gauge Fields: Lattice sites are connected by segments called *(directed) string bits.* An example is shown in Fig. 26.5(a). The gauge fields are assumed to reside on these links, which transport color information between lattice sites; thus they are analogs of the gluon operators in a continuum theory. The string bits (*links*) are lattice versions of the path-dependent gauge group representations discussed for the continuum theory in Section 26.5.1. As suggested by Problem 26.4 and Example 26.2, each string bit has a color at one end and an anticolor at the other. A string bit is described by a state $|n\mu kld\rangle$, where n and $n + \mu$ denote the lattice sites the bit connects, k is the color index, l is the anticolor index, and d indicates the direction of the segment. A string bit is assumed to have a mass (since the gauge field carries energy) but no internal degrees of freedom (they cannot vibrate); only the colored ends of the strings attached to the lattice sites are relevant for the quantum theory.

In the continuum limit of zero lattice spacing the lattice must be invariant under color SU(3). This property can be built directly into the lattice, even for finite spacing, by forbidding a quark on the lattice with no attached gauge fields to move between sites. This is illustrated in Fig. 26.5(b), where a red quark at site A moves over time to site B with no other change. This is consistent with global color invariance but it violates *local color invariance* because site A changes from red to colorless while site B changes from colorless to red. However, a string representing a color gauge field with color on one end and anticolor on the other can combine with colored quarks and antiquarks to form a *colorless object* that transforms as a color SU(3) scalar; Fig. 26.5(c) illustrates for a $q\bar{q}$ pair.

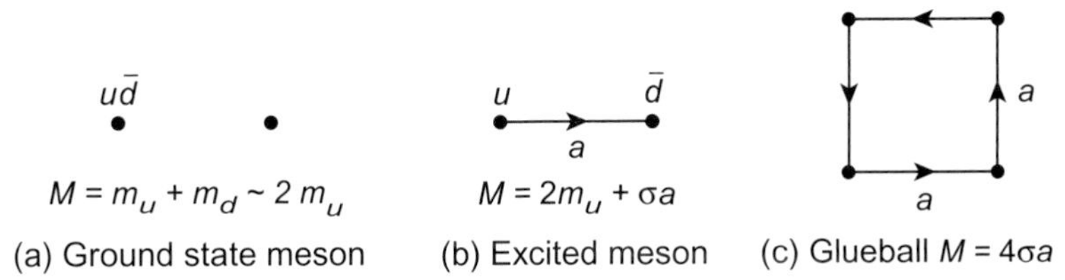

Fig. 26.6 Some static, gauge-invariant states in lattice QCD. Mass of the up quark is m_u, mass of the down quark is m_d, the string tension is σ, the lattice spacing is a, and the total mass is M.

Example 26.4 Figure 26.5(c) represents a lattice model of a meson. The colors of the gauge string ends and of the quarks and antiquarks on the two sites cancel, leaving a color singlet state that can propagate without violating local color SU(3) symmetry.

One can adapt the continuum operators defined in Section 26.5.1 to corresponding lattice operators and construct a complete lattice gauge formulation of QCD [45, 133, 213, 214]. We conclude our introduction with a brief summary of the lattice picture for color SU(3) scalars that could be physical QCD states in the continuum limit.

Static Gauge-Invariant States: In a static picture (no kinetic energy terms), the simplest gauge-invariant states on the lattice are the following.

1. *Ground state mesons,* corresponding to a quark and antiquark on the same lattice site in a color singlet state with no string bits.
2. *Ground state baryons,* corresponding to three quarks on the same lattice site in a color singlet state with no string bits.
3. *Excited meson and baryon states,* corresponding to quarks, antiquarks, and string bits in color singlet configurations.
4. *Glueballs,* constructed from pure string bits (no quarks) arranged in a closed loop.
5. *Exotic states,* constructed from four or more quarks and antiquarks in a color singlet configuration, with no string bits in the ground state.

Some of these lattice states are illustrated in Fig. 26.6. The states are locally gauge invariant but not Poincaré invariant, because the lattice violates translational and rotational symmetry. Only in the continuum limit does the lattice approach full Poincaré invariance.

Background and Further Reading

For a simple overview of many ideas in this chapter see Ryder [174], or Cheng and Li [41]. For a more mathematical overview, see Nakahara [152]. For physics applications of fiber bundles Singer [182] gives an overview, Pires [164] provides a concise outline, and Nash and Sen [156] or Nakahara [152] may be consulted for more mathematical treatments.

Problems

26.1 For coordinates (x^1, x^2) and metric $g = \text{diag}(g_{11}, g_{22})$, the Gaussian curvature is

$$K = \frac{1}{2g_{11}g_{22}}\left\{-\frac{\partial^2 g_{22}}{(\partial x^1)^2} - \frac{\partial^2 g_{11}}{(\partial x^2)^2} + \frac{1}{2g_{11}}\left[\frac{\partial g_{11}}{\partial x^1}\frac{\partial g_{22}}{\partial x^1} + \left(\frac{\partial g_{11}}{\partial x^2}\right)^2\right]\right.$$
$$\left. + \frac{1}{2g_{22}}\left[\frac{\partial g_{11}}{\partial x^2}\frac{\partial g_{22}}{\partial x^2} + \left(\frac{\partial g_{22}}{\partial x^1}\right)^2\right]\right\}.$$

For a sphere with coordinates defined in the following figure,

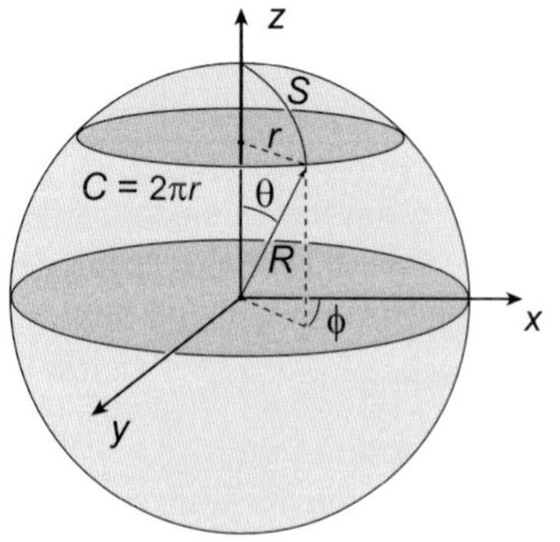

show that a Gaussian curvature $K = R^{-2}$ is obtained using spherical coordinates $(x^1, x^2) = (S, \phi)$ or cylindrical coordinates $(x^1, x^2) = (r, \phi)$.***

26.2 Consider the holonomic basis defined in Box 26.1. Using that the tangent vector for a curve can be written $t = t^\mu e_\mu = (dx^\mu/d\lambda)\, e_\mu$, show that

$$ds^2 = ds \cdot ds \equiv g_{\mu\nu}dx^\mu dx^\nu = (e_\mu \cdot e_\nu)dx^\mu dx^\nu.$$

Thus, $g_{\mu\nu} = e_\mu(x) \cdot e_\nu(x)$ and the components of the metric tensor are defined by the scalar products of the coordinate-dependent basis vectors. ***

26.3 The *Lie bracket* of vector fields A and B is defined as their commutator, $[A, B] = AB - BA$. The Lie bracket of two basis vectors vanishes for a coordinate basis but not for a non-coordinate basis. For the 2D plane parameterized with cartesian (x, y) or polar (r, θ) coordinates and the bases

$$(e_1, e_2)_1 = \left(\frac{\partial}{\partial x}, \frac{\partial}{\partial y}\right) \qquad (e_1, e_2)_2 = \left(\frac{\partial}{\partial r}, \frac{\partial}{\partial \theta}\right) \qquad (e_1, e_2)_3 = \left(\frac{\partial}{\partial r}, \frac{1}{r}\frac{\partial}{\partial \theta}\right),$$

show that $(e_1, e_2)_1$ and $(e_1, e_2)_2$ are coordinate bases, but that $(e_1, e_2)_3$ is not.

26.4 Prove the result of Eq. (26.10) that a path-dependent representation of a gauge group is sensitive to a gauge transformation only at the endpoints of the path. ***

26.5 Demonstrate that for a closed path $\text{Tr}\, U_\gamma(x, x)$ is gauge invariant, where $U_\gamma(x_0, x_1)$ is defined by Eq. (26.9).

26.6 In Eq. (26.7), invert $A^\mu(x) = A^\mu_i(x)\tau^i$ to obtain $A^\mu_i(x) \equiv 2\mathrm{Tr}\,(\tau_i A^\mu)$.

26.7 Prove that $dV/d\lambda$ for a vector field $V(\lambda)$ evaluated along a curve $x^\mu(\lambda)$ parameterized by λ leads to Eq. (26.4) for the absolute derivative of a vector field. *Hint*: Begin with Eq. (26.1), and assume that a position-dependent basis $e_\mu(\lambda)$ is defined along the curve and that its derivative can be expanded as $\partial_\nu e_\mu = \Gamma^\alpha_{\mu\nu} e_\alpha$. ***

27 Geometrical Phases

This chapter addresses phases that are geometrical in origin and that may have quite surprising consequences. We shall illustrate first with the Aharonov–Bohm effect and then with the Berry phase, which is in some sense a generalization of the Aharonov–Bohm effect in real space to the configuration space of a dynamical system. In later chapters we will see that the geometrical Berry phase can also have important topological implications when evaluated over a closed path in the Hilbert space of a quantum problem, because of the *Chern theorem* discussed in Section 28.4.2, and related concepts.

27.1 The Aharonov–Bohm Effect

The Aharonov–Bohm effect [4] is important in quantum mechanics for at least two reasons.

1. It is one of the simplest and most elegant manifestations of geometrical effects in a quantum-mechanical system.
2. It demonstrates that the electromagnetic 4-vector potential can have measurable consequences in a quantum theory, even for regions having no electric or magnetic fields.

The second point contrasts with classical electromagnetism, where electric and magnetic fields are the essential objects and vector potentials are only a mathematical convenience.

27.1.1 Experimental Setup

The experimental setup for the Aharonov–Bohm effect is illustrated in Fig. 27.1(a). It is an electron scattering, two-slit interference experiment, with the added twist that a thin solenoid is placed in the region behind the screen and between the two classical paths that electrons passing through the slits would follow to reach a point on the screen. The solenoid is long and thin, and confines the magnetic field to regions that the electrons should not pass through. Nevertheless, experiments show that the solenoid alters the interference pattern for the electrons scattering through the two slits [37]. We now examine this effect and show that it is a consequence of the gauge symmetry of the electromagnetic field and the non-trivial topology of the scattering plane caused by the presence of the solenoid [174].

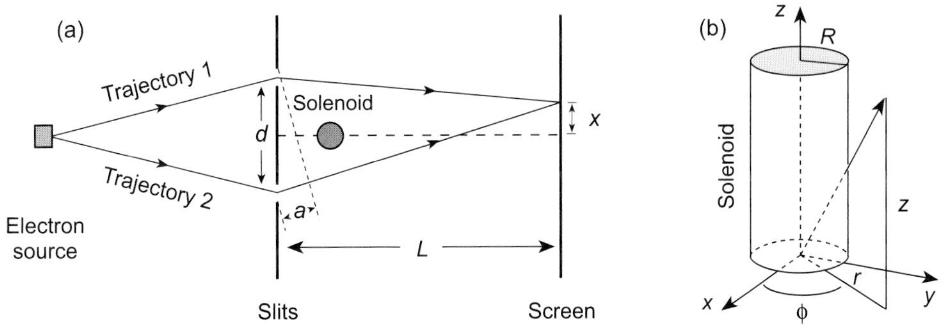

Fig. 27.1 (a) Aharonov–Bohm experiment. (b) Cylindrical coordinates for the Aharonov–Bohm effect.

27.1.2 Analysis of Magnetic Fields

From the geometry illustrated in Fig. 27.1(a), the phase difference between waves of de Broglie wavelength λ associated with paths 1 and 2 is $\delta = 2\pi a/\lambda$, implying that $x \simeq L\lambda\delta/2\pi d$, where we have approximated $a \simeq (x/L)d$ for $x \ll L$. Introduce the cylindrical coordinate system illustrated in Fig. 27.1(b) and a vector potential of the form

$$\begin{aligned} &\text{Inside solenoid:} \quad A_r = A_z = 0 \qquad A_\phi = \frac{Br}{2}, \\ &\text{Outside solenoid:} \quad A_r = A_z = 0 \qquad A_\phi = \frac{BR^2}{2r}. \end{aligned} \tag{27.1}$$

As shown in Problem 27.1, the components of the magnetic field $\boldsymbol{B} = \boldsymbol{\nabla} \times \boldsymbol{A}$ are then

$$\begin{aligned} &\text{Inside solenoid:} \quad B_r = 0 \qquad B_\phi = 0 \qquad B_z = B, \\ &\text{Outside solenoid:} \quad B_r = 0 \qquad B_\phi = 0 \qquad B_z = 0. \end{aligned} \tag{27.2}$$

Thus, the magnetic field $\boldsymbol{B}$ vanishes outside the solenoid but the vector potential A_ϕ does not, falling off as $1/r$ for $r > R$ (Fig. 27.2). Experimentally, the solenoid can be made of a very thin metallic whisker and shielding can ensure that the electrons never enter the solenoid and therefore never pass through regions where the magnetic field is non-zero. Classically, a particle of charge q in an electromagnetic field experiences a Lorentz force, $\boldsymbol{F} = q(\boldsymbol{E} + \boldsymbol{v} \times \boldsymbol{B})$, where $\boldsymbol{v}$ is the velocity, $\boldsymbol{E}$ is the electric field, and $\boldsymbol{B}$ is the magnetic field. Thus classically electrons cannot be influenced since they experience no Lorentz force. However, the electrons pass through regions in which *the magnetic field is zero but the vector potential is finite.* This has non-trivial implications for quantum theory.

27.1.3 Phase of the Electron Wavefunction

The free-electron wavefunction is given by $\psi = |\psi| e^{i\alpha}$, where $\alpha \equiv \boldsymbol{p} \cdot \boldsymbol{r}/\hbar$. Gauge invariance of the interaction between the electron and the electromagnetic field requires the minimal substitution $\boldsymbol{p} \to \boldsymbol{p} - e\boldsymbol{A}$, where $\boldsymbol{A}$ is the 3-vector potential (see Section 16.3), which alters the electron phase locally to,

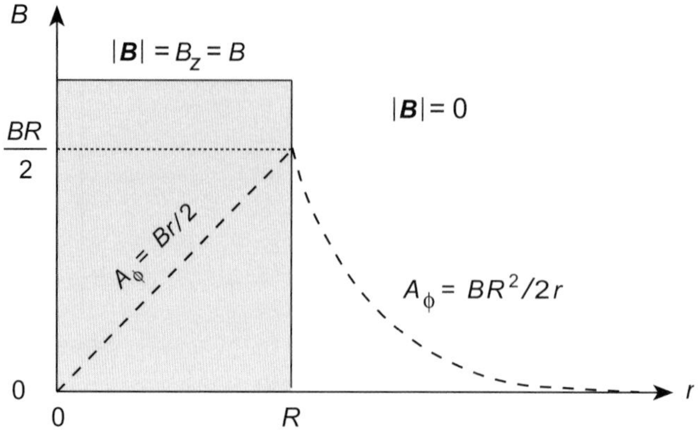

Fig. 27.2 Magnetic field B and vector potential A_ϕ for a solenoid of radius R in Fig. 27.1(a).

$$\alpha \rightarrow \frac{1}{\hbar}(\boldsymbol{p}\cdot\boldsymbol{r} - e\boldsymbol{A}\cdot\boldsymbol{r}) = \alpha - \frac{e}{\hbar}\boldsymbol{A}\cdot\boldsymbol{r}.$$

Then the change in phase over a trajectory τ is given by the integral

$$\Delta\alpha = -\frac{e}{\hbar}\int_\tau \boldsymbol{A}\cdot d\boldsymbol{r}, \tag{27.3}$$

and the shift in phase between trajectories 1 and 2 produced by the vector potential is

$$\Delta\delta = \frac{e}{\hbar}\left(\int_1 \boldsymbol{A}\cdot d\boldsymbol{r} - \int_2 \boldsymbol{A}\cdot d\boldsymbol{r}\right) = \frac{e}{\hbar}\oint \boldsymbol{A}\cdot d\boldsymbol{r}. \tag{27.4}$$

Now apply *Stokes' theorem,* which is summarized in Box 27.1:

$$\oint_C \boldsymbol{A}\cdot d\boldsymbol{r} = \int_S (\boldsymbol{\nabla}\times\boldsymbol{A})\cdot\boldsymbol{n}\, dS \equiv \int_S (\boldsymbol{\nabla}\times\boldsymbol{A})\cdot d\boldsymbol{S}, \tag{27.5}$$

where $\boldsymbol{S}$ is the surface enclosed by the contour C and $\boldsymbol{n}$ is an outward normal to the surface, to obtain (Problem 27.2)

$$\Delta\delta = \frac{e}{\hbar}\Phi \qquad \Phi \equiv \pi R^2 B, \tag{27.6}$$

where Φ is the magnetic flux through the solenoid. Thus, the presence of the solenoid shifts the interference pattern of Fig. 27.1(a) by

$$\Delta x = \frac{L\lambda}{2\pi d}\Delta\delta = \frac{L\lambda}{2\pi d}\frac{e}{\hbar}\Phi, \tag{27.7}$$

even though the electrons are never permitted to enter regions of finite magnetic field $\boldsymbol{B}$.

27.1.4 Topological Origin of the Aharonov–Bohm Effect

The magnetic field is given by $\boldsymbol{B} = \boldsymbol{\nabla}\times\boldsymbol{A}$. Outside the solenoid we must have an electromagnetic vacuum with $\boldsymbol{B} = 0$. This is satisfied by the choice $\boldsymbol{A} = 0$, but because of

Box 27.1 Stokes' Theorem

For a 3D vector field $\boldsymbol{A}$, *Stokes' theorem* relates a surface integral of the curl vector field $\boldsymbol{\nabla} \times \boldsymbol{A}$ to a line integral around the (smooth) boundary of that surface:

$$\oint_C \boldsymbol{A} \cdot d\boldsymbol{r} = \int_S (\boldsymbol{\nabla} \times \boldsymbol{A}) \cdot \boldsymbol{n}\, dS \equiv \int_S (\boldsymbol{\nabla} \times \boldsymbol{A}) \cdot d\boldsymbol{S},$$

where S is the 2D surface enclosed by the 1D boundary C, the outward normal to the surface is $\boldsymbol{n}$, the curl $\boldsymbol{\nabla} \times \boldsymbol{A}$ is given in cartesian coordinates by

$$\boldsymbol{\nabla} \times \boldsymbol{A} = \begin{vmatrix} \boldsymbol{i} & \boldsymbol{j} & \boldsymbol{k} \\ \dfrac{\partial}{\partial x} & \dfrac{\partial}{\partial y} & \dfrac{\partial}{\partial z} \\ A_x & A_y & A_z \end{vmatrix} = \left(\frac{\partial A_z}{\partial y} - \frac{\partial A_y}{\partial z} \right) \boldsymbol{i} + \left(\frac{\partial A_x}{\partial z} - \frac{\partial A_z}{\partial x} \right) \boldsymbol{j} + \left(\frac{\partial A_y}{\partial x} - \frac{\partial A_x}{\partial y} \right) \boldsymbol{k}.$$

Here the orientation of the surface and the direction of the integration path around the boundary are related by a *right-hand rule*: if the thumb of your right hand is pointed in the direction of a unit normal vector near the edge of the surface and you curl your fingers, the direction that your fingers point indicates the direction of integration around the boundary curve. The physical content of Stokes' theorem is easy to visualize, as illustrated in the following figure.

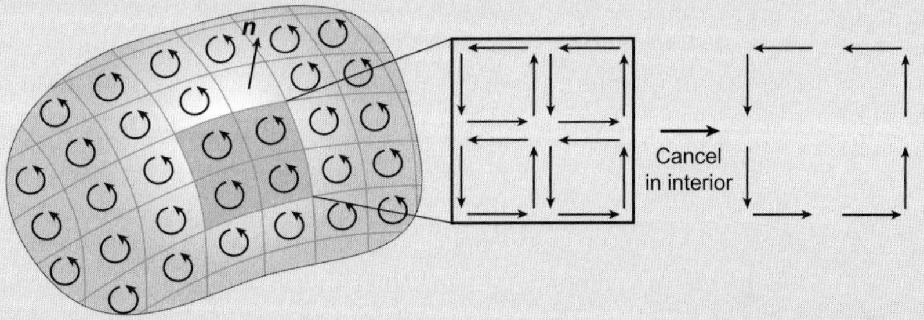

For the representative plaquettes highlighted in darker gray the circulating currents cancel in the interior, leaving only a net contribution from the boundary. Generalizing to the whole surface in the limit that the plaquette size tends to zero, we expect all interior contributions to the surface integral to cancel, leaving only the contributions from the boundary. That is the essence of Stokes' theorem.

gauge invariance the magnetic field is unchanged by the transformation $A_\mu \to A_\mu + \partial_\mu \chi \equiv A'_\mu$, where χ is an arbitrary scalar function (see Section 14.1.2). Thus, the 4-vector potential $A_\mu = \partial_\mu \chi$, which is a gauge transform of the trivial vacuum $A_\mu = 0$, also satisfies the required boundary condition. It follows that in the cylindrical coordinates of Fig. 27.1(b),

$$\boldsymbol{A} = (A_r, A_\phi, A_z) = \boldsymbol{\nabla}\chi \qquad (27.8)$$

is the most general vector potential giving a vanishing electromagnetic field. As shown in Problem 27.3, the scalar function χ is then given by

$$\chi = \frac{1}{2} B R^2 \phi. \qquad (27.9)$$

Therefore, χ increases by $\pi B R^2$ as ϕ increases by 2π and *is not single valued.* This is possible because *the vacuum configuration of the Aharonov–Bohm experiment is not simply connected.*

From Fig. 27.1, the space for the Aharonov–Bohm experiment is a plane pierced by a hole representing the solenoid, which is $\mathbb{R} \times S^1$ topologically (see Example 24.9). From Section 16.5, electromagnetism is invariant under local phase rotations $e^{-ie\chi}$. Thus, the gauge group is U(1), which is S^1 topologically, and we must consider mappings between $\mathbb{R} \times S^1$ and S^1. But $\mathbb{R}$ is simply connected so the only part of this mapping that matters is $S^1 \to S^1$, which is characterized by an infinity of integer winding numbers. (This reasoning is a bit slapdash; Problem 27.4 requests a more formal proof.) Hence, *the gauge function* χ *is multivalued* because the mappings cannot all be deformed to the trivial mapping χ = constant, which would give [by Eq. (27.8)] $A_\mu = 0$ and no Aharonov–Bohm effect.

27.2 The Berry Phase

The Aharonov–Bohm effect leads to rather unexpected results because of a topologically non-trivial map between the manifold of the electromagnetic gauge group and the manifold corresponding to the real-space experimental setup. As we shall now demonstrate, similar more abstract but no less observable effects may occur in the configuration space of a classical or quantum-mechanical system. The signatures of these effects are often termed *geometrical phases,* and have their origin in the failure of a physical system to return completely to its original state after traversing a closed path in its parameter space. This is called *holonomy,* as discussed further in Box 27.2. One way to see the potential effect of quantum holonomies is to examine critically a standard assumption in many treatments of quantum systems: that dynamical degrees of freedom operating on very different timescales may be separated approximately by solving for fast degrees of freedom at fixed values of parameters describing slow degrees of freedom.

27.2.1 Fast and Slow Degrees of Freedom

Assume that a Hamiltonian describes a system in terms of some "slow" degrees of freedom $(\boldsymbol{P}, \boldsymbol{R})$ and some "fast" degrees of freedom $(\boldsymbol{p}, \boldsymbol{r})$, and that the full Hamiltonian is

$$H(\boldsymbol{P}, \boldsymbol{R}, \boldsymbol{p}, \boldsymbol{r}) = H_0(\boldsymbol{P}, \boldsymbol{R}) + h(\boldsymbol{p}, \boldsymbol{r}, \boldsymbol{R}) = -\frac{\hbar^2}{2M}\frac{\partial^2}{\partial \boldsymbol{R}^2} + h(\boldsymbol{p}, \boldsymbol{r}, \boldsymbol{R}), \tag{27.10}$$

where M is the mass and where the first term on the right depends only on $\boldsymbol{P}$ and $\boldsymbol{R}$. A standard solution of such a problem utilizes an approximate separation between fast and slow degrees of freedom (the *adiabatic* or *Born–Oppenheimer approximation*).

1. Fix the value of the slow coordinate $\boldsymbol{R}$.
2. Diagonalize the Schrödinger equation in the fast coordinates for fixed $\boldsymbol{R}$,

$$h(\boldsymbol{R}) \left| n, \boldsymbol{R} \right\rangle = \epsilon_n(\boldsymbol{R}) \left| n, \boldsymbol{R} \right\rangle, \tag{27.11}$$

which yields eigenfunctions and eigenvalues that depend on the parameter $\boldsymbol{R}$.

Box 27.2 **Holonomy**

Holonomy is the failure of a parameterized system to return to its initial physical state after traversing a closed loop in parameter space.[a] Specifically, a holonomy occurs when a physical system parameterized by a set of variables x_i is taken around a closed loop in the parameter space $\boldsymbol{x}$ and one or more measurable physical quantities q characterizing the system fails to return to its original value.

Parallel Transport on Curved Surfaces

A standard example of holonomy is the failure of a vector to return to its original orientation following parallel transport in a closed path on a 2D sphere, as in Fig. 26.3(b). In this case the parameters are the two angles specifying location on the sphere and the physical quantity that fails to return to its original value is the direction the vector points. A quantitative measure of this holonomy is the difference in initial and final orientation angles, which is proportional to the solid angle subtended by the closed path on the sphere, as seen from the center of the sphere. For the closed path in Fig. 26.3(b) the holonomy is $\frac{\pi}{2}$. Holonomies are independent of specific parameterizations of the manifold, depending only on its intrinsic geometry. Such geometric holonomies can exhibit *global change without local change,* whereby observables fail to return to their original values after a closed loop in parameter space, even though there is *no local intrinsic rate of change* as parameters are varied.

Quantum Holonomies

We shall be interested particularly in holonomies associated with variation of parameters for a quantum system, which typically alters phases of wavefunctions. It was long thought that these phase changes would be unobservable but Berry showed that they can have measurable consequences [22]. This is the celebrated *Berry phase,* which is discussed in Section 27.2.

Geometrical and Topological Phases

Holonomies of interest to us are typically *geometrical* in nature, as in the examples given above. However, under certain conditions holonomies may also acquire a *topological meaning.* The distinction lies in whether the quantities that measure the holonomy change continuously when conditions are varied, in which case we speak of geometrical phases, or whether they change only by discrete amounts under this variation, in which case we speak of topological phases. The difference between geometrical and topological holonomies is discussed further in Section 27.4.

[a] This is the definition used by mathematicians and many physicists. Some physicists term this instead *anholonomy.*

3. Write the full wavefunction as the product

$$\left|\psi\right\rangle = \Phi(\boldsymbol{R})\left|n, \boldsymbol{R}\right\rangle, \tag{27.12}$$

where $\Phi(\boldsymbol{R})$ is the wavefunction for the slow degrees of freedom.

4. Use the basis (27.12) to diagonalize additional terms in the Hamiltonian that mix fast and slow degrees of freedom.

Such a separation of degrees of freedom operating on very different timescales is used either explicitly or implicitly in most treatments of dynamical problems.

27.2.2 The Berry Connection

Let us examine in more detail the implications of the approximate separation of slow and fast degrees of freedom described in the preceding section. Inserting Eqs. (27.12) and (27.10) into the time-independent Schrödinger equation $H\left|\psi\right\rangle = E\left|\psi\right\rangle$ gives

$$H\left|\psi\right\rangle = \frac{-\hbar^2}{2M}\frac{\partial^2}{\partial \boldsymbol{R}^2}\Phi(\boldsymbol{R})\left|n,\boldsymbol{R}\right\rangle + h(\boldsymbol{p},\boldsymbol{r},\boldsymbol{R})\Phi(\boldsymbol{R})\left|n,\boldsymbol{R}\right\rangle = E\Phi(\boldsymbol{R})\left|n,\boldsymbol{R}\right\rangle. \quad (27.13)$$

As shown in Problem 27.7, this equation and Eq. (27.11) imply that

$$\begin{aligned}\left\langle n,\boldsymbol{R}\right| H\left|\psi\right\rangle = \frac{-\hbar^2}{2M}\Bigg[\Phi(\boldsymbol{R})\left\langle n,\boldsymbol{R}\left|\frac{\partial^2}{\partial \boldsymbol{R}^2}\right|n,\boldsymbol{R}\right\rangle + \frac{\partial^2\Phi(\boldsymbol{R})}{\partial \boldsymbol{R}^2} \\ +2\left(\frac{\partial\Phi(\boldsymbol{R})}{\partial \boldsymbol{R}}\right)\left\langle n,\boldsymbol{R}\left|\frac{\partial}{\partial \boldsymbol{R}}\right|n,\boldsymbol{R}\right\rangle\Bigg] + \epsilon_n(\boldsymbol{R})\Phi(\boldsymbol{R}) = E\Phi(\boldsymbol{R}). \quad (27.14)\end{aligned}$$

Let us rewrite the first term inside the square brackets of Eq. (27.14) by noting that

$$\begin{aligned}\left\langle n,\boldsymbol{R}\left|\frac{\partial^2}{\partial \boldsymbol{R}^2}\right|n,\boldsymbol{R}\right\rangle &= \left\langle n,\boldsymbol{R}\left|\frac{\partial}{\partial \boldsymbol{R}}\left(\frac{\partial}{\partial \boldsymbol{R}}\right|n,\boldsymbol{R}\right\rangle\right) \\ &= \sum_{n'}\left\langle n,\boldsymbol{R}\left|\frac{\partial}{\partial \boldsymbol{R}}\right|n',\boldsymbol{R}\right\rangle\left\langle n',\boldsymbol{R}\left|\frac{\partial}{\partial \boldsymbol{R}}\right|n,\boldsymbol{R}\right\rangle = \left|\left\langle n,\boldsymbol{R}\left|\frac{\partial}{\partial \boldsymbol{R}}\right|n,\boldsymbol{R}\right\rangle\right|^2,\end{aligned}$$

where in the last step we have inserted $1 = \sum_{n'}\left|n',\boldsymbol{R}\right\rangle\left\langle n',\boldsymbol{R}\right|$ and assumed that coupling between different values of n is negligible. Then Eq. (27.14) is replaced by

$$\begin{aligned}\left\langle n,\boldsymbol{R}\right| H\left|\psi\right\rangle = \frac{-\hbar^2}{2M}\Bigg[\Phi(\boldsymbol{R})\left|\left\langle n,\boldsymbol{R}\left|\frac{\partial}{\partial \boldsymbol{R}}\right|n,\boldsymbol{R}\right\rangle\right|^2 + \frac{\partial^2\Phi(\boldsymbol{R})}{\partial \boldsymbol{R}^2} \\ +2\left(\frac{\partial\Phi(\boldsymbol{R})}{\partial \boldsymbol{R}}\right)\left\langle n,\boldsymbol{R}\left|\frac{\partial}{\partial \boldsymbol{R}}\right|n,\boldsymbol{R}\right\rangle\Bigg] + \epsilon_n(\boldsymbol{R})\Phi(\boldsymbol{R}) = E\Phi(\boldsymbol{R}). \quad (27.15)\end{aligned}$$

Now we introduce the definitions

$$V_n(\boldsymbol{R}) \equiv \epsilon_n(\boldsymbol{R}) \qquad \boldsymbol{A}_n(\boldsymbol{R}) \equiv i\hbar\left\langle n,\boldsymbol{R}\left|\boldsymbol{\nabla}_{\boldsymbol{R}}\right|n,\boldsymbol{R}\right\rangle, \quad (27.16)$$

where $\boldsymbol{\nabla}_{\boldsymbol{R}} \equiv \partial/\partial\boldsymbol{R}$. Upon inserting these definitions, Eq. (27.15) becomes (Problem 27.8),

$$\left[\frac{1}{2M}\left(\frac{\hbar}{i}\boldsymbol{\nabla}_{\boldsymbol{R}} - \boldsymbol{A}_n(\boldsymbol{R})\right)^2 + V_n(\boldsymbol{R})\right]\Phi(\boldsymbol{R}) = E\Phi(\boldsymbol{R}), \quad (27.17)$$

which has the form of an electromagnetic gauge coupling by the minimal substitution prescription to a vector potential $\boldsymbol{A}_n(\boldsymbol{R})$ and a scalar potential $V_n(\boldsymbol{R})$ (see Section 16.3), even though *no actual electric or magnetic fields have been introduced.* Therefore, the equation governing the slow variable function $\Phi(\boldsymbol{R})$ is (27.17), with the definitions (27.16).

The fast degrees of freedom have induced a "vector potential" $\boldsymbol{A}_n(\boldsymbol{R})$ coupled to the slow degrees of freedom, even though no actual electromagnetic field is anywhere to be found.

The quantity $\boldsymbol{A}_n(\boldsymbol{R})$ of Eq. (27.16) is an example of a *Berry connection* or *Berry potential.*[1]

27.2.3 Trading the Connection for a Phase

Gauge symmetries are phase symmetries and the effect of a Berry connection can be moved from the Hamiltonian to the wavefunction in Eq. (27.17) by a suitable choice of phase. Let

$$|\tilde{n}, \boldsymbol{R}\rangle \equiv \exp\left(i\gamma_n(\boldsymbol{R})\right)|n, \boldsymbol{R}\rangle, \tag{27.18}$$

where the phase $\gamma_n(\boldsymbol{R})$ is defined by

$$\gamma_n(\boldsymbol{R}) = \frac{1}{\hbar}\int_{\boldsymbol{R}_0}^{\boldsymbol{R}} \boldsymbol{A}_n(\boldsymbol{R})d\boldsymbol{R}, \tag{27.19}$$

with the integration along a path in the $\boldsymbol{R}$-space starting at $\boldsymbol{R}_0$ and ending at $\boldsymbol{R}$. As shown in Problem 27.9, the phase transformed vector potential vanishes, $A_{\tilde{n}}(\boldsymbol{R}) \equiv i\hbar\langle\tilde{n}, \boldsymbol{R}|\boldsymbol{\nabla}_{\boldsymbol{R}}|\tilde{n}, \boldsymbol{R}\rangle = 0$. Thus, under the transformation

$$|n, \boldsymbol{R}\rangle \;\rightarrow\; |\tilde{n}, \boldsymbol{R}\rangle \equiv \exp\left(\frac{i}{\hbar}\int_{\boldsymbol{R}_0}^{\boldsymbol{R}} A_n(\boldsymbol{R})d\boldsymbol{R}\right)|n, \boldsymbol{R}\rangle,$$

we find that $(\hbar/i)\boldsymbol{\nabla}_{\boldsymbol{R}} - A_n(\boldsymbol{R}) \;\rightarrow\; (\hbar/i)\boldsymbol{\nabla}_{\boldsymbol{R}}$ and the Berry connection $A_n(\boldsymbol{R})$ is eliminated from the Hamiltonian of Eq. (27.17) in favor of an additional phase,

$$\exp(i\gamma_n) = \exp\left(\frac{i}{\hbar}\int_{\boldsymbol{R}_0}^{\boldsymbol{R}} A_n(\boldsymbol{R})d\boldsymbol{R}\right) \tag{27.20}$$

for the wavefunction.

27.2.4 Berry Phases

Physical quantities are *invariant under gauge transformations* in electromagnetism. Because of the parallels between Berry connections and electromagnetism, we conjecture

[1] "Berry connection" references a formal similarity to connection coefficients prescribing parallel transport in curved manifolds (Section 26.1.2). "Berry potential" derives from the similarity of $\boldsymbol{A}_n(\boldsymbol{R})$ to the electromagnetic vector potential (Section 14.1.1). Note that we will often use natural units and drop the $\hbar$ in Eq. (27.16).

that physical effects associated with the Berry connection should involve "gauge-invariant" quantities also. The Berry connection $\boldsymbol{A}_n(\boldsymbol{R})$ itself is not gauge invariant, transforming as

$$\boldsymbol{A}_n(\boldsymbol{R}) \rightarrow \boldsymbol{A}_n(\boldsymbol{R}) - \boldsymbol{\nabla}_{\boldsymbol{R}}\chi(\boldsymbol{R}) \tag{27.21}$$

under a local phase rotation $|n, \boldsymbol{R}\rangle \rightarrow e^{i\chi(\boldsymbol{R})}|n, \boldsymbol{R}\rangle$, where $\chi(\boldsymbol{R})$ is a scalar function (Problem 27.10). However, let us define the *Berry phase* γ_n by

$$\gamma_n \equiv \oint_C \boldsymbol{A}_n(\boldsymbol{R}) \cdot d\boldsymbol{R} = i \oint_C \langle n, \boldsymbol{R}| \boldsymbol{\nabla}_{\boldsymbol{R}} |n, \boldsymbol{R}\rangle \, d\boldsymbol{R}, \tag{27.22}$$

where C represents the contour of a closed path in parameter space (identification of endpoints $\boldsymbol{R}$ and $\boldsymbol{R}_0$; note that changing the direction on the closed path changes the sign of γ_n). Then it is shown in Problem 27.12 that γ_n is *gauge independent* (up to a multiple of 2π, as befits a phase). The Berry phase (27.22) is also often written as

$$\gamma_n = -\text{Im} \oint_C \langle n, \boldsymbol{R}| \boldsymbol{\nabla}_{\boldsymbol{R}} |n, \boldsymbol{R}\rangle \, d\boldsymbol{R}, \tag{27.23}$$

where Im denotes the imaginary part. Because $\langle n, \boldsymbol{R}| \boldsymbol{\nabla}_{\boldsymbol{R}} |n, \boldsymbol{R}\rangle$ is pure imaginary, γ_n is real. As emphasized in Box 27.3, the Berry phase is a *geometrical effect* that is distinct from the normal dynamical phase associated with time evolution of a wavefunction.

27.2.5 Berry Curvature

Let us exploit further the analogy of Berry phases with electromagnetism by introducing a rank-2 antisymmetric tensor analogous to the field tensor $F^{\mu\nu}$ defined in Eq. (14.14) that we shall call the *Berry curvature tensor,*

$$\Omega^n_{\mu\nu} \equiv \partial_\mu A^n_\nu(\boldsymbol{R}) - \partial_\nu A^n_\mu(\boldsymbol{R}), \tag{27.24}$$

where $\partial_\alpha \equiv \partial/\partial R_\alpha$ and the Berry connection $A^n(\boldsymbol{R})$ is defined in Eq. (27.16). In d dimensions a rank-2 antisymmetric tensor has $\frac{1}{2}d(d-1)$ independent components. For the special case $d = 3$ the tensor has three components, which is the same number of components as a 3D vector. Hence for a 3D parameter space the Berry curvature tensor $\Omega^n_{\mu\nu}$ can be replaced by a *Berry curvature vector* $\boldsymbol{\Omega}$, with the standard mapping between tensor components $\Omega^n_{\mu\nu}$ and vector components Ω^n_λ in 3D being $\Omega^n_{\mu\nu} = \epsilon_{\mu\nu\lambda}\Omega^n_\lambda$, where $\epsilon_{\mu\nu\lambda}$ is the completely antisymmetric or Levi-Cevita tensor.[2] Then the Berry curvature vector $\boldsymbol{\Omega}_n(\boldsymbol{R})$ in 3D is

$$\boldsymbol{\Omega}_n(\boldsymbol{R}) \equiv \boldsymbol{\nabla}_{\boldsymbol{R}} \times \boldsymbol{A}_n(\boldsymbol{R}). \tag{27.25}$$

This has the same form as the definition of a magnetic field $\boldsymbol{B}$ in terms of a vector potential $\boldsymbol{A}$ that was given in Eq. (14.3) and we may think of the Berry curvature as a "magnetic field," but defined in the parameter space rather than in spacetime. Just as for the Berry phase and the electromagnetic field tensor $F^{\mu\nu}$, the Berry curvature is gauge independent. In contrast to the Berry connection, which becomes gauge invariant only after integrating

[2] Strictly, $\boldsymbol{\Omega}$ is a *pseudovector* or *axial vector* (just as for an actual magnetic field). For brevity we will term it a vector.

Box 27.3 **Geometrical Origin of the Berry Phase**

The Berry phase is *geometrical in origin* and is distinct from the ordinary dynamical phase that any time-evolving quantum system accumulates by virtue of the time-dependent Schrödinger equation. [a] If the preceding derivation is repeated using the time-dependent Schrödinger equation, one obtains for the wavefunction in adiabatic approximation for a closed path in parameter space,

$$\left|\psi(t)\right\rangle = e^{i(\gamma_n(t)+\xi(t))}\left|n, \boldsymbol{R}(t)\right\rangle,$$

where the *dynamical phase* is

$$\xi(t) = \frac{1}{\hbar}\int_0^t dt' E_n(\boldsymbol{R}(t'))$$

and the *geometrical or Berry phase* is

$$\begin{aligned}\gamma_n(t) &\equiv i\int_0^t dt'\left\langle n, \boldsymbol{R}(t')\middle|\frac{\partial}{\partial t'}\middle|n, \boldsymbol{R}(t')\right\rangle \\ &= i\oint_C d\boldsymbol{R}\cdot\langle n, \boldsymbol{R}|\,\boldsymbol{\nabla}_{\boldsymbol{R}}\,|n, \boldsymbol{R}\rangle \\ &= \oint_C d\boldsymbol{R}\cdot\boldsymbol{A}_n(\boldsymbol{R}),\end{aligned}$$

where C is a closed contour and the geometrical phase depends only on the path followed and not on the time dependence of that path because the Berry connection $\boldsymbol{A}_n(\boldsymbol{R})$ is a function of the coordinates $\boldsymbol{R}(t)$, but is not an explicit function of time.

[a] As a reminder of the difference between geometrical and dynamical information, consider a trip in a car. The distance measured by the odometer depends on the route followed (geometrical information), but it does not depend on how the speedometer varies (dynamical information).

it over a closed path to yield the Berry phase, the Berry curvature is a local gauge-invariant measure of geometric properties.

By applying Stokes' theorem (Box 27.1) to Eq. (27.22), the Berry phase γ_n may be written as an integral of the Berry curvature over the surface $\boldsymbol{S}$ bounded by the contour C,

$$\begin{aligned}\gamma_n &= \oint_C d\boldsymbol{R}\cdot\boldsymbol{A}_n(\boldsymbol{R}) \\ &= \int_S (\boldsymbol{\nabla}_{\boldsymbol{R}}\times\boldsymbol{A}_n(\boldsymbol{R}))\cdot d\boldsymbol{S} \\ &= \int_S \boldsymbol{\Omega}_n(\boldsymbol{R})\cdot d\boldsymbol{S} \qquad (\text{mod } 2\pi),\end{aligned} \tag{27.26}$$

where mod 2π indicates that γ_n is a phase that is defined up to multiples of 2π. Thus the Berry phase may be viewed as the flux of Berry curvature through the surface $\boldsymbol{S}$ bounded by the path C. As a consequence, the Berry phase is sometimes termed the *Berry flux.* By analogy with electromagnetism, we may conjecture that the Berry connection is not an

Table 27.1. Berry phase analogies [38]

Electromagnetism	Quantum holonomy
Vector potential $\boldsymbol{A}(\boldsymbol{r})$	Berry connection $\boldsymbol{A}(\boldsymbol{R})$
Magnetic field $\boldsymbol{B}(\boldsymbol{r})$	Berry curvature $\boldsymbol{\Omega}(\boldsymbol{R})$
Magnetic monopole	Degenerate point
Magnetic charge	Berry index
Magnetic flux $\Phi(C)$	Berry phase $\gamma(C)$

observable since it is gauge dependent, but that the gauge-invariant Berry phase and gauge-invariant Berry curvature may be associated with measurable effects. The Berry curvature can also be expressed in terms of energy eigenstates $|n, \boldsymbol{R}\rangle$ as

$$\boldsymbol{\Omega}_n(\boldsymbol{R}) \equiv \text{Im} \sum_{m \neq n} \frac{\langle n, \boldsymbol{R}| \boldsymbol{\nabla}_{\boldsymbol{R}} H(\boldsymbol{R}) \,|m, \boldsymbol{R}\rangle \times \langle m, \boldsymbol{R}| \boldsymbol{\nabla}_{\boldsymbol{R}} H(\boldsymbol{R}) \,|n, \boldsymbol{R}\rangle}{[E_n(\boldsymbol{R}) - E_m(\boldsymbol{R})]^2}, \tag{27.27}$$

where $\times$ indicates the 3-vector cross product and $H(\boldsymbol{R})$ is the Hamiltonian. This form is often useful in practical calculations; for example, in Problem 27.6.

This discussion suggests close analogies between classical electromagnetism described in Section 14.1 and quantum holonomies exemplified by Berry phases, as summarized in Table 27.1. These are *analogies only*: Berry phase effects do not require the presence of an actual electromagnetic field. The appearance of the magnetic monopoles and magnetic charges of Table 27.1 in actual electromagnetism would contradict the Maxwell equation (14.1c). There is no solid evidence for fundamental magnetic monopoles so Maxwell's equations are safe (for now), but mathematically analogous particles may occur as *emergent phenomena* in condensed matter systems. These are quasiparticles created by many-body interactions that resemble magnetic monopoles abstractly, but do not imply a fundamental violation of Maxwell's equations.

27.3 An Electron in a Magnetic Field

In evaluating the physical significance of a Berry phase the essential question is whether it vanishes along any closed path in parameter space. If so, it represents a phase choice for the wavefunction that has no observable significance. There are many instances where the geometric phase does not vanish and leads to observable consequences. We examine one such case in Example 27.1, and will see others in Chs. 28 and 29.

Example 27.1 The Hamiltonian for an electron interacting with a magnetic field $\boldsymbol{B}$ is

$$H = -\frac{1}{2}\mu\boldsymbol{\sigma} \cdot \boldsymbol{B} = -\frac{1}{2}\mu \left(\sigma_x B_x + \sigma_y B_y + \sigma_z B_z\right), \tag{27.28}$$

where the σ_i are Pauli matrices. The eigenvalues for $H|\psi\rangle = E|\psi\rangle$ are (Problem 27.5)

$$E_{\pm}(B) = \pm\frac{1}{2}\mu\sqrt{B_x^2 + B_y^2 + B_z^2} = \pm\frac{1}{2}\mu B, \tag{27.29}$$

where $B \equiv |\boldsymbol{B}|$, and the corresponding eigenvectors are given by

$$|\psi_+\rangle = \begin{pmatrix} \cos\frac{\theta}{2}e^{-i\phi/2} \\ \sin\frac{\theta}{2}e^{i\phi/2} \end{pmatrix} \qquad |\psi_-\rangle = \begin{pmatrix} -\sin\frac{\theta}{2}e^{-i\phi/2} \\ \cos\frac{\theta}{2}e^{i\phi/2} \end{pmatrix}, \tag{27.30}$$

where the angles θ and ϕ are related to the field components by

$$B_x = B\sin\theta\cos\phi \qquad B_y = B\sin\theta\sin\phi \qquad B_z = B\cos\theta.$$

After some evaluation, Eq. (27.27) gives

$$\Omega_-(\boldsymbol{B}) = -\frac{1}{2}\left(\frac{\boldsymbol{B}}{B^3}\right) \qquad \Omega_+(\boldsymbol{B}) = \frac{1}{2}\left(\frac{\boldsymbol{B}}{B^3}\right), \tag{27.31}$$

and the Berry phase is (Problem 27.6)

$$\gamma_m(C) = \mp\pi(1-\cos\theta), \tag{27.32}$$

where $\pi(1-\cos\theta)$ is half the solid angle enclosed by the path C as seen from the origin, $\boldsymbol{n}$ is a unit vector orthogonal to the surface, and $\boldsymbol{n}\cdot\boldsymbol{B} = B$. These results are equivalent to those that would result from placing a monopole of magnetic charge $\pm\frac{1}{2}$ at the origin of parameter space.

Example 27.1 shows that a non-trivial Berry phase is associated with electronic motion in a magnetic field, suggesting that geometrical phases can have physical consequences.

27.4 Topological Implications of Berry Phases

As illustrated in Figs. 26.3 and 27.3(a), parallel transport over a closed path on a 2-sphere rotates vectors. The holonomy (difference in initial and final vector orientations) measures curvature enclosed by the path. This is a *geometrical holonomy* because it results solely from geometry of the manifold. On the other hand, the Möbius strip in Fig. 27.3(b) is *intrinsically flat* but *non-orientable,* because it has only one side.[3] A vector perpendicular to the surface that is parallel transported continuously around the strip will point in the opposite direction after once around [arrows 1–5 in Fig. 27.3(b)], but will return to the original orientation after twice around. Thus there are two discrete outcomes for a vector transported on a closed path around the strip an integer number of times: the starting

[3] *Orientability* for a surface is a topological invariant. A thin rectangular strip has two sides and could be oriented with one or the other side up; it is *orientable.* For a Möbius strip a continuous path can be traced over all of the surface without crossing an edge, so it has only one side and is *non-orientable.* But a Möbius strip is made by putting a twist in a flat strip and gluing the ends together, so it has no intrinsic curvature.

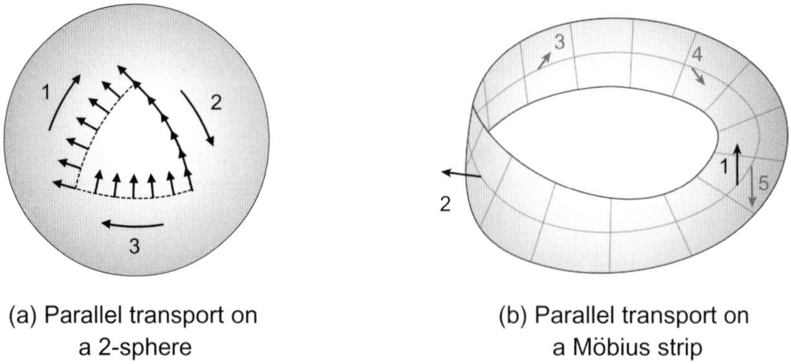

Fig. 27.3 (a) *Geometrical holonomy* associated with curvature of a sphere. (b) *Topological holonomy* associated with orientability of a Möbius strip.

orientation (+1) for an even number of trips, or the reversed orientation (−1) for an odd number of trips. This change in orientation is a (discrete) *topological holonomy* because it results from topological properties of the manifold. For the geometrical holonomy discussed above the rotation of the vector depends on the path and takes continuous values, since it is proportional to the enclosed curvature. Conversely, the topological holonomy of the Möbius strip takes discrete values and is unaffected by smooth deformations of the path that do not change the number of complete trips around the strip.

Berry phases and Berry curvature take continuous values so they are geometrical in nature. However, as we shall discuss in Section 28.4, the *Chern theorem* asserts that the integral of the Berry curvature over a closed manifold is quantized in units of 2π, and the *Chern numbers* resulting from that quantization are *topological invariants taking integer values.* Thus, quantities calculated from geometrical Berry phases may in some cases acquire a topological significance. Indeed, we shall see in Chs. 28 and 29 that topological invariants deriving from integrals of Berry curvature in electronic configuration space are essential to the understanding of topological matter.

Background and Further Reading

For an introduction to the Aharonov–Bohm effect see Feynman, Leighton, and Sands [62], Nakahara [152], and Ryder [174]. For geometrical and topological phases, see Dittrich and Reuter [50], and Nakahara [152]. The original paper on Berry phases is Ref. [22] and Berry [23] documents some historical anticipations. A resource letter on geometric phases in physics may be found in Anandan, Christian, and Wanelik [10]. Batterman [20] gives a wide-ranging discussion of physical manifestations for geometric phases. Vanderbilt [200] discusses Berry phases in electronic structure, while El-Batanouny [54], and Girvin and Yang [75] give an overview of Berry phases in condensed matter applications like the quantum Hall effect and topological matter that we shall expand upon in Chs. 28 and 29.

Problems

27.1 Show that the vector potentials given in Eq. (27.1) imply the magnetic fields given in Eq. (27.2) by evaluating $\boldsymbol{B} = \boldsymbol{\nabla} \times \boldsymbol{A}$ in the cylindrical coordinates of Fig. 27.1(b).

27.2 Use Stokes' theorem [Eq. (27.5)] to prove that Eq. (27.4) leads to Eq. (27.6). ***

27.3 Demonstrate that in the Aharonov–Bohm effect the scalar function χ corresponding to the vector potential $\boldsymbol{A} = \boldsymbol{\nabla}\chi$ outside the solenoid is given by Eq. (27.9).

27.4 Use the results of Table 24.1 to show formally that for the Aharonov–Bohm effect the mapping discussed in Section 27.1.4 from the electromagnetic gauge group manifold U(1) to the plane pierced by the solenoid is non-trivial because it admits an infinity of topological sectors characterized by an integer winding number.

27.5 Show that solution of the eigenvalue problem $H\left|\psi\right\rangle = E\left|\psi\right\rangle$ for the Hamiltonian (27.28) gives the eigenvalues (27.29) and the eigenfunctions (27.30). ***

27.6 Prove that for a spin-$\frac{1}{2}$ particle in a magnetic field, Eqs. (27.29) and (27.30) imply the Berry phase (27.32). *Hint*: Use Eq. (27.27) to find the Berry curvature in Eq. (27.31) and then use (27.26) to solve for the Berry phase. ***

27.7 Show that Eqs. (27.13) and (27.11) imply Eq. (27.14). *Hint*: Write

$$\frac{\partial^2}{\partial \boldsymbol{R}^2}\Phi(\boldsymbol{R})\left|n,\boldsymbol{R}\right\rangle = \frac{\partial}{\partial \boldsymbol{R}}\left(\frac{\partial}{\partial \boldsymbol{R}}(\Phi(\boldsymbol{R})\left|n,\boldsymbol{R}\right\rangle)\right)$$

and evaluate the resulting derivatives of products. ***

27.8 Prove using the definitions (27.16) that Eq. (27.15) is equivalent to Eq. (27.17), which has the form of a gauge coupling to a vector potential $\boldsymbol{A}_n(\boldsymbol{R})$.

27.9 Prove that the Berry connection can be eliminated from the Hamiltonian in favor of an added phase for the wavefunction given by Eq. (27.20). *Hint*: Evaluate $\partial\left|\tilde{n},\boldsymbol{R}\right\rangle/\partial\boldsymbol{R}$ from Eq. (27.18) and multiply the result from the left by $i\hbar\left\langle\tilde{n},\boldsymbol{R}\right|$. ***

27.10 Show that under a local gauge transformation $\left|n,\boldsymbol{R}\right\rangle \rightarrow e^{i\chi(\boldsymbol{R})}\left|n,\boldsymbol{R}\right\rangle$, the Berry connection $\boldsymbol{A}_n(\boldsymbol{R})$ is not invariant, transforming instead as in Eq. (27.21).

27.11 Show that the Berry curvature (27.24) can also be written as

$$\Omega^n_{\mu\nu}(\boldsymbol{R}) = i\left(\langle\partial_\mu n(\boldsymbol{R})|\partial_\nu n(\boldsymbol{R})\rangle - \langle\partial_\nu n(\boldsymbol{R})|\partial_\mu n(\boldsymbol{R})\rangle\right),$$

where $\partial_\alpha \equiv \partial/\partial R_\alpha$ and the definition (27.16) of the Berry connection A^n_μ was used.

27.12 Demonstrate that both the Berry phase γ_n and the Berry curvature $\boldsymbol{\Omega}_n(\boldsymbol{R})$ are invariant under a local gauge transformation. ***

28 Topology of the Quantum Hall Effect

The *integer quantum Hall effect* and the *fractional quantum Hall effect* represent extraordinary physics that appears when a strong magnetic field is applied to a low-density electron gas confined in two dimensions at very low temperature. As we shall see, these effects can be interpreted in terms of quantum numbers that are *topological,* and hint at the existence of whole new classes of *topological matter,* which will be the subject of Ch. 29. These topological classifications introduce the new idea that important phases of matter may have a topological characterization even if they cannot be distinguished on the basis of traditional symmetry or broken symmetry, though we will find that symmetry and broken symmetry still play a role in determining whether topological states can occur. We begin by recalling the properties of the classical Hall effect.[1]

28.1 The Classical Hall Effect

The *classical Hall effect* is illustrated in Fig. 28.1. Since the Lorentz and Hall forces cancel in steady state, the current is entirely in the x (longitudinal) direction and the Hall field is perpendicular to the current (in the direction $\boldsymbol{j} \times \boldsymbol{B}$) if the sample is uniform.

28.1.1 Hall Effect Measurements

We may analyze the situation in the classical limit governed by the *Lorentz force,*

$$\boldsymbol{F} = q\left(\boldsymbol{E} + \frac{1}{c}\boldsymbol{v} \times \boldsymbol{B}\right), \tag{28.1}$$

where q is the charge, $\boldsymbol{B}$ is the applied magnetic field (taken here to be in the z direction), $\boldsymbol{E}$ is the electric field, and $\boldsymbol{v}$ is the electron velocity. As indicated in Fig. 28.1(d), in a typical classical Hall experiment the transverse voltage $V_{\rm H}$ and the longitudinal voltage $V_{\rm L}$ are measured. If the sample is approximated as 2D with length L and transverse width w [see Fig. 28.1(d)], and Ohm's law in the form $j_x = \sigma E_x$ is assumed (see Box 28.1), you are asked to show in Problem 28.1(d) that

$$E_y = \frac{V_{\rm H}}{w} \qquad I = w j_x \qquad R_{\rm H} \equiv \frac{V_{\rm H}}{I} = -\frac{B}{ecn_e} \qquad R_{\rm L} \equiv \frac{V_{\rm L}}{I} = \frac{L}{w\sigma}, \tag{28.2}$$

[1] The presentation in this chapter assumes a familiarity with the Bloch theorem and Brillouin zones described in Ch. 5, the basic principles of topology reviewed in Ch. 24, and the Berry phase discussed in Ch. 27.

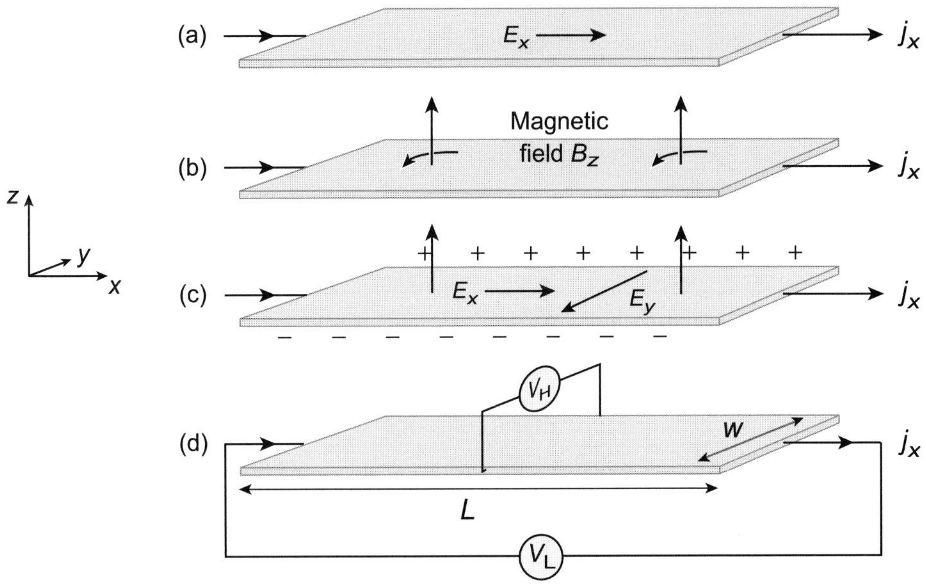

Fig. 28.1 Classical Hall effect. (a) An electric field E_x causes a current density j_x in a thin rectangular sample in the x direction. (b) A uniform magnetic field B_z is placed on the sample in the positive z direction. Curved arrows indicate the response of the electrons to the magnetic field, causing a deflection in the negative y direction. (c) Electrons accumulate on one edge and a positive ion excess on the other edge, producing a transverse electric field E_y (*Hall field*) that just cancels the force produced by the magnetic field. Thus in equilibrium current flows only in the x direction. (d) Typically the longitudinal voltage V_L and the transverse (Hall) voltage V_H are measured.

where V_H is the *Hall voltage* (associated with the induced electric field E_y), I is the total current, j_x is the current density in the x direction, $q = -e$ was assumed for electrons, R_H is the *Hall resistance,* and R_L is the *longitudinal resistance.* The sign of the Hall resistance depends on the sign of the charge carriers and its magnitude is inversely proportional to their density. Hence a measurement of the classical Hall resistance for a sample determines the sign and the density of the charge carriers.

28.1.2 Quantization of the Hall Effect

A richer set of possibilities than for the classical Hall effect is found in high-precision experiments at low temperatures and high magnetic fields. These *quantum Hall effects* are associated with quantization of the Hall conductance in terms of a fundamental quantum of flux, and can be separated into two sets of phenomena:

1. the *integer quantum Hall effect (IQHE)*, and
2. the *fractional quantum Hall effect (FQHE),*

which entail related physics but arise for different microscopic reasons. We shall describe the IQHE and the FQHE shortly, but first we discuss quantized electrons confined to 2D in strong magnetic fields, since this lies at the heart of the quantum Hall effect.

Box 28.1 Resistance and Resistivities

In the simplest cases the current and the electric field may be assumed collinear and an elementary form of *Ohm's law* holds: $V = IR$, where V is voltage, I is current, and R is resistance. This can also be expressed in terms of the electric field E and current density j as $j = \sigma E$ where σ is the scalar *conductivity*, or the inverted expression $E = \rho j$, where the scalar *resistivity* ρ is given by $\rho = \sigma^{-1}$.

Resistivity and Conductivity Tensors

In more complex situations such as in the Hall effect where there are both electric and magnetic fields, the resistivities and related quantities become tensors. Restricting to two dimensions for the quantum Hall problem, the resistivity tensor ρ and conductivity tensor σ may be expressed as the matrices

$$\rho = \begin{pmatrix} \rho_{11} & \rho_{12} \\ \rho_{21} & \rho_{22} \end{pmatrix} = \begin{pmatrix} \rho_{xx} & \rho_{xy} \\ \rho_{yx} & \rho_{yy} \end{pmatrix} \qquad \sigma = \begin{pmatrix} \sigma_{11} & \sigma_{12} \\ \sigma_{21} & \sigma_{22} \end{pmatrix} = \begin{pmatrix} \sigma_{xx} & \sigma_{xy} \\ \sigma_{yx} & \sigma_{yy} \end{pmatrix},$$

where the components are $\rho_{ij} = E_i/j_j$ and $\sigma_{ij} = j_i/E_j$, with ρ and σ related by matrix inversion, $\rho = \sigma^{-1}$. Then $E_i = \rho_{ij} j_j$ and $j_i = \sigma_{ij} E_j$, or

$$\begin{pmatrix} E_x \\ E_y \end{pmatrix} = \begin{pmatrix} \rho_{xx} & \rho_{xy} \\ \rho_{yx} & \rho_{yy} \end{pmatrix} \begin{pmatrix} j_x \\ j_y \end{pmatrix} \qquad \begin{pmatrix} j_x \\ j_y \end{pmatrix} = \begin{pmatrix} \sigma_{xx} & \sigma_{xy} \\ \sigma_{yx} & \sigma_{yy} \end{pmatrix} \begin{pmatrix} E_x \\ E_y \end{pmatrix},$$

written out explicitly.

Symmetries of the Tensors

For the quantum Hall experiment rotational invariance about the z-axis implies that $\rho_{yy} = \rho_{xx}$ and $\rho_{yx} = -\rho_{xy}$, so the preceding equations simplify to

$$\begin{pmatrix} E_x \\ E_y \end{pmatrix} = \begin{pmatrix} \rho_{xx} & \rho_{xy} \\ -\rho_{xy} & \rho_{xx} \end{pmatrix} \begin{pmatrix} j_x \\ j_y \end{pmatrix} \qquad \begin{pmatrix} j_x \\ j_y \end{pmatrix} = \frac{1}{\rho_{xx}^2 + \rho_{xy}^2} \begin{pmatrix} \rho_{xx} & -\rho_{xy} \\ \rho_{xy} & \rho_{xx} \end{pmatrix} \begin{pmatrix} E_x \\ E_y \end{pmatrix}.$$

From Problem 28.7, for the 2D Hall setup the resistance R and the resistivity ρ have the same units, and upon comparing with Fig. 28.1 the Hall resistance is $R_{\text{H}} = R_{xy} = \rho_{xy}$ and the longitudinal resistance is $R_{\text{L}} = R_{xx} = (L/w)\rho_{xx}$. Notice that if ρ_{xy} vanishes the resistivity matrix becomes diagonal and we revert to the simple Ohm's law description with a scalar resistivity. It is the off-diagonal components of the resistivity (or conductivity) tensor that are responsible for the Hall effect.

28.2 Landau Levels for Non-Relativistic Electrons

Let us consider the quantum-mechanical description of electrons in a 2D gas with a magnetic field oriented perpendicular to the 2D plane, assuming electrons to be non-relativistic so that the Schrödinger equation applies. Since we shall be concerned with strong magnetic fields that can align spins, the states will be assumed to all be of one electron spin polarization (magnetic substate) and the constant energy contributed by each spin through the Zeeman interaction will be omitted. We shall also neglect initially the effect of the Coulomb interaction between electrons relative to the effect of the strong magnetic field.

28.2.1 Hamiltonian and Schrödinger Equation

By the minimal prescription to ensure gauge invariance given in Section 16.3, the effect of coupling an electromagnetic field may be included by modifying the momentum operator $\boldsymbol{p} = (p_x, p_y)$ in the Hamiltonian to include a term depending on the vector potential $\boldsymbol{A}$,

$$H = \frac{1}{2m}\left(\frac{\hbar}{i}\boldsymbol{\nabla} + \frac{e}{c}\boldsymbol{A}\right)^2, \tag{28.3}$$

where the curl of the vector potential $\boldsymbol{\nabla} \times \boldsymbol{A}$ yields the magnetic field $\boldsymbol{B}$:

$$\boldsymbol{B} = \boldsymbol{\nabla} \times \boldsymbol{A} \qquad \boldsymbol{B} \equiv (B_x,\ B_y,\ B_z) = (0, 0, B).$$

Two common gauge choices leading to this magnetic field are the *Landau gauge* (often convenient for rectangular geometry), where

$$\boldsymbol{A} = (0,\ Bx,\ 0) \qquad \text{(Landau gauge)}, \tag{28.4}$$

and the *symmetric gauge* (often convenient for spherical geometry), where

$$\boldsymbol{A} = \frac{1}{2}B(-y,\ x,\ 0) \qquad \text{(symmetric gauge)}. \tag{28.5}$$

The electromagnetic field is invariant under gauge transformations so no physical observable depends on the gauge chosen, but the forms of the vector potential and of the wavefunction (which are *not observables*) do. As shown in Problem 28.2, in Landau gauge with the *cyclotron frequency* ω_c and the *magnetic length* Λ defined by

$$\omega_c \equiv \frac{eB}{mc} \qquad \Lambda \equiv \sqrt{\frac{\hbar c}{eB}}, \tag{28.6}$$

the wavefunction resulting from solving $H\psi = \epsilon\psi$ is a plane wave in the y direction and of localized harmonic oscillator form in the x direction, and is labeled by two quantum numbers (n, k),

$$\psi_{nk}(x, y) = H_n\left(\frac{x}{\Lambda} - \Lambda k\right) e^{-(x-x_k)^2/2\Lambda^2} e^{iky}, \tag{28.7}$$

where H_n is a Hermite polynomial and $x_k = \Lambda^2 k$. The corresponding energy is

$$\epsilon_n = \hbar\omega_c\left(n + \frac{1}{2}\right), \tag{28.8}$$

which is independent of the quantum number k, depending only on the principal Landau quantum number $n = 0, 1, 2, \ldots, \infty$. From Eq. (28.7), for each value of k the wavefunction is localized in x near $x = x_k = \Lambda^2 k$, but extended in y. (This depends on our gauge choice, but remember that the wavefunction is not an observable; see Problem 28.4.)

28.2.2 Landau Levels and Density of States

Energy levels labeled by n in Eq. (28.8) are termed *Landau levels*.[2] The degeneracy at a given n is determined by how many k values correspond to that n. Since ϵ_n is independent of k, in the absence of electron scattering the level density $g(\epsilon)$ is a series of δ-functions at the energies ϵ corresponding to discrete values of n, weighted by a degeneracy factor N,

$$g(\epsilon) = N \sum_n \delta \left[\epsilon - \hbar\omega_c \left(n + \frac{1}{2} \right) \right], \tag{28.9}$$

where N is the number of electrons that can occupy a Landau level and is given for a sample of width w, length L, and magnetic length Λ by

$$N = \frac{BLw}{hc/e} = \frac{Lw}{2\pi\Lambda^2}. \tag{28.10}$$

Equation (28.10) has a simple physical interpretation.

For a sample of area $A = Lw$, the magnetic flux through the sample is BLw, so N is the magnetic flux measured in units of the quantum of magnetic flux, hc/e.

Equation (28.10) implies that the number of states per unit area in each Landau level is

$$n_B = \frac{B}{hc/e} = \frac{1}{2\pi\Lambda^2}, \tag{28.11}$$

which is independent of the Landau level but increases linearly with the strength of the magnetic field. Thus, for large magnetic fields the Landau levels will be highly degenerate. Let the actual number density of electrons be n_e and define a *filling factor* ν by

$$\nu \equiv \frac{n_e}{n_B} = \frac{n_e hc}{eB}. \tag{28.12}$$

If ν is an integer, for the ground state the lowest ν Landau levels will be filled and the rest empty, while if ν is fractional the last level will be partially filled. Thus, if ν equals an integer M there will be M filled levels and an energy gap between the $n = M$ and $n = M+1$ Landau levels that inhibits excitation.

28.3 The Integer Quantum Hall Effect

Let us examine the Hall effect for a dilute 2D electron gas at low temperature with a strong magnetic field, where we might expect quantum effects to become important.

[2] Landau levels are a general consequence of quantizing the motion of charged particles in a magnetic field. They may be thought of as arising from a charged particle moving in a circle with a radius determined by the quantization condition that an integral number of de Broglie wavelengths fit around the circle. The radius of the circle is termed the *cyclotron radius* and the center of the circle is termed the *guiding center*. The high degeneracy associated with the k quantum number is a consequence of there being many different possible locations for the center of a circle of given radius.

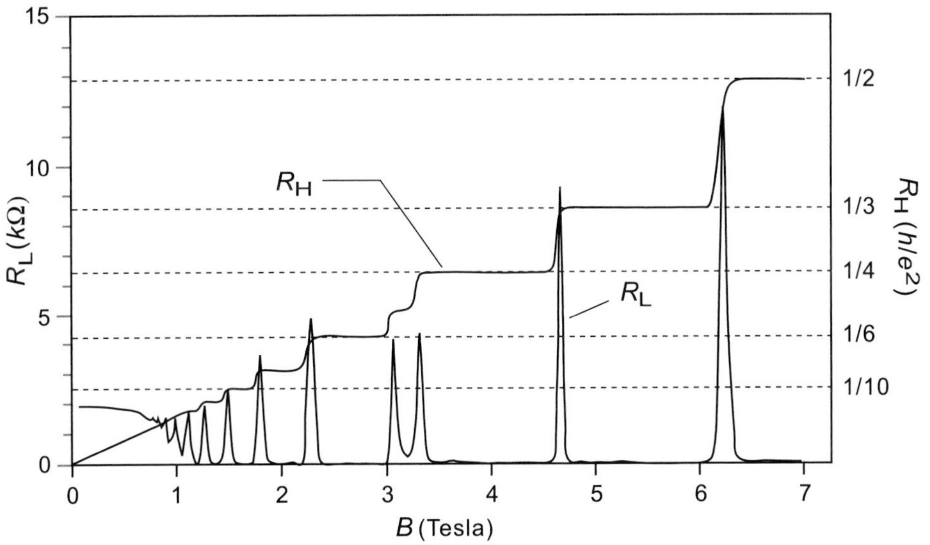

Fig. 28.2 Data illustrating the integer quantum Hall effect [204]. The Hall resistance R_{H} (upper curve) varies stepwise with changes in magnetic field B. Step height is given by the physical constant h/e^2 divided by an integer [see Eq. (28.13)]. The lower curve with multiple peaks is the longitudinal resistance R_{L}, which vanishes over each plateau in R_{H}.

From Eq. (28.2), the classical expectation is that R_{H} should increase linearly with magnetic field, while R_{L} should be constant at a non-zero value. Experimental data are shown in Fig. 28.2 [204]. For very low magnetic fields the classical predictions hold, but they are violated strongly for large fields. Above $B \sim 1$ T the Hall resistance R_{H} exhibits plateaus of height[3]

$$R_{\mathrm{H}} = \rho_{yx} = \frac{V_y}{I_x} = \frac{1}{n}\frac{h}{e^2}, \tag{28.13}$$

where n is an integer, and the longitudinal resistance R_{L} peaks at the transitions between R_{H} plateaus and drops to essentially zero over the range of each plateau. The phenomena displayed in Fig. 28.2 constitute the *integer quantum Hall effect (IQHE)*. Let us see if we can explain this non-classical behavior using the quantum theory developed in Section 28.2.

28.3.1 Understanding the Integer Quantum Hall Effect

The integer quantum Hall effect is related solely to the filling of Landau levels and our discussion of Landau levels (supplemented by impurity scattering and topological considerations to be discussed below) provides the basis for understanding its essential features. From Eq. (28.13) and Fig. 28.2, the Hall resistance plateaus occur at precisely

[3] Our discussion will sometimes refer to Hall resistance R_{H} and sometimes to Hall conductance $\sigma_{\mathrm{H}} = R_{\mathrm{H}}^{-1}$.

quantized values $R_H = \rho_{xy} = h/e^2 n$, with n an integer. Comparing with $\rho_{xy} = B/n_e ec$ from Eq. (28.2),

$$n = \frac{hcn_e}{Be} = \frac{n_e}{n_B} = \nu,$$

where ν is the Landau filling factor defined in Eq. (28.12).

The integer n in Eq. (28.13) labeling R_H plateaus in Fig. 28.2 is the Landau filling factor ν, and the plateau heights correspond to exact filling of Landau levels.

From Eq. (28.10), the number of electrons that can be placed in each Landau level depends on the magnetic field strength. As the magnetic field is changed, a match between the capacity of the Landau levels and the number of electrons in the sample at certain values of the magnetic field causes an integer number of Landau levels to be filled exactly. The special values of the field at which this occurs are those at which the ratio of the number of electrons per unit area to the number of units of flux h/e^2 [the filling factor ν defined in Eq. (28.12)] is an integer.

Furthermore, when the electrons fill ν Landau levels exactly there is an energy gap $\Delta E = \hbar\omega_c$ between filled and empty states. A decrease in the area A of the sample decreases the number of flux quanta N_ϕ that pierce it, requiring that electrons be promoted to the next Landau level [see Eq. (28.10)]. But an energy price $\hbar\omega_c$ is required and the system resists compression. As described in Box 28.2, such a system is said to be an *incompressible quantum fluid.* If we apply a small electric field to this state the electrons cannot move because they have no place to go, as in an insulator. Thus $\rho_{xx} = 0$, as required by data.[4]

Therefore, Landau level filling and the associated "shell closures" might plausibly account for the IQHE, except for two essential details that we have yet to explain.

1. Why are the jumps in R_H followed by broad plateaus in the Hall resistance of Fig. 28.2?
2. Why do those plateaus correspond to remarkably constant Hall resistance, quantized to a precision of one part in a billion?[5]

The first suggests that not all of the states within a Landau level contribute to the conductance. The second suggests a fundamental principle responsible for quantization of the Hall resistance that renders it insensitive to details. We tackle first the issue of why

[4] A vanishing longitudinal resistivity ρ_{xx} has a different meaning in a system with tensor resistivity than in one with scalar resistivity. In the latter, conductivity and resistivity are related by $\sigma_{xx} = 1/\rho_{xx}$ and they cannot both be zero, but for the quantum Hall system σ and ρ are *matrix inverses* of each other and from Box 28.1,

$$\sigma_{xx} = \frac{\rho_{xx}}{\rho_{xx}^2 + \rho_{xy}^2} \qquad \sigma_{xy} = \frac{-\rho_{xy}}{\rho_{xx}^2 + \rho_{xy}^2}.$$

Now if $\rho_{xy} = 0$ we get the familiar $\sigma_{xx} = 1/\rho_{xx}$, but if $\rho_{xy} \neq 0$ *both* σ_{xx} and ρ_{xx} can be zero. Then $\sigma_{xx} = 0$ means no current is flowing in the longitudinal direction (as in an insulator), but at the same time $\rho_{xx} = 0$ means that no energy is being dissipated (as in a perfect conductor).

[5] Because of this precision, the quantum Hall effect is used to maintain the international standard for electrical resistance. Furthermore, the speed of light c is a defined quantity, so the precise measurement of h/e^2 in the integer quantum Hall effect is equivalent to measuring precisely the dimensionless *fine-structure constant* $\alpha_{CGS} \equiv e^2/\hbar c \sim 1/137$, which defines the fundamental coupling strength of quantum electrodynamics.

Box 28.2 Incompressible Quantum States

In a bound, quantized, fermionic system the lowest-energy state (the ground state) is formed by filling single-particle energy levels from the bottom, subject to the Pauli principle. The energy of the highest-energy filled level is called the *Fermi energy*.

Gapped and Ungapped States

Two qualitatively different possibilities may be identified (see also Box 5.1):

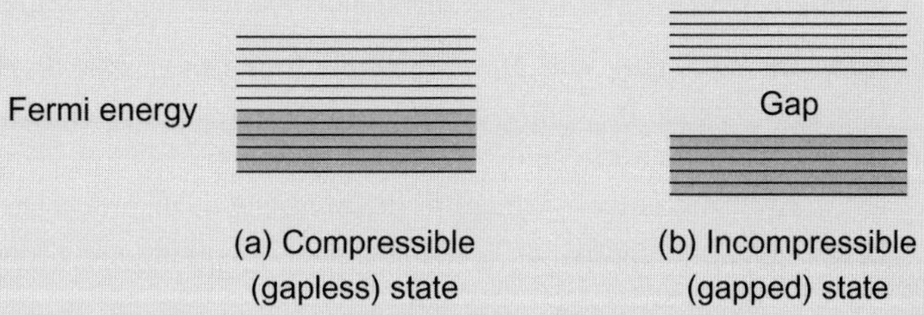

with occupied levels shaded in darker gray. In (a) there is no energy gap near the Fermi energy; this state is *gapless*. In (b) there is a gap between the highest occupied and lowest unoccupied levels; this state is *gapped*.

Compressible and Incompressible States

Now consider how the material corresponding to the states in (a) or (b) responds to applied pressure. A reduction of the volume requires that particles be excited (uncertainty principle). In case (a) there is no gap to excitation, so only a relatively small effort is required to compress the state. We say that such an ungapped state is *compressible*. In case (b) there is a gap to excitation that must be overcome to compress the material. We say that such a gapped state is *incompressible*. Thus "incompressible" is an indirect reference to a state that is gapped, meaning that a larger than average energy price must be paid to compress the sample.

there are plateaus by taking a detour to consider the influence of impurities on the transport properties of an electron gas. Then we shall consider the issue of what protects the Hall resistance quantization so that these plateaus are so remarkably flat.

28.3.2 Disorder and the Integer Quantum Hall State

The integer quantum Hall effect can occur for samples that contain impurities, so they are subject to disorder. Let us now consider how this affects the quantum Hall state. We begin by observing that electronic states in a sample may be classified into two broad categories, according to their spatial localization.

1. *Extended states,* where the wavefunction falls off slowly with distance.
2. *Localized states,* where the wavefunction is finite only in a small region.

Extended states can carry current and thus increase the conductance, while localized states interfere with charge transport and increase the resistance. The very flat plateaus of the integer quantum Hall effect exhibited in Fig. 28.2 cannot occur in a perfect crystal, for which the system is translationally invariant, electrons can propagate as Bloch waves, and all states are extended states (see Section 5.4). Thus, the Hall resistance in a perfect crystal should change monotonically with the magnetic field, as for the classical Hall effect, but contrary to the behavior observed in Fig. 28.2. In fact, increasing the impurities is found to increase the strength of the integer quantum Hall effect, up to a point. Another name for impurities is "dirt," so it seems contrary to intuition that adding dirt should cause integers to emerge from a pristine physical system. We shall see that the answer to this puzzle is topology, which cares little for fine details like chemical composition.

Impurity Scattering and Mobility Gaps: Impurities cause an increase in localization and thus in resistance by scattering the charge carriers.[6] Systems disordered by impurities are notoriously difficult to analyze because neither translational invariance nor the concept of an energy band is well defined in the presence of disorder. However, it will be sufficient here to apply qualitative insights. The most important is the following.

> In systems with disorder some states are spatially extended and some are localized, and the extended states lie very near in energy to the unperturbed levels.

As illustrated in Fig. 28.3(a), in a clean system the density of states corresponds to δ-functions at the location of the filled Landau levels, weighted by a degeneracy factor depending on the strength of the magnetic field. Impurity scattering will modify the energy of the Landau levels, breaking the degeneracy and broadening the level density distribution, as in Fig. 28.3(b). Furthermore, it is reasonable to expect that the effect of impurity scattering is to localize a state, and that the stronger the impurity scattering the more the energy will be perturbed from the clean limit and the more the state will be localized. For example, think of the extreme case where the interaction is so strong that the impurity captures the electron into a bound state. Thus in the broadened peaks of Fig. 28.3(b), states near the centers of peaks are least disturbed from the clean limit and are the most extended, while states in the wings of the peak are most disturbed by the impurity scattering and are the most localized, as illustrated in Fig. 28.3(c). This produces *mobility gaps,* corresponding to ranges of states that do not support electron transport [the lighter gray regions of Fig. 28.3(c)].

Normally even a small amount of impurity scattering destroys all states carrying current in 2D systems. However, a magnetic field breaks time-reversal symmetry and interferes with localization processes. Detailed studies show that current-carrying states remain even for samples with impurities in strong magnetic fields, and that they are concentrated near the energies of the unperturbed Landau levels, as our simple reasoning would suggest.

Origin of the Hall Plateaus: Figure 28.3(c) explains the Hall plateaus. Only states near the centers of the peaks (near the unperturbed Landau energy) are delocalized and

[6] A simple picture of localization by impurity scattering will be adequate for the present discussion but the full topic is an involved one. Readers wishing a deeper understanding are urged to begin with *Anderson localization* [11]; an introduction may be found in Phillips [163].

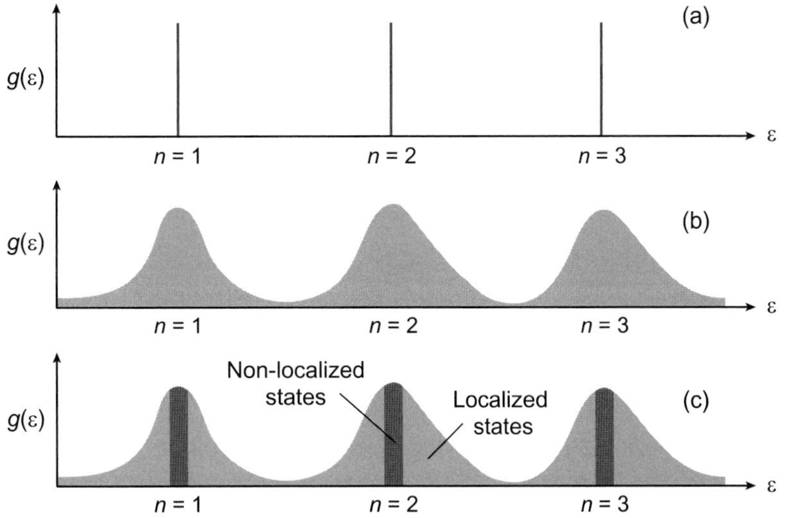

Fig. 28.3 Level densities $g(\epsilon)$ for a 2D electron gas in a magnetic field as a function of energy ϵ. (a) No impurity scattering (clean limit): Landau level filling corresponds to δ-functions at integer values of the principal Landau quantum number n, as in Eq. (28.9). (b) Impurity scattering broadens the distributions. (c) The peaks broadened by impurity scattering consist of localized states in the wings and non-localized states near the centroids.

carry current. Thus the current jumps discontinuously as the chemical potential is tuned (by varying particle number or the magnetic field) through the center of each Landau level, and remains constant as it is tuned through the localized states in the wings of the peak because the chemical potential is then passing through states that cannot carry current. Hence the conductance does not change, producing the Hall plateaus. The existence of the plateaus implies that the range of extended states must be very narrow. In summary, for the integer quantum Hall effect the crucial points are

1. quantization of Landau levels in the magnetic field,
2. the presence of a random scattering potential because of impurities, and
3. because of mobility gaps created by impurity scattering, each Landau level contributes a fixed amount to the Hall conductance.

Thus, conductance is quantized because it counts filled Landau levels.

28.3.3 Edge States and Conduction

Let us now look more carefully at the nature of the currents carried by the delocalized states in the integer quantum Hall effect.

Transport by Edge Currents: For a confined 2D electron gas with a magnetic field orthogonal to the 2D plane, the electrons move in circular (cyclotron) orbits because of the Lorentz force. As illustrated in Fig. 28.4, this means that no net current flows in the central region because the orbits there are localized and the *current vanishes in the bulk.* How is

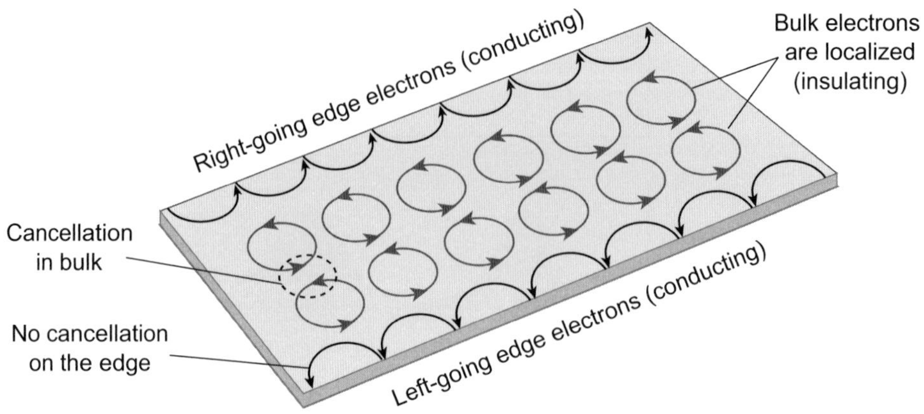

Fig. 28.4 Cartoon illustrating conduction by edge states for a 2D electron gas in a perpendicular magnetic field. The currents are *chiral* (have a "handedness") because only right-going electrons exist at the upper edge and only left-going electrons exist at the lower edge, if the width is such that quantum tunneling between the two conducting edges is negligible.

the current carried, then? Figure 28.4 suggests an answer. Edge states do not suffer this localization because of confinement by the sides of the sample. Thus boundary electrons move by "skipping" along the edges and they carry the current. This edge current is *chiral* (it has a "handedness") because the current is right-going on one edge and left-going on the opposite edge. The preceding argument is classical and Fig. 28.4 is schematic at best, but the physics that they suggest is still basically correct. In a full quantum treatment the solutions are modified by the presence of a confining potential representing the edge of the Hall apparatus, which leads to a wavefunction delocalization that produces the chiral edge currents. The energy levels in the first few Landau levels are illustrated in Fig. 28.5 [207]. This shows that states are generally gapped in the bulk (interior of the sample), where an energy cost $\Delta E = \hbar\omega$ must be paid to excite the system, but are gapless near the edges, where excitations of vanishing energy are possible in the thermodynamic limit of many particles.

Chiral Protection of Edge Currents: The chiral nature of the edge states is the reason why they remain extended, even in the presence of impurities (Figs. 28.3 and 28.4). Scattering between time-reversed states of opposite momentum ("backscattering") results in insulator character due to wavefunction localization for typical low-dimensional systems. If the scattering potential does not depend on spin, chiral 2D states are protected from such localization because scattering of a particle carrying definite chirality into a time-reversed state would switch the chirality. This cannot happen if the available states are all of one chirality, which is the case if the right-going current on one edge and the left-going current on the opposite edge in Fig. 28.4 cannot interact by quantum tunneling.

Protection of the Plateau Quantization: Let us now address the other question posed earlier: why are the Hall resistance plateaus so flat? The impressive precision of Hall

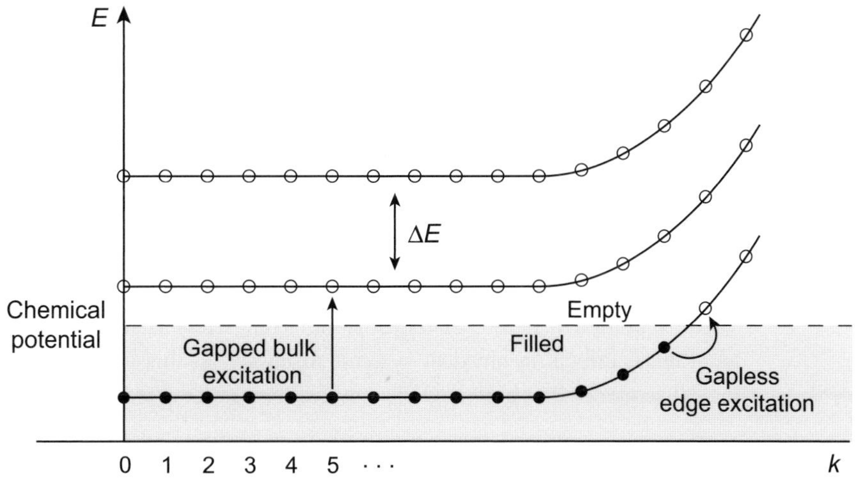

Fig. 28.5 First three Landau levels with a smooth confining potential on the right [207]. States are labeled by k and indicated by circles. Filled states are shaded, empty states are unshaded, and the chemical potential is indicated by the dashed line. Bulk excitations are inhibited by a gap $\Delta E = \hbar\omega_c$, but the gap for edge excitations tends to zero in the thermodynamic limit.

quantization in integral multiples of e^2/h and its insensitivity to different geometries, electron densities, and impurity concentrations argue that a fundamental principle is in play. The indication that current is carried by edge states hints that the current may be topological in nature, and that the remarkable precision of Hall quantization may result from conservation of a *topological quantum number*. Indeed David Thouless, a pioneer in understanding topological matter, credited his initial interest in whether the integer quantum Hall effect could be a consequence of topology to a question posed by Hans Dehmelt: How could the quantum Hall effect lead to such precise and reproducible measurements when so little was known about either the details of the devices or the theory of the effect at the time the question was first asked [193]? In the next section, we shall develop a topological interpretation of the integer quantum Hall effect that explains the remarkable stability of the Hall plateaus in terms of conserved topological quantum numbers.

28.4 Topology and Integer Quantum Hall Effects

Topology has long been used to classify defects such as vortices, dislocations, and disclinations for real-space configurations in condensed matter physics.[7] These are sometimes

[7] Vortices are swirls such as the flow of water into a drain. Dislocations are points in a regular crystal structure where an atom has been displaced from its expected position in the periodic lattice. A disclination is a line defect in a crystal that violates rotational symmetry.

termed *topological textures*. A major advance was achieved when Thouless, Kohmoto, Nightingale, and den Nijs (TKNN) [195] showed in 1982 that a derived topological quantity now called the *TKNN invariant* (which is equivalent to a *Chern index*, see Section 28.4.2) is related to integers defining the quantum Hall conductance steps. This accounted for the Hall conductance quantization and for why the IQHE is so robust.[8] In retrospect, this was the first inkling of *topological phases* in quantum matter [65, 102, 127, 128], which will be the subject of Ch. 29. Unlike the real-space topological textures alluded to above, topological phases of matter are characterized by a more abstract topological non-triviality in the Hilbert space of quantum states.

The topological nature of the quantum Hall effect is less obvious than many of the topological effects in physical systems discussed to this point. To uncover it requires a logical sequence that pulls together several topics discussed in earlier chapters.

1. The integer quantum Hall effect can be described in terms of a Hamiltonian depending on two angular variables.
2. Adiabatic transport in this parameter space around a closed path can produce a non-trivial Berry phase (Ch. 27).
3. By arguments similar to those of Ch. 26 concerning parallel transport in curved spaces, the Berry phase may be used to define a measure of *curvature* in the parameter space.
4. By a generalization of the *Gauss–Bonnet theorem* (Box 28.3) called the *Chern theorem,* this *geometrical property* of the manifold (curvature) may be linked to a *topological invariant* of the manifold called the *Chern class.*
5. The integer *Chern number* labeling the Chern class is a topological index that can be related to the integer quantum Hall quantization of conductance.

Let us begin this conjuring of the topology hiding in the integer quantum Hall effect by recalling the properties of Berry phases and of parallel transport in curved manifolds.

28.4.1 Berry Phases and Adiabatic Curvature

In Ch. 27 a Berry geometrical phase was introduced that describes the phase evolution of a vector in a complex vector space as a set of parameters is varied. We now wish to show that the description of the integer quantum Hall effect involves Berry phases where the vector space corresponds to Bloch wavefunctions and the parameter space corresponds to wavevectors $\boldsymbol{k}$ in the Brillouin zone (see Section 5.4). Assume a set of electronic bands with lattice momentum $\boldsymbol{k}$ in the 2D Brillouin zone, where the wavefunctions in a band are of the Bloch form (5.9): $\psi_{\boldsymbol{k}} = e^{i\boldsymbol{k}\cdot\boldsymbol{x}} u_{\boldsymbol{k}}(\boldsymbol{x})$. We assume also that the electrons are non-interacting, and that the Fermi energy lies in a gap between filled and unfilled Landau levels. If a state is transported adiabatically around a closed loop in parameter space (k_x, k_y) using the time-dependent Schrödinger equation, the phase mismatch associated with parallel transport will be given by the Berry phase, implying a Berry connection,

[8] Much of the foundation for a topological explanation of the IQHE is provided by the "Laughlin gauge argument" [135]. It is rather technical and we will not discuss it directly, given our level of presentation. An introduction may be found in Phillips [163] and a more technical discussion in Tong [197].

$$A_i(\boldsymbol{k}) = -i \left\langle u_{\boldsymbol{k}} \left| \frac{\partial}{\partial k_i} \right| u_{\boldsymbol{k}} \right\rangle \tag{28.14}$$

and a Berry curvature Ω given by [see Eq. (27.24)],

$$\Omega = \partial_y A_x - \partial_x A_y = 2 \text{ Im } \langle \partial_y u_{\boldsymbol{k}} | \partial_x u_{\boldsymbol{k}} \rangle, \tag{28.15}$$

where Im denotes the imaginary part and $\partial_\alpha \equiv \partial/\partial k^\alpha$. That we can assign a Berry curvature to a vector space of quantum states defined in the Brillouin zone is a deep and beautiful insight but the Berry curvature takes continuous values and thus is not topological, so it alone cannot explain the stability of the integral Hall effect. Now we must take one final step, which is to link this geometry associated with the Berry phase to a topological conservation law that arises because the Brillouin zone is a closed manifold that is a circle in 1D, a 2-torus in 2D, and a 3-torus in 3D, and these have non-trivial connectivity. (Strictly, we should use the *magnetic Brillouin zone*; however, our conclusions will not be substantially affected by using the normal Brillouin zone as the basis of our discussion.)

28.4.2 Chern Numbers

The geometrical curvature just established in the quantum Hall parameter space through Berry phases can be related to a topological invariant called a *Chern number*. Let us see how.

The Gauss–Bonnet Equation: To understand Chern numbers, their topological significance, and their connection with geometry, we begin with a simpler but related problem embodied in the *Gauss–Bonnet theorem,* which is described in Box 28.3. The content of the Gauss–Bonnet theorem may be expressed concisely by the *Gauss–Bonnet equation*

$$\frac{1}{2\pi} \int_S K \, dA = 2(1 - g). \tag{28.16}$$

The Gauss–Bonnet theorem relates the geometrical properties (carried by the local curvature K) and topological properties (carried by the genus g, which is a topological invariant since connectedness is invariant under homeomorphisms; see Section 24.3.2).

Example 28.1 Suppose that we deform the surface S through a small, smooth and continuous deformation that does not puncture the surface so that its genus is unchanged. Then the curvature K on the left side of Eq. (28.16) may change locally but the *right side cannot change* because it corresponds to an even integer and integers are incapable of changing continuously by small amounts. Thus the integral of the curvature on the left side cannot change either and is invariant under smooth deformations. Figure 28.6 illustrates.

Thus the integral over curvature for the surface S is *topologically protected* against small changes in local curvature K.

The Gauss–Bonnet–Chern Equation: The topological protection demonstrated in Example 28.1 raises the issue of whether topology could account for precise quantization of the IQHE conductance, but Eq. (28.16) deals with the geometry and topology of a spatial

Box 28.3 The Gauss–Bonnet Theorem

For a 2D surface there is a remarkable relationship between its geometry and its topology called the *Gauss–Bonnet theorem.*[a] This theorem can be expressed as the *Gauss–Bonnet equation,*

$$\frac{1}{2\pi}\int_S K dA = 2(1-g),$$

where the integral is over a closed surface S, the local curvature of the surface is K, and g is the genus of the surface (the number of "holes" or "handles" characterizing its topology). The right side of the Gauss–Bonnet equation can also be expressed in terms of the *Euler characteristic,* $\chi = 2 - 2g$. As examples, consider the manifolds

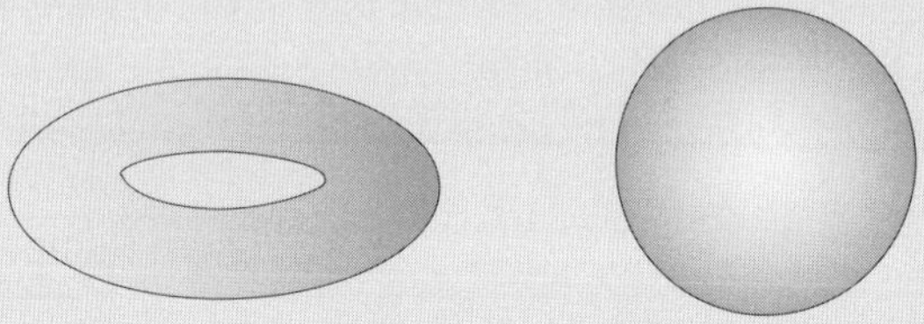

The genus of the torus is one and the genus of the sphere is zero. Inserting these values in the formula above, we find that for the sphere the Gauss–Bonnet theorem leads to the usual formula for the area (Problem 28.5), but for the torus the integral on the left side vanishes. This is because the torus has regions of positive and negative curvature K that cancel exactly in their contribution to $\int_S K dA$, which is required by the Gauss–Bonnet theorem since $2(1-g)$ vanishes for $g = 1$.

[a] The content of the Gauss–Bonnet theorem was known to Karl Friedrich Gauss but he never published it. A special case of the theorem was published by Pierre Ossian Bonnet in 1848.

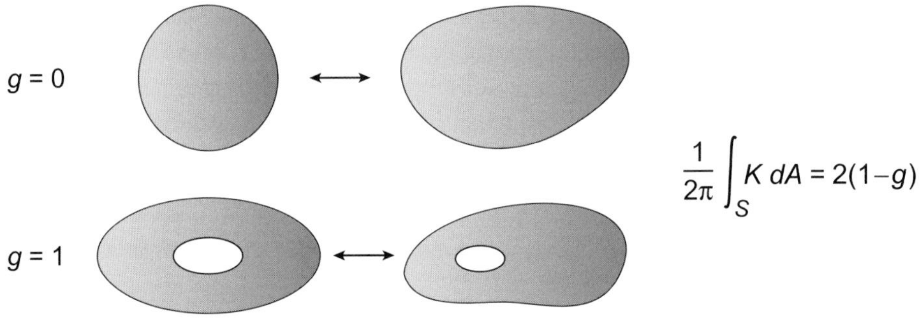

Fig. 28.6 Smooth deformations that do not change the genus *g* cannot change the Gauss–Bonnet topological invariant defined by the curvature *K* integrated over the surface in Eq. (28.16).

manifold, not with quantum states. However, in 1944, Shiing-Shen Chern generalized the Gauss–Bonnet equation to the *Gauss–Bonnet–Chern equation,*

$$\frac{1}{2\pi}\int_S K dA = C_n. \tag{28.17}$$

This resembles the 2D Gauss–Bonnet formula (28.16), with the following exceptions.

1. The Gauss–Bonnet–Chern formula is valid for a Riemannian manifold of dimension $2n$, where n is a positive integer.
2. The *Chern number* (or *Chern index*) C_n on the right side labels a *Chern class* and is an integer, but not necessarily an even one.[9]
3. The curvature K is the *Berry curvature* associated with a manifold of states defined on S.
4. Unlike for the right side of the Gauss–Bonnet equation (28.16), the Chern index C_n in Eq. (28.17) is not determined by the genus of the surface S but rather by the *topology associated with a manifold of states defined over S*.

For brevity we shall refer to the general implications of Eq. (28.17) as the *Chern theorem*.

Chern Theorem: The integral of Berry curvature over any closed manifold is quantized in terms of Chern numbers that take integer values.

A proof of the Chern theorem is given in Box 28.4. Now we outline how Eq. (28.17) may be used to construct a topological description of the integer quantum Hall effect.

Topological Quantization of Hall Conductance: Restricting to two spatial dimensions so $n = 1$ and substituting the Berry curvature Ω for K in Eq. (28.17), the Hall conductance $\sigma_{\rm H}$ contributed by a single non-degenerate electronic band may be expressed as[10]

$$\sigma_{\rm H} = \frac{e^2}{2\pi h}\int_S \Omega\, dS = C_1\frac{e^2}{h} \equiv C\frac{e^2}{h}, \tag{28.18}$$

and if multiple bands contribute $\sigma_{\rm H}$ will be a sum of terms (28.18), one for each filled band.

1. The surface S in Eq. (28.18) is a torus parameterized by variables ϕ and θ, which are periodic and so are angular variables. This arises because the Brillouin zone (Section 5.3) can be regarded as a closed 2-torus for the 2D electron gas by virtue of periodic boundary conditions in the k_x and k_y directions [the momentum vectors (k_x, k_y), $(k_x + 2\pi, k_y)$, and $(k_x, k_y + 2\pi)$ are equivalent], as illustrated in Box 28.5.
2. The curvature Ω is the Berry curvature from Eq. (27.24), evaluated for a set of quantum states defined on S.
3. The coefficient $C \equiv C_1$ is the *first Chern number* for the band, which can take any integer value.
4. The topological character of the Berry phase winding under gauge transformations and the relationship to Chern numbers are illustrated in Fig. 28.7.

   a. The 2D Brillouin zone is topologically a 2-torus (see Box 28.5), as in Fig. 28.7(a).
   b. By gauge invariance the Berry phase must match mod 2π at $\lambda = 0$ and $\lambda = 2\pi$, so over a cycle in λ the angle $\beta(\lambda)$ must evolve by $2\pi m$ for some integer m, as in Fig. 28.7(b).

[9] Chern classes arise naturally in the discipline of *algebraic topology*. The index n labels different classes of Chern numbers C_n. For 2D manifolds the relevant topological invariant is an index of the first Chern class (*first Chern number*, C_1), which takes values from the set of integers Z. For a more general and more mathematical discussion of Chern classes, see Frankel [63] or Nakahara [152].

[10] The full proof of this result is rather technically involved and will not be presented here. A pedagogical overview of the proof may be found in Tong [197].

Box 28.4 Proof of the Chern Theorem

In the Gauss–Bonnet theorem g is an integer because it counts holes for the surface, so the right side of Eq. (28.16) must be an even integer. But the Chern number in Eq. (28.17) does not count holes, so what restricts it to integer values? To answer that question, let us outline a simple derivation of the Chern theorem [200].

Spherical Manifold

Consider a 2D spherical manifold on which we draw a closed loop P,

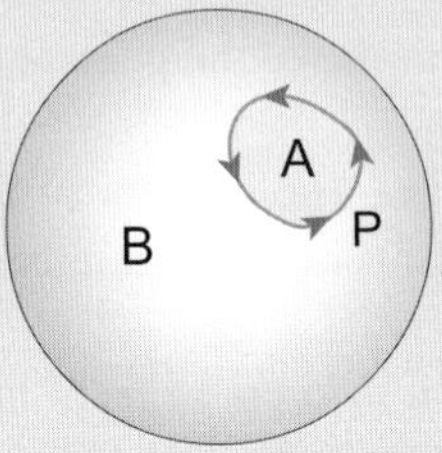

which divides the sphere into two patches, A and B, that together cover the entire sphere. Now apply Stokes' theorem (Box 27.1) twice to the loop P, once considering the enclosed area to be A and once considering it to be B (which we can do because the sphere is compact so the loop bounds both A and B). By the rules for signs and loop directions in Box 27.1, if P traverses A in the forward direction it traverses B in the reverse direction. Applying Stokes' theorem to both A and B gives

$$\int_{\mathrm{A}} \mathbf{\Omega} \cdot d\boldsymbol{S} = \gamma \qquad \int_{\mathrm{B}} \mathbf{\Omega} \cdot d\boldsymbol{S} = -\gamma,$$

where γ is the Berry phase for the loop P traversed in the forward direction. Up to a sign the Berry phase must agree for the two calculations since they are evaluated for the same loop, but *only modulo* 2π. Thus, adding the two results from above, the integral of the Berry curvature over the entire sphere is

$$\oint \mathbf{\Omega} \cdot d\boldsymbol{S} = \int_{\mathrm{A}} \mathbf{\Omega} \cdot d\boldsymbol{S} + \int_{\mathrm{B}} \mathbf{\Omega} \cdot d\boldsymbol{S} = 0 \bmod 2\pi = 2\pi C,$$

where $C = 0, 1, 2, \ldots$ is an integer. This is the Chern theorem (28.17) for K equal to the Berry curvature Ω.

Compact 2D Manifolds

A similar strategy applies for proving the Chern theorem on any compact and orientable 2D manifold.

1. Decompose into an atlas of local charts that cover the manifold (Section 24.5.1), such that smooth, continuous functions can be defined.
2. Apply Stokes' theorem to loops for each chart individually.
3. Sum the results over all charts

As for the sphere above, one side of the resulting equation will be $\oint_S \mathbf{\Omega} \cdot d\boldsymbol{S}$ and the other side will be the sum of Berry phases along the boundaries, which must cancel modulo 2π, giving the Chern theorem of Eq. (28.17).

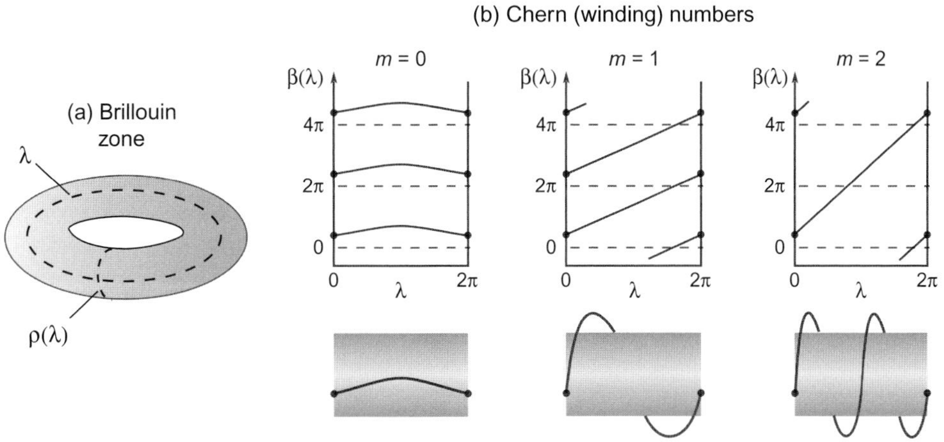

Fig. 28.7 Chern numbers C associated with the Berry phase in the 2D Brillouin zone. (a) The toroidal Brillouin zone parameterized by the periodic angular variables λ and $\rho(\lambda)$ (see Box 28.5). (b) Winding numbers m arising from allowed gauge transformations $|u_k\rangle \to \exp\,(i\beta(\lambda))\;|u_k\rangle$ that correspond to Chern numbers $C = m = (0, 1, 2)$. The points $\lambda = 0$ and $\lambda = 2\pi$ are identified, so each cylinder in (b) is actually the 2-torus in (a).

From Eq. (28.17), the integral of Berry curvature over a closed manifold is 2π times the Chern number, so the Chern number C is equal to the winding number m. Hence the Chern number may be interpreted as a winding number along ρ as λ evolves once around the torus in Fig. 28.7(a).

Thus the Berry curvature takes continuous values and is not topological, but constraints imposed by a closed loop in parameter space do admit an integer winding number arising from a topological classification of the possible gauge transformations on the loop. Equivalently, the space of complex wavefunctions may be parameterized by a sphere as in Box 29.4 and C may be viewed as the integer number of times the torus is wrapped around the sphere in the map between the Brillouin zone parameter space and the vector space of complex wavefunctions. Then, since the Chern number C is an integer, Eq. (28.18) explains the quantization of Hall conductance in units of e^2/h that is exhibited in Fig. 28.2.

Topological Character of the Chern Number: By arguments similar to those given above for the Gauss–Bonnet theorem, the Chern number is topological in that it cannot be modified in small steps. If the Hamiltonian is altered locally by a small amount we expect that the local curvature K may change by an accordingly small amount. But the right side of Eq. (28.17) is an integer, which cannot change in small increments. Hence the integer quantum Hall conductance should have plateaus for which the conductance is constant because it is protected topologically against small changes, punctuated by sudden jumps corresponding to large changes (quantum phase transitions) in the Hamiltonian, exactly as displayed in Fig. 28.2. These large changes correspond to the ground state crossing other eigenstates in the low-lying spectrum in response to changes in magnetic field or electron density. At such a *level crossing* in a quantum Hall system the curvature Ω diverges and the Chern number becomes undefined. Beyond the level crossing the old ground state is

Box 28.5

Topology of the Brillouin Zone

Because of the periodicity implied by Bloch's theorem (Section 5.4), the Brillouin zone BZ has a non-trivial topology with physical implications (see Fig. 28.7).

Topology of 1D Brillouin Zones

For a 1D crystal the band energy depends on a single momentum k. Taking the BZ to range from $k = -\pi/a$ to $+\pi/a$, where a is the lattice spacing, the momentum dependence of a single-band energy can be displayed as in Fig. (a) below.

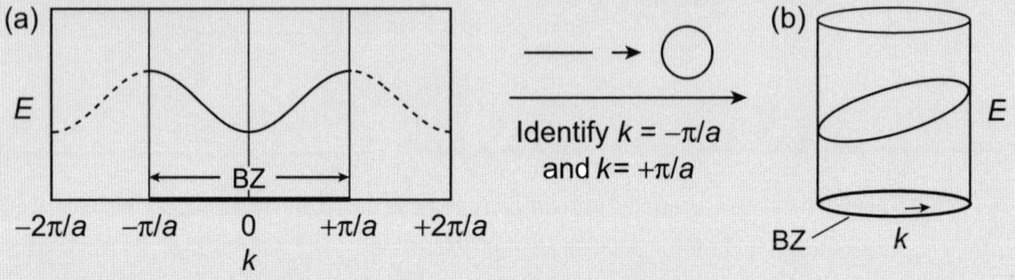

But Fig. (a) fails to indicate clearly that the right and left sides of the BZ at $k = \pm\pi/a$ are *equivalent*. This can be remedied by identifying $k = +\pi/a$ and $k = -\pi/a$, which converts the 1D Brillouin zone from an interval on the real number line to a circle and the corresponding band-energy plot then becomes cylindrical, as in Fig. (b).

Topology of 2D and 3D Brillouin Zones

The Brillouin zone for a 2D crystal is displayed in Fig. (c) below,

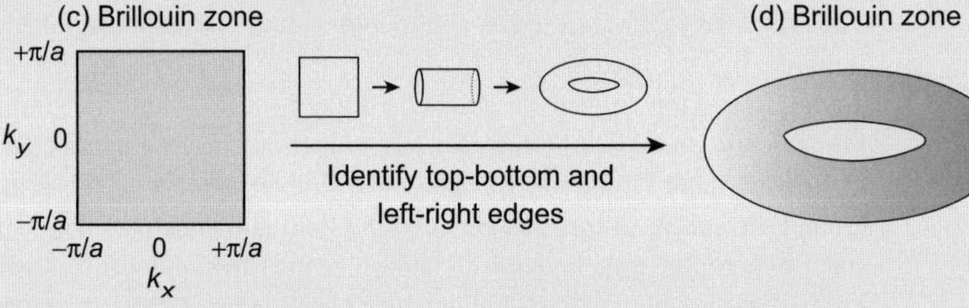

Periodicity in k_y implies identification of top and bottom edges, converting the square to a cylinder. Periodicity in k_x identifies left and right edges, which joins the ends of the cylinder, giving the 2-torus of Fig. (d). By like reasoning, the 3D Brillouin zone is a 3-torus. For the 2D Brillouin zone there are different types of closed paths:

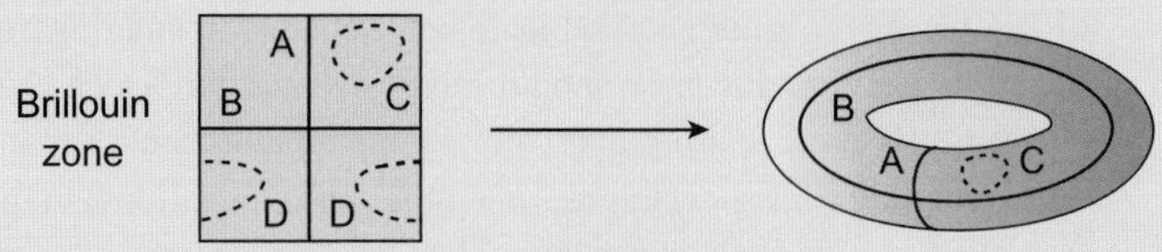

Loops such as C and D can be shrunk continuously to a point (D is closed because of the periodic boundary condition); loops such as A and B cannot. Thus A, B, and C cannot be deformed into each other and lie in distinct topological sectors.

Box 28.6 **The Hofstadter Butterfly**

A graphical example of integer quantum Hall phases known as the *Hofstadter butterfly* is shown in the following figure.

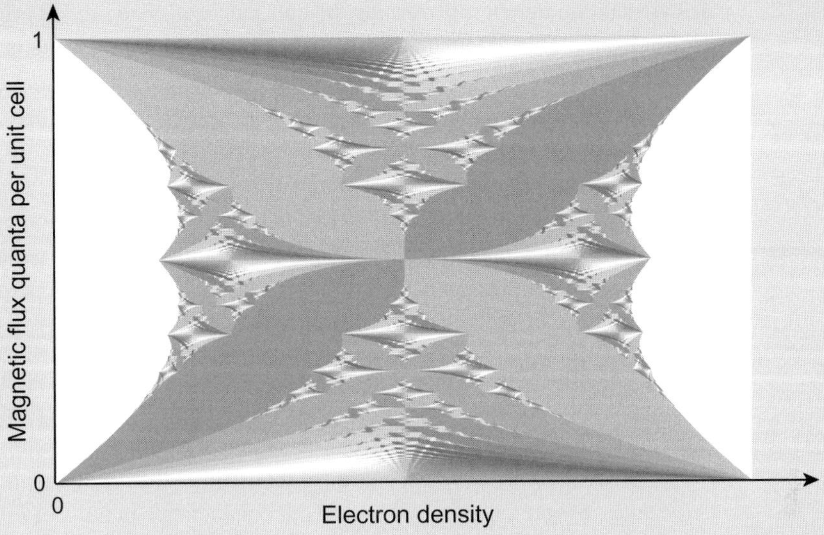

Shades of gray indicate different phases characterized by quantized Hall conductance and associated Chern numbers (they are easier to distinguish in color but are obvious even in grayscale). The pattern is fractal (self-similar). Chern numbers take any integer value, so there are infinitely many distinct phases. White areas to the left and right correspond to Chern number zero. This diagram corresponds to no disorder. Much of the fractal structure disappears in the presence of disorder.

replaced by a new ground state corresponding to the crossing level and the Chern number becomes well defined again, but with a value generally different from its value before the phase transition. Thus a state labeled by a Chern number cannot change by incremental steps into a state labeled by a different Chern number. It can only do so by a *quantum phase transition* that radically alters the topological properties of the state.

Example 28.2 A spectacular example of quantum Hall phases called the *Hofstadter butterfly* is discussed in Box 28.6. Typically, conventional thermodynamical systems exhibit a few well-defined phases. Here an *infinite number of phases occur,* labeled by different values of the first Chern number $C \equiv C_1$.

That a simple two-dimensional model can lead to the complex phase diagram displayed in Box 28.6 highlights the topological richness of the integer quantum Hall effect.

Bulk–Boundary Correspondence: In the quantum Hall effect and in topics to be discussed in Ch. 29, it becomes important to distinguish the interior of samples, which will be termed the *bulk,* from the surface (in 3D), edges (in 2D), or ends (in 1D). In the quantum Hall effect the strong magnetic field confines the motion of the electrons in the bulk, but at the same time forces them into delocalized (that is, ungapped or metallic) edge states. This relationship is called *bulk–boundary correspondence,* and will be found to be a general feature of topology in the Brillouin zone. Typically bulk–boundary correspondence implies that topological quantum numbers governing the bulk can be used to predict properties of the boundary. For example, in 2D insulators described by Chern numbers the net number of metallic states on an edge $N_+ - N_-$, where N_+ is the number of right-moving edge states and N_- is the number of left-moving edge states, is equal to the Chern number C of the bulk [14].

Thus the integer quantum Hall effect suggests that topological considerations can upend the simple insulator–metal classification from non-topological band theory outlined in Box 5.1: a 2D electron gas in a strong magnetic field can exhibit gapped insulating behavior in the bulk but ungapped metallic conduction along the edge. In general it may be shown that the surface conduction occurs through a discrete number of open edge-state channels, with the number of such channels linked to the Chern number.

Topological Quantum Matter: Let us close this section by summarizing concisely the topological roots of the integer quantum Hall effect. Topological aspects of the integer quantum Hall effect arise from a map between a 2-torus Brillouin zone and a space of complex wavefunctions. The topological invariant is the *first Chern class* of the mapping, which takes integer values equal to the Hall conductance in units of e^2/h, and is determined through Berry curvature evaluated for Brillouin zone wavefunctions. As we shall elaborate in Ch. 29, the quantum Hall effect is the harbinger of a new class of materials called *topological matter,* which is not characterized by symmetry or broken symmetry in the normal sense, and that owes its unusual stability to forms of topological protection generalizing that found here for the integer quantum Hall effect.

28.5 The Fractional Quantum Hall Effect

The fractional quantum Hall effect (FQHE) was discovered in 1982 [198], when it was found that for very pure samples at higher magnetic fields and lower temperatures the quantum Hall effect can occur for a *fractional value of* $\frac{1}{3}(e^2/h)$, which corresponds to a filling fraction $\nu = \frac{1}{3}$. Laughlin gave an explanation in terms of a new state of matter formed by electrons interacting strongly within a single Landau level [136]. Later work showed that the FQHE could occur for a series of fractional values $\frac{1}{m}(e^2/h)$, with m an even or odd integer.

28.5.1 Properties of the Fractional Quantum Hall State

Let us consider a theoretical understanding of this quantum Hall effect with a filling factor ν that is fractional.

Fractional Filling with an Energy Gap: A fractional filling factor in itself is not difficult to explain. If the lowest Landau level is filled partially with non-interacting electrons having a fractional occupation ν, the conductance will be $\nu e^2/h$.

1. In the absence of significant electron–electron interaction there will be no energy gap at the Fermi surface and adding electrons will cost essentially zero energy. This would destroy the plateau structure of the Hall conductance.
2. However, interactions will scatter the electrons into empty Landau states, leading to a finite longitudinal resistance, in contradiction to the observation that longitudinal resistance vanishes in the plateau regions.

Thus a theoretical understanding of the fractional quantum Hall effect requires that the ground state correspond to a fractionally filled Landau level, and that this state have an *energy gap* with respect to electronic excitation.

As noted in Box 28.2, a state having an energy gap resists compression, so such a state is also termed *incompressible.* Ground states with an energy gap and negligible resistance to transport of electrons were known before the discovery of the fractional quantum Hall effect. However, none fit the bill for explaining the FQHE.

Example 28.3 A superconductor has an energy gap and zero resistance to electron transport, but normal superconductors are destroyed by strong magnetic fields. Thus, it is unlikely that the FQH state could be a superconducting condensate of the usual type.

A new kind of quantum state is required involving strong interactions among the electrons in the partially filled Landau level that lift the enormous degeneracy of possible ground states, producing a unique ground state corresponding to an incompressible (gapped) electronic liquid for a Landau level filling factor that is fractional.

Contrasting Integer and Fractional Quantum Hall Effects: The integer quantum Hall effect and the fractional quantum Hall effect both result from quantization of 2D electron gases in strong magnetic fields, but their mechanisms differ fundamentally. The IQHE involves weakly interacting electrons but the FQHE is a consequence of strong Coulomb interactions between the degenerate electrons in the Landau levels. This problem of strongly correlated electrons in a magnetic field is complex, and was solved only by making an educated guess for the form of the wavefunction. The inherent difficulty lies in some features of the FQHE that make it different from most other many-body problems [123].

1. Observed filling fractions are less than one, suggesting that the FQHE corresponds to partial filling of the lowest Landau level (LLL). In the absence of interactions the ground state is then highly degenerate because there are many ways to fill the LLL partially that give the same energy. Thus, it is not at all obvious which linear combination of these states will define the physical ground state when interactions are turned on.
2. In the high magnetic fields characteristic of FQHE experiments the problem reduces approximately to just Coulomb interactions among electrons in the LLL. Thus, the fractional quantum Hall effect appears to contain no parameters since the Coulomb interaction just sets an energy scale. A theory without parameters is attractive because it leads to bold and falsifiable predictions, but with no parameters to adjust it becomes difficult to gain intuition about the physics primarily responsible for the unknown solution.

3. It is common in many-body physics to view a complicated collective state as resulting from an instability of a "normal state" that would be obtained if the interactions were removed. For example, conventional superconductivity is a consequence of the Cooper instability in a Fermi liquid parent state (see Boxes 32.1 and 32.2), and much understanding of conventional superconductors derives from this instability. In contrast, turning off the interactions for the FQHE state does not lead to a normal state that could be viewed as the parent of the fractional quantum Hall state.

Thus the FQHE solution cannot be obtained by creeping up on it in small steps. It must be constructed (or guessed) in one fell swoop.

28.5.2 Fractionally Charged Quasiparticles

If the magnetic field is tuned slightly away from a FQHE fraction, quasiparticle excitations called *anyons* appear carrying a charge $\frac{1}{3}e, \frac{1}{2}e, \ldots$ that is a rational fraction of the electronic charge e. They obey *fractional statistics,* as elaborated further in Section 29.8.1. These fractionally charged quasiparticles result from complicated many-body interactions among the electrons that are responsible for producing the fractional quantum Hall effect.

28.5.3 Nature of the Edge States

Electrons interact weakly in the IQHE and the edge states carrying the current may be described by the *Fermi liquid theory* of Box 32.1. However, the edge currents responsible for charge transport in the FQHE are essentially 1D with strong correlations (they may be viewed as moving on 1D quantum wires), and correlated 1D systems tend to behave as the *Luttinger liquids* described in Box 28.7. Thus, in the FQHE the edge excitations may act approximately as a fluid of Luttinger liquid quasiparticles carrying fractional charge. Since the edge excitations in a FQHE liquid with $\nu = 1/m$ move only in one direction, they are formally equivalent to to the right-moving (or left-moving) branch of a Luttinger liquid. Because of their chiral nature, these edge currents in the FQHE are often termed *chiral Luttinger liquids.* This difference in edge states is illustrated in the figures of Box 28.7.

28.5.4 Topology and Fractional Quantum Hall States

Chern numbers do not play a direct role in the fractional quantum Hall effect. However, there are examples of more general topological considerations which do.

1. The fractional quantum Hall effect can be understood in terms of topological order described by a *Chern–Simons field theory.*
2. As indicated in Section 28.5.2, quasiparticles called *anyons* can appear that carry fractional quantum numbers and obey fractional statistics.

A discussion of Chern–Simons theory would exceed our scope, but anyons and fractional statistics will be revisited in Section 29.8.

Box 28.7

Luttinger Liquids

Luttinger liquids can occur in one-dimensional correlated electron systems and differ fundamentally from the more usual Fermi liquids described in Box 32.1 [163]. In 1D even weak interactions can destabilize Fermi liquid behavior, producing a Luttinger liquid phase where the low-energy excitations are collective density waves moving freely to the right or left ("right movers" and "left movers," respectively), rather than electrons dressed by weak interactions. Luttinger liquids are plausible in 1D but it is unclear whether they occur in higher dimensions.

Holons and Spinons

One remarkable property of Luttinger liquids is that electrons of the non-interacting system become two distinct bosonic quasiparticles in the interacting system, with one (termed a *holon*) carrying the charge and one (termed a *spinon*) carrying the spin of the electron. This implies that electronic spin and electronic charge can propagate independently in the interacting system described by a Luttinger liquid.

Comparison with Fermi Liquids

In Fermi liquids there is a one to one correspondence between non-interacting fermions and quasifermions of the interacting system; Luttinger liquids differ fundamentally because of their 1:2 relationship between electrons and quasiparticles. Energy–momentum relations for (a) a 1D Fermi liquid and (b) a 1D chiral Luttinger liquid are illustrated in the following figure.

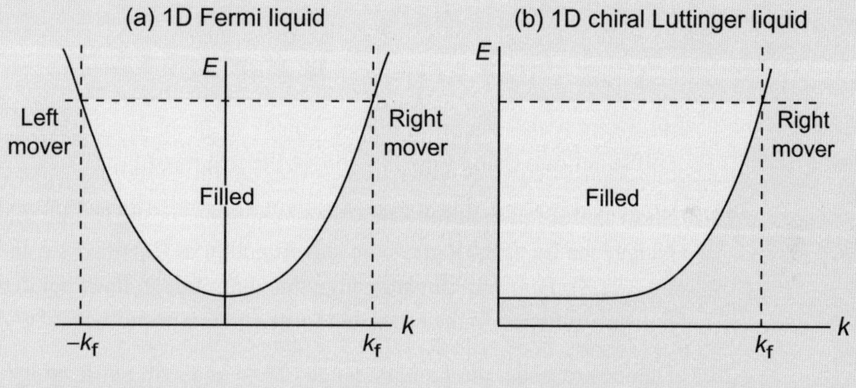

In the Luttinger state the edge current moves in only one direction (chosen as right here). Edge currents can have both right and left movers in a Fermi liquid [207].

Background and Further Reading

An accessible overview of the quantum Hall effect is given in Laughlin's 1998 Nobel Prize lecture [137], while more technical presentations may be found in Tong [197] and Phillips [163]. An overview of topological quantum numbers in non-relativistic physics is presented in Thouless [193]. Avron, Osadchy, and Seiler [17] give a pedagogical discussion of the quantum Hall effect from a topological perspective. A more advanced discussion of the quantum Hall effect and other topological properties in condensed matter physics may be found in El-Batanouny [54], and in Girvin and Yang [75]. The relationship of Berry phases, Berry curvature, and Chern classes is discussed pedagogically in Ramirez and Skinner [168], and at a more advanced level in Vanderbilt [200] and El-Batanouny [54].

Problems

28.1 (a) Starting from the classical Lorentz force equation (28.1), show that $E_y = v_x B_z/c$ in Fig. 28.1. *Hint*: Forces in the y direction should cancel. (b) The electron velocity $\boldsymbol{v}$ depends on interactions with the lattice. In the classical *Drude model* of electron transport the force is given by $\boldsymbol{F} = m(d\boldsymbol{v}/dt + \boldsymbol{v}/\tau)$, where m is the effective electron mass and τ is the mean time between electron–ion collisions. Assuming steady-state electron flow and defining the *cyclotron frequency* ω_c by Eq. (28.6), show that the classical equations of motion for electrons in the Drude model approximation are

$$v_x = -\frac{e\tau}{m} E_x - \omega_c \tau v_y \qquad v_y = -\frac{e\tau}{m} E_y + \omega_c \tau v_x \qquad v_z = -\frac{e\tau}{m} E_z .$$

(c) Use the results from parts (a) and (b) to show that in Fig. 28.1, $E_y = -(eB\tau/mc)E_x$. *Hint*: At equilibrium, $v_y = 0$. (d) The current density $\boldsymbol{j}$ in the Drude model is

$$\boldsymbol{j} = \frac{e^2 n_e}{m} \tau \boldsymbol{E} = \sigma \boldsymbol{E} \qquad \sigma \equiv \frac{e^2 n_e \tau}{m},$$

where n_e is the electron number density, σ is the conductance, and τ is the mean collision time. Using Fig. 28.1, prove the results of Eq. (28.2). ***

28.2 Show that solution of the Schrödinger equation for the Hamiltonian (28.3) in Landau gauge [see Eq. (28.4)] gives the wavefunction of Eq. (28.7), with an energy given by Eq. (28.8). *Hint*: In Landau gauge the Hamiltonian does not depend on y, so write the wavefunction in the separable form $\psi(x, y) = \phi(x)e^{iky}$. ***

28.3 Demonstrate that the Landau gauge (28.4) and symmetric gauge (28.5) both lead to the same magnetic field $\boldsymbol{B} = (B_x, B_y, B_z) = (0, 0, B)$, where $B = |\boldsymbol{B}|$.

28.4 Show that under a local gauge transformation (14.5) the vector potential $\boldsymbol{A}$ and the form of the wavefunction (28.7) are changed, but no observable is affected.

28.5 Evaluate the formula for the Gauss–Bonnet theorem in Box 28.3 for a 2-sphere and show that this leads to the usual relation for the area of a sphere. *Hint*: The local curvature for a 2-surface is the Gaussian curvature.

28.6 Evaluate the Berry flux $\int_S \Omega_{\mu\nu} dS$, with $\Omega_{\mu\nu}$ given by Eq. (27.24), for the 2D Brillouin zone parameterized in Fig. 28.7(a) by the angular variables λ and ρ. *Hint*: The second term of the integral vanishes because $A_\lambda(1) = A_\lambda(0)$, and from Eq. (27.22) the Berry phase for a cycle in the ρ direction is $\gamma^\rho = \int_0^1 A_\rho d\rho$. ***

28.7 (a) Show for a 2D Hall bar of length L and width w that $j = \sigma E$ (where j is the current density, E is the electric field, and σ is the conductivity) is equivalent to the simple form of Ohm's law, $V = IR$. *Hint*: The resistance is $R = \rho L/w$, where $\rho = 1/\sigma$ is the resistivity. (b) Show that for Fig. 28.1 the resistance components $R_{ij} = V_i/I_j$ and resistivity components $\rho_{ij} = E_i/j_j$ are related by $R_{yx} = \rho_{yx}$ and $R_{xx} = (L/w)\rho_{xx}$. (c) Verify the final forms of the ρ and σ tensors in Box 28.1.

29 Topological Matter

A burgeoning subfield of condensed matter physics and materials science concerns itself with *topological states of matter,* where a quantum many-body system may exhibit non-trivial topology within its function space that has observable consequences. The *topological matter* in which the states labeled by these quantum numbers occur typically enjoys a degree of stability ensured by *topological protection,* which follows from the difficulty of changing dynamically a quantum number that derives from topological and not dynamical quantization. In addition to its obvious intrinsic interest, such matter could be of large technical importance for applications in quantum computing because of the stability conveyed by topological protection. This chapter provides an overview of these ideas. It assumes that the reader has prior acquaintance with the description in Ch. 5 of electrons on periodic lattices, the Berry phase described in Section 27.2, the introduction to topology given in Ch. 24, and the topological interpretation of the integer quantum Hall effect expounded in Section 28.4.

29.1 Topology and the Many-Body Paradigm

Traditional theoretical many-body physics in disciplines such as condensed matter and nuclear structure physics has rested on two broad principles: (1) *adiabatic continuity* and (2) *spontaneous symmetry breaking.* Let us review these concepts.

29.1.1 Adiabatic Continuity

Adiabatic continuity is the idea that a complicated system of fundamental particles (electrons, for example) can be replaced by a simpler system of effective particles (quasi-particles), with the predicted observables being essentially the same, and that the transition from the complicated system to the simpler system can be affected in an infinitesimal series of steps; we say that one system can be *adiabatically deformed* into the other, meaning that the deformation is sufficiently slow that the system remains in its ground state at all times. At the quantum level this means that the *matrix elements*, which determine the values of observables, are equivalent in the two systems, even though the operators and wavefunctions appearing in the matrix elements (which are *not* observables) are likely quite different.

Example 29.1 Perhaps the most successful application of adiabatic continuity is the *Fermi liquid theory* of Box 32.1, in which a possibly strongly interacting fermionic system can be adiabatically deformed into a weakly interacting set of fermionic quasiparticles. Then the low-energy properties of strongly interacting fermions can be mapped to the properties of weakly interacting quasifermions with renormalized interactions.

Fermi liquid theory is highly successful where it is applicable, but many phenomena are not amenable to a description using adiabatic continuity, typically because of *phase transitions.* For example, a superconductor cannot be obtained by adiabatic deformation of weakly interacting fermions because the superconducting phase differs so radically from the normal phase. This is where the second principle of spontaneously broken symmetry enters.

29.1.2 Spontaneous Symmetry Breaking

As has been discussed in Chs. 17 and 23, a powerful approach to the description of phase transitions is to view the system before the phase transition as having a symmetry and the system after the phase transition as having lost that symmetry (or vice versa).

Example 29.2 Consider the superconducting transition again. Above the critical temperature T_c the system has finite resistance to charge transport, conserves electron number and therefore charge exactly (gauge invariance), and its bulk can be permeated by a magnetic field. As the temperature is lowered through the superconducting transition temperature T_c the system undergoes a phase transition to a fundamentally new state (the superconductor), where

1. charge flows without resistance,
2. an energy gap opens in the spectrum at the Fermi surface, and
3. magnetic fields are expelled from the bulk of the sample (Meissner effect).

In a mean-field approximation such as the BCS model introduced in Section 22.4, the superconducting state has indefinite electron number and fails to conserve charge (Box 22.4).[1] Therefore, in mean-field approximation the superconducting transition *breaks*

[1] Do not be confused by the (standard) jargon in use here. Charge is *conserved exactly* in any physical state. The statement that the superconducting state violates charge symmetry concerns an *approximation of the exact solution* such as BCS, which captures simply much of the superconducting state's character at the expense of requiring charge to be conserved only on average. However, the *exact* solution would conserve charge exactly, not just on average. From the discussion of spontaneous symmetry breaking in Ch. 17 and restoration of symmetry by projection in Ch. 22, formation of a pairing condensate implies a transition to a phase having an infinity of degenerate classical ground states [compare Fig. 17.3(b)]. A mean-field approximation for the quantum ground state like BCS selects *one* of these degenerate states as the ground state, which breaks gauge symmetry since a preferred direction has been selected in an intrinsically isotropic space; thus the (approximate) BCS state does not conserve charge. However, the correct solution can be obtained by using the generator coordinate and projection methods described in Ch. 22 to project from a superposition of the degenerate classical ground states a quantum state conserving particle number. This state has the advantage of correctly conserving charge, but the disadvantage of being more complicated than the approximate BCS state.

spontaneously the gauge symmetry: the higher-temperature state respects gauge symmetry and conserves charge but the lower-temperature superconducting state does not.

Landau introduced the concept of an *order parameter* to measure symmetry. This permits an operational characterization of the phase because the order parameter takes qualitatively different values in the two phases: typically zero in the high-temperature symmetric phase and non-zero in the low-temperature phase with broken symmetry.

Example 29.3 A suitable order parameter for a superconductor is the *pairing gap,* which measures the magnitude of the energy gap at the Fermi surface in the single-particle spectrum. The gap order parameter vanishes above the critical temperature T_c for the superconducting transition and takes a finite value below T_c.

The highly-successful Ginzburg–Landau phenomenological theory (see [54] or [75] for a description) is built on this conception of phase transitions and is used widely to describe low-temperature, broken symmetry phases such as superconductors or ferromagnets.

29.1.3 Beyond the Landau Picture

In broad overview, much of traditional many-body physics employs adiabatic continuity in the guise of Fermi liquid theory to describe systems in higher-temperature phases, and order parameters implemented through Ginzburg–Landau phenomenology to describe spontaneously broken symmetry in low-temperature phases. However, the quantum Hall effect discussed in Ch. 28 calls into question the completeness of this paradigm as a framework for understanding many-body fermionic systems.

1. The fractional quantum Hall state is strongly correlated, is not a Fermi liquid, and cannot be obtained by adiabatic variation of weakly interacting electrons in a magnetic field.
2. Different fractional quantum Hall and integer quantum Hall states have the *same symmetry* in the Landau sense, so they cannot be distinguished by order parameters associated with spontaneously broken symmetry.

We conclude that the traditional approach to many-body systems could be overlooking some important features.

Quantum Hall effects imply that some properties of quantum many-body systems are *topological in origin* and do not derive from the traditional paradigms of adiabatic continuation and Landau broken symmetry.

That is, the quantum Hall effect suggests that the properties of some materials are determined primarily by global topological invariants rather than by local symmetries. For example, in a topological material the electronic properties can be determined only by examining the complete set of states for an electronic band. A useful analogy is the

Möbius strip of Fig. 27.3(b), which has topological properties that cannot be discerned by examining any local region of the strip. They are only clear upon examining the behavior of normal vectors under parallel transport an integer number of times around the entire strip. Stated another way, because the Möbius strip retains its topological properties under smooth, continuous deformations, the "twist" that makes the topology unique cannot be localized to any point and must be thought of as a global property of the strip. Therein lies much of the practical interest in topological materials: if a property is not stored locally, it is hard for a localized perturbation such as the effect of an impurity to change that property. Thus material properties derived from topological conservation laws are unusually robust.

Fermi liquid theory and the characterization of phases through symmetry that is broken spontaneously are concepts pioneered in large part by Russian physicist Lev Landau. Therefore, it is sometimes said that understanding the quantum Hall effect – and subsequently discovered topological phenomena in other contexts to be discussed in this chapter – represents a systematic effort to transcend the Landau paradigm that has ruled our understanding of many-body systems for well over half a century, specifically by introducing a topological view of matter. Let us now consider this possibility.

29.2 Berry Phases and Brillouin Zones

Of particular interest for our discussion will be the Berry phase of Section 27.2, evaluated for a closed path in the Brillouin zone of a periodic crystal (Section 5.3). An electron executing a loop P in the Brillouin zone acquires a Berry phase (27.22) given by

$$\gamma_P \equiv \frac{1}{\hbar} \oint_P \boldsymbol{A} \cdot d\boldsymbol{k}, \tag{29.1}$$

where $\boldsymbol{A}$ is the Berry connection, as illustrated in Fig. 29.1. If the integration is along the boundaries of the Brillouin zone, the corresponding Berry phase is given by

$$\gamma_{\mathrm{BZ}} = \frac{1}{\hbar} \oint_{\mathrm{BZ}} \boldsymbol{A} \cdot d\boldsymbol{k},$$

where the integration path is indicated by the dashed lines in Fig. 29.1 and the sign is determined by whether the closed path is traversed in a clockwise or counterclockwise sense (because this reverses the sign of $d\boldsymbol{k}$). If traversing the path around the edges of the BZ clockwise gives a phase γ_{BZ}, a counterclockwise path gives $-\gamma_{\mathrm{BZ}}$. But because of the periodic boundary conditions opposite edges of the BZ are physically equivalent states (see Box 28.5), so clockwise and counterclockwise paths must cause equivalent changes of the wavefunction phase. That is possible only if γ_{BZ} is an integer multiple of 2π,

$$\gamma_{\mathrm{BZ}} = \frac{1}{\hbar} \oint_{\mathrm{BZ}} \boldsymbol{A} \cdot d\boldsymbol{k} = 0, 2\pi, 4\pi, \ldots \tag{29.2}$$

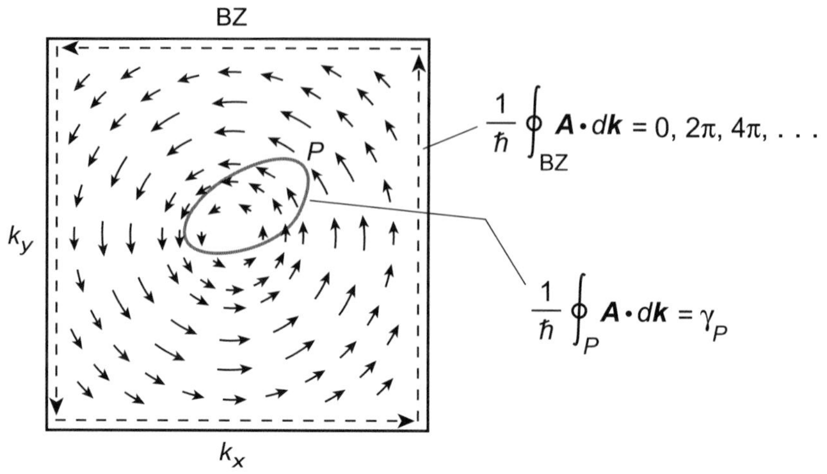

Fig. 29.1 Berry connection vector field $\boldsymbol{A}$ [see Eq. (27.16)] indicated by arrows in the 2D Brillouin zone (BZ) [168]. If an electron is accelerated and decelerated such that it follows a closed path labeled P in the momentum space, its wavefunction acquires a Berry phase given by Eq. (29.1), with a sign indicating whether the path P is clockwise or counterclockwise. On the other hand, for a closed path that runs along the BZ boundaries (the dashed lines), the corresponding phase is required to be a multiple of 2π, as indicated in Eq. (29.2).

It is often convenient to use Stokes' theorem (Box 27.1) to transform this condition on the Berry phase by converting the contour integral along the BZ boundary into a surface integral evaluated over the full Brillouin zone [see Eq. (27.26)]. This gives

$$\frac{1}{2\pi\hbar^2}\int_{BZ} \Omega\, d^2k = C_1, \tag{29.3}$$

where $\boldsymbol{\Omega} = \boldsymbol{\nabla} \times \boldsymbol{A}$ is the Berry curvature defined in Eq. (27.25) and C_1 is a (first) Chern number that takes integer values. The Chern number in a band insulator can be computed for each band so it will generally carry a band index n. It is topological in that if the Hamiltonian is deformed adiabatically (for example, by changing the interatomic spacing slowly and smoothly) without allowing the gap between band n and any other band to close, the Berry curvature varies continuously but the Chern number (integral of the Berry curvature over the BZ) is an integer and cannot change continuously. The only way that the Chern number can change is if the gap of the nth band closes and reopens, implying a quantum phase transition to a new phase characterized by a new Chern number.

29.3 Topological States and Symmetry

Topological states of matter are distinguished by topological quantum numbers and not by traditional symmetry considerations. However, in addressing whether topological states are possible in a given system, the discrete symmetries of time reversal and spatial inversion can be crucial. In particular, an important question is that of the symmetry conditions that permit non-vanishing Berry curvature and Chern numbers. For a band labeled by an index n, the Berry curvature $\boldsymbol{\Omega}_n(\boldsymbol{k})$ in the Brillouin zone has the following behavior [54, 200].

1. Symmetry under spatial inversion implies that $\boldsymbol{\Omega}_n(\boldsymbol{k}) = \boldsymbol{\Omega}_n(-\boldsymbol{k})$.
2. Symmetry under time reversal implies that $\boldsymbol{\Omega}_n(\boldsymbol{k}) = -\boldsymbol{\Omega}_n(-\boldsymbol{k})$, meaning that integrals of the Berry curvature over the BZ vanish.
3. Symmetry under both spatial inversion and time reversal implies that $\boldsymbol{\Omega}_n(\boldsymbol{k}) = 0$.
4. Point group symmetries may impose additional constraints that we will not discuss here.

The second and third items in this list suggest the following.

> Topological conservation laws associated with non-zero Chern numbers require that *time-reversal symmetry must be broken.*

In the integer quantum Hall effect time-reversal symmetry is broken by the applied magnetic field,[2] leading to topological quantization of Hall conductance in terms of Chern numbers. But is it possible to engineer materials that mimic the effect of an external magnetic field because of their internal structure, even in the absence of an applied magnetic field, and/or that acquire stability through topological invariants other than a Chern number? In the remainder of this chapter we provide an introduction to possibilities.

29.4 Topological Insulators

Topological aspects of the integer quantum Hall effect described in Section 28.3 are enabled by an applied magnetic field that breaks time-reversal symmetry. However, it has been shown that the crucial aspect of the IQHE leading to properties like insulating bulk and metallic edge states is not the magnetic field itself but rather the non-trivial topology of occupied bands that it induces, thus suggesting the possibility of topological matter that requires no external magnetic field and is symmetric under time reversal. Most materials are topologically trivial but some exhibit striking topological properties, even in the absence of applied fields. Estimates based on systematic analysis of symmetries and band structure suggest that at least 10%, perhaps considerably more, of known materials might exhibit intrinsic topological properties under appropriate conditions. For example, topological states in 3D, 2D, and 1D materials have been found called variously *topological insulators* or *quantum spin Hall states* [65, 102, 112, 127, 128] that are distinct topologically from all other known states of matter, including the quantum Hall states of Ch. 28 [165]. Topological insulators have a bandgap and behave like a normal insulator in their interiors ("in the bulk"), but have conducting states on their edges or surfaces that

[2] In a quantum Hall experiment the magnetic field forces electrons into trajectories with a handedness set by the direction of the magnetic field. Playing a movie of these trajectories backward without reversing the sign of the applied magnetic field exhibits motion inconsistent with the Lorentz force law (16.32).

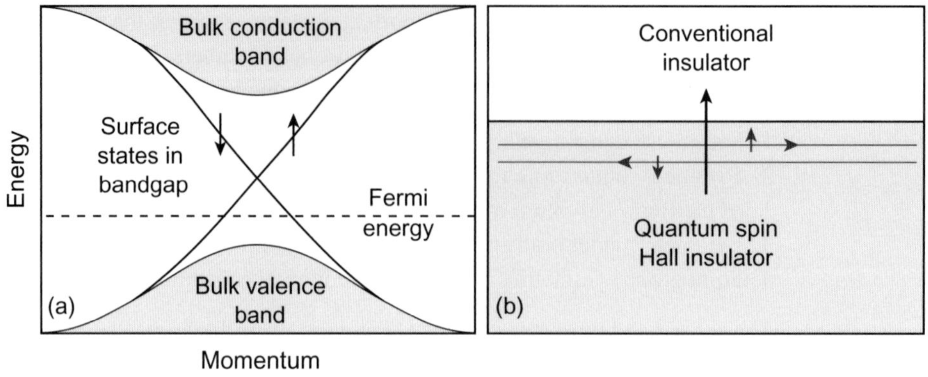

Fig. 29.2 (a) Schematic topological insulator. The Fermi energy lies within the bulk bandgap, which is crossed by robust surface states that are topologically protected (arrows indicate spin polarization). (b) Interface between a topological insulator and conventional insulator, indicating that metallic surface states as in (a) are expected generally at such a boundary.

are topologically protected from perturbations like impurity scattering.[3] These conducting surface states occur in the bandgap separating valence and conducting bands, as illustrated in Fig. 29.2(a).

Such surface states for bulk insulators of topological origin are expected on quite general grounds. To see this, let us place a topological insulator adjacent to a conventional insulator (which could be empty space), as in Fig. 29.2(b). Now imagine interpolating on a path indicated by the vertical arrow from the topological insulator into the conventional insulator. At some point on this path the energy gap must vanish because the phase transition between topological and conventional insulators can occur only if the energy gap goes to zero. Thus, there must be low-energy electronic states in the surface region where the gap closes.

29.4.1 The Quantum Spin Hall Effect

In the quantum Hall systems described in Ch. 28, charge is transported by edge currents because the magnetic field breaks time-reversal symmetry. However, topological edge currents can exist in a material that preserves time-reversal symmetry with no external magnetic field if electron spin and momentum are strongly coupled. In the simplest example this allows the two electron spin states to have *non-zero but opposite Chern numbers,* as illustrated in the quantum spin Hall (QSH) system of Fig. 29.3. This system has no external magnetic field and is invariant under time reversal, which changes $\boldsymbol{k}$ to $-\boldsymbol{k}$ and spin up to spin down. The surface electrons have their spin locked

[3] For convenience we will sometimes call these "edge states" for both 2D and 3D materials. This topological protection is of large intrinsic interest because a topological material could allow surface spin currents with no dissipation. This in turn has technical implications because of potential applications in *spintronics,* which uses electron spin to construct devices in a way analogous to the use of electron charge in conventional solid-state electronics. Normal insulators can have conducting surface states too, but they typically are not protected by symmetries or topology and are destroyed easily by impurities.

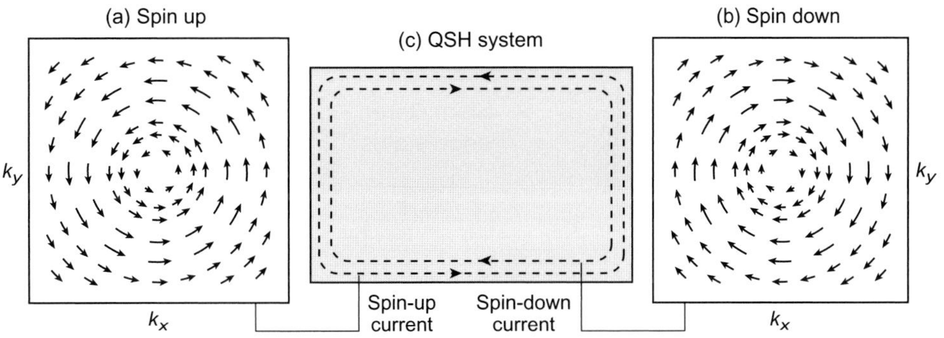

Fig. 29.3 Berry connection (arrows indicate the vector field) in the Brillouin zone for a quantum spin Hall system [168]. (a) Connection winds counterclockwise for spin-up electrons, giving Chern number $C = +1$. (b) Connection winds clockwise for spin-down electrons, giving Chern number $C = -1$. (c) Quantum spin Hall topological insulator, which has $C = 0$ and is time-reversal invariant.

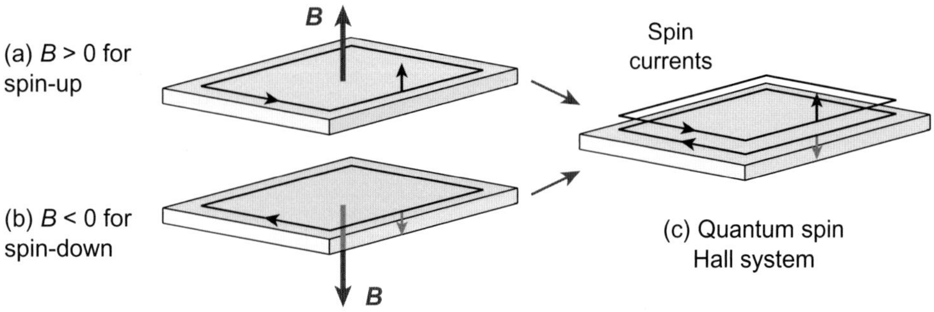

Fig. 29.4 Schematic illustration of the quantum spin Hall system in Fig. 29.3 as a superposition of two quantum Hall systems with canceling magnetic fields [54, 150].

perpendicular to their momentum and are said to be *spin-filtered,* because a momentum in a particular direction is associated with only one spin state. They also are termed *helical,* because of the similarity to states of only one helicity (projection of spin on momentum). The only other nearby states have opposite spin, so backscattering is strongly suppressed and surface conduction is highly metallic.

Conceptually, the QSH system of Fig. 29.3 may be thought of as a superposition of two quantum Hall systems with oppositely directed magnetic fields, and the charge current replaced with a spin current, as illustrated in Fig. 29.4. In this example a left-moving spin-up electron is converted into a right-moving spin-down electron under reversal of the direction of time, which implies that time-reversal symmetry is maintained overall if bands for spin-up and spin-down electrons have opposite Chern numbers. The edge currents accomplish net transfer of spin since spin-up and spin-down currents flow in opposite directions. Thus the quantum spin Hall effect has much in common with the quantum Hall effect, but with no external magnetic field, and for a current of spins rather

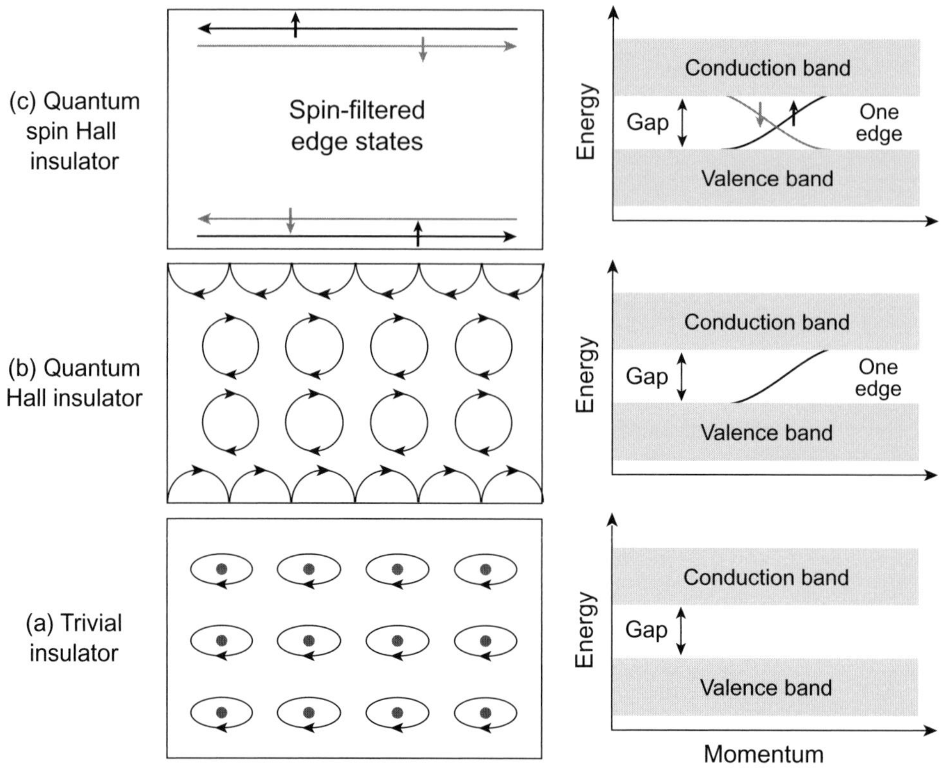

Fig. 29.5 Bulk and boundary states for 2D insulators. (a) Trivial (conventional) insulator, with no robust edge currents. (b) Quantum Hall system (with perpendicular magnetic field). The edge states support a chiral charge current; see Figs. 28.4 and 28.5. (c) Quantum spin Hall system with spin-filtered (helical) edge states and a spin edge current.

than charge. Figure 29.5 compares the bulk and boundary states for several topologically distinct insulators.

Physically, quantum spin Hall states can be realized through spin–orbit coupling in atomic orbitals.[4] Recall that for an electron with orbital angular momentum $\boldsymbol{\ell}$ and intrinsic spin $\boldsymbol{s}$ the spin–orbit interaction is proportional to $\boldsymbol{\ell} \cdot \boldsymbol{s}$, which changes sign for opposite spin projections, as would an interaction with a magnetic field. Thus spin–orbit coupling produces an effective internal magnetic field acting on electrons, which can in turn lead to a finite Berry curvature and non-zero Chern number for each spin.

29.4.2 The $\mathbf{Z}_2$ Topological Index

For quantum Hall systems the Chern topological invariant $C \equiv C_1$ is responsible for the integer quantum Hall effect. It can be non-zero because the applied magnetic field breaks

[4] Spin–orbit coupling is a consequence of Lorentz invariance (Chs. 13 and 14). It is a Zeeman-like interaction caused by two special relativity effects: (1) the magnetic moment associated with intrinsic spin of a Dirac electron, and (2) the apparent magnetic field seen by an electron in motion relative to a nuclear electrostatic field. Hence spin–orbit coupling tends to be largest in heavy atoms where electron velocities are highest.

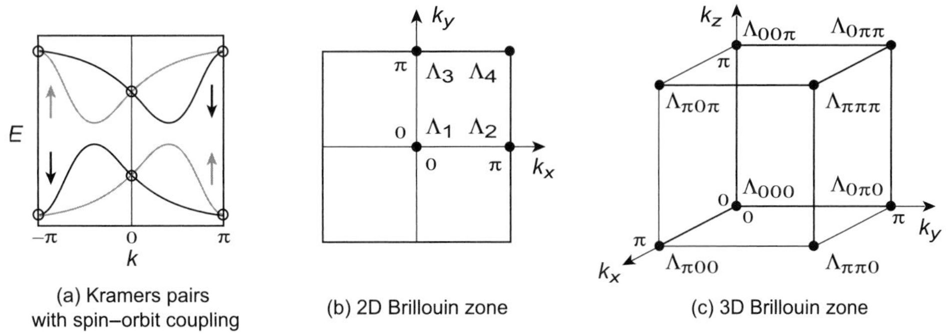

Fig. 29.6 Time-reversal invariant momenta (TRIM) and Kramers degeneracy [54] (see Box 29.1). (a) Dispersion for two Kramers pairs of bands with spin–orbit interaction. TRIM points are at 0 and π (which is equivalent to $-\pi$ by time reversal). Arrows indicate spin projection in the absence of spin–orbit interaction. At TRIM, bands are Kramers degenerate (open circles); at other momenta spin–orbit coupling breaks the degeneracy. (b) TRIM (dots) in the 2D square Brillouin zone. (c) TRIM (dots) in the 3D Brillouin zone.

time-reversal symmetry. For quantum *spin* Hall systems there is no magnetic field, the spin-up and spin-down Chern numbers effectively cancel (see Figs. 29.3 and 29.4), and the net Chern number is zero because of the time-reversal invariance; yet the QSH ground state is distinct topologically from a normal insulator. Kane and Mele proposed a classification that distinguishes a non-topological normal insulator from a topological QSH ground state [127]. The corresponding topological index ν may be viewed as an element of Z_2, the group of integers under addition mod 2, and is a measure of whether there are phase-twisted boundary conditions [analogous mathematically to the two choices in Fig. 26.4(b–d) for joining the ends of a flat strip into a cylinder or a Möbius band].

2D Topological Insulators: The primary distinction between the insulating phases labeled by the Z_2 index lies in the behavior of the edge states.

1. In the QSH phase, for each energy in the bulk gap there is a corresponding single time-reversed pair of eigenstates on each edge (*Kramers doublet*; see the discussion of time-reversal invariance and TRIM points in Box 29.1 and Fig. 29.6). These states are stable against small perturbations because time-reversal symmetry prevents mixing of Kramers doublets and single-particle elastic backscattering is forbidden.
2. The edge currents are not chiral, as they are in the integer quantum Hall effect,[5] but they are still insensitive to disorder caused by impurity scattering because their directionality is correlated with their spin on each edge, which suppresses backscattering.
3. For the simple insulator there is an even number of Kramers pairs at each energy; this permits elastic backscattering, so such edge states can be localized easily.

[5] For *Chern insulators* (2D insulators with non-zero Chern number), chiral edge currents propagate without reflection because there are no counter-propagating states to scatter into. However, time-reversing an edge state would produce a counter-propagating state, so insulators symmetric under time reversal cannot be Chern insulators.

Box 29.1 Time-Reversal Symmetry and Kramers Degeneracy

Ordinary symmetries are implemented by unitary operators. For example, if Π inverts the spatial coordinates its action on a wavefunction is $\Pi\, \psi(\boldsymbol{r}) = \psi(-\boldsymbol{r})$. Then, if the Hamiltonian H commutes with Π, simultaneous eigenvectors of H and Π can be constructed, labeled by a *parity* ± 1 that is the eigenvalue of Π.

Time-Reversal Symmetry

In contrast to ordinary symmetries, time reversal is implemented by an *antiunitary operator* Θ that must satisfy

$$\Theta(c_1\,|\alpha\rangle + c_2|\beta\rangle) = c_1^*\Theta\,|\alpha\rangle + c_2^*\Theta|\beta\rangle \qquad \langle\tilde{\beta}|\tilde{\alpha}\rangle = \langle\beta|\alpha\rangle^*,$$

where c_1 and c_2 are scalars and the tilde indicates a time-reversed state, and must satisfy $\Theta^2 = -1$ when operating on fermions. It is common to choose for fermions $\Theta = i\sigma_2\mathcal{K}$, where σ_2 is the second Pauli matrix and $\mathcal{K}$ complex conjugates constants standing to its right: $\mathcal{K}\, c\,|\alpha\rangle = c^*\,|\alpha\rangle$. A Hamiltonian is said to have time-reversal symmetry if it commutes with Θ, but the implications differ from those for commutation with a unitary operator. For example, states cannot be classified simply under time reversal as was done for spatial inversion above.

Kramers Degeneracy

Time-reversal invariance has a fundamental consequence for fermionic systems.

> ***Kramers' Theorem:*** For spin-$\frac{1}{2}$ systems that are symmetric under time-reversal, all eigenstates are (at least) doubly degenerate.

Corresponding doubly degenerate pairs are called *Kramers doublets.* If spin–orbit coupling is small, members of a Kramers doublet are typically spin-up and spin-down partners, but more generally they are time-reversed pairs.

Time Reversal and Bloch Wavefunctions

In a crystal, time-reversal symmetry implies that for Bloch wavefunctions $\left|\psi_{n\boldsymbol{k}}\right\rangle$, the cell-periodic wavefunctions $|u_{n\boldsymbol{k}}\rangle$ of Eq. (5.9), and the $\boldsymbol{k}$-dependent Hamiltonian $H_{\boldsymbol{k}}$,

$$\Theta\left|\psi_{n\boldsymbol{k}}\right\rangle = e^{i\phi}\left|\psi_{n,-\boldsymbol{k}}\right\rangle \qquad \Theta\,|u_{n\boldsymbol{k}}\rangle = e^{i\phi}\,|u_{n,-\boldsymbol{k}}\rangle \qquad \Theta H_{\boldsymbol{k}}\Theta^{-1} = H_{-\boldsymbol{k}},$$

where n is a band index, ϕ is a phase angle that may depend on n and $\boldsymbol{k}$, and $H_{\boldsymbol{k}}\,|u_{n\boldsymbol{k}}\rangle = E_{n\boldsymbol{k}}\,|u_{n\boldsymbol{k}}\rangle$. Thus the Bloch wavefunction at momentum $\boldsymbol{k}$ is degenerate with a time-reversed partner at $-\boldsymbol{k}$ for each band.

Time-Reversal Invariant Momenta (TRIM)

If $\boldsymbol{k} = -\boldsymbol{k}$ up to a reciprocal lattice vector, the labels $\boldsymbol{k}$ and $-\boldsymbol{k}$ are equivalent, $\Theta H_{\boldsymbol{k}}\Theta^{-1} = H_{\boldsymbol{k}}$ so that $H_{\boldsymbol{k}}$ commutes with time reversal, and all states are doubly (Kramers) degenerate at that $\boldsymbol{k}$. These special points are called *time-reversal invariant momenta (TRIM)*; they occur when k_i is 0 or π, so there are two TRIM points in 1D, four in 2D, and eight in 3D; see Fig. 29.6.

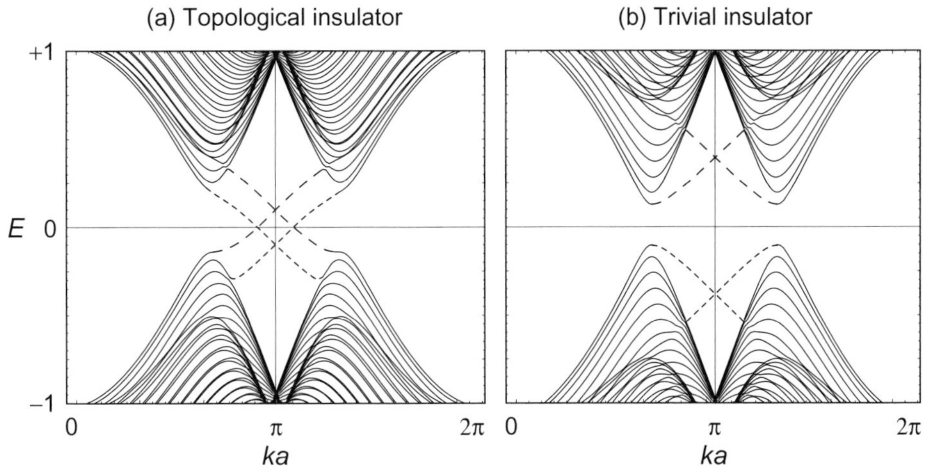

Fig. 29.7 Energy of edge-state pairs (dashed curves) as a function of momentum k for a 1D zigzag strip of graphene in the Kane–Mele model [127]. (a) Topological insulator. (b) Trivial insulator. Pairs of edge states cross at the TRIM point $ka = \pi$ because of time-reversal invariance (Kramers' theorem), but the Kramers pairing differs fundamentally between (a) and (b), as explained more schematically in Fig. 29.8.

The nature of edge-state pairs is illustrated in Fig. 29.7, where we can see the following.

1. The edge-state configurations in Figs. 29.7(a) and 29.7(b) cannot be deformed continuously into each other without gap closures, so they are topologically distinct.
2. Edge states in the bulk bandgap of Fig. 29.7(a) cannot be removed by smoothly deforming the bands, so this is a bulk insulator with a conducting surface that is protected.
3. If any edge states in Fig. 29.7(b) should lie in the bandgap they can be moved out of the bandgap by smooth deformations, so any conducting surface states are not protected.

These topological alternatives are illustrated more schematically in Fig. 29.8. Time-reversal invariance requires that the bands must be doubly degenerate at the TRIM points Λ_a and Λ_b (Kramers' theorem; see Box 29.1 and Fig. 29.6).

This Kramers degeneracy condition can be satisfied in two distinct ways: (1) in Fig. 29.8(a) the Kramers pairs "switch partners" between Λ_a and Λ_b, leading to a characteristic zigzag pattern; (2) in Fig. 29.8(b) the partners do not switch.

The patterns in Figs. 29.8(a) and 29.8(b) have different values of Z_2.

1. In the "switched" pattern of Fig. 29.8(a), a horizontal line intersects an edge state an odd number of times and for the Z_2 invariant ν we have ν = (odd integer) mod 2 = 1. In this case there will always be an edge state at the Fermi energy and the edge is metallic.[6]

[6] This Z_2 classification is stable against continuous deformations. For example, if the curve in Fig. 29.8(a) passing through the dashed horizontal line is distorted continuously into an S shape by a smooth variation of the Hamiltonian it could cross the dashed line three times, but that still is an odd number and $\nu = 3 \bmod 2 = 1$.

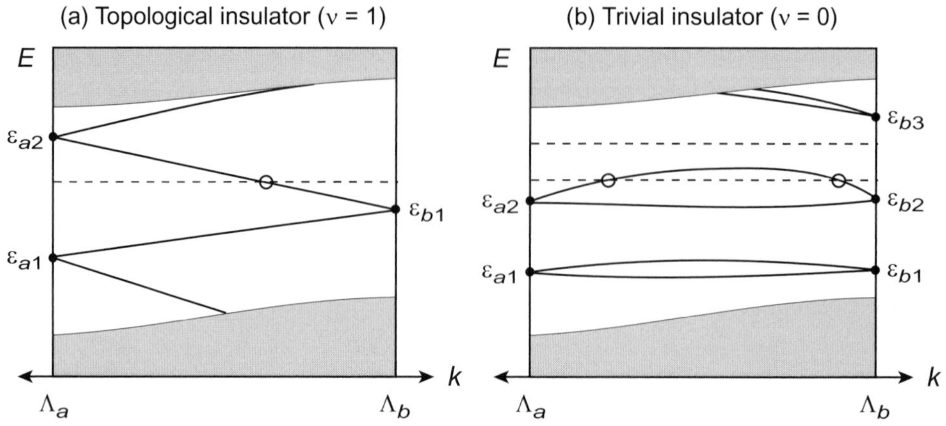

Fig. 29.8 Two ways of connecting TRIM points Λ_a and Λ_b for edge states in 2D or 3D time-reversal invariant systems so that the Kramers degeneracy condition (Box 29.1) is satisfied at the TRIM points [64]. The bulk valence and conduction bands are shaded and the bulk gap is unshaded. The curves are discrete surface or edge states localized near a surface. (a) A *topological insulator,* where a dashed horizontal line (for example, a Fermi energy) drawn in the bulk gap intersects an edge state an odd number of times (one in this case, indicated by the open circle). (b) A *trivial (conventional) insulator,* where a dashed horizontal line in the bulk gap intersects an edge state an even number of times (zero for the upper dashed line and two for the lower dashed line in this case).

2. In the "unswitched" pattern of Fig. 29.8(b), a horizontal line in the bulk gap intersects an edge state an even number of times and $\nu =$ (even integer) mod 2 $= 0$. In this case the Fermi energy might intersect edge states so that the surface is metallic [lower dashed line in Fig. 29.8(b)], but small changes in the Hamiltonian could place the Fermi energy in an insulating gap [upper dashed line in Fig. 29.8(b)]. Thus, if there are metallic surface states in the trivial insulator they are not robust.

The Kramers degeneracy pattern of Fig. 29.8(a) cannot be distorted continuously into the Kramers degeneracy pattern of Fig. 29.8(b) without gap closures, because at each infinitesimal step the Kramers pairs are uniquely pinned at both Λ_a and Λ_b. Hence the conducting surface states of the topological insulator are protected and robust.

Z_2 Classification for 2D: We conclude that insulator states differ by the number of edge-state pairs at a single edge, modulo 2, which sorts all 2D time-reversal invariant insulators into classes labeled by the Z_2 invariant ν.

1. The *trivial class* ($\nu = 0$), for which there is an even number of edge-state pairs.
2. The *topological class* ($\nu = 1$), for which there is an odd number of edge-state pairs.

These classes differ fundamentally in their physical attributes. For example, disorder can suppress edge-state conduction for $\nu = 0$ insulators but conducting edge states for

Generally *any* smoothly distorted curve in the gap of Fig. 29.8(a) that satisfies the Kramers theorem at the endpoints Λ_a and Λ_b and does not close the gap will cross a horizontal line in the gap an odd number of times. Thus the $\nu =$ (odd integer) mod 2 $= 1$ classification is topological and is unaffected by smooth deformations.

$\nu = 1$ insulators are protected topologically. Systems within each class are *related by a homeomorphism*: one member can be distorted into another by a continuous deformation that does not close the gap. Conversely, $\nu = 0$ and $\nu = 1$ states cannot be distorted into each other without a gap closure and a corresponding topological phase transition.

Example 29.4 Topological insulator states often are a consequence of band inversion. As an example, consider the two-band mixing Hamiltonian [54]

$$H = \begin{pmatrix} d_3 & d_1 - id_2 \\ d_1 + id_2 & -d_3 \end{pmatrix} \qquad \begin{cases} d_1 \equiv a_0 \sin k_x \\ d_2 \equiv a_0 \sin k_y \\ d_3 \equiv m + 2 - \cos k_x - \cos k_y \end{cases} \tag{29.4}$$

where a_0 and m are parameters. As shown in Problem 29.6, diagonalization of this Hamiltonian gives for the eigenvalues $E_\pm$ and wavefunctions $|\chi_\pm\rangle$

$$\begin{gathered} E_\pm = \pm\sqrt{(m + 2 - \cos k_x - \cos k_y)^2 + a_0^2(\sin^2 k_x + \sin^2 k_y)}, \\ |X_+\rangle = A\,|\chi_1\rangle + B\,|\chi_2\rangle \qquad |X_-\rangle = C\,|\chi_1\rangle + D\,|\chi_2\rangle, \end{gathered} \tag{29.5}$$

where $|\chi_1\rangle$ and $|\chi_2\rangle$ are wavefunctions for the unperturbed bands. Eigenvalues of this Hamiltonian are plotted in Figs. 29.9 and 29.10 for several values of m, and corresponding squared wavefunction amplitudes A^2, B^2, C^2, and D^2 are displayed in Fig. 29.10.

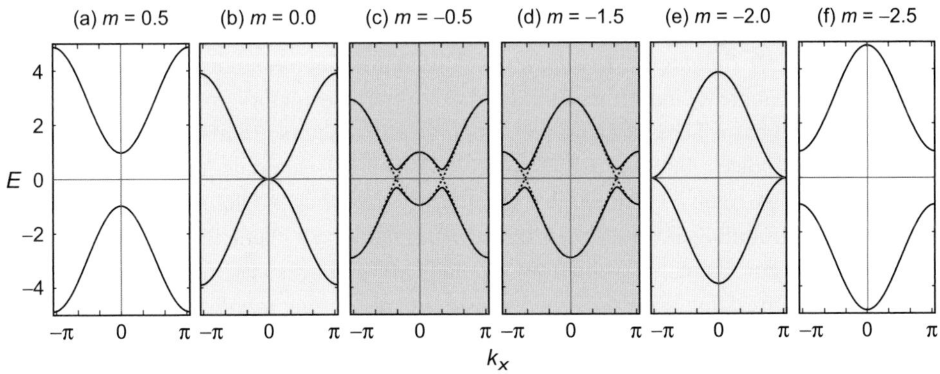

Fig. 29.9 Band-mixing model of Example 29.4 for different values of m assuming $a_0 = 0.2$ and $k_y = 0$. (a) Trivial insulator phase. (b) Phase transition at gap closure. (c)–(d) Topological insulator phase with "twist" caused by band inversion. (e) Phase transition at gap closure. (f) Another trivial insulator phase. Dotted curves ignore the off-diagonal terms causing band mixing in Eq. (29.4). A white background indicates a trivial insulator phase, a light gray background indicates a phase transition (gap closure), and a dark gray background indicates a topological phase.

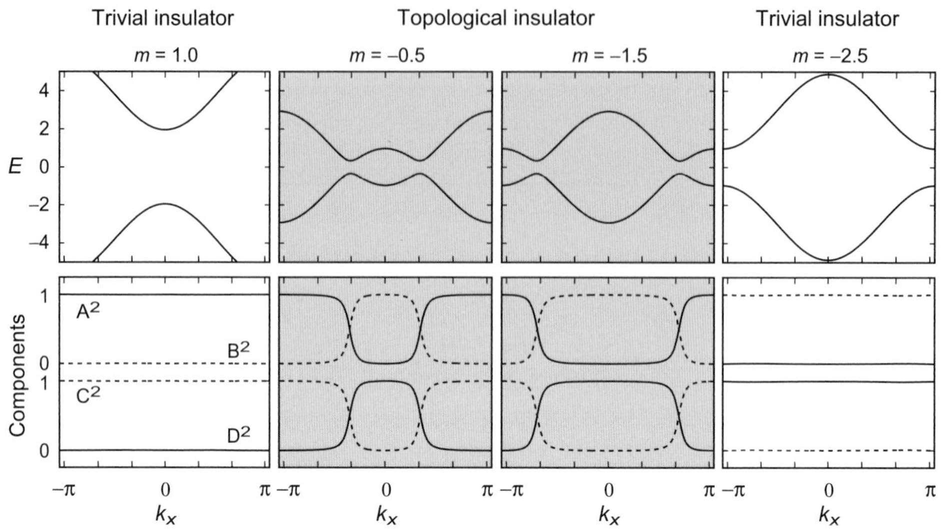

Fig. 29.10 Band-mixing model of Example 29.4 with $a_0 = 0.2$ and $k_y = 0$. Top panels: eigenvalues. Bottom panels: square of coefficients in wavefunctions (29.5). Trivial phases are indicated by a white background and the topological phase by a gray background. The band inversions shown for $m = -0.5$ and $m = -1.5$ are associated with the topological phase.

In Figs. 29.9 and 29.10 we see that the insulator phase diagram is controlled by the parameter m, with onset of the topological phase when m becomes sufficiently negative to invert the original order of the bands for a range of momenta. In typical topological insulators such a band inversion can occur because of strong spin–orbit interaction.

3D Topological Insulators: Two-dimensional topological insulators are characterized by a single Z_2 index that distinguishes ordinary insulators from a topological quantum spin Hall phase. Three-dimensional topological insulators are also characterized by Z_2 topological indices, but four are required [64]. For 3D materials, three Z_2 invariants ν_1, ν_2, and ν_3 are relatively susceptible to impurity scattering and define weak topological insulators; the fourth Z_2 invariant ν_0 is more robust and $\nu_0 = 1$ characterizes strong topological insulators. For 3D materials the four topological Z_2 indices $(\nu_0, \nu_1, \nu_2, \nu_3)$ distinguish among *trivial insulator, weak topological insulator,* and *strong topological insulator* phases.

29.5 Weyl Semimetals

Weyl semimetals (semimetals are defined in Box 5.1) are 3D materials but they may exhibit topological states that can be understood in terms of 2D Chern topological invariants, as illustrated in Figs. 29.11 and 29.12.

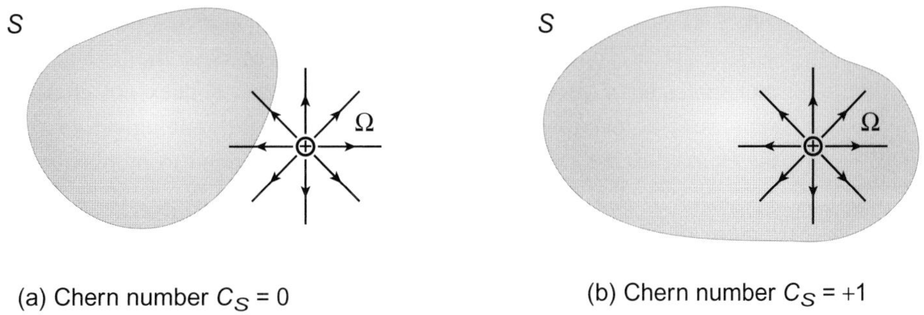

Fig. 29.11 Enclosure of a Berry monopole source ⊕ by a 2D surface S in a 3D Brillouin zone. The Berry curvature $\mathbf{\Omega}$ may be displayed as a vector field flowing from or into the monopole source, which is called a *Weyl point* or a *Weyl node.* (a) The surface S does not enclose the Berry monopole so the Chern number is $C_S = 0$. (b) The surface has been distorted to enclose the monopole, causing the Chern number to jump discontinuously to $C_S = +1$.

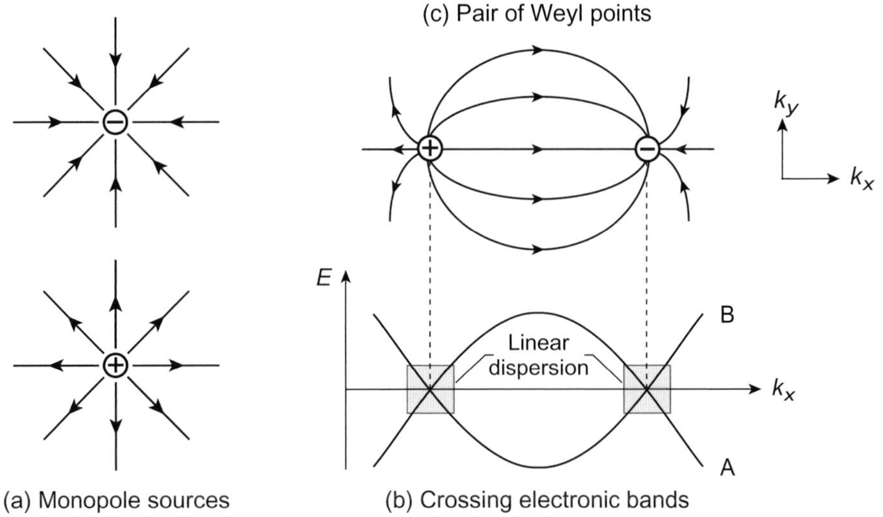

Fig. 29.12 Berry curvature $\mathbf{\Omega}$ plotted as vector fields for a Weyl semimetal [168]. (a) A Weyl semimetal derives from monopole Berry flux sources located in momentum space at *Weyl points.* A 2D surface that encloses a Weyl point acquires a Chern number $C_S = \pm 1$, as in Fig. 29.11. (b) At each Weyl point (in momentum space), two electronic bands A and B having different orbital angular momentum content cross. Near the Weyl points the dispersion E versus k_x is linear. (c) In a Weyl semimetal, Weyl points come in pairs with opposite topological charge located at the momenta of the band crossings in (b).

29.5.1 A Topological Conservation Law

By reasoning similar to that for the Chern theorem in Section 28.4.2, there is a Chern number C_S associated with the 2D surface S in Fig. 29.11 that is quantized according to

$$\frac{1}{2\pi\hbar^2}\int_S \mathbf{\Omega}\cdot d\mathbf{A} = C_S, \tag{29.6}$$

where $\boldsymbol{\Omega}$ is the Berry curvature and the integer C_S describes a net flux through the surface [168]. Thus $C_S = +1$ in Fig. 29.11(b) because the surface S encloses a Berry monopole, but $C_S = 0$ in Fig. 29.11(a) because S does not enclose any Berry flux sources. The locations in momentum space of the topological charge associated with the Berry flux are called *Weyl points*. They correspond to monopoles of Berry flux that are topological analogs of electrical charge. They always occur in pairs of opposite *topological charge* C_S. The surface integral over S appearing in Eq. (29.6) then mirrors *Gauss' law* in electromagnetism.[7] By familiar arguments, the integer C_S cannot change in small increments. The only way to modify C_S is through a discontinuous jump caused by distorting S to include or exclude a source of Berry flux, as in Fig. 29.11(b). Generally, C_S for a surface S sums algebraically the topological charge of the Weyl points enclosed by S, which is invariant under smooth distortions of S that do not change the number of enclosed Weyl points.

29.5.2 Realization of a Weyl Semimetal

Are there actual materials that might exhibit Weyl semimetal behavior? A way in which a pair of Weyl points could be realized is illustrated in Fig. 29.12. Two bands A and B of electronic states having different angular momenta cross each other at two different momenta as a consequence of strong spin–orbit coupling, with Weyl points forming at the two momenta where the bands just touch. If band A is occupied in the ground state at small k_x, at the first band crossing with increasing k_x it becomes favorable energetically to occupy band B and the orbital character of the ground state wavefunction changes abruptly. These conditions can create the pair of Weyl points illustrated in Fig. 29.12(c).

Weyl Hamiltonian: As discussed in Box 29.2, band crossing as in Fig. 29.12(b) is unusual in normal semiconductors and metals because of *avoided crossing*. But in Weyl semimetals the degenerate bands at the Weyl points are protected by quantization of the Chern number [Eq. (29.6)], which prevents the band hybridization that would destroy the degeneracy and open a spectral gap. The degeneracy can be removed only by a perturbation large enough to bring the two Weyl points in Fig. 29.12(c) together so that they annihilate each other.

> Weyl semimetals are gapless electronic systems with opening of gaps prohibited by topological protection of the Weyl points.

As illustrated in Fig. 29.12(b), the dispersion near the Weyl points is approximately linear. A Dirac "Hamiltonian" was defined in Eq. (14.44). Setting the mass in it to zero suggests a corresponding Weyl Hamiltonian,

$$H(\boldsymbol{k}) = c\chi \boldsymbol{k} \cdot \sigma, \tag{29.7}$$

[7] Recall that Gauss' law relates the distribution of electrical charge to the corresponding electric field. In integral form it states that the flux of the electric field out of an arbitrary closed surface is proportional to the electric charge enclosed by that surface, independent of the distribution of the charge within the surface.

Box 29.2 Band Degeneracies

Traditionally, degeneracy of electronic bands is understood in terms of symmetry, with the degeneracy at a given $\boldsymbol{k}$ given by the dimensionality of the corresponding irreducible representation of the symmetry group. However, Weyl semimetals have points of band degeneracy corresponding to gap closures that originate in topology, not symmetry, with topological protection from symmetry breaking perturbations that could lift degeneracies.

Avoided Crossings

In band theory it is unusual for bands to create degeneracies by crossing. If two non-interacting bands cross, interaction terms in the Hamiltonian will mix the bands into two hybrid bands (symmetric and antisymmetric linear combinations), with one pushed up in energy and one pushed down in energy:

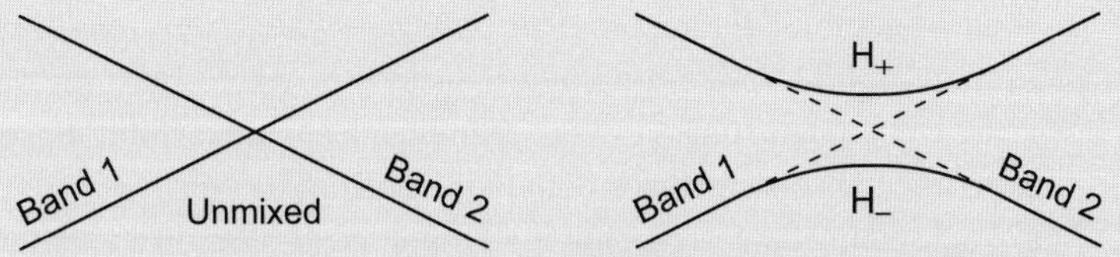

(see Example 29.4). Thus the actual bands H_+ and H_- appear to "repel" each other and never become degenerate. This is called an *avoided crossing.*

Essential and Accidental Degeneracies

Quantum degeneracies may be sorted into two basic categories: (1) *essential* and (2) *accidental.* Essential degeneracies typically are a consequence of symmetry (see Problem 29.1). Accidental degeneracies result from special features of a specific system. In a band theory essential degeneracies are typically lifted by symmetry breaking perturbations. This leads to avoided crossings as described above, which opens a gap in the band spectrum. However, features of particular systems like topological protection may interfere with band hybridization, leaving the degeneracy associated with crossing bands and no energy gap.

where $\boldsymbol{k}$ is the 3-momentum, $\boldsymbol{\sigma}$ is spin, c is the speed of light, and $\chi = \pm 1$ is the chirality. This is similar to the effective Hamiltonian in the linear dispersion region for two crossing bands in Fig. 29.12(b) if c is replaced by the Fermi velocity v_{f}, which is why the electrons in the present non-relativistic context are called *Weyl fermions.*[8] Now that neutrinos have been shown to have finite mass through observation of neutrino flavor oscillations, it seems that *no known elementary fermions are massless.* Thus Weyl semimetals may provide an example where emergent "elementary particles" in a many-body system exhibit a property that has never been observed for actual elementary particles.

[8] As discussed in Section 14.5.2, $H(\boldsymbol{k})$ can be interpreted as a Hamiltonian only for weak, slowly varying fields. Recall from Section 14.4 that Weyl fermion solutions behave like massless Dirac solutions, but each corresponds to a single handedness (chirality). That is, *for massless fermions* we may view the Dirac equation as being composed of two independent Weyl equations describing opposite chiralities.

On account of Eq. (29.7) with c replaced by v_f, a chirality χ may be assigned to the Weyl points of Fig. 29.12 and the topological charge for a Weyl point is often termed the *chiral charge.* We adopt a convention that a source of Berry flux $\oplus$ corresponds to $\chi = +1$ and a sink $\ominus$ corresponds to $\chi = -1$. By considering the boundary of the Brillouin zone to be an enclosing 2-surface, one finds that the net topological charge $\sum C_S$ must vanish for the entire Brillouin zone. Thus, the Brillouin zone can contain only an even number of Weyl-point pairs, with the members of each pair having opposite topological charges.

Gapless Surface States: The Weyl nodes in a Weyl semimetal correspond to points of band degeneracy that occur in the *bulk of the 3D material.* However, the chiral charge associated with each Weyl node provides topological protection for gapless surface states that appear on the boundary of a bulk sample. These states manifest as *Fermi arcs* that connect the projections of bulk Weyl nodes in the surface Brillouin zone, which is an example of the topological *bulk–boundary correspondence* that was introduced in Section 28.4.2. Unlike topological insulators where only the surface states are thought to be very interesting, a Weyl semimetal should exhibit novel structure with various experimental consequences both in the bulk and on the surface. Experimental discovery of a Weyl semimetal corresponding to topological states in a tantalum arsenide single crystal, with the predicted Fermi arcs, has been reported [113, 225].

29.6 Majorana Modes

In many-body physics composite emergent states can imitate the properties of various elementary particles.[9] For example, the Higgs boson was first suggested by collective phenomena in condensed matter before it was proposed as an elementary particle in the Standard Model of particle physics. Neutrinos may correspond to the elementary Majorana particles discussed in Section 14.6. Neutrino oscillation experiments indicate that neutrinos have mass so they are not Weyl particles, but ongoing experiments have yielded no evidence yet whether neutrinos are Dirac or Majorana particles (see Section 14.6.2). Meanwhile, much interest has been directed at the possibility of an emergent mode in condensed matter having the properties of Majorana particles, which would be of fundamental interest but also could have significant practical implications for quantum computing applications.

The electrons in condensed matter applications cannot be Majorana particles because their electrical charge distinguishes particle from antiparticle. However, in interacting

[9] Formally, an elementary particle exhibits no internal structure when probed on any energy (momentum transfer) scale. *Practically,* whether a particle is "elementary" is a question of characteristic energy scale for the physics being investigated. The atom ^{4}He is a composite of elementary fermions (electrons and quarks), virtual bosons (virtual photons and gluons), virtual particles and antiparticles from vacuum fluctuations, orbital and spin angular momentum, and binding energy. However, a low-energy probe can see none of this and ^{4}He behaves in all respects as a spinless elementary boson. Generally any quasiparticle will act like an elementary particle if the energy scale is low enough that its internal structure is not relevant.

systems quasiparticles corresponding to collective excitations of electrons can have exotic properties quite different from those of their constituent electrons. For example, we saw in Box 28.7 that for Luttinger liquids the charge and spin of individual electrons in the non-interacting system can be separated in the interacting system, with one type of quasiparticle carrying the spin and another type carrying the charge. This raises the issue of whether effective Majorana particles might arise as quasiparticles in interacting condensed matter systems. To be viewed as a Majorana particle in the condensed matter context a quasiparticle must satisfy two conditions. (1) It must obey a Dirac-like wave equation. (2) It must behave as its own antiparticle. Neither of these conditions is normally fulfilled but, as we now discuss, they could occur in certain correlated electronic systems.

29.6.1 The Dirac Equation in Condensed Matter

Particles in condensed matter are non-relativistic and normally satisfy the Schrödinger equation, which gives a parabolic energy dispersion (energy proportional to the square of the momentum), not the linear dispersion (energy proportional to momentum) expected for a relativistic Dirac solution.[10] However, for special circumstances such as near a sharp band crossing a non-relativistic Hamiltonian can exhibit a linear dependence on the momentum at low energy, leading to a wave equation of the Dirac form. We have come across several examples.

1. We saw in Fig. 20.2 that monolayer graphene exhibits linear dispersion (Dirac cones).
2. As was seen in Section 29.5, for Weyl semimetals the low-energy electrons behave as solutions of the Weyl equations (Dirac equation for massless fermions).
3. The topological superconductors of Section 29.7 can support gapless fermions on spatial boundaries, or near defects like vortices, that are described by Dirac-like equations.

In such linear dispersion approximations the electron Fermi velocity (highest electron velocity in the system) typically plays the part of light velocity in the actual Dirac equation.

29.6.2 Quasiparticles and Anti-Quasiparticles

The possibility of identifying a quasiparticle with its antiparticle that is a necessary condition for a quasiparticle to behave as a Majorana particle can arise within the context of superconductivity. In quantum field theory we may view the absence of a particle as a *hole,* with quantum numbers that cause a particle and corresponding hole to act as if they are a particle and its corresponding antiparticle. Superconductors correspond to a condensate of Cooper pairs (Box 32.2). As shown in Sections 22.4.2 and 32.1, superconducting states have particle–hole symmetry, which means that when the BCS wavefunction (22.31) is expressed in the quasiparticle basis (22.29), we can say the following.

[10] Non-relativistically, energy is $E = p^2/2m + V$, while relativistically $E^2 = (p^2 + m^2)^{1/2}$. Neglecting V gives $E \propto p^2$ for non-relativistic particles and assuming $p \gg m$ gives $E \propto p$ for relativistic particles.

Zero-energy Bogoliubov quasiparticles are equal mixtures of particles and holes. Thus the hermitian conjugate of this state is the same as the original state, meaning that the quasiparticle and anti-quasiparticle are equivalent.

Therefore, superconductors have an effective charge conjugation (that is, particle–hole) symmetry that can cause its quasiparticles to behave as Majorana fermions.

29.7 Topological Superconductors

Topological superconductors are characterized by a non-trivial topological structure for their Cooper pairs [178, 179].

29.7.1 Topological Majorana Fermions

In vortex cores and on sample surfaces, Majorana fermions can appear as zero-energy Bogoliubov quasiparticles associated with the superconductor. Note, however, that any superconductor contains Majorana particles in the loose sense defined above (since it contains zero-energy particle–hole pairs), but in strict usage a Majorana fermion is a *Dirac particle* that is its own antiparticle. Thus Majorana fermions can occur only in superconductors that exhibit *linear dispersion* at the Fermi surface: $\epsilon \sim k$, where ϵ is the energy and k the momentum. This is not the case for most conventional and unconventional superconductors, which are solutions of Schrödinger equations exhibiting quadratic dispersion, $\epsilon \sim k^2$.

The Majorana zero-energy state in a superconductor is protected by bulk topological non-triviality of the Hilbert space. As a consequence, it is resistant to extrinsic perturbations such as impurities and crystal imperfection. Therein lies the great interest in Majorana fermions as a basis for fault-tolerant quantum computing (*topological quantum computation*), because they are expected to be highly resistant to environmental decoherence of the quantum wavefunction. We will take this up further in Section 29.9.

29.7.2 Fractionalization of Electrons

The exotic behavior of Majorana fermions in a superconductor may be thought of as resulting from the electron being split into two parts.[11] An electron field is complex but a Majorana field is real. Thus, two Majorana fields can be combined into a single electron

[11] We do not mean that individual electrons have been split! We mean that the many-body state (electrons plus correlations) behaves *as if* it is composed of two kinds of quasiparticles, one carrying the charge and the other carrying the spin. But if the interactions that produce the many-body state were switched off we would be left with only free electrons, each carrying the full electronic spin and full electronic charge.

(or electron hole) field, and the formation of a Majorana quasiparticle in a superconductor is functionally equivalent to a *fractionalization* of an electron and its quantum numbers into two quasiparticles. Majorana fermions in topological superconductors may exhibit *non-abelian statistics* (exchange operations on two particles do not commute), which differs from normal Fermi–Dirac or Bose–Einstein statistics. Implications for quantum statistics are taken up in Section 29.8 and for quantum computing in Section 29.9.

29.8 Fractional Statistics

Quantum statistics deals with how wavefunctions are affected by exchange of identical particles. In three dimensions there are two possible outcomes for exchange on a state of two identical particles (see Ch. 4): (1) the wavefunction is unchanged, or (2) it is multiplied by a factor of -1. In *two dimensions* the effect of a transposition can depend on exactly how the operation is carried out. This more complicated realization of quantum statistics is in fact a topological property of such phases of matter. A question of fundamental interest, and of practical importance for quantum computing, is how this topological information might be accessed and manipulated experimentally.

29.8.1 Anyon Statistics

The wavefunction for a quantum system with two indistinguishable particles can be denoted $|\psi_1\psi_2\rangle$. Under exchange of states for the two particles, $|\psi_1\psi_2\rangle \rightarrow |\psi_2\psi_1\rangle$ we should obtain the same two-particle state, up to a phase factor,

$$|\psi_1\psi_2\rangle = e^{i\theta}|\psi_2\psi_1\rangle. \tag{29.8}$$

In a space with three or more dimensions the choices for the phase factor θ are governed by the *spin–statistics theorem* (see Box 29.3), which permits two possibilities.

1. *Fermi–Dirac statistics,* where $\theta = \pi$ so that $e^{i\pi} = -1$ and elementary particles are *fermions* with half-integer spins.
2. *Bose–Einstein statistics,* where $\theta = 2\pi$ so that $e^{i\pi} = +1$ and elementary particles are *bosons* with integer spins.

However, in two spatial dimensions these restrictions no longer apply and quasiparticles are at least theoretically possible for which θ can take continuous values. For the 2D case with non-degenerate states the phase θ in Eq. (29.8) is shorthand for a prescription that when quasiparticle 1 and quasiparticle 2 are interchanged in a manner such that each makes a counterclockwise half revolution around the other, the wavefunction of the two-particle system is multiplied by a complex phase factor $e^{i\theta}$, or multiplied by a factor $e^{-i\theta}$ if the half revolution is instead clockwise. Notice that this prescription makes sense only in two spatial dimensions, where clockwise and counterclockwise have well-defined meaning.

Box 29.3 The Spin–Statistics Theorem

In (Lorentz-invariant) relativistic quantum field theory, fields at two spacetime points separated by a spacelike interval (see Section 13.3.3) are required by causality to either commute or anticommute. Consistency requires the choice of

- *anticommutation* for particles of half-integer spin (fermions) and
- *commutation* for particles of integer spins (bosons).

This is called the *spin–statistics theorem.* It is valid for free and interacting fields in three or more spatial dimensions that (1) are Lorentz invariant, (2) have positive energies for all particles, and (3) have positive norms for all Hilbert space states.

Such quasiparticles are called *anyons.* The symmetry of these anyon exchange operations may be encoded in the *braid group,* which we now discuss.

29.8.2 The Braid Group

One way to explore the statistics and information content of a topological system is to move its quasiparticles slowly (*adiabatically,* so that the motion does not excite the system) until they have returned to their original position, up to possible permutations with other quasiparticles. To visualize this procedure we may employ a standard tool from relativity called a *spacetime diagram,* in which the trajectory of a particle (called the *worldline* of the particle) is plotted versus both space and time. Adopting a convention that the time axis is oriented vertically and spatial axes are oriented horizontally, for two spatial dimensions Fig. 29.13 illustrates for two particles and Fig. 29.14 illustrates worldline entanglement for two pairs of particles. A collection of (possibly crossed) strands with *fixed endpoints* as in Figs. 29.13 and 29.14 is called a *braid.* Rules for valid braids and operations are illustrated in Fig. 29.15, and a natural braid multiplication is illustrated in Fig. 29.16.

> Two braids with the same number of strands may be multiplied by joining the ends of the strands for one braid to the beginning of the strands for the other braid to give the product braid, and then simplifying (if necessary) by pushing, pulling, or stretching the strands while keeping the endpoints fixed.

Using braid multiplication one finds the following.

1. Multiplication of two n-strand braids gives another n-strand braid.
2. A braid with no strands that cross each other acts as a unit braid.
3. Each braid has an inverse braid; the product of the braid and its inverse is the unit braid.
4. Braid multiplication is associative.

Therefore, the set of all braids having n strands (all *n-braids*) forms a group that may be labeled B_n, with equivalence classes of n-braids as the group elements and the braid multiplication defined in Fig. 29.16 as the group multiplication (composition) operation.

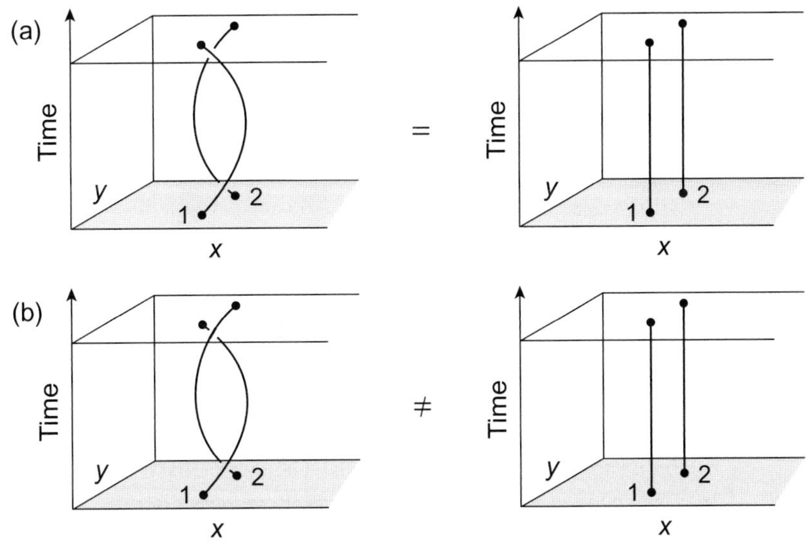

Fig. 29.13 Braiding of two quasiparticle worldlines in two spatial and one time dimension (2 + 1 spacetime) [170]. (a) The worldline for particle 1 crosses above that for particle 2 twice, so these worldlines can be deformed continuously into the unentangled configuration on the right. (b) The worldline for particle 1 crosses over the worldline for particle 2 and then later crosses under it. These worldlines cannot be deformed continuously into the configuration on the right while keeping the endpoints fixed.

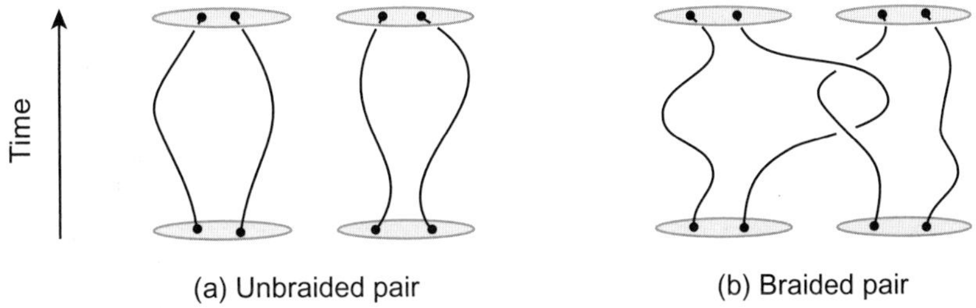

Fig. 29.14 Worldlines for pairs of particles in a spacetime with two spatial and one time coordinate [177]. The braided pair in (b) cannot be deformed continuously into the unbraided pair in (a) since the endpoints of the worldlines are fixed by boundary conditions. Thus time evolution in (a) and (b) correspond to different histories.

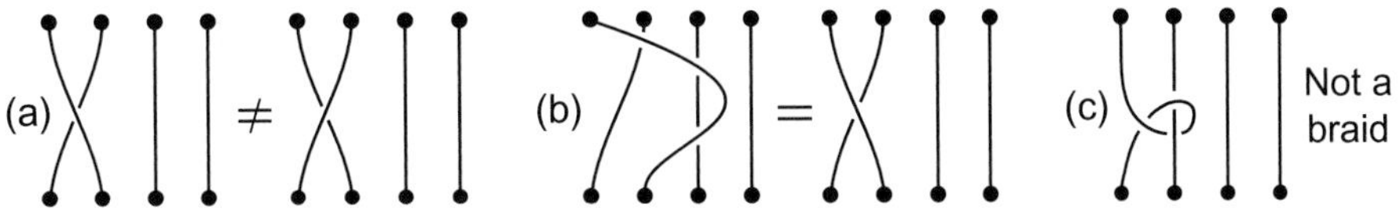

Fig. 29.15 Braid etiquette. (a) Strands cannot pass through each other so whether a strand passes over or under another strand matters; these two braids are not equivalent. (b) Strands can be stretched, pulled, or pushed while keeping endpoints fixed, but never cut. Two braids that look the same after stretching, pushing, or pulling the strands as needed are equivalent. (c) Not a braid: braids cannot have strands that reverse direction and produce a knot.

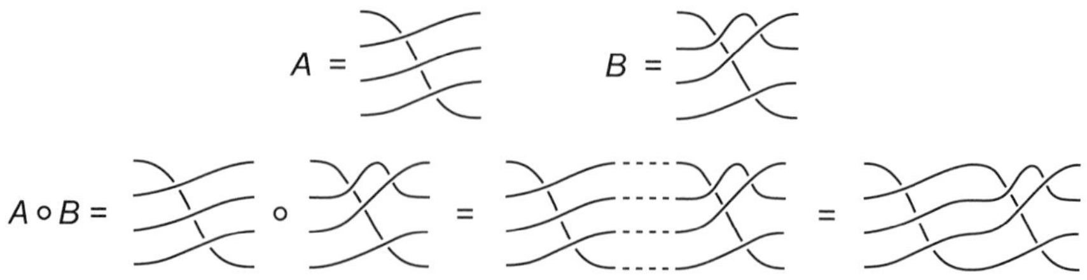

Fig. 29.16 Multiplication (composition) of two braids A and B, with $\circ$ indicating the composition operation. As notational shorthand we may write $A \circ B \equiv AB$. For convenience, when multiplying braids we display the stands horizontally rather than vertically, and omit the dots at the ends of strands indicating explicitly that the endpoints must be kept fixed.

29.8.3 Abelian and Non-Abelian Anyons

Anyons can be classified as *abelian* or *non-abelian,* depending upon their *braid-group representations*. For fermions and bosons in 3D space the statistics operators are irreps of the permutation group S_n for n indistinguishable particles, but in 2D space the anyonic statistics operators are irreducible representations of the *braid group* B_n. If a system exhibits *degeneracy* such that more than one distinct state has the same particle configuration, exchange of particles can cause a transition to a different state with the same configuration of particles, not just a change of an overall phase factor. Thus, particle exchange may be viewed more generally in 2D space as a *linear transformation on the subspace of degenerate states.* If there is no degeneracy, the subspace is 1D and the linear transformations (multiplication by a phase factor) commute; these are called *abelian anyons.* However, if there is degeneracy the subspace is of higher dimension and the linear transformations need not commute among themselves (just as arithmetic multiplication commutes but a general matrix multiplication does not); these are called *non-abelian anyons.*

29.9 Quantum Computers and Topological Matter

Considerable interest has attached to the possibilities inherent in a *quantum computer* [177]. Suppose that we have at our disposal a controllable quantum system that can be initialized in some known state $|\psi_0\rangle$ and evolved by a unitary transformation $U(t)$ to a final state $U(t)|\psi_0\rangle$. We assume sufficient control over the Hamiltonian of the system that $U(t)$ can be any unitary transformation that is desired, and that a method has been devised to read out the state of the system at the end of the evolution. This overall process of (1) initialization, (2) evolution, and (3) measurement is called *quantum computation.* Quantum computing is a subfield of *quantum information processing,* which includes also subfields like *quantum cryptography* and *quantum teleportation.*

29.9.1 Qubits and Quantum Information

The basic information unit of quantum information processing is the *quantum bit* or *qubit,* which is a two-level quantum system with states $|0\rangle$ and $|1\rangle$. A classical bit can be in one of its two states, conveniently labeled 0 or 1. Unlike a classical bit, a qubit is a quantum object that can be in a linear superposition state $\alpha\,|0\rangle + \beta\,|1\rangle$. By the rules of quantum measurement, when read out a qubit will be found to be either in the definite state $|0\rangle$ or in the definite state $|1\rangle$, with the probability for each depending on the quantum state of the qubit. A useful representation of a qubit called a *Bloch sphere* is discussed in Box 29.4.

Box 29.4 The Bloch Sphere

In a quantum system having two basis states $|0\rangle$ and $|1\rangle$, a general wavefunction can be expressed as $|\psi\rangle = \alpha\,|0\rangle + \beta\,|1\rangle$, where α and β are complex numbers, requiring four real numbers to specify them. However, normalization to unit probability gives the constraint $|\alpha|^2 + |\beta|^2 = 1$ and the overall phase is not observable, so one coefficient may be chosen real. Thus, the most general state has two degrees of freedom and may be parameterized as

$$|\psi\rangle = \cos\left(\frac{\theta}{2}\right)|0\rangle + e^{i\phi}\sin\left(\frac{\theta}{2}\right)|1\rangle,$$

which may be visualized using the *Bloch sphere* illustrated in the following figure,

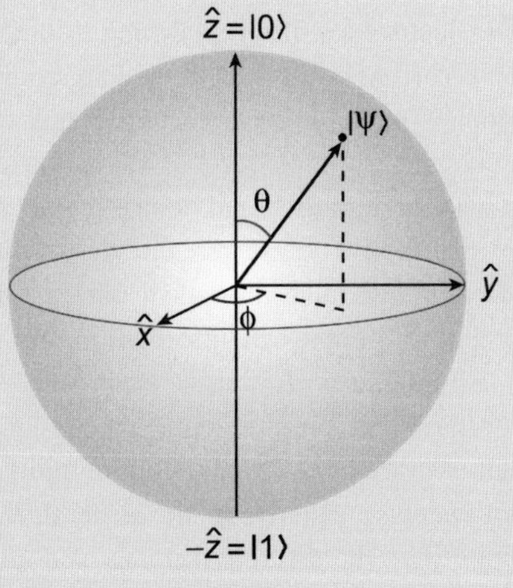

where $\hat{x}$, $\hat{y}$, and $\hat{z}$ are unit vectors. The spherical surface represents a 2D space of possible pure states for the system. The two basis states are denoted by the unit vectors $\hat{z} = |0\rangle$ and $-\hat{z} = |1\rangle$. Thus any pure state is a linear combination of these states and is represented by a point on the sphere parameterized by the spherical coordinates (θ, ϕ), as illustrated. The Bloch sphere representation is useful in quantum computing, where the poles of the Bloch sphere represent the two allowed states of a classical bit and the full spherical surface represents the superposition states encoded by a quantum bit (*qubit*).

In quantum computing the qubits and their interactions carry the content of the computation and the "computer code" may be viewed as the unitary operations $U(t)$ that evolve the quantum system. It may then be shown that because of (1) linear superposition, (2) unitary evolution, and (3) the exponentially large Hilbert spaces of entangled states,[12] a quantum computer can be *much* more powerful than an ordinary classical digital computer:

- a single qubit is in any linear superposition of two distinct quantum states,
- a two-qubit state is a superposition of four such states,
- a three-qubit state is a superposition of eight such states, and in general
- an n-qubit state is an arbitrary superposition of 2^n distinct states.

In contrast, a classical bit can be in only *one* of these 2^n different quantum states at any given time.

> An n-qubit quantum computer could in principle encode and process *exponentially more information* than a corresponding n-bit classical computer.

Obviously an exponential increase in computing power would enable solution of problems that lie beyond the capabilities of even the most powerful classical supercomputers.

29.9.2 The Problem of Decoherence

Many challenging technical problems must be solved to implement the attractive theoretical scheme sketched in the preceding paragraphs; the problem of overriding importance is the inevitable occurrence of errors during a computation. The information in the quantum computation is carried by a coherent wavefunction that is highly susceptible to *decoherence* by interaction with the environment. A many-qubit quantum computer will be able to maintain a coherent superposition of states for only a finite length of time. Any external interaction, whether unintentional or a deliberate measurement, can collapse the wavefunction, thereby decohering the system and destroying the encoded quantum information.

Classical computers also must deal with errors, but there the problem is surmounted by keeping multiple copies of information and checking these copies for discrepancies during the computation. A similar approach is much more difficult for a quantum computer because the measurement or copying of a quantum state in the course of the computation may alter the quantum-computational information, by basic uncertainty principle arguments. Furthermore, the errors that are possible in quantum computation can be more subtle than the discrete flipping of a qubit such as $|1\rangle \to |0\rangle$ that would be the analog of a typical classical digital computer error: it can be a continuous phase error such as $\alpha\,|0\rangle + \beta\,|1\rangle \to \alpha\,|0\rangle + \beta e^{i\phi}\,|1\rangle$ for arbitrary values of ϕ.

In spite of these difficulties, it has been shown that quantum error correction is possible in principle, implying that *fault tolerant* quantum computation is (perhaps) within reach.

[12] A set of quantum particles is entangled when the properties of any one particle cannot be described independent of all the other particles, even if the particles are widely separated. The possibility of entangled states is a fundamental property that distinguishes quantum from classical systems.

However, quantum error correction algorithms may be noisy and may themselves introduce errors. Thus, quantum computation is feasible only if the intrinsic error rate is small. A typical estimate is that a quantum computer must be able to perform reliably $\sim 10^3$ consecutive flawless operations to make known quantum error correction schemes practical.

29.9.3 Topological Quantum Computation

The discussion of the preceding section implies that a quantum computer requires extremely strong noise suppression. However, the overall considerations of this chapter suggest an alternative: instead of making the system noiseless we could construct it such that it does not hear most of any noise that is present and thus is immune to the usual sources of quantum decoherence [177]. Most quantum systems exhibit *local interactions,* meaning that a typical decohering process affects the quantum computer only over a region corresponding to a few atoms. However, if a physical system has topological degrees of freedom they are *global properties* that are *insensitive to local perturbations.* If we could build a quantum computer that encodes its information in topological degrees of freedom, that information would be automatically shielded against corruption by local interactions. Thus fault tolerant quantum computing might be implemented through *topological protection,* rendering the quantum computer largely immune to noise because it cannot hear it.

Until recently such considerations were faced with the practical issue that, with the exception of the quantum Hall effect, which requires strong magnetic fields and low temperatures that are problematic for building devices, few condensed matter systems were expected to have topological degrees of freedom. However, as this chapter has documented there has been an explosion in recent years of evidence for a broad variety of topological matter. Thus there is some optimism that a topological quantum computer could be within our technical grasp using materials that are known or conjectured to exist, but many issues remain to be resolved and the timescale for such development is uncertain. To date (2021), demonstrations of quantum computing have used technologies that are not topological.

Background and Further Reading

A pedagogical discussion of topological properties in condensed matter physics may be found in Ramirez and Skinner [168], and Castelvecchi [34]; more advanced discussions are given in Girvin and Yang [75] and El-Batanouny [54]. Wilczek has given a synthesis of topological concepts in elementary particle and condensed matter physics [212]. An overview of topological quantum numbers in non-relativistic physics is given by Thouless [193]. Topological insulators are reviewed in Asbóth, Oroszlány, and Pályi [14], Bernevig [21], Hasan and Kane [107], and Qi and Zhang [165]. Majorana fermions are discussed in Elliott and Franz [57] and Sato and Fujimoto [179], while Aguado and Kouwenhoven [3], and Alicea [9] discuss topologically protected Majorana qubits for quantum information applications. Topological superconductors are reviewed in Sato and Ando [178]. Read [170] gives an accessible overview of topological phases of matter, non-abelian statistics, the braid group, and quantum computation. Sarma, Freedman, and Nayak [177] review topological quantum computing and Stanescu [183] gives an introduction to both topological matter and quantum computation.

Problems

29.1 The normal (not accidental; see Box 29.2) degeneracies in a quantum system result from symmetry. Show that if a Hamiltonian H is invariant under transformation by a unitary symmetry operator S, then (a) H commutes with S, and (b) if $|\alpha\rangle$ is an eigenstate of H with energy E, then the symmetry transformed state $|\beta\rangle = S\,|\alpha\rangle$ is also an eigenstate with energy E. Thus, if $|\alpha\rangle$ and $|\beta\rangle$ are linearly independent they define states that are degenerate in energy because they are related by symmetry.

29.2 Construct the braid group products

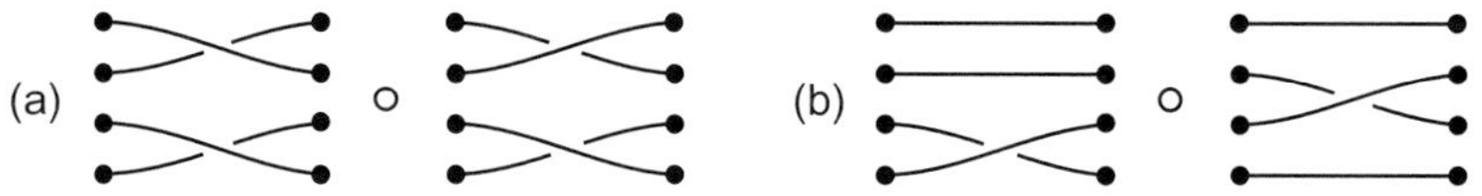

using the algorithm of Fig. 29.16. ***

29.3 Show that the two braids

are mutual inverses under braid multiplication.

29.4 For states $|\alpha\rangle$ and $|\beta\rangle$, define time-reversed states by $|\tilde{\alpha}\rangle = \Theta\,|\alpha\rangle$ and $|\tilde{\beta}\rangle = \Theta|\beta\rangle$. Show that the overlap of the time-reversed states is given by $\langle\tilde{\beta}|\tilde{\alpha}\rangle = \langle\beta|\alpha\rangle^*$. *Hint*: Expand $|\tilde{\alpha}\rangle$ and $|\tilde{\beta}\rangle$ in complete sets of states using $\sum_a |a\rangle\langle a| = 1$, and use that the complex conjugation operator has no effect on a ket, $\mathscr{K}\,|\alpha\rangle = |\alpha\rangle$. ***

29.5 Demonstrate that the operator $\Theta = i\sigma_2\mathscr{K}$ defined in Box 29.1 meets all the criteria given there for a fermion time-reversal operator: it preserves the norm of a wavefunction, is antilinear, and satisfies $\Theta^2 = -1$.

29.6 Reproduce the plots in Fig. 29.10 by deriving formulas for the eigenvalues and eigenfunctions of the Hamiltonian (29.4). *Hint*: See the solution of Problem 14.14. ***

29.7 Prove Kramers' theorem (Box 29.1) for spin-$\frac{1}{2}$ fermions by showing that a state and its time reverse are degenerate and orthogonal, so they are two independent states with the same energy. *Hint*: Use that the time-reversal operator obeys $\Theta^2 = -1$, and that $\langle\tilde{\beta}|\tilde{\alpha}\rangle = \langle\beta|\alpha\rangle^*$ from Problem 29.4. ***

PART IV

A VARIETY OF PHYSICAL APPLICATIONS

30 Angular Momentum Recoupling

The basics of tensor methods for angular momentum operators were introduced in Section 6.4. In this chapter we expand that discussion to more ambitious cases of coupling three or four angular momenta. This topic is more complex than many in this book, often involving long equations with many indices. However, it is an essential issue for those fields like atomic or nuclear physics where one commonly deals with many-body states of good total angular momentum, with more than one way to couple particles to form those states (for example, L–S and J–J coupling schemes). This chapter is optional if your interests lie in fields like condensed matter or particle physics where many-body states of good total angular momentum are less important.

30.1 Recoupling of Three Angular Momenta

If we have more than two angular momenta, the manner in which they can be coupled to a good total angular momentum is not unique. For example, three angular momenta j_1, j_2, and j_3 could be coupled in either of the following ways to a total angular momentum J,

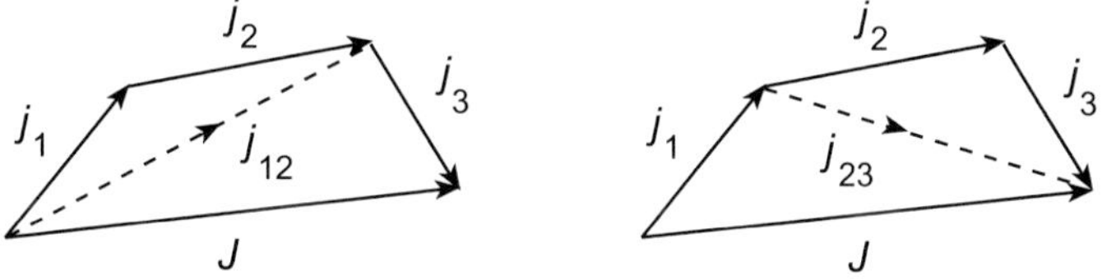

The explicit forms for these state vectors are

$$|(j_1 j_2)_{j_{12}} j_3, JM\rangle = \sum_{m_1 m_2} \sum_{m_3 m_{12}} \langle j_1\, m_1\, j_2\, m_2 |\, j_{12}\, m_{12}\rangle \times \langle j_{12}\, m_{12}\, j_3\, m_3 |\, J\, M\rangle\, |j_1 m_1\rangle\, |j_2 m_2\rangle\, |j_3 m_3\rangle\,, \qquad (30.1)$$

$$|j_1 (j_2 j_3)_{j_{23}}, JM\rangle = \sum_{m_1 m_2 m_3} \sum_{m_{23}} \langle j_2\, m_2\, j_3\, m_3 |\, j_{23}\, m_{23}\rangle \times \langle j_1\, m_1\, j_{23}\, m_{23} |\, J\, M\rangle\, |j_1 m_1\rangle\, |j_2 m_2\rangle\, |j_3 m_3\rangle\,. \qquad (30.2)$$

Since these provide only different labelings of the same problem, the two coupling schemes are connected by a unitary transformation,

$$|j_1 (j_2 j_3)_{j_{23}}, JM\rangle = \sum_{j_{12}} \langle (j_1 j_2)_{j_{12}} j_3, JM | j_1 (j_2 j_3)_{j_{23}}, JM\rangle |(j_1 j_2)_{j_{12}} j_3, JM\rangle. \qquad (30.3)$$

The transformation coefficient $\langle (j_1 j_2)_{j_{12}} j_3, JM | j_1 (j_2 j_3)_{j_{23}}, JM\rangle$ actually is independent of M, by arguments similar to those leading to the Wigner–Eckart theorem (see Edmonds [51]). Expanding the wavefunctions with Clebsch–Gordan coefficients and invoking their orthogonality, we find explicitly that

$$\langle (j_1 j_2)_{j_{12}} j_3, J | j_1 (j_2 j_3)_{j_{23}}, J\rangle = \sum_{m_1 m_2 m_3} \sum_{m_{12}} \sum_{m_{23}} \langle j_{12}\, m_{12}\, j_3\, m_3 |\, J\, M\rangle \langle j_1\, m_1\, j_2\, m_2 |\, j_{12}\, m_{12}\rangle$$
$$\times \langle j_2\, m_2\, j_3\, m_3 |\, j_{23}\, m_{23}\rangle \langle j_1\, m_1\, j_{23}\, m_{23} |\, J\, M\rangle . \tag{30.4}$$

The transformation coefficients among the other possible couplings of three angular momenta may be deduced from (30.3) by changing the order of the angular momenta in the Clebsch–Gordan coefficients, using Eqs. (6.49).

30.1.1 6*J* Coefficients

To exploit the symmetry of the recoupling we may define the 6J *coefficient* by

$$\begin{aligned} \begin{Bmatrix} j_1 & j_2 & j_{12} \\ j_3 & J & j_{23} \end{Bmatrix} &\equiv \frac{(-1)^{j_1+j_2+j_3+J}}{\sqrt{(2j_{12}+1)(2j_{23}+1)}} \langle (j_1 j_2)_{j_{12}} j_3, J | j_1 (j_2 j_3)_{j_{23}}, J\rangle \\ &= \frac{(-1)^{j_1+j_2+j_3+J}}{\sqrt{(2j_{12}+1)(2j_{23}+1)}} \sum_{m_1 m_2} \langle j_1\, m_1\, j_2\, m_2 |\, j_{12}, m_1+m_2\rangle \\ &\quad \times \langle j_{12}, m_1+m_2, j_3, M-m_1-m_2 |\, JM\rangle \\ &\quad \times \langle j_2 m_2 j_3, M-m_1-m_2 |\, j_{23}, M-m_1\rangle \langle j_1 m_1 j_{23}, M-m_1 |\, JM\rangle , \end{aligned} \tag{30.5}$$

where $m_1 + m_2 = M$ was used to reduce the summation indices. The 6J symbol is

1. invariant under any permutation of columns, and
2. invariant under interchange of upper and lower arguments in any two columns.

The transformation between coupling schemes is then of the form

$$|j_1 (j_2 j_3)_{j_{23}}, JM\rangle = \sum_{j_{12}} \frac{(-1)^{j_1+j_2+j_3+J}}{\sqrt{(2j_{12}+1)(2j_{23}+1)}} \times \begin{Bmatrix} j_1 & j_2 & j_{12} \\ j_3 & J & J_{23} \end{Bmatrix} |(j_1 j_2)_{j_{12}} j_3, JM\rangle. \tag{30.6}$$

This may be inverted using the unitarity of the transformation. For example,

$$|(j_1 j_2)_{j_{12}} j_3, JM\rangle = \sum_{j_{23}} \langle j_1 (j_2 j_3)_{j_{23}}, J | (j_1 j_2)_{j_{12}} j_3, J\rangle\, |j_1 (j_2 j_3)_{j_{23}}, JM\rangle. \tag{30.7}$$

30.1.2 Racah Coefficients

The recoupling transformation is also commonly defined in terms of the *Racah coefficient* W, which is related to the 6J symbol by

$$\begin{Bmatrix} j_1 & j_2 & j_3 \\ l_1 & l_2 & l_3 \end{Bmatrix} = (-1)^{j_1+j_2+l_1+l_2} W\,(j_1 j_2 l_2 l_1; j_3 l_3)\,. \tag{30.8}$$

Table 30.1. Some Racah coefficients $W(j_1, j_2, j_3, j_4; j_5, j_6)$ [29]

$$W(a, a+\tfrac{1}{2}, b, b+\tfrac{1}{2}; \tfrac{1}{2}, c) = (-1)^{a+b-c}\left[\frac{(a+b+c+2)(a+b-c+1)}{(2a+1)(2a+2)(2b+1)(2b+2)}\right]^{1/2}$$

$$W(a, a+\tfrac{1}{2}, b, b-\tfrac{1}{2}; \tfrac{1}{2}, c) = (-1)^{a+b-c}\left[\frac{(a-b+c+1)(c-a+b)}{2b(2a+1)(2a+2)(2b+1)}\right]^{1/2}$$

$$W(a, a+1, b, b+1; 1, c) = (-1)^{a+b-c}\left[\frac{(a+b+c+3)(a+b+c+2)(a+b-c+2)(a+b-c+1)}{4(2a+3)(a+1)(2a+1)(2b+3)(2b+1)(b+1)}\right]^{1/2}$$

$$W(a, a+1, b, b; 1, c) = (-1)^{a+b-c}\left[\frac{(a+b+c+2)(a-b+c+1)(a+b-c+1)(c-a+b)}{4b(2a+3)(a+1)(2a+1)(b+1)(2b+1)}\right]^{1/2}$$

$$W(a, a+1, b, b-1; 1, c) = (-1)^{a+b-c}\left[\frac{(c-a+b)(c-a+b-1)(a-b+c+2)(a-b+c+1)}{4b(2a+3)(a+1)(2a+1)(2b-1)(2b+1)}\right]^{1/2}$$

$$W(a, a, b, b; 1, c) = (-1)^{a+b-c-1}\left[\frac{a(a+1)+b(b+1)-c(c+1)}{[4ab(a+1)(2a+1)(b+1)(2b+1)]^{1/2}}\right]$$

Some explicit formulas for Racah coefficients are given in Table 30.1; they may be related to $6J$ coefficients using Eq. (30.8).

30.2 Matrix Elements of Tensor Products

As an illustration of the importance of the $6J$ coefficients, let us use them to evaluate matrix elements of tensor products. Suppose we have a tensor that may be expressed as the coupled product of two tensors

$$T_{KQ}(kk') = \sum_q T_{kq} T_{k',Q-q} \langle kqk'(Q-q)|\, KQ\rangle . \tag{30.9}$$

The reduced matrix element between states of good angular momentum is

$$\begin{aligned}\langle J \| \boldsymbol{T}_K \| J'\rangle &= \sum_Q \sqrt{2J+1}\,\langle J'\,(M-Q)\,K\,Q|\,J\,M\rangle\langle JM|\,T_{KQ}\,|J'(M-Q)\rangle \\ &= \sum_{qQ} \sqrt{2J+1}\,\langle J'\,(M-Q)\,K\,Q|\,J\,M\rangle\langle k\,q\,k'\,(Q-q)|\,K\,Q\rangle \\ &\quad\times \langle JM|\,T_{kq}T_{k',Q-q}\,|J'(M-Q)\rangle ,\end{aligned} \tag{30.10}$$

where we may write

$$\begin{aligned}\langle JM|\,T_{kq}T_{k',Q-q}\,|J'(M-Q)\rangle &= \sum_{J''}\langle JM|\,T_{kq}\,|J''(M-q)\rangle\langle J''(M-q)|\,T_{k',Q-q}\,|J'(M-q)\rangle \\ &= \sum_{J''}\langle J''(M-q)kq|\,JM\rangle\langle J'(M-Q)k'(Q-q)|\,J''(M-q)\rangle\langle J\,\|\,\boldsymbol{T}_k\,\|\,J'\rangle\langle J''\,\|\,\boldsymbol{T}_{k'}\,\|\,J'\rangle.\end{aligned}$$

Inserting this into (30.10) and comparing with the definition of the Racah coefficient [see Eqs. (30.5)–(30.8)] gives

$$\langle J \| \boldsymbol{T}_K \| J' \rangle = \sum_{J''} (-1)^{K-k-k'} \sqrt{2K+1}\, W(JJ'kk'; KJ'') \times \langle J \| \boldsymbol{T}_k \| J'' \rangle \langle J'' \| \boldsymbol{T}_{k'} \| J' \rangle. \tag{30.11}$$

As another example, consider a matrix element $\langle j_1 j_2 J | T_{kq}(1) | j_1' j_2' J' \rangle$, where the tensor operator $T_{kq}(1)$ operates only on the part of the system associated with the angular momenta labeled by the subscript 1. The reduced matrix element is (Problem 30.5)

$$\langle j_1 j_2 J \| \boldsymbol{T}_k(1) \| j_1' j_2' J' \rangle = (-1)^{J'+k+j_1+j_2} \sqrt{(2J+1)(2J'+1)} \times \begin{Bmatrix} j_1 & j_1' & k \\ J' & J & j_2 \end{Bmatrix} \langle j_1 \| \boldsymbol{T}_k(1) \| j_1' \rangle \delta_{j_2 j_2'}. \tag{30.12}$$

Example 30.1 and Problem 30.3 illustrate using this formula.

Example 30.1 Using Eq. (30.12) to calculate the reduced matrix element of a spherical harmonic between l–s coupled states gives

$$\langle lsj \| \boldsymbol{Y}_L \| l'sj' \rangle = (-1)^{j-1/2} \sqrt{\frac{(2L+1)}{4\pi(2l+1)}} \sqrt{(2j'+1)(2j+1)} \begin{Bmatrix} j & j' & L \\ -\frac{1}{2} & \frac{1}{2} & 0 \end{Bmatrix}, \tag{30.13}$$

as you are asked to show in Problem 30.6.

30.3 Recoupling of Four Angular Momenta

Four angular momenta may be coupled pairwise and then the pairs coupled to the total angular momentum. There are several ways to do this. For example, one possibility is

$$\boldsymbol{j}_1 + \boldsymbol{j}_3 = \boldsymbol{j}_{13} \qquad \boldsymbol{j}_2 + \boldsymbol{j}_4 = \boldsymbol{j}_{24} \qquad \boldsymbol{j}_{13} + \boldsymbol{j}_{24} = \boldsymbol{J}$$

and another is

$$\boldsymbol{j}_1 + \boldsymbol{j}_2 = \boldsymbol{j}_{12} \qquad \boldsymbol{j}_3 + \boldsymbol{j}_4 = \boldsymbol{j}_{34} \qquad \boldsymbol{j}_{12} + \boldsymbol{j}_{34} = \boldsymbol{J}.$$

How are these choices related?

30.3.1 9*J* Coefficients

The unitary transformation connecting these different coupling schemes may be used to introduce the $9J$ *symbol* (so called because it contains nine angular momenta j) through

$$\langle (j_1 j_3)_{j_{13}} (j_2 j_4)_{j_{24}}; J | (j_1 j_2)_{j_{12}} (j_3 j_4)_{j_{34}}; J \rangle \equiv \sqrt{(2j_{13}+1)} \times \sqrt{(2j_{24}+1)(2j_{12}+1)(2j_{34}+1)} \begin{Bmatrix} j_1 & j_2 & j_{12} \\ j_3 & j_4 & j_{34} \\ j_{13} & j_{24} & J \end{Bmatrix}. \tag{30.14}$$

The $9J$ symbol in curly brackets exhibits a high degree of symmetry. An even permutation of rows or columns, or a transposition of rows and columns, leaves the symbol invariant, while an odd permutation of adjacent rows or columns multiplies a $9J$ symbol by a phase $(-1)^p$, where $p = \sum j's$ is a sum over all nine entries in the symbol. The $9J$ coefficients can be expressed in terms of $3J$ symbols,

$$\begin{Bmatrix} j_1 & j_2 & j_{12} \\ j_3 & j_4 & j_{34} \\ j_{13} & j_{24} & J \end{Bmatrix} = \sum_{\text{all } m,M} \begin{pmatrix} j_1 & j_2 & j_{12} \\ m_1 & m_2 & m_{12} \end{pmatrix} \begin{pmatrix} j_3 & j_4 & j_{34} \\ m_3 & m_4 & m_{34} \end{pmatrix} \times \begin{pmatrix} j_{13} & j_{24} & J \\ m_{13} & m_{24} & M \end{pmatrix} \begin{pmatrix} j_1 & j_3 & j_{13} \\ m_1 & m_3 & m_{13} \end{pmatrix} \times \begin{pmatrix} j_2 & j_4 & j_{24} \\ m_2 & m_4 & m_{24} \end{pmatrix} \begin{pmatrix} j_{12} & j_{34} & J \\ m_{12} & m_{34} & M \end{pmatrix}, \tag{30.15}$$

or equivalently in terms of $6J$ symbols as

$$\begin{Bmatrix} j_1 & j_2 & j_{12} \\ j_3 & j_4 & j_{34} \\ j_{13} & j_{24} & J \end{Bmatrix} = \sum_j (-1)^{2j}(2j+1) \begin{Bmatrix} j_1 & j_3 & j_{13} \\ j_{24} & J & j \end{Bmatrix} \times \begin{Bmatrix} j_2 & j_4 & j_{24} \\ j_3 & j & j_{34} \end{Bmatrix} \begin{Bmatrix} j_{12} & j_{34} & J \\ j & j_1 & j_2 \end{Bmatrix}. \tag{30.16}$$

30.3.2 Transformation Between *L–S* and *J–J* Coupling

An important use of $9J$ coefficients occurs in the transformation between L–S and J–J coupling in atomic or nuclear physics. In particular, we have

$$|(l_1 l_2)L(s_1 s_2)S; JM\rangle = \sum_{j_1 j_2} \sqrt{(2S+1)(2L+1)} \times \sqrt{(2j_1+1)(2j_2+1)} \begin{Bmatrix} l_1 & l_2 & L \\ s_1 & s_2 & S \\ j_1 & j_2 & J \end{Bmatrix} |(l_1 s_1)j_1(l_2 s_2)j_2; JM\rangle \tag{30.17}$$

and the inverse relation

$$|(l_1 s_1)j_1(l_2 s_2)j_2; JM\rangle = \sum_{LS} \sqrt{(2S+1)(2L+1)} \times \sqrt{(2j_1+1)(2j_2+1)} \begin{Bmatrix} s_1 & l_1 & j_1 \\ s_2 & l_2 & j_2 \\ S & L & J \end{Bmatrix} |(l_1 l_2)L(s_1 s_2)S; JM\rangle \tag{30.18}$$

for such transformations.

30.3.3 Matrix Element of an Independent Tensor Product

Another important case in which the $9J$ symbol appears is in the reduced matrix element of a tensor product operator when each of the product tensors operates on a separate part of the system. For a coupled tensor

$$T_{KQ}(k_1k_2) = \sum_{q_1q_2} \langle k_1q_1k_2q_2|\, KQ\rangle\, R_{k_1q_1}(1)S_{k_2q_2}(2), \tag{30.19}$$

the matrix elements are

$$\begin{aligned}\langle j_1j_2JM|\, T_{KQ}(k_1k_2)\, |j_1'j_2'J'(M-Q)\rangle &= \sum_{m_1m_2}\sum_{m_1'm_2'}\sum_{q_1q_2} \langle j_1\, m_1\, j_2\, m_2|\; J\, M\rangle \\ &\times \langle j_1'm_1'j_2'm_2'|\; J'(M-Q)\rangle\, \langle k_1q_1k_2q_2|\; KQ\rangle \\ &\times \langle j_1JM|\, R_{k_1q_1}(1)\, |j_1'J'(M-Q)\rangle\, \langle j_2JM|\, S_{k_2q_2}(2)\, |j_2'J'(M-Q)\rangle\,.\end{aligned}$$

Introducing reduced matrix elements on both sides by the Wigner–Eckart theorem gives an expression that involves a sum over a product of six Clebsch–Gordan coefficients, which may be replaced by a $9J$ symbol to give

$$\begin{aligned}\langle j_1j_2J\,\|\,\boldsymbol{T}_K(k_1k_2)\,\|\,j_1'j_2'J'\rangle &= \sqrt{(2J'+1)(2K+1)(2j_1+1)(2j_2+1)} \\ &\times \begin{Bmatrix} J & J' & K \\ j_1 & j_1' & k_1 \\ j_2 & j_2' & k_2 \end{Bmatrix} \langle j_1\,\|\,\boldsymbol{R}_{k_1}(1)\,\|\,j_1'\rangle\langle j_2\,\|\,\boldsymbol{S}_{k_2}(2)\,\|\,j_2'\rangle,\end{aligned} \tag{30.20}$$

as the final result.

30.3.4 Matrix Element of a Scalar Product

A $9J$ symbol is proportional to a $6J$ symbol if one of the angular momentum arguments is zero. For example,

$$\begin{Bmatrix} a & b & c \\ d & e & f \\ g & h & 0 \end{Bmatrix} = \frac{(-1)^{b+d+g+c}}{\sqrt{(2c+1)(2g+1)}} \begin{Bmatrix} a & b & c \\ e & d & g \end{Bmatrix} \delta_{cf}\delta_{gh}, \tag{30.21}$$

and a $9J$ symbol with a zero in some other position can be evaluated from this expression using the symmetries of the $9J$ coefficients given in Section 30.3.1. There are two important cases for which the expression (30.20) reduces to one with a $6J$ symbol. (1) If either R or S is a scalar (k_1 or k_2 is zero), we recover the previous example of a tensor operating on only one part of a system. (2) If the operator is a *scalar product*,

$$\boldsymbol{R}_k \cdot \boldsymbol{S}_k \equiv (-1)^k\sqrt{2k+1}\, T_{00}(kk) = \sum_q (-1)^q R_{kq} S_{k-q}, \tag{30.22}$$

then $K = 0$ and

$$\langle j_1 j_2 J \,\|\, \boldsymbol{R}_k(1)\cdot \boldsymbol{S}_k(2) \,\|\, j_1' j_2' J'\rangle = (-1)^{j_1'+j_2+J}\delta_{JJ'}\sqrt{2J+1} \times \begin{Bmatrix} j_1 & j_1' & k \\ j_2' & j_2 & J \end{Bmatrix} \langle j_1 \,\|\, \boldsymbol{R}_k \,\|\, j_1'\rangle\langle j_2 \,\|\, \boldsymbol{S}_k \,\|\, j_2'\rangle. \tag{30.23}$$

For example, from Eq. (30.23) the matrix element of a spin–orbit operator $\boldsymbol{l}\cdot\boldsymbol{s}$ between coupled angular momentum states $|lsj\rangle$ is [29]

$$\begin{aligned}\langle lsj|\,\boldsymbol{l}\cdot\boldsymbol{s}\,|l's'j'\rangle &= \delta_{ss'}\delta_{ll'}\delta_{jj'}(-1)^{j-l-s}\sqrt{l(l+1)s(s+1)(2l+1)(2s+1)}\,W(llss;1j)\\ &= \frac{1}{2}\left[j(j+1)-l(l+1)-s(s+1)\right].\end{aligned}$$

In this particular case it would be easier and faster to use the vector-coupling model and evaluate the matrix elements of $2\,\boldsymbol{l}\cdot\boldsymbol{s} = \boldsymbol{j}^2 - \boldsymbol{l}^2 - \boldsymbol{s}^2$, but the tensor method discussed here is more general and therefore more flexible in application to complex problems.

Background and Further Reading

Tensor operators for SO(3) and the associated angular momentum algebra are discussed extensively in Brink and Satchler [29], deShalit and Feshbach [48], Edmonds [51], Rose [173], Varshalovich, Moskalev, and Khersonskii [201], and Wybourne [224].

Problems

30.1 Prove that for $6J$ symbols

$$\begin{Bmatrix} a & b & c \\ d & e & 0 \end{Bmatrix} = \frac{(-1)^{a+b+c}}{\sqrt{(2a+1)(2b+1)}}\delta_{ae}\delta_{bd}.$$

Note that since the $6J$ symbol is invariant under interchange of columns or the interchange of upper and lower arguments in any two columns, this formula can be used to evaluate $6J$ coefficients with a zero in any position. *Hint*: A $6J$ recoupling coefficient can be expressed in terms of sums over products of $3J$ coefficients [48],

$$\begin{Bmatrix} j_1 & j_2 & j_3 \\ \ell_1 & \ell_2 & \ell_3 \end{Bmatrix} = \sum (-1)^{-l_3-j_3-j_1-j_2-\ell_1-\ell_2-m_1-m_1'} \times \begin{pmatrix} j_1 & j_2 & j_3 \\ m_1 & m_2 & -m_3 \end{pmatrix}\begin{pmatrix} \ell_1 & \ell_2 & j_3 \\ m_1' & m_2' & m_3 \end{pmatrix}\begin{pmatrix} j_2 & \ell_1 & \ell_3 \\ m_2 & m_1' & -m_3' \end{pmatrix}\begin{pmatrix} \ell_2 & j_1 & \ell_3 \\ m_2' & m_1 & m_3' \end{pmatrix},$$

where the sum is over all m and m'.

30.2 Prove that

$$\begin{Bmatrix} j_1 & j_2 & J \\ j_1' & j_2' & J' \\ k & k' & 0 \end{Bmatrix} = \frac{(-1)^{j_2+j_1'+k+J}}{\sqrt{(2J+1)(2k+1)}} \begin{Bmatrix} j_1 & j_2 & J \\ j_2' & j_1' & k \end{Bmatrix} \delta_{kk'}\delta_{JJ'}.$$

Hint: Use the results of Problem 30.1.

30.3 For orbital angular momentum L, spin angular momentum S, total angular momentum J, and projection M of the total angular momentum, use the Wigner–Eckart theorem to express $\langle LSJM|\, L_z + 2S_z \,|LSJM\rangle$ in terms of reduced matrix elements. Evaluate the reduced matrix elements to show that $\langle LSJM|\, L_z + 2S_z \,|LSJM\rangle = Mg$, where the *Landé g-factor* is

$$g \equiv 1 + \frac{J(J+1) + S(S+1) - L(L+1)}{2J(J+1)}.$$

Hint: Consult Example 6.9, and note that L operates only on the orbital part and S only on the spin part of the wavefunction, so Eq. (30.12) is appropriate. ***

30.4 Two-body matrix elements for particles moving in central potentials are important in many areas of physics. These typically involve the matrix elements of Legendre polynomials, which can be written using the spherical harmonic addition theorem as

$$P_k(\cos\theta_{12}) = \frac{4\pi}{2k+1}\sum_q Y^*_{kq}(\theta_1,\phi_1)Y_{kq}(\theta_2,\phi_2) \equiv \boldsymbol{C}^{(1)}_k \cdot \boldsymbol{C}^{(2)}_k.$$

Calculate the reduced matrix element $\langle l_1 l_2 L \,\|\, \boldsymbol{C}_k(1)\cdot \boldsymbol{C}_k(2) \,\|\, l_1' l_2' L'\rangle$.

30.5 Evaluate the matrix element $\langle j_1 j_2 J|\, T_{kq}(1) \,\big|j_1' j_2' J'\big\rangle$, where the tensor operator $T_{kq}(1)$ operates only on the part of the system associated with the angular momenta labeled by the subscript 1, to obtain Eq. (30.12). ***

30.6 Use Eq. (30.12) to evaluate the reduced matrix element of a spherical harmonic between states that are l–s coupled to good total angular momentum j; thus obtain Eq. (30.13). *Hint*: The reduced matrix element of a spherical harmonic $\langle l|\, \boldsymbol{Y}_L \,|l'\rangle$ between states of good l was evaluated previously in Eq. (6.74). ***

30.7 Find the reduced matrix element $\langle j_1 j_2 J \,\|\, \boldsymbol{J}_1 \cdot \boldsymbol{J}_2 \,\|\, j_1' j_2' J\rangle$ where $\boldsymbol{J} = \boldsymbol{j}_1 + \boldsymbol{j}_2 = \boldsymbol{j}_1' + \boldsymbol{j}_2'$ and where $\boldsymbol{J}_1$ and $\boldsymbol{J}_2$ are angular momentum operators operating on the indices 1 and 2, respectively, with $\langle J \,\|\, \boldsymbol{J} \,\|\, J'\rangle = \delta_{JJ'}$.

30.8 Use tensor methods to evaluate the reduced matrix element of the spin–orbit interaction $\langle J \,\|\, \boldsymbol{l}\cdot\boldsymbol{s} \,\|\, J'\rangle$ between states of good angular momentum $\boldsymbol{J} = \boldsymbol{l} + \boldsymbol{s}$.

31 Nuclear Fermion Dynamical Symmetry

Dynamical symmetry applied to nuclear structure physics has a long history tracing back to Wigner supermultiplet theory, SU(2) quasispin models, and the Elliott SU(3) model described in Ch. 10. These models have had broad conceptual influence but more limited practical application in the full context of nuclear structure physics because the conditions for their application are realized only in some nuclei. In this chapter we introduce more modern applications of dynamical symmetry to nuclear structure that use the properties of Lie algebras and Lie groups to describe the systematics of nuclear structure across the full range of nuclear species. Our primary emphasis will be on the *Fermion Dynamical Symmetry Model (FDSM),* which represents a truncation of the full Hilbert space of the nuclear structure problem to a collective subspace where emergent states live by using the principles of dynamical symmetry outlined in Ch. 20. The methods to be discussed here are based generally on the nuclear shell model, which was introduced in Fig. 10.1 and is illustrated more completely in Fig. 31.1 and Box 31.1. The central thesis of the FDSM is that collective modes can be described efficiently in a basis of shell model fermion pairs coupled to angular momentum zero (S pairs) and angular momentum two (D pairs). The FDSM is an ambitious extension of the *Ginocchio model,* so we begin there.

31.1 The Ginocchio Model

Ginocchio [74] introduced a model where the total angular momentum $\boldsymbol{j}$ of a shell model particle is decomposed into a *pseudo-orbital* angular momentum $\boldsymbol{k}$, where $k = |\boldsymbol{k}|$ is integer, and a *pseudospin* angular momentum $\boldsymbol{i}$, where $i = |\boldsymbol{i}|$ is half-integer,

$$\boldsymbol{j} = \boldsymbol{k} + \boldsymbol{i}. \tag{31.1}$$

The single-nucleon creation operator $b^\dagger_{km_k im_i}$ in the k–i scheme is then related to the usual shell model nucleon creation operator by,[1]

$$a^\dagger_{jm} = \sum_{m_i, m_k} \langle k\, m_k\, i\, m_i |\, j\, m\rangle\, b^\dagger_{km_k im_i}, \tag{31.2}$$

where $\langle k\, m_k\, i\, m_i |\, j\, m\rangle$ is a Clebsch–Gordan coefficient.

[1] Clebsch–Gordan coefficients are elements of a unitary transformation so Eq. (31.2) does not change any physics. However, it suggests new ways to approximate and solve the interacting many-fermion problem.

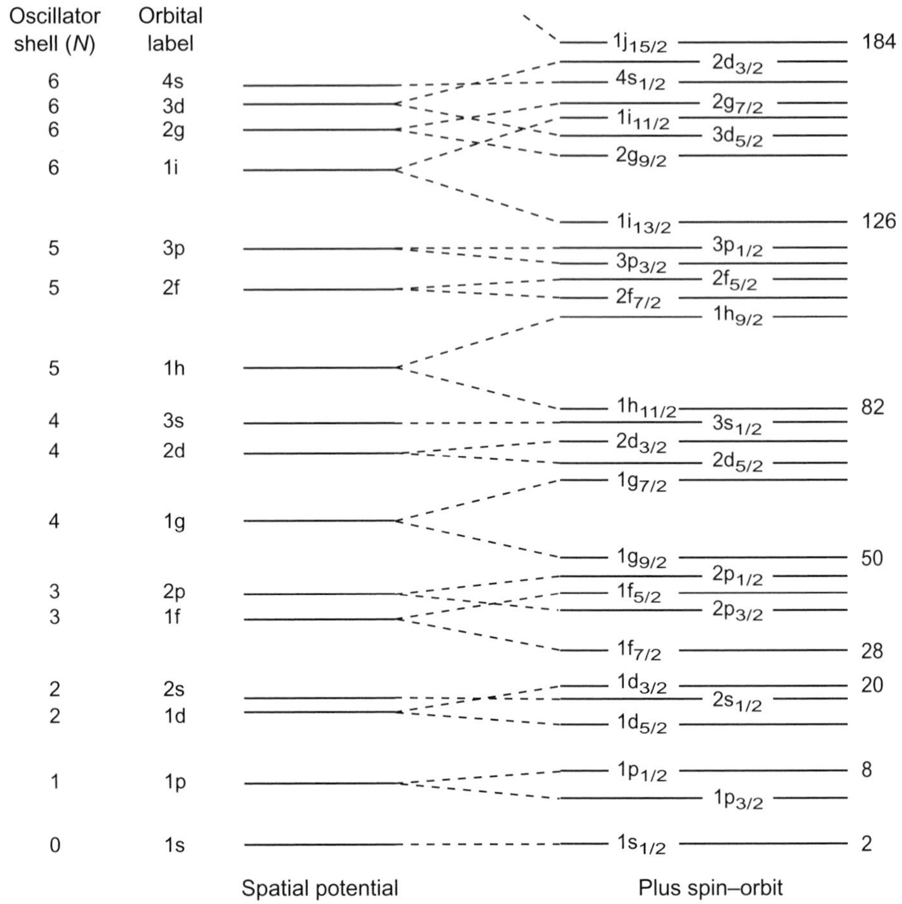

Fig. 31.1 The nuclear shell model. Energy levels labeled "Spatial potential" correspond to those in the realistic potential of Fig. 10.1(a). Energy levels labeled "Plus spin–orbit" correspond to splitting of orbital degeneracies by spin–orbit coupling. Numbers on the right give "magic numbers" corresponding to shell-gap closures. Since neutron and proton single-particle potentials differ due to issues like Coulomb interaction between protons, separate shell structures for neutrons and protons are typically used except in the lightest nuclei.

Dynamical Symmetry of the k–i Coupling Scheme: There are two alternatives for constructing S and D fermion pairs that lead to a dynamical symmetry group structure [74].

- The *k-active scheme*: choose $k = 1$ and couple i to zero for the pair.
- The *i-active scheme*: choose $i = \frac{3}{2}$ and couple k to zero for the pair.

These schemes, along with the $J = 0$ pairing scheme implemented in quasispin models (see Problem 31.1) that will be discussed further below, are illustrated in Fig. 31.2. Commutation of the single-particle and two-particle creation and annihilation operators closes an SO(8) Lie algebra for the i-active scheme and an Sp(6) Lie algebra for the

Box 31.1

The Nuclear Shell Model

The starting point for most dynamical symmetry applications in nuclear structure physics is the *nuclear shell model,* which is a mean-field, single-particle description that is illustrated in Fig. 10.1 for the first few shells, and is extended to shells for heavier nuclei in Fig. 31.1.

Normal-Parity and Abnormal-Parity Orbitals

An important characteristic of nuclear shells is that for heavier nuclei a given shell contains several *normal-parity orbitals* all of the same parity, and a single *abnormal-parity orbital* of opposite parity that has been lowered in energy by spin–orbit interaction from the shell above (it is sometimes called the *intruder level*).

> ***Example:*** In Fig. 31.1 the shell lying between magic numbers 82 and 126 contains five odd-parity orbitals that arise from the $N = 5$ oscillator shell, and one even-parity $i_{13/2}$ orbital that arises in the $N = 6$ oscillator shell but has been lowered strongly in energy by the spin–orbit interaction.

Collective emergent states correspond to strongly mixed shell model orbitals. For heavier nuclei this mixing is essentially different for normal-parity and abnormal-parity orbitals within a shell: there are multiple normal-parity orbitals that can mix strongly, but only one abnormal-parity orbital, which cannot mix easily because it has been shifted in energy from its like-parity siblings by the spin–orbit interaction. Thus, to first approximation we expect normal-parity orbitals within a shell to be able to hybridize into highly collective configurations, but the single abnormal-parity orbital will be largely unmixed and retain its shell model character.

Truncation of the Shell Model Space

Shell model spaces for complex nuclei are huge, so solutions require truncation to a more manageable subspace. We are particularly interested in truncations to a *collective subspace* that separates out the part of the Hilbert space primarily responsible for (collective) emergent modes. Typical truncation schemes use energy as a criterion, retaining only lower-energy single-particle states in the basis. In this chapter we explore an alternative truncation in terms of dynamical symmetries.

k-active scheme, with the subgroup chains (dynamical symmetries) illustrated in Fig. 31.3. These dynamical symmetries will be explained in more detail below.

Dynamical Symmetry Truncation of Hilbert Space: If a Hamiltonian is constructed from rotationally invariant [SO(3) scalar] combinations of the SO(8) and Sp(6) generators, the (S, D) subspace of angular momentum $J = 0$ and $J = 2$ fermion pairs is decoupled from the remainder of the shell model space, thus implementing by group-theoretical means the collective subspace truncation illustrated schematically in Fig. 20.1(a). This

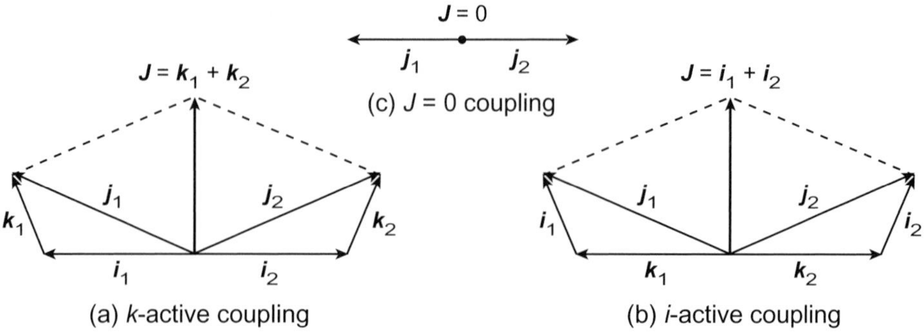

Fig. 31.2 The (a) k-active and (b) i-active coupling schemes of the Ginocchio model, and (c) the $J = 0$ coupling scheme of a quasispin pairing model.

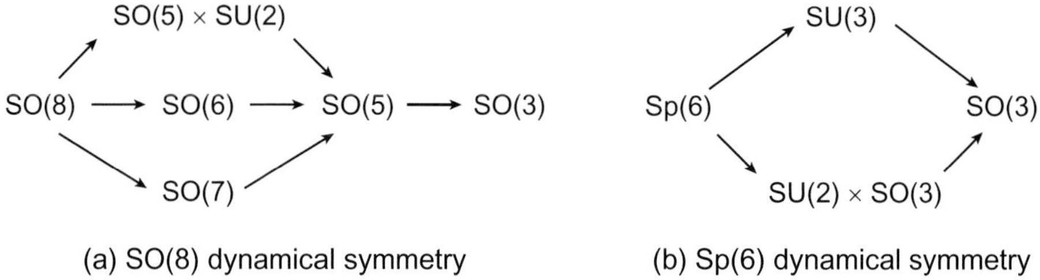

Fig. 31.3 Dynamical symmetries of the Ginocchio model that end in the SO(3) group representing conservation of angular momentum. (a) The three SO(8) subgroup chains corresponding to i-active coupling. (b) The two Sp(6) subgroup chains corresponding to k-active coupling.

truncation and decoupling is the basis for the dynamical symmetry models discussed in this chapter.

31.2 The Fermion Dynamical Symmetry Model

The Fermion Dynamical Symmetry Model (FDSM) [215, 216] is a theoretical framework for nuclear structure physics that has been described in an extensive review [218]. The overall model is rather technical since it is shaped by specialized knowledge from nuclear structure physics. In particular, the equations are complicated by the angular momentum algebra (Chs. 6 and 30) that is required because nuclear physics experiments often measure states of good angular momentum. In this discussion we shall provide a high-level overview of the FDSM as an illustration of the dynamical symmetry approach that was introduced in Ch. 20. This overview will emphasize the role of Lie algebras, Lie groups, and fermion dynamical symmetry in the physical description of nuclei, while omitting many of the more technical details (which the interested reader may find in the review [218] and references cited there).

31.2.1 Dynamical Symmetry Generators

The Fermion Dynamical Symmetry Model may be viewed as an extension of the Ginocchio model described in Section 31.1 that

1. provides a firm basis for relating the Ginocchio model to the nuclear shell model,
2. establishes a framework enabling extensive comparisons with data, and
3. cures some issues (both real and imagined) that were thought originally to restrict the utility of the Ginocchio model.

For the physical reasons discussed in Box 31.1, the FDSM adopts the k–i coupling scheme of the Ginocchio model illustrated in Fig. 31.2(a,b) for the normal-parity orbitals of nuclear shells, but employs the $J = 0$ pair coupling scheme of Fig. 31.2(c) for abnormal-parity orbitals. In the k–i basis the FDSM generators associated with normal-parity orbitals are

$$S^\dagger = \sqrt{\frac{\Omega_{ki}}{2}} \left[b^\dagger_{ki} b^\dagger_{ki}\right]^{00}_{00} \qquad \text{(any } k, i\text{)}, \tag{31.3a}$$

$$D^\dagger_\mu = \begin{cases} \sqrt{\frac{\Omega_{1i}}{2}} \left[b^\dagger_{1i} b^\dagger_{1i}\right]^{20}_{\mu 0} & (k\text{-active}), \\ \sqrt{\frac{\Omega_{k3/2}}{2}} \left[b^\dagger_{k3/2} b^\dagger_{k3/2}\right]^{02}_{0\mu} & (i\text{-active}), \end{cases} \tag{31.3b}$$

$$P^r_\mu = \begin{cases} \sqrt{\frac{\Omega_{1i}}{2}} \left[b^\dagger_{1i} \tilde{b}_{1i}\right]^{r0}_{\mu 0} & (k\text{-active}, r = 0, 1, 2), \\ \sqrt{\frac{\Omega_{k3/2}}{2}} \left[b^\dagger_{k3/2} \tilde{b}_{k3/2}\right]^{0r}_{0\mu} & (i\text{-active}, r = 0, 1, 2, 3), \end{cases} \tag{31.3c}$$

where the corresponding pair annihilation operators are $S = (S^\dagger)^\dagger$ and $D = (D^\dagger)^\dagger$, and

$$\Omega_{ki} \equiv \frac{1}{2}(2k+1)(2i+1) \qquad \tilde{b}_{km_k im_i} \equiv (-1)^{k-m_k+i-m_i} b_{km_k im_i}, \tag{31.4}$$

where Ω_{ki} is the pair degeneracy and where a notation like $[b^\dagger_{ki}\tilde{b}_{ki}]^{KI}_{m_K m_I}$ means that the k values are angular momentum coupled to total (K, m_K) and the i values are angular momentum coupled to total (I, m_I), using Clebsch–Gordan coefficients or $3J$ symbols.

For the abnormal-parity orbitals the collective degrees of freedom are assumed to be dominated by monopole pairing in light of the arguments in Box 31.1, implying the quasispin SU(2) symmetry explored in Problems 31.1 through 31.6, with generators

$$\mathbb{S}^\dagger = \sqrt{\frac{\Omega_{j0}}{2}} \left[a^\dagger_{j0} a^\dagger_{j0}\right]^0_0 \qquad \mathbb{S} = (\mathbb{S}^\dagger)^\dagger \qquad \mathbb{S}_0 = \frac{n_0 - \Omega_{j0}}{2}, \tag{31.5}$$

where the number of particles in the abnormal-parity orbital n_0 is given by

$$n_0 = 2\sqrt{\frac{\Omega_{j0}}{2}} \left[a^\dagger_{j0} \tilde{a}_{j0}\right]^0_0. \tag{31.6}$$

We shall use the italicized symbol $SU(2)$ to denote this quasispin symmetry group for the abnormal-parity orbitals, to distinguish it from SU(2) symmetries for normal-parity orbitals.

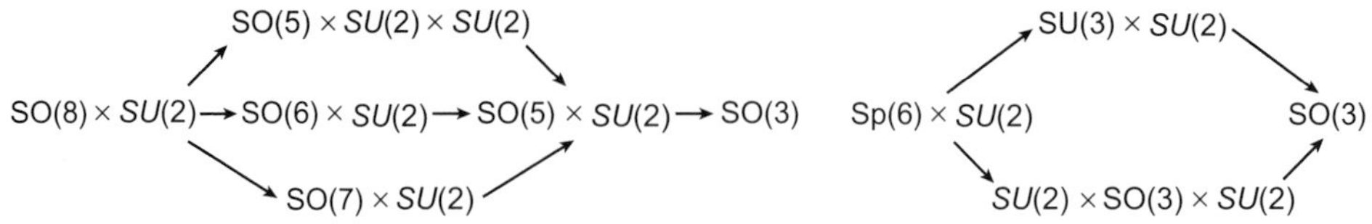

Fig. 31.4 The five primary dynamical symmetry subgroup chains of the FDSM. Italicized *SU*(2) symbols denote quasispin symmetry for abnormal-parity orbitals. For brevity the left pattern will be termed the SO(8) symmetry and the right pattern the Sp(6) symmetry.

31.2.2 The FDSM Dynamical Symmetries

In Eq. (31.3b) the index μ can take five values and in Eq. (31.3c) it can take $2r + 1$ values. Thus from Eqs. (31.3) there are 21 generators for k-active coupling,

$$G_{\mathrm{Sp}(6)} = \{S, S^\dagger, D_\mu, D^\dagger_\mu, P^0, P^1_\mu, P^2_\mu\},$$

and commutation of these generators closes an Sp(6) Lie algebra. Likewise, in Eqs. (31.3) there are 28 generators for i-active coupling,

$$G_{\mathrm{SO}(8)} = \{S, S^\dagger, D_\mu, D^\dagger_\mu, P^0, P^1_\mu, P^2_\mu, P^3_\mu\},$$

and commutation of these generators closes an SO(8) Lie algebra [218].[2] The FDSM dynamical symmetry subgroup chains that describe the low-energy nuclear structure of heavy nuclei are displayed in Fig. 31.4. These are basically the Ginocchio chains of Fig. 31.3, but modified to account consistently for the role of normal-parity and abnormal-parity orbitals in the nuclear shell model. All subgroup chains end in the group SO(3) corresponding to conservation of angular momentum.

Utilizing the microscopic dynamical symmetry methods introduced in Ch. 20, physical matrix elements may be determined for the FDSM dynamical symmetry chains of Fig. 31.4 and these may be used to attach physical interpretations to the dynamical symmetries. As is common in nuclear structure physics, the physical nature of collective states will often be expressed in the geometrical language of rotors and vibrators. Of the five dynamical symmetry chains sketched in Fig. 31.4, two have the character of collective rotational states, two have the character of collective vibrations, and one has the character of a critical transitional symmetry. Let us describe them briefly.

The SU(3) Axially Symmetric Rotor: The Sp(6) ⊃ SU(3) dynamical symmetry limit has the matrix elements and character of a collective, axially symmetric rotor.

The SO(6) γ-Soft Rotor: The SO(8) ⊃ SO(6) dynamical symmetry limit has matrix elements suggesting an interpretation as a rotor, but one soft against deviations from axial symmetry. In nuclear physics this is known as a γ-soft rotor.

[2] In the table of classical Lie algebras given in Appendix D, the Sp(6) algebra corresponds to Sp($2L$) with $L = 3$ and the SO(8) algebra corresponds to SO($2L$) with $L = 4$.

The SU(2) Spherical Vibrator: The Sp(6) $\supset$ SU(2) $\times$ SO(3) dynamical symmetry has the matrix elements of a spherically symmetric vibrational state.

The SO(5) Spherical Vibrator: The SO(8) $\supset$ SO(5) $\times$ SU(2) dynamical symmetry has the characteristics of a spherical vibrator, but with differences in microscopic detail relative to the SU(2) spherical vibrator.

The SO(7) Critical Dynamical Symmetry: The SO(8) $\supset$ SO(7) dynamical symmetry is a kind of soft vibrator, but with very unusual properties. Historically it was the first discovered example of the *critical dynamical symmetries* discussed in Box 20.3.[3]

31.2.3 FDSM Irreducible Representations

Many techniques introduced in earlier chapters may be used to determine the FDSM irrep structure, in particular the method of Young diagrams as illustrated for the Elliott SU(3) model in Section 10.3. Once the irreps are constructed, general dynamical symmetry methods may be used to deduce matrix elements of physical relevance. This topic is considered in much more depth in the review [218], but here we will illustrate with a few examples for the Sp(6) $\supset$ U(3) $\supset$ SU(3) dynamical symmetry. Let us assume n_1 valence fermions, corresponding to $N_1 = \frac{1}{2}n_1$ fermion pairs, participating in the SU(3) dynamical symmetry.

Allowed Partitions: The states and quantum numbers may be obtained from the allowed partitions $[h_1, h_2, h_3]$ of U(3) Young diagrams with symmetric pairs $\square\!\square$ as building blocks, where $h_1 + h_2 + h_3 = n_1 = 2N_1$ is the number of particles participating in the SU(3) symmetry, h_i $(i = 1, 2, 3)$ is the number of boxes in row i of the diagram, and $h_1 \geq h_2 \geq h_3$. Each valid Young diagram then is an SU(3) irrep with SU(3) quantum numbers (λ, μ),

$$\lambda \equiv h_1 - h_2 \qquad \mu \equiv h_2 - h_3. \tag{31.7}$$

The allowed values of the SU(3) quantum numbers are[4]

$$(\lambda, \mu) = \begin{cases} (2N_1, 0), (2N_1 - 4, 2), \dots, (0, N_1) \text{ or } (2, 2N_1 - 1) \\ (2N_1 - 6, 0), (2N_1 - 10, 2), \dots, (0, N_1 - 3) \text{ or } (2, 2N_1 - 4) \\ (2N_1 - 12, 0), (2N_1 - 16, 2), \dots, (0, N_1 - 6) \text{ or } (2, 2N_1 - 7), \dots \end{cases} \tag{31.8}$$

and for a given SU(3) irrep the possible angular momenta J are given by

$$J = \begin{cases} K_{\rm m}, K_{\rm m} - 2, K_{\rm m} - 4, \dots, 0 \text{ or } 1 & (\kappa = 0), \\ \kappa, \kappa + 1, \kappa + 2, \dots, \kappa + K_{\rm m} & (\kappa \neq 0), \end{cases} \tag{31.9}$$

[3] The SO(7) dynamical symmetry was proposed as a new nuclear collective mode by the FDSM and a survey of nuclear data gave compelling evidence for modes with the predicted properties [35]. The quantum critical nature of the SO(7) phase was then elucidated in FDSM coherent state calculations [229, 230], and later work in various fields found additional examples of critical dynamical symmetries, as summarized in Box 20.3.

[4] The sequences in Eqs. (31.8)–(31.10) are understood to terminate if a quantity becomes negative.

where we define

$$K_0 \equiv \min(\lambda, \mu) \qquad K_\mathrm{m} \equiv \max(\lambda, \mu) \qquad \kappa = K_0, K_0 - 2, K_0 - 4, \ldots, 0 \text{ or } 1 \tag{31.10}$$

and there are $2J + 1$ values of M for each J.

Eigenstates and Spectrum: An irreducible basis for states with N_1 unbroken pairs in the SU(3) dynamical symmetry chain may be denoted by $|N_1(\lambda\mu)\kappa JM\rangle$, where $(\lambda\mu)$ labels SU(3) irreps according to Eq. (31.8), κ is is defined in Eq. (31.10), J is the total angular momentum of the state, and M labels the $2J+1$ projection quantum numbers corresponding to J. Assuming a minimal spherically symmetric Hamiltonian the eigenvalues are

$$E(N_1(\lambda\mu)\kappa J) = E_0(N_1) - \beta C(\lambda, \mu) + \alpha J(J + 1), \tag{31.11}$$

where we take E_0 to be constant, α and β depend on the effective interactions, and

$$C(\lambda, \mu) = 2C_{\mathrm{su(3)}}(\lambda, \mu) = \lambda^2 + \mu^2 + \lambda\mu + 3\lambda + 3\mu, \tag{31.12}$$

with $C_{\mathrm{su(3)}}$ the quadratic SU(3) Casimir operator.

Example 31.1 Spectra for SU(3) irreps corresponding to $N_1 = 4$ pairs (8 fermions) are illustrated in Fig. 31.5, where Eqs. (31.7)–(31.12) were used. You are asked to work these out in Problem 31.7. The spectral pattern of Fig. 31.5 is common in many heavy nuclei and the FDSM accounts quantitatively for low-lying spectra in such nuclei.

General Matrix Elements: Once the representation structure and spectrum have been determined, standard dynamical symmetry methods may be used to calculate both diagonal and non-diagonal matrix elements for more general physical observables within the SU(3) dynamical symmetry basis $|N_1(\lambda\mu)\kappa JM\rangle$. For example, the reduced electric quadrupole transition rates $B(E2)$ for transitions between the states of the irrep $(2N_1, 0)$ of Eq. (31.8) and Fig. 31.5 (which may be measured in γ-ray spectroscopy experiments) are given by [218]

$$B(E2, J + 2 \to J) = \frac{3}{4} C^2 \frac{(J + 1)(J + 2)}{(2J + 3)(2J + 5)} (n_1 - J)(n_1 + J + 3), \tag{31.13}$$

where the single parameter C depends on the effective charge for electromagnetic transitions and may be determined empirically for a given set of nuclei. The result (31.13) can be obtained analytically because the quadrupole operator in the matrix element that must be evaluated is a generator of SU(3). This is expected to be a common feature of dynamical symmetries. If the transition were not caused by an operator that is a generator of SU(3), the transition would take us out of an SU(3) multiplet and break the symmetry.

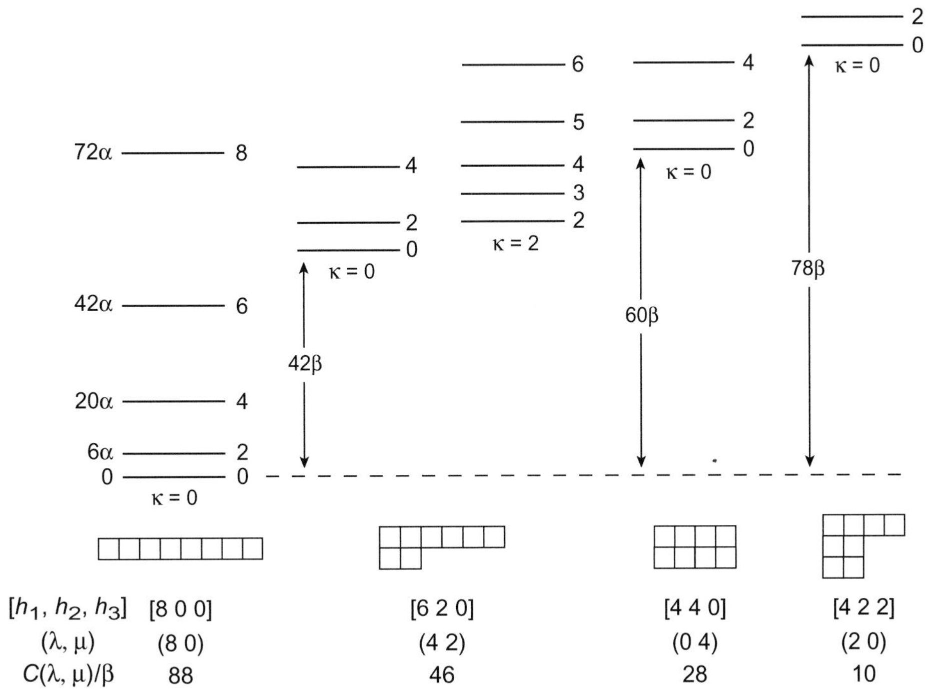

Fig. 31.5 Spectrum resulting from the eigenvalues (31.11) for some FDSM SU(3) irreducible representations for 8 fermions labeled by Young diagrams, U(3) partitions $[h_1, h_2, h_3]$, SU(3) quantum numbers (λ, μ), and Casimir energy $C(\lambda, \mu)$ [218]. States in the spectrum are labeled by angular momentum J and have energy $\alpha J(J+1)$ relative to the bandhead (these energies are displayed for the [800] band).

> A dynamical symmetry implies an irrep structure and diagonal matrix elements like an energy spectrum, but it also implies that *transition operators must be proportional to group generators* if the symmetry is not to be broken by dynamics.

This implies that both diagonal and non-diagonal matrix elements for a dynamical symmetry may be evaluated analytically using the group properties.

31.2.4 Quantitative FDSM Calculations

The FDSM leads to a systematic global classification of nuclear structure, but also accounts *quantitatively* for experimental quantities. Many examples are summarized in the review [218] and references cited there. Here we restrict discussion to a few representative cases.

Electric Quadrupole Transition Rates: A quantitative description of electric quadrupole transition rates in heavy rare earth nuclei is illustrated in Fig. 31.6. The dashed curves are a bosonic approximation that neglects the influence of the Pauli principle on the structure of collective states. The bosonic approximation is good only for a small number of valence particles, which minimizes the Pauli effect. Near the middle of the shells (dashed vertical

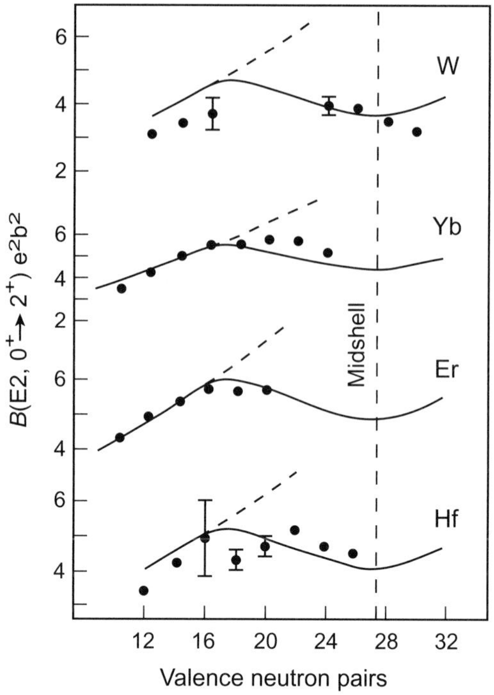

Fig. 31.6 Reduced electric quadrupole transition rates $B(\mathrm{E2}, 0^+ \rightarrow 2^+)$ for ground state rotational bands in 29 rare earth isotopes [217]. Units are e^2b^2, where e is the charge unit and b (barns) $= 10^{-24}$ cm. Dots are data, solid curves are FDSM calculations, and dashed curves are a bosonic approximation neglecting the Pauli principle. Midshell for neutron pairs is indicated by the vertical dashed line. There is a single adjustable parameter: the same electromagnetic coupling strength C^2 in Eq. (31.13) applied to all 29 isotopes.

line in Fig. 31.6) the data and the FDSM calculations exhibit a strong suppression reflecting the competition between the correlations, which would increase quadrupole collectivity, and the Pauli effect, which blocks occupation of some orbitals that would otherwise lead to optimal quadrupole correlation.

This essential tension between occupation of states that would enhance correlation and Pauli blocking of those orbitals is a fundamental property of strongly correlated fermionic states termed the *dynamical Pauli effect* [60, 92]. It manifests itself in the *Pauli blocking of FDSM SU(3) irreducible representations* that would be permitted to contribute in a bosonic approximation. Hence, the bosonic approximation predicts that for SU(3) symmetry optimal collectivity would occur near half filling of the valence neutron shell, whereas the FDSM predicts maximum SU(3) collectivity near $\sim \frac{1}{3}$ shell filling and a local minimum in the collective quadrupole strength near midshell because of the SU(3) dynamical Pauli effect. It is obvious from Fig. 31.6 that the data support the FDSM prediction, confirming the role of the SU(3) dynamical Pauli effect in fermionic collective states.[5]

[5] Suppression of collectivity by the Pauli principle depends on the dynamical symmetry (physical nature of the collective mode). For the SO(6) fermion dynamical symmetry maximum quadrupole collectivity is predicted

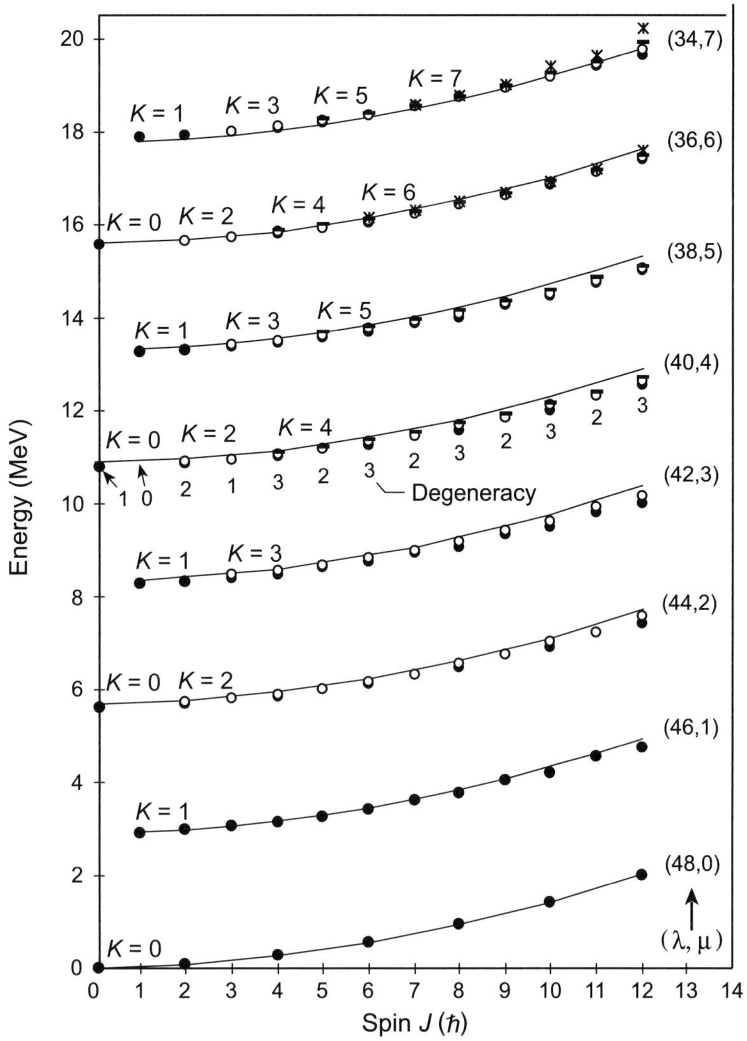

Fig. 31.7 Spectrum corresponding to coupled rotations of neutrons and protons in ^{168}Er. Symbols indicate numerical Projected Shell Model calculations [188, 189], with states labeled by spin *J* and projection *K*. Curves are analytical FDSM results labeled by SU(3) quantum numbers (λ, μ). Many states are degenerate, as indicated explicitly for the (40, 4) band. Reproduced with permission from APS: *Physical Review Letters*, Scissors-Mode Vibrations and the Emergence of SU(3) Symmetry from the Projected Deformed Mean Field, Y. Sun, C.-L. Wu, K. Bhatt, M. W. Guidry, and D. H. Feng, **80**, 672 (1998). Copyright (1998) by the American Physical Society.

Numerical Corroboration of Emergent FDSM Symmetries: Figure 31.7 displays evidence from a numerical solution of the Projected Shell Model (PSM) [105] for the

to occur at midshell, as supported by data. The reason for the difference between the SU(3) and SO(6) fermion dynamical symmetries is that they have different irreducible representation structure so the Pauli effect acts differently on them in collective states. This is discussed more extensively in Refs. [60, 92, 218].

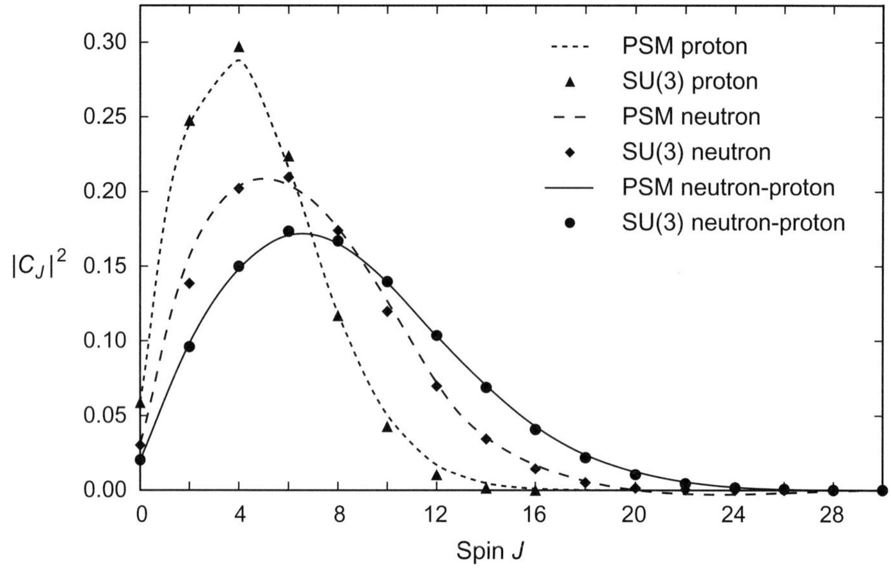

Fig. 31.8 Wavefunction amplitudes for numerical PSM and analytical SU(3) calculations [189]. Points are PSM results and curves are FDSM results. Reproduced with permission from Elsevier: *Nuclear Physics*, SU(3) symmetry and scissors mode vibrations in nuclei, Y. Sun, C.-L Wu, K. Bhatt, and M. W. Guidry, **A703**, 130 (2002). Copyright Elsevier (2002).

existence of effective FDSM SU(3) symmetries in heavy nuclei. The regular pattern of bands up to 20 MeV of excitation and $12\hbar$ in spin found in PSM calculations and the agreement of these hundreds of states with the analytical FDSM predictions is remarkable. The PSM result is a numerical diagonalization that knows nothing of SU(3) symmetry. Conversely, the analytical SU(3) calculation knows nothing of the numerical details corresponding to the PSM calculation; it asserts only that the complicated PSM numerical manipulations should conspire to produce an emergent state described by SU(3) symmetry in which the complex quasiparticle interactions are absorbed into a few effective interaction parameters and the resulting state is described by a simple spectrum and pattern of physical matrix elements, as observed. That this is not just a remarkably fortuitous accident is supported by Fig. 31.8, which compares wavefunctions for the numerical PSM calculation and an SU(3) dynamical symmetry calculation. The agreement between SU(3) and PSM calculations confirms that indeed a collective SU(3) symmetry consistent with the predictions of the FDSM has emerged from the purely numerical PSM calculations.

31.3 The Interacting Boson Model

The Interacting Boson Model (IBM) [116] is an approach similar in spirit to the FDSM. This largely phenomenological model pioneered the systematic use of dynamical symmetry to describe nuclear rotational and vibrational states, but assumed nuclear states to be described by pairs of nucleons obeying bosonic rather than fermionic statistics. Of course matter is composed of fermions, not bosons, but often correlations between fermions lead to collective states that can be approximated by interacting bosons. We have in fact seen evidence earlier of exactly when such an approximation is expected to be valid: if available fermionic states are only sparsely populated, the Pauli principle is less important and the distinction between fermions and bosons is minimized. As illustrated in Fig. 31.6, for small numbers of valence particles the FDSM calculation (solid curves) and the bosonic approximation neglecting Pauli effects (dashed curves) agree well with each other and with data. However, as valence spaces acquire increasing fermionic population the bosonic approximation diverges rapidly from the FDSM calculation and from data in Fig. 31.6.

Fermionic dynamical symmetries have a broader range of applicability than bosonic dynamical symmetries, and at a general level interacting bosons are the limit of interacting fermions if the Pauli principle is ignored. Thus we will not dwell on bosonic dynamical symmetries here but note that many dynamical symmetry concepts and group-theoretical techniques employed here for fermionic systems were developed originally within the context of bosonic systems, and that the IBM and FDSM are expected to give similar results for applications where the Pauli effect is of minimal importance. Readers interested in bosonic dynamical symmetries are referred to the prolific literature on the subject, starting with [116] and references therein.

Background and Further Reading

The Fermion Dynamical Symmetry Model is discussed in depth by Wu, Feng, and Guidry [218]. A complete introduction to the Interacting Boson Model may found in Iachello and Arima [116]. The relationship of the boson approximation to fermion dynamical symmetries is discussed in Ref. [218].

Problems

31.1 Shell models are important in various fields of physics. Consider a shell of fermions consisting of $(2j + 1)$ degenerate levels of angular momentum j, with each level labeled by a projection quantum number m (this is often called a "single j-shell model"). Define the fermion operators for a fixed angular momentum j as

$$s_+^{(m)} = a_m^\dagger a_{-m}^\dagger \qquad s_-^{(m)} = a_{-m}\, a_m \qquad s_0^{(m)} = \frac{1}{2}(a_m^\dagger a_m + a_{-m}^\dagger a_{-m} - 1),$$

where $a_m^\dagger$ and a_m are the usual fermion creation and annihilation operators, respectively, obeying anticommutation relations of the form (3.23). Show that these operators (called *quasispin* or *pseudospin* operators) close under

$$[\, s_+^{(m)}, s_-^{(m)} \,] = 2s_0^{(m)} \qquad [\, s_0^{(m)}, s_\pm^{(m)} \,] = \pm s_\pm^{(m)},$$

which is the SU(2) commutator algebra. Quasispin models are of physical relevance for fermion shells exhibiting a strong tendency for pairs of particles to couple to zero total angular momentum. Such models and their generalizations have found substantial application in nuclear, atomic, and condensed matter physics. ***

31.2 For the quasispin model of Problem 31.1, find the eigenvalues of $s_0^{(m)}$ for the levels labeled by m. Show that the system has a total quasispin $\boldsymbol{S}$ that is the vector sum of quasispins for each level m, which is the mathematical analog of total spin.

31.3 Show that for a Hamiltonian of the form

$$H = -G \sum_{m,m'>0} a^\dagger_{m'} a^\dagger_{-m'} a_{-m} a_m,$$

the energy eigenvalues for the quasispin model of Problem 31.1 are given by

$$E = G\left(S(S+1) + \frac{1}{4}(N-\Omega)^2 + \frac{1}{2}(N-\Omega)\right),$$

where N is half the particle number, $\Omega = \frac{1}{2}(2j+1)$ is half the shell degeneracy, and $S(S+1)$ is the eigenvalue of the operator

$$S^2 = \frac{1}{2}\sum_{m>0}\left(s_+^{(m)}s_-^{(m)} + s_-^{(m)}s_+^{(m)} + s_0^{(m)}s_0^{(m)}\right),$$

corresponding to the total quasispin. ***

31.4 *Seniority quantum numbers* typically measure how many fermions are in some sense "not paired" with another fermion. For the quasispin model of Problem 31.3, define the *Racah seniority* v through

$$S = \frac{1}{2}(\Omega - v) \qquad v = \begin{cases} 0, 2, 4, \ldots, N & \text{(for } N \text{ even)} \\ 1, 3, 5, \ldots, N & \text{(for } N \text{ odd).}\end{cases}$$

Show that (i) the absolute value of the energy depends on N and v but the spectrum for fixed N is a function only of v (degenerate in all other quantum numbers); (ii) the ground state corresponds to having all the "quasispins" aligned if G is positive and this is a $v = 0$ state; (iii) for an even number of particles the first excited state is $v = 2$, and its energy is independent of the number of particles. ***

31.5 Consider a system described by the quasispin model of Problem 31.3 for $G > 0$ with two identical fermions in a single j-shell. Show that the allowed seniorities are $v = 0, 2$, the allowed angular momenta are $J = 0, 2, \ldots, 2j-1$, and that

$$\begin{array}{ccc} v = 2 & \text{———————} & J = 2, 4, \ldots, 2j-1 \\ v = 0 & \text{———————} & J = 0 \end{array}$$

is the schematic form of the spectrum.

31.6 Prove that the binding energy of the ground state for the quasispin model in Problem 31.3 is linear in the number of pairs of particles N for small N.

31.7 Use Eqs. (31.7)–(31.12) to verify the irreducible representations, quantum numbers, and spectrum of Fig. 31.5. ***

32 Superconductivity and Superfluidity

This chapter considers some applications of Lie algebras, dynamical symmetries, and generalized coherent states to superconductivity (SC) and superfluidity (SF) in various many-body systems. The theory of conventional SC is based on the Bardeen–Cooper–Schrieffer or BCS formalism [18] and its improvements. In recent decades many *unconventional superconductors* have been discovered, with properties such as anomalously high SC transition temperatures that confound BCS expectations. We shall illustrate that dynamical symmetries can unify descriptions of conventional and unconventional superconductivity and superfluidity across many fields, and that such a symmetry based theory can give quantitative descriptions rivaling those of more conventional methods within those fields.

32.1 Conventional Superconductors

A *Fermi liquid* is a set of interacting fermions that can be transformed into non-interacting effective fermions in a series of small steps (Box 32.1). The final fermions, which may be greatly modified from the original fermions but still obey fermionic statistics, are termed *quasiparticles* (Section 22.4.2). The traditional view of superconductivity is based on the *Cooper instability* of a Fermi liquid, described in Box 32.2. The simple Cooper model illustrates the instability that leads to formation of conventional SC, but a superconducting state is a collective state involving many fermions. A many-body state exhibiting the physics of Cooper pairing is described by the BCS wavefunction of Eq. (22.24),

$$|\Psi_{\rm BCS}\rangle = \prod_{\boldsymbol{k}} (u_{\boldsymbol{k}} + v_{\boldsymbol{k}} c^{\dagger}_{\boldsymbol{k}\uparrow} c^{\dagger}_{-\boldsymbol{k}\downarrow})\, |0\rangle\,, \tag{32.1}$$

which is a coherent superposition of singlet (spin-0) Cooper-like pairs. We shall consider a *conventional superconductor* to be one described approximately by the BCS wavefunction (32.1). *Unconventional superconductors* typically do not have Fermi liquid parent states. This implies that a different approach must be used to describe unconventional superconductors than that developed over decades to describe conventional superconductivity.

32.2 Unconventional Superconductors

We shall take as a formal microscopic definition that the pairing interaction leading to the SC state is spherically symmetric in conventional superconductors and leads to a

Box 32.1 Fermi Liquids

Imagine a set of non-interacting fermions for which interactions are turned on slowly (adiabatically). Initially the ground state is a non-interacting Fermi sea. As the interaction is slowly increased the non-interacting ground state evolves into a ground state of interacting fermions. If each "bare" fermion of the non-interacting system can be continuously deformed to a "dressed" particle of the interacting system by adiabatically increasing the interaction strength from zero to its final value, (1) the single-particle states of the interacting system may be expected to be in one to one correspondence with those of the non-interacting system, and (2) the dressed particles of the interacting system may be expected to continue to obey Fermi–Dirac statistics. The interacting system is then termed a *Fermi liquid,* its dressed particles are termed *(Landau) quasiparticles,* and the one to one correspondence between the initial particles and final quasiparticles is called *adiabatic continuity.*

Adiabatic Continuity for Strongly Interacting Systems

Adiabatic continuity seems plausible if the interactions are weak. The utility of the Fermi liquid description is that even if the final interactions are quite strong, it is still possible that the final states can be reached by a continuous deformation of the initial states so that states remain in one to one correspondence and the Pauli principle continues to be obeyed by the quasiparticles. Then even a strongly interacting fermionic system may have a relatively simple description in terms of weakly interacting quasifermions. This will be true if (1) much of the interaction has been absorbed into the structure of the quasifermions, and (2) the quasifermions *continue to obey the Pauli principle,* which greatly restricts the range of possibilities for processes involving particles near the Fermi surface.

Validity of the Quasiparticle Concept

The quasiparticle concept is valid only for low temperatures and energies near the Fermi energy. In many interacting systems the large suppression of the scattering rate by the Pauli principle for states in the vicinity of the Fermi surface is sufficient to satisfy the Fermi liquid criteria. We then may view the system as being composed of weakly interacting fermionic quasiparticles, with single-particle states that can be connected to the states of the non-interacting fermionic system by a continuous, adiabatic deformation, even if the bare interaction between particles is large.

Deviations from Fermi Liquid Behavior

A great deal of understanding in condensed matter and nuclear physics relies on the Fermi liquid concept. Many systems are at least approximately Fermi liquids, but some important ones are not. The fractional quantum Hall state of Section 28.5 and the Luttinger liquid of Box 28.7 are two examples. Another is the Mott insulating parent state of cuprate superconductors (see Box 32.3). These states require approaches more sophisticated than adiabatic continuity and Fermi liquid theory.

BCS-like wavefunction (32.1), but may have a more complex geometry in unconventional superconductors. Let us concentrate on the cuprate high-temperature superconductors as exemplary of unconventional SC. It is rather uniformly agreed that cuprate SC forms from a parent state that is an antiferromagnetic (AF) *Mott insulator* (see Box 32.3). Evidence

Box 32.2

Cooper Pairs and the Cooper Instability

The Cooper model consists of two fermions with opposite momenta and spin projections outside a filled and inert Fermi sea; the following figure illustrates.

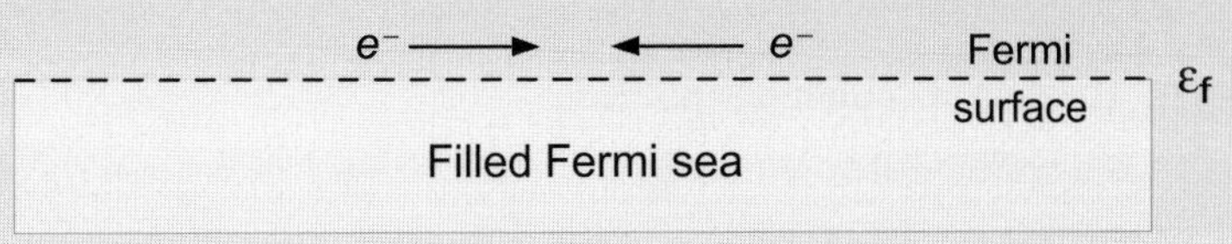

The Fermi sea of electrons blocks occupation of levels below the Fermi energy by the Pauli effect. The two fermions are assumed to interact with each other through a constant attractive interaction over a limited energy range of width $\hbar\omega_{\rm D}$ at the Fermi surface, where $\omega_{\rm D}$ is the Debye frequency characterizing vibrations of the crystal lattice. If they do not interact, the minimum energy of the two particles is $2\epsilon_{\rm f}$, since all lower-energy states are blocked by the filled Fermi sea.

Cooper Pairs

Solving the two-particle Schrödinger equation assuming weak coupling of strength λ then implies a bound state of the interacting system with binding energy

$$E_{\rm b} = 2\hbar\omega_{\rm D}e^{-2/\lambda}.$$

This state, which can form for *any finite attractive* λ, is called a *Cooper pair.*

The Cooper Instability

Thus a Fermi liquid state (Box 32.1) can become unstable to condensing pairs of fermions for *any weakly attractive pairing interaction*. This is called the *Cooper instability.* The Fermi sea is inert but essential to this result: for two particles interacting in empty 3D space there normally is an energy threshold for particle binding and there are no bound states for vanishingly small attractive interactions.

The Instability Is Non-Perturbative

The expression given above for the binding energy $E_{\rm b}$ of the Cooper pair is not an analytic function of the potential λ near $\lambda = 0$, since it tends to infinity as $\lambda \to 0$. This implies that the Cooper pairing state cannot be reached by a perturbative expansion around the normal state.

suggests that SC in the cuprates is a result of Cooper pairing, but the SC state cannot result from the normal Cooper instability because an AF Mott insulator is not a Fermi liquid. We will now show that fermion dynamical symmetries based on non-abelian Lie algebras generalize the Cooper instability to a corresponding instability in doped Mott insulators, and can describe *both* conventional and unconventional superconductors and superfluids in a unified and quantitative way.

Box 32.3

Mott Insulators

Normal insulators result from an energy gap above a filled band, which inhibits excitation of electrons to a conduction band and suppresses charge transport (see Box 5.1). But an insulating state can arise for a quite different reason.

The Role of Coulomb Repulsion

If a system exhibits very strong onsite Coulomb repulsion between electrons, configurations with two electrons on the same lattice site will be highly disfavored energetically. Suppose that the highest occupied band is half filled. In the absence of Coulomb repulsion this state should be metallic, since it corresponds to a partially filled conduction band. However, the presence of strong onsite repulsion will favor a classical ground state in which each lattice site is occupied by exactly one electron from the highest band. Thus an electron in the half filled band cannot move easily because it must hop to another site that is already occupied by an electron, which is disfavored by the Coulomb repulsion.

Charge transport suppressed by on site Coulomb repulsion

Hence, in the presence of strong Coulomb repulsion the half filled band produces an insulating rather than metallic state.

> Materials that are insulating, not because of band structure but because onsite Coulomb repulsion suppresses charge transport, are termed *Mott insulators.*

Mott insulators differ fundamentally from normal bandgap insulators because the mechanism suppressing charge transport is different.

Mott Insulators Are Not Fermi Liquids

Mott insulator states are not in one to one correspondence with states of a non-interacting Fermi gas because configurations with spin-up and spin-down electrons on the same site occur in the non-interacting system but are strongly suppressed in the Mott insulator. Thus *Mott insulators are not Fermi liquids* (see Box 32.1).

32.3 The SU(4) Model of Non-Abelian Superconductors

In Ch. 5, wavefunctions on periodic lattices were classified with respect to translational, rotational, and discrete symmetries. In this chapter we ask whether there are additional properties of lattice wavefunctions associated with symmetries of the Hamiltonian itself (dynamical symmetries) that can provide new insight in the physics of the solid state. As already illustrated for graphene in a magnetic field in Section 20.2, and for complex nuclei in Section 31.2, we shall see that such methods can lead to enormous simplifications of complex many-body wavefunctions, and to physical insight that would be difficult to obtain by other means. As elaborated in Section 20.1.4, this is because the method allows *dynamics* to be constrained. This section will introduce methods that use Lie algebras and dynamical symmetries of fermionic systems to obtain exact many-body solutions within symmetry truncated Hilbert spaces [218]. It will then be demonstrated that such methods can be used to obtain an accurate many-body solution for a lattice model in which antiferromagnetism and *d*-wave singlet superconductivity compete on an equal footing.

The minimal model incorporating both AF and *d*-wave SC, while conserving charge and spin, has an SU(4) symmetry [93, 94, 190, 222]. These results may be expressed in simple and compact form because of the symmetry, but even the minimal (analytically solvable) theory represents a rich many-body solution capable of modeling extremely complex behavior. Remarkably, the commutation relations implied by the SU(4) symmetry automatically enforce a no-double-occupancy (Mott insulator) condition on the lattice in the copper–oxygen plane. Therefore, the microscopic SU(4) symmetry itself implies that ground states are AF Mott insulators at zero doping [corresponding to a dynamical symmetry associated with an SO(4) subgroup of SU(4)], that with increased doping the system evolves to a ground state having large fluctuations in both pairing and antiferromagnetism with pseudogap properties [corresponding to an SO(5) subgroup], and that beyond a critical doping point the system makes a transition to a ground state exhibiting pure *d*-wave superconducting order [corresponding to an SU(2) subgroup]. Thus SU(4) dynamical symmetry leads to a sophisticated yet tractable model in which antiferromagnetism, *d*-wave superconductivity, and a Mott insulator constraint enter as intrinsic microscopic properties of an exact many-body solution. We now describe the construction of this model.

32.3.1 The SU(4) Algebra

Let us begin by introducing creation and annihilation operators for the relevant physical modes. Since the phase diagram exhibits SC states, our first consideration is the types of pairs that can form on a cuprate lattice. Because Coulomb repulsion may be significant for the valence orbitals that contribute to the electronic structure, we shall allow two kinds of fermionic pairs: (1) *onsite pairs,* and (2) *bondwise pairs,* as illustrated in Fig. 32.1.

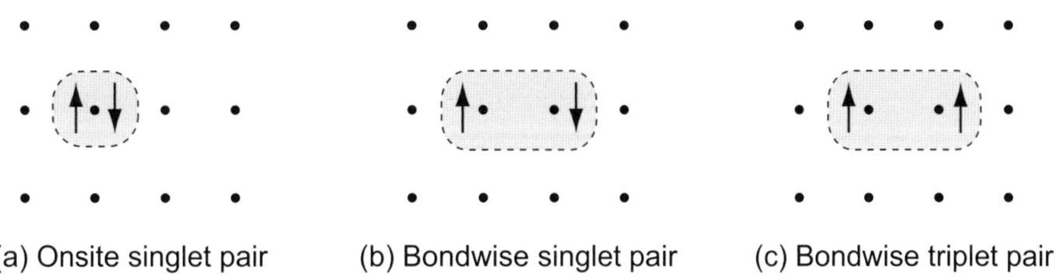

Fig. 32.1 (a) Onsite singlet pair. (b) Nearest-neighbor bondwise singlet pair. (c) Nearest-neighbor bondwise triplet pair. Bondwise pairs are important if Coulomb repulsion disfavors onsite pairing. For simplicity only nearest-neighbor bondwise pairs will be considered explicitly. The pairs defined in (b) and (c) may then be viewed as effective pairs that are renormalized by any contributions from next-nearest-neighbor pairing, and so on.

Generators of the Algebra: An N-dimensional basis has a minimal closed algebra SO($2N$) if all possible bilinear particle–hole and pair operators are taken as generators [126]. The simplest basis for cuprate superconductors may be regarded as four-dimensional, since electrons can exist in four basic states: (1) on even sites with spin up, (2) on even sites with spin down, (3) on odd sites with spin up, and (4) on odd sites with spin down, where the sites of the 2D lattice are divided into two sets labeled even and odd. Thus, the minimal Lie algebra for a set of particle–hole and pairing generators that can describe a cuprate system is SO($2N$) = SO(8) [corresponding to the Dynkin classification SO($2L$) with $L = 4$ from Appendix D, which has $L(2L-1) = 28$ generators]. Let us investigate this 28-element algebra. First we introduce the 16 operators

$$\vec{S} = \left(\frac{S_{12} + S_{21}}{2}, -i\,\frac{S_{12} - S_{21}}{2}, \frac{S_{11} - S_{22}}{2} \right), \tag{32.2a}$$

$$\vec{Q} = \left(\frac{Q_{12} + Q_{21}}{2}, -i\,\frac{Q_{12} - Q_{21}}{2}, \frac{Q_{11} - Q_{22}}{2} \right), \tag{32.2b}$$

$$\vec{\pi}^\dagger = \left(i\,\frac{q^\dagger_{11} - q^\dagger_{22}}{2}, \frac{q^\dagger_{11} + q^\dagger_{22}}{2}, -i\,\frac{q^\dagger_{12} + q^\dagger_{21}}{2} \right), \tag{32.2c}$$

$$\vec{\pi} = \left(-i\,\frac{q_{11} - q_{22}}{2}, \frac{q_{11} + q_{22}}{2}, i\,\frac{q_{12} + q_{21}}{2} \right), \tag{32.2d}$$

$$p^\dagger = p^\dagger_{12} \qquad p = p_{12} \qquad \hat{n} = \sum_{k,i} c^\dagger_{\boldsymbol{k},i} c_{\boldsymbol{k},i} = S_{11} + S_{22} + \Omega, \tag{32.2e}$$

$$Q_+ = Q_{11} + Q_{22} = \sum_k \left(c^\dagger_{\boldsymbol{k}+\boldsymbol{Q}\uparrow} c_{\boldsymbol{k}\uparrow} + c^\dagger_{\boldsymbol{k}+\boldsymbol{Q}\downarrow} c_{\boldsymbol{k}\downarrow} \right), \tag{32.2f}$$

where we have defined[1]

$$\begin{gathered} p^\dagger = \sum_{\boldsymbol{k}} g(\boldsymbol{k}) c^\dagger_{\boldsymbol{k}\uparrow} c^\dagger_{-\boldsymbol{k}\downarrow} \qquad p = (p^\dagger)^\dagger \qquad q^\dagger_{ij} = \sum_{\boldsymbol{k}} g(\boldsymbol{k}) c^\dagger_{\boldsymbol{k}+\boldsymbol{Q},i} c^\dagger_{-\boldsymbol{k},j} \qquad q = (q^\dagger)^\dagger, \\ Q_{ij} = \sum_{\boldsymbol{k}} c^\dagger_{\boldsymbol{k}+\boldsymbol{Q},i} c_{\boldsymbol{k},j} \qquad S_{ij} = \sum_{\boldsymbol{k}} c^\dagger_{\boldsymbol{k},i} c_{\boldsymbol{k},j} - \frac{\Omega}{2}\,\delta_{ij}, \end{gathered} \tag{32.3}$$

Ω is the degeneracy, and $\boldsymbol{Q} = (Q_x, Q_y) = (\pi, \pi)$ is an AF ordering vector in k-space. The operators appearing in Eqs. (32.2) may be given simple physical interpretations.

- The vector $\vec{S}$ is the electron spin operator.
- The vector $\vec{Q}$ is the staggered magnetization operator characterizing antiferromagnetism.
- $p^\dagger$ (p) is a creation (annihilation) operator for bondwise singlet pairs.
- $\vec{\pi}^\dagger$ ($\vec{\pi}$) is a vector of creation (annihilation) operators for bondwise spin-triplet pairs.
- $\hat{n}$ is the electron number operator.
- Q_+ is a commensurate charge density wave operator.

[1] The operators in Eqs. (32.2) and (32.3) are assumed to receive contributions from a single band near the Fermi surface, which is a good initial approximation for the cuprates and is sufficient to illustrate the method. Generalization to include contributions from multiple bands is given in Refs. [98, 191]. This generalization is important if more than one band lies near the Fermi surface, as in application of SU(4) symmetry to the iron-based high-temperature superconductors that is discussed in Refs. [95, 98].

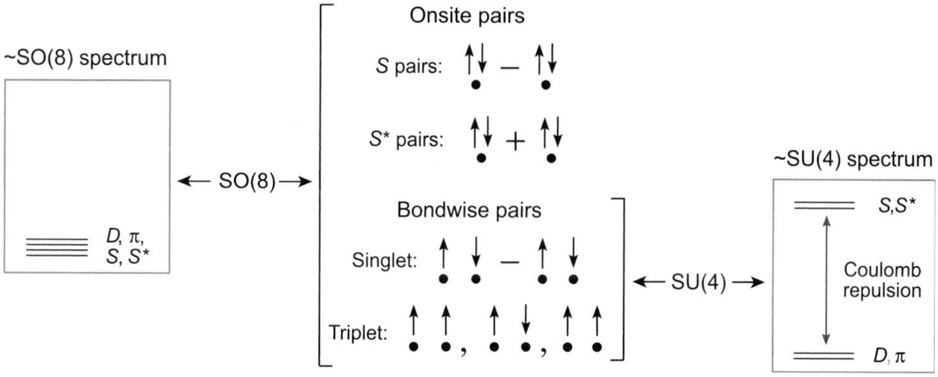

Fig. 32.2 Schematic difference between bondwise (D, π) and onsite (S, S^*) pair energies (see Fig. 32.1). If onsite repulsion is weak the pairing states are approximately degenerate, which yields an SO(8) symmetry generated by the 28 operators (32.2) and (32.4). If it is strong the onsite pairs are pushed up in energy, reducing the SO(8) symmetry to an effective SU(4) low-energy symmetry generated by the operators (32.2).

The 16 operators (32.2) are consistent with bondwise pairing [Fig. 32.1(b)]. Let us introduce 12 additional operators:

$$\begin{aligned}
\bar{p}_{12}^{\dagger} &= \sum_{\mathbf{r}\in A}\left(c_{\mathbf{r}\uparrow}^{\dagger}c_{\mathbf{r}\downarrow}^{\dagger} - c_{\bar{\mathbf{r}}\downarrow}^{\dagger}c_{\bar{\mathbf{r}}\uparrow}^{\dagger}\right) &\qquad \bar{q}_{12}^{\dagger} &= \pm\sum_{\mathbf{r}\in A}\left(c_{\mathbf{r}\uparrow}^{\dagger}c_{\mathbf{r}\downarrow}^{\dagger} + c_{\bar{\mathbf{r}}\downarrow}^{\dagger}c_{\bar{\mathbf{r}}\uparrow}^{\dagger}\right),\\
\bar{S}_{ij} &= \sum_{\mathbf{r}\in A}\left(c_{\mathbf{r},i}^{\dagger}c_{\bar{\mathbf{r}},j} - c_{\mathbf{r},j}c_{\bar{\mathbf{r}},i}^{\dagger}\right) &\qquad \bar{Q}_{ij} &= \pm\sum_{\mathbf{r}\in A}\left(c_{\mathbf{r},i}^{\dagger}c_{\bar{\mathbf{r}},j} + c_{\mathbf{r},j}c_{\bar{\mathbf{r}},i}^{\dagger}\right),\\
\bar{p}_{12} &= (\bar{p}_{12}^{\dagger})^{\dagger} &\qquad \bar{q}_{12} &= (\bar{q}_{12}^{\dagger})^{\dagger}.
\end{aligned} \tag{32.4}$$

Equation (32.4) contains spin-singlet pairs created by $\bar{p}_{12}^{\dagger}$ and $\bar{q}_{12}^{\dagger}$, which we shall term S and S^* pairs, respectively. Unlike the bondwise pairs associated with Eq. (32.2), these pairs are *onsite* [Fig. 32.1(a)], where the two electrons (or two holes) occupy the same site, with equal probability to appear anywhere in the lattice coherently.

The 28 operators in Eqs. (32.2) and (32.4) close an SO(8) Lie algebra under commutation and are the basis of a general description of conventional and unconventional superconductors. However, the 16 operators in Eqs. (32.2) close a U(4) $\supset$ U(1) $\times$ SU(4) subalgebra of SO(8), where the U(1) factor is generated by Q_+ and SU(4) is generated by the other 15 operators in Eq. (32.2). Our primary interest here will be in the SU(4) subalgebra. The relationship between the full SO(8) algebra and the SU(4) subalgebra is illustrated in Fig. 32.2.

Since the Cooper pair condensate that forms the superconductor is known to be dominated by singlet pairs, it might be thought that the triplet pair operators in the above expressions are superfluous in a minimal model. However, as illustrated in Fig. 32.3, interaction of singlet pairs with the antiferromagnetic operators will inevitably produce triplet pairs. Thus, even if one starts with only singlet pairs, in the presence of antiferromagnetism singlet and triplet pairs will interconvert and both will be present.

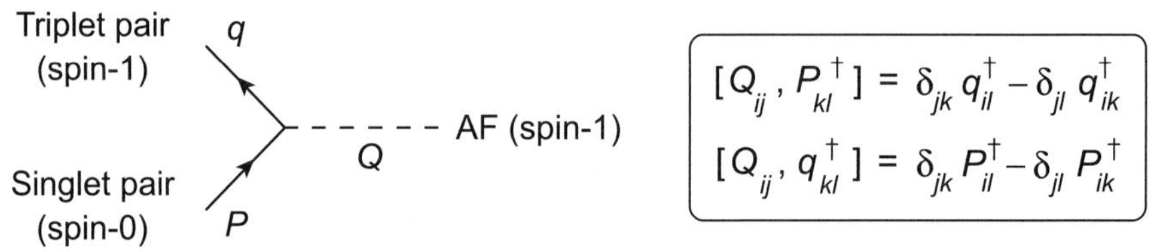

Fig. 32.3 Scattering of singlet pairs by AF operators will necessarily produce triplet pairs, even if none existed before. Thus a physically and mathematically complete Hilbert subspace must contain both kinds of pairs in the presence of antiferromagnetism.

Closure of the SU(4) Lie Algebra: Inserting the AF ordering vector $\boldsymbol{Q} = (Q_x, Q_y) = (\pi, \pi)$ appropriate for the cuprates and calculating all commutators for the 16 operators in Eqs. (32.2), the set is closed under commutation if the pairing formfactor $g(\boldsymbol{k})$ satisfies:

$$g(\boldsymbol{k}) = g(-\boldsymbol{k}) \qquad g(\boldsymbol{k}+\boldsymbol{Q}) = \pm g(\boldsymbol{k}) \qquad |g(\boldsymbol{k})| = 1. \tag{32.5}$$

If these conditions are met (we shall elaborate on their physical meaning later), the operators defined in Eq. (32.2) close a U(4) ⊃ U(1) × SU(4) Lie algebra, where the U(1) factor is generated by the commensurate charge density wave operator Q_+ that commutes with the other generators [93, 94]. The direct product structure U(1) × SU(4) means that we may take cuprate states to be described by an SU(4) symmetry generated by pairing, antiferromagnetic correlations, and global charge and spin conservation, with the U(1) charge density wave sector treated independently. There are three independent subgroup chains of this SU(4) group,

$$\mathrm{SU(4)} \begin{cases} \supset \mathrm{SO(4)} \times \mathrm{U(1)} \supset \mathrm{SU(2)_s} \times \mathrm{U(1)} & \text{(32.6a)} \\ \supset \mathrm{SO(5)} \supset \mathrm{SU(2)_s} \times \mathrm{U(1)} & \text{(32.6b)} \\ \supset \mathrm{SU(2)_p} \times \mathrm{SU(2)_s} \supset \mathrm{SU(2)_s} \times \mathrm{U(1)} & \text{(32.6c)} \end{cases}$$

that end in the subgroup $\mathrm{SU(2)_s} \times \mathrm{U(1)}$ representing spin and charge conservation, with the U(1) subgroup being generated by the charge operator $M = \frac{1}{2}(S_{11} + S_{22})$ and the $\mathrm{SU(2)_s}$ subgroup being generated by the three spin operators S_1, S_2, and S_3. Notice that the commutation algebra of the operators (32.2) *would not close* if we omitted the six triplet pair operators (32.2c) and (32.2d). The specific reason is that commuting singlet pair operators with the AF operators gives triplet pair operators and commuting triplet pair operators with the AF operators gives singlet pairs. Thus the SU(4) Lie algebra is the *algebraic embodiment* of the *physical argument* made in connection with Fig. 32.3 that for a pair basis in the presence of antiferromagnetism the Hilbert space must contain both singlet and triplet pair states.

32.3.2 The SU(4) Collective Subspace

In keeping with the general discussion in Ch. 20 of dynamical symmetry as a systematic means to truncate the full Hilbert space to a manageable collective subspace, the dynamical symmetry chains just obtained permit the definition of a collective subspace in which symmetry can be used to obtain analytic solutions relevant for cuprate high-temperature

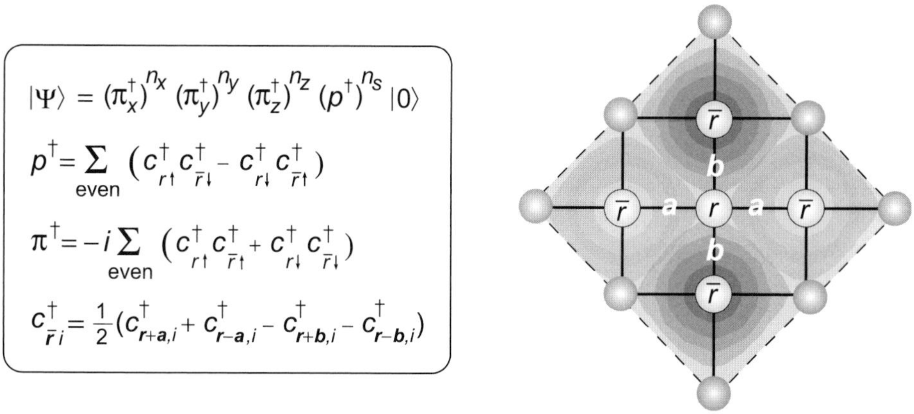
$|\Psi\rangle = (\pi_x^\dagger)^{n_x} (\pi_y^\dagger)^{n_y} (\pi_z^\dagger)^{n_z} (p^\dagger)^{n_s} |0\rangle$
$p^\dagger = \sum_{\text{even}} (c_{r\uparrow}^\dagger c_{\bar{r}\downarrow}^\dagger - c_{r\downarrow}^\dagger c_{\bar{r}\uparrow}^\dagger)$
$\pi^\dagger = -i \sum_{\text{even}} (c_{r\uparrow}^\dagger c_{\bar{r}\uparrow}^\dagger + c_{r\downarrow}^\dagger c_{\bar{r}\downarrow}^\dagger)$
$c_{\bar{r}i}^\dagger = \frac{1}{2}(c_{r+\mathbf{a},i}^\dagger + c_{r-\mathbf{a},i}^\dagger - c_{r+\mathbf{b},i}^\dagger - c_{r-\mathbf{b},i}^\dagger)$
r
b
a

Fig. 32.4 Linear superpositions in coordinate space implied by the wavefunction (32.7) [98]. The inset figure is discussed more extensively in connection with Fig. 32.7. Reproduced with permission from Springer Nature: *Frontiers of Physics*, Fermion Dynamical Symmetry and Strongly-Correlated Electrons: A Comprehensive Model of High-Temperature Superconductivity, M. Guidry, Y. Sun, and L.-A. Wu, **15**, 43301 (2020).

superconductors. The group SU(4) is rank-3 and the irreps may be labeled by three weight space quantum numbers, $(\sigma_1, \sigma_2, \sigma_3)$.[2] We assume a collective subspace illustrated in Fig. 20.1(a) that is spanned by the vectors

$$|\Psi\rangle = |n_x n_y n_z n_s\rangle = (\pi_x^\dagger)^{n_x} (\pi_y^\dagger)^{n_y} (\pi_z^\dagger)^{n_z} (p^\dagger)^{n_s} |0\rangle \tag{32.7}$$

[compare Eq. (20.16)], where $n_x + n_y + n_z + n_s$ is the total number of fermion pairs. If there are no unpaired particles the collective subspace is associated with "fully stretched," and therefore maximally collective, irreducible representations of the form

$$(\sigma_1, \sigma_2, \sigma_3) = \left(\frac{\Omega}{2}, 0, 0\right). \tag{32.8}$$

More general representations having broken pairs are described in Ref. [190], but here we restrict to no unpaired particles to keep the discussion simple. A subspace associated with such maximally collective configurations is an obvious candidate for describing the lowest-energy states of the system. This wavefunction has a rich structure associated with a coherent superposition of pairs, as illustrated in Fig. 32.4.

32.3.3 The Dynamical Symmetry Hamiltonian

SU(4) has a quadratic Casimir operator ,

$$C_{\text{su}(4)} = \vec{\pi}^\dagger \cdot \vec{\pi} + p^\dagger p + \vec{S} \cdot \vec{S} + \vec{Q} \cdot \vec{Q} + M(M-4), \tag{32.9}$$

[2] We employ an isomorphism between the groups SU(4) and SO(6) to label irreducible representations using SO(6) quantum numbers [74].

and the corresponding expectation value evaluated in the irreducible representations (32.8) is a constant,

$$\langle C_{\mathrm{su}(4)} \rangle = \frac{\Omega}{2}\left(\frac{\Omega}{2} + 4\right). \tag{32.10}$$

The most general two-body Hamiltonian within the collective pair space consists of a linear combination of (lowest-order) Casimir operators C_{g} for all subgroups g in the subgroup chains (32.6) of SU(4):[3]

$$\begin{aligned} C_{\mathrm{so}(5)} &= \vec{\pi}^\dagger \cdot \vec{\pi} + \vec{S} \cdot \vec{S} + M(M-3) \qquad C_{\mathrm{so}(4)} = \vec{Q} \cdot \vec{Q} + \vec{S} \cdot \vec{S}, \\ C_{\mathrm{su}(2)_\mathrm{p}} &= p^\dagger p + M(M-1) \qquad C_{\mathrm{su}(2)_\mathrm{s}} = \vec{S} \cdot \vec{S} \qquad C_{\mathrm{u}(1)} = M \text{ and } M^2. \end{aligned} \tag{32.11}$$

Using the constant SU(4) Casimir expectation value (32.10) to eliminate terms in $\vec{\pi}^\dagger \cdot \vec{\pi}$, the most general SU(4) Hamiltonian restricted to two-body interactions is [93, 97]

$$H = H_0 - \tilde{G}_0\,[(1-\sigma)p^\dagger p + \sigma \vec{Q} \cdot \vec{Q}] + g'\vec{S} \cdot \vec{S}, \tag{32.12}$$

where H_0, $\tilde{G}_0$, and g' are effective interaction parameters, $p^\dagger$ creates singlet pairs, $\vec{Q}$ is the staggered magnetization, $\vec{S}$ is spin, $\tilde{G}_0 = \chi(x) + G_0(x)$, and where σ, given by

$$\sigma = \sigma(x) = \frac{\chi(x)}{\chi(x) + G_0(x)}, \tag{32.13}$$

[with $G_0(x)$ and $\chi(x)$ the effective SC and AF coupling strengths, respectively] governs the relative strength of antiferromagnetic and pairing interactions [190, 191]. In these expressions doping is characterized by a parameter

$$x = 1 - \frac{n}{\Omega}, \tag{32.14}$$

for an n-electron system, with Ω the maximum number of doped holes (or doped electrons for electron-doped compounds) that can form coherent pairs, assuming the normal state at half filling ($n = \Omega$, implying $M = 0$) to be the vacuum. Since $\Omega - n$ is the hole number when $n < \Omega$, positive x represents the case of hole-doping, with $x = 0$ corresponding to half filling (no doping) and $x = 1$ to maximal hole-doping. Negative x ($n > \Omega$) is then the relative doping fraction for electron-doping.

32.3.4 The SU(4) Dynamical Symmetry Limits

Each of the dynamical symmetry limits given in Eqs. (32.6) defines a basis in which the Hamiltonian (32.12) is diagonal. Thus, they have exact solutions [86, 93, 222] that may be constructed using the methods developed in Ref. [218]. These solutions result from special choices of the parameter σ in the Hamiltonian (32.12), and are summarized in Table 32.1.

[3] Larger groups may have more than one Casimir invariant. We shall use the unqualified term "Casimir" to refer to the lowest-order such invariants (which are usually quadratic in the group generators). In the present discussion, quadratic Casimirs are associated microscopically with two-body interactions and higher-order Casimirs are associated with three-body and higher interactions. The restriction of the Hamiltonian to lowest-order Casimir operators is then a physical restriction to one-body and two-body interactions.

Table 32.1. SU(4) dynamical symmetries

σ	Symmetry	Physical interpretation
0	SO(4)	Antiferromagnetic Mott insulator
$\frac{1}{2}$	SO(5)	Critical dynamical symmetry
1	SU(2)	*d*-wave singlet superconductor

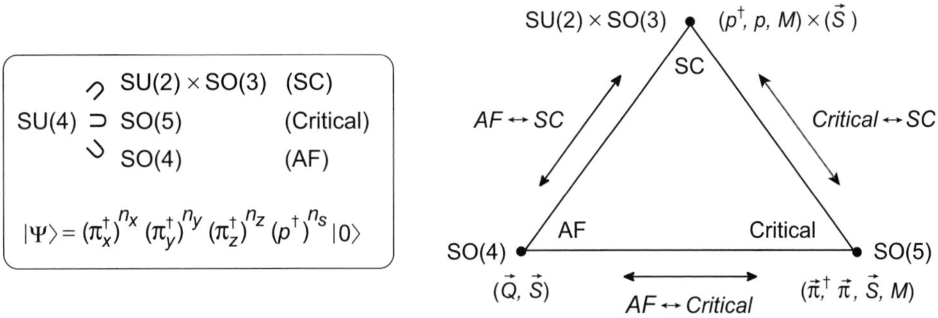

Fig. 32.5 Relationships among the SU(4) dynamical symmetries summarized in Table 32.1.

1. The SO(4) limit corresponds to choosing $\sigma = 1$ in Eq. (32.12). It represents a collective Mott insulator, AF state defined by the subgroup chain of Eq. (32.6a).
2. The $SU(2)_p$ limit [SU(2) limit for brevity] corresponds to choosing $\sigma = 0$ in Eq. (32.12). It represents a collective *d*-wave SC state defined by the subgroup chain of Eq. (32.6c).
3. The SO(5) limit corresponds to choosing $\sigma = \frac{1}{2}$ in Eq. (32.12), which leads to a Hamiltonian in which pairing and antiferromagnetism enter with equal weight. The SO(5) limit represents a *critical dynamical symmetry* (see Box 20.3) that interpolates dynamically between the SC and AF phases. It is defined by the subgroup chain of Eq. (32.6b).

The detailed mathematical properties of these symmetry limits are derived and discussed in the cited references and their schematic relationships are displayed in Fig. 32.5. We now discuss the justification for the above identifications and the important physical properties of the exact solutions in these symmetry limits.

32.3.5 The SO(4) Dynamical Symmetry Limit

The dynamical symmetry subgroup chain $SU(4) \supset SO(4) \times U(1) \supset SU(2)_s \times U(1)$ of Eq. (32.6a), which is termed the *SO(4) dynamical symmetry* for brevity, is the symmetry limit of Eq. (32.12) when $\sigma = 1$. As we now argue, it corresponds to a collective state having long-range antiferromagnetic order and Mott insulator transport properties. The SO(4) subgroup is locally isomorphic to the product group $SU(2)_F \times SU(2)_G$ that is generated by the linear combinations

$$\vec{F} = \frac{\vec{Q} + \vec{S}}{2} \qquad \vec{G} = \frac{\vec{Q} - \vec{S}}{2}, \tag{32.15}$$

of the original SO(4) generators $\vec{Q}$ and $\vec{S}$ (Problem 32.2). The ground state corresponds to $\frac{1}{2}n$ spin-up electrons on the even sites ($F = \frac{1}{2}N$) and $\frac{1}{2}n$ spin-down electrons on odd sites ($G = \frac{1}{2}N$), or vice versa. Thus it has maximal staggered magnetization,

$$Q = \frac{1}{2}\Omega(1 - x) = \frac{n}{2}, \tag{32.16}$$

and a large energy gap associated with the antiferromagnetic correlation $\vec{Q} \cdot \vec{Q}$,

$$\Delta E = 2\chi(1 - x)\Omega. \tag{32.17}$$

In addition, the pairing gap,

$$\Delta = \frac{1}{2}G_0\Omega\sqrt{x(1 - x)}, \tag{32.18}$$

is small near half filling ($x = 0$), and we shall demonstrate below that the SU(4) symmetry requires the lattice to have no double occupancy at half filling and thus to be a Mott insulator. We conclude that near half filling these SO(4) states are identified naturally with a collective, antiferromagnetic, Mott insulating state. This identification will be corroborated by the ground state energy surface evaluated in this limit (see Fig. 32.9).

32.3.6 The SU(2) Dynamical Symmetry Limit

The subgroup chain SU(4) $\supset$ SU(2)$_\text{p}$ $\times$ SU(2)$_\text{s}$ $\supset$ SU(2)$_\text{s}$ $\times$ U(1) of Eq. (32.6c), which is termed the *SU(2) dynamical symmetry* for brevity, corresponds to the $\sigma = 0$ symmetry limit of Eq. (32.12). The ground state has a large pairing gap $\Delta E = G_0\Omega$, the pairing correlation Δ is the largest among the three symmetry limits, and the staggered magnetization Q vanishes in the ground state:

$$\Delta = \frac{1}{2}G_0\Omega\sqrt{1 - x^2} \qquad Q = 0. \tag{32.19}$$

Thus we interpret this dynamical symmetry as a d-wave pair condensate associated with spin-singlet superconducting order. This identification will be strengthened below by examination of the ground state energy surface evaluated in this limit (see Fig. 32.9).

32.3.7 The SO(5) Dynamical Symmetry Limit

The subgroup chain SU(4) $\supset$ SO(5) $\supset$ SU(2)$_\text{s}$ $\times$ U(1) of Eq. (32.6b), which is termed the *SO(5) dynamical symmetry* for brevity, is a *critical dynamical symmetry* (Box 20.3) that results when $\sigma = \frac{1}{2}$ in Eq. (32.12). Many SO(5) states having different numbers λ of triplet pairs can mix into the ground state when x is small because the excitation energy in the SO(5) limit is $\Delta E = \lambda G_0\Omega x$. In particular, near half band filling ($x \sim 0$) the low-lying states are highly degenerate in λ and mixing components with different numbers of π pairs in the ground state costs no energy. The π pairs must be responsible for the antiferromagnetism in this phase, since within the model space only π pairs carry spin if

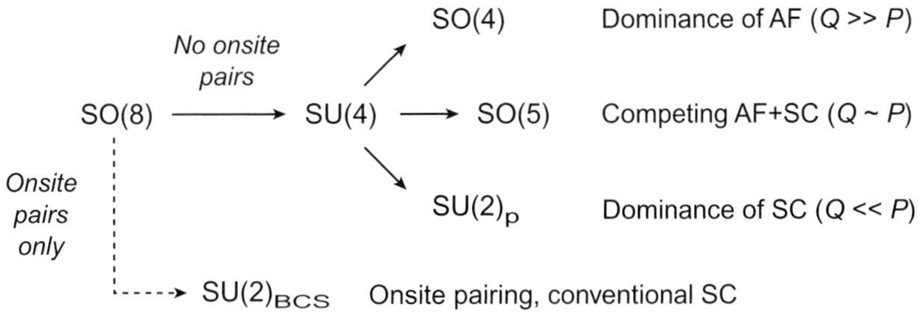

Fig. 32.6 Relationship among SO(8), SU(4), and conventional BCS SU(2) symmetry. Antiferromagnetic correlations are denoted by Q and antiferromagnetic states by AF; pairing correlations are denoted by P and superconducting states by SC.

there are no unbroken pairs. Thus the SO(5) ground state has large fluctuation in AF (and SC) order,[4] and acts as a quantum critical *doorway state* between AF and SC order. This identification will be strengthened by the energy surface evaluated in this limit (Fig. 32.9).

32.3.8 Conventional and Unconventional Superconductors

The dynamical symmetry chains deriving from SO(8) imply a unified view of conventional and unconventional superconductivity that is illustrated in Fig. 32.6. Both the $SU(2)_p$ and $SU(2)_{BCS}$ subgroups have pseudospin pair generators, but differ in that the pairs are onsite for $SU(2)_{BCS}$ and bondwise for $SU(2)_p$, implying different orbital formfactors for pairing in the two cases. The symmetry $SU(2)_{BCS}$ with conventional formfactor is favored only if onsite Coulomb repulsion, and any collective modes like AF competing with pairing, can be neglected. In the presence of strong onsite Coulomb repulsion and a competing long-range order such as antiferromagnetism the $SO(8) \supset SU(4)$ dynamical symmetry is favored, giving birth to unconventional superconductors.

32.4 Some Implications of SU(4) Symmetry

The SU(4) dynamical symmetry model outlined in Section 32.3 has a number of implications for understanding unconventional superconductors in general, and high-temperature cuprate and iron-based superconductors in particular. Extensive overviews are available [98, 99]. Here we restrict to summarizing a few representative examples.

32.4.1 No Double Occupancy

Onsite Coulomb repulsion inhibits cuprate charge transport, leading to Mott insulator normal states at half filling. This is often imposed by requiring no double occupancy

[4] It is an essential feature of the SU(4) dynamical symmetry that the antiferromagnetic order and superconducting order are not independent because both are generated by subsets of the SU(4) generators.

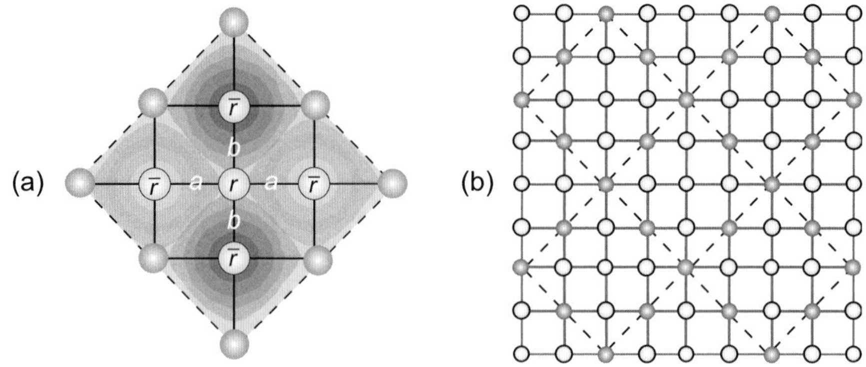

Fig. 32.7 (a) Schematic SU(4) hole pair on the real-space lattice. (b) Tiling of the Cu–O plane by the hole pairs of (a) that leads to the Mott insulator no double occupancy condition.

of lattice sites in a basis. The SU(4) model automatically imposes no double lattice occupancy for configurations contributing to emergent states through a *fundamental symmetry constraint*.

The SU(4) symmetry responsible for the emergent states describing cuprate superconductivity would be broken by double occupancy of lattice sites.

A schematic hole pair is sketched in Fig. 32.7(a). Lighter-shaded balls in the interior are sites where electron holes form a pair: one hole at $\mathbf{r}$, the other with equal probability of $\frac{1}{4}$ at the four neighboring sites ($\bar{\mathbf{r}} = \mathbf{r} \pm \mathbf{a}$ and $\mathbf{r} \pm \mathbf{b}$). Darker balls on the outer boundary (connected by dashed lines) represent sites where the presence of holes would imply average double occupancy of some lattice sites. The no double occupancy restriction follows geometrically from tiling the plane with such pairs, as illustrated in Fig. 32.7(b). Each diamond outlined by dashed boundaries corresponds to one unit pair from (a). Furthermore, by simple counting the lattice can ensure no double occupancy [and thus preserve SU(4) symmetry] only if it is not more than $\frac{1}{4}$ occupied by electron holes. This provides a natural explanation for the experimental finding that cuprate superconductivity is generally not observed beyond hole-doping fractions of about 25% [see Fig. 32.8(a)].

32.4.2 Quantitative Gap and Phase Diagrams

Cuprate superconductors exhibit a characteristic temperature–doping phase diagram. An SU(4) calculation of the phase diagram for hole-doped cuprates is compared with data in Fig. 32.8(a). The pseudogap temperature is T^* and the SC transition temperature is T_c. The AF correlations vanish, leaving a pure singlet d-wave condensate, above the critical doping P_q. Dominant correlations in each region are indicated by italic labels. The two curves for the pseudogap temperature are calculations assuming that momentum k is either resolved (T^*_{max}) or not (T^*_{avg}) [191]. Circles indicate data that resolve k; squares indicate data that do not. The SU(4) theory is seen to reproduce the experimental cuprate phase diagram

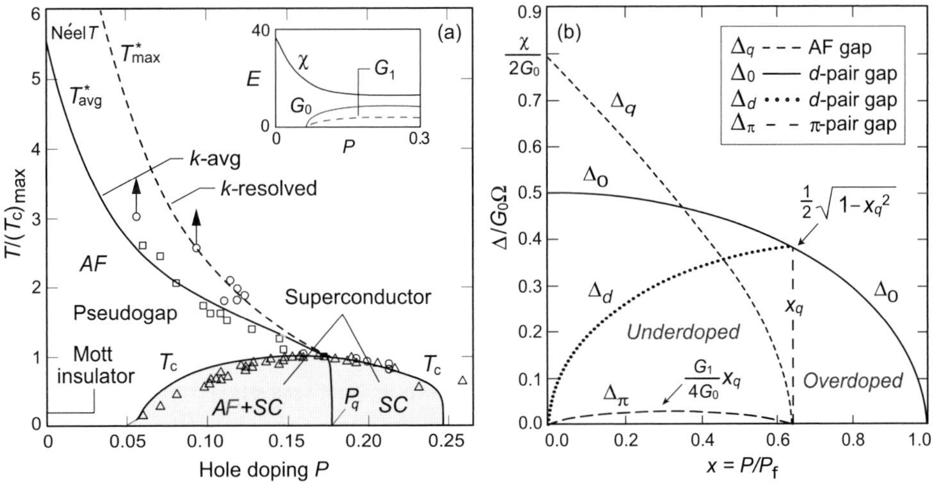

Fig. 32.8 (a) SU(4) cuprate temperature T and doping P phase diagram compared with data. Open triangles and open squares from Ref. [46]; open circles from Ref. [33]. Three parameters were fit to global cuprate data [190]: AF coupling strength χ, singlet pairing strength G_0, and triplet pairing strength G_1, as shown in the inset as a function of doping. (b) Generic gap diagram (correlation energies versus doping), as predicted by the SU(4) model at $T = 0$.

quantitatively. Physically it would be expected that the effective coupling strengths have a smooth dependence on doping and we have assumed the dependence shown in the inset of Fig. 32.8(a) for χ, G_0, and G_1. However, as discussed further below, even if the parameters were held constant the basic features of Fig. 32.8(a) would survive.

The phase diagram reflects the interplay of various correlation energies. An SU(4) gap diagram for hole-doped cuprates is shown in Fig. 32.8(b). Energies are scaled by $G_0\Omega$ and the doping parameter x is scaled by the maximum allowed SU(4) doping fraction $P_f = 0.25$ [94]. The critical SU(4) doping parameter is $x_q = 0.64$, corresponding to a critical physical doping parameter $P_q = x_q P_f \sim 0.16$ in Fig. 32.8(b). Interaction strengths were assumed not to depend on doping. This is physically unrealistic and a better fit to data would follow from allowing some doping dependence of parameters, as in Fig. 32.8(a). However, this minimal calculation shows that the basic features of the gap and phase diagrams follow directly from SU(4) dynamical symmetry, with parameters affecting only details [98].

32.4.3 Coherent State Energy Surfaces

The coherent state methods described in Ch. 21 yield SU(4) solutions that are not restricted to dynamical symmetry limits, but it is instructive to examine coherent state energy surfaces for those limits (which have exact solutions), as illustrated in Fig. 32.9(a–c). In these figures the horizontal axis measures AF order in terms of an order parameter β, curves are labeled by lattice occupation fractions with half filling corresponding to a value of one, and σ is the ratio of AF coupling to the sum of AF and pairing coupling strengths.

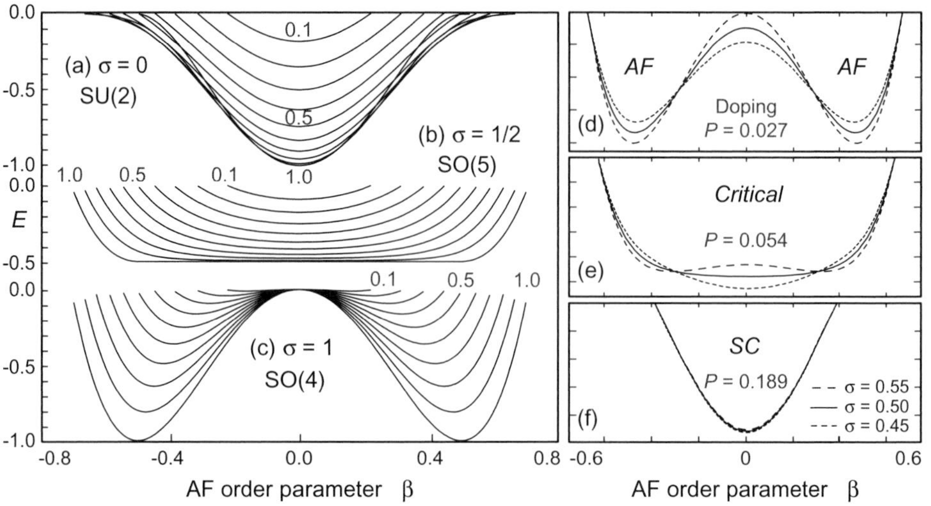

Fig. 32.9 (a–c) Coherent state energy surfaces for symmetry limits of the SU(4) Hamiltonian [98, 222]. (d–f) Effect of altering the ratio σ for three values of doping in the cuprates. In (d) and (f) the system is in the stable minima associated with AF and SC, respectively, and changing σ by 10% hardly alters the location of the energy minima, but in (b) the energy surface is critical and the perturbation can flip the nature of the ground state between SC and AF minima. Reproduced with permission from Springer Nature: *Frontiers of Physics*, Fermion Dynamical Symmetry and Strongly-Correlated Electrons: A Comprehensive Model of High-Temperature Superconductivity, M. Guidry, Y. Sun, and L.-A. Wu, **15**, 43301 (2020).

Energy Surfaces in the SU(2) Limit: The energy surface corresponding to the SU(2) dynamical symmetry limit is displayed in Fig. 32.9(a). The minimum energy occurs at $\beta = 0$ for all values of n and Δ reaches its maximum value of $\Delta_{\text{max}} = \frac{1}{2}\Omega(1 - x^2)^{1/2}$, indicating a state having superconducting order but vanishing antiferromagnetic order.

Energy Surfaces in the SO(4) Limit: The energy surface corresponding to the SO(4) dynamical symmetry limit is displayed in Fig. 32.9(c). The point $\beta = 0$ is unstable and an infinitesimal fluctuation will drive the system to the energy minima at finite $\beta = \pm\frac{1}{2}(n/\Omega)^{1/2}$, which corresponds to $\Delta = 0$ and a state having purely AF order.

Energy Surfaces in the SO(5) Limit: The energy surface corresponding to the SO(5) dynamical symmetry limit is displayed in Fig. 32.9(b). When n is near Ω (half filling), the energy surface is almost flat for broad ranges of β, suggesting a phase very soft against fluctuations in the AF (and SC) order parameters. However, as n/Ω decreases, fluctuations are reduced and the energy surface tends to the SU(2) (superconducting) limit.

Dynamical Criticality: An important consequence of the energy surface for SO(5) dynamical symmetry is illustrated in Figs. 32.9(d–f). The locations of minima in the AF region (d) and the SC region (f) are stable under perturbations (parameterized by changing the coupling ratio σ), but the effect in the critical SO(5) region (e) can be dramatic,

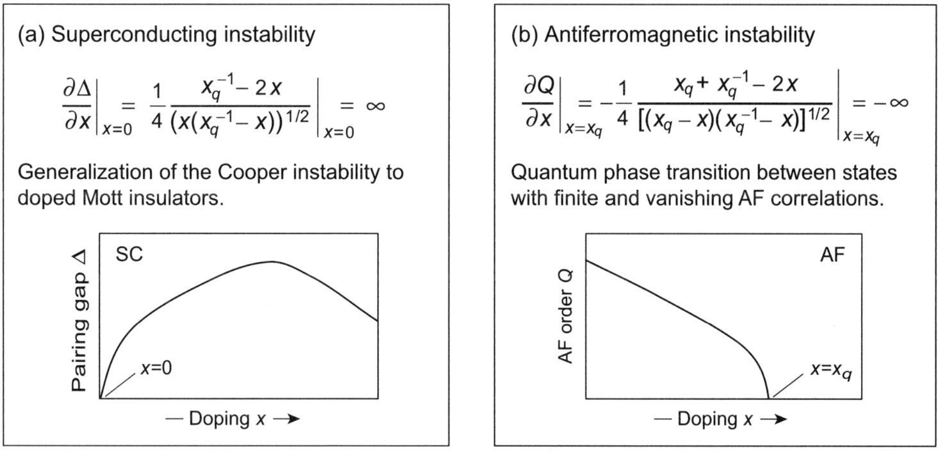

(a) Superconducting instability
$\left.\frac{\partial \Delta}{\partial x}\right|_{x=0} = \frac{1}{4} \left.\frac{x_q^{-1} - 2x}{(x(x_q^{-1} - x))^{1/2}}\right|_{x=0} = \infty$
Generalization of the Cooper instability to doped Mott insulators.
Pairing gap Δ
SC
x=0
— Doping x →
(b) Antiferromagnetic instability
$\left.\frac{\partial Q}{\partial x}\right|_{x=x_q} = -\frac{1}{4} \left.\frac{x_q + x_q^{-1} - 2x}{[(x_q - x)(x_q^{-1} - x)]^{1/2}}\right|_{x=x_q} = -\infty$
Quantum phase transition between states with finite and vanishing AF correlations.
AF order Q
AF
x=x_q
— Doping x →

Fig. 32.10 Two SU(4) instabilities that govern the behavior of high-temperature superconductors through quantum phase transitions. (a) Generalized Cooper instability. (b) AF instability.

even causing switches between SC-favoring and AF-favoring energy surfaces because of the critical SO(5) energy surface. It has been proposed that this instability may be responsible for many of the inhomogeneity effects that have been reported for underdoped cuprates [97].

32.4.4 Fundamental SU(4) Instabilities

The relationship between dynamical symmetry chains imposed by the highest SU(4) symmetry has important implications for quantum phase transitions, as elaborated in Section 23.7. Two instabilities that are key to understanding the behavior of cuprate superconductors are illustrated in Fig. 32.10. (1) The *generalized Cooper instability* explains the tendency of cuprate Mott insulator states at half band filling to become superconductors with only small hole doping. (2) The *AF instability* explains how cuprate superconductors can exhibit a rather universal phase diagram with strong SC over a broad range of doping, and at the same time display marked local inhomogeneity at lower doping.

The Generalized Cooper Instability: The pairing instability of cuprate superconductors has proven difficult to understand in traditional terms. Conventional SC develops from a Fermi liquid normal state but cuprate superconductors emerge from an AF Mott insulator state that harbors a secret propensity to superconductivity: these compounds can be turned from AF Mott insulators into high-temperature superconductors by hole doping at a modest 3–5% level. Emergent SU(4) symmetry provides a natural explanation, as illustrated in Fig. 32.10(a). From the $T = 0$ coherent state solution for the pairing gap Δ [98],

$$\left.\frac{\partial\Delta}{\partial x}\right|_{x=0} = \frac{1}{4}\left.\frac{x_q^{-1} - 2x}{(x(x_q^{-1} - x))^{1/2}}\right|_{x=0} = \infty, \tag{32.20}$$

displaying explicitly a fundamental pairing instability signaled by divergence of $\partial\Delta/\partial x$ at doping $x = 0$.

> SU(4) symmetry implies that the cuprate ground state at half filling is an AF Mott insulator that is *fundamentally unstable against condensing hole pairs* under infinitesimal hole doping if there is a finite attractive pairing interaction.

This instability is hinted at in the solution of Problem 32.1, which shows that even the pure SU(4) ⊃ SO(4) AF Mott insulator has finite pairing correlations unless the doping x vanishes identically.

The AF Instability: SU(4) symmetry implies a second fundamental instability near the critical doping point x_q that is also decisive for the behavior of cuprate superconductors. From the $T = 0$ coherent state solution for the AF correlation Q,

$$\left.\frac{\partial Q}{\partial x}\right|_{x=x_q} = -\frac{1}{4}\left.\frac{x_q + x_q^{-1} - 2x}{[(x_q - x)(x_q^{-1} - x)]^{1/2}}\right|_{x=x_q} = -\infty, \tag{32.21}$$

and a small change in doping will cause a large change in AF correlations near $x = x_q$, as illustrated in Fig. 32.10(b). The AF instability also is a direct consequence of SU(4) symmetry, which requires that Q vanish for doping $x \geq x_q$ and be finite for $0 < x < x_q$. An important consequence of the AF instability is illustrated in Fig. 32.8(b), where at a critical doping $x = x_q \sim 0.65$ (corresponding to $P \sim 0.16$) there is a quantum phase transition from a weaker superconducting state with AF perturbations for $x < x_q$ to a pure singlet d-wave SC state for $x > x_q$. Data strongly support this picture that the cuprates are superconducting at low temperature for $x < x_q$ but become more robust superconductors for $x > x_q$. As suggested by Fig. 32.8, this instability also is an essential aspect of pseudogap behavior, since x_q marks the upper doping boundary of fluctuations responsible for the pseudogap.

32.4.5 Origin of High Critical Temperatures

Unconventional superconductors often have transition temperatures that are higher than expected. The traditional view is that the critical temperature for onset of SC depends largely on the magnitude of the pairing interaction. However, the dynamical symmetry solutions discussed in this chapter suggest that for unconventional SC there is another emergent factor of central importance that is illustrated in Fig. 32.11 [98, 99]. If the superconductor is unconventional it is likely a sign of some order competing with pairing. If the competing order is *related to the SC* because they correspond to different subgroups of a higher symmetry like SU(4) (implying that they compete for the same Hilbert subspace), the competing order parent state can "precondition" the SC phase transition. Thus it can occur at higher T because the low-entropy competing order ground state can be rotated collectively into the SC state in group space, as illustrated schematically in

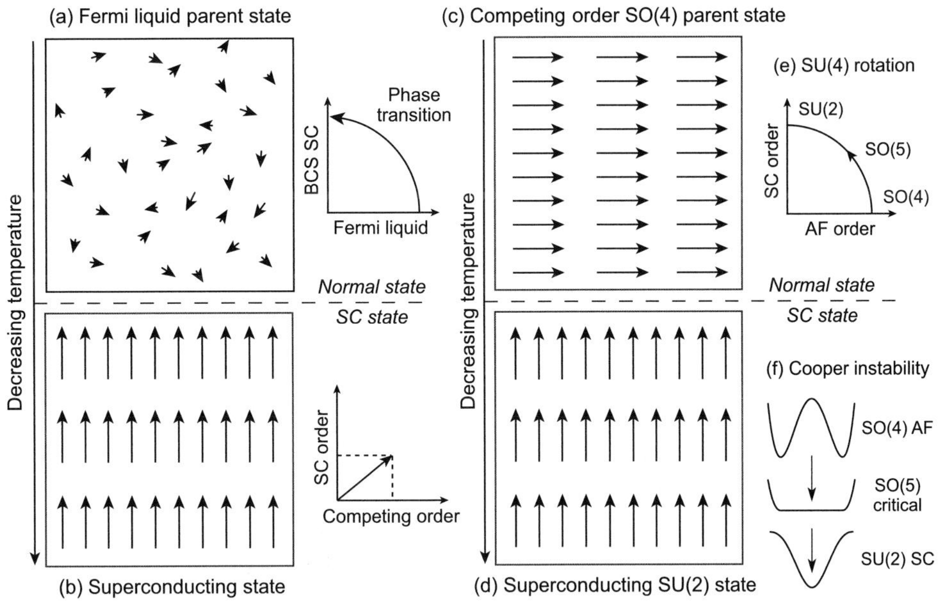

Fig. 32.11 Possible origin of high critical temperatures in the cuprates [98]. (a)–(b) Formation of a BCS superconductor by lowering the temperature of a Fermi liquid through T_c. Direction of vectors indicates relative strength of competing order (x) and SC (y); length indicates total SU(4) strength. The SC transition converts a high-entropy state (a) into a highly ordered state (b), implying a low T_c. (c)–(d) Formation of SC from a parent state having order that competes with SC but is related to SC by symmetry. This requires imposing SC order (d) on a state (c) already highly ordered, which can occur at a higher T_c because it is a collective rotation in the group space between two low-entropy states. (e) Collective rotation in SU(4) group space. (f) SU(4) generalized Cooper instability. Reproduced with permission from Springer Nature: *Frontiers of Physics*, Fermion Dynamical Symmetry and Strongly-Correlated Electrons: A Comprehensive Model of High-Temperature Superconductivity, M. Guidry, Y. Sun, and L.-A. Wu, **15**, 43301 (2020).

Fig. 32.11. The superconducting transition between Figs. 32.11(c) and 32.11(d) can occur *spontaneously* if there is no barrier to the SU(4) rotation. The SU(4) $\supset$ SO(5) critical dynamical symmetry exhibits such a property, as illustrated in Figs. 32.11(e–f).

1. At low doping there are degenerate AF and SC ground states with effectively no energy barrier separating them [see the curves in Fig. 32.9(b) for $n/\Omega = (1 - x) \sim 1$].
2. Thus AF and SC phases can be connected by infinitesimal rotations through nearly degenerate SO(5) "doorway states" exhibiting a broad range of AF and SC components.

This entropy argument is equivalently an information argument. Figure 32.11(d) is obtained from Fig. 32.11(c) by collectively rotating all arrows. This can be specified in terms of a single rotation angle, which requires minimal information. Conversely, in Fig. 32.11(a) there is no order in the parent state and each arrow must be lengthened and oriented separately to give Fig. 32.11(b), which requires much more information. Thus the entropy reduction needed to condense the SC state from the parent state is much greater

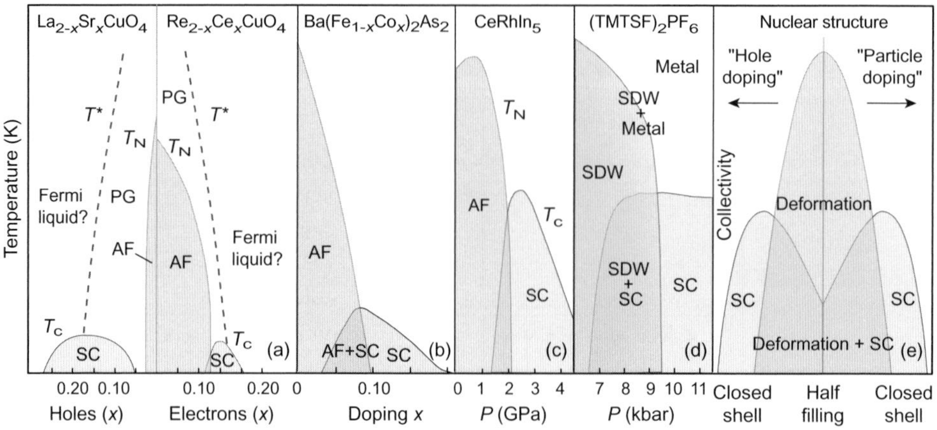

Fig. 32.12 Universality of superconductivity and superfluidity [91, 98]. (a) Phase diagram for hole- and electron-doped cuprates [13]. Superconducting (SC), antiferromagnetic (AF), and pseudogap (PG) regions are labeled, as are Néel (T_N), SC critical (T_c), and PG (T^*) temperatures. (b) Phase diagram for Fe-based SC [59]. (c) Heavy-fermion phase diagram [132]. (d) Phase diagram for an organic superconductor (SDW denotes spin density waves) [129]. (e) Generic correlation energy diagram for nuclear structure [91]. Reproduced with permission from Springer Nature: *Frontiers of Physics*, Fermion Dynamical Symmetry and Strongly-Correlated Electrons: A Comprehensive Model of High-Temperature Superconductivity, M. Guidry, Y. Sun, and L.-A. Wu, **15**, 43301 (2020).

in Figs. 32.11(a–b) than in Figs. 32.11(c–d). The information argument also distinguishes independent competing modes from those related by a higher symmetry. In the former case a phase transition requires deconstructing one mode and using the pieces to construct the other mode. In the latter case *the higher symmetry encodes the relationship between the two modes.* Hence only minimal additional information is required to produce the SC state from the competing order state, because they arise from the same collective Hilbert subspace and correspond to subgroups of the same highest symmetry.

32.4.6 Universality of Dynamical Symmetry States

In Section 20.3 we illustrated the remarkable universality of collective states generated by dynamical symmetries through the very similar energy surface diagrams of Fig. 20.10. In Fig. 32.12 another aspect of that universality is illustrated by displaying remarkably similar phase diagrams for superconductors and superfluids across a broad range of disciplines. This universality occurs because the systems have completely different microscopic structure but share a similar symmetry dictated truncation of their Hilbert space for their emergent modes [91, 98].

Background and Further Reading

A comprehensive review of the SU(4) model with references to the primary literature is given in Refs. [98] and [99]. This chapter has relied heavily on those reviews.

Problems

32.1 Highlight the propensity of cuprate antiferromagnetic Mott insulator states to condense a superconductor in the presence of small hole doping by showing that even the AF Mott insulator limit of SU(4) symmetry implies non-zero pairing correlations in the ground state unless the hole doping x is identically zero. *Hint*: From Ref. [190], SU(4) symmetry requires that $Q^2 + \Delta^2 + \Pi^2 = \frac{1}{4}(1 - x^2)$, where Δ is the singlet pair correlation, Π is the triplet pair correlation, and Q is the AF correlation, but for the SU(4) $\supset$ SO(4) AF symmetry limit $Q^2 = \frac{1}{4}(1 - x)^2$. ***

32.2 Show that the AF Mott insulator symmetry SU(4) $\supset$ SO(4) described in Section 32.3.5 is locally isomorphic to SU(2) $\times$ SU(2), if new generators are defined by Eq. (32.15).

33 Current Algebra

Box 19.1 introduced the Fermi current–current theory of weak interactions. In the interest of simplicity it was illustrated there for leptonic weak currents. For hadronic weak currents we might expect that the strong interactions would renormalize such matrix elements substantially from their leptonic values. However, the hadronic matrix elements are found to be much less renormalized than might be expected. As we shall see, this is because of symmetries that partially protect the currents from renormalization by strong interactions. In elementary particle physics these ideas go under the rubric of *current algebra.*

33.1 The CVC and PCAC Hypotheses

From Box 19.1, a matrix element of the leptonic weak current takes the form $\langle e| l^\alpha |\nu_e\rangle \sim \bar{u}_e\gamma^\alpha(1-\gamma_5)u_\nu$. It is found that classical β-decay is described by matrix elements *analogous to the leptonic form,*

$$\langle p| h^\alpha |n\rangle \sim \bar{u}_p\gamma^\alpha(g_V - g_A\gamma_5)u_n, \tag{33.1}$$

with $g_V \sim 0.98$ and $g_A \sim 1.24$. The vector coefficient g_V is nearly the same as for the leptonic current and the axial vector coefficient g_A is only $\sim$ 25% larger than for the leptonic current. Feynman and Gell-Mann [61] explained $g_V \sim 1$ by noting that in QED the electric charge is exactly conserved, even for strongly interacting particles, because the electromagnetic 4-current is a conserved current. This suggests that a conserved current may be responsible for the protection of hadronic vector matrix elements, with a logical candidate being a conserved isospin current. This is called the *conserved vector current (CVC) hypothesis*. Likewise, the deviation of the hadronic axial vector coefficient g_A by only about 25% from its leptonic value was hypothesized to be due to a *partially conserved axial vector current (PCAC)*.

33.1.1 Current Algebra and Chiral Symmetry

Restricting to u and d quarks, we assume a triplet of *conserved* vector isospin currents

$$\partial_\alpha V_i^\alpha = 0 \qquad V_i^\alpha(x) \equiv \frac{1}{2}\bar{q}(x)\gamma^\alpha\tau_i q(x) \qquad q = \begin{pmatrix} u \\ d \end{pmatrix}, \tag{33.2}$$

and a triplet of *partially conserved* axial vector currents

$$\partial_\alpha A_i^\alpha \simeq 0 \qquad A_i^\alpha(x) \equiv \frac{1}{2}\bar{q}(x)\gamma^\alpha\gamma_5\tau_i q(x) \qquad q = \begin{pmatrix} u \\ d \end{pmatrix}, \tag{33.3}$$

where the τ_i $(i = 1, 2, 3)$ are isospin operators. Charges associated with these vector and axial currents may be constructed:

$$Q_i(t) \equiv \int d^3x\, V_i^0(\mathbf{x}, t) = \frac{1}{2}\int d^3x\, q^\dagger(x)\tau_i q(x), \tag{33.4}$$

$$Q_{i5}(t) \equiv \int d^3x\, A_i^0(\mathbf{x}, t) = \frac{1}{2}\int d^3x\, q^\dagger(x)\gamma_5\tau_i q(x), \tag{33.5}$$

which obey the commutation relations (Problem 33.2)

$$[\,Q_i, Q_j\,] = i\epsilon_{ijk}Q_k \qquad [\,Q_i, Q_{j5}\,] = i\epsilon_{ijk}Q_{k5} \qquad [\,Q_{i5}, Q_{j5}\,] = i\epsilon_{ijk}Q_k. \tag{33.6}$$

Then the left-handed charge $Q_{\rm L}^i$ and the right-handed charge $Q_{\rm R}^i$ defined by

$$Q_{\rm L}^i \equiv Q_-^i = \frac{Q_i - Q_{i5}}{2} \qquad Q_{\rm R}^i \equiv Q_+^i = \frac{Q_i + Q_{i5}}{2}, \tag{33.7}$$

satisfy the commutation algebra

$$[\,Q_{\rm L}^i, Q_{\rm L}^j\,] = i\epsilon_{ijk}Q_{\rm L}^k \qquad [\,Q_{\rm R}^i, Q_{\rm R}^j\,] = i\epsilon_{ijk}Q_{\rm R}^k \qquad [\,Q_{\rm L}^i, Q_{\rm R}^j\,] = 0, \tag{33.8}$$

which may be recognized immediately as the Lie algebra SU(2) × SU(2). This algebra of the charges associated with the vector and axial currents is called the *chiral* $\mathrm{SU(2)_L} \times \mathrm{SU(2)_R}$ *algebra*.[1] The algebra is said to be chiral because the generators of the independent $\mathrm{SU(2)_L}$ and $\mathrm{SU(2)_R}$ algebras are related by a parity transformation (Problem 33.3).

33.1.2 The Partially Conserved Axial Current

Conservation of the axial current (33.3) would imply the existence of a massless pseudoscalar, isovector particle. Since the pion carries these quantum numbers and is the lowest-mass hadron, it is assumed to be the "massless" particle associated with the axial current. However, in reality pions have a small but finite mass, which leads to the *partially conserved axial current (PCAC) hypothesis*.

PCAC Hypothesis: The divergence of the axial vector current $\partial_\mu A_i^\mu(x)$, which measures the degree to which the current is not conserved, is related to the pion mass m_π through

$$\partial_\mu A_i^\mu(x) = f_\pi m_\pi^2 \pi_i(x) \neq 0, \tag{33.9}$$

where $\pi_i(x)$ is a pseudoscalar field operator and f_π is a pion coupling constant.

[1] As discussed in Section 19.2.3, if u, d, and s quark flavors are included a similar derivation leads to a chiral $\mathrm{SU(3)_L} \times \mathrm{SU(3)_R}$ algebra. We shall use the simpler $\mathrm{SU(2)_L} \times \mathrm{SU(2)_R}$ algebra to illustrate our discussion here.

We may expect that a realistic Hamiltonian describing the strong interactions is of the form $H = H_0 + \epsilon H_1$, where H_0 is chiral invariant and H_1 is not, but ϵ should be small because $\epsilon \sim \mathcal{O}\left(m_\pi^2\right)$ and m_π is small compared with the characteristic mass scale for hadrons.

33.2 The Linear σ-Model

The *linear* σ-*model* [32, 47, 67] illustrates many of the ideas we have just introduced concerning PCAC. The Lagrangian density is $\mathscr{L} = \mathscr{L}_0 + \mathscr{L}'$, where $\mathscr{L}' = \epsilon\sigma$ and

$$\begin{aligned} \mathscr{L}_0 &= \overline{N}(i\partial\!\!\!/ + g(\sigma + i\boldsymbol{\tau}\cdot\boldsymbol{\pi}\gamma_5))N \\ &\quad + \frac{1}{2}\left(\partial_\mu\sigma\right)^2 + \frac{1}{2}\left(\partial_\mu\boldsymbol{\pi}\right)^2 - \frac{1}{4}\lambda(\sigma^2 + \boldsymbol{\pi}\cdot\boldsymbol{\pi} - v^2)^2, \end{aligned} \tag{33.10}$$

with N an isodoublet nucleon field, $\boldsymbol{\pi}$ an isotriplet pion field, and σ a scalar meson field. The scalar products in Eq. (33.10) are with respect to components in the isospace, $\boldsymbol{\pi}\cdot\boldsymbol{\pi} = \pi_1^2 + \pi_2^2 + \pi_3^2$ and $\boldsymbol{\tau}\cdot\boldsymbol{\pi} = \tau_1\pi_1 + \tau_2\pi_2 + \tau_3\pi_3$, and the constants λ and ϵ are assumed to be positive. The term $\mathscr{L}_0$ is invariant under isospin and chiral transformations, while $\mathscr{L}'$ is isospin invariant but breaks chiral symmetry. To investigate this model, let us treat the $\boldsymbol{\pi}$ and σ fields as classical through replacements such as $\langle\sigma^4\rangle \simeq \langle\sigma\rangle^4 \equiv \sigma^4$ (where $\langle\ \rangle$ denotes the quantum expectation value) and introduce a meson potential energy

$$\begin{aligned} V(\pi,\sigma) &\equiv \frac{1}{4}\lambda\left(\sigma^2 + \boldsymbol{\pi}\cdot\boldsymbol{\pi} - v^2\right)^2 - \epsilon\sigma \\ &= \frac{1}{4}\lambda\left(\sigma^2 + \boldsymbol{\pi}\cdot\boldsymbol{\pi}\right)^2 + \frac{1}{2}\mu^2\left(\sigma^2 + \boldsymbol{\pi}\cdot\boldsymbol{\pi}\right) - \epsilon\sigma, \end{aligned} \tag{33.11}$$

where $\mu^2 \equiv -\lambda v^2$ and a constant term was dropped. This gives a Lagrangian density

$$\mathscr{L} = \overline{N}(i\partial\!\!\!/ + g(\sigma + i\boldsymbol{\tau}\cdot\boldsymbol{\pi}\gamma_5))N + \frac{1}{2}\left(\partial_\mu\sigma\right)^2 + \frac{1}{2}\left(\partial_\mu\boldsymbol{\pi}\right)^2 - V(\pi,\sigma). \tag{33.12}$$

Let us explore the particle spectrum expected for this theory.

33.2.1 The Particle Spectrum

First neglect the chiral symmetry breaking term in the potential (33.11) by setting $\epsilon = 0$. From the discussion in Section 17.5, we expect two qualitatively different cases, depending on the sign of μ^2. If $\mu^2 > 0$ the symmetry will be implemented in the Wigner mode, as illustrated in Fig. 33.1(a). The classical vacuum is the symmetric minimum at $\boldsymbol{\pi} = \sigma = 0$, low-energy excitations correspond to oscillations around this symmetric minimum, and the particle spectrum can be read off from the terms quadratic in the fields in $\mathscr{L}$,

$$\mathscr{L} \simeq \overline{N}\left(i\partial\!\!\!/ + g(\sigma + i\boldsymbol{\tau}\cdot\boldsymbol{\pi}\gamma_5)\right)N + \frac{1}{2}\left(\partial_\mu\sigma\right)^2 + \frac{1}{2}\left(\partial_\mu\boldsymbol{\pi}\right)^2 - \frac{1}{2}m_{\sigma,\pi}^2(\sigma^2 + \boldsymbol{\pi}\cdot\boldsymbol{\pi}).$$

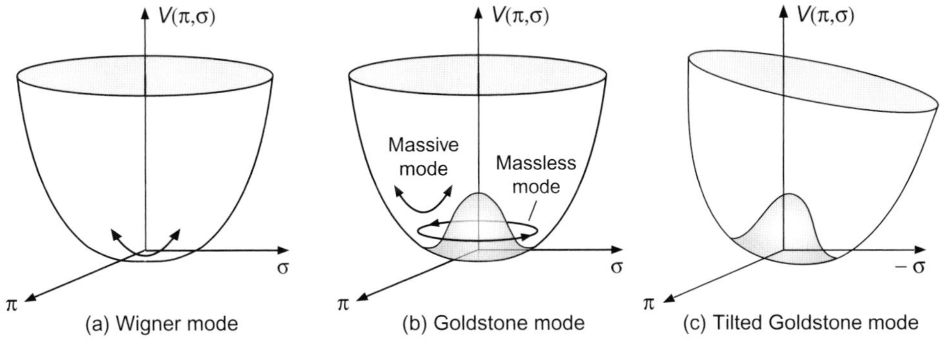

Fig. 33.1 Potential $V(\pi, \sigma)$ of the linear σ-model. (a) Symmetry realized in the Wigner mode ($\mu^2 > 0$). (b) Symmetry broken spontaneously in the Goldstone mode ($\mu^2 < 0, \epsilon = 0$). (c) Spontaneous symmetry breaking in the Goldstone mode ($\mu^2 < 0$), along with an *explicit* breaking of chiral symmetry ($\epsilon \neq 0$).

This corresponds to a spectrum of massless nucleons (a nucleon mass term $m\,\overline{N}N$ would break chiral symmetry), and π and σ mesons with masses given by

$$m_\pi^2 = m_\sigma^2 = \mu^2 = -\lambda v^2 \equiv m_{\sigma,\pi}^2. \tag{33.13}$$

But this is in poor agreement with the phenomenology of elementary particles: real nucleons have significant mass and no scalar particle is observed that is degenerate in mass with the pseudoscalar pions. Therefore, let us consider the symmetry to be broken spontaneously. Following the examples in Ch. 17, we choose $\mu^2 < 0$ in Eq. (33.11) to give the potential in Fig. 33.1(b), and expand the solution around the new classical vacuum that is no longer at the origin. The ring of minima in Fig. 33.1(b) satisfies $\sigma^2 + \boldsymbol{\pi}\cdot\boldsymbol{\pi} = v^2$ and we may choose the minimum on the negative σ-axis to be the classical vacuum.[2] Introducing a new variable $\sigma' \equiv \sigma + v$, the Lagrangian density becomes

$$\begin{aligned}\mathscr{L} &= \overline{N}\left(i\partial\!\!\!/ - gv + g(\sigma' + i\boldsymbol{\tau}\cdot\boldsymbol{\pi}\gamma_5)\right)N + \frac{1}{2}\left(\partial_\mu\sigma'\right)^2 + \frac{1}{2}\left(\partial_\mu\boldsymbol{\pi}\right)^2 \\ &\quad - \frac{1}{4}\lambda(\sigma'^2 + \boldsymbol{\pi}\cdot\boldsymbol{\pi})^2 - \lambda v^2\sigma'^2 + \lambda v\sigma'(\sigma'^2 + \boldsymbol{\pi}\cdot\boldsymbol{\pi}).\end{aligned} \tag{33.14}$$

Identifying masses by picking out terms quadratic in the fields gives

- a nucleon isodoublet with a mass $m_{\mathrm{N}} = gv$,
- an isoscalar σ-meson with mass $m_\sigma = \sqrt{2\lambda}\,v = \sqrt{-2\mu^2}$, and
- an isotriplet π-meson of mass $m_\pi = 0$.

We may interpret σ as massive radial oscillations of the field, and π as massless excitations (no restoring force) for motion around the bottom of the valley in Fig. 33.1(b). This is a far more realistic spectrum.

[2] Choosing the vacuum state to lie on the σ-axis ($\pi = 0$) is motivated physically because σ is a Lorentz scalar but π is a Lorentz pseudoscalar. Allowing the vacuum to contain a scalar field but not pseudoscalar fields ensures that the vacuum state conserves parity, as required by experimental data.

1. The nucleons N are massive.
2. The scalar σ and the pseudoscalar π are no longer necessarily degenerate.
3. The pion is the Goldstone boson associated with the spontaneous breaking of chiral SU(2) × SU(2) symmetry, so it is identically massless.

The predicted scalar particle could be problematic since there is little evidence for low-lying scalars in the experimental spectrum, but the σ mass can be adjusted independently to push it to higher energy. In reality $m_\pi c^2 \sim 140$ MeV, which is small for an elementary particle, so the pion is "almost massless." However, it is possible to do even better for the pion mass if the chiral symmetry is also broken *explicitly.*

33.2.2 Explicit Breaking of Chiral Symmetry

If we set $\epsilon \neq 0$ for the linear term that breaks chiral symmetry explicitly in Eq. (33.11) the potential of Fig. 33.1(b) is tilted, as illustrated in Fig. 33.1(c). Now the classical ground state is unique and not continuously degenerate. The pion, which was the zero-mass Goldstone boson corresponding to unimpeded circular motion in Fig. 33.1(b), acquires a small mass and is only an "approximate Goldstone boson" because the explicit symmetry breaking implies a small restoring force for angular motion away from the unique minimum. If ϵ is assumed small and the Lagrangian density is expanded around the new minimum,

1. the shift in nucleon and σ masses is of order ϵ and thus negligible,
2. the pion acquires a small mass $m_\pi^2 \sim \epsilon/v$, and
3. the axial current is only partially conserved, $\partial_\mu A_i^\mu = \epsilon\pi_i$.

The small but non-zero pion mass and partial conservation of the axial current are now in improved agreement with weak interaction phenomenology.

Background and Further Reading

Pedagogical introductions to current algebra, chiral symmetry, and the linear σ-model that have influenced our presentation may be found in Refs. [32, 85].

Problems

33.1 Derive the commutator $[\,Q_i, Q_j\,] = i\epsilon_{ijk}Q_k$ for the charge defined in Eq. (33.4). *Hint*: Use the charge (33.4) to write the commutator, displaying explicit matrix indices

$$[\,Q_a(t), Q_b(t)\,] = \int d^3x\, d^3x'\; t^a_{ij}\, t^b_{kl} [\,q_i^\dagger(x) q_j(x), q_k^\dagger(x') q_l(x')\,],$$

where $t^c \equiv \frac{1}{2}\tau^c$. Then use that the fermion operators anticommute. ***

33.2 Prove that the charges (33.7) obey the commutators in Eq. (33.8).

33.3 Show that the generators of the algebra (33.8) are related by parity. *Hint*: For a Dirac wavefunction the action of parity is $P\psi(\boldsymbol{x},t)P^{-1} = \gamma_0\psi(-\boldsymbol{x},t)$, up to a phase. ***

33.4 Verify that the potential $V(\pi,\sigma)$ can be written as Eq. (33.11), and that if $\epsilon = 0$ and the symmetry is implemented in the Wigner mode the masses for the π and σ fields are degenerate and given by Eq. (33.13). *Hint*: A constant term may be discarded from a Lagrangian density without affecting physical quantities.

34 Grand Unified Theories

In Ch. 19 we saw that the weak, electromagnetic, and strong interactions are all described by local gauge theories and, if gravity is neglected, the fundamental interactions correspond to a gauge symmetry $SU(2)_w \times U(1)_y \times SU(3)_c$ called the Standard Model. Above the electroweak scale of about 100 GeV the symmetry is assumed to be unbroken and the weak and electromagnetic interactions are unified, but below the electroweak scale the symmetry is broken according to the pattern

$$SU(2)_w \times U(1)_y \times SU(3)_c \longrightarrow SU(3)_c \times U(1)_{QED},$$

where the arrow indicates that the Higgs field breaks the electroweak gauge group $SU(2)_w \times U(1)_y$ down to $U(1)_{QED}$ below the electroweak scale, so that in our low-energy world we obtain the following.

1. The weak and electromagnetic interactions differ from each other fundamentally in strength, range, and conserved symmetries.
2. The intermediate vector bosons $W^\pm$ and Z^0 are massive.
3. The photon is massless and the electromagnetic charge is conserved exactly by virtue of the remaining U(1) gauge symmetry of QED.
4. The color gauge symmetry $SU(3)_c$ of the strong interactions is conserved exactly and color is confined, so physical states are color SU(3) singlets.

This invites speculation that perhaps this symmetry breaking at the electroweak scale is just the last step in a hierarchy of symmetry breaking originating in a more comprehensive gauge symmetry that unifies all interactions at some very large mass scale under a single gauge coupling constant. Theories that attempt to unify the presently known non-gravitational fundamental interactions in a larger gauge group $\mathbb{G}$ that contains the Standard Model symmetry $SU(2)_w \times U(1)_y \times SU(3)_c$ as a subgroup are called *grand unified theories* or GUTs. We shall term the conjectured group $\mathbb{G}$ the *GUT gauge group*.

34.1 Evolution of Fundamental Coupling Constants

Two general observations motivate further investigation of GUTs. (1) If the GUT gauge group operates on the familiar quarks and leptons of the low-energy world that are listed in Fig. 19.1, candidates for $\mathbb{G}$ may be found that accommodate assignment of known particles to low-dimensional representations. (2) The momentum-dependent coupling strengths of the strong, weak, and electromagnetic interactions evolve with the square of momentum transfer Q^2 and they become comparable in magnitude at an energy of $\sim 10^{16}$ GeV.

Box 34.1 Anomalies in Quantum Field Theory

Anomalies imply that a theory is rendered non-renormalizable by the quantization process itself (see Section 17.1). Specifically, they occur when a symmetry that is respected at the classical level is no longer respected when the field is quantized. Anomalies are an advanced topic that we shall not discuss except to note that they appear only for certain gauge group representations. Therefore, construction of non-anomalous (and therefore renormalizable) gauge theories can imply restrictions on the permissible gauge groups and representations for the matter fields. An example of an anomaly-free representation in the SU(5) grand unified theory is given in Section 34.3. For a more extensive introduction to anomalies in the Standard Model, see Ryder [174].

34.2 Minimal Criteria for a Grand Unified Group

We assume that the GUT group $\mathbb{G}$ operates on the familiar quarks and leptons, which are grouped into generations as in Fig. 19.1. Gauge couplings preserve chirality and it is convenient to express the generations of Fig. 19.1 entirely as left-handed fields. Utilizing the results of Problem 34.1, we may rewrite the three Standard Model generations as

$$\begin{aligned} \text{I:}\quad &\left(\nu_e,\, e^-,\, e^+,\, u_i,\, d_i,\, \bar{u}_i,\, \bar{d}_i\right)_{\rm L}, \\ \text{II:}\quad &\left(\nu_\mu,\, \mu^-,\, \mu^+,\, c_i,\, s_i,\, \bar{c}_i,\, \bar{s}_i\right)_{\rm L}, \\ \text{III:}\quad &\left(\nu_\tau,\, \tau^-,\, \tau^+,\, t_i,\, b_i,\, \bar{t}_i,\, \bar{b}_i\right)_{\rm L}, \end{aligned} \tag{34.1}$$

where Roman numerals label generations, L denotes left-handed components, bars indicate charge conjugates, and $i = 1, 2, 3$ is a quark color index. Thus each generation consists of 15 two-component fields, three for each (colored) quark and one for each lepton.[1] The GUT gauge group $\mathbb{G}$ may be expected to meet a few minimal criteria.

1. It should be *simple* (Section 2.13), so that there is only one gauge coupling strength.
2. It must contain $\mathrm{SU(2)_w \times U(1)_y \times SU(3)_c}$ as a subgroup, which requires it to be at least of rank $1 + 1 + 2 = 4$.
3. It must accommodate complex representations (see Section 8.5 and the solution of Problem 8.6) since the generations (34.1) contain fields and their charge conjugates.
4. The representations of $\mathbb{G}$ to which particles are assigned must satisfy technical conditions ensuring freedom from anomalies, which are discussed in Box 34.1.

Various candidate GUT groups meeting these requirements have been suggested, with two popular ones being SU(5) and SO(10).

[1] In the Standard Model neutrinos are assumed identically massless and only the left-handed components enter the weak interactions. We now know that neutrinos have finite but tiny masses, but we will keep things simple and assume massless neutrinos for purposes of this discussion.

34.3 The SU(5) Grand Unified Theory

The simplest theory meeting the conditions of the preceding section is the SU(5) model of Georgi and Glashow [69]. Although this theory suffers the misfortune of being contradicted by data (it predicts a proton lifetime too short to be consistent with experimental limits), it gets many things right and we study it here as a prototype of a grand unified theory [41, 85, 166].

SU(5) Gauge Bosons: Gauge bosons transform as the adjoint representation of the gauge group, which is of dimension $N^2 - 1$ for SU(N). Thus an SU(5) GUT requires 24 gauge bosons. We presume that 12 are already familiar from the Standard Model: eight gluons, three intermediate vector bosons, and one photon. The remaining 12 gauge bosons are termed *leptoquark bosons,* denoted by X and Y. The gauge bosons X and Y carry both weak isospin and color quantum numbers, so they can mediate transitions between leptons and quarks.

Particle Assignments: The fundamental and conjugate representations of SU(5) are the **5** and $\overline{\mathbf{5}}$, respectively, and each fermion generation of the Standard Model fits neatly into a *direct sum* of SU(5) representations: $\overline{\mathbf{5}} \oplus \mathbf{10}$, where the **10** comes from the antisymmetric part of $\mathbf{5} \otimes \mathbf{5}$. We may classify these SU(5) multiplets according to their $\mathrm{SU(3)_c}$ and $\mathrm{SU(2)_w}$ subgroup dimensionalities using the notation (Dim SU(3), Dim SU(2)). Then for the first generation of Standard Model particles,

$$\begin{aligned} \mathbf{5} &= (\mathbf{3},\mathbf{1}) + (\mathbf{1},\mathbf{2}) = d_\mathrm{R} + (\bar{\nu}_\mathrm{e}, e^+)_\mathrm{R}\,, \\ \overline{\mathbf{5}} &= (\overline{\mathbf{3}},\mathbf{1}) + (\mathbf{1},\mathbf{2}) = \bar{d}_\mathrm{L} + (\nu_\mathrm{e}, e^-)_\mathrm{L}\,, \\ \mathbf{10} &= (\mathbf{3},\mathbf{2}) + (\overline{\mathbf{3}},\mathbf{1}) + (\mathbf{1},\mathbf{1}) = (u,d)_\mathrm{L} + \bar{u}_\mathrm{L} + e^+_\mathrm{L}. \end{aligned} \tag{34.2}$$

Generations II and III have analogous assignments.

Example 34.1 The SU(5) representation $\overline{\mathbf{5}}$ contains the left-handed components of ν_e and e^- [a weak isospin doublet but a color singlet, since ν_e and e^- do not see color], and left-handed components of the antidown quark [a $\overline{\mathbf{3}}$ under SU(3) color but an $\mathrm{SU(2)_w}$ singlet, since only right-handed antiparticles couple to the charged weak current].

Example 34.2 The **10** of SU(5) contains the left-handed e^+, which is a color and weak singlet; the left-handed $\bar{u}$ quark, which transforms as a color $\overline{\mathbf{3}}$ but a weak singlet; and the left-handed u and d quarks, which constitute a weak isospin doublet and transform as $\mathrm{SU(3)_c}$ triplets.

You are asked to show in Problem 34.3 that the $\mathbf{5} \oplus \mathbf{10}$ representation of SU(5) is anomaly free (Box 34.1), because anomalies associated with the **5** and **10** are each finite, but they cancel exactly in the direct sum.

Quantization of Electrical Charge in SU(5): The magnitudes of the electrical charge carried by protons and electrons are known experimentally to be equivalent to extremely high precision. The Standard Model gives no explanation for this fact, which is fundamental to the existence of stable atoms. The SU(5) GUT fermion assignments account for this by reproducing the factors of $\frac{1}{3}$ between the quark charges found in Table 9.1 and leptonic charges, thereby explaining the equality of proton and positron charges. The SU(5) generators are traceless as for SU(2) and SU(3), and because the photon is one of the SU(5) gauge bosons the electrical charge is a traceless diagonal matrix acting on the **5** and $\bar{\mathbf{5}}$. This implies that the sum of the charges in an SU(5) multiplet must vanish.

Example 34.3 Neglecting chirality labels, the $\bar{\mathbf{5}}$ in Eq. (34.2) has the particle content $\bar{d}$, ν_e, and e^-. Therefore, the sum of the charges must satisfy $3Q(\bar{d}) + Q(\nu_e) + Q(e) = 0$, where the factor of 3 is for the color degeneracy of the $\bar{d}$ quark. Since $Q(\nu_e) = 0$, this implies that $Q(e) = -3Q(\bar{d}) = -1$, where we have used $Q(\bar{d}) = +\frac{1}{3}$ from Table 9.1 evaluated for the antiquark (see table footnote). This indicates that the third-integer charges for quarks in Table 9.1, which were suggested originally by the phenomenological flavor symmetry, are actually a consequence of *color gauge symmetry*.

The Standard Model imposes no conditions on electrical charge because it is associated with an *abelian* U(1) factor. Conversely, the constraints imposed by the *non-abelian* SU(5) commutator algebra imply that *charge must be quantized.*

Symmetry Breaking Hierarchy: Leptons and quarks have been assigned to the same SU(5) representations and the leptoquark bosons X and Y carry both weak isospin and color quantum numbers, so transitions that fail to conserve baryon number and lepton number are possible in the SU(5) theory. The poster child for such processes is decay of the proton, which is forbidden if baryon number is conserved. On dimensional grounds the mean life for proton decay to lighter subatomic particles in the SU(5) theory is

$$\tau_p \simeq \frac{1}{\alpha_G^2} \frac{M_x^4}{M_p^5}, \tag{34.3}$$

where M_p is the proton mass, M_x is the (unknown) leptoquark boson mass scale, and α_G is the (unknown and momentum-dependent) gauge coupling strength for SU(5). Since the lifetime of the proton is known to be greater than 10^{34} yr, reasonable choices of α_G in Eq. (34.3) require that $M_x \sim 10^{15}$ GeV or larger to avoid obvious conflict with experiments. A mass scale that large can be acquired only by spontaneous symmetry breaking (Ch. 17). Thus the SU(5) GUT requires two levels of spontaneous symmetry breaking.

1. On a mass scale $M_x \sim 10^{15}$ GeV a 24-dimensional representation of real scalar fields breaks the SU(5) symmetry down to the Standard Model by the Higgs mechanism,

$$\mathrm{SU}(5) \rightarrow \mathrm{SU}(2)_w \times \mathrm{U}(1)_y \times \mathrm{SU}(3)_c,$$

with the 12 X and Y bosons acquiring masses of order 10^{15} GeV, the other 12 gauge bosons remaining massless, and 12 Higgs bosons appearing with masses comparable to the leptoquark bosons. (This is a more complicated version of the simple Higgs mechanisms described in Sections 18.2 and 19.1.3.) The 24 gauge bosons of SU(5) may be classified under the $SU(3)_c$ and $SU(2)_w$ subgroups as

$$\mathbf{24} = \underset{\text{gluons}}{(\mathbf{8},\mathbf{1})} + \underset{W^\pm, Z^0, \gamma}{(\mathbf{1},\mathbf{3}) + (\mathbf{1},\mathbf{1})} + \underset{\bar{X},\bar{Y}}{(\mathbf{3},\mathbf{2})} + \underset{X,Y}{(\bar{\mathbf{3}},\mathbf{2})},$$

where a notation (Dim $SU(3)_c$, Dim $SU(2)_w$) is used. Thus the massive X and Y bosons can mediate interactions that interconvert leptons and quarks because they transform as triplets under color SU(3) and doublets under weak isospin SU(2).

2. On a mass scale of ~ 100 GeV the standard electroweak symmetry breaking occurs:

$$SU(2)_w \times U(1)_y \times SU(3)_c \rightarrow SU(3)_c \times U(1)_{QED}.$$

For the SU(5) theory this is accomplished using a set of 10 fields transforming as a complex **5** under SU(5). Three of the original scalars give the weak vector bosons $W^\pm$ and Z^0 mass by the Higgs mechanism, leaving seven physical Higgs fields.

We note in passing that dealing with the very large difference between these symmetry breaking scales in perturbation theory leads to a serious issue associated with fine-tuning of parameters called the *gauge hierarchy problem.* Introductions to the problem may be found in the literature [41, 166]. We will not pursue this technical issue here, except to note below that constraints imposed by a conjectured symmetry between fermions and bosons might be relevant to resolving this issue.

34.4 Beyond Simple GUTs

The simplest GUT is the SU(5) theory just outlined, but it is ruled out by prediction of a proton decay lifetime that is several orders of magnitude shorter than the present experimental limit. More generally, there is as yet no evidence for the baryon nonconservation implied by such theories. Extensions of SU(5), and GUTs based on larger groups like SO(10) [which contains SU(5) as a subgroup], can push the proton decay lifetime higher at the expense of more complicated formalisms and larger theoretical ambiguity.

Considerable attention has been focused on GUTs based on supersymmetries, which are symmetries for which bosons and fermions appear in the same group representations (see Box 34.2). These theories predict supersymmetric fermion partners for known elementary bosons and supersymmetric boson partners for known elementary fermions. Supersymmetries are also crucial to *superstring theory* and *M-theory,* which attempt to unify all four fundamental interactions, and undiscovered supersymmetric particles are a popular candidate for the dark matter required by modern cosmology.

If supersymmetry existed, it might also help with the gauge hierarchy problem alluded to above: if every fermion has a boson partner and every boson has a fermion partner, then in perturbation theory there could be large cancellations in the contributions from each fermion–boson pair, thus at least partially alleviating the parameter fine-tuning problem.

Box 34.2 Supersymmetry

Many ideas in modern physics invoke a property of the Universe called *supersymmetry* that is conjectured theoretically, but for which there is not yet any evidence.

Fermions and Bosons

Elementary particles can be divided into two classes that exhibit fundamentally different quantum statistics. *Fermions* carry half-integer spins and obey *Fermi–Dirac statistics,* implying wavefunctions that are antisymmetric with respect to exchange of identical fermions. The most notable implication of fermionic statistics is the *Pauli principle*: no two identical fermions can occupy the same quantum state. *Bosons* carry integer spins and obey *Bose–Einstein statistics,* implying wavefunctions that are symmetric with respect to exchange of identical bosons. The most notable implication of bosonic statistics is boson condensation: many identical bosons can occupy the same quantum state. We have no fundamental explanation for why, but all known elementary particles are classified uniquely as either bosons or fermions.

Symmetries Relating Fermions and Bosons

Normal quantum symmetries relate bosons to bosons or fermions to fermions, but *not bosons to fermions.* For example, in color SU(3) a group multiplet contains either fermions (quarks) or bosons (gluons), but not both. There are attractive theoretical reasons to believe that the Universe will eventually be found to exhibit a fundamental symmetry that relates fermions to bosons called *supersymmetry.* Under supersymmetry each fermion has a partner boson of the same mass and each boson has a partner fermion of the same mass.

> Electrons are spin-$\frac{1}{2}$ fermions with a definite mass. In supersymmetry the electron has a partner called a *selectron* of the same mass, but having a spin of zero and obeying boson statistics.

Thus supersymmetric fermions and bosons can occur in the same (super)symmetry multiplets.

Experimental Evidence for Supersymmetry

The problem with this rather beautiful idea is that no experiment has ever seen a supersymmetric partner of any known particle. However, supersymmetry could be broken, with the masses for supersymmetric partners pushed up to high values that could explain the lack of observational evidence. The Large Hadron Collider (LHC) has searched for supersymmetry at energies of hundreds of GeV where the simplest implementations of broken supersymmetry suggest that there could be new supersymmetric particles. As of this writing in 2021 there is no evidence for such new particles in these LHC experiments (or in high-energy cosmic rays), which bodes poorly for at least the simplest implementations of supersymmetry.

This is because in perturbative quantum field theory diagrams corresponding to closed loops (for example, as in Fig. 19.3) contribute with opposite sign for a fermion loop relative to a boson loop. Thus if each fermion is related by (super)symmetry to a corresponding boson, each such fermion–boson pair could give canceling contributions to matrix elements corresponding to loop diagrams, by virtue of the symmetry constraint. However, as discussed in Box 34.2 there is as yet no experimental evidence for the supersymmetries required for these approaches to explaining our Universe.

Background and Further Reading

The prototypical GUT is the SU(5) model of Georgi and Glashow [69]. Textbook discussions of GUTs may be found in Cheng and Li [41], Guidry [85], and Quigg [166].

Problems

34.1 Prove that in Eq. (34.1) we may take as independent fermion fields $\psi_{\rm L}$ and $\psi^{\rm c}_{\rm L}$, instead of $\psi_{\rm L}$ and $\psi_{\rm R}$, because the charge conjugate of a right-handed fermion field is left-handed.***

34.2 Repeat Example 34.3 to relate the quark and leptonic charges for the **5** and **10** representations of Eq. (34.2).

34.3 The anomaly $A(R)$ of a fermion representation R is given by [41]

$$\mathrm{Tr}\left(\{T^a(R), T^b(R)\}\, T^c(R)\right) = \frac{1}{2}\, d^{abc} A(R),$$

where $T^a(R)$ is a representation matrix, the d^{abc} are totally symmetric constants defined through the anticommutator of the generators as in Eq. (11.17), and $A(R)$ is normalized to unity for the fundamental representation and is independent of the generators. Prove that the SU(5) representation $\bar{\mathbf{5}} \oplus \mathbf{10}$ is anomaly free because the anomalies of the $\bar{\mathbf{5}}$ and $\mathbf{10}$ are non-zero, but they exactly cancel each other. *Hint*: Charge Q is a generator so evaluate the above expression for $T^a = T^b = T^c = Q$. ***

Appendix A Second Quantization

Many examples discussed in this book represent quantum *many-body systems* containing a large number of *identical* particles that can be divided into two broad categories according to their exchange properties: *bosons* and *fermions*. Constructing a total wavefunction having proper exchange characteristics for a many-particle system can be a cumbersome task. The method of *second quantization* or *occupation number representation* using *Dirac notation* greatly facilitates this endeavor and is used extensively in this book. This Appendix provides a concise review of such methods for the non-relativistic many-body problem.

A.1 Symmetrized Many-Particle Wavefunctions

In quantum mechanics particles of the same type are *identical particles* since they are *indistinguishable observationally.* For the wavefunction of a quantum system with N identical particles $\Psi(x_1, x_2, \ldots, x_N)$, with $x_i \equiv (\mathbf{r}_i, s_i)$ denoting position ($\mathbf{r}_i$) and spin projection (s_i) coordinates of particle i, indistinguishability implies that

$$|\Psi(x_1, \ldots, x_i, \ldots, x_j, \ldots, x_N)|^2 = |\Psi(x_1, \ldots, x_j, \ldots, x_i, \ldots, x_N)|^2.$$

This identity requires that the probabilities for observations associated with the wavefunction be the same for configurations that differ in the exchange of coordinates between any two particles i and j. Let us see how to ensure this requirement of exchange symmetry for a many-body wavefunction. We assume an N-particle quantum system with a Hamiltonian

$$H = H_0 + H_M \qquad H_0 = \sum_{i=1}^{N} h(x_i) \qquad h(x_i) = -\frac{\hbar^2}{2m}\nabla_i^2 + V(x_i), \tag{A.1}$$

where H_0 represents the non-interacting part of the Hamiltonian, which is a *one-body operator* since it involves single-particle coordinates only, and H_M is a sum of terms involving the coordinates of more than one particle that represents interactions among the particles. For most problems it is a sufficient initial approximation to ignore three-body and higher interactions, and we assume H_M to be a *two-body operator* in this discussion.

It is convenient to write the many-particle wavefunction in a basis corresponding to eigenfunctions of H_0. To do so, one first solves the eigenvalue equation $h(x)\phi_a(x) = \epsilon_a \phi_a(x)$ for the one-body operator $h(x)$, where a represents a set of quantum numbers that uniquely characterize the single-particle state ϕ_a. For identical particles a viable *many-particle wavefunction* is *not* simply a product $\phi_{a_1}(x_1)\phi_{a_2}(x_2)\cdots\phi_{a_N}(x_N)$ of single-particle wavefunctions, but rather is a superposition of such product states corresponding to all possible permutations of the N coordinates $x_1, x_2, \ldots, x_N$ that fulfill the exchange symmetry.

A.1.1 Bosonic and Fermionic Wavefunctions

For boson systems the wavefunction must be *symmetric* under exchange of coordinates. The symmetrized wavefunction for bosons can be written

$$\Psi^{(S)}(x_1, x_2, \ldots, x_N) = \frac{1}{\sqrt{N!}\sqrt{\prod_a n_a!}} \sum_{P \in S_N} P\, \phi_{a_1}(x_1)\phi_{a_2}(x_2)\cdots\phi_{a_N}(x_N), \qquad \text{(A.2)}$$

with the superscript S standing for "symmetric." The summation runs over all permutations P of the N coordinates $x_1, x_2, \ldots, x_N$, with S_N indicating the set of permutations in the permutation group (Section 2.3.4). The product of factorials $n_a!$ runs over all states in the single-particle basis with n_a the number of particles in the single-particle state a.

For fermion systems the wavefunction must be *antisymmetric* under exchange of coordinates. The antisymmetrized wavefunction for fermions can be written

$$\Psi^{(A)}(x_1, x_2, \ldots, x_N) = \frac{1}{\sqrt{N!}} \sum_{P \in S_N} \mathrm{sgn}(P) \times P\, \phi_{a_1}(x_1)\phi_{a_2}(x_2)\cdots\phi_{a_N}(x_N), \qquad \text{(A.3)}$$

with the superscript A standing for "antisymmetric." There are two important differences compared to the bosonic case in Eq. (A.2). (1) For fermions n_a can only be 0 or 1, leading to $\prod n_a! = 1$. (2) There is a new factor $\mathrm{sgn}(P)$ in the summation, which denotes the sign of the permutation P. A permutation has a positive (negative) sign if it corresponds to an even (odd) number of two-particle exchanges (*transpositions*).

Example A.1 A permutation can be written as a product of transpositions P_{ij} that interchange the indices of the coordinates at positions i and j. For example, $P_{13}(123) = (321)$, which interchanges the indices of the coordinates at the positions 1 and 3, and $P_{12}(321) = (231)$. Then successive application of P_{13} and P_{12} gives $P_{12}P_{13}(123) = (231)$. Thus this permutation can be written in terms of an *even* number of transpositions (two) and the sign of the permutation is *positive*. Alternative sequences of transposition can give the same result. For example, applying four transpositions $P_{23}P_{13}P_{12}P_{23}$ on (123) also gives (231), but four is an even number and the sign is again positive. The general rule is that the way to express a permutation in terms of transpositions is not unique but the evenness/oddness is unique so the sign $\pm$ for $\mathrm{sgn}(P)$ in Eq. (A.3) is unambiguous.

A.1.2 Slater Determinants

The fermionic wavefunction (A.3) can be expressed conveniently as a *Slater determinant,*

$$\Psi^{(A)}(x_1, x_2, \ldots, x_N) = \frac{1}{\sqrt{N!}} \begin{vmatrix} \phi_{a_1}(x_1) & \phi_{a_2}(x_1) & \ldots & \phi_{a_N}(x_1) \\ \phi_{a_1}(x_2) & \phi_{a_2}(x_2) & \ldots & \phi_{a_N}(x_2) \\ \vdots & \vdots & \ddots & \vdots \\ \phi_{a_1}(x_N) & \phi_{a_2}(x_N) & \ldots & \phi_{a_N}(x_N) \end{vmatrix}, \qquad \text{(A.4)}$$

which ensures fidelity to the *Pauli principle.* It is easily checked that the determinant vanishes if it contains identical rows and/or identical columns. The physical meaning for a vanishing determinant with two identical columns is that it is not possible to put more than one identical fermion into a given single-particle state. The determinant also vanishes for two identical rows, meaning that it is not possible to have two identical fermions at the same x (remember that x denotes position and spin projection of a single-particle state). The meaning of a Slater determinant can be appreciated from a simple example.

Example A.2 For two fermions with coordinates x_1 and x_2, the *Hartree product* state is

$$\Psi(x_1, x_2) = \phi_1(x_1)\phi_2(x_2).$$

This state is not satisfactory because the wavefunction is not antisymmetric under exchange of the coordinates of the two fermions. A correct wavefunction should have the property $\Psi(x_1, x_2) = -\Psi(x_2, x_1) = -\phi_1(x_2)\phi_2(x_1)$, which is satisfied if we write

$$\Psi(x_1, x_2) = \frac{1}{\sqrt{2}}[\phi_1(x_1)\phi_2(x_2) - \phi_1(x_2)\phi_2(x_1)] = \frac{1}{\sqrt{2}} \begin{vmatrix} \phi_1(x_1) & \phi_2(x_1) \\ \phi_1(x_2) & \phi_2(x_2) \end{vmatrix},$$

where $1/\sqrt{2}$ is a normalization factor. This equation (1) is now antisymmetric, (2) no longer distinguishes between identical fermions, and (3) satisfies the Pauli principle.

A single Slater determinant is often a poor approximation to the many-particle wavefunction because it is formed using the solution of the non-interacting H_0 in the Hamiltonian. In more accurate treatments a linear combination of Slater determinants must be constructed and mixed by the two-body operator H_M in Eq. (A.1).

A.2 Dirac Notation

Dirac notation, invented by Paul Dirac in 1939, is ubiquitous in modern quantum mechanics. As one can see from many examples in the present book, this powerful and concise notation allows physics to be discussed while accounting for implicitly, but without having to display explicitly, mathematical details such as the integral and differential calculus appearing in Schrödinger's wave mechanics.

A.2.1 Bras, Kets, and Bra-Ket Pairs

A *ket* $|\,\rangle$ denotes a vector in a complex vector space (*Hilbert space*) representing a state of a quantum system. For example, $|x\rangle$ can represent a state for a particle at position x, $|p\rangle$ can represent a state for a particle having momentum p, and $|J, M\rangle$ can represent a state with total angular momentum J and projection M. The spin operator $\hat{\sigma}_z$ on a two-dimensional spinor space has eigenvalues $\pm\frac{1}{2}$ with eigenspinors (ψ_+, ψ_-), which may be expressed in Dirac notation as $\psi_+ = |+\rangle$ and $\psi_- = |-\rangle$. More generally, an expression $|\Psi\rangle$ represents a quantum system in the state Ψ and is termed a *state vector*. A *bra* $\langle\,|$ is the *hermitian conjugate* of a state represented by a corresponding ket, with the bra and ket states related by

$$\langle\Psi|^\dagger = |\Psi\rangle \qquad |\Psi\rangle^\dagger = \langle\Psi|, \tag{A.5}$$

where $\dagger$ indicates *hermitian conjugation* (transposition of matrices and complex conjugation of their entries).

Inner Product: A bra-ket combination $\langle\Psi_2|\,\Psi_1\rangle$ represents the Hilbert space inner product of state vectors $|\Psi_1\rangle$ and $|\Psi_2\rangle$, which may be viewed as the projection of $|\Psi_1\rangle$ onto $|\Psi_2\rangle$ and corresponds to the overlap integral in ordinary quantum notation. In Dirac notation a bra may be used to specify detailed physical meaning of a ket. For example, the bra-ket $\langle x|\,\Psi\rangle$ specifies the state $|\Psi\rangle$ represented in coordinate space x, and $|\langle x = 0.68|\,\Psi\rangle|^2$ expresses the probability for the particle in state $|\Psi\rangle$ to be found at $x = 0.68$.

Completeness: A ket-bra combination $|\,\rangle\langle\,|$ is a *projection operator.* If $|e_i\rangle$ denotes a basis vector in an N-dimensional space, $|e_i\rangle\langle e_i|\,\Psi\rangle$ is the projection of $|\Psi\rangle$ on the ith axis. Summation over all projection operators in the space gives the *completeness condition,*

$$\sum_{i=1}^{N} |e_i\rangle\langle e_i| = 1, \tag{A.6}$$

where 1 is the unit matrix in the space. Insertion of a complete set of states through the identity (A.6) is a common manipulation.

Direct Products: If a system is composed of two independent subsystems U and V, the Hilbert space of the entire system is the tensor product of the two spaces, $U \otimes V$. If $|A\rangle$ is a ket in U and $|B\rangle$ is a ket in V, the direct product of the two kets is $|A\rangle\,|B\rangle$.

A.2.2 Bras and Kets as Row and Column Vectors

A ket in Dirac notation can be identified with a column vector, and a bra with a row vector. For an N-dimensional vector space with a given orthonormal basis, the inner product can be written as a matrix multiplication of a row vector with a column vector,

$$\langle A|\,B\rangle = a_1^* b_1 + a_2^* b_2 + \cdots + a_N^* b_N = (a_1^*\; a_2^* \cdots a_N^*) \begin{pmatrix} b_1 \\ b_2 \\ \vdots \\ b_N \end{pmatrix},$$

implying that a bra is a $1 \times N$ row matrix and a ket is an $N \times 1$ column matrix in this basis,

$$\langle A| = (a_1^* \, a_2^* \cdots a_N^*) \qquad |B\rangle = \begin{pmatrix} b_1 \\ b_2 \\ \vdots \\ b_N \end{pmatrix}. \tag{A.7}$$

Note that the elements a_i^* in the row matrix Eq. (A.7) must be complex conjugates of a_i, by virtue of Eq. (A.5). It is natural then that the complete set of projection operators in Eq. (A.6) becomes an $N \times N$ identity matrix in a matrix representation.

A.2.3 Linear Operators Acting on Bras and Kets

A linear operator $\hat{T}$ transforms a vector in Hilbert space by operations like rotation or reflection, resulting in another vector in the same space. From Eq. (A.7), in an N-dimensional Hilbert space a ket can be expressed as an $N \times 1$ column matrix. Now letting $\hat{T}$ act on a basis vector $\hat{T}|e_j\rangle$ and utilizing Eq. (A.6),

$$\hat{T}|e_j\rangle = 1 \times \hat{T}|e_j\rangle = \sum_{i=1}^{N} |e_i\rangle\langle e_i|\hat{T}|e_j\rangle = \sum_{i=1}^{N} T_{ij}\,|e_i\rangle \qquad (j = 1, 2, \ldots, N), \tag{A.8}$$

where the matrix elements $T_{ij} \equiv \langle e_i|\hat{T}|e_j\rangle$ correspond to N^2 complex numbers. Thus, linear operations in an N-dimensional Hilbert space can be identified with $N \times N$ matrices. Taking i as row index and j as column index, the corresponding matrix is

$$\mathbf{T} = \begin{pmatrix} T_{11} & T_{12} & \ldots & T_{1N} \\ T_{21} & T_{22} & \ldots & T_{2N} \\ \vdots & \vdots & \ddots & \vdots \\ T_{N1} & T_{N2} & \ldots & T_{NN} \end{pmatrix}. \tag{A.9}$$

Transformations of kets or bras in the Hilbert space then correspond to matrix multiplications, as outlined in Box A.1. This is particularly useful for numerical calculations since matrix multiplications of large dimensions are easily implemented by computers.

A.3 Occupation Number Representation

A quantum state represented by $|\mathscr{S}\rangle$ looks rather abstract. For many discussions it is sufficient to know that this is a vector in the relevant Hilbert space. By applying Dirac notation, one may specify it with respect to a particular basis.

1. Recall from introductory quantum mechanics that wavefunctions are often given as functions of position, $\psi(x)$. Using a bra-ket combination, this may be viewed as the coefficient in the expansion of $|\mathscr{S}\rangle$ in the basis of position eigenfunctions; this may be expressed as $\psi(x) = \langle x|\,\mathscr{S}\rangle$, where $|x\rangle$ is the eigenvector of the position operator $\hat{x}$.

Box A.1 Action of Linear Operators on Bras and Kets

Equations (A.8) and (A.9) involve the action of linear operators on bras and kets, which can be cast generally as matrix multiplications.

Linear Operators Acting on Kets

If $\mathbf{T}$ is a linear operator and $|A\rangle$ a ket-vector, then $\mathbf{T}$ acting from the left on $|A\rangle$ produces another ket-vector $\mathbf{T}|A\rangle$ in the same space. This operation can be realized by matrix multiplication with an $N \times N$ (generally complex) matrix multiplying an $N \times 1$ column matrix, resulting in a new $N \times 1$ column matrix.

Linear Operators Acting on Bras

Operations can be viewed as $\mathbf{T}$ acting on bras $\langle A|$ from the right, resulting in another bra $\langle A|\mathbf{T}$. In an N-dimensional Hilbert space $\langle A|$ is a $1 \times N$ row matrix and matrix multiplication of it by an $N \times N$ matrix results in a new $1 \times N$ row matrix.

Linear Operators Sandwiched by a Bra and Ket

The expression $\langle A|\,\mathbf{T}\,|B\rangle$ can be understood either as $\langle A|(\mathbf{T}|B\rangle)$ ($\mathbf{T}$ acting from the left on $|B\rangle$), or $(\langle A|\mathbf{T})|B\rangle$ ($\mathbf{T}$ acting from the right on $\langle A|$).

> In an N-dimensional Hilbert space, matrix multiplication of a $1\times N$ row matrix, $N\times N$ matrix, and $N \times 1$ column matrix (the order is important!) results in a c-number.

If $\mathbf{T}$ is hermitian ($\mathbf{T} = \mathbf{T}^\dagger$), then $\langle\, A|\mathbf{T}|B\,\rangle$ is a *transition matrix element* corresponding to the amplitude for an oservable probability for transitions between $|A\rangle$ and $|B\rangle$. If the same state vector appears on both bra and ket sides, $\langle A|\,\mathbf{T}\,|A\rangle$, the matrix element is termed an *expectation value* and if $\mathbf{T}$ is hermitian its square represents a static observable property like an energy or magnetic moment.

2. Similarly, the wavefunction in the momentum space $\Phi(p)$ may be viewed as the expansion coefficient of $|\mathscr{S}\rangle$ in the basis of momentum eigenfunctions: $\Phi(p) = \langle p|\,\mathscr{S}\rangle$.

3. Another example encountered when the superposition principle is considered is to expand $|\mathscr{S}\rangle$ in the basis of energy eigenfunctions obtained by solving the Schrödinger equation with the Hamiltonian $\hat{H}$: $c_n = \langle n|\,\mathscr{S}\rangle$, where $|n\rangle$ represents the nth (discrete) wavefunction, $\psi_n(x)$.

These examples express the same vector $|\mathscr{S}\rangle$ in different bases, so they are related by linear transformations. The relationship is evident if we use different expansion coefficients to express the position eigenfunction:

$$\psi(x) = \int \psi(y)\delta(x-y)dy \tag{A.10a}$$

$$= \int \Phi(p)\frac{1}{\sqrt{2\pi\hbar}}e^{ipx/\hbar}dp \tag{A.10b}$$

$$= \sum_n c_n\, \psi_n(x). \tag{A.10c}$$

Thus, $\psi(x)$, $\Phi(p)$, and c_n describe the *same* state vector $|\mathscr{S}\rangle$ but in *different* bases, specified by position (x), momentum (p), and (discrete) energy states (c_n), respectively.

A.3.1 Creation and Annihilation Operators

It can be tiresome to write symmetrized wavefunctions for many-particle systems in terms of single-particle states as in Eqs. (A.10). On the other hand, it is redundant (and physically impossible) to know detailed spatial information for every particle since they are identical. The essential information that we need for many-particle wavefunctions is simple: *labels for the basis states* and *particle occupation numbers for those states*. In fact, anyone who has taken a course in introductory quantum mechanics is familiar with this idea: *spin* cannot be expressed using spatial functions. For example, a spin-$\frac{1}{2}$ particle wavefunction can be expressed as a two-element column matrix

$$\chi = \begin{pmatrix} a \\ b \end{pmatrix} = a\psi_+ + b\psi_- \qquad \psi_+ = \begin{pmatrix} 1 \\ 0 \end{pmatrix} \qquad \psi_- = \begin{pmatrix} 0 \\ 1 \end{pmatrix},$$

with the convention that the *occupation number* 1 for the upper (lower) element of the column matrix represents the spin-up (spin-down) state. In *occupation number representation* a many-particle state is specified by the number of particles occupying each basis single-particle state. Let us illustrate, first for fermions and then for bosons.

Fermion Occupation Numbers: Consider a three-fermion system with single-particle states labeled 1, 3, and 6 being occupied, with all other states empty. This three-particle state can be expressed by a many-particle ket written in occupation number representation as

$$|1, 0, 1, 0, 0, 1, 0, 0, 0, 0, \dots\rangle, \tag{A.11}$$

where the numbers represent occupations and horizontal position of numbers indicates the state. Thus in Eq. (A.11) one particle is in the single-particle state labeled 1, one in state 3, one in state 6, and there are no particles in the remaining states.

Changes in occupation number representation are implemented by *creation* and *annihilation* operators. To illustrate, let us start with a true *vacuum state* having no particles in any single-particle states, $|0\rangle = |0_1, 0_2, 0_3, \dots\rangle$. The subscripts denote the ordinal number of the basis states; they are not essential since the position of the number indicates the state [as in Eq. (A.11)] but we will use them here temporarily to make the arguments more transparent. Now create a fermion in state 1 by operating on the vacuum with a *creation operator* $c_1^\dagger$,

$$c_1^\dagger\,|0\rangle = |1_1, 0_2, 0_3, \dots\rangle, \tag{A.12}$$

where the subscript 1 for the operator indicates that creation occurs specifically in the single-particle state 1. We then create a second fermion in state 2, which results in

$$c_2^\dagger c_1^\dagger\,|0\rangle = c_2^\dagger\,|1_1, 0_2, 0_3, \dots\rangle = -\,|1_1, 1_2, 0_3, \dots\rangle. \tag{A.13}$$

Note the minus sign that appears because state 2 must interchange its position with state 1 to be acted by $c_2^\dagger$. If the two operators are placed instead in opposite order,

$$c_1^\dagger c_2^\dagger |0\rangle = c_1^\dagger |0_1, 1_2, 0_3, \ldots\rangle = +|1_1, 1_2, 0_3, \ldots\rangle. \tag{A.14}$$

This sign difference reflects the antisymmetry of the fermions; from Eqs. (A.13) and (A.14) we have $(c_1^\dagger c_2^\dagger + c_2^\dagger c_1^\dagger)|0\rangle = 0$, or $c_1^\dagger c_2^\dagger = -c_2^\dagger c_1^\dagger$. Thus for two fermion creation operators

$$[c_1^\dagger, c_2^\dagger]_+ \equiv c_1^\dagger c_2^\dagger + c_2^\dagger c_1^\dagger = 0, \tag{A.15}$$

where the subscript + indicates *anticommutation* (*summation* of the two terms).

Fermion antisymmetry manifests itself in anticommutation relations that fermion operators must obey in occupation number representation.

These anticommutation relations carry exactly the same antisymmetrization information as the Slater determinant in Eq. (A.4), but in more concise form.

Boson Occupation Numbers: If we repeat the above considerations for *bosons,* the minus sign in Eq. (A.13) is absent because *bosons commute.* The two terms appearing in Eq. (A.15) then *subtract* from each other for bosons, leading to the *commutation relation*

$$[c_1^\dagger, c_2^\dagger]_- \equiv c_1^\dagger c_2^\dagger - c_2^\dagger c_1^\dagger = 0, \tag{A.16}$$

where the subscript − distinguishes boson commutators in (A.16) from the fermion anticommutators in (A.15). Proceeding in this way we find the following.

In occupation number representation the commutation and anticommutation relations for creation and annihilation operators are

$$[c_i^\dagger, c_j^\dagger]_\mp \equiv c_i^\dagger c_j^\dagger \mp c_j^\dagger c_i^\dagger = 0, \tag{A.17a}$$

$$[c_i, c_j]_\mp \equiv c_i c_j \mp c_j c_i = 0, \tag{A.17b}$$

$$[c_i, c_j^\dagger]_\mp \equiv c_i c_j^\dagger \mp c_j^\dagger c_i = \delta_{ij}, \tag{A.17c}$$

with the minus sign denoting commutation relations for bosons and the plus sign denoting anticommutation relations for fermions.

Note that in the chapters of this book we will typically use a shorthand notation $\{A, B\} \equiv [A, B]_+$ for anticommutators, and $[A, B] \equiv [A, B]_-$ for commutators.

A.3.2 Basis Transformations

Basis transformations are particularly simple within the second-quantized formalism. Let $|a\rangle$ be a set of single-particle kets whose overlaps with the position spin eigenstates $|x\rangle$ are the single-particle wavefunctions $\phi_a(x) \equiv \langle x|\, a\rangle$. Consider a different set of single-particle kets $|a'\rangle$ with corresponding single-particle wavefunctions $\phi'_{a'}(x) \equiv \langle x|\, a'\rangle$. To transform between the two single-particle bases we use Eq. (A.6) to write

$$|a\rangle = 1 \times |a\rangle = \sum_{a'} |a'\rangle \langle a'|\, a\rangle = \sum_{a'} \langle a'|\, a\rangle\, |a'\rangle , \tag{A.18}$$

where the $|a\rangle$ are expressed as a sum over $|a'\rangle$ with coefficients $D_{a',a} \equiv \langle a'|\, a\rangle$. This can also be written in terms of operator transformations. Since

$$c_a^\dagger\, |0\rangle = |a\rangle = \sum_{a'} \langle a'|\, a\rangle\, |a'\rangle = \sum_{a'} \langle a'|\, a\rangle\, c_{a'}^\dagger\, |0\rangle ,$$

we have that

$$c_a^\dagger = \sum_{a'} \langle a'|\, a\rangle\, c_{a'}^\dagger \qquad c_a = \sum_{a'} \langle a|\, a'\rangle\, c_{a'}, \tag{A.19}$$

where $c_a = (c_a^\dagger)^\dagger$ was used. Similarly, one can derive the inverse transformations,

$$c_{a'}^\dagger = \sum_{a} \langle a|\, a'\rangle\, c_a^\dagger \qquad c_{a'} = \sum_{a} \langle a'|\, a\rangle\, c_a . \tag{A.20}$$

The transformation between single-particle wavefunctions $\phi_a(x)$ and $\phi'_{a'}(x)$ is then,

$$\phi_a(x) = \langle x|\, a\rangle = \langle x| \Big(\sum_{a'} \langle a'|\, a\rangle\, |a'\rangle \Big) = \sum_{a'} \langle a'|\, a\rangle \langle x|\, a'\rangle = \sum_{a'} \langle a'|\, a\rangle\, \phi'_{a'}(x),$$

and the inverse transformation is $\phi'_{a'}(x) = \sum_a \langle a|\, a'\rangle\, \phi_a(x)$. In quantum mechanics basis transformations are unitary and the corresponding matrix of transformation coefficients $D_{a',a} \equiv \langle a'|\, a\rangle$ is unitary ($D^\dagger = D^{-1}$). Thus, basis transformation in the second quantization formalism is just basic linear algebra implemented for unitary matrices.

A.3.3 Many-Particle Vector States

An arbitrary many-particle vector state of an N-particle system can be constructed with a set of creation operators in an appropriate order. Starting from a true vacuum (no particles in any single-particle states), $|0\rangle \equiv |0, 0, 0, \ldots\rangle$, with the condition $c_i\, |0\rangle = 0$ for all i, and applying the commutation relations for bosons or fermions in Eq. (A.17), we can construct general many-particle vector states with n_1 particles in single-particle state 1, n_2 particles in state 2, n_3 particles in state 3, …, with $\sum_i n_i = N$. For a system of identical bosons,

$$|n\rangle_{\rm b} = |n_1, n_2, \ldots, n_i, \ldots\rangle_{\rm b} = \frac{1}{\sqrt{n_1!\, n_2! \cdots n_i! \cdots}} \prod_i (c_i^\dagger)^{n_i}\, |0\rangle_{\rm b} . \tag{A.21}$$

For fermions n_i can only be 0 (unoccupied) or 1 (occupied), so $n_i! = 1$ and

$$|n\rangle_{\rm f} = |n_1, n_2, \ldots, n_i, \ldots\rangle_{\rm f} = \prod_i (c_i^\dagger)^{n_i}\, |0\rangle_{\rm f} \qquad (n_i = 0 \text{ or } 1). \tag{A.22}$$

In occupation number representation a particularly important operator is one to count the number of particles of a particular type. Box A.2 describes the second-quantized *particle-number operator* that accomplishes this task.

Box A.2 The Number Operator

Consider applying the operator $c_i^\dagger c_i$ to an arbitrary vector state $|n\rangle$ in Eq. (A.21) or Eq. (A.22), acting specifically on the term n_i such that $c_i^\dagger c_i\,|n\rangle = c_i^\dagger c_i\,|\ldots, n_i, \ldots\rangle$.

1. If $|n\rangle_\mathrm{b}$ is a bosonic vector it is easy to show that

$$c_i^\dagger\,|\ldots, n_i, \ldots\rangle_\mathrm{b} = \sqrt{n_i + 1}\,|\ldots, n_i + 1, \ldots\rangle_\mathrm{b}\,,$$
$$c_i\,|\ldots, n_i, \ldots\rangle_\mathrm{b} = \sqrt{n_i}\,|\ldots, n_i - 1, \ldots\rangle_\mathrm{b}\,.$$

   Applying the above equations successively results in

$$c_i^\dagger c_i\,|\ldots, n_i, \ldots\rangle_\mathrm{b} = \left(\sqrt{n_i}\right)^2\,|\ldots, n_i, \ldots\rangle_\mathrm{b} = n_i\,|\ldots, n_i, \ldots\rangle_\mathrm{b}\,.$$

2. If $|n\rangle_\mathrm{f}$ is a fermionic vector it is easy to show that

$$c_i^\dagger\,|\ldots, n_i, \ldots\rangle_\mathrm{f} = (-1)^{\sum_{j<i} n_j}(1 - n_i)\,|\ldots, 1_i, \ldots\rangle_\mathrm{f}\,,$$
$$c_i\,|\ldots, n_i, \ldots\rangle_\mathrm{f} = (-1)^{\sum_{j<i} n_j}\,n_i\,|\ldots, 0_i, \ldots\rangle_\mathrm{f}\,.$$

   Applying the above equations successively results again in

$$c_i^\dagger c_i\,|\ldots, n_i, \ldots\rangle_\mathrm{f} = n_i\,|\ldots, n_i, \ldots\rangle_\mathrm{f}\,.$$

Thus the operator $\hat{n}_i \equiv c_i^\dagger c_i$ counts the number of particles (either bosons or fermions) in the single-particle state i.

We term $\hat{n}_i$ the *number operator* for state i, and define

$$\hat{N} = \sum_i \hat{n}_i = \sum_i c_i^\dagger c_i$$

to be the *total number operator* satisfying $\hat{N}\,|n\rangle = N\,|n\rangle$, with $N = \sum_i n_i$ being the total number of particles in $|n\rangle$.

A.3.4 One-Body and Two-Body Operators

We have introduced creation and annihilation operators to express vectors for many-particle states. The same can be done for *operators* that act on these states. We show next how to construct one-body and two-body operators for an N-particle system.

One-Body Operators: Examples of *one-body operators* for an N-particle system include those for kinetic energy or total momentum; these are usually given as the sum of N operators $\hat{h}_i$ acting on the ith particle, $\hat{H}_0 = \sum_{i=1}^N \hat{h}_i$. A one-body operator transformed to occupation number representation can be written as

$$\sum_{i=1}^{N} \hat{h}_i \Longrightarrow \sum_{\alpha\alpha'} h_{\alpha\alpha'} c^{\dagger}_{\alpha} c_{\alpha'} \qquad h_{\alpha\alpha'} \equiv \langle \alpha | \, \hat{h} \, | \alpha' \rangle = \int dx \, \phi^{*}_{\alpha}(x) \hat{h}(x) \phi_{\alpha'}(x), \qquad \text{(A.23)}$$

where $h_{\alpha\alpha'}$ is the matrix element in occupation number representation, and it has been evaluated in the coordinate representation with $x \equiv (\mathbf{r}, s)$ denoting position ($\mathbf{r}$) and spin projection (s),

Two-Body Operators: Examples of *two-body operator* include spin–spin interaction or Coulomb interaction for charged particles, which is generally given for an N-particle system by $\hat{H}_M = \sum_1^N \hat{v}_{ij}$ with the restriction $i < j$. In occupation number representation a two-body operator can be expressed as

$$\sum_{i<j=1}^{N} \hat{v}_{ij} \Longrightarrow \frac{1}{2} \sum_{\alpha\beta\alpha'\beta'} v_{\alpha\beta\alpha'\beta'} c^{\dagger}_{\alpha} c^{\dagger}_{\beta} c_{\beta'} c_{\alpha'} \qquad v_{\alpha\beta\alpha'\beta'} \equiv \left\langle \alpha\beta \middle| \, \hat{v} \, \middle| \alpha'\beta' \right\rangle,$$

where $v_{\alpha\beta\alpha'\beta'}$ is the matrix element in the occupation number representation, which can be evaluated explicitly as

$$v_{\alpha\beta\alpha'\beta'} = \iint dx \, dx' \phi^{*}_{\alpha}(x) \phi^{*}_{\beta}(x') \hat{v}(x, x') \phi_{\alpha'}(x) \phi_{\beta'}(x'),$$

with $x \equiv (\mathbf{r}, s)$.

Example A.3 Consider a system of N spinless, non-interacting fermions. The Hamiltonian in occupation number representation is simply $\hat{H}_0 = \sum_{\alpha} E_{\alpha} a^{\dagger}_{\alpha} a_{\alpha}$, where α labels single-particle states with increasingly ordered energy (assuming no degeneracy) $E_1 < E_2 < \cdots < E_{\alpha} < \cdots < E_N$. Since no single-particle state can hold more than one fermion, the ground state wavefunction $|\psi_{\text{gs}}\rangle$ is obtained by filling the first N states,

$$|\psi_{\text{gs}}\rangle = \prod_{\alpha=1}^{N} a^{\dagger}_{\alpha} \, |0\rangle = a^{\dagger}_{1} a^{\dagger}_{2} \ldots a^{\dagger}_{N} \, |0\rangle = | \underbrace{1 \ldots 1}_{N}, 00 \ldots \rangle,$$

with 1 and 0 indicating occupied and unoccupied states, respectively. Thus the ground state has the lowest N single-particle states occupied and all others empty. The energy,

$$E_{\text{gs}} = \sum_{\alpha} E_{\alpha} = E_1 + E_2 + \cdots + E_N,$$

is the sum of all N single-particle energies and the energy of the last occupied single particle, E_N, is the *Fermi energy* of the system. Now let us remove the particle from the state N and excite it to the next unoccupied state $N+1$. This process creates a hole (absence of a particle) below the Fermi energy and a particle above the Fermi energy by *particle–hole excitation.* This excited state wavefunction $|\psi_{\text{es}}\rangle$ can be expressed as

$$|\psi_{\text{es}}\rangle = | \underbrace{1 \ldots 1}_{N-1}, 010 \ldots \rangle = a^{\dagger}_{N+1} a_N |\psi_{\text{gs}}\rangle. \qquad \text{(A.24)}$$

This shows clearly that, with respect to $|\psi_{gs}\rangle$, the particle in the state N is *annihilated* by a_N and a particle in the state $N+1$ is *created* by $a^\dagger_{N+1}$. The energy of this state is thus

$$E_{es} = E_1 + E_2 + \cdots + E_{N-1} + E_{N+1} = E_{gs} + E_{N+1} - E_N > E_{gs},$$

so it is higher in energy than the ground state and is termed an *excited state.*

Therefore, the occupation number representation provides a clear and intuitive picture of the ground state and particle–hole excited states for the many-fermion system.

Appendix B Natural Units

Often it is convenient to define new units where values of fundamental constants such as the speed of light or Planck's constant are set to one. These are called *natural units* because they are suggested by the physics of the problem. For example, the velocity of light c is clearly of fundamental importance in problems where special relativity is applicable. In that context, it is natural to use c to set the scale for velocities. Defining a set of units where c takes unit value is equivalent to making velocity a dimensionless quantity that is measured in units of c, as illustrated below, thus setting a "natural" scale for velocity.

B.1 The Advantage of Natural Units

The introduction of a natural set of units has the virtue of more compact notation, since the constants rescaled to unit value need not be included explicitly in the equations, and the standard "engineering" units like MKS may be restored by dimensional analysis if they are required to obtain numerical results. This Appendix outlines the use of such natural units for various problems encountered in this book.

B.2 Natural Units in Quantum Field Theory

In relativistic quantum field theory it is convenient to define natural units where $\hbar = c = 1$. Using the notation $[a]$ to denote the dimension of a and using $[L]$, $[T]$ and $[M]$ to denote the dimensions of length, time, and mass, respectively, for the speed of light c,

$$[c] = [L][T]^{-1}. \tag{B.1}$$

Setting $c = 1$ then implies that $[L] = [T]$, and since $E^2 = \boldsymbol{p}^2 c^2 + M^2 c^4$,

$$[E] = [M] = [\boldsymbol{p}] = [\boldsymbol{k}], \tag{B.2}$$

where $\boldsymbol{p} = \hbar \boldsymbol{k}$. Furthermore, because

$$[\hbar] = [M][L]^2[T]^{-1}, \tag{B.3}$$

setting $\hbar = c = 1$ and using $[L] = [T]$ yields

$$[M] = [L]^{-1} = [T]^{-1}. \tag{B.4}$$

Thus $[M]$ may be chosen as the single independent dimension of our set of $\hbar = c = 1$ natural units. Useful conversions are

$$\begin{aligned} \hbar c &= 197.3\,\text{MeV fm} \qquad 1\,\text{fm} = \frac{1}{197.3}\,\text{MeV}^{-1} = 5.068\,\text{GeV}^{-1} \\ 1\,\text{fm}^{-1} &= 197.3\,\text{MeV} \qquad 1\,\text{GeV} = 5.068\,\text{fm}^{-1}, \end{aligned} \tag{B.5}$$

where 1 MeV = 10^6 eV, 1 GeV = 10^9 eV, an eV is an electronvolt with 1 eV = 1.602×10^{-19} Joule, and 1 fm = 10^{-13} cm = 10^{-15} m (where fm denotes a fermi or a femtometer). Example B.1 illustrates conversion from natural units to normal engineering units.

Example B.1 The mass of a pion is given by $M_\pi c^2 \sim 140$ MeV, and the reduced Compton wavelength is $\bar{\lambda}_\pi = \hbar/M_\pi c^2$. Then

$$\lambda_\pi = \frac{1}{M_\pi} \simeq (140\,\text{MeV})^{-1}$$

in $\hbar = c = 1$ units. Utilizing Eq. (B.5), this may be converted to

$$\lambda_\pi = \left(\frac{1}{140}\,\text{MeV}^{-1}\right) \times 197.3\,\text{MeV fm} = 1.41\,\text{fm} = 1.41 \times 10^{-15}\,\text{m}$$

in standard units.

Problems 3.19–3.22 give some practice using natural units.

Appendix C Angular Momentum Tables

This Appendix contains tables of quantities relevant for using SO(3) and SU(2) symmetries to describe angular momentum in quantum mechanics. The d-functions tabulated in Table C.1 are defined in Eq. (6.32) and are related to the D-functions (matrix elements of the angular momentum operator) in Eq. (6.31a). The spherical harmonics of Table C.2 are related to the angular momentum D-functions in Eq. (6.40). Clebsch–Gordan (vector-coupling) coefficients are defined in Section 6.3.6 and tabulated in Table C.3. They are related to the $3J$ symbols of Table C.4 in Section 6.3.8.

Table C.1. Some d-functions $d^j_{mn}(\theta)$

$$d^{\frac{1}{2}}_{\frac{1}{2}\frac{1}{2}}(\theta) = \cos\left(\tfrac{\theta}{2}\right) \qquad d^{\frac{1}{2}}_{\frac{1}{2}-\frac{1}{2}}(\theta) = -\sin\left(\tfrac{\theta}{2}\right) \qquad d^1_{11}(\theta) = \tfrac{1}{2}(1+\cos\theta)$$

$$d^1_{10}(\theta) = -\tfrac{1}{\sqrt{2}}\sin\theta \qquad d^1_{1-1}(\theta) = \tfrac{1}{2}(1-\cos\theta) \qquad d^1_{00}(\theta) = \cos\theta$$

$$d^{\frac{3}{2}}_{\frac{3}{2}\frac{3}{2}}(\theta) = \tfrac{1}{2}(1+\cos\theta)\cos\left(\tfrac{\theta}{2}\right) \qquad d^{\frac{3}{2}}_{\frac{3}{2}\frac{1}{2}}(\theta) = -\tfrac{\sqrt{3}}{2}(1+\cos\theta)\sin\left(\tfrac{\theta}{2}\right)$$

$$d^{\frac{3}{2}}_{\frac{3}{2}-\frac{1}{2}}(\theta) = \tfrac{\sqrt{3}}{2}(1-\cos\theta)\cos\left(\tfrac{\theta}{2}\right) \qquad d^{\frac{3}{2}}_{\frac{3}{2}-\frac{3}{2}}(\theta) = -\tfrac{1}{2}(1-\cos\theta)\sin\left(\tfrac{\theta}{2}\right)$$

$$d^{\frac{3}{2}}_{\frac{1}{2}\frac{1}{2}}(\theta) = \tfrac{1}{2}(3\cos\theta-1)\cos\left(\tfrac{\theta}{2}\right) \qquad d^{\frac{3}{2}}_{\frac{1}{2}-\frac{1}{2}}(\theta) = -\tfrac{1}{2}(3\cos\theta+1)\sin\left(\tfrac{\theta}{2}\right)$$

$$d^2_{22}(\theta) = \tfrac{1}{4}(1+\cos\theta)^2 \qquad d^2_{21}(\theta) = -\tfrac{1}{2}(1-\cos\theta)\sin\theta \qquad d^2_{20}(\theta) = \tfrac{\sqrt{6}}{4}\sin^2\theta$$

$$d^2_{2-1}(\theta) = -\tfrac{1}{2}(1-\cos\theta)\sin\theta \qquad d^2_{2-2}(\theta) = \tfrac{1}{4}(1-\cos\theta)^2$$

$$d^2_{11}(\theta) = \tfrac{1}{2}(1-\cos\theta)(2\cos\theta-1) \qquad d^2_{10}(\theta) = -\tfrac{\sqrt{3}}{\sqrt{2}}\sin\theta\cos\theta$$

$$d^2_{1-1}(\theta) = \tfrac{1}{2}(1-\cos\theta)(2\cos\theta+1) \qquad d^2_{00}(\theta) = \tfrac{3}{2}\cos^2\theta - \tfrac{1}{2}$$

Additional values may be inferred from $d^j_{mk}(\theta) = (-1)^{k-m} d^j_{km}(\theta) = d^j_{-k-m}(\theta)$

Table C.2. Some spherical harmonics $Y_{lm}(\theta,\phi)$

$$Y_{10}(\theta,\phi) = \sqrt{\tfrac{3}{4\pi}}\cos\theta \qquad Y_{11}(\theta,\phi) = -\sqrt{\tfrac{3}{8\pi}}\sin\theta e^{i\phi} \qquad Y_{20}(\theta,\phi) = \sqrt{\tfrac{5}{4\pi}}\left(\tfrac{3}{2}\cos^2\theta - \tfrac{1}{2}\right)$$

$$Y_{21}(\theta,\phi) = -\sqrt{\tfrac{15}{8\pi}}\sin\theta\cos\theta e^{i\phi} \qquad Y_{22}(\theta,\phi) = \sqrt{\tfrac{15}{32\pi}}\sin^2\theta e^{2i\phi}$$

Other values may be obtained from the symmetry $Y_{l-m}(\theta,\phi) = (-1)^m Y^*_{lm}(\theta,\phi)$

Table C.3. Some SO(3) Clebsch–Gordan Coefficients $\langle j_1 m_1 j_2 m_2 | JM \rangle$

j_1	j_2	m_1	m_2	J	M	CG	j_1	j_2	m_1	m_2	J	M	CG
1/2	1/2	1/2	1/2	1	1	1	1/2	1/2	1/2	−1/2	1	0	$\sqrt{1/2}$
1/2	1/2	1/2	−1/2	0	0	$\sqrt{1/2}$	1/2	1/2	−1/2	1/2	1	0	$\sqrt{1/2}$
1/2	1/2	−1/2	1/2	0	0	$-\sqrt{1/2}$	1/2	1/2	−1/2	−1/2	1	−1	1
1	1/2	1	1/2	3/2	3/2	1	1	1/2	1	−1/2	3/2	1/2	$\sqrt{1/3}$
1	1/2	1	−1/2	1/2	1/2	$\sqrt{2/3}$	1	1/2	0	1/2	3/2	1/2	$\sqrt{2/3}$
1	1/2	0	1/2	1/2	1/2	$-\sqrt{1/3}$	1	1/2	0	−1/2	3/2	−1/2	$\sqrt{2/3}$
1	1/2	0	−1/2	1/2	−1/2	$\sqrt{1/3}$	1	1/2	−1	1/2	3/2	−1/2	$\sqrt{1/3}$
1	1/2	−1	1/2	1/2	−1/2	$-\sqrt{2/3}$	1	1/2	−1	−1/2	3/2	−3/2	1
1	1	1	1	2	2	1	1	1	1	0	2	1	$\sqrt{1/2}$
1	1	1	0	1	1	$\sqrt{1/2}$	1	1	0	1	2	1	$\sqrt{1/2}$
1	1	0	1	1	1	$-\sqrt{1/2}$	1	1	1	−1	2	0	$\sqrt{1/6}$
1	1	1	−1	1	0	$\sqrt{1/2}$	1	1	1	−1	0	0	$\sqrt{1/3}$
1	1	0	0	2	0	$\sqrt{2/3}$	1	1	0	0	1	0	0
1	1	0	0	0	0	$-\sqrt{1/3}$	1	1	−1	1	2	0	$\sqrt{1/6}$
1	1	−1	1	1	0	$-\sqrt{1/2}$	1	1	−1	1	0	0	$\sqrt{1/3}$
1	1	0	−1	2	−1	$\sqrt{1/2}$	1	1	0	−1	1	−1	$\sqrt{1/2}$
1	1	−1	0	2	−1	$\sqrt{1/2}$	1	1	−1	0	1	−1	$-\sqrt{1/2}$
1	1	−1	−1	2	−2	1							
2	1/2	2	1/2	5/2	5/2	1	2	1/2	1	−1/2	5/2	3/2	$\sqrt{1/5}$
2	1/2	2	−1/2	3/2	3/2	$\sqrt{4/5}$	2	1/2	1	1/2	5/2	3/2	$\sqrt{4/5}$
2	1/2	1	1/2	3/2	3/2	$-\sqrt{1/5}$	2	1/2	1	−1/2	5/2	1/2	$\sqrt{2/5}$
2	1/2	1	−1/2	3/2	1/2	$\sqrt{3/5}$	2	1/2	0	1/2	5/2	1/2	$\sqrt{3/5}$
2	1/2	0	1/2	3/2	1/2	$-\sqrt{2/5}$	2	1/2	0	−1/2	5/2	−1/2	$\sqrt{3/5}$
2	1/2	0	−1/2	3/2	−1/2	$\sqrt{2/5}$	2	1/2	−1	1/2	5/2	−1/2	$\sqrt{2/5}$
2	1/2	−1	1/2	3/2	−1/2	$-\sqrt{3/5}$	2	1/2	−1	−1/2	5/2	−3/2	$\sqrt{4/5}$
2	1/2	−1	−1/2	3/2	−3/2	$\sqrt{1/5}$	2	1/2	−2	1/2	5/2	−3/2	$\sqrt{1/5}$
2	1/2	−2	1/2	3/2	−3/2	$-\sqrt{4/5}$	2	1/2	−2	−1/2	5/2	−5/2	1
3/2	1/2	3/2	1/2	2	2	1	3/2	1/2	3/2	−1/2	2	1	1/2
3/2	1/2	3/2	−1/2	1	1	$\sqrt{3/4}$	3/2	1/2	1/2	1/2	2	1	$\sqrt{3/4}$
3/2	1/2	1/2	1/2	1	1	−1/2	3/2	1/2	1/2	−1/2	2	0	$\sqrt{1/2}$
3/2	1/2	1/2	−1/2	1	0	$\sqrt{1/2}$	3/2	1/2	−1/2	1/2	2	0	$\sqrt{1/2}$
3/2	1/2	−1/2	1/2	1	0	$-\sqrt{1/2}$	3/2	1/2	−1/2	−1/2	2	−1	$\sqrt{3/4}$
3/2	1/2	−1/2	−1/2	1	−1	1/2	3/2	1/2	−3/2	1/2	2	−1	1/2
3/2	1/2	−3/2	1/2	1	−1	$-\sqrt{3/4}$	3/2	1/2	−3/2	−1/2	2	−2	1

Table C.4. Some 3*J* coefficients [29]

$$\begin{pmatrix} a & a+1/2 & 1/2 \\ b & -b-1/2 & 1/2 \end{pmatrix} = (-1)^{a-b-1}\left[\frac{a+b+1}{(2a+1)(2a+2)}\right]^{1/2}$$

$$\begin{pmatrix} a & a & 1 \\ b & -b-1 & 1 \end{pmatrix} = (-1)^{a-b}\left[\frac{(a-b)(a+b+1)}{2a(a+1)(2a+1)}\right]^{1/2}$$

$$\begin{pmatrix} a & a & 1 \\ b & -b & 0 \end{pmatrix} = (-1)^{a-b}\frac{b}{[a(a+1)(2a+1)]^{1/2}}$$

$$\begin{pmatrix} a & a+1 & 1 \\ b & -b-1 & 1 \end{pmatrix} = (-1)^{a-b}\left[\frac{(a+b+1)(a+b+2)}{(2a+1)(2a+2)(2a+3)}\right]^{1/2}$$

$$\begin{pmatrix} a & a+1 & 1 \\ b & -b & 0 \end{pmatrix} = (-1)^{a-b-1}\left[\frac{(a-b+1)(a+b+1)}{(a+1)(2a+1)(2a+3)}\right]^{1/2}$$

$$\begin{pmatrix} a & a+1/2 & 3/2 \\ b & -b-3/2 & 3/2 \end{pmatrix} = (-1)^{a-b-1}\left[\frac{3(a+b+1)(a+b+2)(a-b)}{2a(2a+1)(2a+2)(2a+3)}\right]^{1/2}$$

$$\begin{pmatrix} a & a+1/2 & 3/2 \\ b & -b-1/2 & 1/2 \end{pmatrix} = (-1)^{a-b}(a-3b)\left[\frac{a+b+1}{2a(2a+1)(2a+2)(2a+3)}\right]^{1/2}$$

$$\begin{pmatrix} a & a+3/2 & 3/2 \\ b & -b-3/2 & 3/2 \end{pmatrix} = (-1)^{a-b-1}\left[\frac{(a+b+1)(a+b+2)(a+b+3)}{(2a+1)(2a+2)(2a+3)(2a+4)}\right]^{1/2}$$

$$\begin{pmatrix} a & a+3/2 & 3/2 \\ b & -b-1/2 & 1/2 \end{pmatrix} = (-1)^{a-b}\left[\frac{3(a-b+1)(a+b+1)(a+b+2)}{(2a+1)(2a+2)(2a+3)(2a+4)}\right]^{1/2}$$

$$\begin{pmatrix} a & a & 2 \\ b & -b-2 & 2 \end{pmatrix} = (-1)^{a-b}\left[\frac{3(a+b+1)(a+b+2)(a-b-1)(a-b)}{a(2a+3)(2a+2)(2a+1)(2a-1)}\right]^{1/2}$$

$$\begin{pmatrix} a & a & 2 \\ b & -b-1 & 1 \end{pmatrix} = (-1)^{a-b}(2b+1)\left[\frac{3(a-b)(a+b+1)}{a(2a+3)(2a+2)(2a+1)(2a-1)}\right]^{1/2}$$

$$\begin{pmatrix} a & a & 2 \\ b & -b & 0 \end{pmatrix} = (-1)^{a-b}\frac{3b^2-a(a+1)}{[a(a+1)(2a+3)(2a+1)(2a-1)]^{1/2}}$$

$$\begin{pmatrix} a & a+1 & 2 \\ b & -b-2 & 2 \end{pmatrix} = (-1)^{a-b}\left[\frac{(a+b+1)(a+b+2)(a+b+3)(a-b)}{a(a+1)(2a+4)(2a+3)(2a+1)}\right]^{1/2}$$

$$\begin{pmatrix} a & a+1 & 2 \\ b & -b-1 & 1 \end{pmatrix} = (-1)^{a-b-1}(a-2b)\left[\frac{(a+b+2)(a+b+1)}{a(a+1)(2a+4)(2a+3)(2a+1)}\right]^{1/2}$$

$$\begin{pmatrix} a & a+1 & 2 \\ b & -b & 0 \end{pmatrix} = (-1)^{a-b-1}b\left[\frac{3(a+b+1)(a-b+1)}{a(a+1)(a+2)(2a+3)(2a+1)}\right]^{1/2}$$

$$\begin{pmatrix} a & a+2 & 2 \\ b & -b-2 & 2 \end{pmatrix} = (-1)^{a-b}\left[\frac{(a+b+1)(a+b+2)(a+b+3)(a+b+4)}{(2a+1)(2a+2)(2a+3)(2a+4)(2a+5)}\right]^{1/2}$$

$$\begin{pmatrix} a & a+2 & 2 \\ b & -b-1 & 1 \end{pmatrix} = (-1)^{a-b-1}\left[\frac{(a+b+1)(a+b+2)(a+b+3)(a-b+1)}{(a+1)(a+2)(2a+1)(2a+3)(2a+5)}\right]^{1/2}$$

$$\begin{pmatrix} a & a+2 & 2 \\ b & -b & 0 \end{pmatrix} = (-1)^{a-b}\left[\frac{3(a+b+1)(a+b+2)(a-b+1)(a-b+2)}{(a+1)(2a+5)(2a+4)(2a+3)(2a+1)}\right]^{1/2}$$

Appendix D Lie Algebras

Lie algebras and corresponding Dynkin diagrams derived using the methods of Ch. 7 are displayed in the following table [224].

Table D.1. Lie algebras and Dynkin diagrams

Group label	Cartan label	Rank of algebra	Number of generators	Dynkin diagram
SU($L+1$)	A_L	L	$L(L+2)$	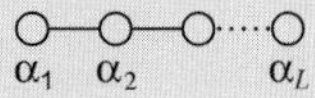
SO($2L+1$) ($L>2$)	B_L	L	$L(2L+1)$	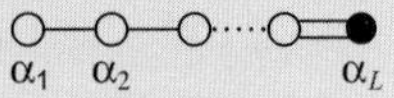
Sp($2L$) ($L\geqslant 3$)	C_L	L	$L(2L+1)$	
SO($2L$) ($L\geqslant 4$)	D_L	L	$L(2L-1)$	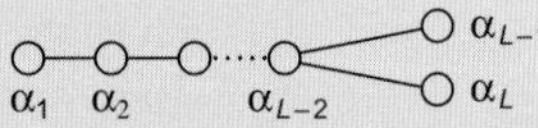
G_2	G_2	2	14	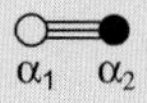
F_4	F_4	4	52	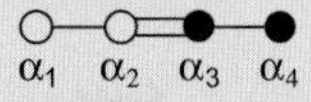
E_6	E_6	6	78	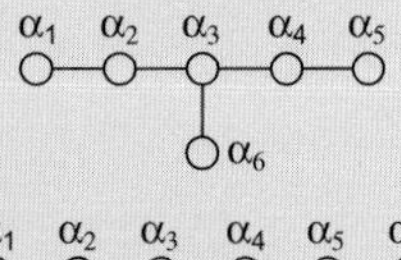
E_7	E_7	7	133	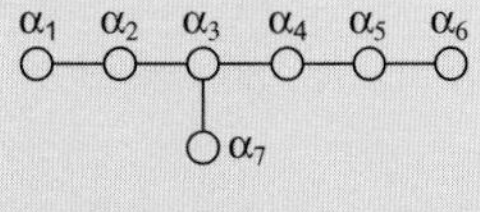
E_8	E_8	8	248	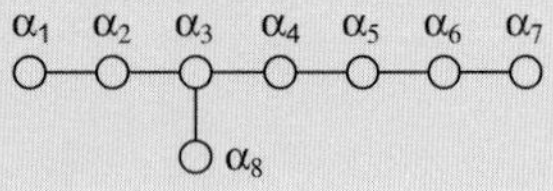

References

[1] Abers, E. S., and Lee, B. W. 1973. *Phys. Rep.*, **C9**, 1.

[2] Actor, A. 1979. *Rev. Mod. Phys.*, **51**, 461.

[3] Aguado, R., and Kouwenhoven, L. P. 2020. *Physics Today*, **73 (6)**, 44.

[4] Aharonov, Y., and Bohm, D. 1959. *Phys. Rev.*, **115**, 484.

[5] Ahmad, I., and Butler, P. A. 1993. *Ann. Rev. Nucl. Part. Sci.*, **43**, 71.

[6] Aitchison, I. J. R. 1982. *An Informal Introduction to Gauge Field Theories*. Cambridge: Cambridge University Press.

[7] Aitchison, I. J. R., and Hey, A. J. G. 1982. *Gauge Theories in Particle Physics*. Bristol: Adam Hilger.

[8] Aivazis, M. 1991. *Group Theory in Physics: Problems & Solutions*. Singapore: World Scientific.

[9] Alicea, J. 2012. *Rep. Prog. Phys.*, **75**, 076501 (arXiv: 1202.1293).

[10] Anandan, J., Christian, J., and Wanelik, K. 1997. *Am. J. Phys.*, **65**, 180 (arXiv: quant–ph/9702011).

[11] Anderson, P. W. 1958. *Phys. Rev.*, **109**, 1492.

[12] Anderson, P. W. 1963. *Phys. Rev.*, **130**, 439.

[13] Armitage, N. P., Fournier, P., and Greene, R. L. 2010. *Rev. Mod. Phys.*, **82**, 2421.

[14] Asbóth, J. K., Oroszlány, L., and Pályi, A. 2016. *A Short Course on Topological Insulators: Band-structure topology and edge states in one and two dimensions*. Heidelberg: Springer.

[15] Ashcroft, N. W., and Mermin, N. D. 1976. *Solid State Physics*. Fort Worth, TX: Saunders College Publishing.

[16] Atiyah, M. 1995. *Rev. Mod. Phys.*, **67**, 977.

[17] Avron, J. E., Osadchy, D., and Seiler, R. 2003. *Phys. Today*, **56 (8)**, 38.

[18] Bardeen, J., Cooper, L. N., and Schrieffer, J. R. 1957. *Phys. Rev.*, **108**, 1175.

[19] Barut, A. O., and Raczka, R. 1986. *Theory of Group Representations and Applications*. Singapore: World Scientific.

[20] Batterman, R. W. 2003. *Stud. Hist. Phil. Sci.*, B34, 527.

[21] Bernevig, B. A. 2013. *Topological Insulators and Topological Superconductors*. NJ: Princeton University Press.

[22] Berry, M. V. 1984. *Proc. R. Soc. London*, **A392**, 45.

[23] Berry, M. V. 1990. *Phys. Today*, **43 (12)**, 34.

[24] Bijker, R., Iachello, F., and Leviatan, A. 1994. *Ann. Phys.*, **236**, 69.

[25] Bitko, D., Rosenbaum, T. F., and Aeppli, G. 1996. *Phys. Rev. Lett.*, **77**, 940.

[26] Bludman, S., and Klein, A. 1962. *Phys. Rev.*, **131**, 2364.

[27] Bohr, A., and Mottelson, B. R. 1969. *Nuclear Structure*, Vol. I. Reading, MA: W. A. Benjamin.

[28] Bohr, A., and Mottelson, B. R. 1975. *Nuclear Structure*, Vol. II. Reading, W. A. Benjamin.

[29] Brink, D. M., and Satchler, G. R. 1968. *Angular Momentum*. Oxford: Clarendon Press.

[30] Butler, P. H. 1981. *Point Group Symmetry Applications: Methods and Tables*. New York: Plenum Press.

[31] Callan, C. G., Dashen, R. F., and Gross, D. J. 1976. *Phys. Lett.*, **63B**, 334.

[32] Campbell, D. K. 1978. In *Nuclear Physics with Heavy Ions and Mesons*, Vol. 2; edited by R. Balian, M. Rho, and G. Ripka. Amsterdam: North-Holland.

[33] Campuzano, J. C., et al. 1999. *Phys. Rev. Lett.*, **83**, 3709.

[34] Castelvecchi, D. 2017. *Nature*, **547**, 272.

[35] Casten, R. F., Wu, C.-L., Feng, D. H., Ginocchio, J. N., and Han, X. L. 1986. *Phys. Rev. Lett.*, **56**, 2578.

[36] Chaikin, P. M., and Lubensky, T. C. 1995. *Principles of Condensed Matter Physics*. Cambridge: Cambridge University Press.

[37] Chambers, R. G. 1960. *Phys. Rev. Lett.*, **5**, 3.

[38] Chang, M. C. 2016. *Lecture notes on topological insulators*. https://phy.ntnu.edu .tw/~changmc/Teach/Topo/latex/02.pdf.

[39] Chen, J. Q., Feng, D. H., and Wu, C.-L. 1986. *Phys. Rev.*, **C34**, 2269.

[40] Chen, Y. S., Sun, Y., and Gao, Z. C. 2008. *Phys. Rev.*, **C77**, 061305(R).

[41] Cheng, T.-P., and Li, L.-F. 1984. *Gauge Theory of Elementary Particle Physics*. Oxford: Clarendon Press.

[42] Close, F. E. 1979. *An Introduction to Quarks and Partons*. New York: Academic Press.

[43] Coleman, S. 1985. *Aspects of Symmetry: Selected Erice Lectures*. Cambridge: Cambridge University Press.

[44] Cooper, L. N. 1956. *Phys. Rev.*, **104**, 1189.

[45] Creutz, M. 1983. *Quarks, Gluons, and Lattices*. Cambridge: Cambridge University Press.

[46] Dai, P., et al. 1999. *Science*, **284**, 1344.

[47] Dashen, R. 1969. *Phys. Rev.*, **183**, 1245.

[48] deShalit, A., and Feshbach, H. 1974. *Theoretical Nuclear Physics Vol. I: Nuclear Structure*. New York: John Wiley and Sons.

[49] deShalit, A., and Talmi, I. 1963. *Nuclear Shell Theory*. New York: Academic Press.

[50] Dittrich, W., and Reuter, M. 2001. *Classical and Quantum Dynamics*. Berlin: Springer.

[51] Edmonds, A. R. 1959. *Angular Momentum in Quantum Mechanics*. Princeton, NJ: Princeton University Press.

[52] Egido, J. L., and Robledo, L. M. 1991. *Nucl. Phys.*, **A524**, 65.

[53] Eisenberg, J., and Greiner, Walter. 1972. *Microscopic Theory of the Nucleus: Vol. 3*. Amsterdam: North–Holland.

[54] El-Batanouny, M. 2020. *Advanced Quantum Condensed Matter Physics*. Cambridge: Cambridge University Press.

[55] Elliott, J. P. 1958. *Proc. R. Soc.*, **A245**, 128, 562.

[56] Elliott, J. P., and Dawber, P. G. 1979. *Symmetry in Physics*. Oxford: Oxford University Press.

[57] Elliott, S. R., and Franz, M. 2015. *Rev. Mod. Phys.*, **87**, 137 (arXiv: 1403.4976).

[58] Englert, F., and Brout, R. 1964. *Phys. Rev. Lett.*, **13**, 321.

[59] Fang, L., et al. 2009. *Phys. Rev.*, **B80**, 140508(R).

[60] Feng, D. H., Wu, C.-L., Guidry, M. W., and Li, Z.-P. 1988. *Phys. Lett.*, **205B**, 156.

[61] Feynman, R. P., and Gell-Mann, M. 1958. *Phys. Rev.*, **109**, 193.

[62] Feynman, R. P., Leighton, R. B., and Sands, M. 1964. *The Feynman Lectures on Physics*. Reading, MA: Addison-Wesley.

[63] Frankel, T. 2012. *The Geometry of Physics*. Cambridge: Cambridge University Press.

[64] Fu, L., and Kane, C. L. 2007. *Phys. Rev.*, **B76**, 045302 (arXiv:cond–mat/0611341).

[65] Fu, L., Kane, C. L., and Mele, E. J. 2007. *Phys. Rev. Lett.*, **98**, 106803.

[66] Gasiorowicz, S. 1966. *Elementary Particle Physics*. New York: Wiley.

[67] Gell-Mann, M., and Levy, M. 1960. *Nuovo Cimento*, **16**, 705.

[68] Georgi, H. 1999. *Lie Algebras in Particle Physics: From Isospin to Unified Theories*, 2nd edition. Boulder, CO: Westview Press.

[69] Georgi, H., and Glashow, S. 1974. *Phys. Rev. Lett.*, **32**, 438.

[70] Gilmore, R. 1972. *Ann. Phys.*, **74**, 391.

[71] Gilmore, R. 1974. *Rev. Mex. de Fis.*, **23**, 142.

[72] Gilmore, R. 1974. *Lie Groups, Lie Algebras, and Some of their Applications*. New York: Wiley.

[73] Gilmore, R. 2008. *Lie Groups, Physics, and Geometry: An Introduction for Physicists, Engineers and Chemists*. Cambridge: Cambridge University Press.

[74] Ginocchio, J. N. 1980. *Ann. Phys.*, **126**, 234.

[75] Girvin, S. M., and Yang, K. 2019. *Modern Condensed Matter Physics*. Cambridge: Cambridge University Press.

[76] Glauber, R. J. 1963. *Phys. Rev.*, **130**, 2529.

[77] Glauber, R. J. 1963. *Phys. Rev.*, **131**, 2766.

[78] Goerbig, M. O. 2011. *Rev. Mod. Phys.*, **83**, 1193.

[79] Goldbart, P. M., and Kamien, R. D. 2019. *Physics Today*, **72**, 46.

[80] Goldstone, J. 1961. *Nuovo Cimento*, **19**, 154.

[81] Goldstone, J., Salam, A., and Weinberg, S. 1962. *Phys. Rev.*, **127**, 965.

[82] Greiner, W., and Müller, B. 1989. *Quantum Mechanics: Symmetries*. Berlin: Springer.

[83] Griffiths, D. J. 1995. *Introduction to Quantum Mechanics*. Englewood Cliffs, NJ: Prentice Hall.

[84] Gross, D. J. 1979. *Phys. Rep.*, **49**, 143.

[85] Guidry, M. W. 1991. *Gauge Field Theories: An Introduction with Applications*. New York: Wiley Interscience.

[86] Guidry, M. W. 1999. *Rev. Mex. Fís.*, **45 S2**, 132.

[87] Guidry, M. W. 2017. *Fortschr. Phys.*, **65**, 160057.

[88] Guidry, M. W. 2019. *Modern General Relativity: Black Holes, Gravitational Waves, and Cosmology*. Cambridge: Cambridge University Press.

[89] Guidry, M. W. 2019. *Stars and Stellar Processes*. Cambridge: Cambridge University Press.

[90] Guidry, M. W., and Billings, J. J. 2018. A Basic Introduction to the Physics of Solar Neutrinos (arXiv: 1812.00035).

[91] Guidry, M. W., and Sun, Y. 2015. *Front. Phys.*, **10**, 107404.

[92] Guidry, M. W., Wu, C. L., and Feng, D. H. 1995. *Ann. Phys.*, **242**, 135.

[93] Guidry, M. W., Wu, L.-A, Sun, Y., and Wu, C.-L. 2001. *Phys. Rev.*, **B63**, 134516.

[94] Guidry, M. W., Sun, Y., and Wu, C.-L. 2004. *Phys. Rev.*, **B70**, 184501.

[95] Guidry, M. W., Sun, Y., and Wu, C.-L. 2009. *Front. Phys.*, **4**, 233 (arXiv: 0908.1147).

[96] Guidry, M. W., Sun, Y., and Wu, C.-L. 2010. *Front. Phys.*, **5**, 171 (arXiv: 0810:3862).

[97] Guidry, M. W., Sun, Y., and Wu, C.-L. 2011. *Chin. Sci. Bull.*, **56**, 367 (arXiv: 0810.5700).

[98] Guidry, M. W., Sun, Y., Wu, L.-A., and Wu, C.-L. 2020. *Front. Phys.*, **15**, 43301 (arXiv: 2003.07994).

[99] Guidry, M. W., Sun, Y., and Wu, L.-A. 2021. *Symmetry*, **13(5)**, 911 (arXiv: 2004.04066).

[100] Guralnik, G. S., Hagen, C. R., and Kibble, T. W. 1964. *Phys. Rev. Lett.*, **13**, 585.

[101] Gürsey, F., and Radicati, L. A. 1964. *Phys. Rev. Lett.*, **13**, 173.

[102] Haldane, F. D. M. 1988. *Phys. Rev. Lett.*, **61**, 2015.

[103] Halzen, F., and Martin, A. D. 1984. *Quarks and Leptons: An Introductory Course in Modern Particle Physics*. New York: Wiley.

[104] Hamermesh, M. 1962. *Group Theory and its Applications to Physical Problems*. Reading, MA: Addison-Wesley.

[105] Hara, K., and Sun, Y. 1995. *Int. J. Mod. Phys.*, **E4**, 637.

[106] Harvey, M. 1968. *Adv. Nucl. Phys.*, **1**, 67.

[107] Hasan, M. Z., and Kane, C. L. 2010. *Rev. Mod. Phys.*, **82**, 3045.

[108] Heine, V. 1993. *Group Theory in Quantum Mechanics*. Mineola, NY: Dover.

[109] Higgs, P. W. 1964. *Phys. Rev. Lett.*, **12**, 132.

[110] Higgs, P. W. 1964. *Phys. Rev. Lett.*, **13**, 508.

[111] Hill, D. L., and Wheeler, J. A. 1953. *Phys. Rev.*, **89**, 1102.

[112] Hsieh, D., et al. 2008. *Nature*, **452**, 970.

[113] Huang, S.-M., et al. 2015. *Nature Commun.*, **6**, 7373.

[114] Hubbard, J. 1963. *Proc. R. Soc. London*, **A276**, 238.

[115] Iachello, F. 2014. *Lie Algebras and Applications*. Heidelberg: Springer.

[116] Iachello, F., and Arima, A. 1987. *The Interacting Boson Model*. Cambridge: Cambridge University Press.

[117] Iachello, F., and Levine, R. D. 1995. *Algebraic Theory of Molecules*. Oxford: Oxford University Press.
[118] Iachello, F., and Truini, P. 1999. *Ann. Phys.*, **276**, 120.
[119] Inui, T., Tanabe, Y., and Onodera, Y. 1996. *Group Theory and Its Applications in Physics*. Berlin: Springer.
[120] Itzykson, C., and Zuber, J.-B. 1980. *Quantum Field Theory*. New York: McGraw–Hill.
[121] Jackiw, R., and Rebbi, C. 1976. *Phys. Rev. Lett.*, **37**, 172.
[122] Jackson, J. D. 1975. *Classical Electrodynamics*. New York: Wiley.
[123] Jain, J. K. 2007. *Composite Fermions*. Cambridge: Cambridge University Press.
[124] Jeevanjee, N. 2015. *An Introduction to Tensors and Group Theory for Physicists*. New York: Birkhäuser Springer.
[125] Jones, H. F. 1990. *Groups, Representations and Physics*. Bristol: Adam Hilger.
[126] Judd, B. R. 1963. *Operator Techniques in Atomic Spectroscopy*. New York: McGraw-Hill.
[127] Kane, C. L., and Mele, E. J. 2005. *Phys. Rev. Lett.*, **95**, 146802.
[128] Kane, C. L., and Mele, E. J. 2005. *Phys. Rev. Lett.*, **95**, 226801.
[129] Kang, K., et al. 2010. *Phys. Rev.*, **B81**, 100509(R).
[130] Kharitonov, M. 2012. *Phys. Rev.*, **B85**, 155439.
[131] Kittel, C. 2004. *Introduction to Solid State Physics*. Wiley.
[132] Knebel, G., Aoki, D., and Floquet, J. 2009. Magnetism and Superconductivity in CeRhIn5 (arXiv: 0911.5223).
[133] Kogut, J. B. 1983. *Rev. Mod. Phys.*, **55**, 775.
[134] Kosniowski, C.İ980. *A First Course in Algebraic Topology*. Cambridge: Cambridge University Press.
[135] Laughlin, R. B. 1981. *Phys. Rev.*, **B23**, 5632.
[136] Laughlin, R. B. 1983. *Phys. Rev. Lett.*, **50**, 1395.
[137] Laughlin, R. B. 1998. *Fractional Quantization (Nobel lecture)*. www.nobelprize .org/nobel_prizes/physics/laureates/1998/laughlin-lecture.html.
[138] Lax, M. 1974. *Symmetry Principles in Solid State and Molecular Physics*. New York: Wiley.
[139] Lee, T. D. 1981. *Particle Physics and Introduction to Field Theory*. New York: Harwood.
[140] Lee, T. D., and Yang, C. N. 1956. *Phys. Rev.*, **104**, 254.
[141] Lichtenberg, D. B. 1978. *Unitary Symmetry and Elementary Particles*. New York: Academic Press.
[142] Lipkin, H. J. 1966. *Lie Groups for Pedestrians*. Amsterdam: North-Holland.
[143] Ludwig, W., and Falter, C. 1996. *Symmetries in Physics: Group Theory Applied to Physical Problems*. Berlin: Springer.
[144] Ma, S.-K. 1976. *Modern Theory of Critical Phenomena*. New York: Benjamin.
[145] Ma, Z.-Q. 2007. *Group Theory for Physicists*. Singapore: World Scientific.
[146] McCarty, G. 1967. *Topology: An Introduction with Applications to Topological Groups*. New York: Dover Publications.

[147] McVoy, K. 1965. *Rev. Mod. Phys.*, **37**, 84.

[148] Mendelson, B. 1990. *Introduction to Topology*. Garden City, NY: Dover.

[149] Mermin, N. D. 1979. *Rev. Mod. Phys.*, **51**, 591.

[150] Murakami, S. 2007. *New J. Phys.*, **9**, 356.

[151] Naber, G. L. 2010. *Topology, Geometry, and Gauge Fields*. Berlin: Springer.

[152] Nakahara, M. 2003. *Geometry, Topology and Physics*. Boca Raton, FL: Taylor and Francis.

[153] Nambu, Y. 1960. *Phys. Rev. Lett.*, **4**, 380.

[154] Nambu, Y., and Jona-Lasinio, G. 1961. *Phys. Rev.*, **122**, 345.

[155] Nambu, Y., and Jona-Lasinio, G. 1961. *Phys. Rev.*, **124**, 246.

[156] Nash, C., and Sen, S. 1983. *Topology and Geometry for Physicists*. New York: Academic Press.

[157] Neto, A. H. C., et al. 2009. *Rev. Mod. Phys.*, **81**, 109.

[158] O'Raifeartaigh, L. 1986. *Group Structure of Gauge Theories*. Cambridge: Cambridge University Press.

[159] Pal, P. B. 2011. *Am. J. Phys.*, **79**, 485 (arXiv: 1006.1718).

[160] Peierls, R. E., and Yoccoz, J. 1957. *Proc. Phys. Soc. (London)*, **A70**, 381.

[161] Perelomov, A. M. 1972. *Commun. Math. Phys.*, **26**, 222.

[162] Pfeuty, P. 1979. *Phys. Lett.*, **A72**, 245.

[163] Phillips, P. 2003. *Advanced Solid State Physics*. Boulder, CO: Westview Press.

[164] Pires, A. S. T. 2019. *A Brief Introduction to Topology and Differential Geometry in Condensed Matter Physics*. Bristol: IOP Press.

[165] Qi, X.-L., and Zhang, S-C. 2010. *Phys. Today*, **63 (1)**, 33.

[166] Quigg, C. 1983. *Gauge Theories of the Strong, Weak, and Electromagnetic Interactions*. Reading, MA: Benjamin/Cummings.

[167] Rajaraman, R. 1982. *Solitons and Instantons: An Introduction to Solitons and Instantons in Quantum Field Theory*. Amsterdam: North-Holland.

[168] Ramirez, A. P., and Skinner, B. 2020. *Phys. Today*, **73 (9)**, 30.

[169] Ramond, P. 2010. *Group Theory: A Physicist's Survey*. Cambridge: Cambridge University Press.

[170] Read, N. 2012. *Phys. Today*, **65 (7)**, 38.

[171] Ring, P., and Schuck, P. 1980. *The Nuclear Many-Body Problem*. New York: Springer.

[172] Rodríguez-Guzmán, R., Schmid, K. W., and Jiménez-Hoyos, C. A. 2012. *Phys. Rev.*, **B85**, 245130.

[173] Rose, M. E. 1957. *Elementary Theory of Angular Momentum*. New York: Wiley.

[174] Ryder, L. H. 1996. *Quantum Field Theory*. Cambridge: Cambridge University Press.

[175] Sachdev, S. 1999. *Quantum Phase Transitions*. Cambridge: Cambridge University Press.

[176] Sachdev, S. 2000. *Science*, **288**, 479.

[177] Sarma, S. Das, Freedman, M., and Nayak, C. 2006. *Phys. Today*, **59 (7)**, 32.

[178] Sato, M., and Ando, Y. 2017. *Rep. Prog. Phys.*, **80**, 076501 (arXiv: 1608.03395).

[179] Sato, M., and Fujimoto, S. 2016. *J. Phys. Soc. Jpn.*, **85**, 072001 (arXiv: 1601.02726).

[180] Schensted, I. V. 1976. *A Course in the Application of Group Theory to Quantum Mechanics*. Peaks Island, ME: NEO Press.

[181] Schrödinger, E. 1926. *Naturwissenschaften*, **14**, 664.

[182] Singer, I. M. 1982. *Physics Today*, **35 (3)**, 41.

[183] Stanescu, T. D. 2017. *Introduction to Topological Quantum Matter and Quantum Computation*. Boca Raton, FL: CRC Press.

[184] Stephenson, G. 2015. *An Introduction to Matrices, Sets and Groups for Science Students*. Mineola, NY: Dover Publications.

[185] Sternberg, S. 1994. *Group Theory and Physics*. Cambridge: Cambridge University Press.

[186] Sudarshan, E. C. G. 1963. *Phys. Rev. Lett.*, **10**, 277.

[187] Sun, Y. 2016. *Phys. Scr.*, **91**, 043005.

[188] Sun, Y., Wu, C.-L, Bhatt, K., Guidry, M. W., and Feng, D. H. 1998. *Phys. Rev. Lett.*, **80**, 672.

[189] Sun, Y., Wu, C.-L, Bhatt, K., and Guidry, M. W. 2002. *Nuc. Phys.*, **A703**, 130.

[190] Sun, Y., Guidry, M. W., and Wu, C.-L. 2006. *Phys. Rev.*, **B73**, 134519.

[191] Sun, Y., Guidry, M. W., and Wu, C.-L. 2008. *Phys. Rev.*, **B78**, 174524 (arXiv: 0705.0818).

[192] Tanabashi, M., and others (Particle Data Group). 2018. Review of Particle Properties. *Phys. Rev.*, **D98**, 030001.

[193] Thouless, D. J. 1972. *The Quantum Mechanics of Many-Body Systems*. New York: Academic Press.

[194] Thouless, D. 1998. *Topological Quantum Numbers in Nonrelativistic Physics*. Singapore: World Scientific.

[195] Thouless, D. J., Kohmoto, M., Nightingale, M. P., and den Nijs, M. 1982. *Phys. Rev. Lett.*, **49**, 405.

[196] Tinkham, M. 2003. *Group Theory and Quantum Mechanics*. Mineola, NY: Dover.

[197] Tong, D. 2016. The Quantum Hall Effect: TIFR Infosys Lectures (arXiv: 1606.06687).

[198] Tsui, D. C., Stormer, H. L., and Gossard, A. C. 1982. *Phys. Rev. Lett.*, **48**, 1599.

[199] Tung, W.-K. 1985. *Group Theory in Physics*. Singapore: World Scientific.

[200] Vanderbilt, D. 2018. *Berry Phases in Electronic Structure Theory*. Cambridge: Cambridge University Press.

[201] Varshalovich, D. A., Moskalev, A. N., and Kersonskii, V. K. 1988. *Quantum Theory of Angular Momentum*. Singapore: World Scientific.

[202] Vergados, J. D. 1968. *Nucl. Phys.*, **A111**, 681.

[203] Vojta, M. 2003. *Rep. Prog. Phys.*, **66**, 2069.

[204] von Klitzing, K., Dorda, G., and Pepper, M. 1980. *Phys. Rev. Lett.*, **45**, 494.

[205] Ward, R. S. 2005. Solitons and other Extended Field Configurations (arXiv: hepth/0505135).

[206] Warner, S. 2019. *Topology for Beginners*. Get 800 LLC.

[207] Wen, X. G. 2004. *Quantum Field Theory of Many-Body Systems*. Oxford: Oxford University Press.

[208] Wherrett, B. S. 1986. *Group Theory for Atoms, Molecules, and Solids*. Englewood Cliffs, NJ: Prentice Hall International.

[209] Wigner, E. P. 1937. *Phys. Rev.*, **51**, 106.

[210] Wigner, E. P. 1937. *Ann. Math.*, **40**, 39.

[211] Wigner, E. P. 1959. *Group Theory and its Application to the Quantum Mechanics of Atomic Spectra*. New York: Academic Press.

[212] Wilczek, F. Particle Physics and Condensed Matter: The Saga Continues. *Nobel Symposium 156 on New Forms of Matter, Topological Insulators and Superconductors, June 13–15, 2014, Stockholm* (arXiv: 1604.05669).

[213] Wilson, K. G. 1974. *Phys. Rev.*, **D10**, 2445.

[214] Wilson, K. G. 1976. *Phys. Rep.*, **23**, 331.

[215] Wu, C. L, Feng, D. H., Chen, X-.G., Chen, J.-Q., and Guidry, M. W. 1986. *Phys. Lett.*, **B168**, 313.

[216] Wu, C. L, Feng, D. H., Chen, X-.G., Chen, J.-Q., and Guidry, M. W. 1987. *Phys. Rev.*, **C36**, 1157.

[217] Wu, C.-L., Feng, D. H., and Guidry, M. W. 1993. *Ann. Phys.*, **222**, 187.

[218] Wu, C.-L., Feng, D. H., and Guidry, M. W. 1994. *Adv. Nucl. Phys.*, **21**, 227.

[219] Wu, C. S., Ambler, E., Hayward, R. W., Hoppes, D. D., and Hudson, R. P. 1957. *Phys. Rev.*, **105**, 1413.

[220] Wu, F., et al. 2014. *Phys. Rev.*, **B90**, 235432.

[221] Wu, L.-A, and Guidry, M. W. 2016. *Sci. Rep.*, **6**, 22423.

[222] Wu, L.-A, Guidry, M. W., Sun, Y., and Wu, C.-L. 2003. *Phys. Rev.*, **B67**, 014515.

[223] Wu, L.-A, Murphy, M., and Guidry, M. W. 2017. *Phys. Rev.*, **B95**, 115117.

[224] Wybourne, B. G. 1974. *Classical Groups for Physicists*. New York: Wiley Interscience.

[225] Xu, S.-Y., et al. 2015. *Science*, **349**, 613.

[226] Yang, C. N., and Mills, R. L. 1954. *Phys. Rev.*, **96**, 191.

[227] Zee, A. 2010. *Quantum Field Theory in a Nutshell*. Princeton, NJ: Princeton University Press.

[228] Zee, A. 2016. *Group Theory in a Nutshell for Physicists*. Princeton, NJ: Princeton University Press.

[229] Zhang, W.-M., Feng, D. H., and Ginocchio, J. N. 1987. *Phys. Rev. Lett.*, **59**, 2032.

[230] Zhang, W.-M., Feng, D. H., and Ginocchio, J. N. 1988. *Phys. Rev.*, **C37**, 1281.

[231] Zhang, W.-M., Feng, D. H., and Gilmore, R. 1990. *Rev. Mod. Phys.*, **62**, 867.

[232] Ziman, J. M. 1979. *Principles of the Theory of Solids*. Cambridge: Cambridge University Press.

Index